Modern Technologies in Manufacturing (MTeM 2017 - AMaTUC)

MATEC Web of Conferences Volume 137 (2017)

Cluj-Napoca, Romania
12 - 13 October 2017

Editor:

Nicolae Balc

ISBN: 978-1-5108-5252-5

Printed from e-media with permission by:

Curran Associates, Inc.
57 Morehouse Lane
Red Hook, NY 12571

proceedings.com

Some format issues inherent in the e-media version may also appear in this print version.

Printed by Curran Associates, Inc. (2018)

For additional information, please contact EDP Sciences – Web of Conferences at the address below.

EDP Sciences – Web of Conferences
17, Avenue du Hoggar
Parc d'Activité de Courtabœuf
BP 112
F-91944 Les Ulis Cedex A
France

Phone: +33 (0) 1 69 18 75 75
Fax: +33 (0) 1 69 28 84 91

contact-edps@webofconferences.org

Additional copies of this publication are available from:

Curran Associates, Inc.
57 Morehouse Lane
Red Hook, NY 12571 USA
Phone: 845-758-0400
Fax: 845-758-2633
Email: curran@proceedings.com
Web: www.proceedings.com

TABLE OF CONTENTS

Foreword

On behalf of the organizing committee of the 13th International Conference MTeM 2017 - AMaTUC, I would like to express a warm 'Thank You" to all the participants, for their important contribution, they brought to the Conference! I strongly appreciate the contribution of the authors, who submitted valuable papers and agreed to do successive revisions of their papers, following the recommendations received from the reviewers.

The aim of the AMaTUC - MTeM Conference was to bring together experts in Manufacturing Engineering, in order to exchange ideas and make new connections to extend and develop partnerships between the researchers from Romania and other European Countries.

We received more than 100 articles, which were peer reviewed by the International Scientific Committee and they were also cross checked for similarities, for good scientific practice in publishing. Only 83 articles are accepted for publication in the MTeM-AMaTUC Proceedings. These articles are grouped within 8 chapters, which are the topics of the conference. About half of the participants came from abroad, 45 articles having authors from 14 countries, which proves that the MTeM-AMaTUC Conference was really international!

The MTeM-AMaTUC Proceedings 2017 are being published online with open-access in a volume of MATEC Web of Conferences (https://www.matec-conferences.org/forthcoming).

Many thanks to the members of the Scientific Committee, who accepted to do the reviewing free of charge, which spent their time and effort to bring a substantial contribution to the AMaTUC - MTeM 2017 Conference, by providing professional evaluation of the papers and suitable recommendations.

During the AMaTUC-MTeM Conference, other important meeting was held in parallel with the Conference sessions: The Consortium meeting of the Deans and other representatives from the Mechanical and Industrial Engineering Faculties from Romania. This National Consortium meeting was organized within the AMaTUC-MTeM 2017 Conference in Cluj-Napoca, as quite many of the Deans from different faculties from Romania, presented also scientific papers within the MTeM Conference.

I am grateful to the AMaTUC partners from Germany, England and Luxembourg, for their support in boosting the scientific excellence and innovation capacity in additive manufacturing of the Department of Manufacturing Engineering (DME) from TUCN and for their support in organizing this Conference. Bringing over the FabBUS mobile 3D Printing laboratory from Aachen was great, to be explored by the AMaTUC-MTeM Conference participants, in Cluj-Napoca.

My colleagues from the DME have done their best and without their excellent work it would have been impossible to organize this event. Thank you very much to the Organizing Committee!

We have received excellent feedback up to now, from all the participants, saying they were having fruitful scientific activities and they spent enjoyable days, during the AMaTUC-MTeM 2017 Conference!

Professor Nicolae Balc
Chairman of the MTeM-AMaTUC International Conference

AMaTUC
European project on Additive Manufacturing

AMaTUC - Additive Manufacturing at Technical University of Cluj-Napoca
www.amatuc.com

Driving TCM Cluj into the future!

edp sciences

MATEC Web of Conferences

Statement of Peer review

In submitting conference proceedings to MATEC Web of Conferences, I certify to the Publisher that I adhere to the **Policy on Publishing Integrity** of the journal in order to safeguard good scientific practice in publishing.

1. All articles have been subjected to peer review administered by the proceedings editors.
2. Reviews have been conducted by expert referees, who have been requested to provide unbiased and constructive comments aimed, whenever possible, at improving the work.
3. Proceedings editors have taken all reasonable steps to ensure the quality of the materials they publish and their decision to accept or reject a paper for publication has been based only on the merits of the work and the relevance to the journal.

Title, date and place of the conference

Modern Technologies in Manufacturing
MTeM 2017 – AMaTUC
12-13 October 2017, Cluj-Napoca-Romania

Proceedings editor(s):

Nicolae BALC

Date and editor's signature

9th November 2017

17, avenue du Hoggar - PA de Courtabœuf – BP 112 - 91944 Les Ulis Cedex A (France)
Tél. : 33 (0)1 69 18 75 75 - Fax : 33(0)1 69 07 45 17 – www.edpsciences.org

Aspects regarding the surface roughness on a steel part cutted using AWJ technology

Adrian-Paul Basarman[1,*] and *Mircea* Lobonțiu[1]

[1]Technical University of Cluj Napoca, North University Center from Baia Mare, Dr. Victor Babeș Street 62A, 430083, Baia Mare, Romania.

Abstract. The abrasive water jet cutting technology is a new and modern technology which uses an abrasive material carried by water at high pressure to pierce and cut a desired material. It is a technology used to cut a high variety of materials, from rocks to glass, metal or even human tissue. Regarding the metal cutting using AWJ technology, the surface roughness is a very important aspect when it comes to surface quality and precision. This paper presents a series of aspects regarding the surface roughness on a C45 steel part cutted using AWJ technology, presenting the variation of the surface roughness and the aspect resulted after cutting, on the surface of the cut.

1 Introduction

The abrasive water jet cutting technology is a new and modern technology which uses an abrasive material carried by water at high pressure to pierce and cut a desired material. This paper presents a series of aspects regarding the surface roughness on a C45 steel part cutted using AWJ technology, presenting the variation of the surface roughness and the aspect resulted after cutting, on the surface of the cut. The abrasive water jet cutting technology is continuously researched by different authors like Srivastava et. al [1], Popan et. al. [2,3], Ravai Nagy et.al. [4], Vusal et. al. [5] and Herghelegiu et. al. [6].

The quality of the cutted surfaces and also precision are the key elements regarding the study and the research of cutting of the metal parts using abrasive water jet.

For better analyzing and in order to make a connection between the AWJ cutting parameters and the aspect/quality of the cutted surface, a series of roughness measurements were taken in different areas of the cutted surface, divided in three rows that represent three different areas (Primary impact zone, smooth zone and rough zone) [7]. For studying the zones in which the surface deviations are different on the height of the cutted surface, a series of macroscopic analyzes were made at different magnification powers.

A paper regarding the analyzing of the surface roughness of the parts cutted using AWJ cutting technology was done by authors [8] and the current paper is a continuation of the same research.

In the present paper it will be analyzed a single surface cutted using the parameters presented in table 1.

* Corresponding author: adrian.basarman@cunbm.utcluj.ro

Table 1. Cutting parameters.

Parameters	Selected value
Breakthrough time	13 s
Breakthrough pressure	3600 bar
Abrasive material used	GMA Garnet 80 Mesh (300-150 micron)
Quantity of abrasive material	342 g/min
Cutting pressure	3600 bar
Cutting speed	45 mm/min
Interior sapphire nozzle	0.28 mm
Exterior nozzle	0.8 mm

1.1 Abrasive water jet cutting machine

Research has been made using the AWJ cutting machine presented in figure 1a) and the part presented in figure 2a) resulted after extracting it from the plate presented in figure 1b).

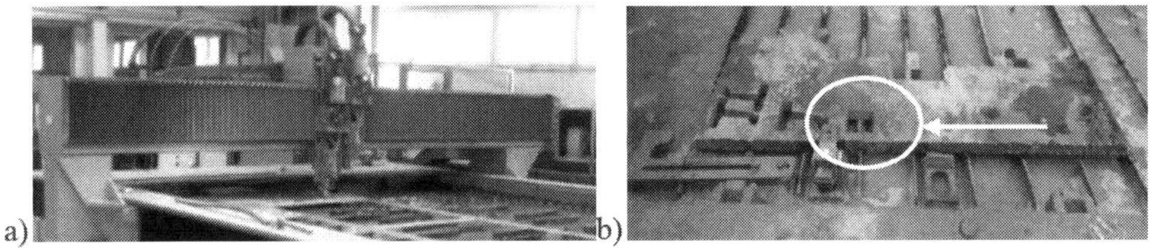

Fig. 1. a) AWJ cutting machine -Bystronic ByJet Pro L-, b) the plate from which the part was cutted.

1.2 Measuring the roughness of the surface cutted using AWJ

The cutted surface was divided in 9 zones, as one can see in figure 2 a). The roughness of the surface was measured using a TR200 roughness tester, as presented in figure 2 b).

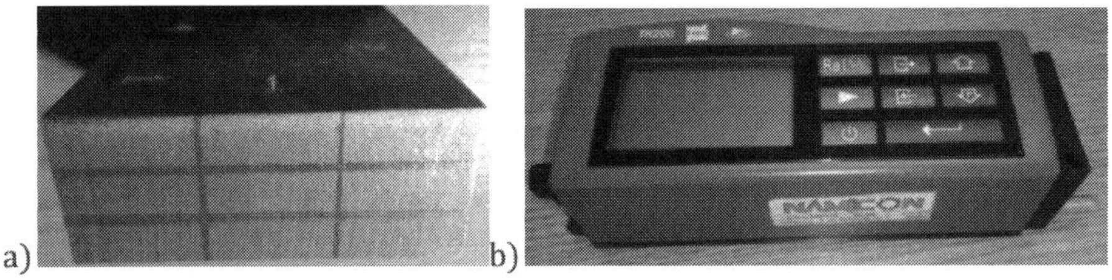

Fig. 2. a) Dividing the surface of the cutted part for measuring, b) TR200 roughness tester.

The dividing was made in such way in order to capture the 3 distinct zones [4] as presented in figure 3.

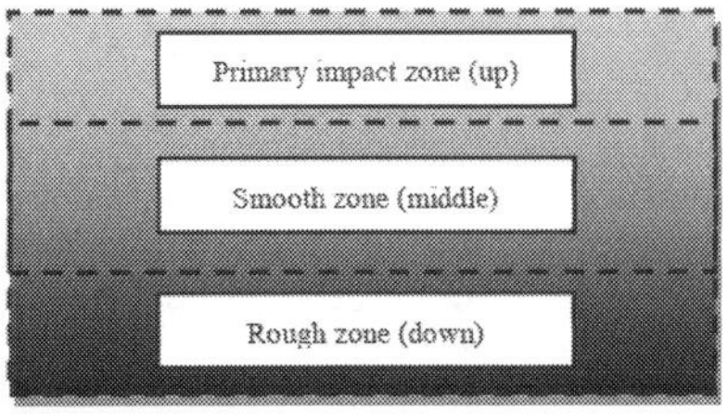

Fig. 3. The zones resulted on an abrasive water jet cutted metal surface.

For better analyzing and understanding of the obtained data, every zone as presented in figure 3 is divided in 3, as seen in figure 2 a) in order to have a number of 3 measurements on every zone. In this way results a number of 9 zones which are numbered as in table 2.

Table 2. Setting and encoding the 9 zones.

Measurements column / Measurement zones	M1	M2	M3
UZ (up zone)	UZM1	UZM2	UZM3
MZ (middle zone)	MZM1	MZM2	MZM3
DZ (down zone)	DZM1	DZM2	DZM3

1.3 Macroscopic analyzing of the surface cutted using abrasive water jet

For start, the surface was cleaned using a solvent that removes the grease applied previously on the surface for anticorrosive protection purpose. That lubricating film makes it impossible to correctly visualize the surface under a microscope.

After removing the lubricating film an optic microscope was used for visualizing the surface and two light sources to prevent the shadowing of surface and assuring a better medium for capturing the images, as it can be seen in figure 4 a) and b).

a) 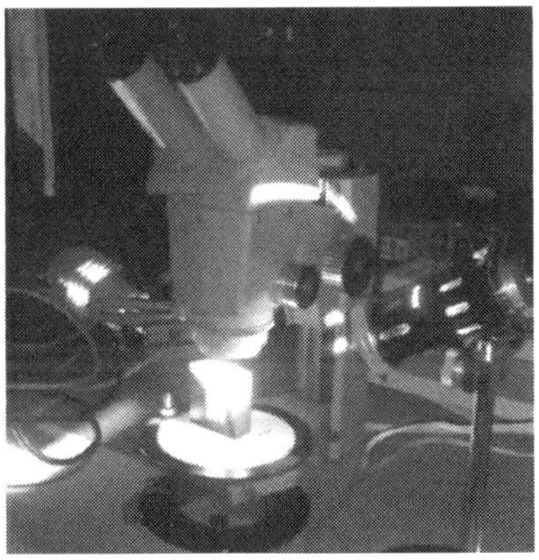 b) 

Fig. 4. The system used for doing the research regarding the surface macrostructure: a) the microscope and the two light sources, b) the metal part analyzed [C45 50x50x20 mm].

To ease the recording and data processing, and also as method for measuring the surface, on the surface that was studied was attached a piece of millimeter paper, as seen in figure 5.

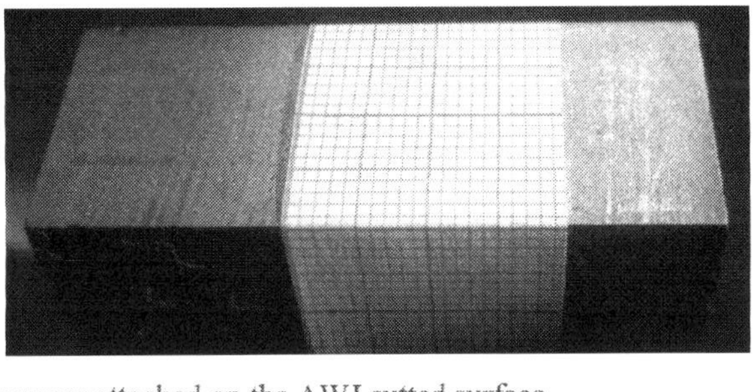

Fig. 5. The millimeter paper attached on the AWJ cutted surface.

DOI: 10.1051/matecconf/201713701001

2 The roughness measurements results

All the measured values are presented in table 3.

Table 3. The measured values corresponded for each of the 9 zones, for both Ra and Rz.

Measurement	**M1**		**M2**		**M3**		**Average**	
Zone	**Ra [µm]**	**Rz [µm]**	**Ra [µm]**	**Rz [µm]**	**Ra [µm]**	**Rz [µm]**	**Ra [µm]**	**Rz [µm]**
UZ	2.597	16.59	3.089	18.84	2.889	16.78	2.858	17.403
MZ	2.831	14.64	3.06	17.54	2.873	16.8	2.903	16..327
DZ	4.105	20.04	3.527	18.87	3.247	16.7	3.626	18.537

For visualizing and a better understanding of the aspect of asperities present on the surface cut using the abrasive water jet cutting technology and, with the help of the roughness tester, was done also a surface profilogram measurement for every zone analyzed.

These profilograms were grouped on measurements so that they can be observed better and, in figures 6, 7 and 8 are presented on columns, the three measurements (M1, M2 and M3), according to tables 2 and 3.

Every figure presents each a set of three profilograms of the surface, the positioning depending of the zone analyzed (UZ, MZ, DZ).

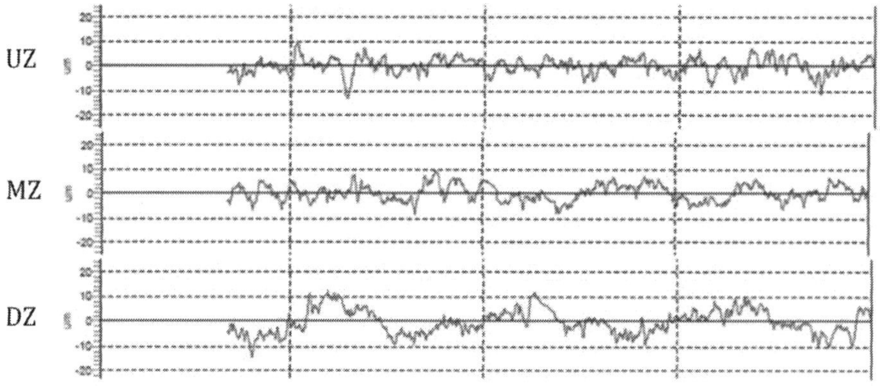

Fig. 6. Profilogram of the measurement M1.

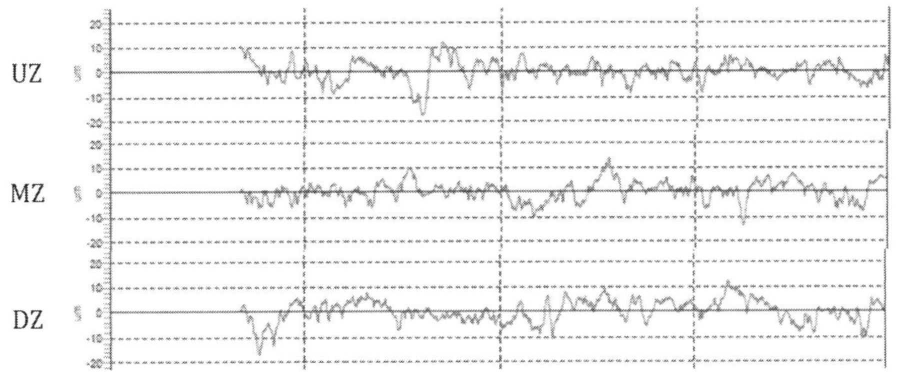

Fig. 7. Profilogram of the measurement M2.

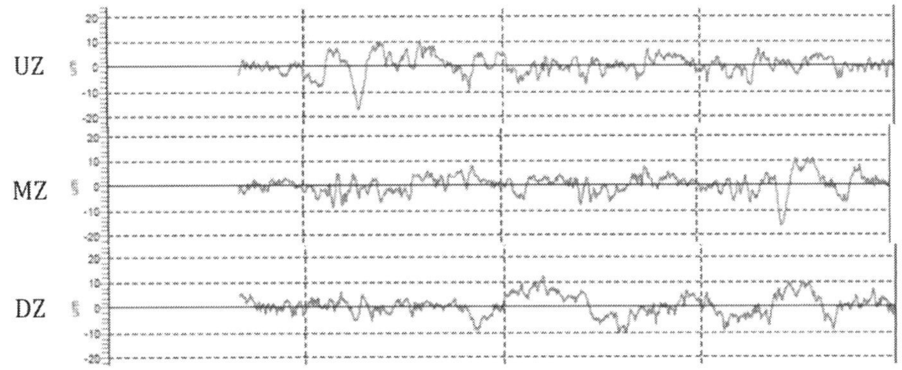

Fig. 8. Profilogram of the measurement M3.

3 The macroscopic analyzing results

For a better analyzing of the abrasive water jet cutted surface, using the optical microscope, the surface was photographed at a magnification power of 10 times bigger and respectively, at a magnification power of 80 times bigger.

Every picture taken were grouped to form a film of the entire surface so, in this case, in the figure 9 is presented, starting from an isometric view of the analyzed surface, from the part cutted using abrasive water jet technology, to a surface 10 times bigger and finally, followed by an 80 times magnification of the same surface.

On the left of each presented surfaces can be seen the millimeter paper which helps for measuring the depth of the zones with different asperities.

After analyzing the surface from figure 9, the depth of each zone with different asperities, as explained in figure 3, were measured and the results can be seen in table 4.

Table 4. The depth of each zone on the AWJ cutted surface.

Measurement zone	Approximated value [mm]
UZ (up zone)	5
MZ (middle zone)	7
DZ (down zone)	8

DOI: 10.1051/matecconf/201713701001

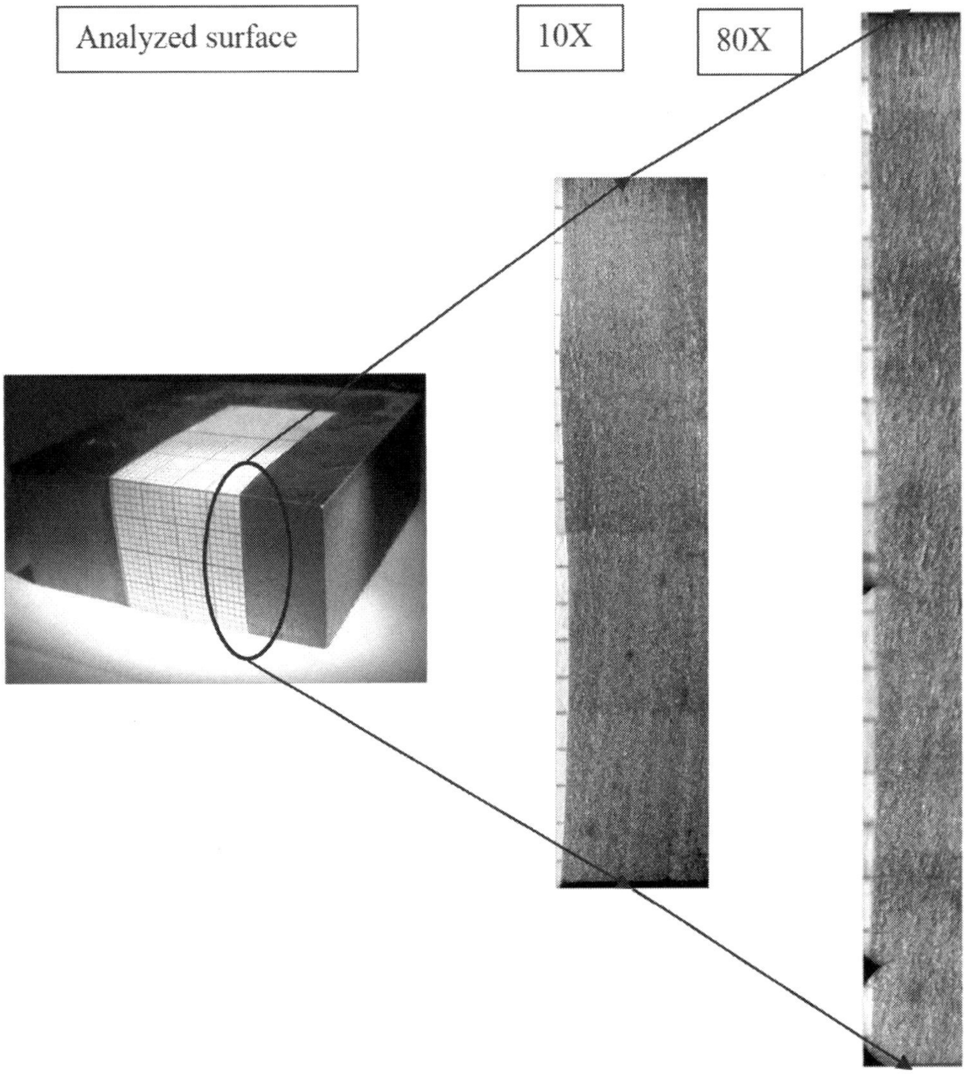

Fig. 9. The macrostructure of the surface cutted using the AWJ cutting technology.

Description of the zones: UZ is the zone from the top part of the analyzed surface which presents the lowest value of the measured roughness; MZ is the middle zone of the metal part where the surface presents a roughness a bit bigger than in the UZ zone but, smaller values that in the DZ zone; and, the DZ zone is the bottom area of the AWJ cutted surface which presents the biggest values for roughness and is the zone with the most surface irregularities.

4 Conclusions

After analyzing the research and results, a series of conclusions emerged:

1) The surface cutted using AWJ cutting technology has asperities with different dimensions in different zones of the surface resulted from cutting.

2) The surface cutted using the AWJ cutting technology presents 3 distinct zones (UZ, MZ, DZ) which present an increasing of the roughness of the surface, from up to down.

3) The depth of the zones is at 5mm for UZ, 7 mm for MZ and 8 mm for DZ, on a 20mm C45 metal part cutted using AWJ cutting technology.

4) Ra roughness recorded for the 3 zones has a min of 2,597 μm in UZM1, 3,006 μm in the middle zone MZM2 and the maximum of 4,105 μm in DZM1.

5) For the first 5mm of the cut the surface has a 2,858 μm Ra roughness, for the next 7 mm a 2,903 μm Ra roughness and for the last 8mm has a 3,626 μm Ra roughness.

6) The surface presents and increasing of roughness from top to bottom, as one can observe also from the visual aspect of the part, seen in the images from figure 9.

References

1. A.K. Srivastava, A. Nag, A.R. Dixit, S. Tiwari, et al., J. Manuf. Process **28**, 11-20 (2017)
2. A. Popan, N. Bâlc, A. Carean, A. Luca, et al. Applied Mechanics and Materials. Trans Tech Publ **760**, 409-414 (2015)
3. A. Popan, N. Bâlc, I. Popan, MATEC Web of Conferences. EDP Sciences, **121**, 02010 (2017)
4. S. Ravai-Nagy, N. Medan, Hidraulica, **4**, 12-17 (2016).
5. B. Vusal, M. Gunel. *An Investigation on the Suitability of Abrasive Waterjet Cutting for Hybrid Manufacturing of Gears in Automotive Industry*, Master of Science Thesis in master program: Production Engineering and Management in the Department of Production Engineering, Royal Institute of Technology, (2015)
6. E. Herghelegiu, M.C. Radu, C. Schnakovszky, C.N. Tampu, MATEC Web of Conferences. EDP Sciences, **94**, 03007 (2017)
7. J. Valíček, S. Hloch, D. Kozak *Study of surface topography created by abrasive waterjet cutting*. Slavonski Brod: Strojarski fakultet u Slavonskom Brodu Sveučilišta J.J. Strossmayera u Osijeku, ISBN 978-953-6048-48-9, (2011)
8. A.P. Basarman, M. Lobonțiu, N. Medan, Scientific Bulletin, Series C: Fascicle Mechanics, Tribology, Machine Manufacturing Technology **30,** 10-16 (2016).

Taxonomy of responsibility allocation in Human-Machine Systems with different levels of automation

Marcin Butlewski[1,*]
[1]Poznan University of Technology, Department of Ergonomics and Quality Engineering, Poland

Abstract. The paper analyses the available taxonomies of responsibility allocation in Human-Machine Systems. The basic assumptions for the division of responsibility in the system and the measures of system wellness are described, as well as the appropriateness of the assignment of responsibility in the system. Finally, a taxonomy of responsibility allocation, consisting of a combination of types of decision resources in the system, was proposed.

1 Introduction

In the anthropotechnical system, man and machine have extremely differing resource characteristics, which results in the different needs of each of these elements in relation to the requirements of the system. The human is particularly unreliable since, unlike the machine, he is not able to maintain for a long period of time and without interruption the high functional parameters required by the work, such as power, force, speed and sensitivity. Therefore, replacing human labor using machines (mechanization) and automated machines (automation) has become the main strategy in the design of Human-Machine (H-M) systems [1]. Reducing human labor can involve manual activities – reducing physical effort and the associated risk of musculoskeletal disorders among different industry sectors [2, 3], however, currently it increasingly involves mental effort, the perception of information and the need for decision-making. The allocation of tasks (or responsibility for tasks due to the over simplification of the issue of the division of labor) between a human and a technical entity seems to be the key factor in determining the effectiveness and reliability of the H-M system. Increasingly complex H-M systems require a more sophisticated way of addressing the problem of responsibility allocation – therefore, this paper proposes taxonomy of responsibility allocation in Human-Machine systems.

2 Criteria for the allocation of functions as a step towards responsibility allocation

For years, the human seemed irreplaceable in areas of work that required the performance of non-standardized procedures, flexibility and unprogrammed action. In turn, the human

* Corresponding author: marcin.butlewski@put.poznan.pl

"loses" with the machine due to his limited reaction time to signals and variable effectiveness with increasing fatigue. The human is also not capable of continuous work, especially when working in unfavorable environmental conditions, e.g. in a significant level of noise, an unfavorable microclimate. The human cannot generate forces rivaling those of machines and fails in situations of repeated and monotonous movements, however, in the end the human ensures the relatively high level of reliability of the system, as its overseer [4, p. 234]. This statement is only seemingly a paradox, because human reliability will be determined by his environment and the working conditions in which he will be put to the test, and one of the factors determining human effectiveness is monotony of work, resulting from the division of tasks in the work system.

With regard to the allocation of functions in the H-M system, there are a number of principles that determine what tasks should or should not be performed by the human – for example, in terms of the allocation of motor functions, the human should only perform those that do not overload him and simultaneously produce the appropriate amount of activity and motor skill [5]. Other decision-making rules developed in this area indicate the need to consider:

- who will be better at performing the given task – the so-called prevailing abilities,
- the good of the whole system, that is, the allocation should be carried out in such a way
- that the measurable values achieved by the system reach an acceptable maximum,
- the harmonization of information in the human-environment system so that the human is not overloaded to the point of impairing the effective performance of work,
- operator satisfaction, and thus adapting the technical system to fit the individual needs of the human.

The obvious difference in the effectiveness of the human and machine during the performance of certain tasks has contributed to the development of lists of the allocation of functions in the H-M system. Task allocation strategies based on the MABA-MABA approach (from the acronym Men Are Better At - Machines Are Better At) were initiated by Fitts [6,7]. This approach, known as the Fitts list, defines the functional division between the operator and machine in the H-M system. In the face of criticism, it has evolved from a strategy of resource complementarity (man and machine mutually complement each other in performing tasks), to a leftover allocation strategy (where all functions that can be automated should be relegated to machines, and the remaining functions should be left to the operators), and finally to a dynamic task allocation (DTA) strategy [8], although even this approach does not seem to exhaust the possibilities of task delegation in H-M systems. Table 1 presents the attributes of tasks, and the suggested relations related to task allocation, however, it should be noted that the division is not exclusive, but rather cooperative [9].

Table 1. Decision table on the allocation of functions in the designed anthropotechnical system [10].

Task attributes	Human (H)	Technical Device (TD)
Flexibility – ease of adaptation to variable conditions	Well suited to a wide range of variable tasks.	Less flexible – specialized.
Adaptation to changing requirements	Good adaptation to unexpected changes in requirements – however, may only work for a limited time in an overload situation.	Work in situations for which the TD is not designed is not possible.
Teaching and training	Easy assimilation of a limited amount of knowledge – improving performance (to a certain level) is the natural domain of man.	Unlimited knowledge – improving performance only within the capacity of the system – requires its control.

Task attributes	Human (H)	Technical Device (TD)
Unstructured tasks in uncertain conditions	In most cases, has a large ability to perform such tasks.	Not adapted to this type of activity – or a significant effort is required to achieve a TD's ability to perform such tasks to a certain extent.
Predictability of outcome	Human activity is not determined and predictable.	Actions with known effectiveness — to a large extent predictable.
Ability to receive and recognize information	Limited number of senses with a high sensitivity results in a limited number of recognizable patterns.	TD can be adapted to recognize any (predictable) stimuli. High performance in this area is expensive.
Monitoring	Very poorly adapted for long-term and routine monitoring due to time-limited vigilance.	Can perform long-term monitoring and recognition of differences.
Incomplete and noisy information	Can act with information shortages or low quality information.	Initially the machine fared quite poorly in such situations, however, work on the development of artificial intelligence led to the creation of decision-making algorithms operating under conditions of uncertainty [11].
Strategic, tactical, organizational planning and decision-making	Well-adjusted when realistic decision models and appropriate conditions (e.g. stress reduction) are ensured.	
Induction and generalization	Can make decisions based on inductive reasoning and generalizing in new situations.	Machines have quite a small capacity for generalization and inductive reasoning.
Memory	A limited short-term memory, and long-term memory is exposed to losses and changes in recorded events.	Memory unrestricted beyond the imposed design limits (what and how long to remember) and phenomena resulting from damage to technical objects.
Continuity of performance	Performance is a function of stress, mental and physical fatigue.	Performance is expressed by mathematical-statistical relationships.
Repeated and monotonous tasks	Human is not suitable for these tasks, although through training can achieve a high level of performance temporarily.	The machine is well suited to this type of task.
Physical abilities	Biologically limited.	Compatible with the design requirements.
Speed	Biologically limited.	An established function obtained through the technical system – considerable speed is possible.
Accuracy	High motor skills, but accuracy is dependent on time spent on the action/training.	Accuracy can be designed as needed, but high accuracy is associated with a considerable cost.
Resistance to environmental factors	Highly resistant in certain areas, beyond which a considerable effort is required to enable a person to work.	Can work under almost all conditions for which the machine was designed.

Determining the allocation of functions in the event that the human and the technical equipment are equal in terms of their available features requires additional analysis and criteria. Helpful in this respect may be determining positions based on Price's diagram of allocation decisions between man and machine [12].

3 Existing typologies in the field of responsibility allocation

Due to the need to create a typology of task allocation in H-M systems that would allow the valuation of the division of functions, incremental scales, such as the 10-point Sheridan-

Verplanck scale, are introduced. It determines the extent that human labor is assisted by automated machinery, ranging from level 1, which means a total lack of assistance, up to level 10, meaning full automation of decision-making and execution of tasks [13]:

1. The system offers no assistance – the human must make all decisions and actions.
2. The system offers a complete set of decision/action alternatives, or
3. Narrows the selection down to a few options (based on the set limitations), or
4. Suggests one alternative, and
5. The system executes that suggestion if the human approves, or
6. The system allows the human a restricted time to veto before automatic execution, or
7. The system makes decisions and executes tasks automatically, then necessarily informs humans, and
8. Informs the human only if asked, or
9. Informs the human only if it, the computer, decides to.
10. The system decides everything and acts autonomously, ignoring the human.

The presented division, though cited in numerous publications, does not allow to determine the level of assistance the technical system offers the human in all areas of resource-efficient decision-making, in terms of perception, decision-making and task execution.

4 Proposed typology of responsibility allocation

The typology of responsibility allocation can be presented by a ternary division of action attributes – Fig. 1.

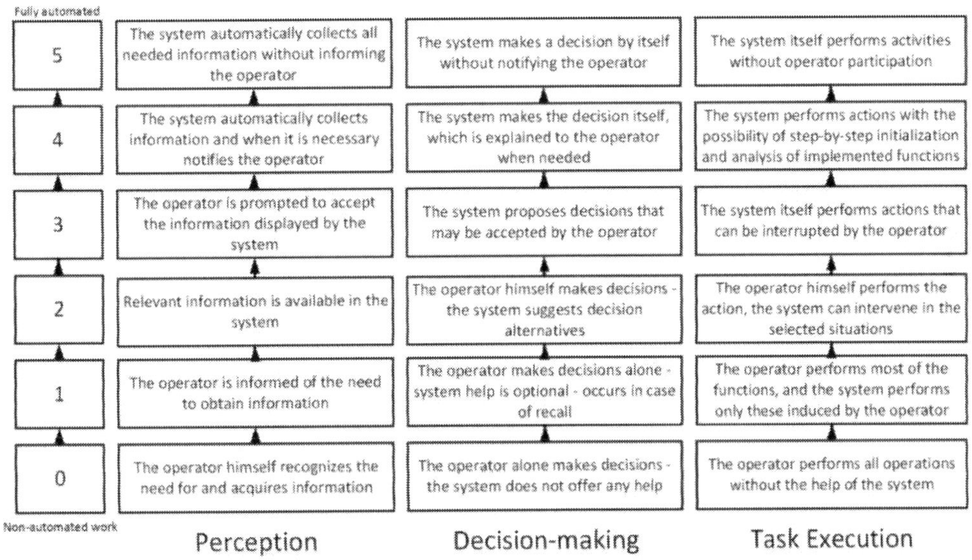

Fig. 1. Levels of automation for main decision resource categories [14].

The extent of the assistance offered to the operator within the H-M system is characterized by three dimensions of his activity (perception X_P, decision-making X_D and task execution X_{TE}) and allows the definition of requirements for different tasks, which may be justified by safety requirements in a specific situation.

For example, in the case of the risk of insufficient implementation resources, the ($X_P.X_D.X_{TE}>3$) strategy can be selected. The functional description of the allocation of tasks may take the form presented in Figure 2.

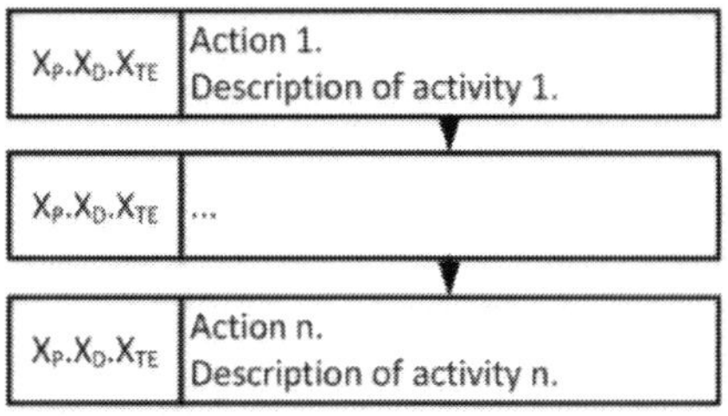

Fig. 2. Proposed structure for describing the levels of automation.

Certain code combinations for major resource categories do not seem to be possible, however, they cannot be excluded under specific conditions of a system's structure (e.g. air traffic control systems, etc.). If the operator makes decisions himself – the system suggests decision-making alternatives (X_P, $X_D(2)$, X_{TE}). These alternatives can be displayed with the percentage probability of a given decision-making alternative, which is characteristic for expert systems.

Mathematical calculations of the expected value for a particular combination of task allocations are also proposed, however, each combination requires the sequential ordering of the values of human and machine abilities to perform a given task and assigning them weights resulting from this order [14]. In determining the desired allocation of functions in the H-M system the Analytic Hierarchy Process (AHP) method is used quite often, and allows for a comparison of importance of a pair of criteria [15]. This approach requires the determination of evaluation criteria, among which the typical ergonomic criteria are found [16-18]:

- criteria related to the operator's well-being,
- criteria related to the operator's effectiveness,
- criteria related to the effectiveness of the system as a whole.

Each of the criteria groups consists of sub-criteria, whose great number cannot be presented in this publication. Some of the criteria will be interdependent and interpenetrating between the mentioned groups. The large number of criteria significantly lengthens the process of comparing them and thus method of support will be needed [19].

5 Conclusion

The allocation of tasks in the system must be dependent on the ability to maintain the performance of tasks over time. This implies the need to designate situations in which the human or machine will assume responsibility for the task being performed (Trading of Authority). This also demonstrates the need for a dynamic functional division. The allocation of responsibilities and the need for its exchange in the H-M system also arises due to situational or compensating resource deficits, among which time and information are the ones most commonly mentioned [20].

These resources are perceived as one because the time available is a factor influencing the ability to acquire and process information, so time will compensate for insufficient information within the system. Systems based on the principle of dynamic sharing of responsibility and the possibility of its exchange are a key element of flexible automation (also known as adaptive automation) [21], whose practical application is observed in intelligent driving systems.

References

1. A.M. Lasota, K. Hankiewicz, In: P. M. Arezes, et al. (Eds.), Occupational Safety and Hygiene, SHO, **138** (2016)
2. A.M. Lasota, K. Hankiewicz, Proceedings of the International Conference on Economics and Management Innovations, **57**, 272 (2016)
3. A. Misztal, M. Butlewski, A. Jasiak, S. Janik, Metalurgija, **54** (2), 429 (2015)
4. A.M. Lasota, K. Hankiewicz, Assessment of risk to work-related musculoskeletal disorders of upper limbs at welding stations, In: P. M. Arezes, et al. (Eds.), Occupational Safety and Hygiene, SHO, 138 (2016)
5. A.M. Lasota, K. Hankiewicz, Proceedings of the 2016 International Conference on Economics and Management Innovations, **57**, 272 (2016)
6. Z. Ratajczak, Niezawodność człowieka w pracy: studium psychologiczne, (Human reliability at work: psychological study) Państwowe Wydawnictwo Naukowe (1988)
7. *** http://www.nop.ciop.pl – accessed 20.02.2017
8. Å. Fasth, & J. Stahre Proceedings of the 3rd CIRP Conference on Assembly Technologies and Systems **1**, (2010)
9. P. M. Fitts, *Human engineering for an effective air-navigation and traffic-control system*, http://www.dtic.mil/get-tr-doc/pdf?AD=ADB815893 (1951)
10. J. Frohm, A. Karsvall, & M. Hassnert, Proceedings of the 8th International Federation of Automatic Control Symposium. Gothenburg, Sweden, **161** (2003)
11. J. M. Hoc, & C. Chauvin, *Cooperative implications of the allocation of functions to humans and machines*, http://jeanmichelhoc.free.fr/pdf/HocChau%202011.pdf (2011)
12. ISO 9355-1:1999 *Ergonomic requirements for the design of displays and control actuators - Part 1: Human interactions with displays and control actuators* (reviewed in 2015)
13. C. J. Hopfe, *Uncertainty and sensitivity analysis in building performance simulation for decision support and design optimization* PhD diss., Eindhoven University, (2009)
14. H. E. Price, Human factors, **27** (1), 33 (1985)
15. R. Parasuraman, T. B. Sheridan, C. D. Wickens, IEEE Transactions on systems, man, and cybernetics-Part A: Systems and Humans, **30** (3), 286 (2000)
16. L. Yong ,W. Jiarui Dong, P. Zhenying, Regional Logistics ijssst.info/Vol-17/No-1/paper12.pdf – accessed 25.02.2017 (2012)
17. Z. G. Wei, A. R. Paauw, P. A. Wieringa, *Subjective evaluation of task allocation: An application of the analytic hierarchy process.* Delft University of Technology, (1996)
18. A. Górny, Posters Extended Abstracts : International Conference, HCI International 2014, Communications in Computer and Information Science, **435**, 449 (2014)
19. A. Górny, MATEC Web of Conferences (the 4th International Conference on Computing and Solutions in Manufacturing Engineering 2016, **94,** (2017)
20. S. Kłos, J. Patalas-Maliszewska, Throughput Analysis of Automatic Production Lines Based on Simulation Methods. Intelligent Data Engineering and automated learning – IDEAL 2015. In.: Jackowski, K., Burduk, R., Walkowiak, K., Wozniak, M., Yin, H. (eds.) Lecture Notes in Computer Science, **9375,** 181 (2015)
21. R. Sika, M. Rogalewicz, Management and Production Engineering Review, **7** (4), 97 (2016)
22. T. Inagaki & Y. Nagai, Journal of Mechanical systems for Transportation and Logistics**, 1** (2), 213 (2008)
23. T. Inagaki, Cognition, Technology & Work, **8** (3), 161 (2006)

Design of the clamping system for the CNC machine tool

Miroslav Císar[1,*], *Ivan* Kuric[1], *Nadežda* Čuboňová[1], and *Matej* Kandera[1]

[1]University of Žilina, Faculty of Mechanical Engineering, Department of Automation and Production Systems, Univerzitná 1, 010 26 Žilina, Slovakia

Abstract. The workpiece clamping and system for automatic workpiece exchange are still topics of great importance. This article describes design of such system implementing interchangeable technological pallets for laboratory training CNC machine tool with confined workspace and limited lengths of feeds. Final design described in article is suitable for implementation in laboratory but also for industrial usage.

1 Introduction

The workpiece clamping is essential part of machining regardless if it is done on conventional machine tools or numerically controlled ones. The time necessary to properly clamp, align, and set up the stock can significantly affect production especially for parts with short cycle times. The approach, relatively common especially in mass production is implementation of technological pallet that host workpieces what allows to parallelize processes of clamping, aligning, and machining itself. Technological pallets are inalienable part of flexible manufacturing systems (FMS) and their implementation, beside time savings and increased potential for automation, it is also capable of promoting further benefits such as greater variability of available mounting systems, increased flexibility in production switching, increased repeatability, and lot more [1].

The systems of technological pallets can utilize various systems for aligning and mounting that should allow fast and simple exchange of individual pallets regardless it is done manually or by automatic pallet exchange system. Each system has its own pros and cons and it is important to choose right one for specific application considering technical and economic aspects [1].

Unfortunately, even pallet systems have downsides and limitations that can negatively affect production, which have to be carefully considered. This article presents design of interchangeable technological pallet system for CNC machine tool Emco Concept Mill 105 (CM 105) – small milling machine with three controlled axis designed specifically for industrial training which is available in laboratory of CNC programing at the Department of Automation and Production Systems, Faculty of mechanical engineering, University of Zilina [1, 2].

* Corresponding author: miroslav.cisar@fstroj.uniza.sk

2 The initial design conditions

The design of proposed pallet system is focused to be compatible with Emco CM 105, what is compact machine tool designed for industrial training and educational purposes. Such machine tools are commonly used at technical high schools and universities. Tools are stored in ten station turret and therefor even unused tools can sometimes overlap to workspace. Slides and load-bearing elements of this machine are made of gray cast iron. The machine tool is already equipped with electronic and mechanic necessities preparation to implement higher level of automation. Standard milling vise with 60mm jaw opening or mill clamps are currently used for workpiece clamping [1].

There is no ready to use commercially available pallet system compatible with this machine tool therefor it was necessary to design custom one. In order to implement system of automatic workpiece exchange, not only manually operated exchange of pallets, it is necessary to add optional accessories such as pneumatic unit, mechanism for automatic door operation and source of compressed air. Further improvement of workplace to implement principles of computer integrated manufacturing (CIM) even to level of FMS would need adding Robotik-Interface hardware and software together with DNC-Interface [2-4].

Proposed design had to take account of the limiting parameters of the machine tool listed in table 1 and also following basic requirements:

- the design shall allow automated and manual workpiece exchange,
- it should allow implementing of various clamping systems,
- overall height should be minimal in order to minimize restrains of workspace,
- the construction should be as simple as possible,
- the pallet system have to withstand maximal possible loads with minimal deformations.

Table 1 The parameters of machine tool relevant to design of pallet system [3].

Travel per axis (X×Y×Z)	[mm]	200×150×250
Effective Z-stoke	[mm]	150
Distance spindle nose – table surface	[mm]	95÷245
Clamping surface (L x W)	[mm]	420×125
Maximum table load	[kg]	10
2 T-slots acc. To DIN 650	[mm]	11
T-slots distance	[mm]	90

3 Design of clamping system

Based on requirements described in chapter above, two distinct designs were created. One of the designs is focused to be fully automatable and fully compatible with other parts of planed FMS. The second design is simpler and designed to be manually operated but it does not require upgrades of systems of machine tool and therefor it can be implemented in case that the project of FMS will not be realized in which case such clamping system still can provide great addition to usability of machine tool in terms of educational potential.

The proposed designs are relatively similar from view of the shape off pallet but with different system of pallet mounting. The common construction of pallet systems uses flat pallets and single piece mounting base that rises pallet by the sum of thicknesses of individual elements. Innovative non-flat shape of pallet with raised ends minimizes workspace limitation due overall thickness of system as it moves mounting and centering elements to the sides. Therefor both proposed designs reduces usable workspace in direction of Z-axis only by thickness of the pallet.

The first design (Fig. 1) is focused to allow fully automated exchange and utilizes zero-point fixture system which is high standard in industry. The zero-point fixtures provide both alignment and mounting of the pallet in wokspace [2, 5-7].

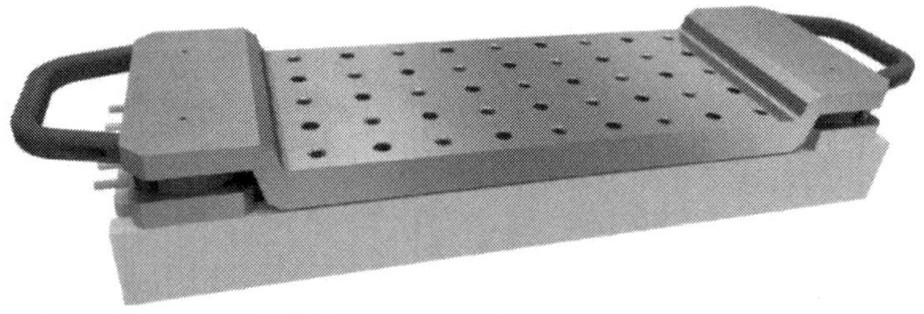

Fig. 1. Pallet system with zero-point fixtures.

The second one (Fig. 2) is simpler and cheaper, it uses manually operated levers with helical surfaces to generate mounting force. The position of pallet on mounting base is ensured by taper pins fitted with housing drilled into pallet [5].

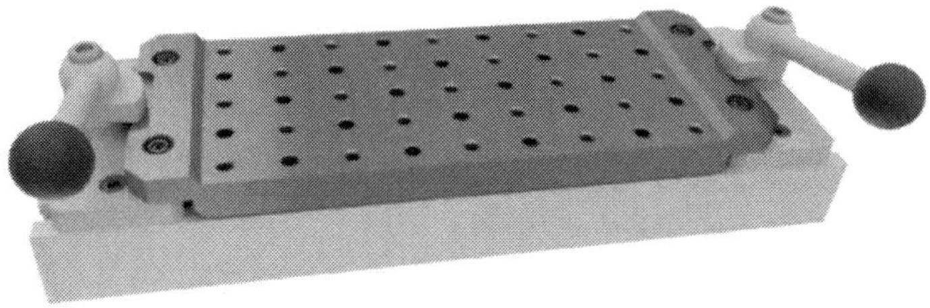

Fig. 2. Pallet system with manually operated levers.

Both proposed designs of clamping system are based on pallet system approach and therefor their construction can be divided into following three construction nodes:
• Mounting base – clamped on worktable, ensures position of mounted pallet in workspace.
• Mounting mechanism – performs safe pallet mounting and releasing.
• Pallet – interchangeable part fitted with geometrical elements necessary for selected clamping system (T-slots, set of holes, threaded holes, etc.)

3.1 Mounting base

Normally, the mounting plate is single piece of metal (baseplate) sheet fitted with mounting mechanism, in this case it was divided to two identical sections in order to minimize height of clamping system and reduce its negative impact to size of effective workspace of machine tool. Both types of baseplates can be mounted on worktable using Allan screws and T-slot nuts as shown on Fig. 3. The proper alignment with direction of axes of machine tool can be reached by using common tool spacer, and dial indicator or a touch probe.

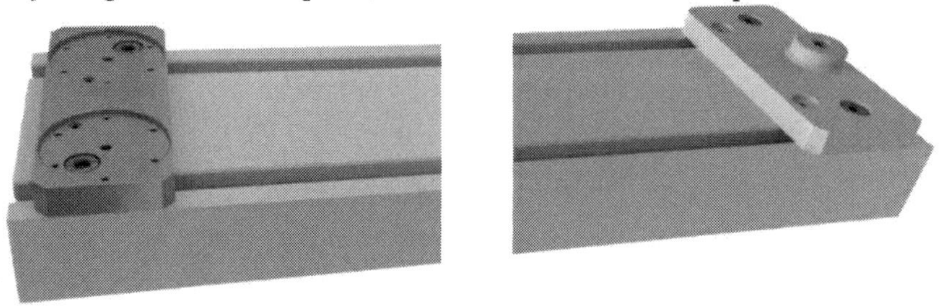

Fig. 3. Divided base plates for zero-point system (left) and lever system (right).

 DOI: 10.1051/matecconf/201713701003

The screws on baseplate for zero-point system (Fig. 3 left) are located in cavity that assure position off zero-point units and thus mounting of such baseplate will take slightly longer time as it is necessary to mount it on worktable and assemble system afterwards. The baseplate also contains holes for distribution of compressed air necessary to operate zero-point fixtures.

The mounting base for the second proposed system (Fig. 3 right) is designed as weldment in order to reduce material requirements. The tapper pins have to be mounted only after the baseplates are mounted to work table and the pallet have to be used to align them properly.

3.2 Mounting mechanism

So-called zero-point fixture system as relatively new trend in fixture market offers easy way to reliably hold pallets where the fixtures have been previously assembled and verified. The clamping system assures that, as the pallets are changed, the fixtures and parts does not need to be realigned to guarantee its correct positioning. Hence, they provide constant zero point with no need of realignment between the pallet and worktable as well as a rapid and safe mounting, exchange, and release of the pallets.

The zero-point fixtures are pneumatically operated what allows to control releasing of pallet by pneumatic unit and standard machine tool interface that is normally used to release workpiece from pneumatic vice as optional equipment provided by Emco.

The first design uses four zero-point fixtures, two on each mounting base as shown on Fig. 4. Each zero-point is mounted to mounting base by six Allen screws and equipped with rubber O-rings to ensure airtight connection to pneumatic system. The fixtures have to be assembled after mounting base plate to worktable.

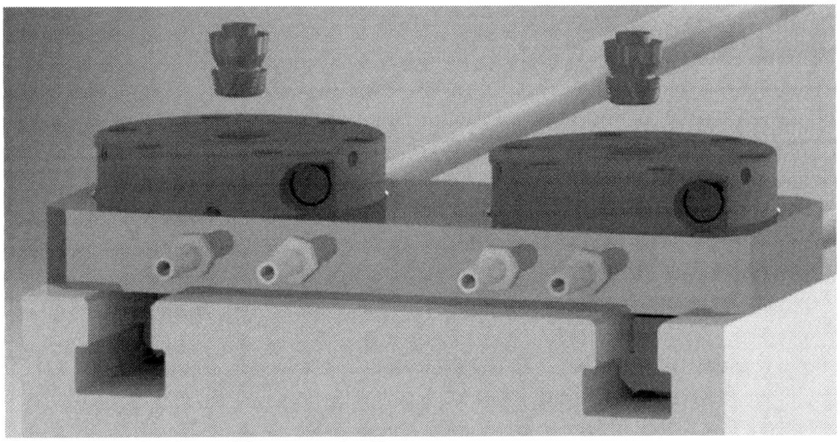

Fig. 4. Mounting mechanism with zero-point fixtures

The pallet stoppers (the pins that fits into zero-point fixture) are mounted to pallet by double ended round collar machine screws with hex recess for Allen wrench. Such mounting allows to set exact position of pallet stoppers on pallet, to align pallet with directions of axes of machine tool, to assemble whole pallet system without necessity to modify standard pallet stoppers.

The second design (Fig. 2) uses more common approach to alignment and mounting of pallet – the taper pins fitted with housing drilled into pallet, that ensures correct position and orientation of pallet, combined with manually operated levers with helical surface that generate clamping force by turning. The levers (Fig. 5 left) are made by Halder Inc. If it will be reasonable to automate releasing of pallet with this system the levers can be replaced by component with same helical shape and size driven by appropriate stepper motors.

DOI: 10.1051/matecconf/201713701003

In this case, the correct position is ensured by four centering pins (Fig. 5 center) mounted to base by Allen screws, paired with bronze housing (Fig. 5 right) fitted to pallet. The centering pins should be mounted to mounting base through housing in pallet in order to set right position and ensure loosen fit that allows comfortable exchange of pallet.

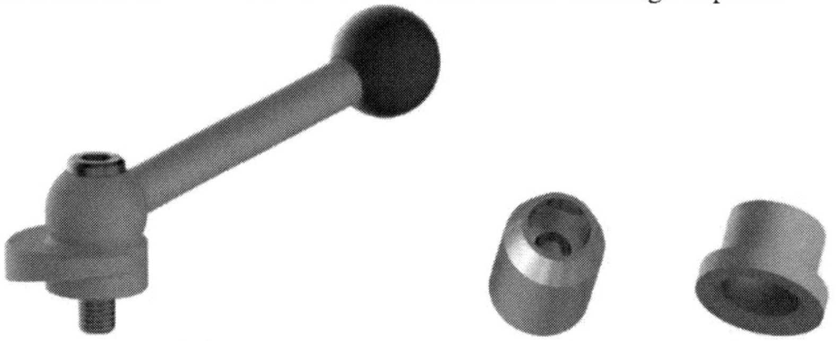

Fig. 5. The mounting lever (left) centering pin (center), and centering housing (right)

3.3 The pallet

Hence, the both proposed designs use mounting base divided to two sections, the general shape of pallet is very similar as shown on Fig. 1 and Fig. 2. The pallet system allow to use various types of clamping systems and mechanisms such as T-slots, grid of holes, grid of threaded holes. For purposes of this article the combination of holes and threaded holes as shown on schematic Fig. 6 was selected as it is compatible with multiple modular fixture systems and it can be used as example of advanced workpiece clamping systems compared to vice commonly used with such machine tools. The machine tool Emco CM 105 is used for educational purposes and the pallet system should widen its educational potential.

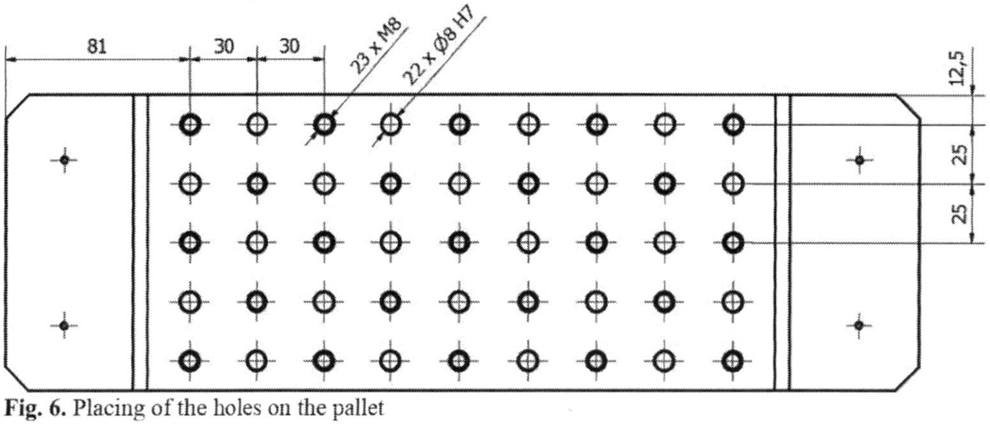

Fig. 6. Placing of the holes on the pallet

In order to allow manipulate pallet by hand the handles have to be added. Simple handle for the first design is on the Fig. 7 (left) and it is also visible mounted on pallet at Fig. 1.

The second design requires different design of handles as the levers would collide with handles. In this case it is possible to use detachable handles (Fig. 7 right) that allow to grab pallet by two Allen screws M8x25 STN 02 1143.52 partially screwed to threaded holes. The handles can be easily fitted over screw heads then pushed to side and secured by lifting pallet. Such system also allow to design implement automatic manipulation of pallets by robot with special effector.

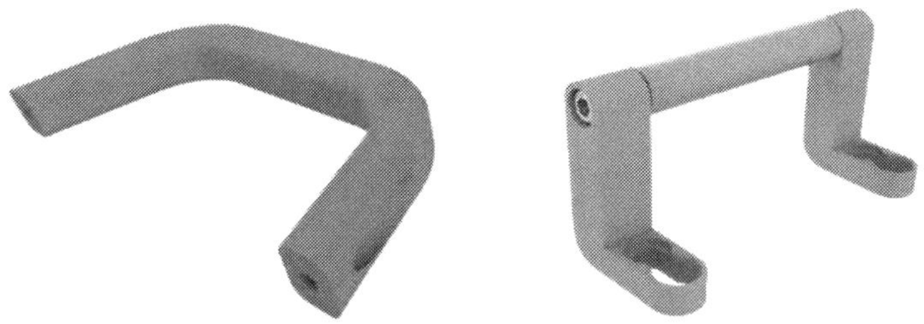

Fig. 7. The fixed handle (left) and the deatachable handle (right)

4 Conclusion

The proposed designs are both suitable for production and implementation on CNC machine tool Emco Concept Mill 105 located in laboratory of CNC programing at the Department of Automation and Production Systems, Faculty of mechanical engineering, University of Zilina. The designs were created in order to implement CIM and to upgrade workplace to FMS. Such upgrade is necessary to create the necessary basis for performing complex multi-criteria diagnostics in controlled environment.

The proposed pallet system can also significantly improve educational value of existing workplace as it can provide real life example of fixtures for mounting workpieces that are used in industry and present its impact to cycle times in production. The proposed designs shows examples of two different approaches to same problem one of which uses modern fixtures what is also great example of technical advancement. The implementation of pallet system with eccentric cam would require additional testing if it is necessary to add safety catch to prevent release of pallet due vibrations.

All models for these designs including renders and drawings were created in Autodesk Inventor 2017 and their dimensions were verified by stress analysis.

The design of pallet with non-flat shape of pallet with raised ends and dividing of the mounting base into two distinct pieces significantly reduces overall height of pallet system and can be utilized not only on training machine tools but also in industrial production especially on small machine tools with limited workspace or in cases when machined workpiece barely fits into workspace.

This article was made under the support of APVV project – APVV-16-0283 Research and development of multi-criteria diagnosis of production machinery and equipment based on the implementation of artificial intelligence methods.

References

1. I. Kuric, AJME Journal, **9** (1), 83 (2011)
2. M. Císar, N. Čuboňová, I. Kuric, MATEC Web of Conferences, **94**, 01003 (2017)
3. EMCO GmbH, *Technical specification Concept MILL 105*, http://www.amsarg.com.ar/Productos/EMCO/spec_cm105_en.pdf
4. M. Císar, I. Kuric, N. Čuboňová, Š. Vlček. ASTRJ, **10**(31), 80-86 (2016)
5. M. Kandera, *Design of clamping system for EMCO Concept Mill 105*, (in Slovak), Master's thesis, University of Zilina (2015)
6. Erwin Halder KG, *Zero-Point Clamping Systems*, http://pdf.directindustry.com/pdf/halder-france/zero-point-clamping-systems/13708-505599-_2.html
7. ZeroClamp GmbH., *General catalogue*, https://adobeindd.com/view/publications/1dfb96fd-fc18-4288-92ee-55e46c791fe4/1/publication-web-resources/pdf/Katalog_A4_neu.pdf

Determination of the Industrial Robot Positioning Performance

Vladimír Tlach[1,*], *Miroslav* Císar[1], *Ivan* Kuric[1] and *Ivan* Zajačko[1]

[1]University of Žilina, Faculty of Mechanical Engineering, Department of Automation and Production Systems, Univerzitná 1, 010 26 Žilina, Slovakia

Abstract. Position performance is one of most important parameters in terms industrial robotics and CNC machine tools. These performance criteria are changed during machine operation and therefore it is necessary to measure device parameters in regular intervals. This measurement should be as fast and easy as possible. In the case of CNC machine tools, the Renishaw Ballbar device is commonly used for such purposes. This device fits both requirements regardless it does not offer same amount of useful information for CNC machine tools and industrial robots that require special approach to measurement and data analysis which is main topic of presented article.

1 Introduction

Industrial robots are, in addition to features such as load capacity, shape and size of the workspace, also specified by performance criteria. These are defined in ISO 9283, which divides the performance criteria into four basic groups: position and path performances, minimum positioning time and static compliance. In the technical documentation, manufacturers of industrial robots most commonly list one-time pose repeatability (RP), which belongs to a group of position performance. [1] This standard, in addition to the definition of individual characteristics, also mentions the recommended conditions for their measurement and evaluation. In the paper [2], the authors pointed to several shortcomings of this standard, including the missing specification of the measuring device for testing individual characteristics. This gives an industrial robot user the ability to decide what measurement device to use. Papers published in this field are presenting various measuring devices in connection with the measurement of the performance characteristics, such as a dial digital indicator [3, 4], laser interferometer [5, 6], laser tracker [7, 8], camera systems [9] and a Ballbar [10, 11].

During the operation of an industrial robot, the performance criteria may change due to various factors. The article [12] mentions five factors that cause them to change. These are environmental, parametric, computational, measurement errors, and errors caused by the robot application itself. At present, industrial robots are also used in applications requiring high repeatability. These are robots with a repeatability of ± 0.02mm, used for precise manipulation or measurement operations. In such applications, it is all the more important to

* Corresponding author: vladimir.tlach@fstroj.uniza.sk

carry out industrial robot measurements at regular intervals. This ensures the long-term monitoring of the industrial robot's condition, which can be used for planning its maintenance or maintenance shutdown. One of the requirements for such a regular measurement is its simplicity and speed. Many of the methods presented in field-specific papers represent a complex process of measurement and another problem may be the high cost of the measuring device.

In the case of CNC machine tools, a Ballbar device is used as a standard tool for measuring performance criteria. This makes possible to carry out a quick and simple measurement for calibration and evaluation of the CNC machine tool. The measurement process itself is based on a change of radius when performing circular paths (or arc portions) around a fixed point determined by the center pivot. By using the attachments, measurements can be made on radii from 50 to 600mm. The data thus obtained are further processed in Ballbar 20 software and used to calculate the total accuracy of the machine in accordance with effective international standards. [13] This paper presents a Ballbar measurement applied to the industrial robot Fanuc LR Mate 200iC. It is an industrial robot of standard design with six controlled axes, placed in the laboratory at the Department of Automation and Production Systems of the University of Žilina.

The purpose of the described measurement is to verify repeatability in positioning the robot end piece in the shortest possible time and in the simplest possible way. Repeatability is the ability of the robot to repeatedly acquire the same position of its TCP point (Tool Center Point). The measurement procedure and the software supplied with Ballbar were specially developed for use on a machine tool in which the circular motion is created by the simultaneous movement of two perpendicular linear axes, which is not possible in case of industrial robots with serial kinematics [5]. For this reason, industrial robot measurement requires a special approach to measurement and data analysis, which is also the main topic of this article. For the purpose of generating circular paths, specialized software under development by the Department of Automation and Production Systems is used.

2 Method of measurement

Measurement was performed on a Fanuc LR Mate 200iC industrial robot in the department's laboratory. The entire measuring assembly is shown in Fig. 1, consisting of the aforementioned robot and the Renishaw Ballbar QC20-W measuring system. The robot end piece is loaded with a weight of 360g (7.2%) during the measurement.

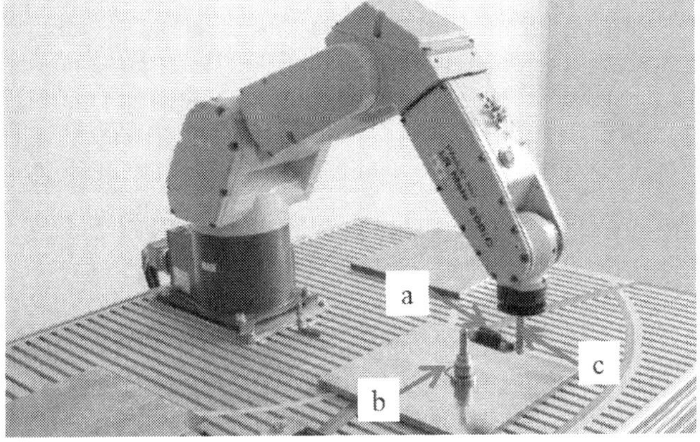

Fig. 1. Measuring assembly: a-Ballbar, b-magnetic center pivot, c-magnetic tool cup.

The measuring assembly is a Ballbar, a magnetic center pivot clamped on a table equipped with a magnetic dish into which the Ballbar ball fits perfectly, and a magnetic tool cup that is clamped onto the robot flange by means of an extension attachment. The magnetic dish of the center stand represents a fixed point of rotation. The measured data are transferred to the PC via the Bluetooth interface. Sensor accuracy of ± 0.5μm (reported at 20°C) and the measuring range of ± 1mm rank Ballbar among very accurate gauges [13]. However, this high accuracy and limited range of measurable deviations may be a problem in measuring the industrial robot performance as the robots does not achieve the same precision as CNC machine tools and therefore this device may not be able to measure their characteristics.

The Ballbar measurement is based on the measurement of the radius deviations between the programmed and the actual radius of the circular path of the robot's end segment. In the Fanuc's programming environment, the circular path can only be created using two semicircles. During the execution of the path thus created, the TCP point of the robot stops at its end point after describing the first semicircle. It then re-starts to describe the second semicircle. A smooth path can be achieved by using the endpoint circumvention function, but in this case, the circular path is deformed to an elliptical one. The circular path made up of two semicircles is therefore inappropriate for Ballbar measurement.

Previous experiments, results of which will be published soon, was performed on a CNC machine tool with a pseudo circular path formed by 512 points connected by linear paths. Comparison of the measurement with such polygonal path and real circular path that uses circular interpolation commands, showed negligible difference similar to the deviation between two consecutive identical measurements. That's why we also chose a pseudo circular path that consists of multiple points when making measurements on the Fanuc robot.

2.1 Preparation for measurement

The first step before the measurement is to program the circular path of the given radius to apply the measurement to. Due to the above reasons, we have chosen a circular path made of multiple points. For the purpose of the experiment, a CGIR (Circular Generator for Industrial Robot) application, Fig. 2a, was created to calculate the coordinates (X and Y) of the individual points on the circle so that the resulting polygon was the most accurate approximation of the measured theoretical circle. In addition, it was used to generate the program to control the Fanuc robot. The generated program's text output is that of an ".LS" extension. If the robot's control system includes the "ASCI Program Loader (R796)" software extension, then the program can be loaded directly. Otherwise, using Fanuc-RoboGuide off-line robot software, the program in that format can be converted to a ".TP" file and then loaded into the robot's control system. The center of calculated circle with coordinates {X = 0; Y = 0; Z = 0}, at the same time represents the start of the user coordinate system (UF). It is defined with respect to the World Coordinate System (WCS) - Fig. 2b. This allows the same program with a measuring circle to be used in any position within the robot workspace. In our case, we have selected three measurement positions, which are changed by the UF coordinate adjustment. For the first, the UF has the coordinates {X = 500; Y = 0; Z = -179}, for the second {X = 200; Y = 494; Z = -179} and for the third measured position {X = 200; Y = -494; Z = -179}.

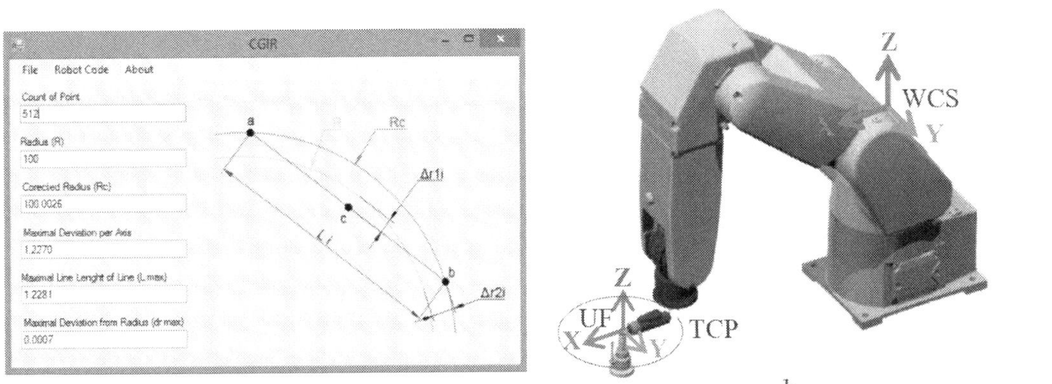

a b

Fig. 2 Application for calculate points on the circle (a) and position of the coordinate systems on the Fanuc LR Mate 200iC robot (b).

For measurement with the Ballbar, a circle with a radius R is required. For the measurement, this circle is replaced by a corrected circle of the radius R_c formed by the n-points. After joining these points with segments, an n - polygon represents the most precise approximation of the desired circle of radius R. Coordinates in the robot control system can be entered up to three decimal places, which is taken into account when calculating X and Y peak coordinates of the said polygon, according to the relations (1) and (2).

$$x_i = \frac{nint(cos(\alpha_i) \cdot R_c \cdot 1000)}{1000} \tag{1}$$

$$y_i = \frac{nint(cos(\alpha_i) \cdot R_c \cdot 1000)}{1000} \tag{2}$$

Where α_i represents the angle between the center flowline of the corrected circle R_c and the counterclockwise X -axis. This angle is determined by the relationship (3).

$$\alpha_i = \frac{\varphi}{n} \cdot i + \varphi \tag{3}$$

Where i ranges from 0 to n, n is the number of points (polygon peaks). φ is the circle's measured angle, that is, if the coordinates of the points on the whole circle are calculated, then $\varphi = 360$.

A numerical method is used to determine the radius of the corrected circle, which, based on geometric conversions, controls the deviation of the polygon from the desired circle using the following formulas.

$$\Delta_{r1i} = \left| R - \frac{\sqrt{(x_i + x_{i+1})^2 + (y_i + y_{i+1})^2}}{2} \right| \tag{4}$$

$$\Delta_{r2i} = \left| R - \sqrt{x_i^2 + y_i^2} \right| \tag{5}$$

$$\Delta_{rmax} = max(\Delta_{r1i} \cup \Delta_{r2i}) \tag{6}$$

From the coordinate points thus determined, the length of the L_i line, joining the two adjacent points on the polygon, is calculated according to the relation (7).

$$L_i = \sqrt{(x_i - x_{i+1})^2 + (y_i - y_{i+1})^2} \tag{7}$$

2.2 Execution of the measurement

For the measurement, the described corrected radius of the R_c circle was used, where the circle itself was made of 512 points with a radius of 100.0026 mm, with the maximum length of the segment L_{max} is 1.2281mm, selected so that the deviation from the actual circle R is as small as 0.0007mm. The corrected radius of the circle described is determined by the calculation so that the resulting polygon is the most accurate approximation of the desired circle. Based on these parameters, two control programs for the Fanuc robot were generated in the application; one for clockwise movement and the other one for the opposite direction. Both of these programs, along with the Magnetic Stand Setup program, were run in automatic mode using remote control. The latter is used in practice to run the robot's automatic cycle. This not only automates the whole process of measurement, but also brings the laboratory tests closer to the real production condition. Thanks to this, we are able to determine the total time needed for measuring - 3 minutes with great precision. This is a very short time that will not significantly affect the manufacturing process of the industrial robot. On the contrary, regular testing is a way of long term monitoring the robot's condition. With sufficient data collected, it is possible to predict a change in the observed characteristics of an industrial robot and thereby plan its maintenance. This may consist of calibration of the controlled robot axes, more detailed measurements of other parameters by other methods (e.g. measurement of the individual joints' play) or other service interventions.

The measurement itself was preceded by a three-hour start-up of the industrial robot, consisting of running circular paths alternating in each of the three measured positions, which reliably mimicked the state in which the robot would be immediately found in the event of production interruption.

3 Measurement results

Performed laboratory tests consisted of two series of measurements. In the first one, 5 repeated measurements were made on each of the three measured positions. One of the findings is the considerable difference between the actual speed of the TCP point of the robot and the speed specified in the control program. In all five measurements, the maximum linear velocity was set for each position, which is 4000mm/s for this robot model. The average measured speed from the five measurements is 77.6mm/s for the first position, 77.5mm/s for the second and 77.9mm/s for the third measured position. The detected velocity differences can be attributed to a very short path between the points of the corrected circle R_c. At the same time, the hypothesis that the stated maximum velocity is only a theoretical speed obtained from the robot's mathematical model and from the maximum angular speeds of the individual robot joints is thereby confirmed. To achieve this high linear velocity, a long enough path would be needed, which, in our opinion, is longer than the robot's workspace allows.

At the same time, this series of measurements suggests a very good repeatability of the given robot, which confirms the results of the measurements mentioned in the article [1] and in the Master's thesis [14], where the repeatability of this industrial robot was measured using the Renishaw XL-80 laser interferometer and the digital indicator. However, the consistency of the Ballbar measurement results needs to be verified by a larger number of measurements.

The second series of measurements was performed on one of the positions (UF = {X = 500; Y = 0; Z = -179}) in order to verify the effect of the CNT parameter on the measurement results. The CNT is a part of the motion command in the robot control program, which defines the size of the programmed bypass point. This parameter can range from 0 to 100. The CNT100 stands for the largest circumvention and vice versa when defining a motion command with CNT0, the TCP point of the robot will stop at individual points. In the first series of five repeated measurements at three different positions, the CNT value was set at 100.

Three measurements were made at CNT100, CNT50 and CNT10, with the speed parameter being the same at all three measurements (4000mm/s). Reduction in the CNT value occurred in two ways. The first is the measurement length or the time required to execute two circular paths in one direction (Fig. 3). The lower the CNT value, the more time was needed to make the one way circular path. At CNT100 it was 17 seconds, at CNT50 100 and CNT10 the path completion in one direction took 226 seconds.

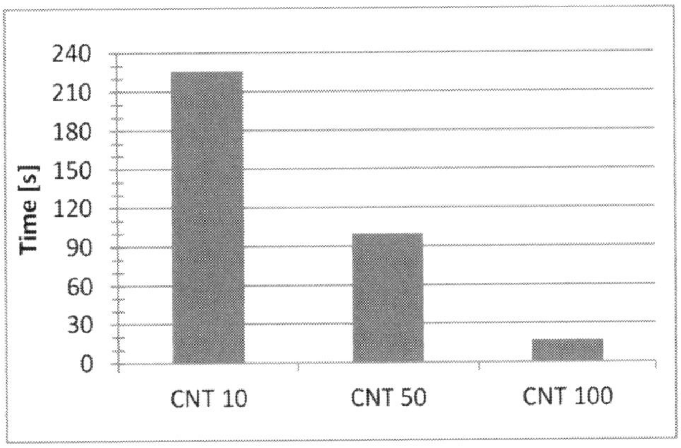

Fig. 3 The effect of the CNT parameter on the time required to execute two circular paths in one direction.

Changes in the CNT value further demonstrated themselves as source of significant vibrations of the robot end piece in course of its running the circular path. They have been expressed as following: the lower the CNT value, the greater the vibrations. The vibrations were evaluated subjectively, at CNT 100, they were not observed and the path run was continuous, but at CNT10, considerable vibrations at the robot's end piece could be observed. In our opinion, this phenomenon is due to the fact that the lower the point circumvention, the closer is the TCP point proximity to the programmed point, thus requiring the reduction of speed and repeated acceleration. With such a large number of points used, the rate of velocity change is very high, which results in the oscillation of the robot's end piece. This phenomenon is also facilitated by the very structure of the robot, which is that of an open kinematic chain.

4 Conclusion

This paper presents the measurement of positioning performance performed on the Fanuc LR Mate 200iC industrial robot using the Renishaw Ballbar QC20-W. A method was developed based on polygonal pseudo circular path, calculated to be the most accurate approximation of the circle required for Ballbar measurement.

Renishaw Ballbar is a popular measuring device for simple and very fast measuring of the working characteristics of CNC machine tools. The proposed measurement method allows this simple and quick measurement to be applied to industrial robot measurements as well. This method can be used in virtually any applications of industrial robots. At the same time, by customizing the control programs of the entire measurement cycle for the automated industrial robot. The article demonstrated the amount of time it takes to interrupt the robot's working cycle in production conditions. The time period of 3 minutes per measured position represents a negligible delay in the production process which can, on the other hand, bring financial savings in terms of industrial robot maintenance as routine testing and monitoring of the robot condition and performance allows precise maintenance planning. However, such planning requires sufficient amount of data collected regularly. This can easily prevents occurrence of collisions, malfunction of the robot, or production of defective products resulting from the deterioration of the working characteristics.

As noted in the introduction, due to high accuracy and the limited measuring range allowed by the Ballbar, it may not be possible for this device to measure the industrial robot characteristics. In our case, none of these problems occurred. Had we gotten close to the Ballbar measurable margin during the measurement, it would have been necessary to adjust the robot's control program so that the measurement could be performed. This can be achieved by subtracting the measured deviations from the radii of the individual programmed points, a suitable method for such a calculation being the decomposition of the measured profile by the Fourier transform and the reverse profile composition of the first 512 harmonics, while the amplitude of the first harmonic component should be adjusted to zero, as it represents the eccentricity, which is considered to be a measurement error and should therefore be excluded.

This article was made under the support of APVV project – APVV-16-0283 Research and development of multi-criteria diagnosis of production machinery and equipment based on the implementation of artificial intelligence methods.

References

1. D. Kumičáková, V. Tlach, M. Císar, *Transactions of the VŠB*, **62**, 39-50 (2016)
2. V. Johnsrud, Improvement of the Positioning Accuracy of Industrial Robots (2014)
3. A. Şirinterlikçi, M. Tiryakioğlu, A. Bird, A. Harris, K. Kweder, *Technology Interface Journal*, **9** (2009)
4. J. Brethé, D. Lefebvre, *International Journal of Factory Automation Robotics and Soft computing*, **2**, 93-101 (2007)
5. M. Slamani, A. Nubiola, I. A. Bonev, *Industrial Robot: An International Journal*, **39**, 57-68 (2012)
6. M. Slamani, I. A. Bonev, *Industrial Robot: An International Journal*. **40**, 441-449 (2013)
7. A. Nubiola, I. A. Bonev, *Robotics and Computer-Integrated Manufacturing*, **29**, 236-245 (2013)
8. W. S. Newman, C. E. Birkhimer, R. J. Horning, A. T. Wilkey, *Robotics and Automation, 2000, ICRA'00, IEEE Int Conference on*. **4**, 3597-3602 (2000)
9. J. Józwik, D. Ostrowski, P. Jarosz, D. Mika, *Advances in Science and Technology Research J.*, **10**, 86-96 (2016)
10. A. Nubiola, M. Slamani, I. A. Bonev, *Precision Engineering*. **37**, 451-460 (2013)
11. M. Slamani, A. Nubiola, I. A. Bonev, *Transactions of the Canadian Society for Mechanical Engineering*, **36**, 83-96 (2012)
12. A. Nubiola, I. A. Bonev, *Precision Engineering*, **38**, 472-480 (2014)
13. *** http://www.renishaw.com/media/pdf/en/bc37e3f237284417baa57889d207cf97.pdf (2013)
14. V. Tlach, Design of workplace for testing the performance characteristics of industrial robot Fanuc LR Mate 200iC, Diploma thesis, University of Žilina (2016)

Verification of static strain and deformation of industrial robot actuator in ANSYS environment

Marek Kočiško[1,*], *Lukáš* Blaško[1], *Petr* Baron[1] and *Dušan* Paulišin[2]

[1] Faculty of Manufacturing Technologies, Technical University of Košice, Štúrova 31, 080 01 Prešov, Slovakia
[2] SPINEA Technologies s.r.o., Volgogradská 13, 080 01 Prešov, Slovakia

Abstract. The article describes the process of experimental analyses in verifying the static strain and deformation of industrial robot actuator. The experiments were carried out in ANSYS analytical system environment. The actuator body behaviour has been verified with respect to its required high accuracy when subjected to radial, axial and torque loads. Due to load, high pressures and consequent partial or total deformation of body, which may result in a deviation in hundreds of millimetres. In case of unsatisfactory results, it is necessary to modify body actuator from structural point of view, either by adding a layer of material to weakened surfaces or by changing the holes layout and functional holes of the actuator. The results were compared with prescribed values of slope of selected material. In first phase of the experiment, forces were applied individually to the body, next one being experimented with their concurrent action.

1 Introduction

Based on the request of partner company Spinea Technologies s.r.o. Prešov which provides the construction of various types of products and technical systems especially in the field of mechatronics and electrical drives, experimental analyses aimed at the static analyses of pressure and deformation in the environment of ANSYS system have been performed at our workplace. The Company Spinea Technologies s.r.o. Prešov provides the development and production of highly precise actuators of the DriveSpin type characterized by outstanding mechanical properties and the use of a compact design AC servo motor. As the only one of its kind in the world, DriveSpin actuators can meet the highest demands of customers in terms of stiffness, transmission precision, transmission of large torques, roll and torsion stiffness, precision and lightness.

The required input parameters for the realization of analysis were provided by the Company Spinea Technologies s.r.o. The analyses focused on an actuator of the 070 series in DSM modification having the special structure of a body which allows for the versatile connection of the whole actuator by means of screws attached to the bottom or the back part when additional equipment is not necessary to attach it.

* Corresponding author: marek.kocisko@tuke.sk

The DSM 070 modular rotary positioning module provide controlled rotary motion and transfer of torque with a high positioning accuracy and precision. The output flange of the module allows capturing both radial and axial forces. Rated output torque is 50 Nm and acceleration/braking output torque is 100 Nm (Table 1). The DSM 070 feature a special design, which allows versatile connections, also without additional devices. Actuators can be combined in many ways using the modular system. The simple design integration ability and small dimensions allow creating kinematic assemblies from DSM modules for end effectors, but also for additional devices and positioners. The selection of a module size depends on the required load carrying capacity and the number of degrees of freedom of the motion axis. The DSM Series is characterized by simple and quick assembly and reduces overall cost. Compact design ensures optimum mounting options and application possibilities, even in confined installation spaces. This actuator is used in applications with request of high torque density, precision and dynamics.

Table 1. The modular actuator DSM is characterized by these technical parameters

DS Actuator			DSM 070
Reduction ratio	I		57, 75
Rated output torque	T_r	Nm	50
Acceleration/braking output torque	T_{max}	Nm	100
Rated input speed of the reduction gear	n_r	min^{-1}	2000
Maximum allowed input speed of the reduction gear	n_{max}	min^{-1}	5000
Tilting stiffness	M_t	Nm/arcmin	35
Torsional stiffness	k_t	Nm/arcmin	7
Maximum lost motion	LM	arcmin	<1.5
Hysteresis	H	arcmin	<1.5
Maximum tilting moment	M_c	Nm	142
Rated radial force	F_{rR}	kN	2.8
Maximum axial force	$F_{a\,max}$	kN	4.1
Allowed temperature range		°C	-10 °C to +40 °C
Reduction gear maximum allowed temperature		°C	65 °C

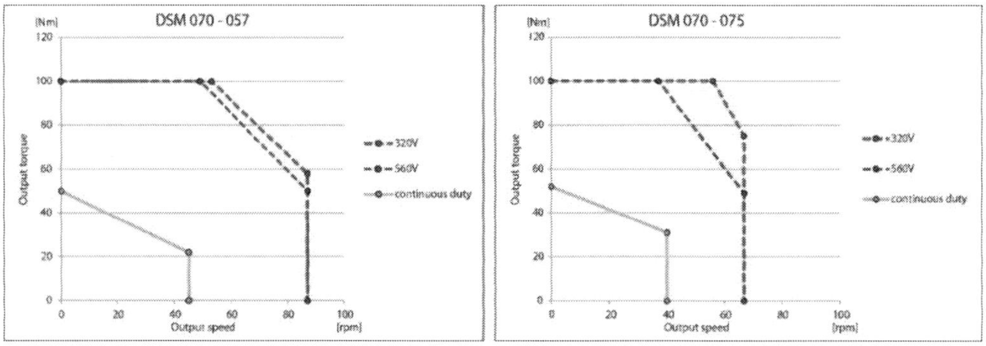

Fig. 1. Performance Characteristics.

Fig. 2. Example of use (connection).

In the following equations, $+i_{out}$ represents input and output rotations in one direction, $-i_{out}$ represents input and output rotations in the opposite direction.

$$i_{out} = \frac{speed_{input}}{speed_{output}}$$

	Input:	Output:	Fixed:
Speed Reduction	**Input:** Input shaft*	**Output:** Output flange	**Fixed:** Case
Speed Acceleration $i_{out} = -i$	**Input:** Output flange	**Output:** Input shaft*	**Fixed:** Case

*Shaft inside the actuator (integrated drive)

Fig. 3. Rotation direction and reduction ratio.

The analyses focused on external stress radial force, axial force and torque. The suitability of the body with respect to its high positioning precision was checked by this experiment. Under stress, high pressure may arise and subsequently the whole or a part of the body may be deformed, which would cause a deviation in hundredths of millimetres. In the event of unsatisfactory results, it is necessary to make structural repairs either by adding a layer of material to the weakened surface or modifying the layout of holes and the holes of the actuator, which would ensure that stress is exerted on the thin surface. The results were compared with the yield strength of the selected material. In the first stage of the experiment, individual forces are applied to the body. In the second stage, the forces are applied simultaneously. By applying all the forces, significant deformation or total deformation along with deviation may occur and be caused by exerting pressure and exceeding the yield strength.

2 The analysis of observed stresses and distortions of DSM 70 actuator body

The analysis of stresses and deformation of the observed actuator body was realized in the environment of ANSYS analytical system. In particular, what was performed is the analysis of maximum stress (von Mises) which is based on the von Mises-Hencky theory known as a shear stress theory or a maximum drawing theory. A safety factor was calculated based on the calculated maximum load bearing capacity and permitted load bearing capacity of the material [2, 5, 6].

$$k = \frac{\sigma_{YieldStress}}{\sigma_{vonMises}} \tag{1}$$

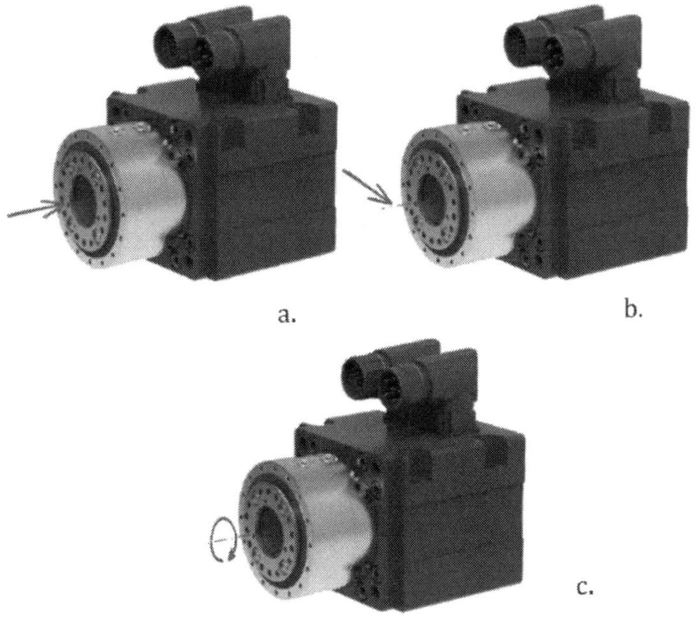

Fig. 4. Effect of axial force (a.) radial force (b.) and torque (c.) on actuator gearbox DSM 070.

The analysis of deformation was performed based on the simulation of total deformation where a resulting maximum value in mm had to be lower than the value of 20 μm (0.02 mm) which was required by the Company Spinea Technologies s.r.o. with respect to the deployment of the DSM module in the manufacturing CNC machines and equipment. The analysis was performed using different means of attaching the body and applying various external forces to the body. Three forces were applied to the body, namely axial force amounting to F = 3.7kN applied to the front part of body, radial force F = 2.6kN and torque Ml = 100Nm (Fig. 4). The given values of forces were simulated using the different means of attaching the body. The first type of analysis assumed the attachment of a back part of body. Subsequently, the body was attached on its bottom (Fig. 5).

Regarding the minimum thickness of walls of the body, which was designed to the value of 4.5 mm, what was chosen is a hollow aluminium alloy profile EN AW-6082 [Al Si1MgMn] with T6 heat treatment whose minimum yield strength Rp0,2 is equal to 250MPa. This value was a starting parameter for comparing the results of the analysis. With respect to its chemical and physical characteristics and advantages over other materials (like steel), an extruded profile made of the above mentioned material was used to analyse stress and deformation of the body of the actuator. It is light, firm, noncorrosive, it may be processed to gain better characteristics and it is used in several industrial fields. At the same time, it is relatively stable, formable and forgeable and it has good tensile strength and good electrical and thermal conductivity [3, 4].

2.1 The stress of the body at a value of 100Nm of torque (attachment on the back of the body)

The body is stressed with torque and it is attached at its back part. The value of stress is shown in Fig. 5 (a). The maximum stress has a value of 9.4991MPa.

The value of permitted stress for the material according to EN AW 6082 is 250MPa. The factor of safety has a value of 26. The comparison of observed and permitted stress (9.4991MPa ≤ 250MPa) confirms that the proposed material is convenient at the maximum torque of 100Nm.

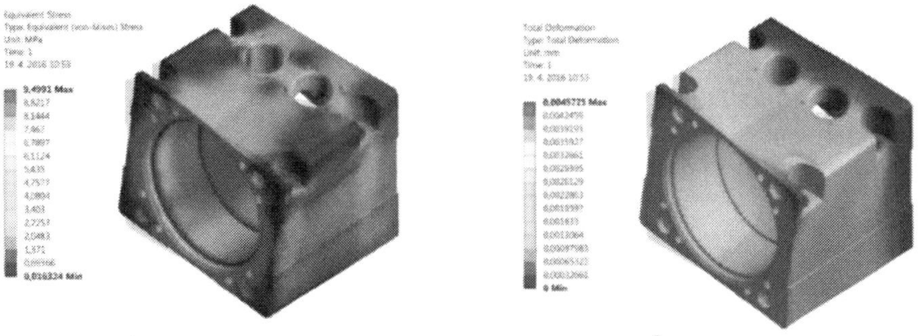

a. b.

Fig. 5. Resulting stress (a.) and deformation (b.) of the actuator body under load exerted by torque (attachment on the back part).

The value of deformation under stress by torque applied to the body (the attachment on the back part) is shown in Fig. 5 (b.). The maximum deformation is 0.0045mm. Based on the assumed use of module in the manufacturing of CNC machines and equipment; we can conclude that the value of deformation is convenient.

2.2 The stress of the body at a value of 100Nm of torque (the attachment on the back of the body)

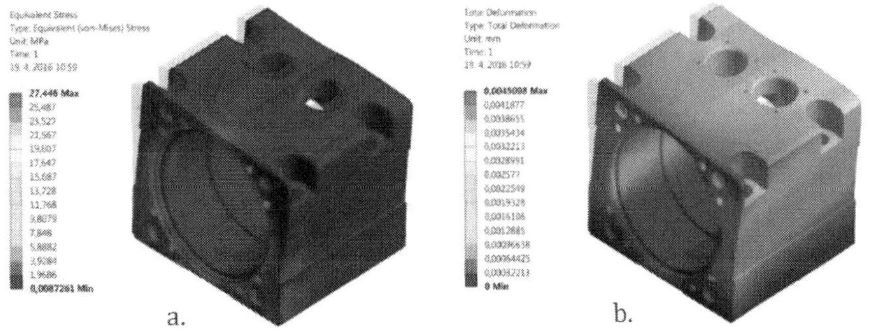

a. b.

Fig. 6. Resulting stress (a.) and deformation (b.) of the actuator body under load exerted by torque (attachment on the bottom part).

In this case, the value of stress is displayed in Fig. 6 (a). The maximum stress has a value of 27.446MPa. In accordance with EN AW 6082, the value of permitted stress for the material is 250MPa. The safety factor has a value of 9. The comparison of identified and permitted stress (27.446MPa ≤ 250MPa) confirms that the proposed material is convenient at the

 DOI: 10.1051/matecconf/201713701005

maximum torque of 100Nm. The value of deformation under load exerted by torque to the body (the attachment on the back part) is shown in Fig. 6 (b.) where the maximum deformation amounts to 0.0045mm. Based on the assumed use of the module in the manufacturing of CNC machines and equipment, the value of deformation is convenient.

2.3 The stress of the body at a value of 3.7kN of axial force (the attachment on the back of the body)

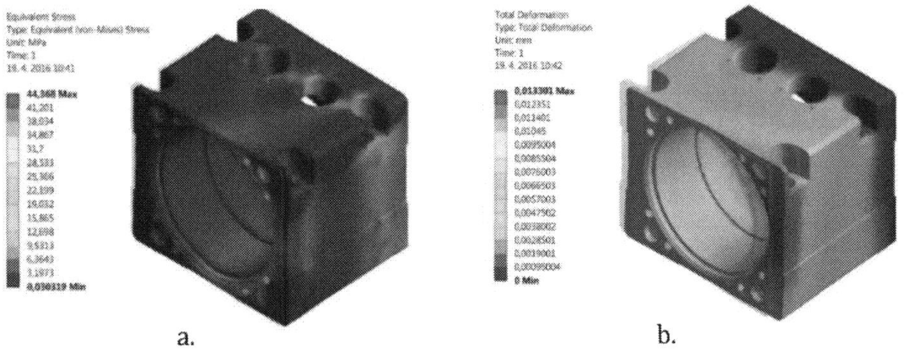

Fig. 7. Resulting stress (a.) and deformation (b.) of the actuator body under load exerted by axial force (attachment on the back part).

The axial force was applied to the body with the attachment at the back (Fig. 7). The value of stress is shown in Fig. 7 (a.). The maximum stress has a value of 44.368MPa. The value of permitted stress for the material according to EN AW 6082 is 250MPa. The factor of safety has a value of 5. The comparison of observed and permitted stress (44.368MPa ≤ 250MPa) confirms that the proposed material is convenient at the maximum axial force of 3.7kN.

The value of deformation under load exerted by axial force to the body (the attachment on the back part) is shown in Fig. 7 (b.). The maximum deformation is 0.0133mm. If the assumed module is put into the manufacturing of CNC machines and equipment, the value of deformation is convenient.

2.4 The stress of the body at a value of 3.7kN of axial force (attachment on the bottom of the body)

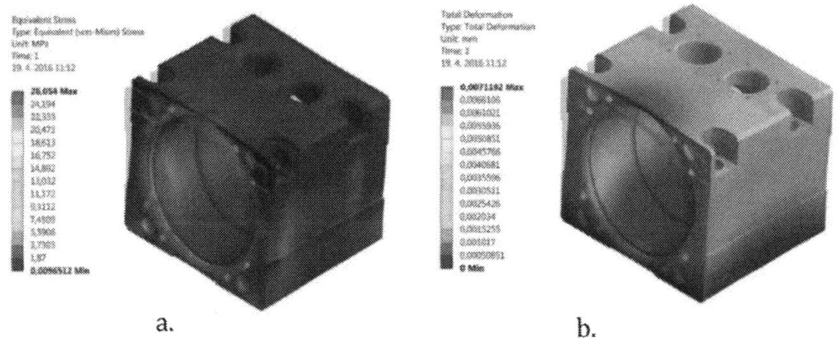

Fig. 8. Resulting stress (a.) and deformation (b.) of the actuator body under load exerted by axial force (attachment on the bottom part).

When axial force was applied to the body with the attachment on the bottom, the value of stress amounted to 26.054MPa. In accordance with EN AW 6082, the value of permitted stress for the material is 250MPa. The safety factor has a value of 9. The comparison of observed and permitted stress (26.054MPa ≤ 250MPa) confirms that the proposed material is convenient at the maximum axial force of 3.7kN.

The value of deformation under load exerted by axial force to the body (attachment on the bottom) is shown in Fig. 8 (b.) and it amounted to 0.0071mm. If the assumed module is put into the manufacturing of CNC machines and equipment, the value of deformation is convenient.

2.5 The stress of the body at a value of 2.6kN of radial force (attachment on the back of the body)

The body of the actuator was subjected to radial force. The body was attached on its back part. The value of stress is shown in Fig. 9 (a.) and maximum stress amounted to 27.175MPa. In accordance with EN AW 6082, the value of permitted stress for the material amounts to 250MPa. The safety factor has a value of 9. The comparison of identified and permitted stress (27.175MPa ≤ 250MPa) confirms that the proposed material is convenient at the maximum radial force of 2.6kN.

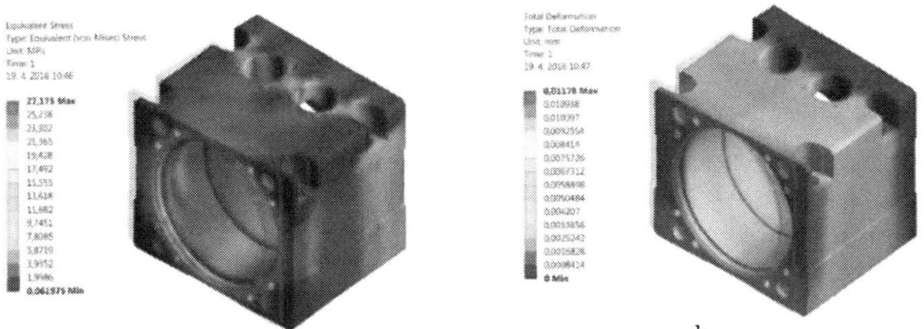

a. b.

Fig. 9. Resulting stress (a.) and deformation (b.) under load exerted by radial force (attachment on the back part).

The value of deformation under load exerted by radial force to the body (attachment on the back) is shown in Fig. 9 (b.) and maximum deformation amounted to 0.0117mm. If the assumed module is put into the manufacturing of CNC machines and equipment, the value of deformation is **convenient**.

2.6 The stress of the body at a value of 2.6kN of radial force (attachment on the bottom of the body)

When radial force was applied to the body and the body was attached on its back, the maximum value of stress amounted to 71.503MPa. In accordance with EN AW 6082, the value of permitted stress for the material amounts to 250MPa. The safety factor has a value of 3. The comparison of identified and permitted stress (71.503MPa ≤ 250MPa) confirms that the proposed material is convenient at the maximum radial force of 2.6kN.

The value of deformation under load exerted by radial force to the body (attachment on the bottom) is shown in Fig. 10 (b.) and maximum deformation amounted to 0.0079mm. If the assumed module is put into the manufacturing of CNC machines and equipment, the value of deformation is convenient.

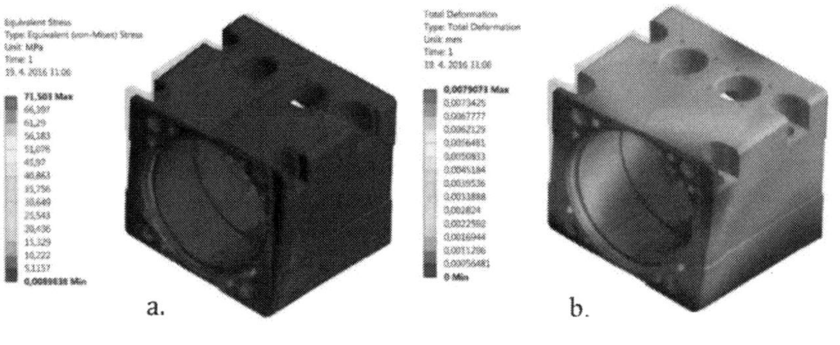

Fig. 10. Resulting stress (a.) and deformation (b.) under load exerted by radial force to the body (attachment on the bottom).

3 Conclusion

The task of the experiment conducted was to predict the behaviour of the body of the actuator under load simultaneously exerted by radial and axial forces along with torque with respect to the selected material of aluminium alloy. During the experiment, the conditions and state were being checked by the analysis in the environment of ANSYS analytical system. The results of experiments demonstrated the structural suitability of the selected material of aluminium alloy [Al Si1MgMn] with obligatory heat treatment in accordance with EN AW-6082. The results were compared with the values of yield strength for the selected material. During potential experiments in future, it would be the best to perform simulation analyses under dynamic load, where more realistic conditions would be created with respect to stress exerted to the body as in practice.

The Ministry of Education, Science, Research and Sport of SR supported this work, contract VEGA No. 1/0492/16, and ITMS project 26220220125.

Agentúra Ministerstva školstva, vedy, výskumu a športu SR pre štrukturálne fondy EÚ

We support research activities in Slovakia/ Project is cofinanced from fonds of EU.
ITMS project code: 26220220125

Európska únia

References

1. M. Staniszewski, S. Legutko, P. Raos, *Tehnicki vjesnik-technical gazette*, **21**,1177-1181, ISSN 1330-3651 (2014)
2. S. Olejarova, J. Dobransky, J. Svetlik, M. Pituk, *Measurement*, **106**, 18-25, ISSN: 0263-2241 (2017)
3. P. Monka, K. Monkova, M. Balara, S. Hloch, J. Rehor, A. Andrej, M. Somsak, Int J of Adv Man Tech, **85**, 2325-2343, ISSN: 0268-3768 (2016)
4. J. Macala, I. Pandova, A. Panda, *Gospodarka surowcami mineralnymi-Mineral Resources Management*, 1, **33**, 125-137, ISSN: 0860-0953 (2017)
5. Š. Salokyová, R. Krehel, *Acta Technica CSAV*, 1, **61**, 45-57, ISSN: 0001-7043 (2016)
6. K. Brezikova, M. Hatala, J. Duplak, D. Mitaľ, S. Radchenko, F. Botko, MM Science J., 1082-1085, ISSN: 18031269 (2016)

System for measurement of interaction forces between wheel and rail for railway vehicles

Ion Manea[1*], *Ioan* Sebesan[2], *Marius* Ene[3], *Mihai Gabriel* Matache[4], *Sorin* Arsene[2]

[1]SC Softronic SRL, R&D Department, Craiova, No.40, Dolj, Romania
[2]Polytechnic University of Bucharest, Sector 6, Splaiul Independentei, No. 313, Romania
[3]Romanian Railway Authority, Calea Grivitei, No. 393, sector 1, Bucharest, Romania
[4]INMA Bucharest, Sector 1, Blv. Ion Ionescu de la Brad, No. 6, Romania

Abstract. Determination of the interaction forces between wheel and rail of railway vehicles is essential for assessment of the vehicle dynamic characteristics from point of view of running safety and rail loading as well as for approval the vehicle and alignment them to the Technical Specification for Interoperability. The direct measurement of transverse and vertical interaction forces using the existing full-disk wheel is practical impossible due to the impossibility of separating the two types of forces. To avoid this impediment it was realized a measuring wheelset fitted with 12 spokes achieved as force transducers for measurement of the vertical forces and 12 spokes achieved as force transducers for measurement of the transverse forces. The measuring wheelset was calibrated as a force transducer and was used to determine the wheel and rail interaction forces for LE-MA 6000kW electric locomotive made by Softronic Craiova. The article presents the measuring wheelset, the calibration principle and the calibration characteristics as well as some time history of the main parameters which characterize the running safety and rail loading, determined in the on-track tests.

1 Introduction

All new or modified railway vehicles shall be certified from the point of view of their dynamic characteristics. The methodology for carrying out the tests to determine the dynamic behaviour and to validate the compliance with international norms in terms of running safety and rail loading is established by norms EN 14363:2005 [1, 2] and UIC 518:2009 [3].

Determination of the running characteristics can be made by tests or by numerical simulation. Determination by tests uses one from the two different types of, complete or partial on-track tests, each of which can be achieved using one of the two different methods, normal measuring method or simplified measuring method.

On-track tests by normal measuring method require assessment of running safety and rail loading of rail vehicle by direct measurement of interaction forces between wheel and rail.

* Corresponding author: ion.manea@softronic.ro

Assessment parameters for running behaviour are either directly measured or derived from other measured parameters. The following parameters generally used for assessment by test of the running behaviour are the following:
• Guiding force, Y, lateral measuring direction;
• Wheel force, Q, vertical measuring direction;
• Sum of guiding forces, ΣY, of a wheelset;
• Quotient, Y/Q, of guiding force/wheel force.

The derived parameters, ΣY, and Y/Q, are the criteria for running safety. They are considered as safety-critical assessment parameters. ΣY is used for the vehicle assessing with regard to the safety against the rail shifting. The quotient Y/Q of the leading wheel is the criterion for safety against the derailment resulting from the climbing of the wheel flange onto the rail.

The guiding force Y and the wheel force Q form the basis for the assessment of lateral and vertical track loading.

On-track tests for assessment of running behaviour are carried out in the four different test zones: in straight track and curves with very large radius, in medium radius curves (600 m $\leq$ R $\leq$ 900 m), in small radius curves (400 m $\leq$ R $\leq$ 600 m), in very small radius curves (250 m $\leq$ R $<$ 400 m).

In order to measure the interaction forces between the wheel and rail, one of the following methods can be used:
• on a standardized railway which correspond in terms of: gauge, cant deficiency, radius curves and running speed, provided with rail sections with built-in force transducers for measurement of vertical and transverse forces;
• using a measuring wheelset equipped with sensors for measurement of the two types of forces, lateral guiding force, Y and vertical wheel force, Q.

Approaching the second measurement method requires the use of a measuring wheelset that very well matches to the wheelset used on the vehicle.

Article presents a measuring wheelset realized for Romanian electric locomotive LE-MA 6000kW, the calibration principle and the calibration characteristics as well as the time history parameters characterizing the wheel and rail interaction forces, determined in the on-track tests, according to the normal measuring method recommended by EN 14363:2005.

2 Measuring wheelset for the measurement of the wheel and rail interaction forces

Measurement of the wheel and rail interaction forces, Y and Q, using a full-disk drive wheels is a very difficult problem because it is practically impossible to delimit areas where can be possible to separate the two types of forces.

To avoid this impediment, a measuring wheelset with spokes was realized to measure the interaction forces with rail of the LE-MA electric locomotive. The spokes of the measuring wheelset were equipped with strain gauges, in full-bridge configuration, being transformed into measuring transducers for Y and Q forces, and were calibrated as measuring transducers as such. Each wheel is equipped with 24 spokes, of which 12 are designed as transducers for transverse forces, Y, and 12 as transducers for vertical forces, Q. The spokes for measurement of Y forces have an elliptical-shapes with the large axis oriented perpendicular to the axis. The spokes for measurement of Q forces are thinner and have a round-shapes.

To select the place for application of the strain gauges on the spokes for transverse forces it was used the ANSYS 15 program.

To increase the sensitivity of the transducers so made, the successive spokes Q were connected in the antiphase and the successive Y spokes were connected in phase. In this way, when running a full wheel length under constant vertical and horizontal loads, the voltage signal supplied by the vertical spokes has an almost sinusoidal shape, and the voltage signal provided by the horizontal spokes has an almost constant shape.

To be used as a transducer for the measurement of independent Y and Q forces, the measuring wheelset was calibrated on a stand allowing successive and independent application of the vertical and transverse forces on each spoke. Loading on the spokes was carried out with a 150kN hydraulic cylinder and the force was measured with an etalon cell.

Figure 1 shows the calibration stand with the wheelset mounted for calibration procedure.

Figure 2 presents a typical calibration chart obtained for Q force applied on spoke 1 of wheel 1. The same calibration charts are obtained for all spokes. It can be seen that the measuring wheelset behaves like a linear transducer.

Fig. 1. Calibration stand with measuring wheelset mounted for the calibration procedure

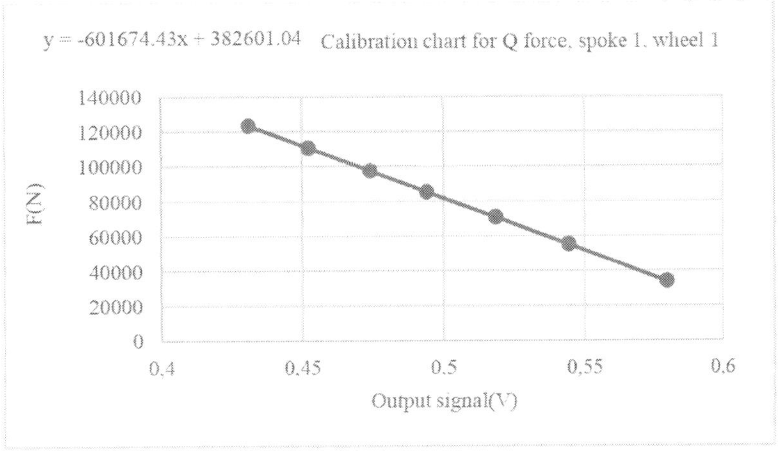

Fig. 2. Calibration chart of measuring wheelset for Q force applied on spoke 1 of wheel 1

The interpolation factors, A and B, were determined for each spoke, in table 1 being represented the interpolation factors for all spokes of the measuring wheelset. From the table it can be seen the need of measurement of angle of the measuring wheelset with the vertical direction for all the moments when the spokes are in the vertical position and they exert the maximum force on the track.

The main characteristics of the measuring wheelset are: gauge: 1 435 mm; diameter: 1250 mm; equivalent conicity: tan γ_e =0.49.

Table 1. Linear interpolation factors, A and B, of the measuring wheelset.

Spoke	Angle	A_{W1_V}	B_{W1_V}	A_{W1_OT}	B_{W1_OT}	A_{W2_V}	B_{W2_V}	A_{W2_OT}	B_{W2_OT}
	(degree)	(kN/V)	(kN)	(kN/V)	(kN)	(kN/V)	(kN)	(kN/V)	(kN)
1	196.9	-601.895	382.649	428.669	234.476	678.128	116.670	440.733	19.485
3	225.7	652.412	-415.721	426.405	233.306	-541.601	-92.720	441.377	19.428
5	255.4	-760.080	483.984	419.512	229.490	635.833	109.542	441.948	19.537
7	284.1	711.484	-454.243	408.377	223.429	-603.775	-104.091	431.795	19.172
9	315.3	-578.033	369.161	402.486	220.075	664.713	114.838	402.063	17.941
11	345	686.146	-440.387	400.337	218.959	-725.561	-126.487	375.194	16.735
13	14.6	-726.165	464.996	400.500	219.062	689.191	120.050	369.188	16.462
15	46.7	651.292	-414.920	400.227	218.971	-704.265	-123.268	381.137	16.954
17	75.7	-695.433	442.849	403.567	220.755	705.218	123.562	404.013	17.968
19	107.1	650.879	-414.818	407.705	223.002	-752.666	-131.244	422.611	18.777
21	137.2	-713.541	453.918	414.363	226.652	662.647	115.550	429.914	19.087
23	165.6	592.151	-377.250	423.745	231.719	-593.637	-102.842	438.076	19.403

3 ON-TRACK determination of wheel and rail interaction forces

Have been carried out complete on-track tests using the normal measuring method for approval of LE-MA 6000kW electric locomotive from point of view of running safety, rail loading and ride characteristics (Figure 3). Assessment from point of view of ride characteristics is not the subject of this article and will not be discussed.

A number of critical problems appear during the measurement of Y and Q forces: the measuring wheelset is in rotation motion; the area is heavily electromagnetically disturbed; the space between the axle of measuring wheelset and the traction motor is very limited and does not allows the positioning of wireless transmission modules, etc. Due to these reasons, it was realized two devices containing each two galvanic isolated modules type MB 38 for strain gauge, in full bridge configuration, powered by supercapacitor battery, mounted on a wheel.

Each module supplies the measuring full bridge strain gauge configuration, mounted on a wheel and it is, in fact, an interface that provides both supplying of strain gauge and protection against electromagnetic disturbances and mechanical shocks.

During the tests, were used the following equipment and software:

• Data acquisition system, including: LAN-XI 3053 B-120, Bruel & Kjaer, 12 input channels, 24 kHz, 24bit, 105789 series and LAN-XI 3160 A-042, series 105628, CE 01.03-346/2015;

Fig. 3. LE-MA 26 equipped with the measuring wheelset during the on-track tests

• Measuring Wheelset type LE: gauge 1 435 mm; diameter: 1250 mm; equivalent conicity: $\tan\gamma_e = 0.49$;
• Locomotive Speed Measuring and Recording Equipment, type IVMS;
• Multifunction calibrator, CALYs75-AOIPsas, series 0306 A11 0994A, CE 0306 A11 0994;
• Piezoelectric accelerometers;
• Microphone (for comments);
• Laptop Dell Precision M6800: 32GB RAM, Processor i7-4930MX;
• PulseLabshop, Bruel & Kjaer, 29B5FCFA series: real-time data acquisition software;
• PulseReflex, Bruel & Kjaer, 29B5FCFA series: post-processing software;
• TestPoint, post-processing software.

During the tests, were determined and calculated the following parameters:
• Guiding force of wheels 1/2: $Y_{1/2}$ (N);
• Vertical wheel forces on 1/2: $Q_{1/2}$ (N);
• Locomotive speed: Speed (km/h);
• Wheel angle: Angle (degrees);
• Accelerations of bogie frame and in the body in vertical and transverse directions;
• Sum of guiding forces, ΣY, on a wheelset;
• Quotient, $Y_{1/2}/Q_{1/2}$, of guiding force / wheel force;

Given the particularities of the Y/Q force measurement system, in the TestPoint programming environment, was developed a program to process the data acquired during the tests in view to determine the above mentioned parameters. The program performs:
• Following the recorded signals, determines the maxima and minima for voltage signals recorded from the vertical spokes, for both wheels R1 and R2;
• For each determined maximum or minimum, read the wheelset angle and determine the spoke who interact with the track. Correspondingly, using the interpolation table 1, determines the parameters A and B for both, vertical and the adjacent horizontal spokes;
• By linear interpolation determines the value of the forces Y and Q;
• Determine the sum of guiding forces, ΣY, on a wheelset and quotients $Y_{1/2}/Q_{1/2}$, of guiding force/wheel force.

Figure 4 shows the equipment used to perform the on-track tests and figure 5 presents a typical time history of the characteristic parameters for running safety and rail loading in successive enrolment in left and right curves on the track of LE-MA 26.

On-track tests recordings are divided into zones of interest and then in sections with lengths imposed by EN 14363-2005, 250m for testing in straight rail and very large curves, 100m for large and small radius curves and 70m for very small radius curves.

From recorded signals are calculated the frequency values y(hj), according to the cumulative curve: y (h1), frequency of cumulative curve h1 = 0,15 %; y(h0), frequency of cumulative curve h0 = 50,0 %; y (h2), frequency of cumulative curve h2 = 99,85 %.

Figure 5 shows the following characteristics:

- Vertical forces on the right wheel, Q1dr (N), black upper trace, ass. with left ordinate;
- Vertical forces on the left wheel, Q2st (N), lt. red upper trace, ass. with left ordinate;
- Sum of guiding forces, ΣY2m (kN), blue trace, associated with left ordinate;
- Quotient, (Y1/Q1), green trace, associated with right ordinate;
- Quotient, (Y2/Q2), lt. magenta trace, associated with right ordinate;
- Position in the current line, Pos(m), red trace, associated with middle ordinate;
- Guiding force of the right wheel, Y1dr (N), black bottom trace, ass. with left ordinate;
- Guiding force of the left wheel, Y2st (N), lt. red bottom trace, ass. with left ordinate;
- Speed (km/h), blue trace, associated with left ordinate.

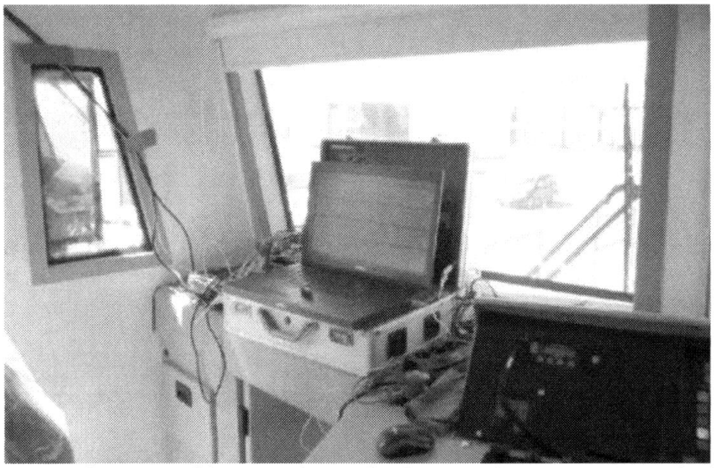

Fig. 4. Measurement equipment used in on-track tests by the normal measurement method

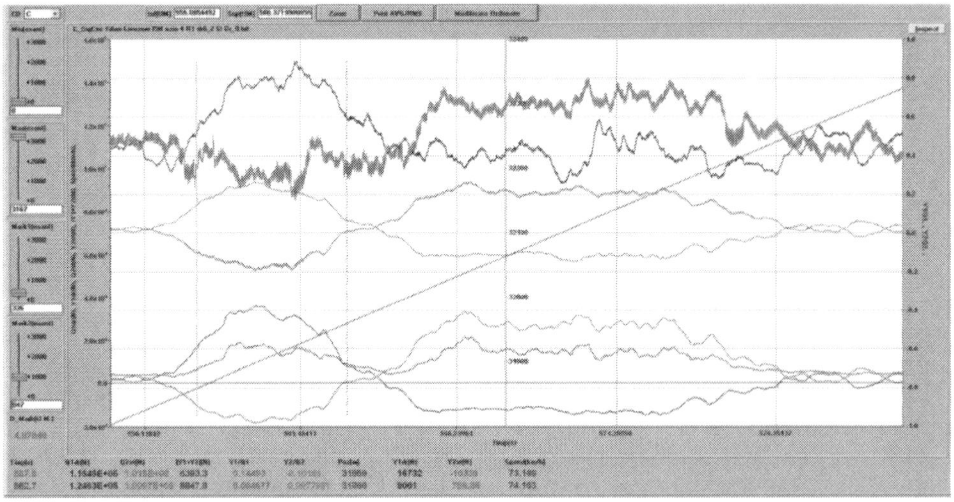

Fig. 5. Time history of characteristic parameters for running safety and rail loading in successive enrolment in left and right curves

Conclusions

The article presents a measuring wheelset realized for determination of the interaction forces between wheel and rail of railway vehicles in view to assessment the running safety and rail loading for electric locomotive LE-MA 6000kW. It is presented the calibration principle and the calibration characteristics as well as the time history of characteristic parameters for wheel-rail interaction forces, determined in the on-track tests, according to EN 14363:2005.

This work was supported by a grant of the Romanian National Authority for Scientific Research and Innovation, CNCS–UEFISCDI, project number PN-III-P2-2.1-PTE-2016-0008, within PNCDI III

References

1. EN 14363:2005 - Railway applications - Testing for the acceptance of running characteristics of railway vehicles. Testing of running behaviour and stationary tests
2. ERA/TD/2012-17/INT rev 3.0 - Running Dynamics. Application of EN 14363:2005 – Modifications and Clarifications
3. UIC 518:2009 - Testing and approval of railway vehicles from the point of view of their dynamic behaviour - Safety - Rail fatigue - Running behaviour

Research Regarding the Influence of Execution, Assembly and Functioning Errors on the Teeth Profile Modification of Spur Gear in Front Plane

Mircea Mera[1,] and *Cristina Stefana* Miron-Borzan[1*]

[1]Technical University of Cluj-Napoca, Manufacturing Engineering Department, 103-105 Muncii Bvd. Cluj-Napoca, Romania

Abstract. The paper presents general aspects related to the influence of execution, assembly and functioning errors on the teeth profile modification from the spur gear in the front plane, in the case when the gear is fixed to the shaft in an asymmetrical position relative to the bearings. The numerical researches were done with FEM.

1 General Considerations

The spur gears have an important place among all the gears used in industry. Until today, many investigations have been presented about the production of the spur gears [1].

Every manufactured gear contains certain types and magnitudes of errors depending on the quality level imposed. Such errors often contribute to the loaded transmission error to affect the meshing dynamics of gears [2].

In the industry, gears have important application areas in the transmission of motion. Many different manufacturing methods have been tried out other than conventional methods [3]. Some researchers [4] manufactured the gear wheel by computerizing the gear tooth profile. Ozek et al. have investigated the possibility of manufacturing spur gears with CNC milling machines by the method of vertical machining [5]. Ozel [6] had investigated the cutting errors in the tooth profile of the spur gears which are manufactured by end mill according to radial cutting method in three-axis CNC milling machine. The results of his study showed that the cutting errors in the involute curve increase according to increments of module, tooth number, pressure angle and cutting angle. The impact of the operating conditions of the gears in the modern era (increasing efforts are being made to reduce the functional dimensions) as well as on the environmental protection (noise reduction and improvement of the medical-social conditions - vibrations) guided the research in the field of gearing to a higher level. The realization of these requirements is possible with the help of the modification of the gear tooth profile, both in the frontal and longitudinal planes. This allow to compensate the elastic deformations of the gear teeth, the execution and mounting errors and to locate the contact patch. Nowadays the design and manufacturing of the gears has as a base the computerised simulation of the meshing, using a Numerical Controlled Machine in manufacturing and for control a 3D Coordinate Measuring System

* Corresponding author: borzan_cristina@ymail.com

(Machine). Numerical simulation is a very effective method for simulation various processes in many domains. A multitude of problems in engineering was solved by Fine Element Analysis [7]. Many authors used numerical simulations [7, 8, 9, 10, 11]. Nistor [8], had investigated the influence of gear geometry such as teeth number and the deformation mechanism by 3D finite-element analysis using FORGE software in terms of teeth forming and forming loads evolution. Based on these simulations, the experimental investigations were carried out to obtain a spur gear form with the good quality, using several billet dimensions. Berri et all [9] have introduced a developed simulator on rotational vibrations of a power transmission spur gear set with one cracked pinion tooth. The simulated outputs depicted precisely the experimental behaviours.

2 Material and Methods

Numerical research was carried out by applying FEM to determine the influence of execution, assembly and functioning errors on the correction depth of tooth gear (flank or correction depth) mounted asymmetrically between bearings. The spur gear is considered to be made of alloy steel, 41Mo Cr10. The cylindrical gears teeth are mainly loaded for bending and contact pressure. The load was made on the peaks of the spur gear teeth to obtain the maximum elastic deformation at bending during operation (deformation which influences the depth value of the gear tooth head modification). The Finite Element Analysis algorithm started from the gear tooth profile that was subjected to profile changes (correction of the head tooth gear). The gear tooth profile was obtained using the MAIN.CPP program. The MAIN.CPP program was developed in [10]. In this software, the 3D meshed model of the gear was made (Figure1), the limit conditions were placed, the forces were loaded and the stresses and the corresponding elastic deformations were determined. The gear model was then exported as DXF files for the ALGOR finite element analysis software. The force loading was made on the peak of the spur gear tooth as linear distributed forces over a band with a certain width due to the contact area between the flanks and they were not concentrated in points. The limitations of Algor's finite element program did not allow another representation. The elastic deformations occurring during the functioning under the load were determined by finite element analysis for each individual gear case. Execution errors, in particular those related to limit deviations for the pinion pitch and for the gear were taken from the Standards, and the assembly and execution errors were taken into account by the way the load was distributed over the tooth width of the gear.

3 Numerical Researches with Finite Element Method

It has been considered a cylindrical spur gear, with z= 30 teeth, module m=4 mm, loaded with 1500 N linear force distributed over the teeth, asymmetrically mounted on a shaft, that sustain onto two bearings.

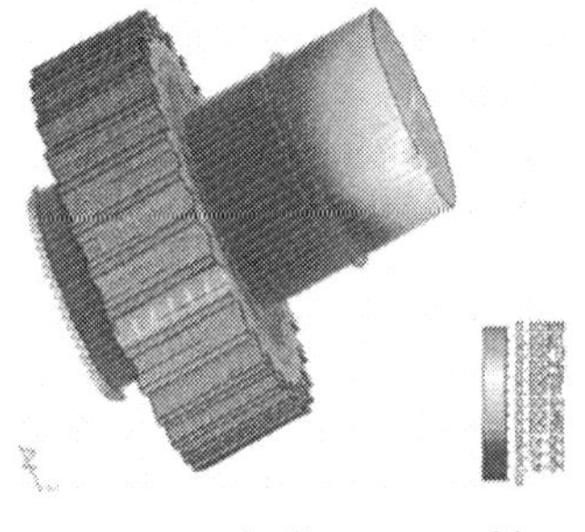

Fig. 1. The elastically deformed gear, asymmetrically mounted between bearings and loaded with linear distributed forces.

There were considered the following situations often encountered in practice classified by the ratio b_{cal}/b, (Figure 2) where: b_{cal} represents the calculation teeth width and b the gear tooth width, in [mm].

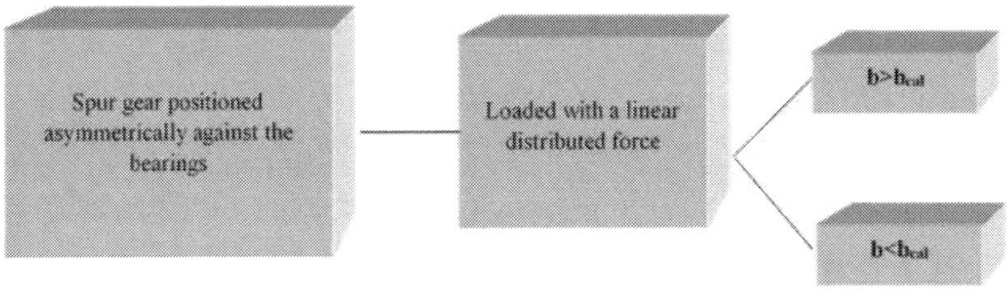

Fig. 2. The analysed algorithm.

In the case of the elastically deformed gear, asymmetrically fixed between bearings and loaded with linearly distributed forces, when $b_{cal} < b$ (the gear is slightly loaded and / or the total deviation of the tooth direction has a high value), the values of the elastic deformations of the gear teeth are shown in Table 1, and the model gear- shaft, elastically deformed in Figure 3.

Table 1. The values for the elastic deformation of the gear tooth in the case of loading with linearly distributed forces, $b_{cal} < b$.

Coordinates of points in the longitudinal plane of the tooth, [mm]	Elastic deformation of the gear according to the applied force direction, [µm]
0	5,606
6,66	4,671
13,32	4,045
19,98	3,296
26.24	2,567
33,3	1,922
40	1,49

Fig. 3. The elastically deformed gear, asymmetrically fixed between bearings and loaded with linearly distributed forces, $b_{cal} < b$.

Analysing the values from Table 1, it can be observed that for the case of asymmetric position of the gear to the bearings, the maximum value was 5,606 µm, on the end portion of the gear from the side where is applied the force that creates the torque moment.

The correction depth of the profile (Δa), determined with relation 1 [10], is 22,6 µm.

$$\Delta_a = f_{\Sigma_{P_b}} + \delta_e \quad (1)$$

where: • δ_e. represents the size of the elastic deformations of the gear teeth under load;
• $f_{\Sigma pb}$ is the mean square deviation of the limit deviations of the pitches of the pinion and gear; it is calculated with:

$$f_{\Sigma_{pb}} = \sqrt{f_{p_{b1}}^2 + f_{p_{b2}}^2} \quad (2)$$

Table 2. The values for the elastic deformation of the gear tooth in the case of loading with linearly distributed loads, $b_{cal} > b$.

Coordinates of points in the longitudinal plane of the tooth, [mm]	Elastic deformation of the gear according to the applied force direction, [µm]
0	4,411
6,66	4,243
13,32	3,786
19,98	3,234
26.24	2,714
33,3	2,253
40	2,07

For the $b_{cal}> b$ case encountered when the gear is heavily loaded and/or the total deviation of the tooth direction has a low value, the gear-shaft model (designed for the asymmetric position of the gear relative to the bearings) is shown in a elastic deformed state, in Figure 4.

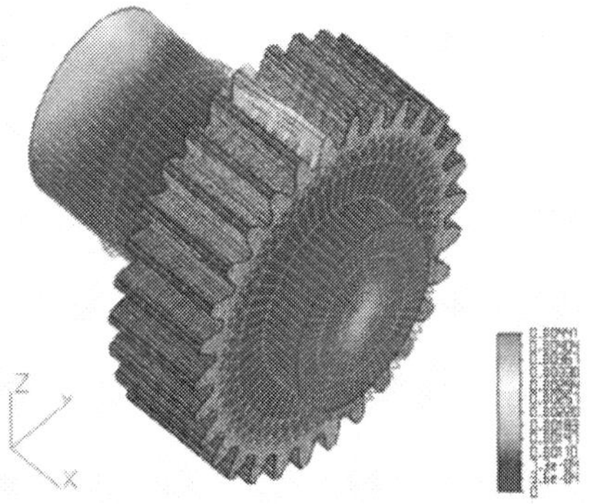

Fig. 4. The elastically deformed gear, asymmetrically fixed between bearings and loaded with linearly distributed forces, $b_{cal} >b$.

The maximum tooth deflection was observed on the end portion of the gear' tooth, from the area where the torque force was applied and it had the value of 4,411 µm. Thus, the recommended depth of head profile correction for this variant is 21,4 µm, calculated with relation (1).

Analysing these values, it results that the elastic deformations for the gears teeth in case of loading with linearly distributed forces are lower for $b_{cal}>b$. It means that the elastic deformations during operating time, in the case of the direction errors of the gear tooth (due to the executions or mounting errors), are lower and/ or the gear is strongly loaded.

3.1 The Load Distribution Calculation on the Tooth Width

The load distribution on the tooth width, when the force has a linear distribution is illustrated in figure 5 and the required formulas for calculate the distribution are presented in the following (3) – (9) relations.

The force linear evolution of the *b* width of the gear tooth can be described as a linear function, such as:

$$p(x) = mx + n \tag{3}$$

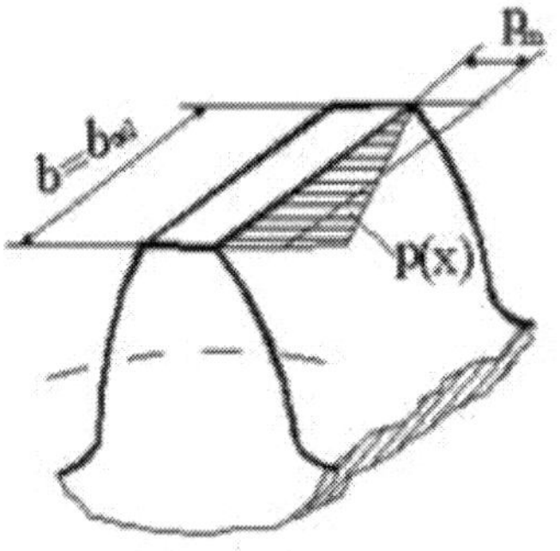

Fig. 5. The forces linear distribution, for the case $b=b_{cal}$.

By imposing the limit conditions, it results:

$$p(0) = p_{max} \Rightarrow n = p_{max}$$
$$p(b) = 0 \Rightarrow mb + n = 0 \Rightarrow m = -\frac{p_{max}}{b} \tag{4}$$

and equation (3) becomes:

$$p(x) = p_{max}(1 - \frac{x}{b}) \tag{5}$$

The F force, which is applied on the tooth, can be calculated by:

$$F = \int_0^b p(x)dx = \int_0^b p_{max}(1 - \frac{x}{b})dx = p_{max}\frac{b}{2} \tag{6}$$

On the b width of the tooth, a number (n) of knocks is established, so (n-1) subintervals are obtained. The force corresponding to each subinterval can be calculated:

$$F_{int} = \frac{2F}{b}(x_1 - x_0)(1 - \frac{x_1 + x_0}{2b}) \tag{7}$$

The following recurrence formula was established:

$$F_k = \frac{F\{2(n-1) - [k + (k-1)]\}}{(n-1)^2} \tag{8}$$

where k represents the subinterval numbers and F_k is the linear variable force on k subinterval.

If is considered the case when $b>b_{cal}$ or $b<b_{cal}$ respectively, the linear variable force on every subinterval is calculated as follows:

$$F_k = \frac{F\{2i - [k + (k-1)]\}}{i^2} \tag{9}$$

where: $b_{cal} = ib/(n-1)$ and *i* represents the number of subintervals on the tooth width.

The table 3 presents the linear distributed forces, calculated with the above presented formulas, applied into the knocks of the finite element network, within the FEM analyses performed for the considered cases.

Table 3. The values for the elastic deformation of the gear tooth in the case of loading with linearly distributed loads, $b_{cal} > b$.

Node number in the finite elements network in which force is applied	The values of the linear force [N], distributed in the network nodes for the cases:		
	$b = b_{cal}$	$b > b_{cal}$	$b < b_{cal}$
	The number of the calculation subinterval, i		
	$i = 6$	$i = 5$	$i = 7$
1	229,165	270	199
2	416,665	480	367,34
3	333,33	360	306,12
4	249,995	240	244,9
5	166,665	120	183
6	83,33	30	122,45
7	20,83		61,22
8			15,3

4 Conclusions

The paper presents a method of determining the value of the modification parameters of the teeth profile for a cylindrical spur gear in the frontal plane. Finite element analysis was used to determine the elastic deformations during operating. The spur gear was considered to be asymmetrically positioned between the bearings, a case commonly encountered in practice. Execution and assembly errors have been highlighted by the force loading distribution on the teeth width of the spur gear (distributed linear forces).

Future researches will be orientated to perform studies which should lead to establish the following things: the influence of the tooth width load distribution on parameters value for gear teeth profile modification; the influence of the gear with respect to the roller bearings, on parameters value for gear teeth profile modification.

This work was supported by a grant of the Romanian National Authority for Scientific Research and Innovation, CNCS/CCCDI–UEFISCDI, project number PN-III-P2-2.1-BG-2016-0216, within PNCDI III.

References

1. C. Özel, Int J. Adv Manuf Tech, **54**, 1–4 (2011)
2. M.Inalpolat, M. Handschuh, A. Kahraman, Int. Gear Conf. Lyon, Gear Techn, 64 (2015)
3. M. Uzun, Adv Manuf Tech, **77**, 5–8 (2015)
4. C.B. Tsay, W.Y. Liu, C. Yi-Chen, J Mater Process Technol **104**, 271 (2000)
5. C. Ozek, C. Ozel, J Mach Eng, **512**, 43 (2002)
6. C. Özel, Int J. Adv Manuf Tech, **59**, 1–4, (2012)
7. P. Bella, P.Buček, M. Ridzoň, M. Mojžiš, L. Parilák, Key Eng. Mater., **716** 708 (2016)
8. L. Nistor, A. Neag, I.Marian, D.Frunza, J. Manuf. Sci. Eng, **138**, 12 (2016)
9. S.M. Berri, J. M. Klosner, Education ASME Turbo Expo 2000: Power for Land, Sea, and Air, Germany, **4**, (2000)
10. M. Mera, *Research on the development of high-performance cylindrical gears*, UTCN (2001)
11. V. A.Ceclan, N.Bâlc, A. V. Miron, C. Borzan, A. Popan, AJME, **9,** 3 (2011)
12. Weck, M. Numerisch gesteuertes Abrichten von Profilschleifscheiben, Industrie Anzeiger, **54**, 12 (1981)
13. M. Mera, C. Gyenge, S. Comşa, O. Mera, *6th*. Int. DAAAM Symposium, Krakow, 227 (1995)
14. M. Mera, C. Gyenge, S. Comşa, Int. DAAAM Symposium Cluj-Napoca (1998)
15. Suhoenkov N. *Modification of the evolving cylindrical spur gears*, (Tehnica, Kiev, 1992)

Decreasing of the manufacturing time for a thermoforming mold by applying the DFM principles

Ancuța Păcurar[1,*], *Răzvan* Păcurar[1], *Beáta* Erőss[1], *Florin* Popister[2], and *Calin Ciprian* Otel[3]

[1]Technical University of Cluj-Napoca, Department of Manufacturing Engineering, Romania
[2]Technical University of Cluj-Napoca, Department of Design and Robotics Engineering, Romania
[3]Technical University of Cluj-Napoca, Department of Industrial Engineering and Management

Abstract. In this paper it is analyzed a product machined at the S.C. ULMA PACKAGING S.A. company, which is a "Thermoforming mold" used in order to obtain plastic containers in which the food or non-food product is packed, making part of a thermoforming machine called TFS 200. The aims of this paper are to determine the optimal technological parameters and to study the effects of the DFM principles and the optimal tool path strategy usage on manufacturing time of the "Thermoforming mold". A redesign of the thermoforming mold is presented based on the failed rules and recommendations given by the DFM program and followed by the analysis of the DFM's benefic effect on the manufacturing time.

1 Introduction

Winning in today's global marketplace requires more than just creating innovative product designs. In order to succeed, the researchers must design products in a right manner from the first time for optimal manufacturing, cost, quality, time, and functionality. DFM (Design for Manufacturing) is a methodology for product development or product improvement projects in which designers and manufacturing engineers work together instead of working separately. The two groups design the product's manufacturing and assembling processes at the same time they design the product itself. As a result of their teamwork and focus on DFM axioms, they create a product that is well-designed, cost-efficient, and easy to produce and maintain [1, 2]. There are many CAD - integrated design for manufacturing programs which help to identify and correct downstream issues early in the design stage, leading to the reduction of the cycle time and, in turn, resulting in high-quality products with lower product development costs. One of these programs is called DFMXpress. In the following, a design for manufacturing analysis will be presented how it was realized by using DFMXpress in order to identify and eliminate the critical design issues of the thermoforming mold, that increase the manufacturing complexity and cost.

* Corresponding author: ancuta.costea@tcm.utcluj.ro

2 Design for manufacturing analysis of the thermoforming mold by using DFMXpres

In this paper it is studied a product manufactured by the Ulma Packaging Company which is called "Thermoforming mold", presented in Figure 1. This thermoforming mold is used in order to obtain plastic containers in which the food or non-food product is packed, Figure 2. The thermoforming mold is part of a thermoforming machine TFS 200, presented in Figure 3, designed to pack food or non-food products.

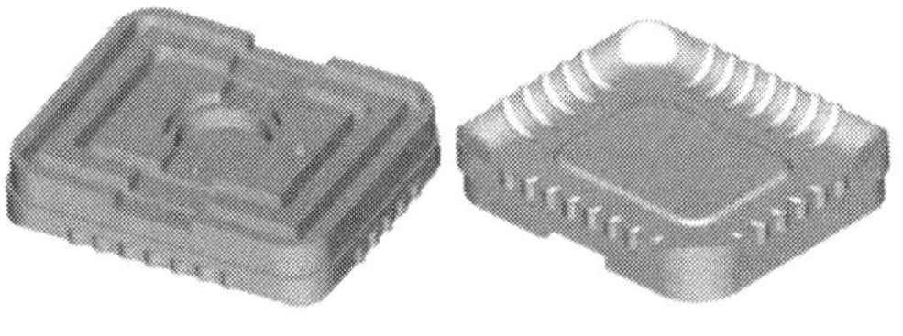

Fig. 1. Analyzed product "thermoforming mold".

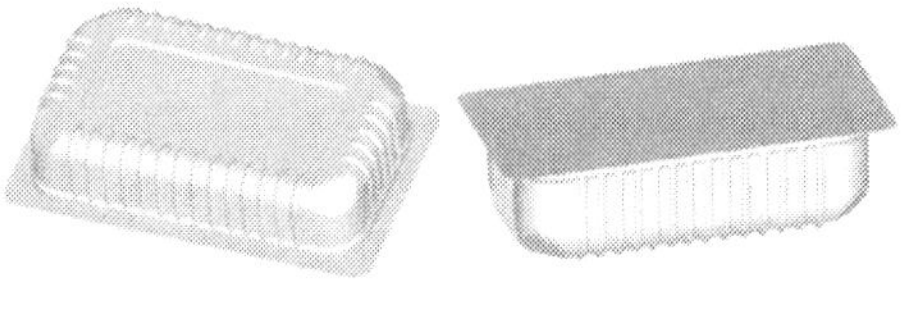

Fig. 2. The obtained plastic food-container after thermoforming process.

Fig. 3. TFS 200 – Thermoforming machine.

An important step of the designing process is the material selection. In this case, the part is made of aluminum alloy EN-AW-5083 H111 which is known for its exceptional performance in extreme environments because it is highly resistant to the chemical attack that might occur in industrial chemical environments. It is very important to take into consideration the fact that the presented part needs to be coated by industrial Teflon in order to eliminate the possibility of sticking of the foil on the preheated thermoforming mold.

First of all, in order to determine the manufacturing time of the "Thermoforming mold", it was necessary to realize the CNC program required for this issue. In this case, EdgeCam 2014 R2 was used to generate the necessary tool paths in order to obtain the final part [3, 4]. Two different tool path strategies were used on specific geometries of the considered part (Figure 4), the aim of this being the determination of the adequate tool path which will guarantee lower machining time.

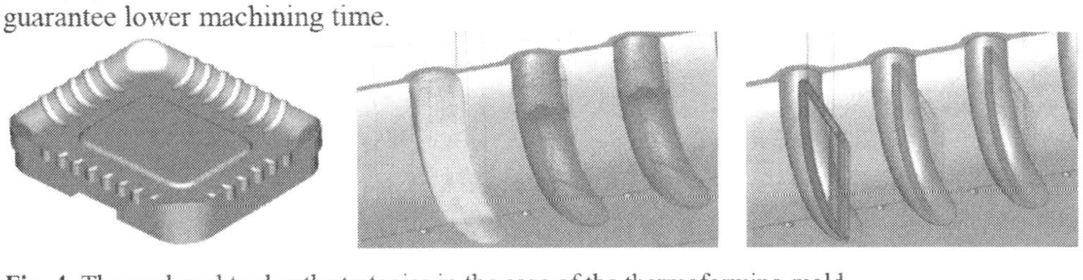

Fig. 4. The analyzed tool path strategies in the case of the thermoforming mold.

Secondly, the thermoforming mold was analyzed from the design for manufacturing point of view using DFMXpress software. After obtaining a redesigned part based on the recommendations of the DFMXpress software, the machining method of the part was changed using less special tools and more tools with higher diameter. These modifications had a significant influence on the machining time.

DFMXpress is a rules-based Solidworks add-in that allows the designer to compare its files against known, good design for manufacturing rules. Within Solidworks, the DFMXpress add-in was opened, the Rule File was selected and the desired values for DFMXpress check were set. For Manufacturing Process the "Mill/Drill only" option was chosen, taking into consideration the fact that the thermoforming mold is a CNC milled part. DFMXpress functionality includes drilled hole checks (hole diameter-to-depth ratio, flat bottoms on holes, perpendicular entry surfaces, holes that intersect internal cavities, partial holes, tolerance checks), milled feature checks (deep pockets and slots, inaccessible features, sharp internal corners, fillets on outside edges) and standard hole size checks (checks hole and available tooling). Therefore, the following DFM rules were verified: Hole Depth/Diameter Ratio, Inaccessible Features, Mill Sharp Internal Corners, Deep Pocket/Slot, Holes with Flat Bottom, Fillets on Outside Edges, Standard Hole Sizes, Partial Hole Rule, Hole Entry/Exit Surface and Hole Intersects Cavity. After running the DFM analysis with the predefined values for the rule parameters, the results can be assessed (Figure 5). It can be observed that 4 rules passed and 6 rules failed. The Partial Hole Rule is respected by the design of the thermoforming mold, which means that the axis of the holes does not intersect any edge of the part or at least 3/4th (75%) of the holes area is within the material. The fact that the Hole Entry/Exit Surface Rule passed means that the entrance and exit surfaces of all the drilled holes are perpendicular to the holes' axes.

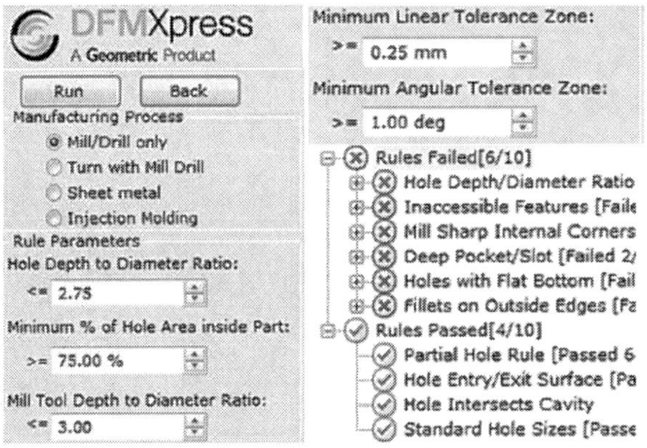

Fig. 5. The obtained DFM results (working with predefined values for Rule Parameters).

It is very important to carefully analyze the failed rules, which represent a suggestion in order to obtain a 3D model which respects the basic DFM axioms. The Hole Depth/Diameter Ratio rule failed due to the fact that the predefined value of 2.75 for this rule parameter was taken into consideration during the DFM analysis. This rule is not respected by the $\phi 1$ mm through holes with the depth of 5 mm (Figure 6), because the Hole Depth to Diameter Ratio in this case is 5 whereas the maximum recommended ratio is 2.75. Although deep holes are difficult to machine and the chip removal becomes difficult to be removed while drilling these types of holes. These through holes are essential due to the fact that these holes guarantee the adherence of the plastic foil to the bottom of the thermoforming mold in which the plastic container gets its shape. The only solution could be the increase of the holes diameter, which is impossible due to the fact that a higher diameter of these holes would leave marks on the surface of the obtained plastic container which cannot be accepted from the esthetical point of view. By taking into consideration the fact that an HSS micro drill with the diameter of $\phi 1$ mm and usable length of 6.8 mm can be used in order to drill these through holes, this rule can be accepted without any modification regarding to the design of the thermoforming mold.

Inaccessible Feature rule failed due to the marked R1 radiuses, presented in the Figure 7. The DFMXpress program considers that the marked features are either not easily accessible for machining (the tool holder assembly is not able to approach the feature and a long slender cutter is required which is not desirable) or require special cutter for machining.

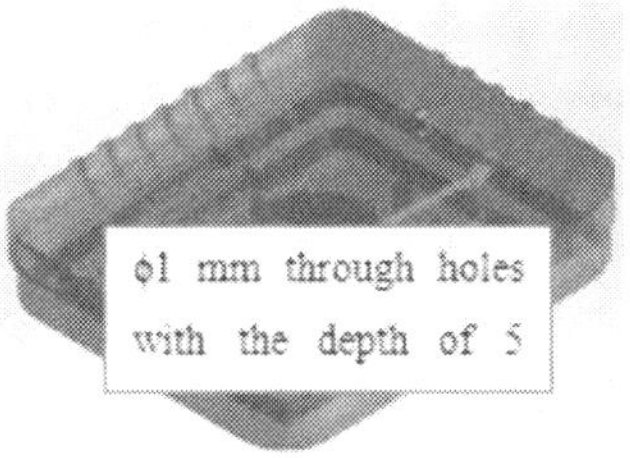

Fig. 6. Failed Hole Depth/Diameter Ratio rule.

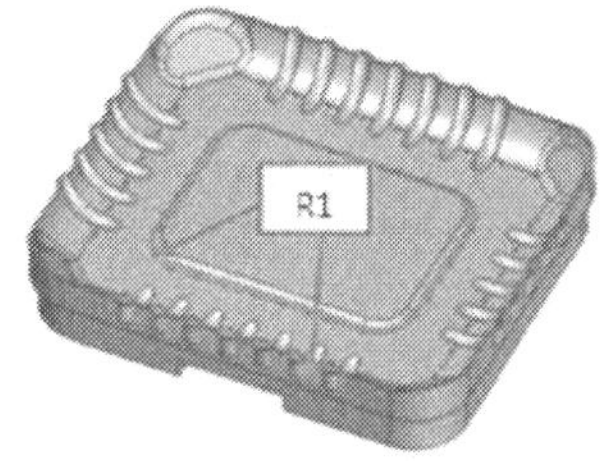

Fig. 7. Failed Inaccessible Feature rule.

3 Analyzing the effect of the DFM analysis on to the manufacturing time

Using the DFMXpress program, a redesign of the thermoforming mold was made based on the failed rules and recommendations given by the DFM program. The R1 radiuses, at the outside and at the bottom of the thermoforming mold, were changed, taking into consideration the fact that the Inaccessible Feature rule failed due to these features (Figure 7). The R1 radius at the bottom of the thermoforming mold was changed into R1.5 radius, therefore the R2 radius of the middle island and the considered R1.5 radius at the part's bottom can be easily machined by using, for both of them, the solid carbide torus cutter with the diameter of ø10 mm and radius of R1.5 mm. This tool will leave a R1.5 radius at the bottom of the thermoforming mold, eliminating therefore the need to use another special tool, the solid carbide torus cutter with the diameter of ø4 mm and radius of R1 mm. Therefore, instead of eliminating the R1 radius from the bottom of the thermoforming mold, it was simpler to modify by increasing it to R1.5 mm, (Figure 8).

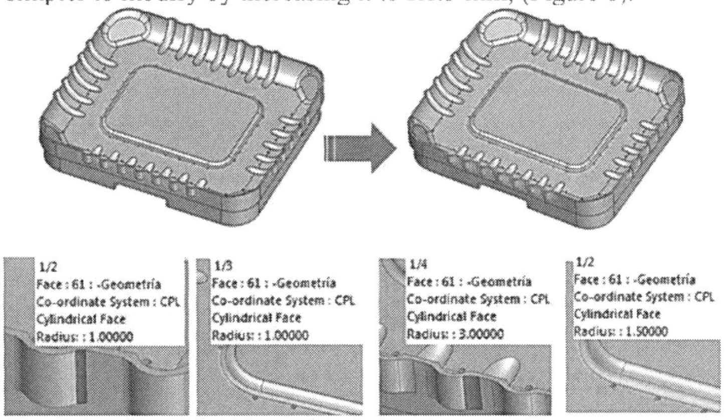

a) b)

Fig. 8. The redesign of the thermoforming mold based on the recommendations of the DFMXpress program: (a) original design; (b) new design.

Initially, these R1 radiuses were machined by using a solid carbide end-mill with the diameter of ϕ2 mm, which can be considered an unproductive machining tool due to its small diameter and low feed rate and cut increment. By increasing the size of these exterior R1 radiuses, it will be easier to be machined with a larger cutter using more productive technological parameters. By taking into consideration the fact that the outside of the

DOI: 10.1051/matecconf/201713701008

thermoforming mold was finished with the ϕ6 solid carbide end-mill, the increase of the R1 value to R3 would eliminate the necessity to use a rest-profiling operation, eliminating therefore the use of the solid carbide end-mill with the diameter of ϕ2 mm at the outside of the thermoforming mold. Due to these modifications, the computer aided manufacturing and the CNC program need to be changed. First of all, the exterior profiling cycle of the R1 radius at the bottom of the thermoforming mold, which initially was realized with the solid carbide torus cutter ø4 R1, was eliminated, due to the fact that the R1 radius was modified into R1.5 one. Therefore, the modified radius will be machined during the interior finishing cycle of the thermoforming mold's cavity, realized by the solid carbide torus cutter ø10 R1.5 [5]. This modification has a negligible influence on to the total machining time, as the whole profiling cycle would take only 34 seconds (Figure 9), but still influences the real-time machining, as the CNC operator does not need to prepare, to measure and to set another special tool, the solid carbide torus cutter ø4 R1.

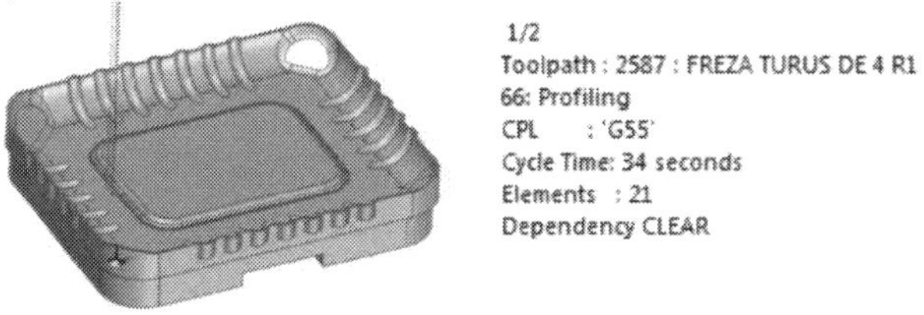

Fig. 9. The solid carbide torus cutter's(ø4 R1) tool path and the machining time for the profiling operation of the R1 radius.

Secondly, the exterior rest-profiling cycle of the R1 radiuses at the outside of the thermoforming mold, which were machined with the solid carbide end-mill ø2, was eliminated, due to the fact that the R1 radiuses were modified into R3 ones [5]. Due to this modification, the R3 radiuses were able to be machined during the exterior profiling cycle, using the solid carbide end-mill with the diameter of ϕ6 mm without changing the machining time regarding to this cutting tool. This modification has a considerable influence on to the total machining time, as the whole rest-profiling cycle by using the ø2solid carbide end-mill would take 32 minutes and 7 seconds (Figure 10). Therefore, by eliminating this rest-profiling cycle, the machining time would be decreased with approximately 32 minutes.

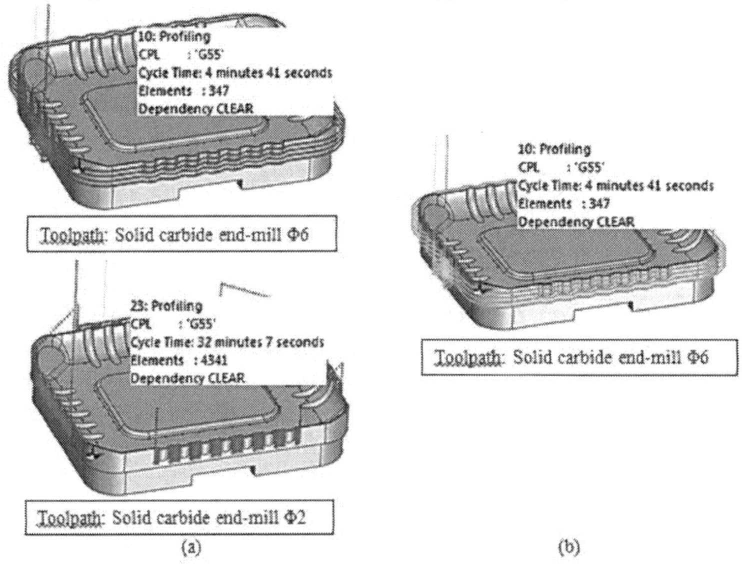

Fig. 10. Machining time for the: (a) R1 radiuses using a profiling and a rest-profiling cycle (original design); (b) R3 radiuses using only a profiling cyle (new design).

As it is presented in the Table 1, it was possible to eliminate the critical design issues and therefore to reduce the machining complexity and costs. Therefore, by modifying the R1 radiuses into R3 ones and the R1 radius into R1.5 one, the manufacturing time was decreased with approximately 33 minutes.

Table 1. The influence of the thermoforming mold's new design (based on DFMXpress recommendations) on to the machining time.

Used cycle		Profiling Cycle	Rest - Profiling Cycle	Profiling Cycle	Total time
Used cutting tool		Φ6solid carbide end-mill	Φ2solid carbide end-mill	Φ4 R1solid carbide torus-cutter	
Original design	R1 – at the outside of the thermoforming mold	4 min 41 s	32 min 7 s	-	37 min 22 s
	R1 – at the bottom of the thermoforming mold	-	-	34 s	
New design	R3 – at the outside of the thermoforming mold	4 min 41 s	-	-	4 min 41 s
	R1.5 – at the bottom of the thermoforming mold	-	-	-	

4 Conclusions

The research presented in this paper was mostly aimed to decrease the manufacturing time of the thermoforming mold by applying optimal tool path strategies and basic DFM principles. A design for manufacturing analysis was realized using DFMXpress in order to identify and eliminate the critical design issues of the thermoforming mold that increase the manufacturing complexity and cost. After running the DFM analysis, the results were assessed, followed by a redesign of the thermoforming mold based on the failed rules and recommendations given by the DFM program. After having a modified design for the thermoforming mold, the benefit effect of the DFM analysis on to the manufacturing time was analyzed. The overall machining time was reduced with approximately 33 minutes for one part, by modifying the R1 radiuses into R3 ones at the outside of the thermoforming mold and the R1 radius into R1.5 one at the bottom of the considered part. Taking into consideration the fact that the required quantity of this thermoforming mold is 25 parts and the manufacturing cost for one hour is equal to 28 euros, it means that the machining time was reduced with 825 minutes (approximately 14 hours) for the whole quantity of parts required to be manufactured. The manufacturing costs were also reduced with 392 euros in total, in this case.

This research was supported by the HORIZON 2020 AMaTUC project (GA 691787) financed from EU.

References

1. D. M. Anderson, Design for Manufacturability; *How to Use Concurrent Engineering to Rapidly Develop Low-Cost,* High-Quality Products for Lean Production, CRC Press, New York, (2014)
2. A. Hoque, P. Halder, M. Parvez, T. Szecsi, J. Computers & Industrial Engineering, **66**, 988 (2013)
3. P. Romero, R. Dorado, F. Diaz, E. Rubio, J. Procedia Engineering, **63**, 523 (2013)
4. C. Zezhong , F. Qiang, J. Computer Aided Design, **43**, 651 (2011)
5. H. Hoffmann Group, *Machining/Clamping technology, Main Catalogue* (2016/2017)

MATEC Web of Conferences 137, 01009 (2017) DOI: 10.1051/matecconf/201713701009
MTeM - AMaTUC 2017

Investigation on standoff distance influence on kerf characteristics in abrasive water jet cutting of composite materials

Ioan Alexandru POPAN[1,*], *Glad* CONTIU[1], and *Ian* CAMPBELL[2]

[1]Technical University of Cluj-Napoca, B-dul. Muncii, 400641, No. 103-105, Cluj-Napoca, Romania
[2]Loughborough University, Leicestershire, LE11 3TU, Loughborough, UK

Abstract. The paper presents an experimental study on processing composite materials using abrasive water jet cutting (AWJC) process, analysing one of the main process parameter, standoff distance (SOD). Carbon Fibre Reinforced Plastics (CFRP) are used in a wide range of industrial applications, like aerospace and automotive. Cutting CFRP using AWJC involves several challenges like high dimensional accuracy and good surface quality. To produce precise parts by using this process must be understand the influence of the process parameters on quality characteristics. On this study was investigated the standoff distance influence on kerf characteristics. The characteristics of the cut surface (kerf geometry, surface roughness and topography) were analyzed.

1 Introduction

Composite materials like Carbon Fibre Reinforced Plastics (CFRP) have wide applications in aerospace, automotive, marine goods, medical fields etc [1, 2]. The main characteristics of CFRPs are: high strength to weight ratio, high modulus, high fracture toughness, corrosion and thermal resistance [3, 4].

Water jet processing uses pressurized water, which acts either by itself at a very high pressure, or in connection with the erosion of the abrasive particles, added into the water jet [5]. Abrasive water jet is used in various machining operations, like cutting [6], milling [7], drilling [8] and turning [9]. This innovative processing solution can be applied in processing for a wide range of materials like aluminium [10], steel [11], stainless-steel [12], composites [13], titanium alloy or even ceramics and diamond [14].

Abrasive water jet cutting (AWJC) process is a proper solution for cutting CFRP because of reduced interface temperature, low mechanical loading, better surface integrity, low tool wear, and small cutting forces [15, 16]. AWJC is a complex process with a large number of parameters, like water pressure, feed rate, abrasive mass flow, standoff distance, etc. [17-20]. The main quality characteristics of the cut parts are: kerf geometry [21], surface roughness [22] and topography [23]. S. Vasanth noticed in his research on titanium alloy processing using AWJC, that abrasive flow rate and standoff distance have a

* Corresponding author: ioan.popan@tcm.utcluj.ro

significant role on determining surface quality. A higher standoff distance (SOD) produces higher surface roughness due to larger and random energy distribution.

Higher SOD allows the waterjet to expand before impingement on the workpiece which may increase the exposure to external drag from the surrounding environment. Thus with the increase in SOD, the jet diverges which results in decreasing the kinetic energy of the jet and thus producing a rough surface [5, 14, 24]. The SOD has effect on kerf geometry, manly on the kerf taper angle and on the top edge radius [25].

The surface cut by AWJC is divided in three zones: initial damage zone, smooth cutting zone and rough cutting zone (figure 1) [26-27]. To understand the influence of the standoff distance on surface quality, each zone must be analysed.

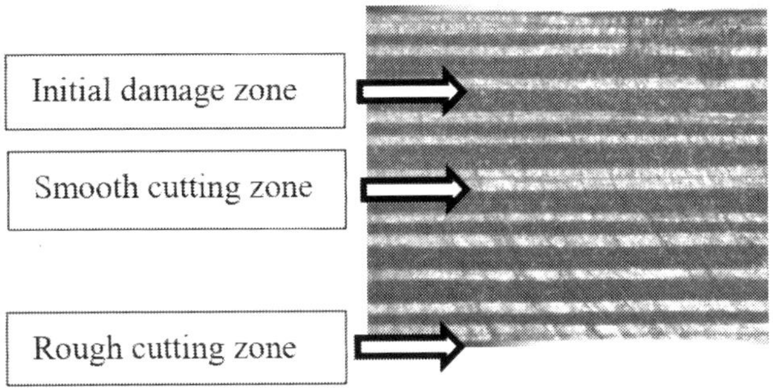

Fig. 1. The main zones of a surface cut by AWJC.

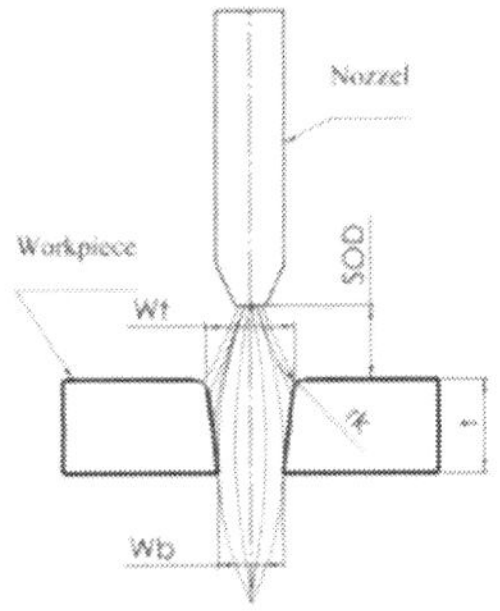

Fig. 2. The kerf geometry.

In this paper the influence of standoff distance (the distance between cutting head and work-piece) on the characteristics of the cut surface is studied.

2 Experimental procedure

In this experimental study, the standoff distance SOD is the variable process parameter. The main characteristics of the cut surface (kerf geometry, surface roughness and topography) were analyzed. Ra / Sa average roughness, Rz / Sz average maximum height of the profile, were the surface roughness parameters studied. The analysed kerf geometry dimensions were: W_t - the top kerf width, W_b - the bottom kerf width and r_k - the top edge radius, presented in figure 2. The constant parameters and the values for SOD used on this study are presented in table 1.

Table 1. Process parameters used in the experiments.

Technological parameter	**Value**
Sample thickness	6 mm
Pressure	355 MPa
Feed rate	100 mm/min
Abrasive type / size	Australian Garnet / 80 mesh
Abrasive mass flow	0.440 kg/min
Nozzle diameter	0.76 mm
Orifice diameter (diamond)	0.35 mm
Standoff distance	**0.5; 1; 2; 3; 4; 5 mm**

The experimental trials were conducted using an Omax 2626, 3-axis abrasive waterjet system (figure 3). It has an OMAX high pressure pump with an output pressure of maximum 400 MPa and the maximum feed rate is 4500 mm/min.

The material used in this study is a multi-layer composite material consisting of carbon fibre reinforced plastics (CFRP): ECOTECH™, made out of HT carbon fiber pre-pregs with a transparent epoxy resin matrix in press molding. One surface layer consists of 3k-fabric style 452 in twill weave 2/2 (fabric pre-preg CE 8201-200-45). The core layer consists of 3k UD-non-crimp fabric 0°/90° (bidirectional) [28].

Fig. 3. The water jet machine Omax 2626.

Fig. 4. The clamping system.

The specimens had the dimensions: 100 x 35 x 6 mm and they were fixed in the clamping system presented in figure 4.

3 Results and discussions

In this section, the influence of standoff distance SOD on kerf geometry, surface roughness and surface topography has been analyzed.

3.1 Influence of the standoff distance on kerf geometry

To measure the dimensions of kerf geometry the PG 2000 microscope was used with 9.5 x zoom. In figure 5 the obtained kerf geometry with different values of SOD (0.5; 1; 2; 3; 4; 5 mm) is presented.

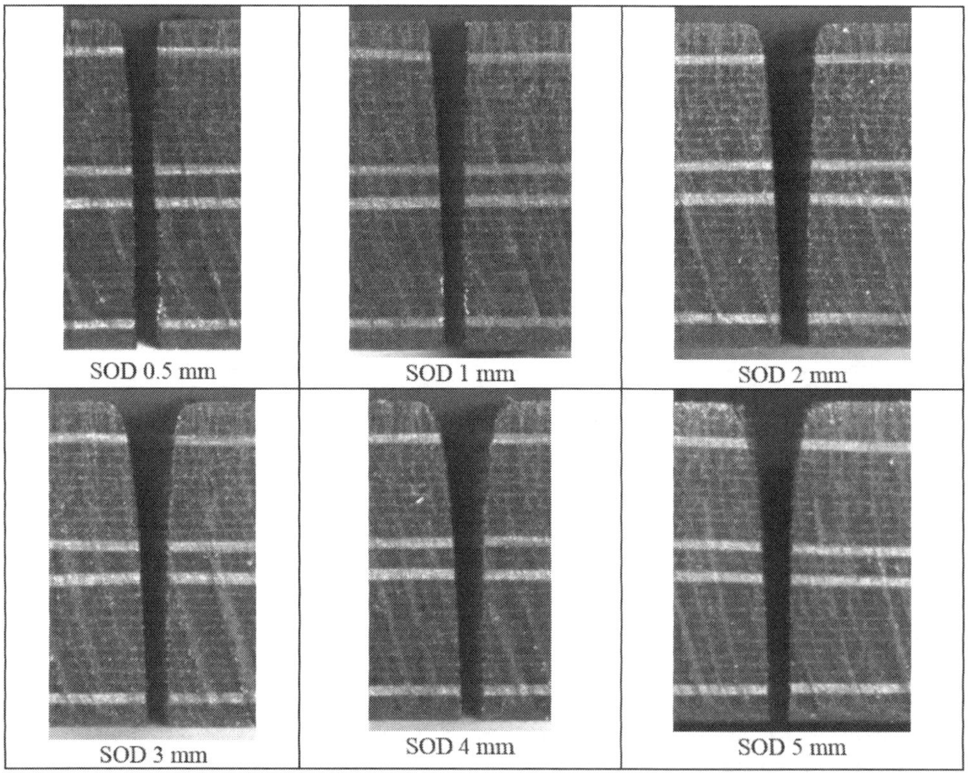

Fig. 5. The kerf geometry obtained with different values of SOD.

The kerf geometry dimensions: W_t - the top kerf width, W_b - the bottom kerf width, r_k - the top edge radius are presented in table 2.

Table 2. The kerf geometry dimensions.

SOD mm	**0.5**	**1**	**2**	**3**	**4**	**5**
W_t mm	0.828	0.963	1.352	1.818	2.016	2.264
W_b mm	0.442	0.457	0.576	0.497	0.546	0.526
r_k mm	0.12	0.18	0.21	0.24	0.32	0.35

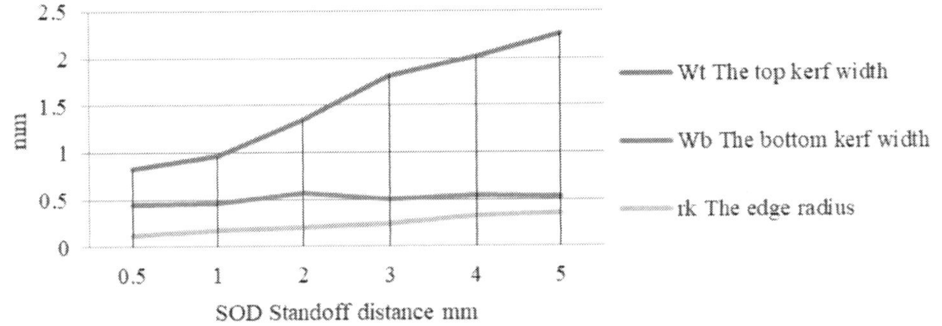

Fig. 6. The influence of the SOD on the kerf geometry dimensions.

Analysing the kerf width variation, figure 6, can be observed that by increasing the SOD form 0.5 up to 5 mm the top kerf width W_t increases from 0.828 up to 2.264 mm (red line) and the bottom kerf width W_b increases from 0.442 up to 0.526 mm (blue line).

In case of the top edge radius r_k represented with green line, by increasing the SOD the edge radius increases form 0.12 up to 0.35 mm.

It can be concluded that, the SOD has a strong influence on the upper zone of the kerf, on the top kerf width and the top edge radius. It doesn't have a notable effect on the bottom kerf width. Similar aspects were observed by D.K. Shanmugam in case of abrasive waterjet machining of alumina ceramics [25].

3.2 Influence of the standoff distance on surface roughness

The NanoFocus 3D microscope µsurf expert was used to measure surface roughness. The surface roughness parameters: R_a; R_z; S_a; S_z are presented in table 3. For this measurement a Gaussian filter 0.8 was used.

Table 3. Surface roughness.

SOD mm	**Initial damage zone**				**Smooth cutting zone**				**Rough cutting zone**			
	Surface roughness µm Gaussian filter 0.8											
	R_a	**R_z**	**S_a**	**S_z**	**R_a**	**R_z**	**S_a**	**S_z**	**R_a**	**R_z**	**S_a**	**S_z**
0.5	1.87	9.82	2.57	28.9	2.80	14.1	4.31	40.6	2.47	11.9	4.09	38.4
1	1.98	10.0	2.84	30.9	2.85	14.5	4.41	42.6	2.52	12.8	4.08	38.3
2	2.21	11.5	3.60	36.1	2.90	14.6	4.54	44.5	2.50	12.1	4.10	40.9
3	2.26	11.5	3.63	36.8	3.05	15.0	4.66	44.7	2.56	13.0	4.09	40.6
4	2.71	12.8	3.80	38.5	3.10	15.6	4.77	48.7	2.55	12.8	4.12	40.9
5	2.74	13.5	4.05	40.1	2.80	13.9	4.30	10.1	2.62	13.3	4.29	41.1

 DOI: 10.1051/matecconf/201713701009

Figure 7 represents the correlation between the standoff distance SOD and the surface roughness for each three main zones of cut surface. The SOD varied between 0.5 and 5 mm, using an abrasive mass flow of 0.45 kg/min and a pressure of 350 MPa.

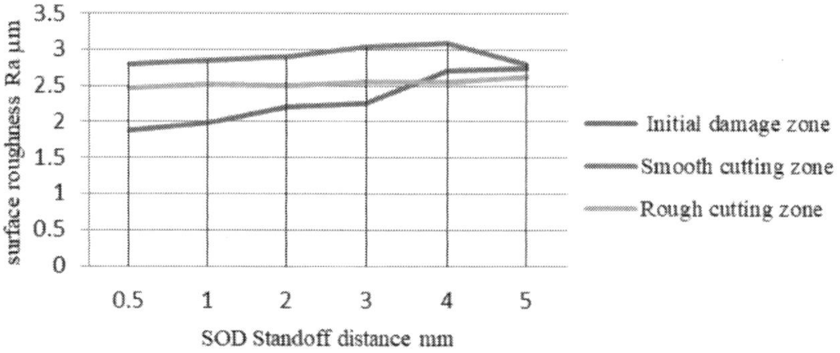

Fig. 7. The influence of the standoff distance on the surface roughness.

By variating the SOD between 0.5 and 5 mm the surface roughness variated between: R_a 1.87 – 3.1 µm, R_z 9.2 - 15.6 µm, S_a 2.57 – 4.77 µm, and S_z 28.9 - 48.7 µm.

Analysing the surface roughness variation on the initial damage zone (represented by the red line) it can be observed that by increasing the SOD form 0.5 up to 5 mm the surface roughness increases R_a from 1.87 to 2.74 µm and R_z 9.82 to 13.5 µm.

In case of surfaces from smooth cutting zone (blue line) and rough cutting zone (green line) the surface roughness variated between R_a 2.8 - 3.1 µm, R_z 9.82 - 13.5 µm and R_a 2.47 - 2.6 µm, R_z 14.1 - 15.6 µm in last zone. On these zones was not observed a influence of SOD on surface quality.

However the standoff distance flow did not have a strong influence on surface quality like other main cutting parameters of the abrasive water jet process, like, pressure, feed rate and abrasive mass flow. It has a low influence on the surface roughness from the initial damage zone. The best surface quality could be obtained by minimizing this parameter, up to 0.5 mm. This value must be correlated with the work-piece surface flatness accuracy so the focusing tube does not break.

3.3 Influence of the standoff distance on surface topography

The same 3D microscope was used for analysing the cut surfaces. In each main zone (initial damage zone, smooth cutting zone and rough cutting zone) a surface was scanned 1.6 x 1.6 mm to obtain the 3D surfaces topography. For surface analysis was used µsoft analysis premium 7.1. The 3D surfaces topography of the cut surfaces are presented in figure 8.

An analysis of the topography showed a smooth surface in case of CFRP processed by AWJC. By variating the SOD from 1 to 5 mm the 3D average roughness S_a between 2.84 and 4.66 µm (figure 8).

The topography of the cut surface presents a random distribution of the valleys, craters and peaks. It is dominated by deep valleys, generated by the curved trajectories of abrasive particles. These valleys have the same direction with the jet travel direction.

On the initial damage zone it can be observed a mixture of valleys and craters. Those craters are generated by abrasive grains from the outlet of the jet. Because the abrasive grains do not have enough kinetic energy to penetrate the entire thickness of material, they remove a small quantity of material from the work-piece, generating valleys and craters.

 DOI: 10.1051/matecconf/201713701009

The surface from the smooth cutting zone exhibits a random distribution of the valleys, craters and peaks. The valleys and craters depth is the smallest from the entire cut surface.

On the rough cutting zone it can be clearly observed that there are deep valleys in the same direction with the jet travel direction.

SOD 1 mm	**SOD 3 mm**	**SOD 5 mm**
	Initial damage zone	
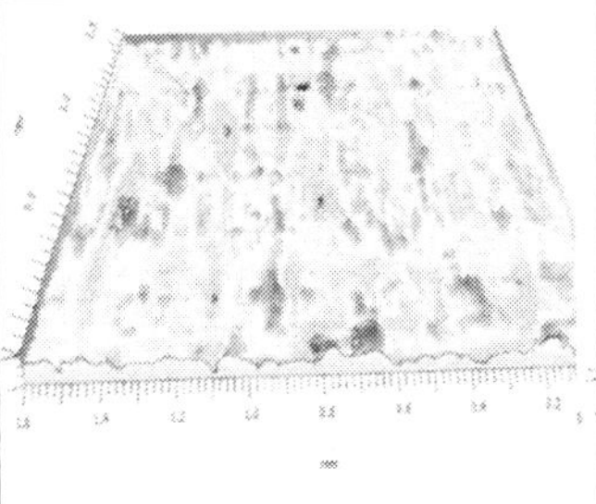	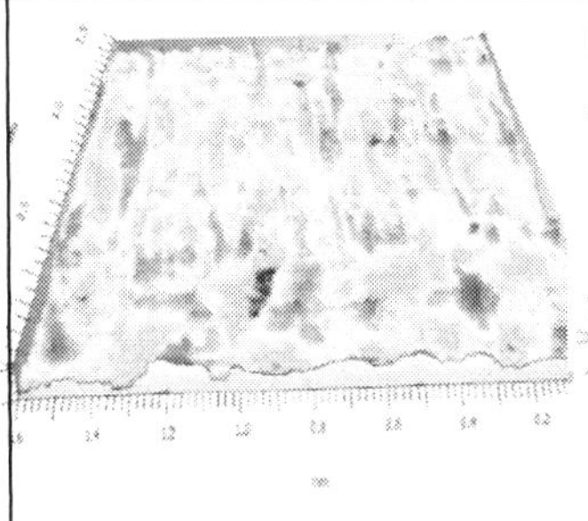	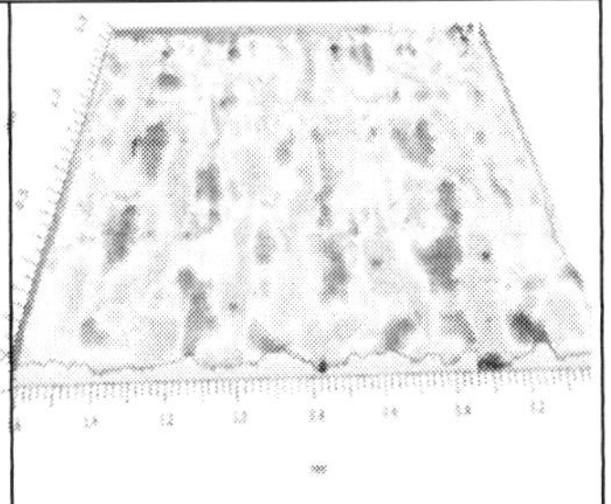
R_a 1.98 µm, S_a 2.84 µm	R_a 2.26 µm, S_a 3.63 µm	R_a 2.74 µm, S_a 4.05 µm
	Smooth cutting zone	
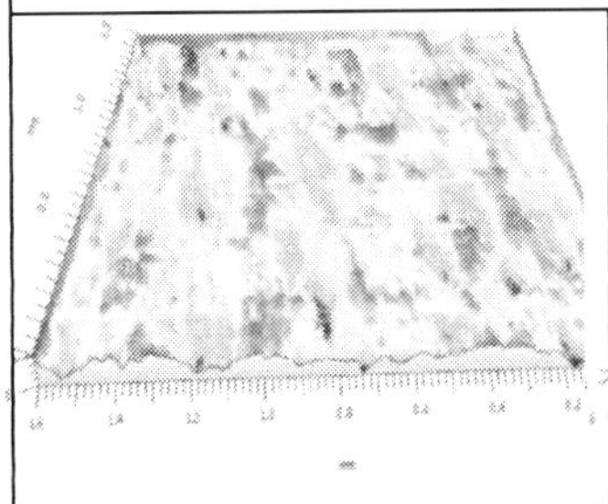	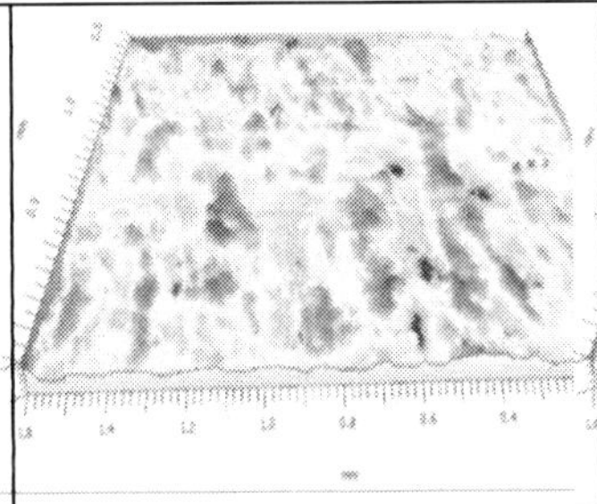	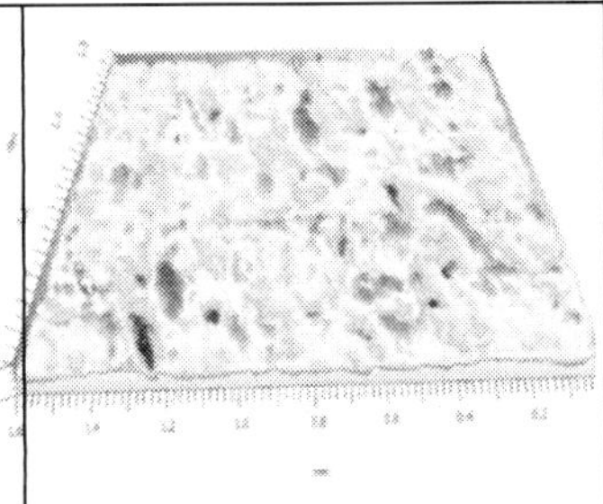
R_a 2.8 µm, S_a 4.41 µm	R_a 2.9 µm, S_a 4.66 µm	R_a 2.8 µm, S_a 4.31µm
	Rough cutting zone	
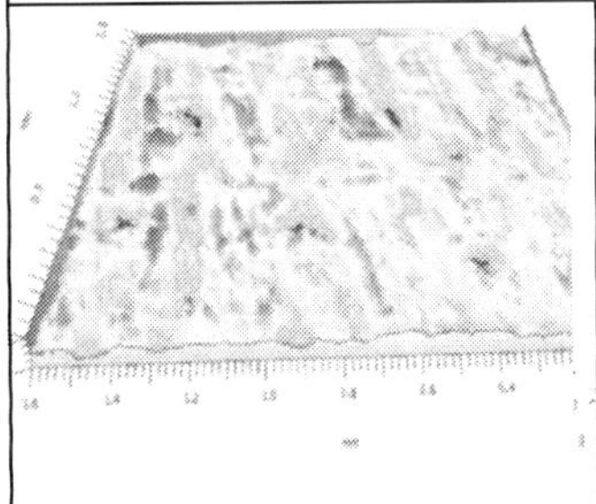		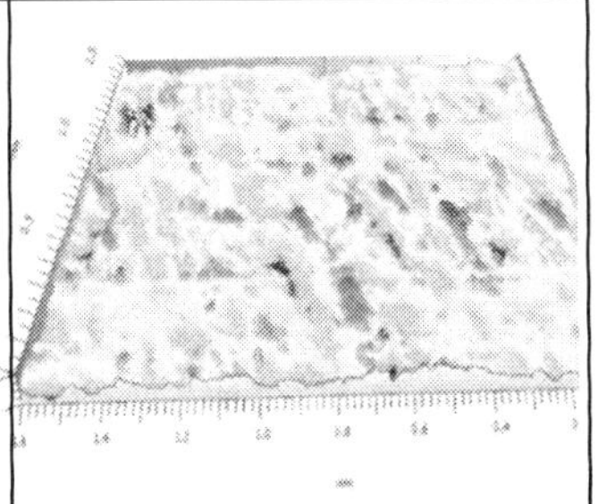
R_a 2.47 µm, S_a 4.08 µm	R_a 2.56 µm, S_a 4.09 µm	R_a 2.62 µm, S_a 4.31 µm

Fig. 8. The 3D surface topography of the cut surfaces.

The 2D surface roughness profiles of the cut surfaces are presented in figure 9 (a–e).

First observation after analysing the 2D roughness profiles is that the peaks and valleys of the profile are not symmetrical and the profile variation is not constant.

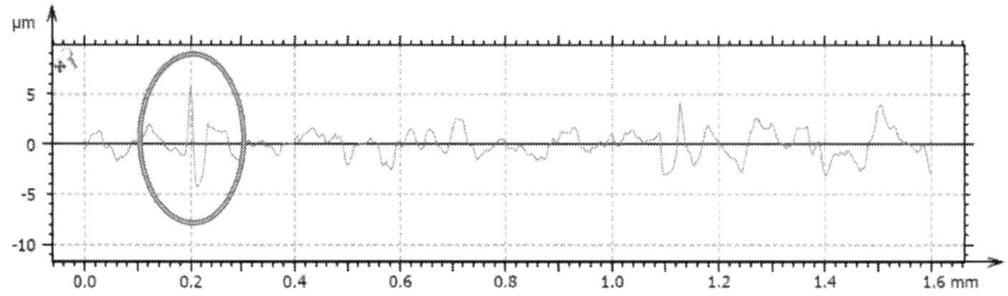

a. Initial damage zone SOD 1 mm.

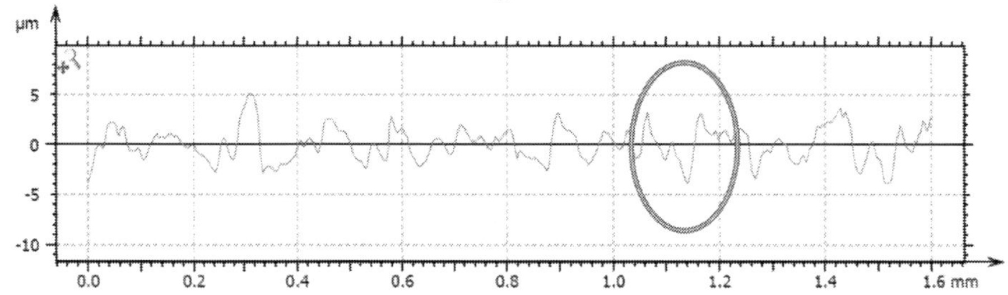

b. Initial damage zone SOD 3 mm.

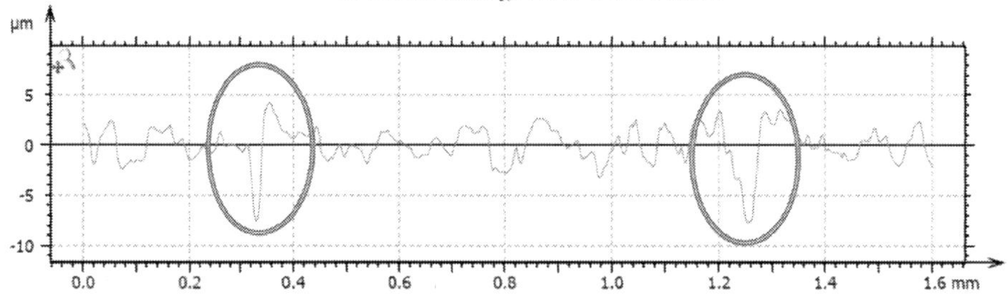

c. Initial damage zone SOD 5 mm.

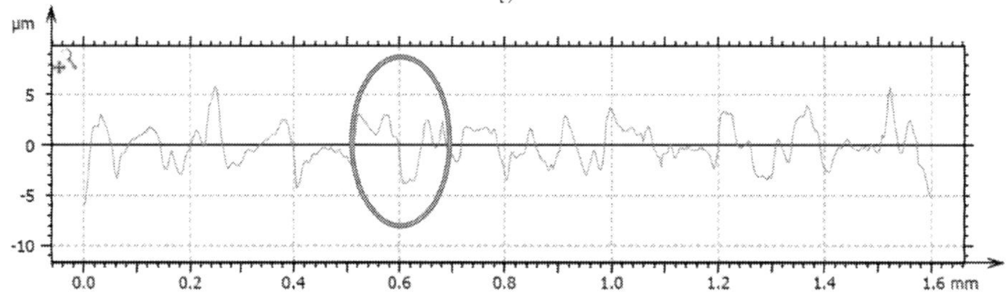

d. Smooth cutting zone SOD 1 mm.

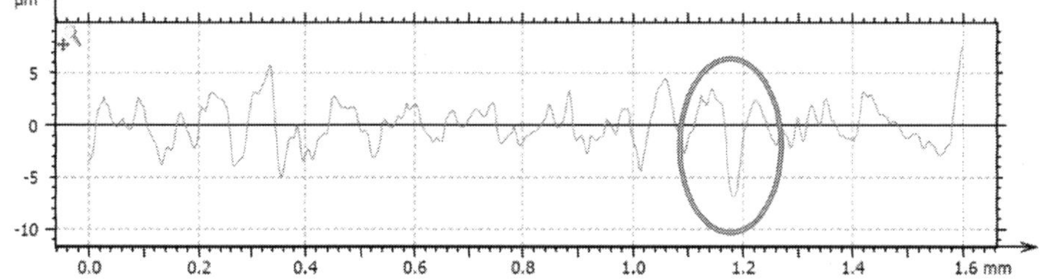

e. Rough cutting zone SOD 1 mm.

Fig. 9. The 2D roughness profiles of the cut surfaces: a. Initial damage zone SOD 1 mm; b. Initial damage zone SOD 3 mm; c. Initial damage zone SOD 5 mm; d. Smooth cutting zone SOD 1 mm; e. Rough cutting zone SOD 1 mm.

On the initial damage zone it can be observed that, by increasing the SOD from 1 to 5 mm, the average maximum height of the profile R_z increases from 10 up to 13.5 µm. For a value of SOD 1mm a maximum valley depths of -4 μm was obtained and the valley depths is increasing up to -8 μm in case of SOD 5 mm (figure 9 a, b and c). The peaks height had a small variation and it is approximately +5 μm.

The SOD has a low influence on surface roughness in the smooth cutting zone and rough cutting zone. A correlation between the SOD variation and the 2D profiles cannot be made.

In the case of the smooth cutting zone, by varying the SOD from 1-5 mm, the average maximum height of the profile R_z varied from 13.9 up to 15.6 μm. The valleys have a maximum depth of approximately -5 μm and approximately +8μm for the peaks height (figure 9 d).

In the case of the rough cutting zone the average maximum height of the profile R_z is between 11.9 and 13.3 μm. In this zone the maximum valleys depth is approximately -7 μm and approximately +7 μm for the peaks height (figure 9 d).

4 Conclusions

The standoff distance is an important parameter in machining of CFRP materials using the abrasive waterjet cutting process.

The specific conclusions from this experimental study are:

• SOD has a strong influence on the kerf geometry, mainly on the upper zone of the kerf. Decreasing the SOD up to 0.5 mm it was noticed that: the top kerf width W_t decreases (up to 0.82 mm), the bottom kerf width W_b decreases (up to 0.442 mm) and the top edge radius r_k decreases also (up to 0.12 mm);

• SOD does not have a strong influence on the surface quality like other main cutting parameters of the AWJC, pressure, feed rate and abrasive mass flow. The main affected zone by the SOD variation is the initial damage zone. On this zone it was observed that by increasing the SOD from 0.5 up to 5 mm the surface roughness increases R_a from 1.87 to 2.74 μm and R_z 9.82 to 13.5 μm. In the case of smooth cutting zone and rough cutting zone surfaces it was not observed any influence of SOD on surface quality.

• Analysing the 3D surfaces topography and 2D roughness profiles it was observed that the peaks, craters and valleys of the surface are asymmetrical and the variation is not constant. On the initial damage zone it can be observed that by increasing the SOD the max. valley depths increase up to 8 μm and the max. peak heights increase up 5μm.

• To improve the CFRP part accuracy and surface roughness, it is recommended to minimize the SOD.

This work was supported by a grant of the Romanian National Authority for Scientific Research and Innovation, CNCS/CCCDI–UEFISCDI, project number PN-III-P2-2.1-BG-2016-0216, within PNCDI III, Technical University of Cluj-Napoca through the research Contract no. 2001/12.07.2017, Internal Competition CICDI-2017 and the H2020 AMaTUC project (GA 691787).

References

1. D. K. Shanmugam, Composites. Part A **39**, 923–929 (2008)
2. I. A. Popan, A. I. Popan, Acta Tehnica Napocensis **60 II**, 251-254 (2017)
3. D. Liu, Y. Tang and WL. Cong, Composite Structures **94,** 1265–1279 (2012)
4. P. Bere, E. Sabau, L. Hancu, A. Popescu, AJME, **11** (4), 128-131 (2013)
5. M. S. Alsoufi, International Journal of Mechanical Engineering and Applications **5**(1), 1-14 (2017)
6. J. Wang, Int J Adv Manuf Technol, **15**, 757–68 (1999)

7. I. A. Popan, N. Balc, A. Carean, A. Luca, A. Miron, ICAMaT 7th International Conference on Advanced Manufacturing Technologies, **760**, 409-414 (2015)
8. M. Ramulu, S. Kraja, 33rd International SAMPE Proceedings, **33**, 1327–1339 (2002)
9. J. Carach, S. Hloch, P. Hlavacek, J. Scucka, P. Martinec, J. Petru, T. Zlamal, M. Zelenak, P. Monka, D. Lehocka, J. Krolczyk, International Journal of Advanced Manufacturing Technology **82**, 1747-1752 (2016)
10. H. Bulea, R Paunescu, A.C. Filip, Applied Mechanics and Materials **760**, 397-402 (2015)
11. A. C. Filip, M. Vasiloni, L. A. Mihail, MATEC Web of Conferences, **94,** (2017)
12. I. A. Popan, N. Balc, A. I. Popan, R. Curta, iCAT, ISSN 1992-5093, (2010)
13. I. A. Popan, N. Balc, A. I. Popan, MSE, MATEC Web Conf.,**121** (2017)
14. M. C. Kong, D.A. Axinte, W. Voice, Journal of Materials Processing Technology **210**, 573–584 (2010)
15. E. Lemma, F. L. Chen, Compos Struct, **57**, 297–303 (2002).
16. M. Ramulu, D. Arola, Composites, **24**, 299–308 (1993)
17. S. Vasanth, T. Muthuramalingam, HPC 2016 – CIRP, Procedia CIRP **46**, 139–142 (2016)
18. I. A. Popan, N. Balc, A. I. Popan, R. Pacurar, The 4th International Conference on Additive Technologies; iCAT, ISSN 1992-5093, (2012)
19. P. Bose1, A. Dhanawade, Journal of the Association of Engineers, India **86**(3-4) (2016)
20. I. A. Popan, N. Balc, A. I. Popan, IMT, Annals Of Oradea University. Fascicle Of Management And Technological Engineering **XIX (IX)**, ISSN 1583-0691, 4.86-4.90 (2010)
21. A. C. Filip, H. Bulea, Advances in Production, Automation and Transportation Systems, ICAPS, 201-205, (2013)
22. A. Miron, N. Balc, I. A. Popan, C. S. Borzan, P. Bere, AJME **11**(2), 87-92 (2013)
23. D. Lehockáa, J. Klichb, J. Foldynab, S. Hloch, ICMEM 2016, Procedia Engineering **149**, 236 – 244 (2016)
24. D. K. Shanmugama, S. H. Masoodb, Journal of Materials Processing Technology, (2003)
25. D. K. Shanmugam, J. Wang, H. Liu, International Journal of Machine Tools & Manufacture **48**, 1527–1534 (2008)
26. P. Löschner, K. Jarosz, P. Niesłony, ICMEM, Procedia Engineering **149**, 276 – 282 (2016)
27. F. L. Chen, E. Siores, Journal of Material Technology **135**, 1-5 (2003)
28. *** http://shop1.r-g.de/en/art/615400 Material specifications

MATEC Web of Conferences **137**, 01010 (2017) DOI: 10.1051/matecconf/201713701010
MTeM - AMaTUC 2017

Forecasting of steel consumption with use of nearest neighbors method

Michał Rogalewicz[1,*], *Robert* Sika[2], *Paweł* Popielarski[2], and *Grzegorz* Wytyk[3]

[1]Poznan University of Technology, Chair of Management and Production, Piotrowo 3 Street, Poland
[2]Poznan University of Technology, Institute of Material Technology, Piotrowo 3 Street, Poland
[3]BiB Systems LLC, Lutycka 11, Poland

Abstract. In the process of building a steel construction, its design is usually commissioned to the design office. Then a quotation is made and the finished offer is delivered to the customer. Its final shape is influenced by steel consumption to a great extent. Correct determination of the potential consumption of this material most often determines the profitability of the project. Because of a long waiting time for a final project from the design office, it is worthwhile to pre-analyze the project's profitability and feasibility using historical data on already realized orders. The paper presents an innovative approach to decision-making support in one of the Polish construction companies. The authors have defined and prioritized the most important factors that differentiate the executed orders and have the greatest impact on steel consumption. These are, among others: height and width of steel structure, number of aisles, type of roof, etc. Then they applied and adapted the method of k-nearest neighbors to the specificity of the discussed problem. The goal was to search a set of historical orders and find the most similar to the analyzed one. On this basis, consumption of steel can be estimated. The method was programmed within the EXPLOR application.

1 Introduction

Development of trade, globalization, as well as diversified needs of clients and development of modern production and computer technologies caused that the modern manufacturing and service companies compete in a global market, where client is the most important. That is why it is important to ensure high quality of products and services, suited to his needs and expectations, as well as supplying them in an acceptable time. In this point of view, improvement of processes inside a company [1,2,3], limitation of wastage [4,5], shortening of a production cycle [6] and changeover times, minimization of distribution costs and improvement of quality of offered services have become the primary goals of organizations. Realization of these goals should be performed in connection with application of newest achievements in science and technology – new technologies on one hand, but also methods of decision-making on the other hand [7,8,9].

* Corresponding author: michal.rogalewicz@put.poznan.pl

Large amounts of data gathered in big databases in scope of a daily functioning of a company are today not always appropriately used for building knowledge about processes and products. Wasting such a detailed source of information, both about a company itself and client requirements, preferences, habits and needs, is a fundamental mistake made in certain organizations. It seems that in the era of Industry 4.0 proper acquisition, processing, analysis and archiving of data gain importance and make for a faster reaction to dynamically changing clients' requirements, shorter time of introduction of a product into market, better estimation of costs of undertakings, as well as obtaining higher client satisfaction [10,11].

To face these challenges, certain methods can be used, allowing analysis of large datasets and/or automatic obtaining of rules and patterns, which are not obvious and trivial for a database user – the so-called Data Mining methods.

This paper presents application of one of the Data Mining methods – the *k*-nearest neighbors method – for a quick, preliminary estimation of steel consumption for building a construction ordered by a client. It facilitates presentation of a preliminary offer to a client, without a need of preparing a detailed project, which could take even several days. Assuming that a company receives several inquiries daily and it orders preparation of a detailed project to a design office, it is important to preliminarily estimate the costs and present them to a client. Only after initial acceptance, actual, detailed project is prepared. The studies were realized in one of Polish construction companies.

2 The *k*-nearest neighbors method

The Data Mining methods fulfill several functions. They enable **describing data/objects**, **classification of objects** (meaning assigning them to classes of a dependent variable, on the basis of values of independent variables describing a given object), **grouping of objects** (creating groups of objects the most similar to each other, on the basis of values of independent variables), **search for associations** (finding interdependencies between data, they are based upon co-appearance of values of particular variables) and **realization of the regression task** (search for a model of relation between independent variables and a dependent variable in a numerical form) [12,13].

An example of a method that allows realization of both classification and regression tasks is the *k*-nearest neighbors method. This paper focuses on using this method for a preliminary estimation of steel consumption for an inquiry of quotation sent by a client, so the regression task is the one being realized. The *k*-nearest neighbors method focuses on finding a determined number of objects located in the closest proximity to a new object and calculating a value of an unknown dependent variable on their basis (usually it is simply an average of a dependent variable value for the selected *k* objects). The distance between objects in a multidimensional space, for the quantitative variables, is defined using one of the several selected metrics, e.g. Euclidean, Manhattan or Chebyshev. The first one is the most commonly used – it represents a standard way of perceiving distance by humans in real world. For the qualitative variables, the "different from" function is used, taking the value of 0 when two objects have the same value of a feature and 1 if the value is different [13,14,15].

The *k*-nearest neighbors method requires normalization of variables, as it is possible that a variable taking high values will totally dominate the distance metric and influence of the other variables. It is also worth ensuring, that there will be no situation, in which many less important variables, for which the studied objects will have a very distant value, will dominate very relevant variables, for which this distance will be very small. To achieve this, the so-called axis stretching in a multidimensional space can be applied, meaning choosing appropriate weights for the more important independent variables [13].

The further part of this paper presents an example of use of the *k*-nearest neighbors method for estimation of steel consumption for the quotation inquiries coming from clients. This estimation is conducted on the basis of already realized, similar projects, archived in a database. The consumption is determined by an a priori defined index. In scope of the performed work, a dedicated software tool EXPLOR was designed. It realizes the following functions: 1C, MultiC and *k*-neighbors. Their development is presented in the further part, on specific examples.

3 Application of the *k*-nearest neighbors method for estimation of steel consumption

3.1 Assumptions for the used method and application

Aim of the work conducted by authors of this paper was development of a method of supporting the decision-making in scope of steel consumption estimation, for building a construction, which is a result of initial discussions with a client. The following assumptions were taken, regarding the used method and the designed software application:
• the application uses a database of projects already realized by the construction company – at the moment of launching the application, the database had approx. 300 records (where a single record contained data from one, specific order realized for a client),
• the most important features, differentiating the projects and having the most important influence on a level of steel consumption are:
- hall width [m],
- hall height, from ground to bottom of roof construction [m],
- number of aisles [1 or 2],
- roof type [penthouse or gable roof],
- roof load [kN/m^2],

• the records regarding halls with a gantry and without it,
• in the case of the *k*-neighbor function, the user was given a choice of number of objects ("neighbors"), on basis of which the estimation is conducted, the Euclidean distance was used,
• the application realizes three tasks, which are presented in a greater detail in chapter 3.2.

3.2 Exemplary use of the application

The authors prepared a software application, realizing three tasks aimed at adjusting value of an independent variable from the already realized orders, to a new offer, on the basis of given values of independent variables. The first two described tasks (1C, MultiC) mostly pertain to rapid comparison of new offers to already realized projects, while the *k*-neighbor method uses a selection of weights of independent values by the authors. Panel of the EXPLOR software is shown in Figure 1 and Figure 2.

The first function of the EXPLOR program (named the 1C) regards use of only one criterion of result adjustment – a specific value of an independent variable, e.g. finding records of an applied value of hall width. This function is particularly useful, when a company must analyze types of already realized projects by a specific, single criterion. The algorithm will often return a large amount of results (in comparison to the other two functions), which may be then used for deepened exploration, e.g. indicate the most profitable orders of the ones selected.

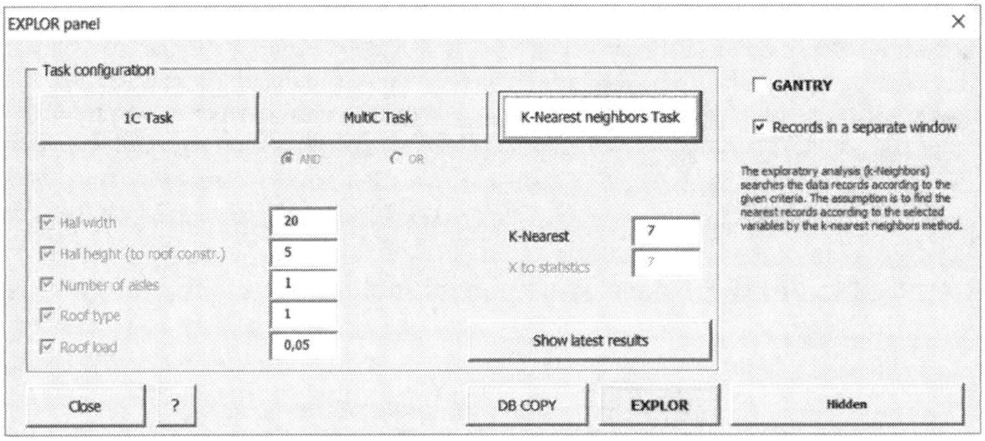

Fig. 1. Main panel of EXPLOR program

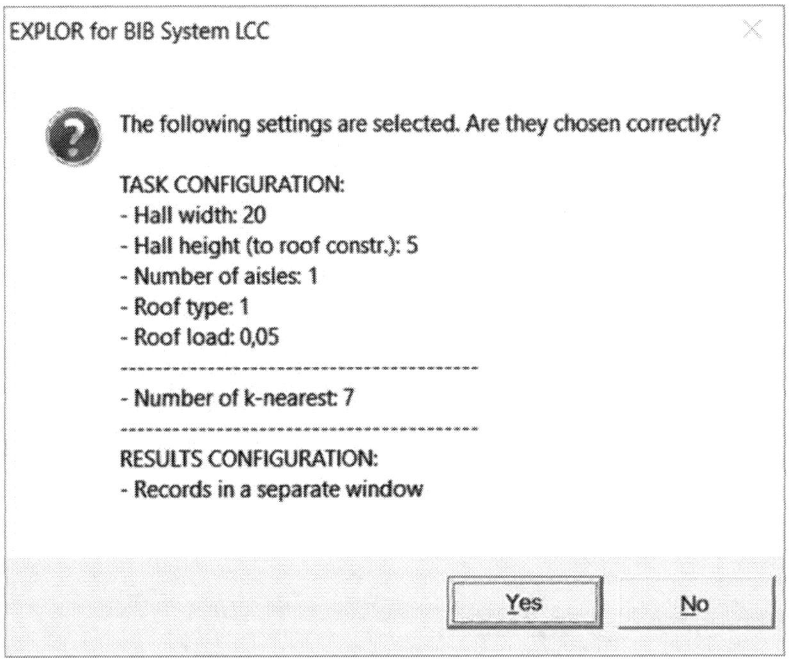

Fig. 2. Settings confirmation in the k-keighbor Task

The second function (named the MultiC) is an expansion to the 1C function. In this case, more than one criterion can be indicated (up to all possible) and logical operators AND and OR can be used. In the first case, results will be returned if there will be records fulfilling all the criteria for the input values. For example, for the following three criteria: hall width (20 m), hall height to bottom of the roof construction (5 m) and amount of aisles (1 aisle), the found records must all contain data about orders concerning 20-meter halls, height equal to 5 meters measured to roof bottom and having only 1 aisle. The second option – OR – returns results when at least one of the indicated criteria is fulfilled.

The last function – *k*-nearest – is the most expanded one. The idea is similar to the MultiC function with the AND operator (the found records will fulfill all the selected criteria for specific values of independent variables), but here the algorithm adjusts values indicated by a user to the closest contained in a database record, using data normalization and axis stretching in a multidimensional space. It is the most useful option in a situation of estimating the steel consumption for quotation inquiries coming from clients. On the basis of experience of the company and the developed know-how, appropriate values of weights were selected for the particular independent variables. The greatest weight is assigned to the most important criterion. Figure 1 shows exemplary settings in the k-nearest function window, while in Figure 2 the settings confirmation in the k-keighbor Task was presented.

As presented, the steel consumption is determined on the basis of 7 nearest neighbors of the case analyzed ("new offer" – see Figure 3). The results will be returned in a separate window and the nearest neighbors are ordered starting from the smallest distance. It is noteworthy, that in this case, the results can be returned even if the found value of the independent variable is not exactly the same as indicated by a user. Figure 3 presents a results window. The analysis result is an average value of the steel consumption index, as well as the standard deviation.

Offer No.	Hall width [m]	Hall height to roof constr. [m]	Number of aisles	Roof type	Roof load [kN/m2]	Location	Steel consumption indicator [kg/m2]	
new offer	**20**	**5**	**1**	**1**	**0,05**	...	**... ? ...**	
1	20	5	1	2	0,05	[illegible]	**28,00**	
45	20	5	1	2	0,1	[illegible]	**26,80**	
186c	20	5	1	2	0,1	[illegible]	**24,00**	
186d	20	5	1	2	0,1	[illegible]	**22,00**	
220a	20	5	1	2	0,15	[illegible]	**27,71**	
47	20	5,5	1	2	0,05	[illegible]	**26,67**	
187d	20	5	1	2	0,25	[illegible]	**25,00**	
							Xaverage	**25,74**
							STD Deviation	**2,18**
				Other results, less matching...				
46	20	6	1	2	0,1	[illegible]	26,00	
48	20	6	1	2	0,1	[illegible]	23,00	
235	24	3,8	1	1	0,15	[illegible]	26,49	
22	20	6	1	2	0,15	[illegible]	23,75	
219b	20	6	1	2	0,15	[illegible]	27,25	
182a	20	4	1	2	0,15	[illegible]	25,12	
148	17,5	6,5	1	1	0,1	[illegible]	31,89	
35b	13	3,8	1	1	0,1	[illegible]	19,66	
34	10	4,5	1	1	0,15	[illegible]	18,15	
4	27,05	5,535	2	1	0,15	[illegible]	32,60	

Fig. 3. Result of work of the *k*-nearest function. The result is 7 closest neighbors.

4 Conclusions

The paper presents an innovative approach to supporting the decision making in the Bidding and Contract Preparation department, on example of a construction company. The authors, together with employees of a company, defined factors having the highest influence on steel consumption and the most differentiating for the steel constructions and ordered them by importance. The possible ways of database exploration were presented, to find already realized orders, which are the most close to an analyzed order, on the basis of values of features defined by the user.

Facing the tight deadlines and difficulties with preparing even the preliminary projects for several or more than a dozen clients at the same time, it was decided to implement the solution proposed by the authors. Quick preparation of an offer using a steel consumption index estimated on the basis of data of already realized orders, allowed the company to perform calculations of profitability of undertakings in virtually no time, as well as to limit costs of project preparation.

The historical data, appropriately processed and analyzed, have turned out to be very useful in daily functioning of a company and building the process knowledge. The tool developed by the authors is universal and may be used by other companies, after modifying the most important independent variables and values of their weights.

References

1. A. Kujawińska, K. Vogt, Man. and Prod. Eng. Rev. **6**, 2 (2015)
2. A. Kujawińska, K. Vogt, A. Hamrol, Adv. in Int. Sys. and Comp., Springer Verlag, **490** (2016)
3. A. Hamrol, Prod. Plan. and Contr., **11**, 8 (2000)
4. M. R. L. Varela, J. Trojanowska, S. Carmo-Silva, N. M. L. Costa, J. Machado, Man. and Prod. Eng. Rev., **8**, 2 (2017)
5. E. Dostatni, J. Diakun, A. Hamrol, W. Mazur,, Ind. Manag. and Data Sys., **113**, 6 (2013)
6. C. Gangala, M. Modi, V. K. Manupati, M. L. R. Varela, J. Machado, J. Trojanowska, In: A. Rocha, A. Correia, H. Adeli, L. Reis, S. Costanzo (eds) Rec. Adv. in Information Sys. and Tech. WorldCIST 2017. Adv. in Int. Sys. and Comp., **571** (2017)
7. J. Trojanowska, M. L. R. Varela, J. Machado, In: A. Rocha, A. Correia, H. Adeli, L. Reis, S. Costanzo (eds) Rec. Adv. in Information Sys. and Tech. WorldCIST 2017. Adv. in Int. Sys. and Comp., **571** (2017)
8. M. Diering, K. Dyczkowski, IEEE Int. Conf. on Ind. Eng. and Eng. Man. (2016)
9. A. Kujawińska, M. Rogalewicz, M. Diering, A. Hamrol, In: A. Rocha, A. Correia, H. Adeli, L. Reis, S. Costanzo (eds) Rec. Adv. in Information Sys. and Tech. WorldCIST 2017. Adv. in Int. Sys. and Comp., **571** (2017)
10. Z. Ignaszak, R. Sika, M. Perzyk, A. Kochański, J. Kozłowski, Arch. of Found. Eng., **145**, 5 (2016)
11. Gorski F., Zawadzki P., Hamrol A., J. of Mach. Eng., **16**, 4 (2016)
12. J. Han, M. Kamber, J. Pei J., *Data Mining: Concepts and techniques*, (Morgan Kaufmann Series in Data Management Systems, Elsevier, 2012)
13. T. Larose, *Discovering knowledge in data: an introduction to Data Mining* (Wiley&sons, 2005)
14. T. Hastie, R. Tibshirani, J. Friedman, *The Elements of statistical learning. Data Mining, inference, and prediction* (Springer, 2009)
15. T. Hill, P. Lewicki, *STATISTICS: methods and applications*. (StatSoft, Tulsa, 2007)

Identification of detailed behaviour of Flexible Manufacturing Cell using Hierarchical Timed Coloured Petri Nets

Sanjib Kumar Saren[1,*], *Florin* Blaga[2], and *Tiberiu* Vesselenyi[1]

[1]Mechatronics Department, University of Oradea, Romania
[2]Industrial Engineering Department, University of Oradea, Romania

Abstract. The goal of this paper is focused on identifying the behavioural structure of a flexible manufacturing cell (FMC) using a model based on Coloured Timed Petri Nets and applying hierarchical techniques. In the paper we explain the modelling procedure of the manufacturing cell with single resource for different types of parts according to proposed decisions on places and transition in order to analyse the overall processing time in the cell. The designed model is based on a manufacturing cell from Faculty of Managerial and Technological Engineering, University of Oradea. In this present work, we present the behaviour of resource places, transitions and evaluate overall performance of the flexible manufacturing cell using CPNTools software.

1 Introduction

The useful nature of Petri Nets (PNs) resides in the power to identify the system behaviour in any kind of manufacturing model related to concurrent structures and asynchronies. Nowadays, the utilization of High Level Petri Nets in modern manufacturing sector for modelling purpose is rising. To establish a compound model using Petri Nets is a common subject now but the identification of the system behaviour in a particular manufacturing model is difficult. Coloured Petri Nets (CPN) provides system analysis and statistical analysis using state space analysis and helps to provide appropriate information about the behaviour in any kind of manufacturing model. The implementation of Hierarchy techniques in manufacturing system is not always necessary but for the design and modelling of complex manufacturing systems sometimes is very useful to build such models. To predict detailed behaviour of the cell or system the modelling and analysing is important. The design of hierarchical models with detailed behaviour with various levels is quite fashion in manufacturing system. Hierarchical Coloured Petri Net (HCPN) application in single model design with relation of sub-nets was first introduced by Huber [1]. Dynamic behaviour of automated manufacturing systems using coloured, timed, object

* Corresponding author: sanjibksaren@gmail.com

oriented PNs was described by Wang [2]. Dynamic tool allocation and performance evaluation of the FMS model was explained by Chen in [3] using object-oriented Coloured Petri Nets. Viswanadham in [4] uses Coloured Petri Nets to model and invariant analysis of Flexible manufacturing cell. In [5] the application of extended Petri Nets in error free hierarchical FMS model with operation of each subsystem using single resource active cycle is described. Blaga in [6] explained the design procedure of a Flexible manufacturing cell (FMC) using CPNTools software. The author in [7] explained the application of CPNTools software to model mechatronics system with implementation of hierarchy and colour. In [8] the application of hierarchical coloured Petri Nets found in production line modelling is described and also the author explained the benefits of CPNTools to model the production system.

In this paper, we briefly explain the Hierarchy modelling techniques of flexible manufacturing cell using Coloured Timed Petri Nets in order to identify the detailed behaviour of the cell and also explain the design procedure of each sub model of the cell and the overall system performance.

2 Flexible manufacturing cell configurations

The proposed, modelled flexible manufacturing cell is operating at the Faculty of Managerial and Technological Engineering, University of Oradea shown in Figure 1. The Flexible manufacturing cell contains one CNC machining centre TMA 55 OP based on a FANUC control system, two ABB IRB 1600 robots equipped with Schunk pneumatic gripping devices, automatic tool changer with a tool capacity in the tool magazine of 40 tools, conveyor transfer system and separate storage area for specific raw material and finished parts and two pallets for transporting the parts in the cell.

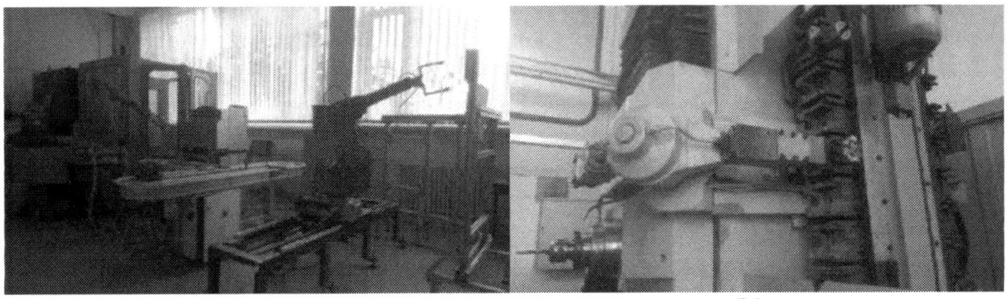

(a) (b)

Fig. 1. (a) Flexible manufacturing cell from the Faculty of Managerial and Technological Engineering, University of Oradea. (b) Automatic tool changer system with tool magazine inside the machine.

3 Proposed Hierarchical Flexible manufacturing cell model

The concept of hierarchical timed Coloured Petri Nets (HTCPN) provides an extended representation of the developed model with colour, time and untimed, hierarchy, priority in transitions etc. Coloured Petri Nets provides a pure graphical model with advanced modelling facilities using hierarchical techniques. CPN Petri nets can be transformed to hierarchical PN using subnet of the model it helps the system to be more transparent with tabular structure which analyses the system more effectively and present the model more detailed in order to better be understand. In this proposed model, we used CPN Tools to construct a single machine cell model with colour, time and hierarchy. The base model is

DOI: 10.1051/matecconf/201713701011

constructed with storage area for raw parts, for loading and unloading purpose we consider two robots, two pallets are considered for transport of raw parts and finished parts through the conveyor and a single machine for machining purpose. The resources are explained in the base model with mention of the places and each place contains meaningful parameters. CPNTools provides the means to construct the hierarchy model. In this cell we developed the base model with three sub models which are respectively: (a) part arrival sub model (b)

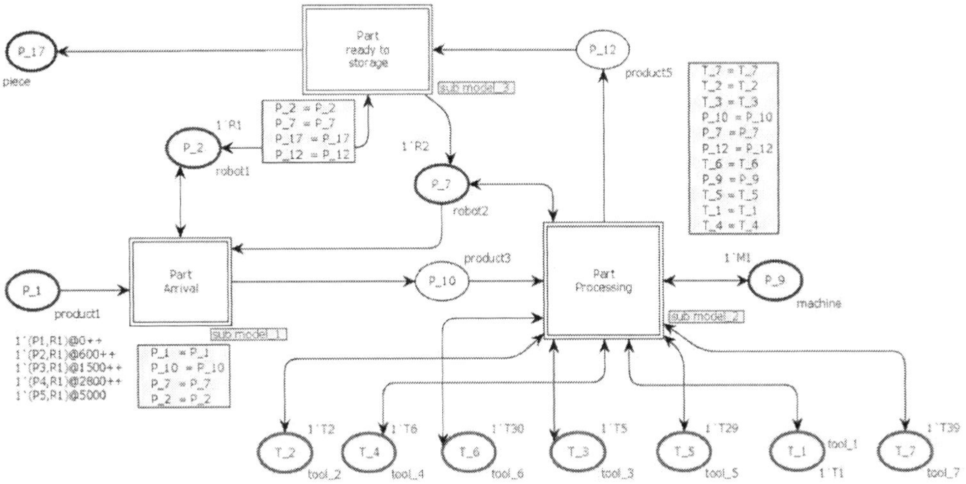

part processing sub model (c) Part complete sub model.

Fig. 2. Hierarchical structure of the flexible manufacturing cell.

Here, we define the colour set for the cell: (1) colset piece = with P1|P2|P3|P4|P5 timed; (2) colset machine = with M1 timed; (3) colset robot1= with R1 timed; (4) colset robot2= with R2 timed; (5) colset pallet1= with B1 timed; (6) colset pallet2= with B2 timed; (7) colset tool_1= with T1; (8) colset tool_2= with T2; (9) colset tool_3= with T5; (10) colset tool_4= with T6; (11) colset tool_5= with T29; (12) colset tool_6= with T30; (13) colset tool_7= with T39. In figure 1 the hierarchical model contains the resource places and blocks of the sub model. In this hierarchical flexible cell, the places and their interpretation play significant role to understand the design of the cell. Place P_1 is Storage area for raw parts which have 5 colour token marking with 1`(P1,R1)++ 1`(P2,R1)++ 1`(P3,R1)++1`(P4,R1)++1`(P5,R1)++, place P_2 and P_7 defines the robots used for loading and unloading purpose both have one colour token and the marking is 1`R1 and 1`R2, place P_9 defines the CNC machine which have one colour token carry marking value 1`M1. All the tools have colour token value one and the places are respectively, T_1, T_2, T_3, T_4, T_5, T_6 and T_7. The marking are for the tools 1`T1, 1`T2, 1`T5, 1`T6, 1`T29, 1`T30, 1`T39. Other place in the model P_10 defines the connection between R1 and M1, P_12 define the connection between R2 and M1 and P_17 defines the final storage area for the finished parts.

3.1 Sub-models of the cell

In this proposed hierarchical flexible cell, arrival time is considered in resource place P_1, for all five different parts mentioned in figure 3. The part arrival time for P1= 0 sec, P2= 600 sec, for P3= 1500 sec, P4=2800 sec, for P5= 5000 sec in the cell. All the parts are processed in the CNC machine and the observed time for each part is collected during experiment. The observed time according to parts are P1 = 336 sec; P2 = 458 sec; P3 = 1158 sec; P4 = 2160 sec and P5 = 2460 sec.

Part arrival sub model can be described as follows: In part arrival section place P_1 contains arrival time for each part. In figure 3, we mentioned the arrival time decision for five different parts in the cell such as P_1 with 1`(P1,R1)@0++1`(P2,R1) @600++1`(P3,R1) @1500++1`(P4,R1) @2800++1`(P5,R1) @5000. Five Colours token appears in the resource place which describes the availability of raw part in storage area. The transitions T1, T2, T3 and T4 contain delay time @+30 seconds to create connection between robot1, pallet1 and robot2 in the sub-model. In Table 2 and 3, we mention related transitions and places for the sub model. In this, sub-model robot1 takes parts from storage area according to arrival time and unload of the part in pallet1 for transport to the next section. Robot1 is referred as an input/output function in this sub-model. Place P_1 and robot2 work as an input function and place P_10 as an output function in the sub-model. The token value represents the status of the resource places such as in this sub-model place P_1 contain token value 5 which represent the number of parts in resource place and P_4

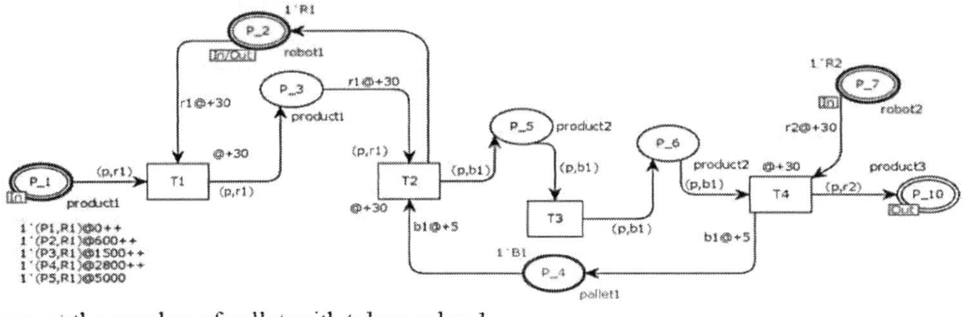

present the number of pallet with token value 1.

Fig. 3. Sub model structure of part arrival for FMC.

In the part processing sub model: the selection of appropriate tool for processing parts is necessary. To construct this section each tool are represented in different resource places, such as 1`T1, 1`T2, 1`T5, 1`T6, 1`T29, 1`T30 and 1`T39.

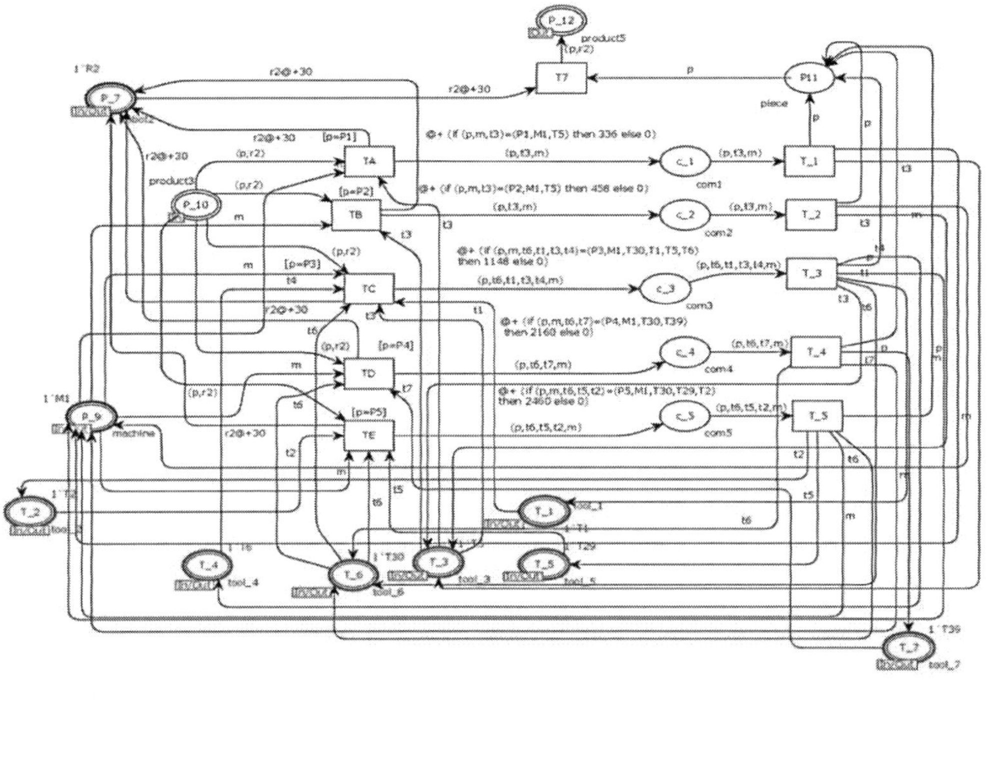

Fig. 4. Sub model structure of part processing for FMC.

The tools are used for different types of operations in the machine. The token value for the tools denoted 1 means only one tool is available in the resource places. The machine, robot2 and all seven tools are defined as an input and output function in this sub model. During design in each transition we define a guard function to identify the exact connection between tool and part. In transition TA, the guard function [p=P1] is activated when information is carried by the arc from place to transitions. So, each transition is designed in such a manner that it can identify the exact part for processing in the machine using the appropriate tools. Also, in this section we have to establish a connection between place and transitions, in the arc we defined variables according to tool and parts. If we closely look to the design of this section, we observe that the processing time is defined in each transition. Each transition defines part processing time for each different part. The processing time is defined for P1=336 sec, P2= 458 sec, P3= 1158 sec, P4= 2160 sec and P5=2460 sec respectively. Each part is processed in the machine using appropriate place such as place c_1 for P1, c_2 for P2, c_3 for P3, c_4 for P4 and c_5 for P5. The manufacturing cell is based on single machine so each time only one transition is activated during operation in the cell. The part storage sub model is defined as followed: place P_2 defined as an input/output function and P_12 defined as an input function in the sub model. P_7 and P_17 are output functions in this sub model. Pallet2 with resource place P_14 contain token valued 1 with marking 1`B2. This section describes the operation when the part is completed robot2 load the part in pallet2 and transfer the part near robot1 and the robot1 unload the part from pallet2 and store the finished part in the storage area. If we closely look at the model we can observe the delay of 30 sec for robot1 and robot2 for loading and unloading purpose.

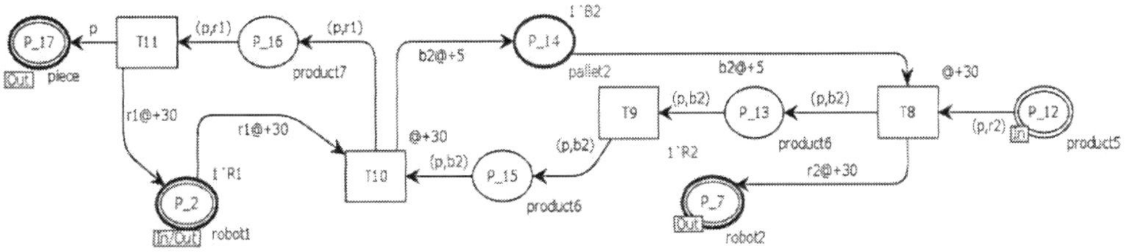

Fig. 5. Sub model structure of end section for FMC.

4 Simulation results

In this flexible manufacturing cell, we are processing five different parts in a single machine according with part arrival time decisions. The part will be released from place P_1 with a specific time. During simulation it is identified that robot1 take the part from resource place and load the part in pallet1 to transfer it through the conveyor. In a specific time later the robot2 takes the part from pallet1 and put it inside the CNC machine, when the part is processed the sub model is activated and the appropriate tool is selected from resource places for the defined part in the cell. Such as, for example, in transition TA, tool_3 is activated for processing part1 in the machine. During simulation, it's identified that the colour token is removed from resource places and performs the operation according to firing on transitions. The part processing is following the arrival time and the part processing time in the cell. After completion of the part robot2 unloads the part from the machine and loads it in pallet2 which carry the part through the conveyor and when robot1 is free it takes the part from pallet2 and loads it in the final storage area. Coloured Petri Nets helps to define simulation process by moving colour token from one place to another. In figure 6, we define the simulation observation after processing all the parts. The cell is modelled using time units so; the time will appear in the mentioned places.

The total time appears in place P_17 with marking value 1`P1@486+++1`P2@1208+++1`P3@2798+++1`P4@5110+++1`P5@7630. Also, in place P_2, P_7 and P_9 the time appears after simulation respectively 7660 sec, 7630 sec after 55 steps. CPNTools provides the number of steps to justify the system identification and provide possible information according to token values and specific time mentioned in resource places and transitions.

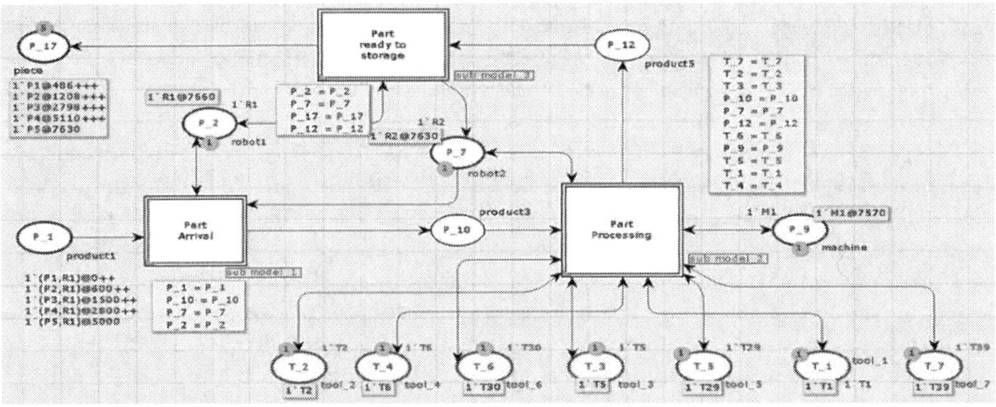

Fig. 6. Final time appears in base model after simulation runs.

5 Conclusions

In this paper we described the modelling techniques and simulation experiment of a flexible manufacturing cell using CPNTools software implement of hierarchical formalism. The contribution of this work was the modelling of flexible cell and the identification of the behaviour of each sub model by simulation of the cell. Each sub model carry significant characteristic to identify the status of the resource places. The implementation of time in hierarchical model provides more realistic results compared with normal Colour Petri Nets. Here, in this cell, the total time observed from the simulation which helps to understand the value for real system.

The paper published has been sponsored under the Erasmus mundus partnership program agreement vide number 2014-0855/001-001 coordinated by and between University of Oradea and City University of London Under Action Plan 2 for the year 2015-2018.

References

1. P. Huber, K. Jensen, R.M. Shapiro, In: Rozenberg G. (eds) Advances in Petri Nets, Lecture Notes in Computer Science, Springer-Verlag (1990).
2. L. C. Wang, S. Y. Wu, Computers and Industrial Engineering, **2, 34,** (1998).
3. J. Chen, F.F. Chen, **2, 21,** (2003).
4. N. Viswanadham, Y. Narahari, Proc. IEEE Int. Conf. Robot. Automat, **4,** (1987).
5. K. P. Valavanis, IEEE Transactions on Systems, Man, and Cybernetics, **1, 20,** (1990).
6. F. Blaga, I. Stanasel, A. Pop, V. Hule and T. Buidos, Annals of the Oradea university, Fascicle of Management and Technological Engineering, **1, 4** (2014).
7. Jadlovský. J, Ilkovič. J, Procedia Engineering, **48,** (2012).
8. Eloundou. J, Baudry. D, Bensrhair. A, Xth International Conference on Integrated Design and Production, CPI (2015).

State space properties of Flexible Manufacturing Cell based on Hierarchical Timed Colored Petri Nets

Sanjib Kumar Saren[1,*], *Florin* Blaga[2], and *Tiberiu* Vesselenyi[1]

[1]Mechatronics Department, University of Oradea, Romania
[2]Industrial Engineering Department, University of Oradea, Romania

Abstract. The subject of this paper is focused on the state space behavioral structure of a flexible manufacturing cell (FMC) using a hierarchical model based on Colored Timed Petri Nets theory. In the paper we discussed various types of properties of the state space in CPNTools and a simulation of the manufacturing cell model with single resource places and transition in order to identify the overall time in the cell. The proposed designed model is based on the manufacturing cell from Faculty of Managerial and Technological Engineering, University of Oradea. In this recent work, we discussed the sequence of resource places and transitions of the flexible manufacturing cell explanation with state space analysis using CPNTools software.

1 Introduction

The implementation of high level Petri Nets provides ideas and applications in various types of problems to improve Flexible manufacturing system (FMS). The applications of Colored Petri Nets are more and more found in manufacturing sector. State space feature is inbuilt in CPNTools software which provides directed graph with a node with each possible reachable marking in the system. State space tools are used to create reachability, boundedness, home properties, and liveness and fairness properties in the system. The application of state space is explained by Christensen in [1] where the use of state space analysis in Colored Petri Nets is described when the state space is finite and also state space allows the user to analyze the system behavior. In [2] the state space method for Colored Petri Nets using time concept and the condensed infinite to finite state space method is presented. In [3] the authors addressed state space analysis in Colored Petri Nets to define each possible behavioral property for nodes and arcs in the concurrent system. Some major features are described in [4] where Colored Petri Nets integrates items such as liveness properties, meaning absence of the deadlocks, boundedness properties meaning finiteness of buffers and resources required for the model and properness which are explained as the ability to recover from failure.

* Corresponding author: sanjibksaren@gmail.com

In this present work, we briefly discussed the Hierarchy model of a flexible manufacturing cell implementing Colored Timed Petri Nets to observe the detailed behavior of the cell and explain the state space analysis of the model with explanation of various properties related to the flexible cell.

2 Hierarchical Flexible manufacturing cell designs

Colored Petri Nets are high level Petri Nets with extension of Place Transitions Petri Nets which allows carrying color token data with values which distinguishes them from other Petri Nets. Temporal behavior of the system can be identified with Colored Petri Nets using delay time and arcs time in the system. Theoretically, the definition of Hierarchical Color Petri Nets (HCPN) is explained elaborately in [5]. A hierarchical CPN has a 9 tuple = (S, SN, SA, PN, PT, PA, FS, FT, and PP) which are explained below:
(a) S is a finite set of pages such that: Each page s ∈ S is a non-hierarchical CPN:
CPN= (Σ, P, T, A, N, C, G, E, I) (The non-hierarchical CPN is defined in [5])
The sets of net elements are pair wise disjoints: ∀ s1, s2 ∈ S: [s1 ≠ s2 ⇒ (P1 ∪ T1 ∪ A1) ∩ (P2 ∪ T2 ∪ A2) = ø].
(b) SN ⊆ T is a set of substitution nodes.
(c) SA is a page assignment function. It is defined from SN into S such that: No page is a subpage of itself: {s0, s1 …sn ∈ S*| n ∈ N+ ∧s0=sn ∧∀k ∈ 1..n: sk ∈ SA(SNsk-1)}= ø.
(d) PN ⊆ P define a set of port nodes.
(e) PT = port type function. It is defined from PN into {in, out, i/o, general}.
(f) PA = it's a port assignment function,
(g) FS ⊆ P is a finite set of fusion sets,
(h) FT = fusion type of function,
(i) PP ∈ SM define multi-set of prime pages.

HCPN are useful to model manufacturing systems with timed constraints and provide high level information about the system. While Coloured Petri Nets used to model the structure and flow in a production system, the Hierarchical Coloured Petri Nets allow the introduction of analysis and information processing. In [6] is described that CPNTools provides to construct hierarchical models. In Figure 1, this manufacturing cell is modeled based on three sub models which are respectively: (a) part arrival (b) part processing and (c) part storage.

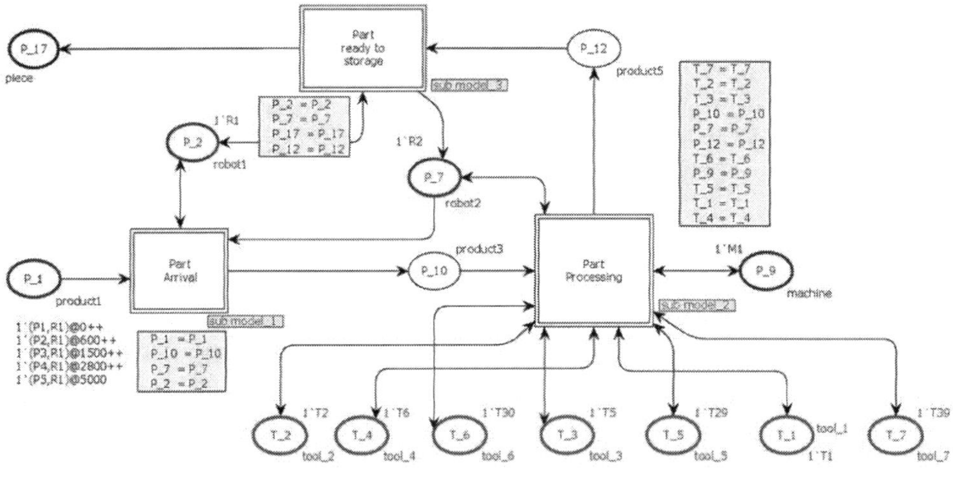

Fig. 1. Hierarchical model of the flexible manufacturing cell.

 DOI: 10.1051/matecconf/201713701012

3 Simulation observation of the model

During the simulation of the hierarchical model of the flexible cell, the number of steps to complete the manufacturing process can be found. During the simulation each sub model according to part arrival in the cell is activated. Each operation is depending on the mentioned color token in resource places and the described time decision in each transition in the cell. The operation steps found in the cell to complete the process is 55 steps. In the followings we provide the information about steps and sequences for the cell. It is clear that in the model we provide the time in transition to execute the operations in the cell and in each case the time involved in transition is observed for all three sub models. The steps provide the behavior of each sub model with respect to color set and variables in the cell. In Figure 2, we represent the observed the time after simulation in the cell and the token value for the respective resources.

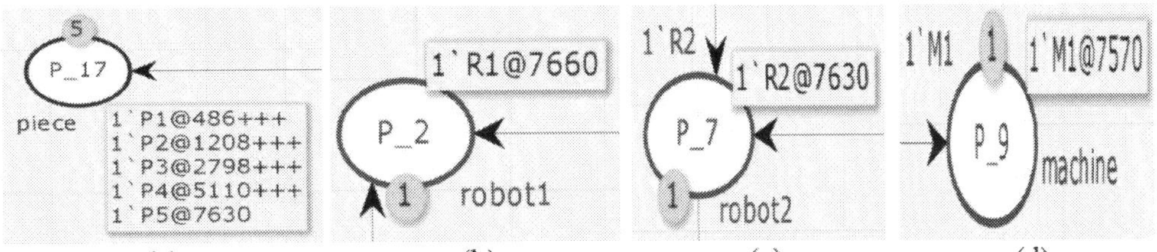

(a) (b) (c) (d)

Fig. 2. Simulation results in (a) Total time in final storage. (b) Time observed for robot1. (c) Time observed for robot2. (d) Time for the machine in hierarchical flexible manufacturing cell model.

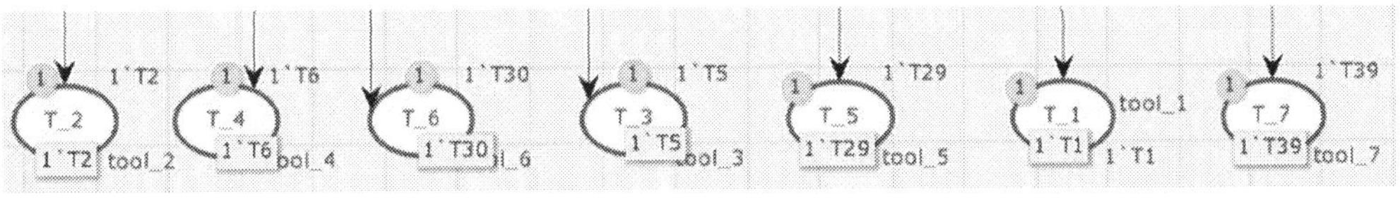

Fig. 3. Tools used for processing of parts containing colour token value after simulation.

The steps and sequences are very important to complete the simulation process for the cell. The model is based on time in each transitions and also resource places. In this hierarchical model during simulation CPNTools provides each sub model operation with the respective to time. The important steps and sequences are mention in Table1.

Table 1. Time and sequences in the cell

Time (Seconds)	Transitions	Sequences
0	T1	(1:sub_model_1) - p = P1 - r1 = R1
486	T11	(1:sub_model_3) - r1 = R1 - p = P1
1208	T11	(1:sub_model_3)- r1 = R1 - p = P2
2798	T11	(1:sub_model_3)- r1 = R1 - p = P3
5110	T11	(1:sub_model_3) - r1 = R1 - p = P4
7630	T11	(1:sub_model_3) - r1 = R1 - p = P5

4 State space analysis of the hierarchical flexible cell

A state space analysis is a report of the analysis for all possible occurrences of the part in the model of the system. State space provides statistic report of the model, boundedness properties, home properties, liveness properties and fairness properties for transition in the model. The CPNTools software features the state space tool which is able to generate strongly connected Component Graph (SCC graph) and reachability properties for the constructed model. To generate and collect a report from the state space module we need to calculate the state space, calculate SCC graph and save the report in a notepad. The display of the nodes with specific options is available in the tool and also it can generate successors and predecessors for the nodes. In Figure 4, we present view of state space tool in CPNTools software to generate the properties for the cell.

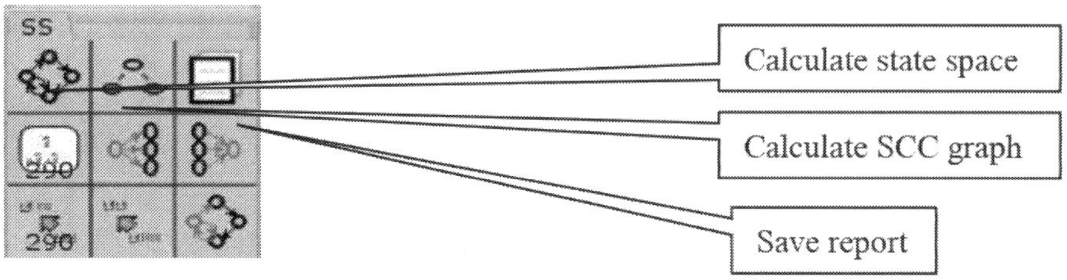

Fig. 4. State space tool in CPNTools software.

The static information of the cell is as follows: in state space, the full status found for the cell, produced 102887 nodes, 204968 arcs and the time took was around 210 sec to generate the state space model. In state space we generate SCC graph and reachability properties. In Figure 5, we show the part of state space properties for the flexible manufacturing cell.

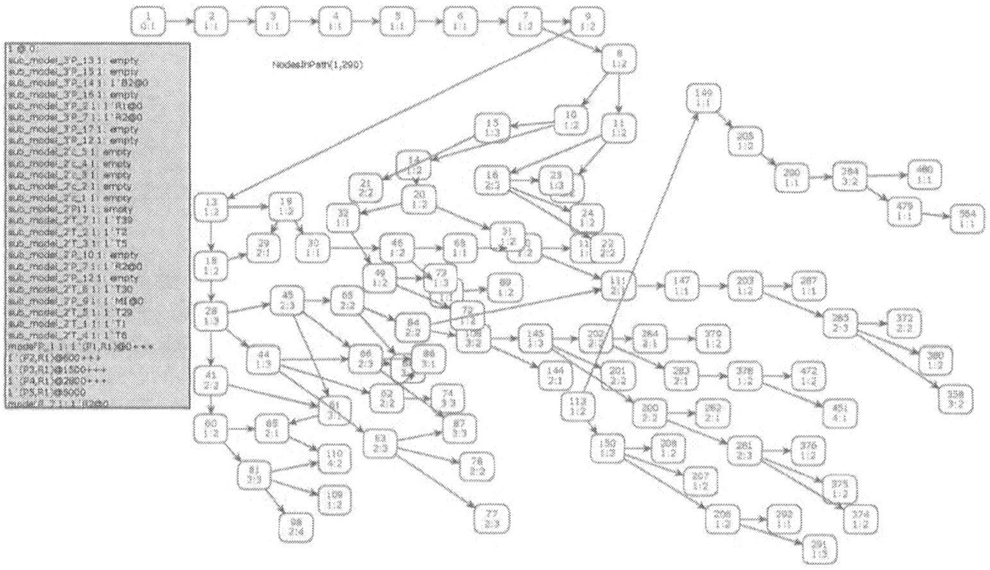

Fig. 5. State space structure view of the hierarchical model based manufacturing cell.

4.1 Boundedness Properties in the cell

We present the boundedness properties of the hierarchical model with the best possible multi set bounds for the cell. The best integer represents all the sections of the cell. In this section, the lower multi set bounds for each place are found empty during generating the state space. Each integer bound presents the marking situation with respect to their places.

Respective upper multi set bounds define the marking of that place with color token value. The best integer with best upper multi set bounds are respectively:

Table 2. Best Integer Bounds for Hierarchical Model

Best Integer Bounds	Best Upper Multi-set Bounds	Best Lower Multi-set Bounds
model'P_10	1`(P1,R2)++1`(P2,R2)++1`(P3,R2)++1`(P4,R2)++1`(P5,R2)	Empty
model'P_1	1`(P1,R1)++1`(P2,R1)++1`(P3,R1)++1`(P4,R1)++1`(P5,R1)	Empty
model'P_12	1`(P1,R2)++1`(P2,R2)++1`(P3,R2)++1`(P4,R2)++1`(P5,R2)	Empty
model'P_17	1`P1++1`P2++1`P3++1`P4++1`P5	Empty
model'P_2	1`R1	Empty
model'P_7	1`R2	Empty
model'P_9	1`M1	Empty
model'T_4	1`T6	Empty
model'T_6	1`T30	Empty
model'T_7	1`T39	Empty
model'T_1	1`T1	Empty
model'T_2	1`T2	Empty
model'T_3	1`T5	Empty
model'T_5	1`T29	Empty
sub_model_1'P_3	1`(P1,R1)++1`(P2,R1)++1`(P3,R1)++1`(P4,R1)++1`(P5,R1)	Empty
sub_model_1'P_4	1`B1	Empty
sub_model_1'P_5	1`(P1,B1)++1`(P2,B1)++1`(P3,B1)++1`(P4,B1)++1`(P5,B1)	Empty
sub_model_1'P_6	1`(P1,B1)++1`(P2,B1)++1`(P3,B1)++1`(P4,B1)++1`(P5,B1)	Empty
sub_model_2'P11	1`P1++1`P2++1`P3++1`P4++1`P5	Empty
sub_model_2'c_1	1`(P1,T5,M1)	Empty
sub_model_2'c_2	1`(P2,T5,M1)	Empty
sub_model_2'c_3	1`(P3,T30,T1,T5,T6,M1)	Empty
sub_model_2'c_4	1`(P4,T30,T39,M1)	Empty
sub_model_2'c_5	1`(P5,T30,T29,T2,M1)	Empty
sub_model_3'P_13	1`(P1,B2)++1`(P2,B2)++1`(P3,B2)++1`(P4,B2)++1`(P5,B2)	Empty
sub_model_3'P_14	1`B2	Empty
sub_model_3'P_15	1`(P1,B2)++1`(P2,B2)++1`(P3,B2)++1`(P4,B2)++1`(P5,B2)	Empty
sub_model_3'P_16	1`(P1,R1)++1`(P2,R1)++1`(P3,R1)++1`(P4,R1)++1`(P5,R1)	Empty

The best Integer bounds for the hierarchy model describe the upper bounds and lower bounds. During generating the state space for the model the values appear in a data sheet. The observed values for integer bounds are respectively: model'P_1(upper bound=5), model'P_17 (upper bound=5), rest of integer have upper bound1 which are model'P_10, model'P_12, model'P_2, model'P_7, model'P_9 , model'T_4, model'T_6 , model'T_7, model'T_1, model'T_2, model'T_3, model'T_5, sub_model_1'P_3, sub_model_1'P_4, sub_model_1'P_5,sub_model_1'P_6,sub_model_2'P11,sub_model_2'c_1,sub_model_2'c_2, sub_model_2'c_3,sub_model_2'c_4,sub_model_2'c_5,sub_model_3'P_13,sub_model_3'P_1

4,sub_model_3'P_15,sub_model_3'P_16 . The lower bounds are found zero for each integer bounds.

4.2 Home, Liveness and Fairness Properties in the cell

The Home Properties of the cell: in this flexible cell, initial marking is not a home marking but in home marking it is possible to return to the initial marking. The marking depends on the design condition of the model. In our case the flexible cell contain no home making which is found during the generation of the state space. So, there is no way to return to the initial making in the cell.*Liveness Properties of the cell*: A liveness property provides information about the enabling of each transition by firing an arbitrary number of transitions in the model. In state space, dead marking means a marking of the net where no transition is enabled. This also provides information about Dead Transition Instances and Live Transition Instances in the state space. In this flexible cell we found five dead marking according to state space generated report. This model, do not contains Dead Transition Instances what means that a transition never become enabled in the cell and also found no Live Transition Instances which means a transition which always will be enabled after firing in the cell. In this, model dead marking were observed during the generation of the state space, the markings are: 290, 99497, 99496, 96100, 96069, and 96068. *Fairness Properties of the cell*: observing the state space we found that no infinite sequence occurred. The state space report provides the order of finite transitions in the cell.

5 Conclusions

In this paper we described the modeling techniques and state space structure of the flexible manufacturing cell using CPNTools software implementing the hierarchical formalism. The contribution of this work is modeling and simulation of flexible cell and identifying the behavior of the hierarchical model. The model carry significant characteristic to identify the status of the places and transitions. Also the state space analysis provides static reports for the model with features which includes the liveness properties, boundedness properties, fairness properties to analyze the simulation process.

The paper published has been sponsored under the Erasmus mundus partnership program agreement vide number 2014-0855/001-001 coordinated by and between University of Oradea and City University of London Under Action Plan 2 for the year 2015-2018.

References

1. C. Soren, L. M. Kristensen, Lscs 01: Petri Net Approaches for Modelling and Validation (2003)
2. K. Jensen, Proceedings of 2nd International Colloquium on Petri Net Technologies for Modelling Communication Based Systems, Berlin (2001)
3. K. Jensen, Formal Methods in System Design, Kluwer Academic Publishers **9** (1-2) 7 (1996)
4. N. Viswanadham, Y. Narahari, Proc. IEEE Int. Conf. Robot. Automat **4** (1987).
5. K. Jensen, *Colored Petri Nets: Basic Concepts, Analysis Methods and Practical Use* **1**, Springer – Verlag, (1992)
6. http://cpntools.org/ accessed on (2015)

Considerations on studying the characteristics of the elastic axle steering system on railway vehicles

Ioan Sebeşan [1,*], *Sorin* Arsene [1], and *Mircea Radu* Sebeşan [2]

[1]Polytechnic University of Bucharest, Sector 6, Splaiul Independenţei, No. 313, Romania
[2]SC Metrorex SA, Sector 1, Dinicu Golescu, No. 38, Romania

Abstract. The new orientation in construction of bogies consists in realization possibility "negotiation the radii of curvature", the axes being placed in an approximately radial position. This reduces wear on the wheel and rail surfaces if the stability of the horizontal transverse movements of the vehicle is not affected. In railway motor vehicles, these requirements are achieved by means of rubber elastic elements which by their own damping capacity also contribute to the reduction of axle movement. The paper presents a series of considerations regarding the elastic constants of the springs from the axle steering system.

1 Introduction

Of particular importance for ensuring a stable circulation and low forces wheel-rail friction it has the axle steering system. This must transmit interaction forces with the tread without impeding the operation of the suspension, while ensuring and a correct position of the axle in the track and towards the vehicle.

At high speeds, it turned out be more advantageous elastic driving of the axles both in the transverse and longitudinal directions.

The new orientation in the construction of bogies consists of making it possible to negotiate of the radii of curvature, that is, the quasi-radial settlement in the curve of the axle of the bogie. This reduces the wear of the wheel and rail contact surfaces, but it is necessary not to cause instability of the as hunting movements.

The tendency of radial settlement of an axle can be done if the rolling profile it is wear and the driving system is sufficiently elastic. An excessive elasticity of the axle steering springs negatively influences the hunting movement of the axle in the sense at it reduces the critical velocity of the hunting movement.

The problems presented are part of the theoretical and experimental studies conducted the coordination of Prof. Eng. Ioan Sebeşan, within the Department of Railway Rolling Stock of the Polytechnic University of Bucharest.

* Corresponding author: ioan_sebesan@yahoo.com

2 Influence of the elasticity of the driving system on the geometric alignment of the axle in the curves

In fig. 1 shows the case of a bogie in which the axle driving arm 1 is connected to the bogie frame 2 by means of a silent-block 3, with possibilities of varying the longitudinal and transverse elasticity by conical tightening of the parts 4.

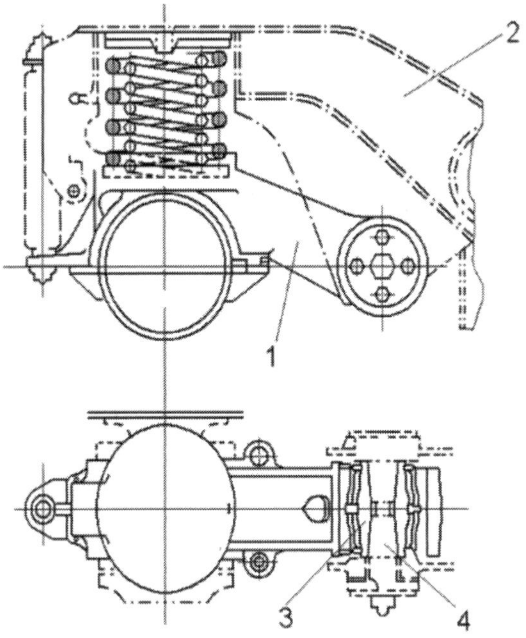

Fig. 1. System with articulated drive arm with a silent block with possibilities for adjusting the elasticity.

The mechanical model of an elastic driving bogie used in the study is shown in fig. 2.

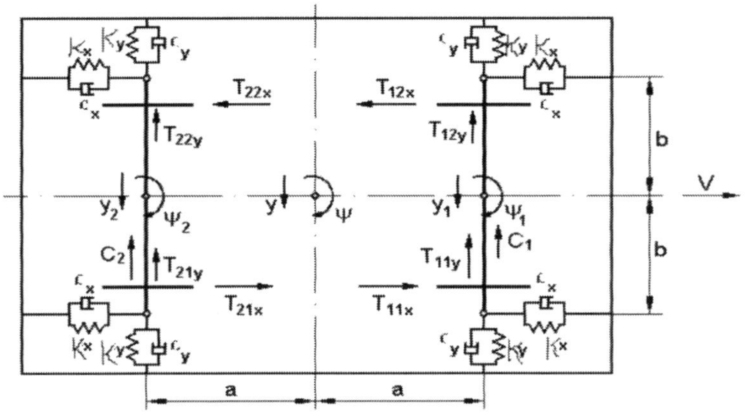

Fig. 2. Mechanical model of the bogie with the forces acting on the elastically driven axles.

For an elastic driving bogie axle, we will analyse the inclusion curve assuming a quasi-static stationary motion regime. A reference work that deals with this problem has been developed by D. E. Newland [1]. It is supposed that under the action of external Fn and contact forces between wheels also rails, bogie is positioned in curve in Fig. 3, axles makes with normals to curve (attack) α_1 and that respectively α_2. Also, it supposed that there are no large sliding of wheels (with profile wear) but pseudo sliding proportional with to contact forces and that forces balance can ensure running of axles without contact on lips (of wheelsets).

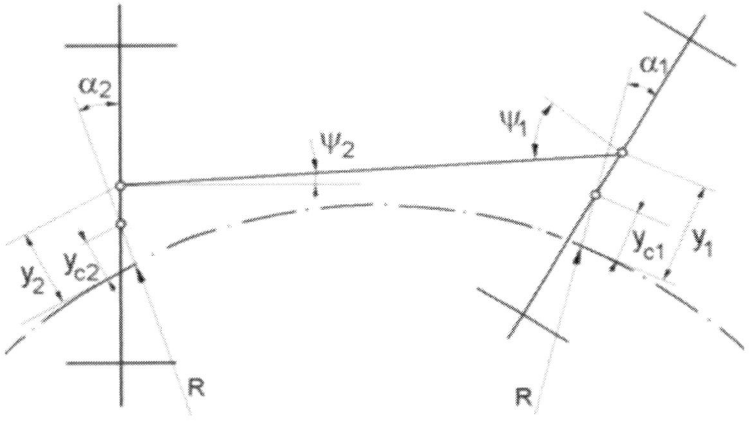

Fig. 3. Driving axle bogie spring curved movements.

To the axis of the track, the axle centers are lagged outside with y_{c1} and, respectively, y_{c2} and the bogie chassis reduced to its longitudinal axis, is lagged with the right axle y_1 and, respectively, y_2. During the movement, the forces of friction, forces of centering and the forces from guidance system, of each axle, must be balanced.

In front of the chassis axles are rotated with ψ_{12}.

On the axes act on the friction forces T_x, T_y, the centering forces C, forces in the guidance system dependent on the elastic constants k_x, k_y as well as an outside force transmitted via bogie system - vehicle bodywork.

The positions of the axles in the curves result from the equilibrium conditions of the forces acting on the two axles and the bogie frame respectively. Friction coefficients between wheels and rails were evaluated on the basis of Kalker's theory [4]. Details of equilibrium equations are presented in detail in the paper [1]

The theoretical study shows that an axle-driven bogie will run without slipping on any curve whose radius is:

$$R \geq \frac{\chi}{2C\mu}\sqrt{1+\frac{a^2}{4\cdot e^2}A^2}; \tag{1}$$

where χ – is the pseudo-slip coefficient $\left(\chi_x = \chi_y = \chi\right)$; μ – coefficient of adhesion; e – semi-track gauge; γ – being the actual conicity of the wheel profile and Q – wheel load.

$$A = 1+\frac{\chi Q}{2c_x b^2}\cdot\frac{e\gamma}{r}\left(1+\frac{4c_x b^2}{c_y a^2}\right); \tag{2}$$

$$C = 1 + \frac{(\chi Q)^2 a}{\left(2c_x b^2\right)^2} \cdot \frac{e\gamma}{r}\left(1 + \frac{4c_x b^2}{c_y a^2}\right); \tag{3}$$

The more elastic the suspension, the smaller the radius R of the curve, the axes tilting to a radial position.

Comparing the displacements of lateral force F_n, is observed that the rear axle is shifted more than the front axle, this movement being independent of the radius of the curve.

The displacement under the effect of force F_n is also independent of the deviation from the track center line, which occurs even if the bogie side not exercise any power and actually indicate the inherent ability to be self-guided bogie in way pseudo sliding forces between wheels and rails.

3 The influence elasticities from conduction system of the axles on the bogie hunting stability

The study stability motion of winding bogie axle with elastic driving axis is based on relationships obtained after linearization phenomenon of winding. Linearization winding phenomenon is realized taking in consideration that the contact forces vary linearly with lateral movement of the axle also, neglecting friction and games of various elements of the bearing structure of the vehicle and neglecting tread irregularities and discontinuities. Not least taking in consideration equivalent conicity wheel profile as constant and proportional to the tangential force pseudo sliding point of contact of the wheel with the rail. [3], [5]

This study aims to determine the velocity at which the stable movement of winding vehicle equipped with elastic driving axle bogies will turn into an unstable movement, to determine the establishment of critical speed, which when exceeded will result in a rapid deterioration walking. In other words, we aimed to determine the maximum speed that can be reached safely by vehicle. Consider the case of general movement of the bogie in winding which the suspension consists in arches having elastic constants k_x, k_y, and the linear characteristic of the shock absorbers (non-viscous), which amortizations constants c_x and c_y (Fig.2).

Center of mass of the bogie is considered located in the axle axles. Determination of critical speed and critical pulse respectively when is neglect the suspended mass of bogie and of amortizations, can be based on the equations of movement for the bogie frame and axles, neglecting balance of axle, the effect of spin and gyroscopic noted: k_x and k_y - the elastic constants in the longitudinal and transverse, m_o - the wheelset (axle) mass, Q - the wheel load, a - the wheelbase bogie, b - the transverse cross midway between the suspension arches e - the midway between the nominal rolling circles, r - the wheel ray (radius), v - the velocity running speed, γ - equivalent conicity.

The model studied was applied to the bogies of towed vehicles by neglecting the suspended bogies and shock absorbers.

For the rail wheel friction coefficient, it can be considered the work of P van Bommel [3] which recommends some approximate values of the pseudo-slip coefficient based on the relations established by Kalker.

In the general motion case, the stability study and the determination of the amplitudes of the repositioning oscillations can be done by applying a numerical method of integrating the motion equations by means of the calculation technique. With the above assumptions, the oscillating system can be reduced to 4 degrees of freedom. In this case, the critical velocity v_c, respectively the critical pulse ω_c, is determined with simple relations derived from the equation of the pulsations of the system [1].

Thus, the characteristic equation of the system is:

$$\begin{vmatrix} T_1(p) & 0 & -4\chi Q & 0 \\ 0 & T_1(p)+4c_y^* & -4c_y^* a & -4\chi Q \\ -4\chi Q\frac{e\gamma}{r} & -4c_y^* a & T_2(p)+4c_y^* a^2 & 0 \\ 0 & 4\chi Q\frac{e\gamma}{r} & 0 & T_2(p)+4c_y^* b^2 \end{vmatrix} = 0; \quad (4)$$

where $c_y^* = \dfrac{c_x c_y b^2}{c_y a^2 + c_x b^2}$; $T_1(p) = 2m_o p^2 + \dfrac{4\chi Q}{v}\text{p}$; $T_2(p) = 2m_o e^2 p^2 + \dfrac{4\chi Q e^2}{v}\text{p}$;

Considering that the stability limit has been reached when $v=v_c$, substituting $p=j\omega_c$ ($j=-1$) results the relations of critical speed determination.

4 Numeric application. Experimental determinations

With the constructional characteristics of the Y32 bogie, which equips the passenger coaches in the CFR park, it results that for the elastic constants of the axle $k_x = 10^7$ N/m and $k_y = 5\cdot10^7$ N/m it results that it runs without slipping of the wheels on the track if the radius of the curve is $R\geq1552$ m.

Studying the stability of the vehicle on straight-line traffic resulted in safe traffic up to critical speed $v_c = 64$ m/s = 230 km/h.

A series of experimental determinations were made at the Romanian Railway Authority (AFER) in Bucharest and Softronic Craiova. On the Făurei ring, the bogie Y32 was experimentally resulting in a stable movement up to 210 km / h.

Experimental determinations have been made at Softronic on special stands for rail vehicle suspensions.

In fig. 4 there is shown the stand for determining the elastic characteristics and dynamic operation of the suspension elastic element.

Fig. 4. Stand for determining elastic features.

5 Conclusions

From the curve flow study is obtained a minimum radius value for not occurring sliding on the wheel - rail contact. In this case, there is a tendency of radial positioning of the first axle. The relationship shown shows the advantages of radial guidance in the case of curves with low radius, but with an appropriate adaptation of the axle guidance system of the vehicle.

Also results of the study have led to the conclusion that, when the speed of about 230 km/h is exceeded, the vehicle's winding movement will become unstable, that leading to inadmissible transverse stresses of the runway and even endangering the safety circulation. The maximum safe running speed of a vehicle shall be less than 10-15% from critical shock velocity if a possible change in time of the elastic characteristics of the axle guidance system is considered.

The presented relations allow to analyze the influence of the various constructive parameters of the bogie on the winding movement and to determine the constructive conditions for expansion at higher winding stability. Although the applied method is based on a series of simplifying hypothesis, it can be used for rapid engineering evaluations of bogie performances.

The calculation model belongs to the author of the paper and it is experimentally validated on a series of high-speed bogies analyzed in the Railway Rolling Stock Department from the Polytechnic University of Bucharest.

Experiments have also been carried out at AFER as well as at Softronic Craiova. They validated the theoretical semi-distances.

This work was supported by a grant of the Romanian National Authority for Scientific Research and Innovation, CNCS–UEFISCDI, project number PN-III-P2-2.1-PTE-2016-0008, within PNCDI III

References

1. D.E. Newland, *Steering a Flexible Railway Track in Curved Track*. (Transaction of ASME 1969)
2. P. van Bommel, *Considerations lineaires concernant le mouvement de lacet d'un vehicule ferroviaire*. (UIC/ORE ,C9 1968)
3. J.J. Kalker, *Three-dimensional elastic bodies in rolling contact.Kluwer* (Academic Publishers, Dortrech E.A. 1990)
4. R. Joly, Rail International, **12** (1971)
5. R Joly, M. Laurent, *Etude de la dynamique transversal d'un vehicule ferroviaire. Banc experimental de Vitry*, (Rapport SNCF, division des essays de materiel 1974)

Selection of methods used for analyzing the standard time needed to complete the assembly process by means of a fuzzy analytic hierarchy process

Robert Cieślak[1]*and *Marcin* Suszyński[2]

[1] State University of Applied Sciences in Konin, Faculty of Engineering
[2] Poznan University of Technology, Faculty of Mechanical Engineering and Management

Abstract. This paper is a review of four main methods of analyzing the standard time needed to complete a given job: chronometric analysis, working-day activity study (working-day photo), work sampling observations, and analysis of basic motions. Thanks to the application of FAHP (Fuzzy Analytic Hierarchy Process) method, the values of weighted coefficients were obtained which describe the importance of selecting the criteria of the choice of methods used to analyze the standard time in the assembly process. In the summary paragraph of this paper, the results of analyzing the criteria are discussed.

1 Analysing the standard time for a given job

1.1 General issues concerning the standard time for a given job

Work can be defined as activities requiring mental and physical efforts, the purpose of which is to produce goods and provide services that meet human needs.

In everyday life an important role is played by gainful work, which is a factor determining people's daily rhythms. From the economic point of view, working time is, besides the achieved results, a measure of the amount of work. Working time also determines the necessary part of work performed by society, which everyone ought to contribute to the production of goods and services which are essential for life and for everyone's development and society.

The Polish employment code establishes that working time ought to be understood as the time during which an employee is at the employer's disposal in the workplace or in another place where an employee does their work [1].

As for as working time is concerned, standards setting its upper limit should be determined. Therefore, the length of working time of an employee can be shorter than or equal to the standard, but it cannot be longer.

* Corresponding author: robertcieslak@wp.pl

In the engineering industry, working time standard is a technically justified amount of time necessary for a specified scope of work to be performed in given technical-organizational conditions in the workplace, by specified number of workers with specified qualifications [2]:

$$t = t_{pz} + n \cdot t_j \tag{1}$$

where:

t – time standard,
t_{pz} – set-up time (preparatory – finishing time),
n – the number of product units in a series manufactured with one setting of a machine tool,
t_j – unit time.

1.2 Research into work practices

Working time analysis involves using methods and experience in analysing and developing work systems in order to facilitate work practices, taking into consideration the following: productivity and needs of people as well as cost-effectiveness of a company.

Analysis and measurements are aimed at, among other things (and provided that measurement and/or control system is acceptable [3]), establishing the real workflow (the stages in a particular work process) and the way the work was performed, revealing time losses resulting from the used method of work (determining production reserves), establishing rational ways and methods of work as well as the necessary time to perform it [4].

2 Methods of working time analysis

In industrial practices, various methods of time-use are applied depending on the purpose and the subject matter of the research. The most commonly used techniques of work standardization include: chronometric analysis, working-day activity study (working-day photo), work sampling observations, and analysis of basic motions – MTM [5].

The results obtained from working time analysis constitute the material for the following:

• working out time standards for typical activities and workers' motions, which later will constitute a source of numerical data to establish work standards,
• analysing methods of work in order to rationalize, that is, searching for better ways of using specific tasks and activities as well as their right sequence in terms of efficiency,
• analysing multi-machine working conditions as one of the ways to use the working time better,
• observing flow production lines to achieve better synchronization of them,
• working out technical standards of time for performing specific operations in case there are no appropriate standards (analytical-measuring method of establishing standards),
• comparing the ability achieved by individual workers or discovering reasons for their inability to achieve 100% of the work standard [4, 6].

2.1 Chronomietrtic analysis (stop-watch study analysis)

It is a method of measuring repetitive operations or their elements (task, activity, workers' motion) designed to determine, on the basis of the specific number of measurements, the appropriate time of their duration and their rational completion at a normal pace of work. Chronometric observations can be carried out continuously or randomly [8].

In order to make labour-intensive chronometric measurements it is necessary that specially prepared (trained) workers should be equipped in appropriate measuring instruments. The most commonly used measuring instruments include: watches and stopwatches, self-recording measuring instruments, photo- and cinematographic equipment, automatic devices for analysis of working time and automatic control devices.

Depending on the number of employees whose work is observed, the chronometric analysis can be carried out for individual workers or a group of them. There are also various types of chronometric analysis, including the continuous, random and cyclical ones. When it comes to the practical application, the following types of chronometric analysis can be enumerated: moving, photographic or selective [5].

Table 1 shows advantages and disadvantages of chronometric analysis.

Table 1. Advantages and disadvantages of chronometric analysis [7, 5].

Advantages	Disadvantages
• the most faithful recording of temporal sequence of events (reliability), • creating a system of information in complete compliance with the real sequence of events, • making sure that information is promptly available, which means that every piece of information obtained in the measurement is available within a designated period of time, • keeping an accurate record of events on the basis of the information derived from authoritative sources (reliable), • simplifying and increasing the pace of preparing information so that it can be used in decision-making processes.	• performing measurements is considerably labour-intensive, • high costs of performing measurements, • high focus of attention on the part of a timekeeper while measuring with the use of a stopwatch without a drag pointer, • the possibility of lengthening the time of making a product by a worker

Figure 1 presents an algorithm showing the procedure in chronometric analysis, worked out on the basis of the published literature [5].

2.2 Working-day activity study (working-day photograph)

Working-day activity study involves continuous observing and performing measurements of time used at the workstation while taking into consideration the time during which employees worked and had breaks. The length of time needed for obtaining one photograph (one self-contained working-day activity study) usually overlaps with a period of one working shift of its part. The purpose of the photograph is to determine the degree to which the working time was used and the value of its real usage which is necessary to work out the standard [10].

The results of working-day activity study are used for improving the organization of production (the organization of work and workstations), for improving material, technical, and organizational production security and they are also the starting point for developing time standards and machine operation norms [14].

In order to obtain the working-day photograph, two basic systems of registering time are used: the tabular and scale systems.

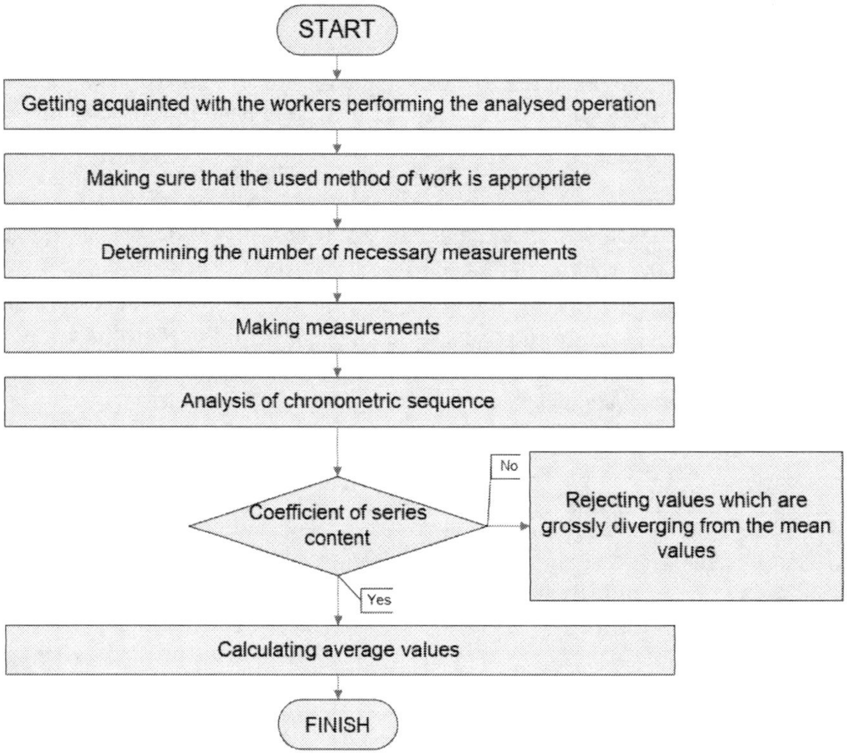

Fig. 1. Procedural algorithm [the author's own study - drawing a diagram].

A worker, machine, a worker and a machine at the same time, and also a team of workers or a team of workers and machines at the same time are usually subject to observations. Depending on the tasks and the kind of objects being observed, the following types of photographs of working time can be distinguished: individual, group, team, a photograph of one worker operating more machines and finally a self-photograph (the study performed by the worker himself/herself [5].

Table 2 shows advantages and disadvantages of working-day activity study.

Table 2. Advantages and disadvantages of working-day activity study [5, 13, 14].

Advantages	Disadvantages
• the possibility of doing research even at six workstations at the same time with one person simultaneously making observations and registering data, • reliability of mean values which are taken into account while determining additions of auxiliary time, • keeping an accurate record of events (the length of time employees work and take breaks). • simplifying and increasing the pace of preparing information so that it can be used in decision-making processes.	• performing measurements is considerably labour-intensive, • high costs of performing measurements, • high focus of attention on the part of a timekeeper while measuring with the use of a stopwatch without a drag pointer, • it does not take account of a random choice of registered real times • the possibility of lengthening the time of making a product by a worker

Figure 2 presents an algorithm showing the procedure in a working-day activity study, worked out on the basis of the published literature [5].

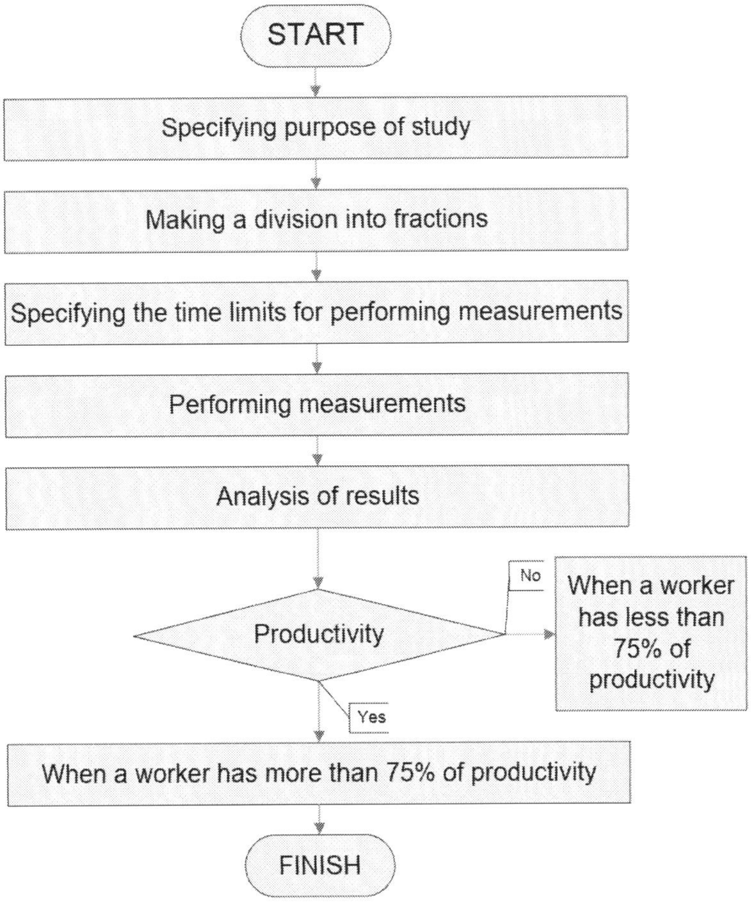

Fig. 2. Procedural algorithm [the author's own study - drawing a diagram].

2.3 Work sampling observations

Work sampling observations have become the basic method designed to analyse working time and they are supplanting the traditional Working-day activity study. It involves establishing the frequency of predefined types of event sequences in one or several work systems, by means of brief observations. These observations can be made on both workers and means of production.

The chronometer (stopwatch) is not used in work sampling observations and, what is more, continuous observations by the observer at workstations are not necessary. The reason for this is that this method involves using the calculus of probability for observations which are made (recorded) at specified intervals.

In order to transfer the results of work sampling analysis, carried out on a sample, onto the whole studied group with a satisfactory reliability, the following conditions have to be fulfilled:

• sample has to be representative, that is, it has to recreate the relations and proportions that occur in the whole studied group,
• any event occurring in the whole studied group has to have an equal probability of being found in the analysed sample [5].

The programme of work sampling study can be divided into several stages:

• preparing for the analysis, which means establishing the purpose and describing types of workflows (event sequences),
• determining the necessary scope of observations,
• determining time points of walking rounds (visits).
• calculating the necessary number of observations,
• making observations,
• the analysis of observation results, including the indirect analysis which is aimed to make sure whether the established number of observation scope is accurate or there is a serious error in the estimation,
• the final analysis.

Table 3 shows advantages and disadvantages of work sampling study.

Table 3. Advantages and disadvantages of work sampling study [8, 9, 14, 15].

Work sampling study	
Strengths	**Weaknesses**
• performing measurements without measuring instruments, • the person carrying out the analysis does not have to continuously observe workstations, • the possibility to observe a number of randomly selected workstations simultaneously, • the possibility to stop and resume making observations periodically, • easy analysis of measurements, • control charts which make it possible to supervise the process of performing measurements, • the possibility to quickly discover errors made while performing measurements, • from 40% to 70% of the time spent on performing time measurements in comparison to other fine-tuned methods of measuring/determining the time consumption of assembly work.	• it is difficult to check a single record in terms of substance because it constitutes a single unrepeatable sequence of activities, • it is impossible to check and compare a report on work sampling study with a report on time measurement, • lack of information on the evaluation of the pace of work, • lack of information on types of activity sequences which comprise less than 1% of all the records, • it is difficult to determine the causes of breaks at work resulting from disruptions, • it is impossible to carry out a detailed analysis of reasons for workers' absences at their workstations.

Figure 3 presents an algorithm showing the procedure in work sampling study, worked out on the basis of the published literature [5].

2.4 Basic motion analysis

Basic motion analysis, also referred to as MTM (Methods Time Measurement), is one of the most commonly used methods of normalizing the length of manual operation time. MTM (Methods Time Measurement) was mainly developed from combining the methodological basics of analyzing motions and their duration time. At the same time, this method is used for improving the flow of work, determining the times of work performance and classifying (determining the level of difficulty). MTM (Methods Time Measurement) assumes that every human manual activity can be divided into basic motions made by the torso and limbs and it also assumes the time consumption of the activity is a sum of duration times of motions which the activity consists of. There are three main groups of motions in this method: 9 hand movements, 2 eye movements and 15 body and leg movements.

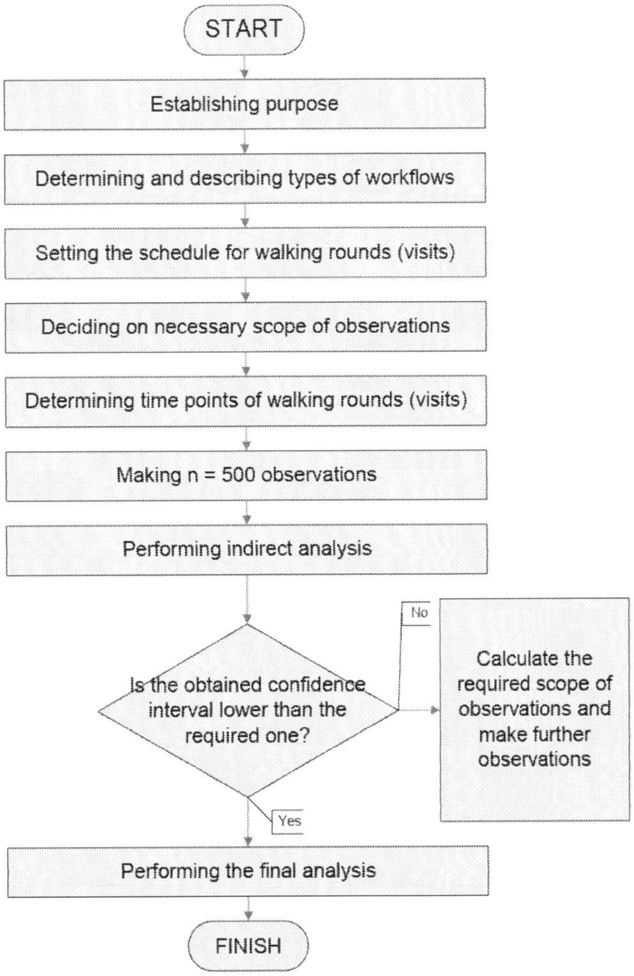

Fig. 3. Procedural algorithm [the drawing made by the author on the basis of literature].

This method does not limit itself to these 26 basic movements. In the first group, the movements are divided into categories, and these, in some cases, can be divided into classes, depending on their character and the conditions in which they are made. Time values are assigned to these categories and classes of movements in special tables of MTM method. The basic time unit in MTM (Methods Time Measurement) method is 1 TMU (Time Measurement Unit) = 0.0006 min.

In practice, the use of this method requires hiring qualified workers who were acquired theoretical knowledge during training courses and who possess a great deal of practical experience.

MTM (Methods Time Measurement) method has its own variations which take account of the level of element integration. The following variations can be distinguished: MTM – 1 method for duration times of operations lasting 0,1-0,5 min and significant repeatability of jobs (multi-series and mass production), MTM – 2, where duration times of operations last 0,5-5 min, and MTM – 3 which is used in special jobs [5].

Tab. 4 shows advantages and disadvantages of MTM method.

Table 4. Advantages and disadvantages of MTM (Methods Time Measurement) method [16].

Advantages	Disadvantages
• the possibility to design the most economical manufacturing process in specified production conditions before launching the production of a given product • improving the work process • numerical and letter symbols make it possible to record performed operations accurately and easily • normative time values in basic time systems are given, depending on parameters (factors) influencing them, such as the length of motion or performance control • operations are easy to analyse • the possibility to determine a precise time needed to perform a job or its elements • it is possible and easy to choose an optimal way of designing or improving a given work process • uniform pace for determined standards • it gives time values before launching production, thus eliminating expensive amendments • improving already existing methods, and procedures constitute additional help and materials for designing new procedures	• it is necessary to provide users of this method with on-site theoretical and practical training courses • there are limited possibilities to analyze the methods and activities of office workers and people operating machinery or equipment and as regards improving and standardizing times of jobs characterized by low pace of production (non-repetitive processes, single and small-scale production) this method is too precise and labour-intensive as a research tool.

3 The analysis of the research into normalizing the working time in the assembly process by means of analytic hierarchy process

3.1 Description of FAHP method

FAHP is a Fuzzy Analytic Hierarchy Process used in developing decision-making models and its purpose is to determine values of weighted coefficients that specify the importance of individual criteria. At the same time, thanks to the use of FAHP, the criteria of lesser importance can be eliminated while selecting auxiliary tools for managing tacit knowledge, [16, 17].

In the analyzed case the choice is made of the most important criteria that a company should take account of while selecting a method of studying the working time in the assembly process. Making an evaluation of objective elements will make it possible to determine the importance of defined criteria of selecting a solution. The preference scale should be defined while determining the relation of relative dominance. According to Nydick i Hill [30], the fuzzy number $\tilde{a}$ = (l, m, u) with a triangular membership function can be assigned to a linguistic variable that describes the relationships. The triangular fuzzy number $\tilde{a}$ = (l, m, u) is defined within the range [*l*, *u*], and its membership function assumes the value which is equal to 1 in point m (fig. 4). The preference scale is presented in tab. 5.

DOI: 10.1051/matecconf/201713701014

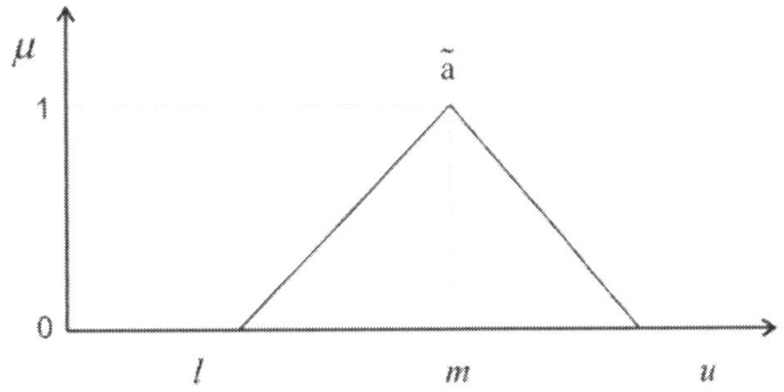

Fig. 4. Triangular fuzzy number ã = (l, m, u).

Table 5. Fuzzy scale used in FAHP method.

Relative dominance	FAHP (Triangular fuzzy scale)	FAHP (Triangular fuzzy scale - inverse)
Equal importance	(1, 1, 1)	(1,1,1)
Moderate importance	(1, 3, 5)	(1/5,1/3,1/1)
Strong importance	(3, 5, 7)	(1/7,1/5, 1/3)
Very strong importance	(5, 7, 9)	(1/9, 1/7, 1/5)
Extreme importance	(7, 9, 9)	(1/9, 1/9, 1/7)

Table 6. The elements of matrix for comparison of criteria selection for analyzing the method of working time study [the author's own study].

Kryteria	NK	NC	CO	ND	ŁP
NK	(1,1,1)	(1,3,5)	(1, 3, 5)	(1/7, 1/5, 1/3)	(3,5,7)
NC	(1/5, 1/3/ 1)	(1,1,1)	(1, 3, 5)	(1/7, 1/5, 1/3)	(1, 3, 5)
CO	(1/5, 1/3/ 1)	(1/5, 1/3/ 1)	(1,1,1)	(1/7, 1/5, 1/3)	(1, 3, 5)
ND	(3, 5, 7)	(3, 5, 7)	(3, 5, 7)	(1,1,1)	(5, 7, 9)
ŁP	(1/7, 1/5, 1/3)	(1/5, 1/3, 1)	(1/5, 1/3, 1)	(1/9, 1/7, 1/5)	(1,1,1)

NK – the lowest cost, NC – the shortest time, CO –observation continuity, ND – the highest accuracy, ŁP – easiness of conducting the study

The geometric mean of evaluations for criterion NK, in accordance with the relationships in FAHP assumes the following value:

$$N\tilde{K} = r\tilde{1} = (\tilde{a}11 \times \tilde{a}12 \times \tilde{a}13 \times \tilde{a}14 \times \tilde{a}15)^{\frac{1}{5}} \tag{2}$$

$$N\tilde{K} = r\tilde{1} = \left(\left(1 \times 1 \times 1 \times \tfrac{1}{7} \times 3\right)^{\frac{1}{5}}, \left(1 \times 3 \times 3 \times \tfrac{1}{5} \times 5\right)^{\frac{1}{5}}, \left(1 \times 5 \times 5 \times \tfrac{1}{3} \times 7\right)^{\frac{1}{5}}\right) \tag{3}$$

$$N\tilde{K} = r\tilde{1} = (0{,}84412; 1{,}55184; 2{,}25519) \tag{4}$$

The values for the other criteria were obtained in an analogous manner to the formula given above:

$$N\tilde{C} = r\tilde{2} = \left(\left(\tfrac{1}{5} \times 1 \times 1 \times \tfrac{1}{7} \times 1\right)^{\frac{1}{5}}, \left(\tfrac{1}{3} \times 1 \times 3 \times \tfrac{1}{5} \times 3\right)^{\frac{1}{5}}, \left(1 \times 1 \times 5 \times \tfrac{1}{3} \times 5\right)^{\frac{1}{5}}\right) \tag{5}$$

$$N\tilde{C} = r\tilde{2} = (0{,}49111; 0{,}90288; 1{,}52814) \tag{6}$$

$$C\tilde{O} = r\tilde{3} = \left(\left(\tfrac{1}{5}\times\tfrac{1}{5}\times 1\times\tfrac{1}{7}\times 1\right)^{\frac{1}{5}}, \left(\tfrac{1}{3}\times\tfrac{1}{3}\times 1\times\tfrac{1}{5}\times 3\right)^{\frac{1}{5}}, \left(1\times 1\times 1\times\tfrac{1}{3}\times 5\right)^{\frac{1}{5}}\right) \quad (7)$$

$$C\tilde{O} = r\tilde{3} = (0{,}35595; 0{,}58181; 1{,}10756) \quad (8)$$

$$N\tilde{D} = r\tilde{4} = \left((3\times 3\times 3\times 1\times 5)^{\frac{1}{5}}, (5\times 5\times 5\times 1\times 7)^{\frac{1}{5}}, (7\times 7\times 7\times 1\times 9)^{\frac{1}{5}}\right) \quad (9)$$

$$N\tilde{D} = r\tilde{4} = (2{,}66726; 3{,}87615; 4{,}98778) \quad (10)$$

$$\L\tilde{P} = r\tilde{5} = \left(\left(\tfrac{1}{7}\times\tfrac{1}{5}\times\tfrac{1}{5}\times\tfrac{1}{9}\times 1\right)^{\frac{1}{5}}, \left(\tfrac{1}{5}\times\tfrac{1}{3}\times\tfrac{1}{3}\times\tfrac{1}{7}\times 1\right)^{\frac{1}{5}}, \left(\tfrac{1}{3}\times 1\times 1\times\tfrac{1}{5}\times 1\right)^{\frac{1}{5}}\right) \quad (11)$$

$$\L\tilde{P} = r\tilde{5} = (0{,}22937; 0{,}31647; 0{,}58181) \quad (12)$$

Consequently, the following was obtained:

$$\tilde{r}1 + \tilde{r}2 + \tilde{r}3 + \tilde{r}4 + \tilde{r}5 = (4{,}58781; 7{,}22915; 10{,}46048) \quad (13)$$

The weight of criterion „NK" was calculated in the following way:

$$\tilde{w}1 = \tilde{r}1 \times (\tilde{r}1 + \tilde{r}2 + \tilde{r}3 + \tilde{r}4 + \tilde{r}5)^{-1} = (0{,}08069; 0{,}21466; 0{,}49156) \quad (14)$$

Then, the weight for the remaining criteria was calculated in an analogous manner to the formula given above:

$$\tilde{w}2 = \tilde{r}2 \times (\tilde{r}1 + \tilde{r}2 + \tilde{r}3 + \tilde{r}4 + \tilde{r}5)^{-1} = (0{,}04694; 0{,}12489; 0{,}33308) \quad (15)$$

$$\tilde{w}3 = \tilde{r}3 \times (\tilde{r}1 + \tilde{r}2 + \tilde{r}3 + \tilde{r}4 + \tilde{r}5)^{-1} = (0{,}03402; 0{,}08048; 0{,}24141) \quad (16)$$

$$\tilde{w}4 = \tilde{r}4 \times (\tilde{r}1 + \tilde{r}2 + \tilde{r}3 + \tilde{r}4 + \tilde{r}5)^{-1} = (0{,}25498; 0{,}53618; 1{,}08718) \quad (17)$$

$$\tilde{w}5 = \tilde{r}5 \times (\tilde{r}1 + \tilde{r}2 + \tilde{r}3 + \tilde{r}4 + \tilde{r}5)^{-1} = (0{,}02192; 0{,}04377; 0{,}12681) \quad (18)$$

The defuzzified values of weights are as follows:

$$\tilde{w}1 = \frac{1}{3}(0{,}08069; 0{,}21466; 0{,}49156) = 0{,}26230 \quad (19)$$

$$\tilde{w}2 = \frac{1}{3}(0{,}04694; 0{,}12489; 0{,}33308) = 0{,}16830 \quad (20)$$

$$\tilde{w}3 = \frac{1}{3}(0{,}03402; 0{,}08048; 0{,}24141) = 0{,}11863 \quad (21)$$

$$\tilde{w}4 = \frac{1}{3}(0{,}25498; 0{,}53618; 1{,}08718) = 0{,}62611 \quad (22)$$

$$\tilde{w}5 = \frac{1}{3}(0{,}02192; 0{,}04377; 0{,}12681) = 0{,}06416 \quad (23)$$

After standardization the weights assume the following values:

$$w1 = 0{,}21161; w2 = 0{,}13578; w3 = 0{,}09570; w4 = 0{,}50513; w5 = 0{,}05176 \quad (24)$$

Thanks to the application of FAHP method, the following importance level of criteria while selecting a method of studying the working time was obtained, respectively:

• criterion importance: **ND – the highest accuracy – value: 0,50513,**
• criterion importance: NK – the lowest cost – value: 0,21161,
• criterion importance: NC – the shortest time – value: 0,13578,
• criterion importance: CO – observation continuity – value: 0,09570,
• criterion importance: ŁPB – easiness of conducting the study – value: 0,05176.

The suggested application of FAHP method makes it possible to make an objective choice of the tool which supports the tacit knowledge management according to the criteria which are difficult to measure. Expressing preferences in the form of fuzzy numbers reflects the uncertainty in formulating judgments in decision-making process. Therefore, a defined order of criteria importance should be taken into consideration while selecting a method of studying the working time.

4 Summary and conclusion

The purpose of this paper is to present the possibility of using Fuzzy Analytic Hierarchy Process (FAHP) in the process of developing a synthetic variable. FAHP (Fuzzy Analytic Hierarchy Process) method takes advantage of experts' opinions in order to determine weighted coefficients specifying the importance of variables. FAHP (Fuzzy Analytic Hierarchy Process) method was used for selecting criteria of the method for studying the working time in the assembly.

The results of the study indicate that, based on the criteria, the highest accuracy is the most important. Then, the lowest cost is the next. The values obtained in the calculations show that the advantage of the highest accuracy over the remaining ones is considerable. The authors of the paper intend to analyze advantages and disadvantages of FAHP (Fuzzy Analytic Hierarchy Process) method in developing more sophisticated technological processes. In order to become convinced that this method can be justifiably used by engineers, the authors intend to analyze more sophisticated technological processes used in assembly work (selecting a technological variant).

References

1. Kodeks Pracy, art. 128
2. A. Giddens, *Socjologia*, PWN, Warszawa (2004)
3. M. Diering, K. Dyczkowski, A. Hamrol, Springer International Publishing Switzerland, **368,** 415 (2015)
4. REFA., *Metody badania prac cz.1 i cz.2*, Cieszyńska Drukarnia Wydawnicza, 12 (1984)
5. Z. Jasiński, *Zarządzanie pracą – organizowanie, planowanie, motywowanie, kontrola*, Agencja Wydawnicza „Placet", Warszawa, 92 (1999)
6. M.R.L. Varela, J. Trojanowska, S. Carmo-Silva, N.M.L. Costa, J. Machado, Management and Production Engineering Review, **8**(2), 69 (2017)
7. H. Mreła, *Technika organizowania pracy*, wyd. Wiedza Powszechna, Warszawa, 289 (1975)
8. R. Wołk, *Podstawy normowania pracy w przemyśle maszynowym*, Wydawnictwa Naukowo – Techniczne, Warszawa, (2012)
9. R. Wołek, Strzelecki T.J., *Badanie metod i normowanie pracy*, Wydawnictwo Politechniki Warszawskiej, Warszawa 39 (1993)
10. H. Mreła, *Metody badania pracy – metody i techniki organizatorskie*, PWE, Warszawa, 84, (1979)
11. V. Libal i zespół (tłumaczył Jeleń W.), *Organizacja i zarządzanie produkcją*, Państwowe Wydawnictwo Ekonomiczne, Warszawa 590 (1976)
12. Z. Mikołajczyk, *Techniki organizatorskie w rozwiązywaniu problemów zarządzania*, PWN, Warszawa (1997)
13. A.P. Muhlemann, J.S. Oakland, K.G. Lockyer, *Zarządzanie produkcja i usługi*, PWN, Warszawa (2001)
14. W. Kieżun, *Podstawy organizacji i zarządzania,* Wydawnictwo Książka i Wiedza, Warszawa **12,** 195 (1980)
15. V. Prabhu, G. Martin Helander and V. Shalin, The International journal of Human Factors in manufacturing **5** (2), 151 (1995)
16. D.-Y. Chang, European Journal of Operational Research, **95** (2), 649 (1996)
17. R.L. Nydick, R.P. Hill, International Journal of Purchasing and Materials Management **28** (2) (1992)

Ergonomic solutions to support forced static positions at work

Marcin Suszyński[1], *Marcin* Butlewski[2,*], and *Róża* Stempowska[2]

[1]Poznan University of Technology, Institute of Mechanical Technology, Poland
[2]Poznan University of Technology, Department of Ergonomics and Quality Engineering, Poland

Abstract. The article analyzes the available ergonomic constructions used for the support of the musculoskeletal system during static, prolonged work performed in forced positions. Possible evaluation methods are presented as well as ergonomic considerations of work performed in inclined positions, where there is no possibility of influencing the working plane. As a result of the presented work, a set of criteria has been proposed and the requirements for methods which can be used to evaluate the technical constructions supporting the worker during tasks performed in forced and static positions.

1 Introduction

Technological developments, and the resulting mechanization and automation of work processes have changed the nature of the demands placed on workers. Professions once characterized by dynamic physical loads, with high energy expenditures, are largely replaced by operator jobs and work requiring precision, where factors such as mental strain, work monotony, repetitive movements, static loads, and ultimately hypokinesia resulting from lack of motion are predominant. The latter two types of loads are to a large extent the main contributing factors to the nature of the work performed by highly qualified specialists such as mechanics, welders [1] or workers in medical professions (medical technicians, dentists, surgeons) [2]. The need to maintain unchanging and unnatural positions at work results in loads on the musculoskeletal system that lead to illnesses known as WRMSDs (work-related musculoskeletal disorders).

The remedial measures employed to prevent static loads are merely of an organizational or technical nature, none of them alone fully address the problem [3]. The problem of static positions is particularly important in those professions where, due to the nature of the work, the position cannot be effectively reorganized in such a way that a worker can assume a natural position at work or interrupt work to perform other tasks in order to relieve the musculoskeletal system. This would be the case for surgical work, where the life and health of the patient (human or animal) is more important than the welfare of the workers (technicians or doctors). The solution to such professions may be a variety of technical measures, however, for various reasons they may not meet the requirements based on which

* Corresponding author: marcin.butlewski@put.poznan.pl

they were created. This article is therefore intended to provide an analysis of the ergonomic criteria for technical solutions that reduce static loads during work.

2 Static positions and their ergonomic evaluation

A static position is defined as any position where load on the musculoskeletal system is maintained for more than 4 seconds. However, a static work position is burdensome due to at least a few more reasons. The most important contributing factors are the following:

- number of body segments involved,
- combination of positions of body segments and the loads (expressed in Nm as the moment) put on reference joints by the body segments' weight and by the forces compressing and stretching these joints,
- influence of external forces resulting from the way the work is performed as well as the individual characteristics of the employee,
- extent to which a position is forced, resulting from the technical and organizational environment of the workplace,
- length of time to maintain a certain body position.

A very important factor when considering the burden of a static position is the length of time required to maintain it. It is evaluated using a parameter known as the percentage of maximum holding time (MHT). A load is considered acceptable if it lasts up to 20% of MHT [4]. MHT is calculated experimentally by determining the amount of time after which the study participant will be unable to continue maintaining a given static position. Along with a prolonged time of maintaining a position, static loads have an impact on the worker's capacity for physical effort – static muscle contractions result in changes in their biochemical composition, leading to increased fatigability and increased load on the musculoskeletal system.

The identification of non-recommended static positions is the subject of ergonomic standards such as ISO 11226, which allows for the evaluation of existing work situations, and EN 1005-4, suited for evaluations during the design/engineering process. In addition, a number of methods exist to assess the positions assumed at work – e.g. by evaluating loads on the whole body, such as OWAS (Ovako Working Posture Analysis System) [5], REBA – Rapid Entire Body Assessment [6], or QEC (Quick Exposure Check) [7]. Some of the methods are dedicated to certain types of loads – such as those on the upper limbs: RULA – Rapid Upper Limb Assessment [8], or certain industries – as is the case for the Ergonomics Assessment Worksheet (EAWS), which is designed for manufacturing industries [9]. The listed methods are only a few selected from a large family of methods used in the risk assessment of WRMSDs, resulting from the manner and circumstances of the work.

Methods that are much more accurate, at least in terms of the classification of the position of the body, are computer simulations, which by modeling reality allow for a precise determination of hazard zones by providing an assessment of: body segment angles, hand locations and force magnitude, spine disc compression. Examples of such programs include 3-D Static Strength Prediction Program (3D SSPP), LifeMOD Biomechanics Modeller or BakPak Lift/Lowering Tasks. These programs are based on biomechanical calculations, and the accuracy of the obtained results is dependent on the extent to which the actual manner of operations was transferred to the virtual model. There are plenty of attempts to make the transition from real work situations to computer simulations [11].

To assess the consequences of static positions as well as the technical solutions (e.g. supports) designed to limit them, one can perform, among others, EMG studies [12] as well as studies employing thermography. Both methods examine muscle activity while assuming a certain position for a specified amount of time. Electromyography (EMG) is an experimental technique related to obtaining, recording, and analyzing specific myoelectrical signals. These signals arise due to physiological changes in the cell membranes of muscle fibers [13]. In addition to basic physiological and biomechanical research, EMG kinesiology can be used as an assessment tool in ergonomics (ergonomic design) research and in studies on the human body's interaction with industrial products and work conditions, which may form the basis for the verification of support prototypes. On the other hand, a study using a thermal imaging camera based on the imaging and analysis of infrared radiation emitted by the examined fragment of the body's surface allows the determination of the human body's interaction with the surface of the support measure. The variable impact of static positions and measures undertaken to reduce them requires a multi-criterial evaluation, by recognizing and interpreting both the actual and the predicted effects of the work performed.

3 Technical solutions to reduce the burden of static positions

Research on improving the ergonomic quality of work performed in forced, inclined and prolonged positions of the body has led to the development of solutions that reduce the risk of such loads. The result of these studies are seats equipped with additional elements (e.g. supports) to relieve the individual body segments or stabilizers supporting the standing position. Examples of such solutions are shown in Figure 1.

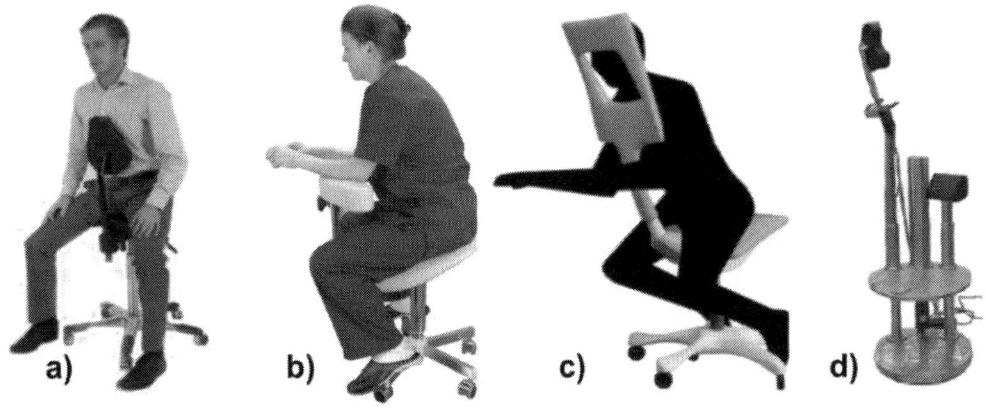

Fig. 1. Solutions supporting the maintenance of an inclined position [14-17].

The presented solutions' aim is to limit static loads, however, their overall ergonomic impact assessment also includes the determination of side effects such as point pressures on soft tissues and constraints on worker mobility associated with the use of the support. In order to assess their validity, their preliminary comparison was made for a surgeon performing laparoscopic procedures.

4 Ergonomic evaluation of selected solutions reducing the impact of static positions

As examples of a forced position at work, three methods of performing surgical tasks during laparoscopic procedures were considered. The most commonly assumed body position when performing such procedures is the standing position. It allows for a free range of movement and a stable position of the body, however, over time, it leads to the development of many ailments, causes fatigue and decreases work performance. Performing laparoscopic surgeries in the upright position involves a frequent tilt of the entire torso and head forward towards the operating plane. One reason for this is difficulties in adjusting the operating table height because of the large amount of space required for maneuvering laparoscopic equipment and long tools. On the other hand, performing laparoscopic surgery while seated requires physicians to maintain a natural lordosis because the use of such supports is not possible. In the seated position there is also the problem of the large distance required between the surgeon's position and the working plane due to the length of laparoscopic instruments. This contributes to excessive forward leaning and a loss of the natural lordosis, which in turn causes numerous ailments and discomfort during the performed work. A comparison was made of the estimated loads during the performance of laparoscopic surgery in three positions: (Figure 2)

A. standing position in close to ideal operating conditions (slight tilt),
B. unsupported sitting position with a tilt typical in this working position,
C. inclined sitting position, obtained when using alternative seating solutions.

Two computer programs, JACK and 3D SSPP, were used for the modeling of musculoskeletal load.

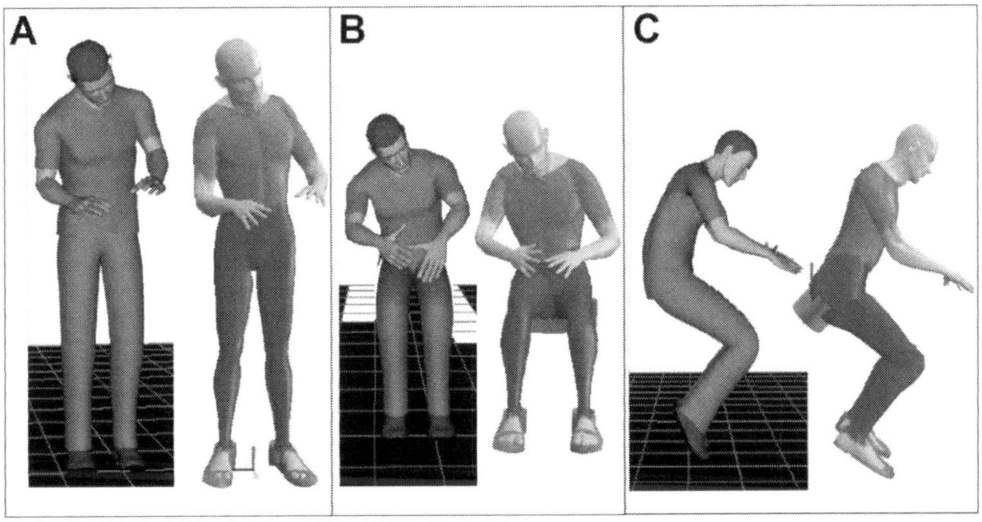

Fig. 2. Analyzed working positions.

The presented avatars represent the above-mentioned positions assumed by the surgeon during laparoscopic procedures. Both the 3D SSPP program and the JACK program are based on NIOSH data, which indicate the limit values for each parameter.

Table 1. Cooperation between positions

Biomechanical parameter \ Position/ Program		Standing (A)		Unsupported sitting (B)		Alternative sitting (C)	
		J-1	3D-1	J-2	3D-2	J-3	3D-3
Compression force on L4/L5 disc [N]		1973	1892,5	1045	2777,7	1205	2050,3
Percentage of the population that is able to transfer the calculated moments for each joint	Wrist	97	92	93	87	100	98
	Elbow	100	100	100	98	100	100
	Shoulder	100	100	99	99	100	100
	Torso	99	97	99	95	100	99
	Hip	97	96	95	100	99	96
	Knee	100	98	96	100	96	99
	Ankle	96	96	99	100	100	97
RULA Result		5		7		3	

Two indicators were used for biomechanical analysis: the compression force on the L4/ L5 intervertebral disc and the percentage of the population that is able to transfer the calculated moments for each joint. The postures of the individual avatars were modeled by defining the position of specific anthropometric points and by assigning angles of the specified joints.

The RULA scores indicate that the highest risk occurs in the sitting position. This level of risk is interpreted as: there is a need to introduce changes to the workstation. The alternative sitting position achieved the lowest average load, which should be interpreted as a sign for further research. It should be noted that the employed biomechanical models were consistent in their percentage results, but the L4/L5 forces indicated by the JACK and 3D SSPP programs differ markedly, despite identically modeled body positions. Both values were marked in the programs as green, which means they do not exceed the NIOSH limit values. The differences between the biomechanical evaluation and RULA scores may arise due to the simplified RULA scales (no exact values are given), as well as the method's greater focus on loads of the upper extremities. Certainly, the greater load will be indicated by the time required to maintain the position. At an incline of only 10°, the acceptable time of holding the static position is 10 minutes [18], and thus much shorter than the time needed for typical activities performed by laparoscopic surgeons.

Based on the data collected and the results obtained, it can be concluded that the most commonly used body position by laparoscopic surgeons contributes to the development of musculoskeletal disorders and changes need to be made to minimize this risk, and a solution in this regard may be the use of dedicated supports for lowering static loads. Solutions of this kind are all the more necessary since the acceptance of the negative impact of work on employees is decreasing [19], which means that only through ergonomic design can one make the appropriate modifications [20-23].

5 Conclusions and recommendations for further research

Ergonomic solutions, designed to limit the problem of static loads in the workplace, must be evaluated from the point of view of a number of factors, among them the main and most consequential is the reduction of the risk of developing WRMSDs. It should be noted, however, that in order for a solution to be considered ergonomic, the appropriate level of worker effectiveness must be ensured, through a proper range of movement and force generated during this movement, as well as a number of utilitarian requirements resulting from the environment in which the solution is used.

The performed analysis revealed the limitations of methods of biomechanical load evaluation that make it difficult to properly evaluate the solutions used to reduce the static loads on employees. The vast majority of methods fail to take into account the transfer of loads to parts of the body other than the limbs. In summary, when evaluating the effectiveness of applications and the ever-increasing number of solutions that support employee positions, one should employ more complex solutions, such as the program AnyBody. Most of common used ergonomic methods does not support alternative position at work.

References

1. A. M. Lasota, & K. Hankiewicz, In 12th International Symposium on Occupational Safety and Hygiene of Portuguese-Society-of-Occupational-Safety-and-Hygiene (SHO), 261-264 (2016)

2. N. Vujica Herzog, R. Vujica Beharic, A. Beharic & B. Buchmeister, International Journal of Simulation Modelling (IJSIMM), **13**(4) (2014)
3. A. Kawecka-Endler, B. Mrugalska, Advances in Social and Organizational Factors, , W. Karwowski Ed., 672-681 (2012)
4. L. Rose, M. Ericson, B. Glimskär, B. Nordgren, and R. Ørtengren, (omputer Applications in Ergonomics, *Occupational Safety and Health,* Mattila, M. and Karwowski, W., Eds., Amsterdam: Elsevier, pp. 461–468 (1992)
5. O. Karhu, Correcting working postures in industry: A practical method for analysis. Applied Ergonomics. **8**(4), 199-201 (1977)
6. S. Hignett, & L. McAtamney, Rapid entire body assessment (REBA). Applied ergonomics, **31**(2), 201-205 (2000)
7. G. Li, P. Buckle, *The development of a practical method for the exposure assessment of risks to work-related musculoskeletal disorders. General report to the HSE* (Contract No. R3408), Robens Centre for Health Ergonomics, European Institute of Health and Medical Sciences, University of Surrey (1998)
8. L. McAtamney, E.N. Nigle Corlett,. RULA, Applied Ergonomics **24** (2), 91-99 (1993)
9. K.G. Schaub, J. Mühlstedt, B. Illmann, Bauer S., L. Fritzsche, T. Wagner, A.C. Bullinger-Hoffmann, R. Bruder, Int. J. Human Factors Modelling and Simulation, **3**, Nos. 3/4 (2012)
10. G. Dahlke, Zeszyty Naukowe Politechniki Poznańskiej. Organizacja i Zarządzanie, **63**, 21-32 (2014)
11. M. Butlewski, M. Sławińska, & M. Niedźwiecki, *Advances in Social & Occupational Ergonomics* 15-23. Springer International Publishing (2017)
12. A. Albayrak, Printer Partners Ipskamp, Enschede, ISBN/EAN: 978-90-5155-050-4 (2008)
13. J. Basmaijan, C. DeLuca, *Muscles alive : their functions revealed by electromyography*, Williams & Wilkins (1985)
14. *** www.kos.ie
15. *** www.jedmed.com
16. *** walyou.com/blog/2010/08/01/unconventional-twist-chair
17. A. Albayrak, M. A. Van Veelen, J. F Prins, C. J. Snijders, , H. de Ridder, & G. Kazemier,. Surgical endoscopy, **21**(10), 1835-1840 (2007)
18. W. Karwowski, & W. S. Marras, Occupational Ergonomics Reference Library-3 Volume Set: The Occupational Ergonomics Handbook (1998)
19. A.M. Lasota, K. Hankiewicz, Proceedings of the 2016 International Conference on Economics and Management Innovations, **57**, 272-278 (2016)
20. A. Górny, Procedia Manufacturing, **3**, 4700-4707 (2015)
21. A. Górny, *Human Factor and Ergonomics in Essential Requirements for the Operation of Technical Equipment*, in: C. Stephanidis (ed.), Posters Extended Abstracts: International Conference, HCI International 2014, Communications in Computer and Information Science, **435**,. 449-454 (2014)
22. A. Kujawińska, K. Vogt, F. Wachowiak, *Ergonomics as Significant Factor of Sustainable Production*, in: P. Golińska, A. Kawa, (ed.), Technology Management for Sustainable Production and Logistics, Book Series: EcoProduction, 193-203 (2015)
23. A. Hamrol, D. Kowalik, A. Kujawinska, Human Factors and Ergonomics in Manufacturing & Service Industries, **21**(2), 156-163 (2011)

Initial Study on the State of the Art Solutions for the Simulation of the Thermoregulatory Sweating Conditions

Ioan Turcin[1, 3*], *Selver* Softic[2], *Udo* Traussnigg[1], and *Adrian* Trif[3]

[1]CAMPUS 02 UAS/Degree Programmes in Automation Technology, 8010 Graz, Austria
[2]CAMPUS 02 UAS/Degree Programmes in IT and Business Informatics, 8010 Graz, Austria
[3]Technical University of Cluj-Napoca, Faculty of Machines Building, 400114 Cluj-Napoca, Romania

Abstract. The paper presents an initial study and basic research of the state of the art systems and solutions for the simulation of the thermoregulatory sweating conditions. The simulation of the thermoregulatory sweating conditions has been treated by many researchers, and has become over the years a major area of interest for many industries (e.g. textile, protective garments, aeronautics, automotive industries and sports). The objective of this paper is to reflect and analyse the existing literature, for this purpose. Three existing systems and two thermoregulatory models will be characterized and critically discussed. The findings support the methodology for future applied scientific research (PhD thesis), which will use existing scientific knowledge to develop a practical application using suitable technology.

1 Introduction

Human body possess a complex thermoregulatory system. Under normal conditions the body keeps, the vital organs such as heart, lung and brain within a narrow temperature range around 37°C. The body produces heat usually through metabolic processes (e.g. muscular activity). During such processes depends on climate circumstances and clothing worn, the body excesses the heat via evaporation, radiation, conduction and respiration [1].

According to the best knowledge of the authors, there is no known, price accessible, measuring equipment for the simulation of the thermoregulatory sweating available on the market for industrial applications, although many manikins and mathematical models have been developed in the last years.

To reproduce a physical, physiological situation or process conveniently, the developed mathematical model has to fit the studied situation. The existing systems are very sophisticated and specific, requiring high levels of training and qualification, which makes them inaccessible for small and medium enterprises (SMEs) with small R&D departments and limited resources. Developing adequate, high qualitative and low-cost equipment,

* Corresponding author: ioan.turcin@campus02.at

which can be adjusted to the demand of SMEs, would enable and motivate these companies to use these simulation systems.

The idea behind the future research (PhD thesis) is to investigate and develop a prototype, using materials or material combinations that are still to be defined, and will lead to accurate thermal conductivity and humidity distribution simulations. Additionally, various advanced manufacturing methods will be investigated, which should allow the integration of sensing and measuring equipment, as well as fluid flow channels, in the to-be-produced prototype at minimal costs (Figure 1). A capable control algorithm based on suitable mathematical model making it easy to implement in future generation of thermal manikins, should accompany the developed element.

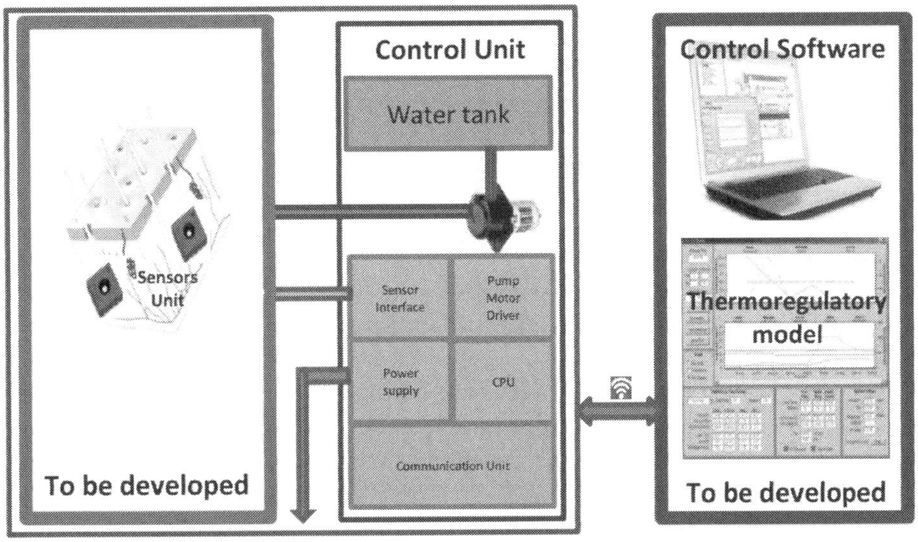

Fig. 1. The block diagramm of the prototype to be developed. The system should consist of three major elements: outer skin prepared by a combination of smart materials (e.g. two of them, sensors included), control unit (including heating, pumping system and needed interfaces) and a modern algorithm for the mathematical model of thermoregulation sweating conditions.

2 Methodology and materials

The intention behind this literature study is to reflect and analyse the developed techniques and state of the art used over the years to simulate the thermoregulatory sweating conditions. There is a large number of published studies about this topic. The focus of the search was on the scientific reflection of the intensely sought-after and found literature, where the information has been collected in a targeted manner. To cover many information and previous works, a variety of online catalogues, databases and open source libraries have been searched (e.g. Google Scholar, Directory of Open Access Journals, Science Direct, Management E-journals (Emerald), Elsevier, The Austrian Library Network and Service). The authors used following keywords for this operation: simulation, thermoregulatory, systems, sweating conditions, manikins and models of thermoregulation. After an intensive review, the materials with following reference number [1], [2], [3], [4], [5], [6] and [7], have been selected for further discussion. This choice criterion in particular was the holistic nature of approaches presented in found works. Only the papers covering all the fields of interest expressed by the searched key words mentioned above have been selected. At the end of selection and evaluation process, only complete and approved systems for simulation of thermoregulatory sweating systems have been considered for further observation and discussion.

3 Discussion

For many years, the developed technique of thermal hot plates or manikins have been used in successful technical applications. There are sweating thermal manikins, where the textile layer, are placed over the thermal manikin and are used for the simulation of the sweating. However, the measurements are too expensive and are not available for example for the SME's of the industry sector (e.g. sweating torso cost around 100.000,00 €). First used thermal manikin was early 1940s, which was a simple one segment copper made manikin introduced by the American army [3].

For a variety of applications in the engineering field, manikins (there are over 100 manikins in the world, which use thermoregulatory systems) are used for the evaluation of clothing thermal, evaporative resistance and similar applications. Advanced technologies have helped in designing of manikins to produce more realistic results than before relating to the human thermal interaction with the environment. In addition, advanced software and computing tools have enabled the combination of the modelling results with the obtained data from manikin tests.

3.1 The manikins

The selection of the manikins SAM (Switzerland), ADAM (USA), X-THERMOMAN (Slovenia) and Wissler (1985) and Fiala (2012) models, is based on the degree of brand awareness, complexity of the system, design of the sweating system used, analysis of the results, different countries of manufacture, etc. The selected manikins and models are introduced in the following paragraphs.

The **SAM** (Fig. 2) is a thermal manikin capable of simulating heavy work conditions, and human movements such as walking and climbing. The purpose was to simulate the human body in terms of heat production, sweat production and movement as closely as possible. The simulation system around SAM manikin allows observation of even small differences in clothing and possesses higher repeatability than those from human tests.

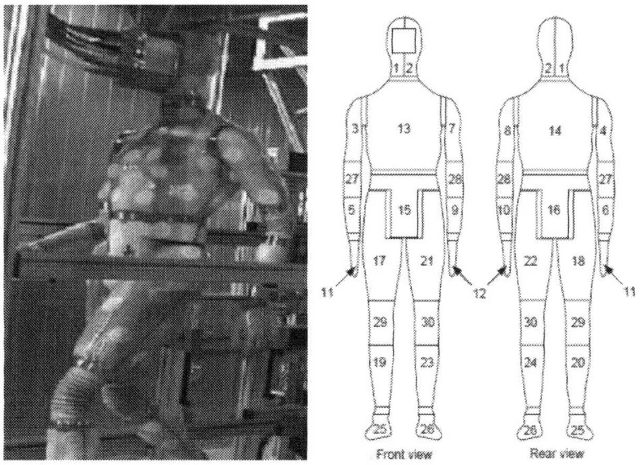

Fig. 2. Manikin SAM with his 30 separately heated sectors [1].

The complete system of SAM consists of 30 separate uniformly heated sectors. Each sector can be heated at a constant average surface temperature (to perform standard ISO TC 38 WG17) or with a constant power (up to 1.2 kW). SAM manikin simulation system allows performing measurements of thermal insulation or supplied with different heating powers to simulate various activities. Sweating system is made of 125 sweat outlets distributed over its surface. It also allows sweat rate regulation from 20 ml/h up to at least 4 litres per hour. In this way, it is possible to simulate various activities and conditions [1].

ADAM (Advanced Automotive Manikin) evaluates transient, non-uniform thermal environments in automobiles, where a sophisticated surface sensor interacts with his environment and is very good linked with a Human Thermal Physiological Model and a Human Thermal Comfort Model (Fig. 3). The manikin does not only respond to thermal inputs such as radiation and convection it also affects the environmental flow field and temperature field [2]. It is a three-dimensional transient finite element model, contains a detailed simulation of human internal thermal physiological systems and thermoregulatory responses.

ADAM consists of on-board power, water, communication, physiological control - designed to respond to transient, non-uniform thermal environments like a human (sweating, breathing) and wears clothes, it is 175 cm tall, has 61 kg, and contains 120 elements (~ 120 cm^2) and 120 zones. It provides a simulated body positioned in a complex thermal environment to measure the transient thermal response with extremely high spatial resolution.

Fig. 3. Thermal manikin ADAM [2].

The finite element model integrated with ADAM provides the manikin with a control algorithm that represents human thermal response. The manikin can be considered as surface sensor that measures the rate of heat loss at each surface zone. The main objective of the ADAM manikin and its model system is to predict the effect of thermal environments on human thermal physiology [2].

X-TERMOMAN, presented manikin in the Figure 4, was the first manikin made at the Faculty of Mechanical Engineering in Maribor in Slovenia.

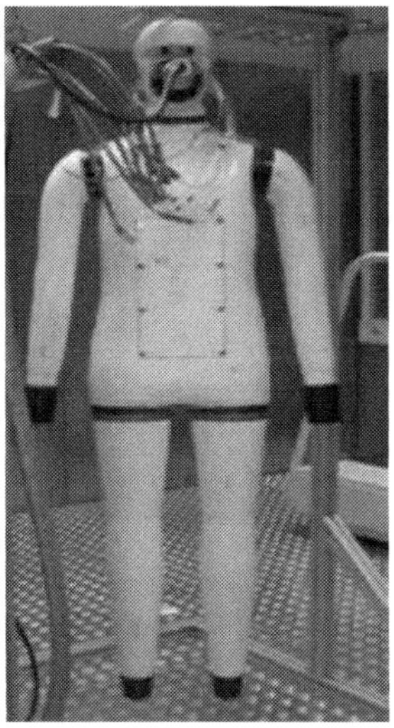

Fig. 4. X-termoman [3].

In order to approximate the desired copy of a human body as close as possible, the required heating power was set to 1000W, where the heaters are electrical (working on resistance principle) and the number of them is optimized in order to get the best possible heat distribution. The X-TERMOMAN was made of aluminium (AlSi9Cu3) which is a cost-efficient replacement for silver that was also protective garment considered as option. The choice fell on aluminium because of the need for thicker walls in order to ensure the self-carrying of the manikin's body [3].

There are significant differences between SAM, ADAM and X-TERMOMAN with respect to the application area, concept, topicality, materials, models of thermoregulation systems and technology used. SAM is capable of simulating movements and produce results, which have a much higher repeatability than those from human tests. On the other hand, ADAM has high spatial resolution, a reasonable response time, and realistic and uniform sweating rates. It is a new tool, which will be useful in predicting the thermoregulatory response of a person in a protective garment, which can also evaluate the effectiveness of personal cooling systems used with these types of suits [2]. Concerning X-TERMOMAN (mostly used for protective garment), it is made of solid blocks of aluminium, selected based on a study of the thermal, mechanical and chemical properties of the material. It includes a self-developed suitable user interface designed to ensure a good overview of all crucial data and to enable simple and controlled testing according to ISO 15831 (2004) [3].

It is to mention, that only ADAM use the new wireless technology for communication between manikin and thermoregulatory model.

Thermal manikins will never really imitate the human and their results are approximate or indicative and can be used for relative tests. Suitable models may overcome limitations of design such as missing thermal capacity, blood flow control and thus a missing core temperature, but all of them reducing the need for expensive human tests. Tests under extreme conditions can be dangerous and they are strictly regulated by law. Using thermal manikins, enables performing the desired tests, without the possible risk of life. Considering the specific requirements and the aim of reducing the overall costs, the techniques are initially expensive to set up, therefore they are relatively exclusive. According to the current findings coming out of the study conducted, there are no standards yet for the materials used, design of the thermoregulatory sweating systems used and the analysis of the results.

3.2 Models of thermoregulation

Some of the existing models of thermoregulation can be divided as follows:

Empirical models: Givoni and Goldman (1972-1973), Pandolf (1986), Malchaire (2000).

Rational models: Wissler (1961-1985), Stolwijk and Hardy 25-node model (1971), Nishi and Gagge 2-node model (1972), Haslam and Parsons, Kraning & Gonzalez (1977), Fiala (1999-2012).

Wissler and Fiala are researchers who realized and described the existence of temperature gradients in the body and have contributed a lot to further development and research of thermoregulation models.

Wissler human thermoregulation model can be used within the areas of physiology, study of thermoregulation, simulation of the human, studying simulated thermoregulatory response and comparing them with real data of human. The continuously development over the past 50 years, was applied to simulate transient thermoregulatory behaviour for numerous environmental heat exchange conditions, types and levels of physical activity.

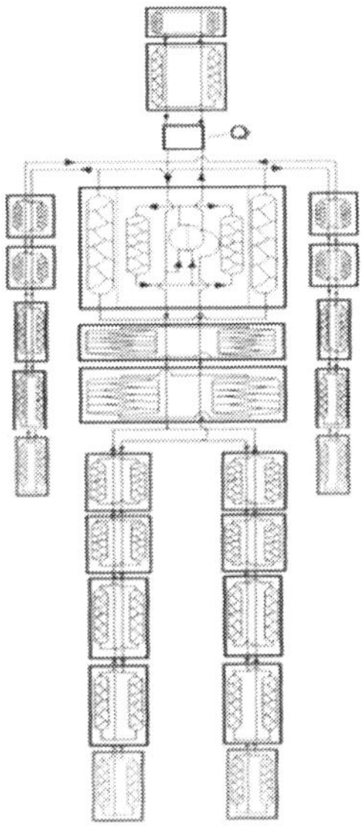

Fig. 5. Graphical representation of the classic Wissler model [4].

Recently, the Wissler human thermoregulation model has been finally translated into C code. The project possesses extensive internal documentation and the model in C is wrapped in Python, to provide a graphical user interface for convenient input and output operations [4].

Fiala considered 1999, that the reasons why regulatory processes did not became widespread, include lack of confidence in their predictive abilities, limited range of applicability, and poor modelling of the heat exchange with the environment. He intended to contribute to research efforts to formulate a more precise, flexible and universal model of the human thermoregulatory system [5]. Over the years, Fiala developed a new version called UTCI model (universal thermal climate index; active system), which is a new version of simplified DTS (Dynamic Thermal Sensation; passive system) models and can be seen in Figure 6.

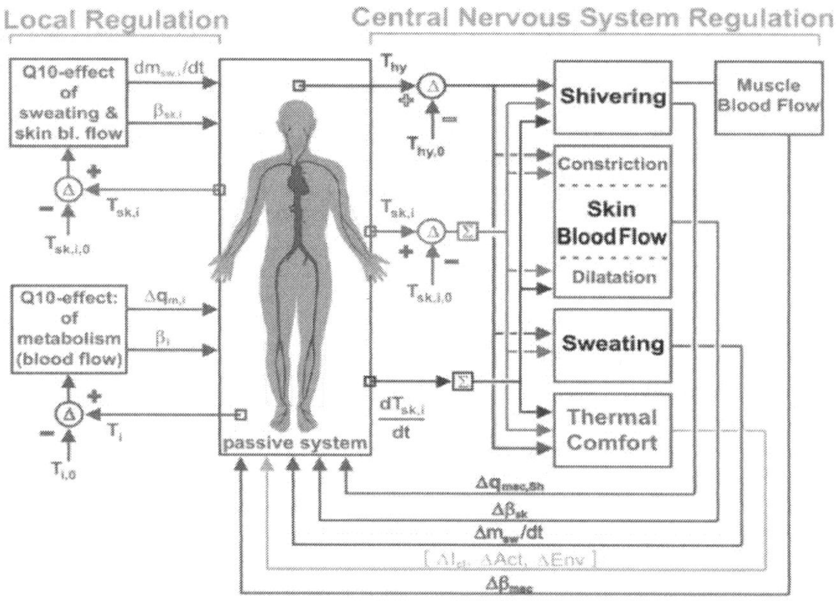

Fig. 6. Fiala's DTS and UTCI block diagram of active system [5].

The entire developed models shares characteristic, they have in common the modelling of the heat exchange within the body and the body with the environment.

Very important difference between the discussed models is in the modelling of the active part that determines physiological responses of the body. A very important distinguishing factor is that the models used different empirical relationships that correlate to diverse physiological responses (sweat rate factors, blood flow rates etc.) [7].

4 Conclusion

All discussed systems are very sophisticated, costly and do not offer the opportunity to focus only on the experiments on thermoregulatory sweating.

The large differences appear to be in apparatus (manikin and climatic chamber), experimental procedures, and methods of calculation used. In applying the different approaches for improving the thermal comfort, a holistic view should be considered and be creative in maximizing the functionality without compromising other important factors such aesthetics, cost, etc.

In order to different requirements and the aim of reducing the overall costs, many of the later manikin designs used plastics or even composite materials instead of metal. The metal manikins are still being manufactured for specific uses such as thermal comfort and heat loss analysis.

The findings of this study have a number of important implications for future practice because it provides a good overview of well-known systems and models and emphasize the missing gap for development of an affordable, general purpose, easily accessible and easy to use system for industrial applications.

The system aimed to be develop by the ongoing PhD thesis will focus on thermoregulatory sweating conditions and employ control algorithms based on developed mathematical models, as well as additive fabrication techniques that will allow it to be flexible, modular and customizable.

This research was supported by the OpTi-DeP Project (no. BG101/2016) financed from the UEFISCDI by the Romanian Government and by the HORIZON 2020 AMaTUC project (GA 691787) financed from EU.

References

1. M.G.M. Richards, N.G. Mattle, Def Tech Inf Cent, Comp Pt Not **ADP012411**, 4-2, 4-4, 4-5 (2002)
2. R.B. Farrington, J.P. Rugh, D. Bharathan, R. Burke, SAE Int, **2004-01-2345**, 1, 2, 5, (2004)
3. I. Pahole, B. Valentan, D. Z. Pavlinić, M. Ficko, J. Balič, T **22(6),** 1623-1626 (2015)
4. K. R. Diller, D. W. Hensley, A. E. Mark, J. R. Abella, G. M. Netscher, E. H. Wissler, J of Biomech Eng **135,** 021005-4, 021005-8 (2013)
5. J. C. Guedes, *Mathematical Modelling of Human Thermoregulation in Hot Environments, PhD Thesis,* (2016)
6. D. Fiala, K. J. Lomas, and M. Stohrer, Int J Biometeorol **45,** 144 (2001)
7. K. Katic, R. Li, W. Zeiler, Elsevier Build & Env **106,** 296 (2016)

Implementation of Lean Manufacturing in Romanian Organisations

Camelia Ioana Ucenic[1,*] and *Claudiu* Ratiu[1]

[1]Technical University Cluj Napoca, Faculty of Machine Building, Romania

Abstract. The implementation of lean manufacturing is one of the main tools for attaining a higher performance level in the Romanian organisations. The study uses data collected at middle management level from organisations located in Transylvania. The organisations come from different fields of activity and are in different stages of lean implementation. The article makes use of quantitative and qualitative instruments for the evaluation of lean in the above mentioned companies. Characteristics as waste reduction, cost, quality, product design have different approaches in the companies. Their way of approach gives useful information regarding the type of company from the point of view of lean manufacturing implementation. This knowledge provides a useful support at different managerial levels in the process of decision making.

1 Introduction

Taiichi Ohno and Shigeo Shingo developed the production systems of Toyota. The book written by Womak – "The Machine which changed the World" promoted their work and made well known the concept of lean production. The application of lean manufacturing principles took a rapid advance in many fields of activity. A lot of companies obtained better results due to lean manufacturing. Airbus, Philips and Rolls Royce are among the brands which implemented lean.

Lean manufacturing is not a universal receipt for success. Not all organisations which tried to apply it reached all the proposed goals. The literature review reveals situations when the company obtained a primary advantage but latter on it was not able to carry on the continuous improvement process. There are many reasons for the failure or for the delay in targets. One of the main reasons for the failure is the insufficient employee education regarding lean manufacturing values. This problem occurred at different managerial levels.

Different lean manufacturing frameworks were proposed. They include various tools, techniques and procedures [1]. Many of them only describe the lean elements. The number of elements which are included in a framework fluctuate from 3 to a maximum 50 elements [2]. The minimum attention given to the stakeholders is a common accuse of these frameworks. The internal stakeholders are the shop floor workers, the executives and the managers. They have different perspectives at a certain moment. The success or the failure of lean implementation depend on the equilibrium point among them.

* Corresponding author: cameliaucenic@gmail.com

2 The principles of lean manufacturing

The main objective of lean is to eliminate the non-value added elements from a process. The elimination of waste at each stage provides competitive advantage for the organisation. Womak and Jones proposed 5 lean principles:
• Identify the value;
• Map the value stream;
• Create the flow after the elimination of waste from the value stream;
• Establish pull and decrease the time to customer;
Look for perfection and make lean a part of corporate culture [3]. A correct implementation of lean requires the involvement of all employees.

Lean is associated more often with manufacturing but are many other fields where was implemented. The service sector benefits a lot after lean implementation. Last but not least are: information technology, software development and maintenance [4].

Swamynathan and Sekkizhar determined that lean means more than a collection of managerial tools. They explained that lean is a viewpoint of continuous improvement that approaches organizations from a systemic perspective [5].

2.1 Traditional organisation versus lean organisation

There are different characteristics which have to be analysed in order to correctly determine the type of an organisation: traditional or lean organisation.

A traditional organisation has the following features:
• The demand is unpredictable
• The producer has the responsibility for the product design
• A low priority is given to the elimination of waste
• The products are standard
• Operations are associated with the most important costs
• The percentage of defective products define the quality performance.
• The supply chain has a moderate capacity to absorb the risk.
A lean organisation is characterised by:
• A predictable demand
• The producer has the responsibility for the product design – same as in traditional organisation
• The elimination of waste has a basic priority
• The products are functional
• Operations are associated with the most important costs– same as in traditional organisation
• The percentage of defective products define the quality performance – same as in traditional organisation
The supply chain has a low capacity to absorb the risk [6].

Many Romanian companies are defined as hybrid organisations. Their demand is unstable and unpredictable. Quality performance has a new way to be measure using the degree of customer satisfaction. The marketing costs are included together with the operation costs in the most important costs. The producer and the customer have the responsibility for the product design. The same as in the traditional organisation, the supply chain has a moderate capacity to absorb the risk [7].

2.2 Correlation between lean principles and their defining elements

The above mentioned five lean principles are explained by the following components:

 DOI: 10.1051/matecconf/201713701017

• Waste reduction
• Order based production
• Customer focus
• Continuous improvement
• Partnership with suppliers.

The waste reduction has as targets the following elements: work in progress, lead time reduction, design for manufacturing, one piece flow and a flat organisational structure. Top management is responsible for the organisational structure and the other goals are tasks for middle management.

Order based production make use of small lot production, pull production, rolling production plans, Kanban system and just in time delivery. Suppliers respond by the just in time delivery. The same practice is considered within the workstations. The other elements are the responsibility of manufacturing department.

To be customer focus means to be in charge with value stream mapping, concurrent engineering, cellular manufacturing, standardisation of parts, and use of small multiple machines. First three elements are associated with the top management level and the last two with the manufacturing.

Continuous improvement implies product simplification, storage reduction, and elimination of buffers, total productive maintenance and new equipment. Top management respond for last two and manufacturing for the other elements.

Supplier partnership has as goals supplier reduction, sole source, long term relationships, supplier proximity and information sharing [8].

3 Methodology

The Kano model was used for this study. The model was presented first time in 1984 and requires a list of potential needs that the process should try to satisfy. Its selection was done because Kano model is recommended as a tool in the process of lean implementation because develops a set of ideas for planning a process or a service. Another reason for model selection was the insufficient employee education in lean attributes [9].

The novelty of this study in comparison with the classical Kano model is the substitution of customers with the internal stakeholders represented by the middle managers. The middle management is responsible for many components which define the lean elements and principles.

The attributes performance of the lean process is not equal for the managers. The performance of some attributes generate higher levels of satisfaction in comparison with other attributes.

A number of 24 middle managers participated at oral interviews. They come from six enterprises located in Cluj-Napoca, Alba-Iulia, Satu-Mare and Bistrita-Nasaud. The companies have implemented lean manufacturing and are in different stages of continuous improvement process.

The number of participant – 24 managers – is in correlation with the prescriptions of literature review. The theory says that 20 to 30 interviews which take place in homogeneous segments are appropriate in order to determine 90 to 95% of all possible process requirements [10].

The requirements are categorised by questionnaire. Each question has two parts:

• The functional form of the question: How do you feel if characteristic "X" is present in the process?
• The dysfunctional form of the question: How do you feel if characteristic "X" is absent in the process?

The answer has to be in one of the next five possibilities:

Table 1. Kano requirements.

Code	Meaning
A	attractive
M	Must be
O	One dimensional
I	indifferent
R	reversed
Q	questionable

Table 2. Kano requirements.

Code	Meaning
If characteristic "X" is good how do I feel?	1. I like it in this way
	2. It must be in this way
	3. I am neutral
	4. I can live with it in this way
	5. I dislike it in this way
If characteristic "X" is poor how do I feel?	6. I like it in this way
	7. It must be in this way
	8. I am neutral
	9. I can live with it in this way
	10. I dislike it in this way

4 Case study: the level of implementation of lean manufacturing in Romanian organisations

The goal of this study was to evaluate how perceive the middle managers from companies from Transylvania the level of implementation of lean manufacturing. As was previously presented one of the main reasons for the failure in lean processes is the insufficient employee education regarding lean manufacturing values.

The selected companies have manufacturing as object of activity. Lean manufacturing is in different stages of implementation. Due to confidentiality reasons neither the companies nor the names of middle managers are not revealed.

The following table presents an example of Kano analysis. The analysis considers nine lean characteristics. It was done using the framework proposed by Agile Logic.

A discrete analysis was done in the second step of the model. Each manager's answer was categorised using the evaluation table, as was presented in table 3. It was calculated the number of total responses for each category and each lean characteristic. The most frequent answer will define the characteristic. The ranking is as following:

• Must-be >Performance>Attractive>Indifferent

• The decrease of work in progress and a small production lot are must be characteristics for lean manufacturing. The category "performance" includes lead time decrease, standardisation of parts and elimination of buffer. The rolling production plan is the only attractive feature.

Table 3. Example of Kano analysis evaluation.

Story Narrative	How would you feel if this was present?	How would you feel if this was absent?	Priority
work in progress decreases	Expect It	Dislike	Must-Be
lead time decreases	Like	Dislike	Performance
organisational structure not flat enough	Don't Care	Live With	Indifferent
design for manufacturing has problems	Dislike	Expect It	Reversal (invalid)
small lot production	Like	Don't Care	Attractive
rolling production plan	Don't Care	Dislike	Must-Be
standardisation of parts	Expect It	Don't Care	Indifferent
elimination of buffer	Expect It	Live With	Indifferent
product simplification	Expect It	Dislike	Must-Be

Table 4. Kano requirements.

Characteristics	**M**	**P**	**A**	**I**	**R**	**Q**	**Total**	**Class**
work in progress decreases	11	3	2	5	1	2	24	M
lead time decreases	4	6	3	4	2	5	24	P
organisational structure not flat enough	5	4	6	9	0	1	24	I
design for manufacturing has problems	3	1	6	4	1	10	24	Q
small lot production	9	4	7	2	1	1	24	M
rolling production plan	4	3	8	4	3	2	24	A
standardisation of parts	3	10	6	4	1	0	24	P
elimination of buffer	2	8	5	4	2	3	24	P
product simplification	3	5	5	7	2	2	24	I

5 Conclusions

The article makes use of quantitative and qualitative instruments for the evaluation of lean in enterprises from Transylvania. The quantitative instrument was Kano model. These firms are in different levels of the lean manufacturing implementation. Twenty four middle managers participated at the interviews. They are considered as internal stakeholders. The model interpretation was done using discrete analysis.

Nine lean characteristics are in the model. Two are must be features, three are performance features, just one is attractive, two indifferent, and one questionable. The model can be developed by adding new lean manufacturing features. New companies have to be tested. A limitation of this study was the fact that all companies are from manufacturing. The model will have a greater value if other types of firms are involved in the study.

Last but not least, the discrete analysis can be replaced with the continuous analysis.

References

1. G.C. Parry, C.E. Turner, Production Planning & Control, **17** (1), 77 (2006)
2. G. Anand, R. Kodali, R., Int. J. Services and Operations Management, **5** (5), 687 (2009)
3. J.P. Womack, D.T. Jones, *Lean Thinking: Banish Waste and Create Wealth in Your Corporation,* New York: Simon & Schuster. (1996)
4. R. Bradley, D. M. Upton. "Lean Principles, Learning, and Software Production: Evidence from Indian Software Services". Harvard Business School Working Paper. No. 08-001 (2007)
5. A. Constancioara, G.L. Florian, O. Brad, Proceedings of the 9th International Management Conference, "Management and Innovation For Competitive Advantage", Romania, 940 (2015)
6. S. Konecka, Electronic Scientific Journal of Logistics, **6** (4), 24 (2010).
7. R. Parthipan, J. Anto, V. Nirmalkannan,. Int. J. of Innovative Research in Science, Engineering and Technology, **4** (2), 91 (2015).
8. R. Alas, Int. J. Services and Operations Management, **4** (4), 481 (2008)
9. ***http://asq.org/learn-about-quality/qfd-quality-function-deployment/overview/kano-model.html
10. L. Phillips: www.businessmeccanica.blogspot.com
11. *** Agil Logic Excel worksheetsheet for Kano analysis

Improving Performance in Supply Chain

Camelia Ioana Ucenic[1,*], *Claudiu* Ratiu[1]

[1]Technical University Cluj Napoca, Faculty of Machine Building, Romania

Abstract. The performance has different meanings and ways of defining and measurement. One cannot evaluate the supply chain of a company without using the concept of performance. There are two ways of measuring it: with qualitative and with quantitative instruments. The quantitative measures include financial and non-financial appraise. This article analysis some of the most used metrics. Resources, flexibility and output are listed among them. The non-financial performance measurements will be analysed from the point of view of the company as well as from the point of view of several of its main customers. The customer evaluation will be used in the second stage for describing the qualitative performance of the analysed supply chain. Customer satisfaction score, customer effort score and net promoter score will be calculated as framework for the qualitative performance of supply chain. The results will provide managerial information for the improvement of this supply chain.

1 Introduction

It is very difficult to provide a general framework for describing and measuring the performance of a supply chain. The industrial and non-industrial systems developed a wide variety of supply chains. Their structure was a linear one at the beginning but in time it was changed. Nowadays can be noticed very often a network type structure. Quantitative [1] and qualitative [2] measures can be used for supply chain performance. Problem identification is the first step of the process. In the second stage, it is necessary to describe the domain of measurement as well as the selected indicators and the relationship among them [3].

The literature review mentions a great amount of indicators from different fields of approaches. There is very complicated to select the most appropriate among them. It is necessary to introduce a set of measurement and selection assumptions in order to increase the quality of analysis. If the set of indicators is too large appears the danger of information dilution. A small number of indicators decreases the quality of performance measurement. The reports resulted after performance analysis are important tools for decision making in the process of planning, monitoring and controlling. The managers can make decisions at all managerial levels and at different time horizons.

There are many studies at the level of supply chain performance and different rankings. For example, Gartner Supply Chain Top 25 ranks every year the companies from specific fields of activity according with their supply chain performance. It makes use of financial and

* Corresponding author: cameliaucenic@gmail.com

DOI: 10.1051/matecconf/201713701018

opinion indicators. An overall weighting of 50-50 was used for the final score. The financial indicators had different weight in the formula: return of assets ROA 25% (net income / total assets), inventory turns 15% (cost of sold goods / inventory) and revenue growth 10% [4].

2 Defining the supply chain performance

The supply chain performance is described as the capacity of the supply chain to deliver with minimum logistic cost the right product at the correct time and in the correct specified location [5]. The delivered products must be in the specified quantities from customers' orders and must have the required quality. The final customer has to receive the products and services at the minimum cost [6].

Other authors define supply chain performance as the overall performance of the processes which are included in the companies' supply chain [7]. Each product has its own unique supply chain. The upstream and downstream relationships between its activities and processes provide the design of the supply chain. The changes in the components of a supply chain will impact its performance level. A correct approach has to define the performance measures as well as the performance measurement systems.

Supply chain performance assessment implies the following aspects:

• defining the indicators: they have to be clear expressed and to allow to be applied for each component of the supply chain,

• perspective of indicators: different components of supply chain may have different approaches regarding an indicator; all parts should agree and decide which is the selected perspective,

• the way of data capturing: the same way of capturing data has to be used in all supply chain components,

• confidentiality: different companies from the same supply chain do not accept to reveal all internal information to the third parties [8].

3 Type of metrics for supply chain performance

Nowadays one of the main challenges when one evaluates the supply chain performance is to decide which the performance metrics are. It has to be specified how are calculated these metrics and how data is obtained. The performance metrics have the following functions: control, improvement and communication [9]. The control function allows to the participants from supply chain to quantity and control the performance of supplies. If the performance level was not reached, the gap between the results and the selected targets can be improved. All stakeholders must know the performance metrics.

The measurement of supply chain performance using a single metric is not recommended. Some authors proposed the usage of three main categories of performance measures: resources, output and flexibility [10]. The objective of performance measure type resources is a high level of efficiency, the goal of output performance measures is a high level of customer service which helps the companies to avoid the transition of customers to other supply chains. The aim of flexibility performance measures is a great ability to answer fast to the demands of a changing environment.

Table 1. Resources, output and flexibility indicators.

Resources	Output	Flexibility
Manufacturing cost	Number of produced units	Increased customer satisfaction
Distribution cost	Number of on-time delivery	Reduction of late orders

Inventory cost	Product quality	Flexibility to demand variation
Return on investment	Customer satisfaction	Flexibility to adapt to new market requirements
Total cost	Sales, profit	Range flexibility (to what extend can change an operation)

Another very common approach of the supply chain performance indicators is the one from financial point of view. It divides the indicators in financial and non-financial. Some of the indicators and their association with the managerial levels are presented in the table 2.

Table 2. Financial and non-financial indicators [11].

Financial	**Non-financial**	**Level**
Rate of return on investment	Delivery lead time	Strategic
Delivery performance	Supplier lead time against industry norm	
Variation against budget	Flexibility of service for customer needs	
Net profit vs. productivity ratio	Customer perceived product value	
Delivery reliability	Accuracy of forecast techniques	Tactical
Supplier cost saving	Purchase order cycle time	
	Order entry methods	
	Product development cycle time	
Cost per operation hour	Capacity utilisation	Operational
Information carrying cost	Supplier rejection rate	
Work in progress inventory	Frequency of delivery	
Scrap level inventory	Quality of delivery documentation	

4 Case study

This article uses the non-financial indices in order to evaluate the performance of the supply chain. The evaluation was done for the supply chain of an automotive company located in Transylvania. The name of the company as well as the name of its customers were not revealed due to confidentiality reasons.

There are many non-financial indicators as was previously described. This study uses the customer satisfaction indicator as measurement for the supply chain performance. Customer satisfaction indicator was selected because a great percentage (89% according with 2011 Customer Experience Impact Report) of customers stop doing business with a firm after they experienced a low customer service. The loss of a customer generates an increase of costs because it is more expensive to attract a new one than retaining the existing ones. The customer satisfaction can be also measured and evaluated through different ways. The customer satisfaction score, customer effort score and net promoter score were chosen for the assessment.

Data collection took place in the interval January – May 2017. The methodology used qualitative and quantitative designs. A structured interview was conducted having as starting point the customer data basis selection given by the firm. The selection considered the main customers from the point of view of their impact in the firm's supply chain. The focus was on the persons who have expert knowledge at supply chain level. The total number of answer was 38. The selection step was 3. Primary and secondary data sources were used in the final evaluation of supply chain performance.

4.1 Customer satisfaction score CSS

The customer satisfaction score is very often used by the companies in their tries to measure and improve firm performance. Different persons rate their satisfaction in different manners in accordance with their culture. Some cultures tend to have a more simple scale to estimate the satisfaction but other have more satisfaction levels. This is the reason which generates different scale ranges for customer satisfaction score. The scale can be: 1-3, 1-5 or 1-10. A large range is not always a better one. For example, the US government employed a 4 star rating [12].

The general way to measure the customer satisfaction score is to address the following question: "How would you rate your global satisfaction of the service you received?" The customer has to rate his degree of satisfaction on the scale 1 to 5: 1 = very unsatisfied, 2= unsatisfied, 3 = neutral, 4 = satisfied and 5 = very satisfied. The composite customer satisfaction score CCSS is calculated as the average result.

This study calculates the CSS on a scale of 0-100%, where the 100 illustrate the total customer satisfaction. Only the number of respondents which rated their satisfaction with score 4 and score 5 are included.

$$SSC = \frac{Number\ of\ satisfied\ customers}{Number\ of\ satisfaction\ survey\ responses} \times 100 \quad (1)$$

The literature review shows that this is a correct approach because only the customers with the two highest possible response rating correctly predict the customer retention [13].

The following questions were included in the survey:

• Q1: How would you rate your global satisfaction regarding our quality of the product P1? P1 is the product with the greatest impact in the volume of sales

$$SSC_1 = \frac{24}{38} \times 100 = 63{,}15\%$$

• Q2: How would you rate your global satisfaction regarding our delivery time for the product P1? (answer shows responsiveness in supply chain)

$$SSC_2 = \frac{17}{38} \times 100 = 44{,}73\%$$

• Q3: How would you rate your global satisfaction regarding our product variety?

$$SSC_3 = \frac{31}{38} \times 100 = 81{,}57\%$$

• Q4: How would you rate your global satisfaction regarding our ability to answer to your changing needs? (answer shows flexibility in supply chain)

$$SSC_4 = \frac{11}{38} \times 100 = 28{,}94\%$$

The results show that the product variety obtained the highest customer satisfaction score. The worst aspect in the supply chain performance is its lack of flexibility, followed by a bad score regarding the responsivity. The SSC for quality of products has the second higher result but it is less than 80%, so still need special attention.

4.2 Customer effort score CES

The customer effort score is important when appear issues which have to be solved. The customer effort score is among the newest performance indicators. It was introduced by CEB (Corporate Executive Board) in 2010. CEB is a global best practice insights and technology company. "CES is 1.8x more predictive of customer loyalty than customer satisfaction (CSAT) measures, plus it is two times more predictive than Net Promoter Score (NPS)" [14].

The problem resolution implies a certain level of effort from the customer point of view. The customer effort score is important because it was demonstrated that customers tend to punish a bad service more often than to recompensed a good service. The clients appreciate more if their problem resolution is as easy as possible. A customer is loyal when his experience was effortless. It is recommended to address the CES question as soon as possible after the customer concluded his experience of interest.

The CES is calculated by asking an answer at the following question: "How easy made for me the company to handle the problem X?" The answer is rated on a scale from 1 to 7, where 1 means very low effort and 7 means a very high effort. The scale for CES: 1 = strongly disagree, 2 = disagree, 3 = somewhat disagree, 4 = neutral, 5 = somewhat agree, 6 = agree, 7 = strongly agree. The scale can be reduced 1 - 5 or 1 – 3 but the results loose part of significance.

The scale 1 – 3 was used. At the first tries with a 7 steps scale, the respondents couldn't decide fast a proper answer. The scale: 1 = difficult, 2 = neutral, 3 = easy.

$$CES = \%Easy - \%Difficult \tag{2}$$

This way of calculation was used because it is done in the same way by the majority of leading American brands. Action plan implementation is easier in this approach instead of calculation CES with the ratio total sum of scores / number of scores.

The following questions were included in the survey:

- Q1: How easy was to make an order using the website?

$$CES_1 = 42\% - 15\% = 27\%$$

- Q2: How easy was to solve your complain in terms of time?

$$CES_2 = 37\% - 26\% = 11\%$$

It is observed a better CES from the point of view of making orders. The long time for solving complains affect the performance of supply chain.

4.3 Net promoter score NPS

The net promoter score is the most common method to evaluate the customer loyalty. NPS together with CES allow to the company to compare the results and to provide a better action plan. The question is if the customer will recommend you to somebody else or not. The scale starts at 0 and goes until 10. The significance of each level is presented in the next table.

Table 3. Net promoter score scale [12].

Code	0	1	2	3	4	5	6	7	8	9	10
Meaning	Not a chance					Neutral					Very likely
Type	Detractors							Passives		Promoters	

The detractors have a score from 0 to 6. They are unsatisfied customers that complain about company, products and services and affect the image and reject other potential customers. The passive customers have score 7 or 8. They are satisfied but don't have loyalty and can be strongly influenced by the competition. They don't damage company image but also don't recommend it to other customers. The promoters have the score 9 and 10. They are happy, satisfied by the company. They keep on buying its products and more than this will recommend it to other people.

$$NPS = \%Promoters - \%Detractors \tag{3}$$

$$NPS = 32\% - 18\% = 14\%$$

A NPS of 14% is not sufficient. It illustrate that many customers are passives. The company performance can be increased by pushing them from "passive" category to "promoters" category. NPS together with other non-financial and financial indicators can help the decision makers in their process to increase the supply chain performance.

5 Conclusions and recommendations

The main objective of this study was to measure and evaluate the supply chain performance of a company from automotive sector using opinion non-financial performance indicators. The company must increase its supply chain performance. This can be done starting from the lowest scores. It must track the negative ratings. Another recommendation is to decrease the time for responding to the customers. The literature review suggest an answer in less than 10 hours for generating scores above 90%.

The percent of resolved issues has to increase. This has to be done in a single interaction, as much as possible. The implemented changes have to be communicated to the clients. The client will feel he is valued for the given input. Another urgent requirement is to reduce the time for complain solving. Customer effort score can be calculated using a scale from 1 to 7, but it is necessary to address the question to persons with more experience in supply chain aspects and with higher positions in the organisational structure.

CES and NPS provide to service leaders a holistic company image: CES gives the service perspective, while NPS gives answers from business perspective. The firm has to decide in the final action plan if wants to reduce the customer effort or to increase the promoters.

Further research can be done. The number of non-financial indicators can be increased and obtain a better evaluation framework. They can be analysed in parallel with financial indicators. A larger set of measures can be proposed in order to increase the performance of supply chain from this company.

References

1. C. Sheperd, H. Gunter, *Behavioral Operations in planning and scheduling*, J.C. Francsoo (Eds), Springer Verlag (2016)
2. F.T.S. Chan et all, Management Decision **41** (7), 635-642 (2003)
3. D. Leonczuk, J. Business Management and Education **14** (1), 103-105 (2016)
4. *** https://www.gartner.com/doc/2493115/gartner-supply-chain-top-#1739401341
5. H. Zhang, S.C. Okoroafo, EMR **4** (1), 38–48 (2015)
6. G.D. Whitten, K.W. Green, P.J. Zelbst, Int J of Operations & Production Management **32** (1), 28–48 (2012)
7. M. Srinivasan, D. Mukherjee, A. Gaur, European Mg. J **29**, 260– 271 (2011)
8. H. Stadtler, C. Kilger, *Supply chain management and advanced planning. Concepts, Models, Software and Case studies. 4th ed.*, Springer-Verlag, Germany (2008)
9. S.A. Melnyk, D.M. Stewart, M. Swink, JOM **22**(3), 209–217 (2004)
10. B.M. Beamon, Int. J. of Operations & Production Management, **19** (3), 275-292 (1999)
11. B. Gunasekaran, C. Patel, R.E. McGaughey, Int. J of Production Economics **87**(3), 333–347 (2004)
12. *** https://www.userlike.com/en/blog/6-proven-methods-for-measuring-your-customer-satisfaction
13. *** https://blog.emolytics.com/customer-experience/customer-satisfaction-score-csat-kpi/
14. M.Dixon, K.Freeman, N.Toman, Harvard Business Review, (2010)

MATEC Web of Conferences **137**, 01019 (2017) DOI: 10.1051/matecconf/201713701019
MTeM - AMaTUC 2017

Analysis of shape correctness of surfaces of diamond burnished components

Gyula Varga[1*] and *Viktoria* Ferencsik[1]

[1]University of Miskolc, Institute of Manufacturing Science, 3515 Miskolc, Egyetemváros, Hungary

Abstract. In these days, diamond burnishing is frequently used for final finishing manufacturing operations of components. Diamond burnishing belongs to cold plastic manufacturing procedures. When using this technique, the following advantages can be obtained: micro-hardness of near surface layer is increasing; surface roughness is improving and the shape correctness is also improving. This paper deals with how the different technological parameters of burnishing, such as the feed rate, the burnishing speed and the burnishing force effect on the cylindricity of the burnished workpiece. The experiment was done on some specimen having outer cylindrical surfaces. The experiments were executed by the Taguchi type Full Factorial Experiment Design method with which empirical formulas can be efficiently created. The measurements of the cylindricity of specimens were done with a circular and position error measuring. From the measured data, special improvement ratios were calculated in order to define the appropriate range of technological parameters which results high improvements. Further aim was to compare the measured values of different cylindricity parameters.

1 Introduction

The development of production engineering and manufacturing is an important part of realization of the tasks determined in the Industry 4.0 [1] in environmentally way [2, 3]. In the last decade, the researches dealing with the shape correctness [4] and surface roughness of the machined surface [5-7], become quicker. More and more papers deal with the examination of residual stresses [8] and changes on the texture [9] taking part on the surface. This is the reason why these features decisively determine the lifetime and reliability of the products [10]. The differently manufactured parts' accuracy is analysed, examined, and compared [11] because parts can be produced with higher accuracy by the newly developed cutting tools [12-14].

During operation, the maximum load of machine tools has effect on the surface of machine elements for a certain depth, so their quality has a great influence mainly on fatigue and so for the operational safety. Burnishing belongs to cold plastic manufacturing which has significant role in forming technology.

* Corresponding author: gyulavarga@uni-miskolc.hu

By its application the surface roughness, the micro-hardness and the corrosion resistance can be increased, just as shape correctness of cylindrical surfaces can be improved [15, 16]. In most cases diamond burnishing can be more effective than conventional chip removal finishing processes as grinding, lapping and polishing. Because of it the working process can be more time and cost effective. Furthermore, environment friendly processing can be implemented because it is chipless so it does not require high volume of coolants and lubricants [17].

The procedure of burnishing is performed for final finishing manufacturing of inner and outer cylindrical surfaces, the main areas of its applications are automotive and aeronautics industries. This study deals with the examination of outer cylindrical surfaces only.

The experiments were performed using the Factorial Experiment Design [18, 19]. Some paper as [20] presented the classification and short description of design of experiment (DoE) methods. The design of experiment is used to determine optimal process parameters in the manufacturing industry, such as for the end milling processes [21].

2 Burnishing of outer cylindrical surfaces

In diamond burnishing the reduction of surface roughness and improving of shape correctness of the surface is caused by the interaction between the tool - which material much harder than the workpiece - and the workpiece while performing rotational and linear movement with sliding friction between them (Fig. 1) [22].

The material of the working part of the burnishing tool can be hardened steel, carbide, mineral ceramics and natural or artificial diamond.

The pressure that is necessary for realizing of cold shaping arrives from the overlapping between the working part of the tool and the surface to be formed of the workpiece. Plastic deformation is realised in 0.01-0.2 mm thick layer in the subsurface of the workpiece because of the static content of the shaping element (burnishing tool) and the outer surface of the workpiece [23-24].

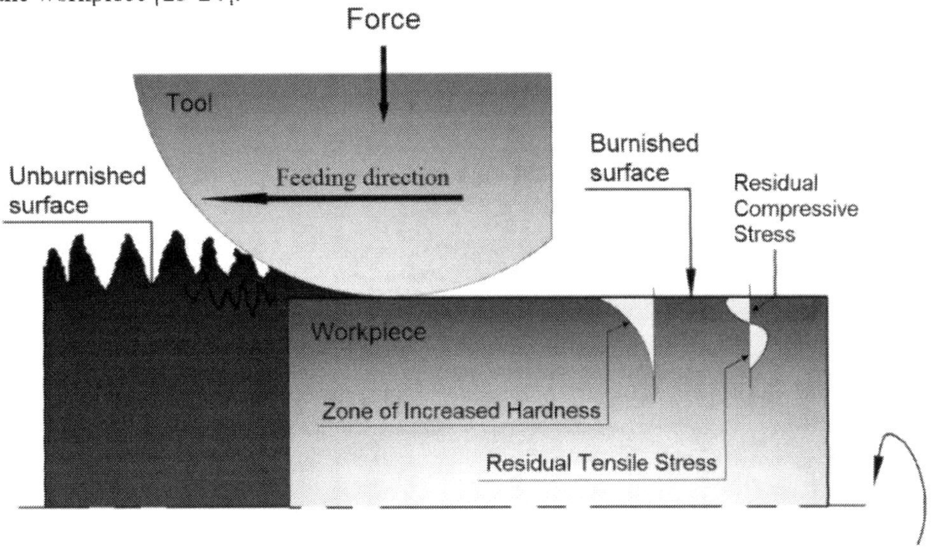

Fig. 1. Schematic illustration of burnishing process [22].

Burnishing of outer cylindrical surfaces can be executed on conventional universal lathes or up-to-date CNC lathes.

3 Experimental conditions

During the experiments, a CNC lathe with flatbed produced by firm OPTIMUM type OPTIturn S600 was used. For the examination, we have chosen hardened steel mark C60 (Rm = 730-1100 MPa, 850-900 HV) as workpiece material and its chemical composition can be found in Table 1 [25]. It was turned to 49 mm diameter dislocated to 5 parts, each of them being 25 mm long (Fig. 2). PCD (polycrystalline diamond) tool was used with 3.5 mm radius and the applied oil kinematic viscosity was 70 mm^2/s.

Table 1. Chemical composition of the burnished specimens [25].

Elements	C	Si	Mn	P	S	Cr	Mo	Ni
Chemical composition (Typical analysis in %)	0.57 0.65	max 0.40	0.60 0.90	max 0.045	max 0.045	max 0.4	max 0.1	0.40

Measurement of the cylindricity was done with a circular and position error measuring machine type Talyrond 365 [26]. In this investigation, inductive sensor was applied for measuring before and after burnishing in 3 mm distance.

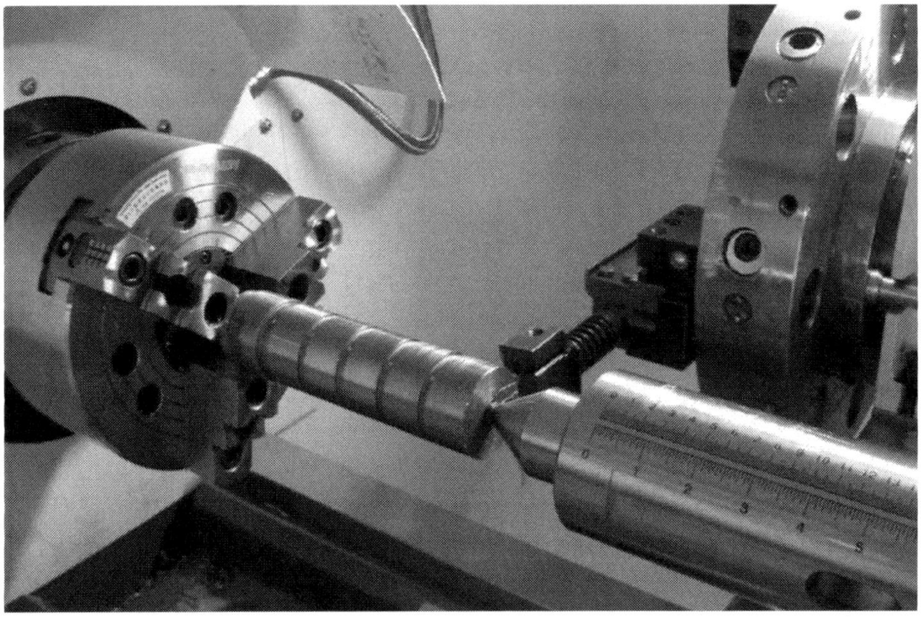

Fig. 2. Execution of diamond burnishing process [27].

In all 16 cylindricity indices were analysed and 4 of them were chosen and examined which mostly determine operating properties. These are: CYLp – peak to reference cylindricity deviation; CYLv – reference to valley cylindricity deviation; CYLt - peak to valley cylindricity deviation and CYLtt – cylinder taper [28], illustrates the rate of conicity as it can be seen in the Fig. 3 [29]. The above mentioned cylindricity parameters are included in standard ISO 12180-1.

Examined parameters: burnishing speed, feed rate and burnishing force which parameter ranges are following:

Burnishing speed: v_{b1} = 40 m/min and v_{b2} = 80 m/min
Feed rate: f_1 = 0.05 mm/rev and f_2 = 0.10 mm/rev
Burnishing force: F_{b1} = 50 N and F_{b2} = 100 N

The matrix of the Taguchi type Factorial Experimental Design [18] can be seen on Table 2, which contains the burnishing parameters in natural dimensions and their transformed values.

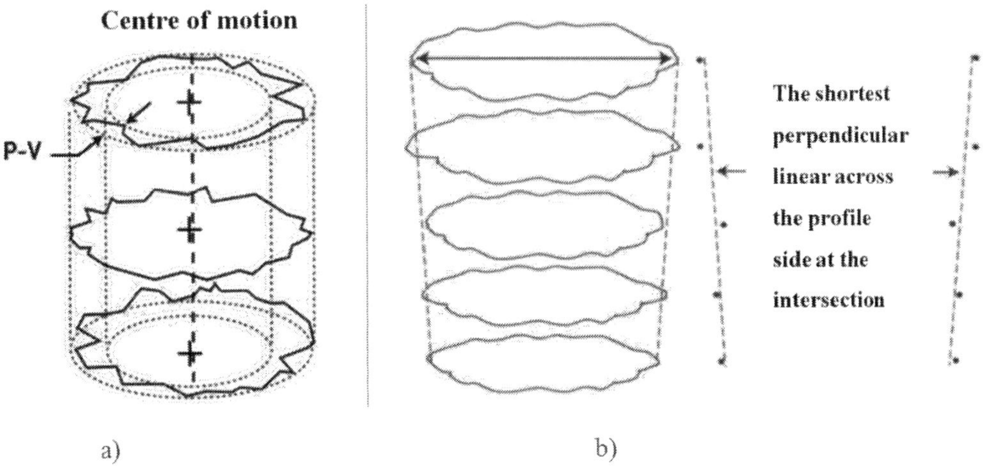

Fig. 3. Cylindricity parameters: CYLp, CYLv, CYLt (a) and CYLtt (b) [25].

Table 2. Applied burnishing parameters.

Sign of experiments	Burnishing parameters			Transformed parameters		
	v_b [m/min]	f [mm/rev]	F_b [N]	X_1	X_2	X_3
1	40	0.05	50	-1	-1	-1
2	80	0.05	50	+1	-1	-1
3	40	0.10	50	-1	+1	-1
4	80	0.10	50	+1	+1	-1
5	40	0.05	100	-1	-1	+1
6	80	0.05	100	+1	-1	+1
7	40	0.10	100	-1	+1	+1
8	80	0.10	100	+1	+1	+1

During examination, the improvement ratios of the different cylindricity parameters were created, the generalized formula of it can be seen in formula (1):

$$\rho_{CYLx} = \frac{CYLx_b - CYLx_g}{CYLx_b} \cdot 100 \quad , \% \tag{1}$$

where:

- ρ_{CYLx} Improvement ratio without dimension to be indicative the change of cylindricity parameter that caused by burnishing process
- CYLx Different cylindricity parameters. *x* (in this experimental) adds value of *p*, *v*, *t*, and *tt*, so it is equal to CYLp, CYLv, CYLt and CYLtt parameters
- $CYLx_g$ Measured cylindricity parameters after grinding
- $CYLx_b$ Measured cylindricity parameters after burnishing

The smallest are the different rates in the negative range, the greater is the improvement.

4 Results

The values of the measurements can be seen in Table 3. with the percental values of the improvement ratios calculated from them.

Table 3. Measured values of the cylindricity parameters.

Sign of the workp.	CYLp [µm]		ρ_{CYLp} [%]	CYLv [µm]		ρ_{CYLv} [%]
	$CYLp_g$	$CYLp_b$		$CYLv_g$	$CYLv_b$	
1	11.33	15.58	27.27	2.71	4.07	33.41
2	19.98	21.13	5.44	3.34	4.07	17.94
3	21.93	18.15	- 20.82	2.53	3.42	26.03
4	10.37	44.14	76.50	3.04	10.97	72.29
5	14.56	15.75	7.56	2.97	4.04	26.49
6	20.63	18.45	- 11.81	2.60	3.58	27.38
7	21.16	18.49	- 14.44	2.57	3.59	28.42
8	16.66	20.34	18.1	2.78	3.66	24.05

Sign of the workp.	CYLt [µm]		ρ_{CYLt} [%]	CYLtt [µm]		ρ_{CYLtt} [%]
	$CYLt_g$	$CYLt_b$		$CYLtt_g$	$CYLtt_b$	
1	14.04	19.92	29.51	- 6.67	18.53	135.99
2	23.33	25.20	7.43	-20.96	-20.48	-2.34
3	24.46	21.56	-13.45	-17.58	-13.14	-33.78
4	13.40	55.12	76.69	-4.35	48.84	108.90
5	17.53	19.78	11.38	-9.81	-14.88	-34.07
6	23.23	22.02	- 5.49	-9.97	-8.67	-14.99
7	23.72	22.09	- 7.37	-17.27	-9.13	-89.15
8	19.44	24.00	19.00	-12.19	-16.37	-25.53

By the use of Factorial Experiment Design empirical formulas (2-5) can be determined. Calculations and demonstrations of the functions were done by using the MathCad software (Fig. 4-5).

$$\rho_{CYL_p} = 331.84 - 5.267 \cdot v_b - 5.211 \cdot 10^3 \cdot f - 2.31 \cdot F_b + 93.195 \cdot v_b \cdot f + 0.035 \cdot v_b \cdot F_b + 37.332 \cdot f \cdot F_b - 0.672 \cdot v_b \cdot f \cdot F_b \quad (2)$$

$$\rho_{CYL_t} = 217.57 - 4.014 \cdot v_b - 2.908 \cdot 10^3 \cdot f - 1.992 \cdot F_b + 64.36 \cdot v_b \cdot f + 0.042 \cdot v_b \cdot F_b + 30.52 \cdot f \cdot F_b - 0.67 \cdot v_b \cdot f \cdot F_b \quad (3)$$

$$\rho_{CYL_t} = 323.3 - 5.212 \cdot v_b - 4.967 \cdot 10^3 \cdot f - 2.331 \cdot F_b + 90.6 \cdot v_b \cdot f + 0.037 \cdot v_b \cdot F_b + 37.276 \cdot f \cdot F_b - 0.69 \cdot v_b \cdot f \cdot F_b \quad (4)$$

$$\rho_{CYL_{tt}} = 1.4037 \cdot 10^3 - 20.331 \cdot v_b - 1.604 \cdot 10^4 \cdot f - 13.573 \cdot F_b + 258.74 \cdot v_b \cdot f + 0.197 \cdot v_b \cdot F_b + 140.464 \cdot f \cdot F_b - 2.365 \cdot v_b \cdot f \cdot F_b \quad (5)$$

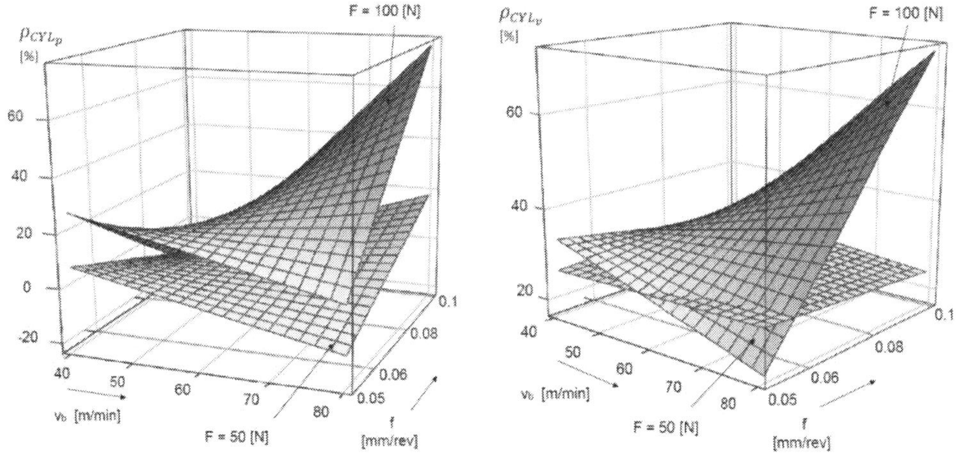

a) b)

Fig. 4. Experimental results in the point of view of peaks (a) and valleys (b) of the cylindricity error.

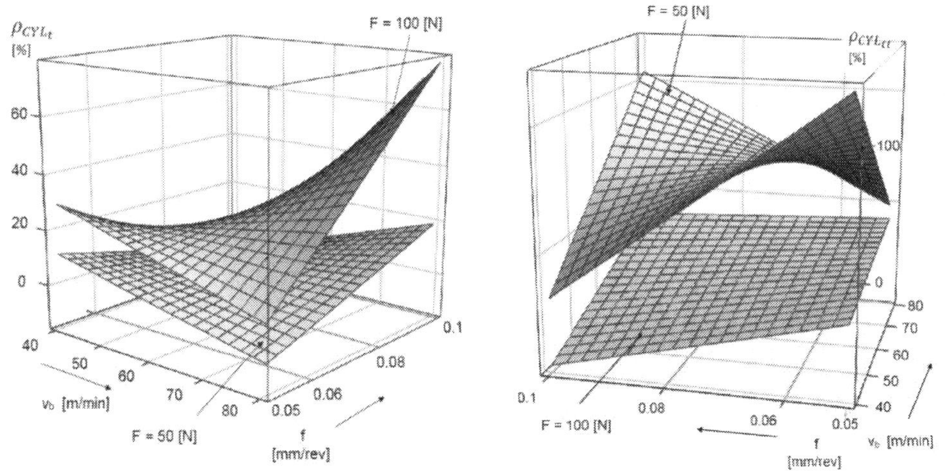

a) b)

Fig. 5. Experimental results in the point of view of largest deviations (a) and conicity of the cylindricity error (b).

Fig. 4-5 show the extent of the improvement of the executed sliding burnishing. It can be stated that the application lower burnishing speed (v_{b1} = 40 m/min) with higher feed rate (f_2 = 0.1 mm/rev) when the burnishing force is lower (F_{b1} = 50 N) show the highest improvement of CYLp and CYLt. When the burnishing force is higher (F_{b2} = 100 N) in that case the improvement is worse by 6.38 % related to the CYLp and 7.07 % related to CYLt. In case of CYLv the higher burnishing speed (v_{b2} = 80 m/min) with lower feed rate (f_1 = 0.05 mm/rev) when the burnishing force is lower (F_{b1} = 50 N) showed the higher improvements. Increasing the burnishing force to the higher value (F_{b2} = 100 N) without changing the value of the other two parameters the improvement ratio of CYLv is reducing by 9.44 %. From the conicity point of view when the feed rate (f_2 = 0.1 mm/rev) and burnishing force (F_{b2} = 100 N) is higher and the burnishing speed is lower (v_{b1} = 40 m/min) show the best result. By increasing the burnishing speed to the higher value, the improvement value is reducing by 63.62 %.

5 Summary

The paper dealt with the experimental analysis of sliding burnishing when the material of the workpiece was hardened steel. Experimental parameters were the burnishing speed, feed rate and burnishing force.

The aim of the experiments was to determine how these parameters have effect to the shape correctness. The experiments were executed and evaluated on the base of Taguchi type Full Factorial Experiment Design. The evaluation was more visible by the use of the dimensionless improvement ratios of different cylindricity parameters (ρ_{CYL_p}, ρ_{CYL_v}, ρ_{CYL_t}, $\rho_{CYL_{tt}}$). After determining the empirical formulas their 3D demonstrations gave to the evaluation a graphical perspective.

On the base of the present research work it can be stated:

• The results of ρ_{CYL_t} (improvement ratio of peak to valley cylindricity deviation) which was calculated from the most frequently used cylindricity parameter (CYLt) are very much similar to the results of ρ_{CYL_p} (improvement ratio of peak to reference cylindricity deviation). According to it the total shape correctness was consisted of peaks (CYLp) in the examined parameter ranges.

• The higher improvement was caused, from the peaks and valleys point of view, by the application of these burnishing parameters:

$v_{b1} = 40$ m/min; $f_2 = 0.1$ mm/rev; $F_{b1} = 50$ N.

• From the point of view of conicity the improvement was served by setting the burnishing parameters to:

$v_{b1} = 40$ m/min; $f_2 = 0.1$ mm/rev; $F_{b2} = 100$ N.

"Project no. NKFI-125117 has been implemented with the support provided from the National Research, Development and Innovation Fund of Hungary, financed under the K_17 funding scheme."
"The described study was carried out as part of the EFOP-3.6.1-16-00011 "Younger and Renewing University – Innovative Knowledge City – institutional development of the University of Miskolc aiming at intelligent specialisation" project implemented in the framework of the Szechenyi 2020 program. The realization of this project is supported by the European Union, co-financed by the European Social Fund."

References

1. P. Tamás, B. Illés, Acad. J. of Manuf. Eng. **14** 4, 119 (2016)
2. P. Tamás, B. Illés, P. Dobos, IOP Conf. Series: Mat. Sci. and Eng. **161**, 1 (2016)
3. M. Rusko, R. Kralikova, Advanced Mat. Research **816-817**, 1225 (2013)
4. M.H. El-Axir, O.M. Othman, A.M. Abodiena, J. Mat. Proc. Tech. **196** 1-3, 120 (2008)
5. J. Kundrák, K. Gyáni, V. Bana, Int. J. of Adv. Manuf. Techn. **38** 1-2, 110 (2008)
6. C. Felhő, J. Kundrák, Key Eng. Materials **496** pp.: 194-199 (2012)
7. G. Varga, Key Eng. Mat. **581** pp.: 403-408 (2014)
8. M. Novák, Manuf. Technology **12** pp.: 66-70 (2012)
9. J. Kundrák, Z. Gácsi, K. Gyáni, V. Bana, G. Tomolya, Int. J. of Adv. Manuf. Techn., **62** 5-8, 457 (2012)
10. W. Zebala, B. Slodki, G. Struzikiewicz, Case Study. Eksploatacja i Niezawodnosc – Maint. and Reliability **15** 4, 421 (2013)
11. J. Kundrák, A.G. Mamalis, A. Markopoulos, Materials and Manuf. Proc. **19** 6, 979 (2004)
12. G. Dezső, J. Herman, F. Szigeti, Int. J. of Eng. **X** 1, 155 (2012)

13. I. Sztankovics, J Kundrák, Manuf. Techn. **14** 1 pp.: 97-104 (2014)
14. J. Beňo, I. Maňková, M. Vrabeľ, D. Kottfer, Measurement **46** 1 pp.: 582-592 (2013)
15. M.R. Stalin John, P. Suresh, D. Raguraman, B.K. Vinayagam, Arab J. SciEng. **39** pp.: 3209–3216 (2014)
16. A. Akkurt, Metal Sci. and Heat Treatm. **53** 3–4 pp.: 145-150 (2011)
17. M. Polankova, F. Manlig, R. Králiková, MM Science Journal. pp.: 691-695, (2015)
18. G. Taguchi, System of experiment design, 1. Experimental design (UNIPUB, Kraus International Publications, White Plains, New York, 1987)
19. L. Fridrik, Chosen chapters from the topics of experimental design of production engineering (Műszaki Könyvkiadó, Budapest, 1987) (In Hungarian)
20. Á. Drégelyi-Kiss, R. Horváth, B. Mikó, Dev. in Mach. Technol. **3**. Ed.: W. Zebala, I. Manková, Cracow University of Technology pp.: 20-34 (2013)
21. M.N. Durakbasa, A. Akdogan, S. Vanli, A. Gunay Bulutsuz, Measurement **68** pp.: 92-99 (2015)
22. A. Akkurt, J. of Materials Eng. and Perform. **20** 6 pp.: 960-968, (2011)
23. L. Luca, S. Neagu-Ventzel, I. Marinescu, Prec. Eng. (Elsevier) **29** pp.: 253-256 (2005)
24. T.A. El-Taweel, M.H. El-Axir, The Int. J. of Adv. Manuf. Techn. **41** pp. 301-310, (2009)
25. G. Varga, Key Eng. Mat. **581** pp.: 403-408 (2014)
26. http://www.steelnumber.com/en/steel_composition_eu.php?name_id=154 (09.01.2017)
27. V. Ferencsik, Sci. Students' Conf., University of Miskolc, pp.: 1-34 (2013) (In Hungarian)
28. D.J. Whitehouse, *Handbook of Surface and Nanometrology*, Second edition, (CRC Press, Taylor and Francis Group, 2011)
29. N.N., Study aid, Taylor Hobson Precision, p.: 34 (In Hungarian) (2013)

Simulation as a designed tool for material flow analysis by means of Witness Horizon

Tomáš Vysocký[1,*], *Ľudmila* Nováková-Marcinčinová[2] and *Ema* Nováková-Marcinčinová[3]

[1]Faculty of manufacturing technologies with a seat in Prešov – Technical University in Košice, Deparment of Computer aid of manufacturing technologies, 08001 Prešov, Slovakia

[2] Faculty of manufacturing technologies with a seat in Prešov – Technical University in Košice, Deparment of Computer aid of manufacturing technologies, 08001 Prešov, Slovakia

[3] Faculty of manufacturing technologies with a seat in Prešov – Technical University in Košice, Deparment of Computer aid of manufacturing technologies, 08001 Prešov, Slovakia

Abstract. In this post, the simulation is analysed, which positively influences the production process. Through simulation, it is possible to identify deficiencies found in different workplaces. Initially, the most familiar simulation software is used, which is used to create different types of simulations. The Witness system is described in more detail by simulation and subsequent results. The basic step before the simulation is to determine the data that enters the simulation. The workplace is composed of three basic workplaces that are contained in a number of machines. There is a conveyor between these workplaces, which ensures the transport of parts throughout the entire production hall. The result of this analysis is to evaluate the efficiency of each workplace and possible suggestions to increase the productivity of the entire process that are displayed using graphs.

1 Introduction

By means of computer simulation we prove create conditions of real experiment. We know follow phenomena, that may arise at exploring different elements. Most often, computer simulation is used in the industry, when is needed prevent mechanical destruction or improve quality of a manufacturing process. Manufacturing process is exposed to all risks, that may arise when its use. The main rules for creating simulation are right set conditions, which influence the whole simulation. By way of example may be mentioned simulation of torpedo with name REMUS, where has been verified this model of six-degrees of freedom [1].

The main elements of this simulation was the torpedo model, which had set center of gravity and coordination system in the middle point of the model. The experiment had to remind underwater environment, so environs had to remind water and all force, which

* Corresponding author: tomas.vysocky@tuke.sk

acting on the solid part. Calculating was created though two methods Euler's method and method Runge-Kutta [1].

The basic tool for creating simulation is hardware and software equipment. For difficult simulation are needed the executive workstation, which have big processing power. In workstation must be mounted a high-capacity RAM, which can be assembled from more elements. Computer processor must contain more cores, compared with normal computer processors. In the case when simulation has a graphical character or result is complicated graphical output, that workstation must contain graphic card, which is determined for calculations. All components must be assembled all right and compatible. Therefore the workstation are bought as one component. In a few cases components are purchased as separate parts and assembled by customer [2].

2 Simulation, simulation software and advantages

In different literatures we can meet with more definitions, what is computer simulation. The most famous definition is it, where is written, simulation is virtual environment, which recalls real world or is same with him. This simple definition showing on simulation, where we need to create 3D model and simulate real conditions. In addition to such simulation is possible simulate different electrical schemes, which are interconnected and are able to start or shut down circuit. The output of the simulation is functional scheme, which is applied into device, which is intended for this.

The main part of simulation is software, which secured of simulation, contain all elements, which are connected with her or can contain library of elements. The most famous software in school environment is Matlab or superstructure in Matlab with same name Simulink. In general, the Matlab is programming environment specializing in scientific or technical numerical calculations, modelling, design of algorithms and so on. Superstructure Simulink is program, which is determined on simulating and modelling dynamical systems, which used algorithm of Matlab.

Environment of this software is similar as environment of basic text editor with little difference that used don't type text, but he inserted blocks, which represent some properties. Superstructure Simulink contains library of blocks, which are used for creation of simulation scheme.

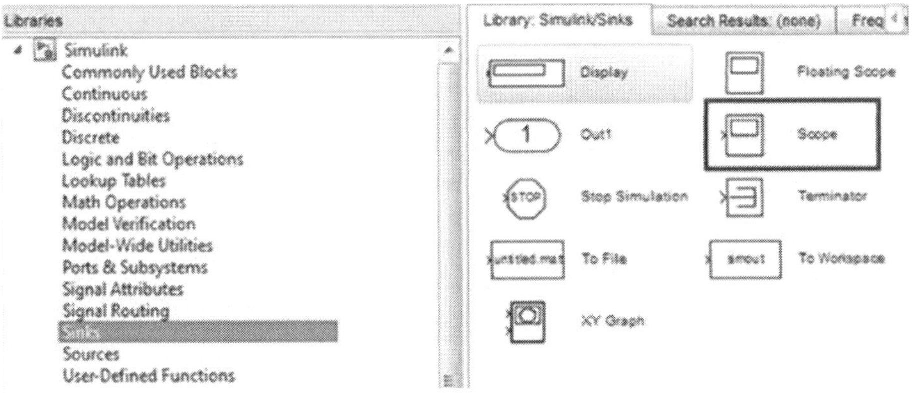

Fig. 1. Library in superstructure Simulink

As can be seen on Fig. 1 library is created very easily and elements aren't graphics intensive. Icons if this elements except graphical site, have functional site. On one or two sites they have small arrows. These arrows represent input or output to next device or element. Outputs of functional scheme can be graphs in different graphical styles, the behaviour of car tachometer, control traffic lights and the like. Outputs of this style has demanding graphic character and it is possible compare them with realistic devices [3].

The following software for simulation of manufacturing process is Flexsim, created by FSP Company. Company was created by Bill Nordgren in 1993. Former name of company was F&H. In 1998, they developed first generation of 3D object-oriented simulation through Taylor ED graphic. First compact software with name Flexsim in version 1.0 was introduced in 2003. In this version of the software included 3D modelling environment and seamless integration of C++ programming language. Animations can be displayed in three views 2D, 3D view and display through virtual reality [4].

FSP Company produces two types of software Flexsim Simulation and Healthcare Simulation. The second software is used for simulation of hospital premises and all things, which are connected with that. Flexsim Simulation is used primarily for simulation of manufacturing process. As already mentioned, the software is based on 3D object-oriented modelling so all elements are inserted through models. Working environment is similar as environment other software's. It contains status bar, working desktop, basic functions and library of elements how is shown on figure 2.

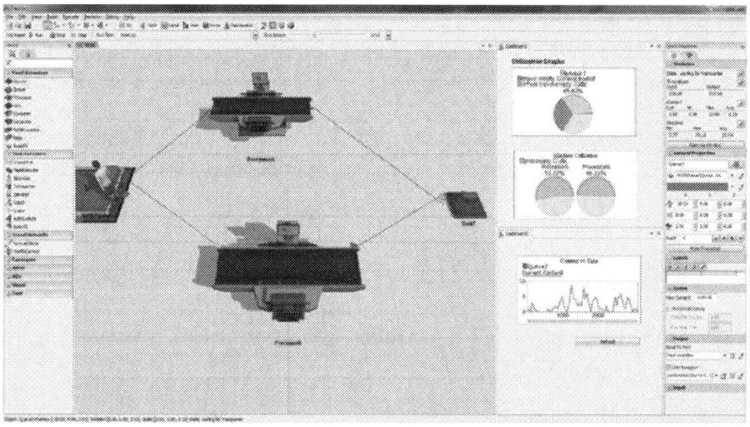

Fig. 2. Environment of Flexsim Simulation.

Modeling in software is simple, here is used drag and drop technique. Elements are inserted from library of elements. Between two elements are connecting lines, which represent connection of elements, for example: warehouse – machining center – conveyor – manipulator – warehouse. This software exempting of the simple controls and optimizations. Advantage of software is option to insert 3D models from CAD systems. Except for models is possible import a drawings, which form the pattern of workplace. Next advantage are integrated scripts, which automatically respond on some elements, so it is not necessary to know programing language on high level.

Similar software as FlexSim is Witness from Lanner Group Company. The simulation software is created on base of object modeling, where predefined elements are inserted into scene. Basic elements represent warehouses, machining centers, manipulators and transport technology. Tools and machines can be found in library of elements, which are implemented in software. Library in addition to elements contain numerical or graphic components how graphs, time and piece units and variables, which are used for analytical part of simulation. This components are perceived how previous elements. One difference between numerical components and machines or warehouses is, that numerical components are not models.

Environment of Witness system is easy and uncluttered. On the top part are basic functions, on the second row are icons, which represent functions as connections, view change, zoom in, zoom out, text format, show and hide grid and more. In the middle part is work place, where is displayed grid. At the bottom part is library of elements, time displayed and icons for control of simulation. All windows can organize your own discretion (Fig. 3).

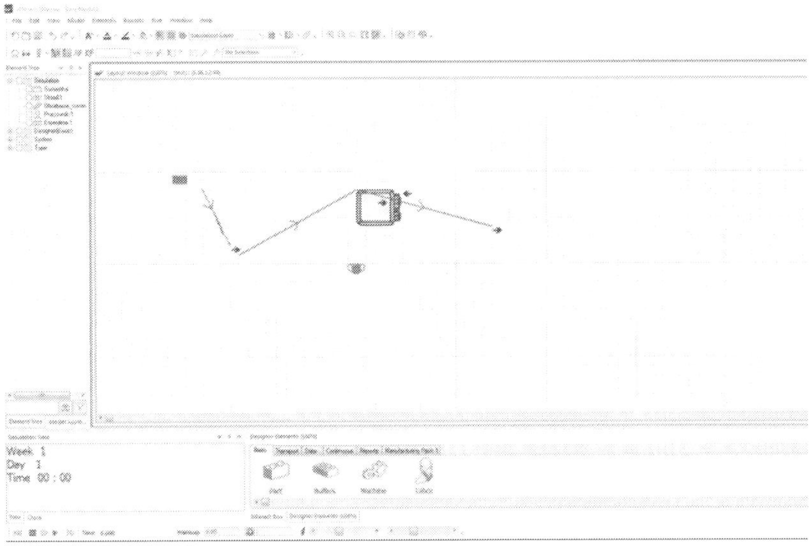

Fig. 3. Environment of Witness Horizon.

Compared to the previous simulation software Witness has better solved rules, where is possible to set for example, change length of the blank on input and output. It is more modular, has a lot of output options, display options and so on.

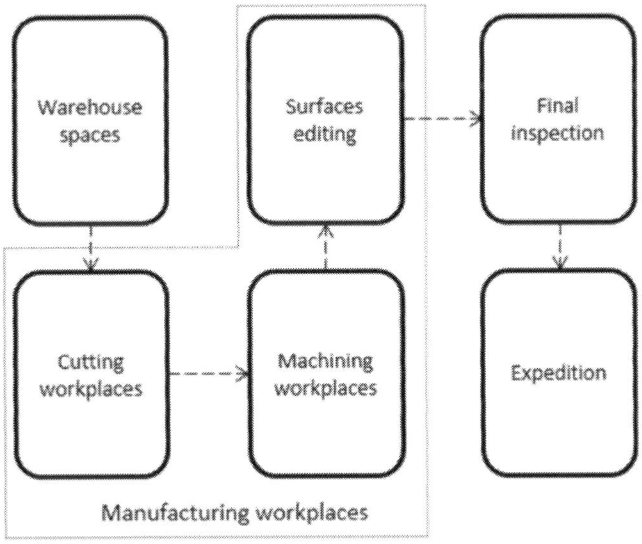

Fig. 4. Basic scheme of manufacturing process.

On Fig. 4 is displayed basic scheme, which points to process blank in its production. First workplace are warehouse spaces, where are saved blanks, which are designated for next processing. Through transport devices, the blanks are transported into manufacturing hall, which contain three cutting workplaces, machining workplace and surfaces editing.

Since those three workplaces are in one manufacturing hall, the best a means of transport is conveyor. At the end of manufacturing process is inspection, where are all parts inspected. If some part contain defects, so they go to waste. The last workplace is expedition, where the packing materials occurs and the temporary storage.

This basic scheme does not contain count of machines and tools, which are used in simulation, but serves as informative character for create real simulation, which contain all parts.

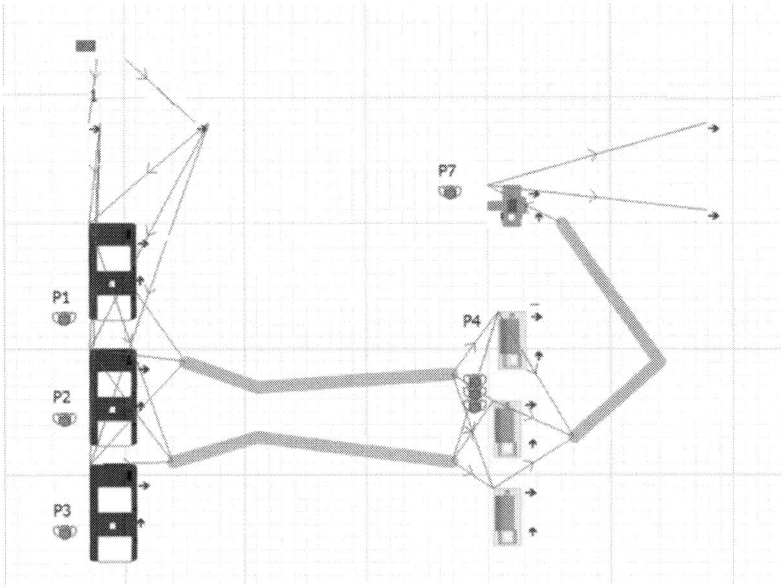

Fig. 5. 2D scheme of workplace.

On Fig. 5 are displayed elements of simulation, which represent real machines and devices. Blue lines in scheme represent connections between elements. Through connections, simple rules as "PUSH to" or "PULL from" are created automatically. Except to these simple rules in simulation are rules, which determine how many percent of components must go into machines or warehouses. In input or output rules are include change of blanks as length, weight and etc. On the start, the blank has the largest size, because he was not in manufacturing process yet. These changes are not include as rules, but as changes on start or end in workplaces. Changes are still identical at the end of first workplace and on the beginning of the second workplace, since the state of the blank between two workplaces is still similar.

In rules of machines are included changes or tool wear, which occurs at a certain time. Except for wear of material, it is need to be reckoned with withdrawals and deposits blanks into and out from machine. These rules are not intended to specifically relationships. They are expressed as settings of machine or work pause of machine.

For machines, which are used for transport of parts the main condition is capacity of machine. The next condition is type of transport as conveyor or rail transport and etc. If you use conveyor, so you need choose type of conveyor, velocity and dimensions. It is possible to choose settings, where length of conveyor is the same of capacity [5].

Important element of simulation are works, who care about a tools change, blanks and collecting of finished pats and moving on transport machines.

3 Flow material in simulation

At the beginning of the simulation are the blanks, which are heading directly into two warehouses with names warehouse1 and warehouse2. Warehouses have capacity 50 pieces of blanks. The blanks enters into warehouses from transport in interval 1 minute about the number 10 pieces. After this, blanks heading directly into cutting workplace. Capacity of one cutting center are 4 blanks. Cutting takes 5 minutes. Cutting workplace contain three machines, so in 5 minutes workplace can produce twelve blanks, who are intended for further processing.

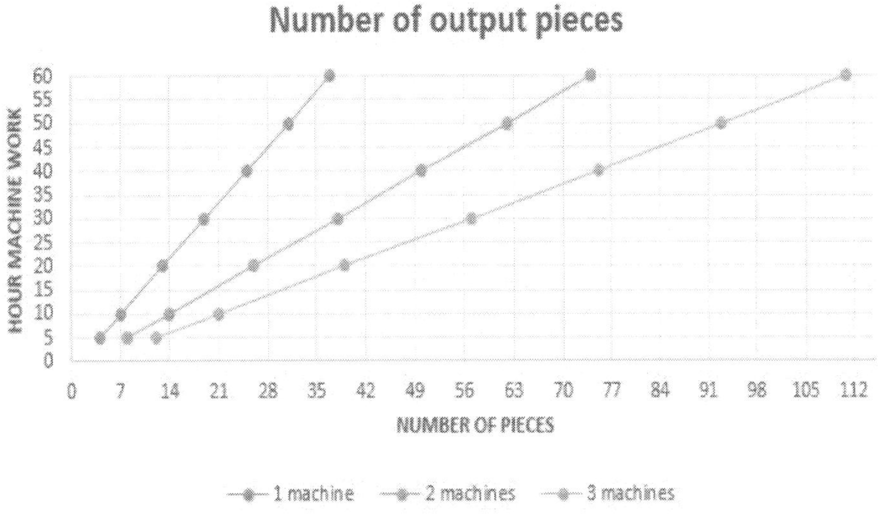

Fig. 6. Number of pieces on cutting workplace.

Fig. 6 describes the effectiveness of the use of multiple machines in the workplace. For one hour, one machine can produce 37 pieces of blanks, which are determined for next processing. If we increase the number of two cutting centers on workplace, so productivity will be 111 pices for one hour. For cutting of blanks is created the loss, which is one respectively three blanks for 5 minutes. The loss for one hour is 33 blanks. It is caused by choosing blanks from centers, moving on conveyors and delivering another blank into cutting center. In cutting center is used one type of tool. In the production of a large number of occurs tool wear. Tool without loss quality of cut can create 111 similar operations. If we use tool for more operations, so on material are created defects, which reduce quality of cut. In that case it is necessary to clean the reservoir of excess material and splinter. Cleaning takes 5 minutes on every workplace. This operation takes place every three hours, also at 8 hour working time to occur 2 times. The total downtime of the machine is in this case 10 minutes.

From first workplace, blanks go through conveyor to another workplace. Capacity of conveyors is max 60 pieces of blanks. If follows that conveyor are equally burdened. It is not possible for one conveyor was idle. Blanks don't have their place on conveyor, but are amassed if the conveyor is not full. The two conveyors operate in parallel. Time of transmission components is 3 minutes from insertion components on conveyor.

The following workplace is machining workplace, where are 3 CNC enters. In these centers occurs components working before final form and surface working. Operations who are carried out on machines takes two and half minutes. Operations are created by two tools (roughing and surface planning). The blank is circular form and it is lathed from one and second sides. Inserting blanks into machine, rotating and removing of part takes 20 seconds. Tool change pending automatically through rotation of tool container. This cycle takes 5 seconds. The analysis shows that all times that are related to wear and tear lasts 35 seconds.

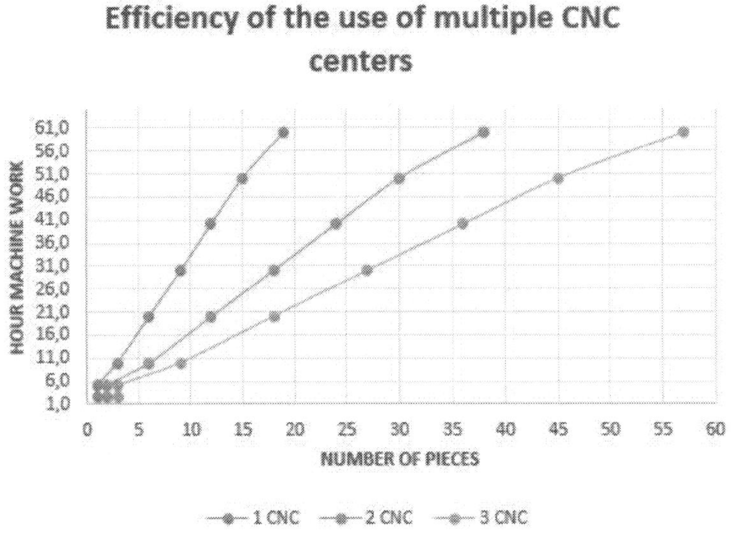

Fig. 7. Number of pieces on machining workplace.

How possible see on Fig. 7, for more complex operations is lower machine productivity. One machining center can produce only 19 pieces of finished products. After calculating working time and pause of machine, which are necessary for good machining time and pause of machine, which are necessary for a good machining of blanks, clean working time of machine is 57 minutes and 45 seconds. When using more CNC machining centers, productivity is 3 times bigger. Working time is the same, since the all machines works in same time. Least effective time is first 5 minutes, when machine needing for production of 2 pieces parts 6 minutes and 30 seconds.

Next stop for finished parts is surface editing. Transport from one to second workplace is through conveyor too. Type of conveyor is similar than conveyor between preceding workplaces. It is roller conveyor. Capacity of conveyor is 60 pieces of parts. Conveyor is still in move. After machining, part is transported to next workplace. It is not possible to be conveyor was full, because parts are automatically sorted from conveyor. Conveyor is full routinely.

Last workplace in manufacturing process is surface editing of parts. In this workplace is only one machine, which able to work with a lot of parts in same time. Container in machine has capacity 100 pieces of parts. Filling one container in average of take 2 hours. These time data are bound on previous workplaces. If be previous workplaces had more machines, so productivity would be bigger. Capacity of conveyor or velocity should be bigger too.

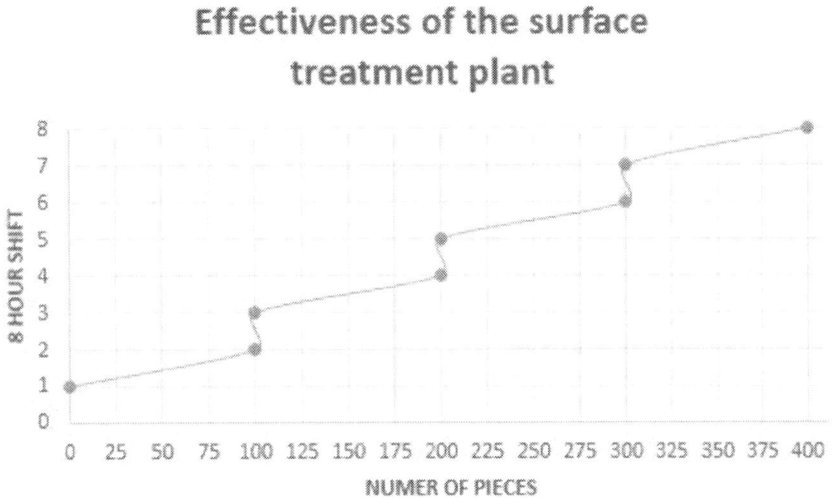

Fig. 8. Productivity of surface editing workplace.

Workplace on surface editing can work with a large number of parts. Hos is displayed on Fig. 8, workplace produce the first parts after 3 hours work. Every next hour has similar data about pieces, because it is caused by processing. Two hours is time, which is necessary for filling of container. This time is measured from beginning manufacturing process until to inserted parts into container. In next hour this data are lower, because work time of machine is 60 minutes. Surface editing takes 50 minutes on all parts. Last 10 minutes is control, which should recognize defects on parts. From two hours waiting on next container, time is reducing on one hour. From picture is possible to evaluate, two hours from beginning of process is machine non-productive. Idle time of machine is on first process, because container is full on the end of last shift. It appears from this that machine can work from begin of day. In this case, machine can produce about 100 pieces of parts more.

Final part in all process is warehousing in temporary storage, which is possible to name a expedition. Even before a certain percentage is moving into storage space with scraps.

4 Conclusion

Simulation of manufacturing process is good tool for data analysis, which taking place in company. Through simulation is possible to watch all areas in company, who are producing some product. In our simulation was created workplace, which contain three manufacturing areas. On these workplaces were monitored a productivity and efficiency of work based on the number of machines. After final calculation was possible detect inefficiency of workplaces.

From the point of productivity best workplace was the last, it was surface editing of products. Workplace includes machines, which can work with a great deal products at the same time. Machines are no productive three hours every day. It is caused by past operations and transport, where product will come after one hour from begin of manufacturing process. These idle time it was possible to eliminate increasing the number of pieces of transport equipment or increase the number of pieces of machinery in the workplace.

Second workplace has balance effectiveness. On the workplace don pt arise idle times in manufacturing process. For the improvement of workplace quality be possible add one or two machines. Second idea is add more conveyor between workplaces or change type and capacity of conveyor.

First workplace has main task in all process. It starting up work on next workplaces. In the case of damaging it is the whole process stopped. Machines, which are on workplace can work with more blanks in same time. In such work takes place more rapidly a tool wear. It is necessary change tool more often. As with the other work efficiency could be increased by adding new equipment or changing technology that is used.

The overall efficiency of the workplace can be improved in various ways. Whether it is a purchase of new machinery and their inclusion in the production or switch to a different technology depending on the requirements of entering as master data in the enterprise. These aspects can simulate and through simulation to create realistic looking to invest in an established company.

References

1. T. Prestero, Verification of a six-degree of freedom simulation model for the REMUS autonomous underwater vehicle, MIT, California, USAm (2001)
2. D. ATHOW, http://www.techradar.com/news/computing/best-workstation-the-best-enterprise-desktop-for-you-1271279 (2017)
3. M. Tothová, J. Piteľ, SISY 2014, IEEE, 209-214, ISBN: 978-1-4799-5997-6 (2014)
4. W. B. Nordgren, WSC '03, *Proceedings of the 35th conference on Winter simulation: driving innovation* (2003)
5. M. Andrejiova, A. Grincova, D. Marasova, Wear. **368**, 400-407, http://www.sciencedirect.com/science/article/pii/S0043164816304616 (2016),

Additive manufacturing in maxillofacial reconstruction

Luciana Laura Dincă[1,*], *Alexandra* Banu[1] and *Aurelian* Vişan[1]

[1]Politehnica University of Bucharest, Department of Manufacturing Technology, Splaiul Independenţei no. 313, sector 6, Bucharest, RO-060042, Romania

Abstract. In this paper the benefits of using additive manufacturing technologies in maxillofacial reconstruction are highlighted. Based on a real clinical case, the paper describes the manufacture of an implant prototype replacing the right zygomatic bone and a part of maxilla using additive manufacturing technologies. The face is the most expressive part of the human body that makes us unique. It was shown that the maxillofacial prostheses help to improve the psychological state of patients affected by, because low self esteem feeling appears commonly to this patients with the facial defects. The aim of this paper is to show how using additive manufacturing technologies methods within this research, the producing a surgical model will help surgeon to improve the pre-operative planning. For this we used additive manufacturing technologies such as Stereolitography to achieve the biomodel and FDM-fused deposition modelling to obtain a prototype model because these technologies make it possible to obtain prosthesis according to the physical and mechanical requirements of the region of implantation.

1 Introduction

Facial defects create psychological effects resulting in a low self-confidence due to their appearance. Because of many factors such as: complexity of the face's bones, the vital adjacent structures, the possibility of infection and because of the uniqueness of the each patient, the manufacturing of maxillofacial implant is a challenge for engineers to design a accurate implant but also is a challenge for doctors to help the patients to look like before [1].

In medicine additive manufacturing technologies are used in different way: for the production of end-use implants, models and moulds usually are used SLS and FDM technologies, for the production of models, moulds and surgical guides are used SLA technology, and SLM and EBM for the production of end-use implants [3].

A method of producing a precise model using 3D printing which can facilitate pre-operative planning is shown in a D. J. Thomas et al. study [4], in which they developed a method of obtaining 3D printed model of mandible highlighting that the surgical procedures reduce risk and also the patient and hospital expenses.

* Corresponding author: lucianalaura_shamieh@yahoo.com

Another study which highlights the benefits of producing the models is that developed by Anni Suomalainen et al. [5] in which had been shown that models facilitate and improve the treatment planning.

The use of CAD / CAM systems have made it possible to design and manufacture customized implants. CAS-computer assisted surgery technique, brings together the use of several technologies such as: Conversion of scans from Computed Tomography, CT; Nuclear Magnetic Resonance, MRI; Rapid Prototyping, RP; 3D CAD design; Finite Element Analysis, FEA; Rapid Manufacturing, RM; Reverse Engineering; Robotics. The most important advantage in CAS techniques is the precision of the patterns resulting from the use of this technique. Specialized imaging techniques from CT, MRI in 3D models can be easily used by doctors to see the defect from any point, so CAS techniques help the doctor to better understand the anatomy involved by viewing and simulating [6].

In maxillofacial reconstruction, these technologies have gained a special place due to the possibility to obtain a perfect implant from patient data.

The main objective of our paper is to highlight the benefits of using additive manufacturing in maxillofacial reconstruction. This study started from a tomographic images of the patient 46 years old male affected by maxillofacial cancer, after which zygomatic bone and a part of maxilla were removed. In the implants manufacturing area, additive manufacturing technologies opens a lot of possibility. The paper is focusing on the steps of obtaining biomodel and the implant prototype using these technologies.

2 Experimental details

2.1 Materials

To fabricate the biomodel, patient data was required. The curable material used to obtain the biomodel in the SLA- stereolithography method was an acrylate.

To obtain the implant prototype we used FDM-fused deposition modeling method and material was an plastic ABS -filament.

2.2 Method

The biomodel and the prototype were made after CT images obtained by Toshiba MRI equipment under the following conditions: 6,47mA, 920 slices and 0.2 mm layer thickness. We used SLA -stereolithography method involving the use of an uv beam to solidify a liquid photopolymer resin in the printer construction drum. The stereolithography equipment which we used to obtain the biomodel was from SLA-250 series. This equipment allows the construction of complicated conceptual models. Due to the high precision and speed of high quality workpiece formation, the SLA-250 machines are successfully used in the medical field.

Manufacturing of implant prototype was performed with the FDM-fused deposition modeling method. We used Zmorph equipment to obtain the implant prototype.

3 Results and discussions

3.1 Design prototype- general principes

To elaborate this paper Mimics software was used, which is a program that helps to convert DICOM images in 3D model and also to design the implant. MIMICS Materialise is a useful viewing and segmenting tool for scanned images from CT or MRI and 3D rendering of images, and more Mimics can be used to diagnose and plan the surgery [7].

The development and design of customized implants is important in order to minimize the possibility of achieving an implant that does not fit the implantation region [8]. There are different methods for designing the custom implant, but the design technique should be taken into account to provide more accurate results and fitting the implants to the place of implantation as much as possible.

As a result of the Moiduddin K, et.al. (2016) study [9], the "mirror" technique offers more precision in the case of medium (20 to 40 mm) and large (41 to 53 mm) sized tumors compared to the "anatomical reconstruction" technique. For small sized tumors (<20 mm) for the reconstruction of maxillofacial implants it is recommended to use the anatomical reconstruction technique. Having the virtual 3D model resulted from conversion of DICOM images into 3D model help us to design implant prototype using the mirror imaging technique, after chose correctly sagittal plane, because of nonsimetry of the patient head. Because additive process offers no limits to design issues or complexities, using additive manufacturing technologies such as SLA - stereolithography and FDM - fused deposition modeling we have obtained the biomodel and the implant prototype.

3.2 The steps in obtaining the implant prototype

At the beginning, the patient went for CT scan to obtain the latest facial morphology. The CT scan shows that in the right side the zygomatic bone and a part of maxilla area are missing (figure 1).

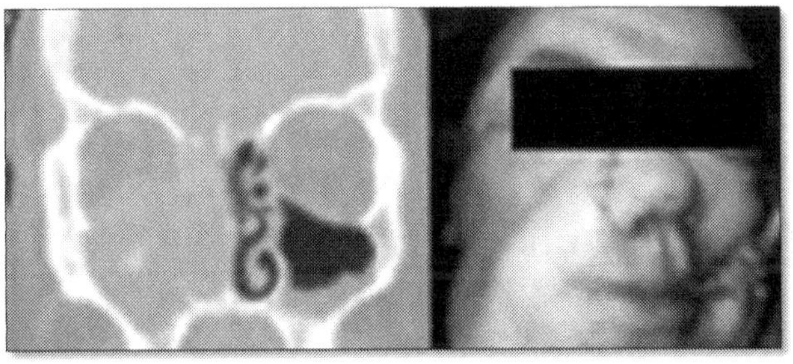

Fig. 1. CT scan confirming the defect and clinical appearance of the patient.

The CT data DICOM was imported into medical imaging software MIMICS, for 3D reconstruction (figure 2).

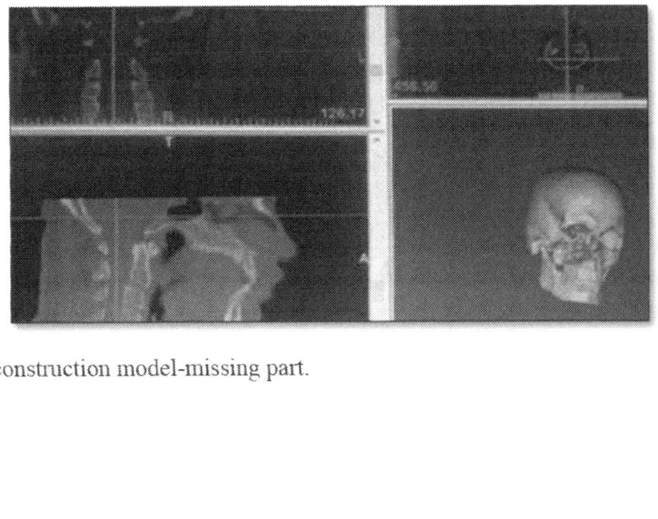

Fig. 2. The reconstruction model-missing part.

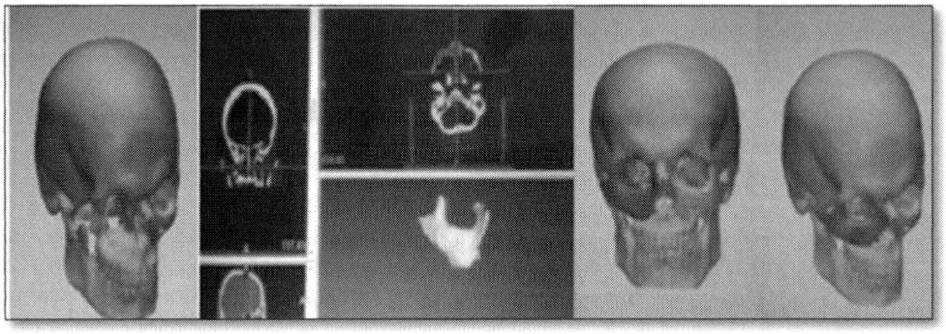

Fig. 3. The implant prototype design steps-the virtual implant.

After this, the next step was to save into stl format and fabricate them, using additive manufacturing technologies. To create the biomodel we used stereolithography, which is a forming additive manufacturing which product part layer by layer using photopolymerization. It is well known that one of the advantages of this technologies is its speed. For biomodel printing time was 16 hours, because of the complexity and because additional support structures were necessary to be built in the beginning of the process. The difficulty we have met when we had to remove the additional structures after process was finised (figure 4).

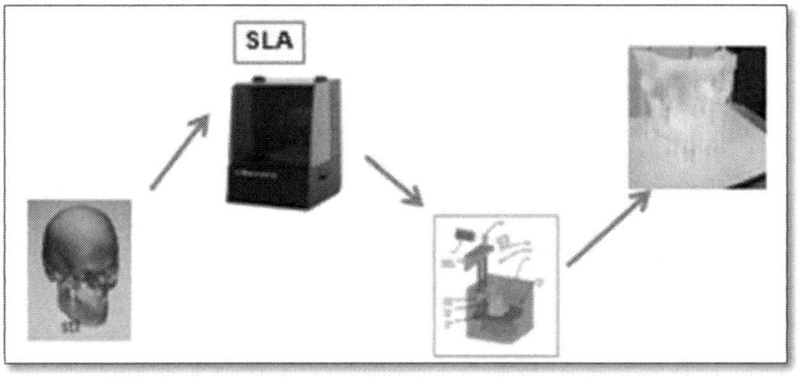

Fig. 4. The SLA manufacturing process of biomodel.

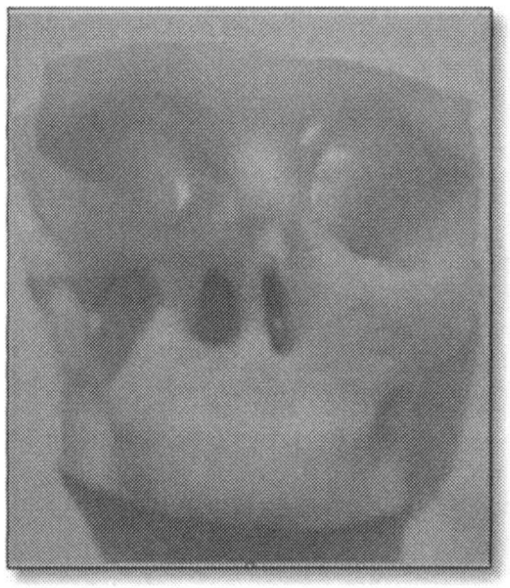

Fig. 5. The patient biomodel manufactured by SLA.

To design the implant we used mirror technique. On the sagittal plane, half of the face, covering the defect area was removed and mirrored the left side.

In the FDM-fused deposition modeling method case from stl format the plastic melt layer after layer, in the implant designed stl shape (figure 6). The thickness layer was

0.2mm. For our implant, were necessary additional vertical support structures, because of the irregular shape.

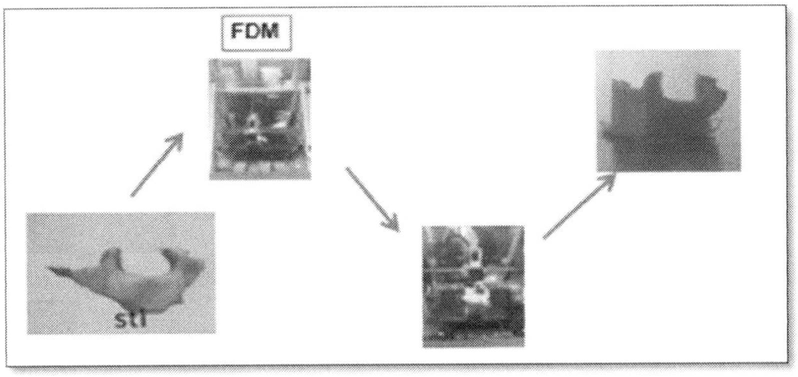

Fig. 6. The FDM manufacturing process of implant prototype.

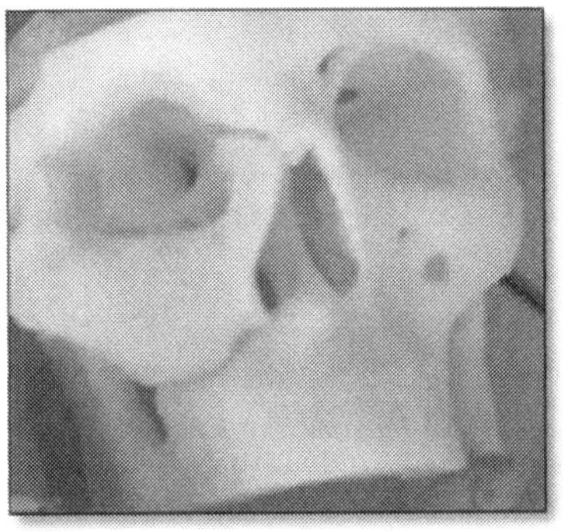

Fig. 7. The implant prototype and biomodel manufactured by additive manufacturing technologies.

Accurate life-sized 3D implant prototype was successfully created for our patient, using additive manufacturing technologies. Fabricating the implant prototype helps us to understand how will be fixed. So the implant model can facilitate the pre-operative planning and make it posible the selection of the most appropriate implant. The next implant will be made of titanium, using also the additive manufacturing technologies, that is SLM- selective laser melting. The titanium implant that we aim to achieve will have a role in restoring the aesthetic appearance, has a support function for attaching the muscle, a protective role for environmental factors, and will also have a support function for a silicone prosthesis, giving the patient a natural look.

4 Conclusions

Additive manufacturing technologies provided the benefits in obtaining the accurate implants. Using these technologies helps us to improve our 3D understanding of anatomy, and how the implant will be fixed. The opportunity to see the model from different angles in a natural way is a new way in medical diagnosis and treatment. This medical model will help surgeon to improve their pre-operative planning of the patient.

References

1. J. Parthasarathy, Annals of Maxillofacial Surgery **4** (1), 9 (2014)
2. K. Kitsakis, Z. Moza, V. Iakovakis, N. Mastorakis, J. Kechagias, An Investigation of Dimensional Accuracy of Multi-Jet Modeling Parts MMMAS 2015 International Conference on Mathematical Models and Methods in Applied Sciences (151-157). Agios Nikolaos, Crete, Greece (2015)
3. A. Thompson, D. McNally, I. Maskery and R. K. Leach, Int. J. Metrol. Qual. Eng. **8**, 17 (2017)
4. D. J. Thomas et al. Int J Adv Manuf Technol **71**, 1643 (2014)
5. A. Suomalainen et al., J Clin Exp Dent. **7** (5), e605 (2015)
6. H. Kawachi, et al., Bull Tokyo Dent Coll. **51** (1), 35 (2010)
7. C. Sadhasivam, S. K. Gunda., International Journal of Science and Research **4** (7), (2015)
8. K. Moiduddin, et al., Technol Health Care **24** (3), 377 (2016)
9. K. Moiduddin, et al., International Journal of Advanced Biotechnology and Research **7** (1), 323- (2016)

Concept Development of a New Lumbar Intervertebral Disk Implant

Cristian-Vasile Doicin[1], *Mihaela-Elena* Ulmeanu[1,*], *Alexa-Sorina* Frîncu[1], and *Vlad-Cristian* Enache[1]

[1]University POLITEHNICA of Bucharest, Splaiul Independenței, No. 313, Sector 6, Bucharest, Romania

Abstract. Worldwide spinal cord injury incidence is rising, due to spikes in traffic incidences, violence and an increase in ageing population, prone to injuries. To satisfy an expanding market, a wide variety of spinal implantable devices are available. The current study develops a new concept for a lumbar intervertebral disk implant which addresses the disadvantages of current commercialised devices. The proposed intervertebral disk implant concept limits the anatomical movements of the trunk, replicating the functions of a natural intervertebral disk. Three concept variations were designed and evaluated using FEA simulations for three main operating hypotheses: Compression of vertebrae in normal up-right position; Lifting weights at a correct and at an incorrect angle; Shock in the lumbar region. Von Misses stress, URES: Resultant Displacement and ESTRN: Equivalent Strain studies were used to evaluate the capability of the intervertebral disk implant concept to withstand the design load. Concept 3 failed URES and ESTRN simulations for the hypotheses which involved lifting weights and delivering a shock in the lumbar region. Simulation revealed optimum results for Concept 1, which was selected for further research.

1 Introduction

Worldwide, according to World Health Organization (WHO), there are between 250 and 500 thousands of people suffering from spinal cord injuries (SCI) and up to 90% of these injuries are traumatic, thus they require the implication of a surgical procedure. The main causes for the traumatic SCI are road traffic crashes, falls or violence. WHO also states that spinal cord injuries "carry substantial individual and societal costs" [1]. Furthermore, a study of the literature that discusses the incidence and prevalence of SCI worldwide, presents that the main causes are car crashes and falling in elderly population [2]. According to the study "Global prevalence and incidence of traumatic spinal cord injury" [3] the United States of America has the highest (906 per million) prevalence of SCI, the lowest values being registered in the Rhone-Alpes region, France (250 per million) and Helsinki, Finland (280 per million). Even if the epidemiology of SCI seems to rise worldwide, as of 2007, in Romania an average of 23 cases per million were registered, and

* Corresponding author: mihaela.ulmeanu@upb.ro

179,312 cases per year [3]. According to the report "Spinal Implants and Surgical Devices Market (...) Global Forecast to 2021", spinal implants and spinal surgeries devices market will reach USD 17.27 Billion by 2021 [4]. The first place for both prevalence of SCI and market share is taken by North America, followed closely by Europe. Considering the rising number of spinal cord injuries, traumatic or non-traumatic, the need to conduct research and development for new concepts of spinal implants is necessary, with focus on decreasing surgery costs, speeding up the healing process, along with ensuring a better patient compliance and an increase in the quality of life.

A variety of both standardised and bespoke spinal implants are available on the market, each catering specific patient needs and performing targeted functions [5, 6]. In order to align the spine in the correct position, the mainly-used spinal implants at the moment, without the requirement of special customized features, are: rods, hooks, screws, plates and cages [6, 7]. All of these show limitations in terms of patient compliance and anatomy adaptability [8, 9, 10]. Also, several types of complex spinal implantable devices have been developed so far, which include: general surgery systems, access and minimally invasive access systems, anterior column plating, interbody and vertebral body replacement systems, spine augmentation systems [4, 6, 11]. Current inter vertebral disk (IVD) bearings tend to be axially rigid and are not designed to resist bending or rotational moments, allowing motion to occur in excess of that permitted by the natural disc [9, 12]. Ball and socket bearings do not completely replicate the functions of the natural IVD, as they tend to increase the hypermobility of the joint, which can lead to several types of complications [7, 13, 14]. Current spinal surgery procedures which require implantation of a medical device, show a variety of complications, amongst the most important are [3, 5, 7, 13, 14]: Regarding spinal fusions, a non-union which undergoes a second surgery; Hardware fractures, like screws getting misplaced; Implant migration; Transitional Syndrome; Paralysis; Death. Manufacturing methods and materials used to obtain such implant systems are available in a wide variety, each with proven advantages [5, 6, 15, 16]. The unrivalled geometric freedom of design provided by AM technologies has contributed significantly to its adoption rate in the medical industry, especially in implantology [6, 17, 18]. Manufactured by metal sintering, the most common used AM spinal implants are fixed, with internal porous structure for light weight and osseointegration properties [8, 15, 18, 19, 20]. In this regard, the current research paper proposes the development of a new intervertebral disk implant concept which brings together the geometry freedom of additive manufacturing with the benefits of standardised medical products.

The main innovations relative to the current state of knowledge are: 1. Improve osseointegration of the intervertebral disk implant through design and manufacture of gripping elements; 2. Diminish the strains in the joints by designing specific gliding surfaces; 3. Limitation of movement within the admissible anatomical and biomechanical limits through innovative geometrical features of the implant joint and elastic core; 4. Modular construction of interlocking joints, which aims at streamlining a possible scale manufacturing of the implant.

2 Concept development of a new lumbar intervertebral disk implant

2.1 Concept development and CAD modelling

Product design and development is a complex process, comprised of numerous stages, as follows: Reason statement; Product (implant) planning and mission statement; Customers' needs portfolio identification; Opportunities, similar products and customers definition; Restrictions identification to developing the implant prototype; Decision matrix compilation for implant concept selection; Mission statement for selected concept development; Identification of customers' needs for the lumbar implant; Setting the specifications of the lumbar implant; Generating implant concepts; Clarifying the problem and defining the general function of the implant; Collection of conceptual solutions for the development of main functions (External research); Generating new conceptual solutions; Concept screening; Concept scoring, Lumbar implant concept selection. All the above mentioned stages were deployed [21] and three variations of the chosen lumbar intervertebral disk implant concept were modelled using SolidWorks Premium 2016 Design Software (Figure 1).

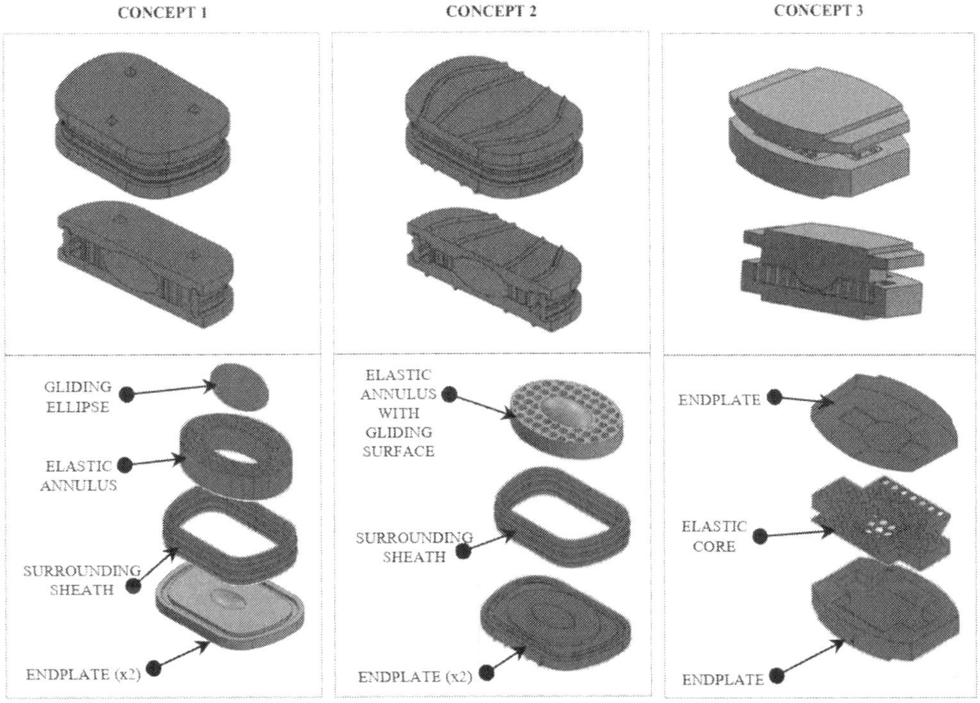

Fig. 1. Main components of the lumbar intervertebral disk implant concept variations.

The IVD implant consists of a nucleus and a surrounding annulus ring. The annulus is assumed to provide shock absorbance. These elements are defined by the study of literature and the former analysis of different aspects. Mean disc height in the lumbar segment was defined between 10 to 12 mm, length varies from 20 to 40 mm and width from 20 to 30 mm [9, 10, 12-14]. The three variations of the chosen concept and their main components are presented in Figure 1. The endplates and gliding ellipse of the final lumbar intervertebral disk implant will be manufactured from a titanium alloy (Ti-6Al-4V) and the elastic annulus and surrounding sheath from polyurethane (11671). The mechanical properties of both materials were considered those given by SolidWorks Premium 2016 Design Software build in material library.

2.2 Simulation Development

Simulation was undertaken using SolidWorks software package defining three types of studies for each of the three hypotheses, as follows: Von Misses stress; URES: Resultant

Displacement and ESTRN: Equivalent Strain. For all simulations the bottom endplate was considered fixed and a global contact set was defined for all three CAD concepts. Bonded contact sets were defined for components of the concept variations presented in figure 1. All initial charts were set to show an automatic deformed shape without a superimposed model. Number formats were set to scientific and the legend had an automatically defined minimum and maximum values. Meshing was undertaken using the characteristics given in Table 1. No errors were found when meshing the assembly.

Table 1. Mesh characteristics of the lumbar intervertebral disk implant concept variations.

Mesh characteristics	Concept 1	Concept 2	Concept 3
Mesh type	Solid Mesh	Solid Mesh	Solid Mesh
Mesher Used:	Standard mesh	Sandard mesh	Sandard mesh
Jacobian points [Points]	4	4	4
Element Size [mm]	2.04721	2.08788	1.96323
Tolerance [mm]	0.102	0.104	0.098
Mesh Quality	High	High	High
Total Nodes	36679	281368	27697
Total Elements	120677	188094	16840
Maximum Aspect Ratio	72.508	167.74	15.952
% of elements with Aspect Ratio < 3	70.1	98.1	90.4
% of elements with Aspect Ratio > 10	1.64	0.1	0.0297
% of distorted elements (Jacobian)	0	0	0
Time to complete mesh [s]:	35	22	11

Load distributions were defined taking into consideration several hypotheses. Three working hypotheses were taken into consideration and analysed in order to understand how the forces act upon the intervertebral disk implant in operation and how well it responds to deformations and stresses. The three hypotheses are: 1. Normal position – compression; 2. Lifting weights: a) correctly, b) wrong (at an incorrect angle); 3. Shock in the lumbar region. The design loads for each hypothesis was set according to literature and own calculations.

Hypothesis 1 *– Compression of vertebrae in normal up-right position*

In the upright standing position the compressive force which acts on the intervertebral disk implant is the weight (W) applied in the centre of gravity of the disc of the upper body [9, 10]. The weight is calculated using relation (1).

$$W = m \cdot g \quad [N] \tag{1}$$

Where, m is the upper body mass, and was averaged according to [9, 10] at 38 kg
g – gravitational acceleration, and was considered 10 m/s^2

Thus, using relation (1) the compressive force which acts on the intervertebral disk implant in a normal standing position was set at 380 N.

Hypothesis 2 *– Lifting weights at a correct and at an incorrect angle*

Implant displacement is commonly encountered when lifting weights [9] and can lead to serious surrounding tissue damage, pain, hernia and even paralysis [13, 14]. These complications can be avoided if a correct position of the upper body is maintained during lifting. An angle of 80^0 of the upper body is considered to ensure safe conditions when lifting weights [9, 10]. By changing the position when lifting to an angle of 65^0 significantly higher forces will be registered [9, 10], usually leading to implant damage and spinal injury. The forces which act on an intervertebral disk wile lifting are presented in Figure 2.

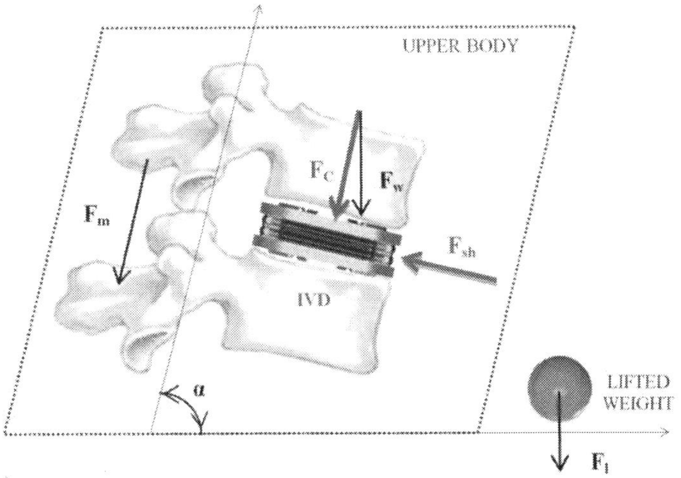

Fig. 2. Loads distribution on an intervertebral disk while lifting weights.

Two main forces are identified within the load distribution for a loading action, namely the compressive force (F_c) and the shear force (F_{sh}). The compressive force which acts on the intervertebral disk implant will be calculated for both cases of hypothesis 2 using relation (2). The shear force was computed using formula (3).

$$F_c = F_m + F_{wx} + F_{lx} \quad [N] \tag{2}$$

Where, F_m is the force applied by the muscles to maintain position $[N]$
F_w – the weight of the upper body $[N]$
F_l – the weight of the lifted object $[N]$
F_c – compressive force exercised on the disc $[N]$
F_{sh} – shear force $[N]$
α – angle of the upper body $[^0]$

$$F_{sh} = F_{wy} + F_{ly} \quad [N] \tag{3}$$

Thus, in case of *hypothesis 2a* the loads distribution was calculated for a lifting angle of 80^0 using the following input values: $F_m = 3020$ N [9]; $F_w = 380$ N; $F_l = 300$ N; $\alpha = 80^0$. The mass of the lifted load was considered to be 30 kg.

$$F_{c/2a} = 3020 + 380 \cdot \sin(80^0) + 300 \cdot \sin(80^0) \Rightarrow F_{c/2a} = 3690\ N$$

$$F_{sh/2a} = 380 \cdot \cos(80^0) + 300 \cdot \cos(80^0) \Rightarrow F_{sh/2a} = 118\ N$$

For *hypothesis 2b* the loads distribution was calculated for a lifting angle of 65^0 using all the other input values as for hypothesis 2a.

$$F_{c/2b} = 3020 + 380 \cdot \sin(65^0) + 300 \cdot \sin(65^0) \Rightarrow F_{c/2b} = 4866\ N$$

$$F_{sh/2b} = 380 \cdot \cos(65^0) + 300 \cdot \cos(65^0) \Rightarrow F_{sh/2b} = 287\ N$$

Hypothesis 3 – Shock in the lumbar region

An impact force on 5000 N was considered for simulation of the falling scenario, as this is the compression tolerance of the lumbar spine given by literature [22].

2.3 Rapid prototyping

Rapid prototyping was done for concept validation and component fitting. In preparation for 3D printing, an *STL file is compiled for each implant component from the CAD models. Based on the *.STL files the 3D printer will manufacture the final part. Involving only general component fitting and concept validation a Fused Deposition Model (FDM) technology was chosen to 3D print the lumbar intervertebral disk implant prototype. Prototype function, medium accuracy, high printing speed and low material cost lead to the selection of this additive manufacturing (AM) technology. 3D printing was undertaken using a 3D Kreator equipment with PLA filament for all print jobs. The filament diameter is 1.75 mm and the material density 1.2 g/cm^3. A 1kg spool of PLA filament was purchased at a price of 30 €. Prototype 3D printing was done following the main FDM manufacturing stages, which are summarized in Table 2.

Table 2. Specific FDM manufacturing stages using Simplify 3D software and 3D Kreator printer.

FDM Stage	Intermediary stages
S1. Print job preparation	*IS 1.1.* PLA material profile loading *IS 1.2.* Part orientation, positioning and scaling on the build platform *IS 1.3.* Support structure generation *IS 1.4.* Process parameters setting for implant shape and functionality
S2. 3D printing using FDM technology	*IS 2.1.* Equipment preparation: heat print platform at 60°C, heat print head at 215°C, print head positioning at start point *IS 2.2.* FDM 3D printing *IS 2.3.* Heating of the printing platform at 80°C for easy part removal
S3. Post-processing	*IS 3.1.* Part removal *IS 3.2.* Support structure deburring *IS 3.3.* Part cleaning

Process parameters were determined using the Simplify 3D software tool, which offers a wide variety of customisable printing features. Print preparation started with selection of the appropriate material printing profile, which in this case is standard PLA. By doing so, the 3D printer automatically loads process parameter values for working with standard PLA material. Using these predefined parameters does not guarantee a successful print job. Thus, using their experience, the authors set the 3D printing process parameters in correspondence with the size, shape geometry and functional surfaces of the lumbar intervertebral disk implant prototype. As there are over one hundred parameters available for modification in the software, a selection of the most important ones is given in Table 3.

After careful preparation of the process settings, the tool-paths for 3D printing and the G-code files were generated. Four build platforms were designed based on abovementioned settings. To avoid any failed parts, the lumbar implant prototype was manufactured in four print jobs (Figure 3), as follows: 1. one gliding ellipse, 2. two end plates; 3. one sheath; 4. one elastic annulus. In order to streamline the optimisation of the build platforms, several types of features are presented in the print preview, marked with different colours. Some of the different features are: travel, outer perimeter, inner perimeter, gap fill, solid layers, infill, bridge, support, dense support, raft, brim, prime pillar, and ooze shield (Figure 3).

Table 3. Process parameters settings in Simplify 3D for the intervertebral disk implant prototype print jobs.

Nr. Crt.	Parameter		Units	Values
1	Retraction	Retraction distance	[mm]	1
2		Retraction speed	[mm/s]	40
3	Temperature	Primary Extruder (start printing/ during printing)	[^{0}C]	215/ 205
4		Platform temperature (start printing/ during printing)	[^{0}C]	60/ 0
5	Movement behaviour	Maximum allowed detour factor	-	1
6	Speeds	Default Printing speed	[mm/s]	60
7		X/Y Axis movement speed	[mm/s]	80
8		Solid Infill underspeed	[%]	80
9	Wipe	Wipe distance	[mm]	2
10	Layer settings	Primary layer height	[mm]	0.15
11	Infill	Interior fill percentage	[%]	50
12		Minimum infill length	[mm]	2
13	Primary extrude toolhead	Extrusion multiplier	-	1
14	Support material generation	Support infill percentage	[%]	30
15		Dense infill percentage	[%]	70
16		Extra inflation distance	[mm]	1.4

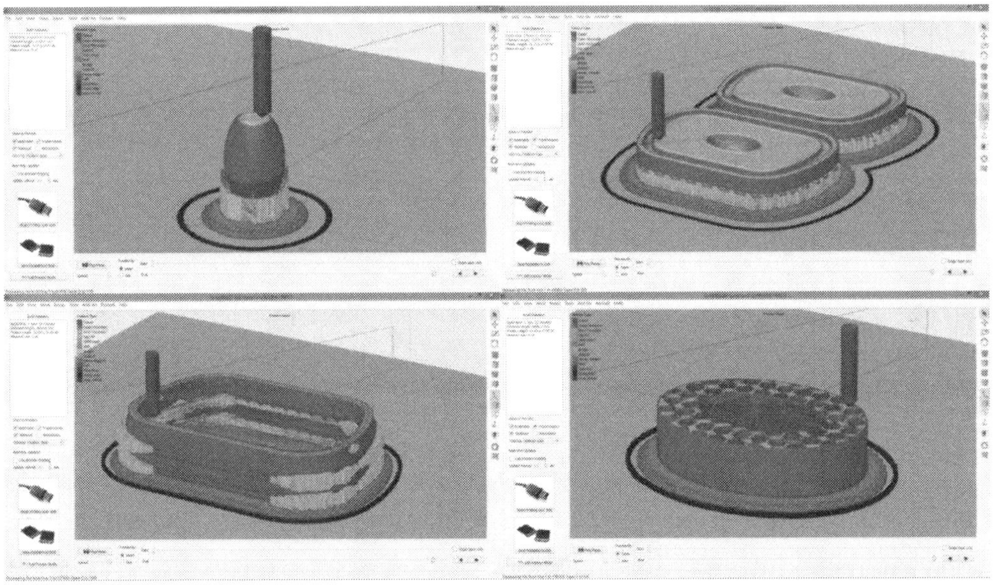

Fig. 3. Build platforms for the intervertebral disk implant prototype designed using Simplify 3D.

Build characteristics for the intervertebral disk implant prototype components are summarised in Table 4.

Table 4. Build characteristics for the intervertebral disk implant prototype components.

	GLIDING ELLIPSE	ENDPLATES	SHEATH	ELASTIC ANNULUS
Build time [hours:min]	0:29	2:21	0:45	1:12
Filament length [mm]	1263.3	11479.7	4736.5	4624.7
Plastic weight [g]	3.65	33.13	13.67	13.35
Material cost [units]	0.11	0.99	0.41	0.40

All five parts were successfully printed and post-processed (Figure 4). Post-processing of the parts included support structure removal and deburring of any stranded filaments. The process lasted approximately ten minutes for all four print jobs. The lumbar intervertebral disk implant prototype components fitted together without any additional post-processing steps, thus the conception process of CAD models was validated. Furthermore, in order to validate the working principle and the surgical technique, evaluation of the prototype was undertaken by one neurosurgeon and one orthopaedic surgeon. Positive feedback from surgeons included remarks regarding the ease of implantation in terms of the surgical instruments used in the theatre. The design of the implant requires no additional tools, standard available surgical instruments being used for conducting the surgical procedure. Its' minimalistic and modular design makes it appropriate for implantation through two available surgical approaches, namely: anterior using an incision on the abdomen wall and a transperitoneal access; anterolateral using an abdominal retroperitoneal access and an extended posterior approach [14]. Some improvements to the designed concepts were suggested: additional fixation is needed, so the contact surface of the endplates should be equipped with more gripping elements; if custom manufacturing is available, the contact surfaces of the endplates should follow the anatomical profile of the patients' corresponding vertebral bodies; for extra support, the elastic annulus shape should follow the sheath shape, by an offset of 2-3 mm; further research should be undertaken to establish the optimum geometry of the elastic annulus and how different mesh structures influence the range of motion offered by the intervertebral disk implant.

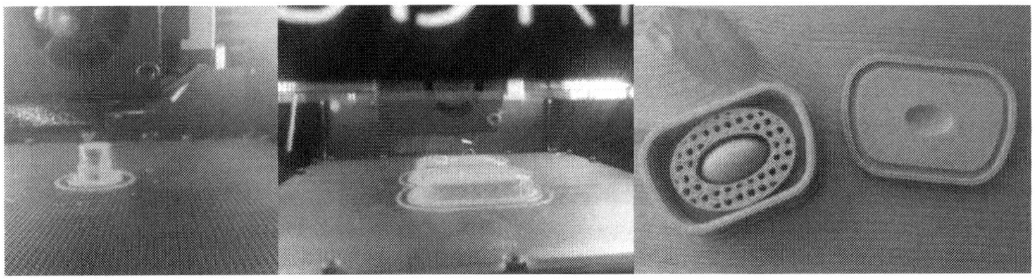

Fig. 4. 3D printing using FDM from PLA and first lumbar intervertebral disk implant prototype.

3 Results interpretation

Simulation results were analysed considering the following stages: Mesh compilation; Study simulation with integrated solver; First evaluation of simulation results; Optimise chart characteristics; Generation of the simulation reports; Interpretation of results; Simulation conclusions regarding the lumbar intervertebral disk implant concept model. Mesh was compiled with the characteristics presented in Table 1 and no errors were given by the mesh solver. Several studies were undertaken for each of the three defined hypotheses, as follows: Von Misses stress; URES: Resultant Displacement; ESTRN: Equivalent Strain. Simulations were run after mesh compilation and a first evaluation of the results led to the optimisation of chart plots characteristics. The deformation scale was adjusted from automatic to 500 units for all plots and no superimposed model was used. Value units were changed from N/m2 to MPa and a floating number format was set. Simulation results for Concept 1 are summarised in Table 5. Similar simulations were undertaken for concepts 2 and 3. Simulation results for all three concepts are summarized in Table 6.

DOI: 10.1051/matecconf/201713702002

Table 5. Concept 1 load distribution simulation analysis.

Case	Von Misses Stress	URES: Resultant Displacement	ESTRN: Equivalent Strain
1	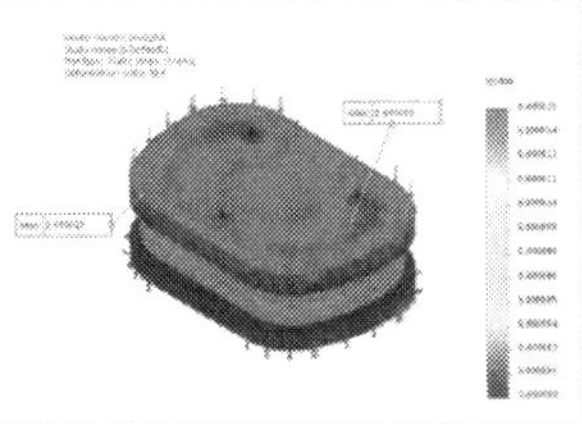	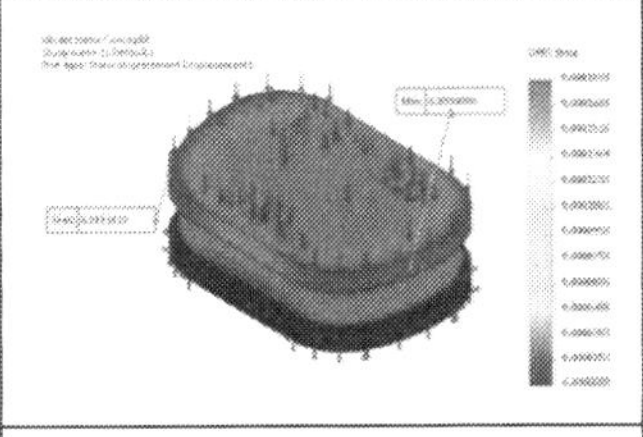	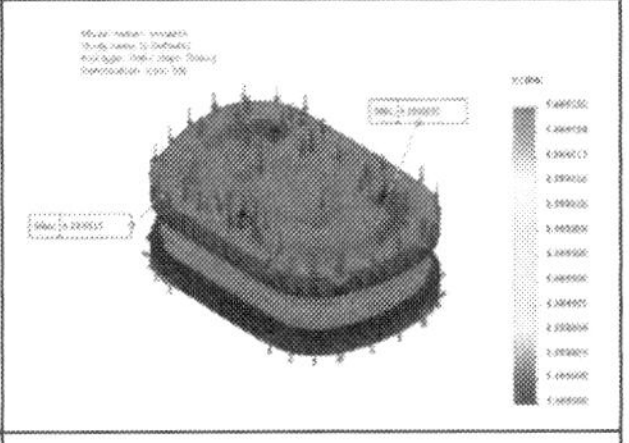
2a	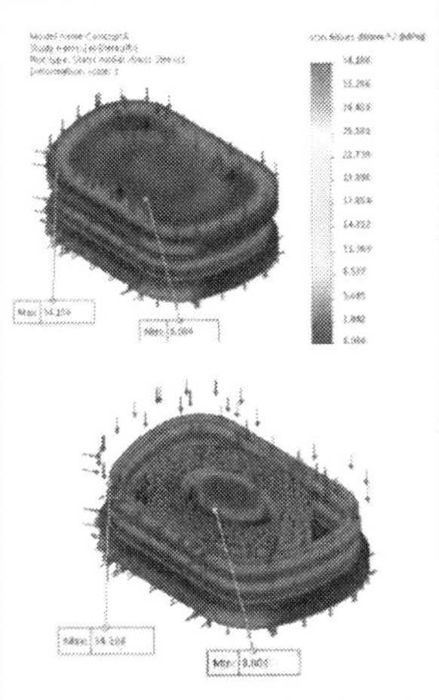	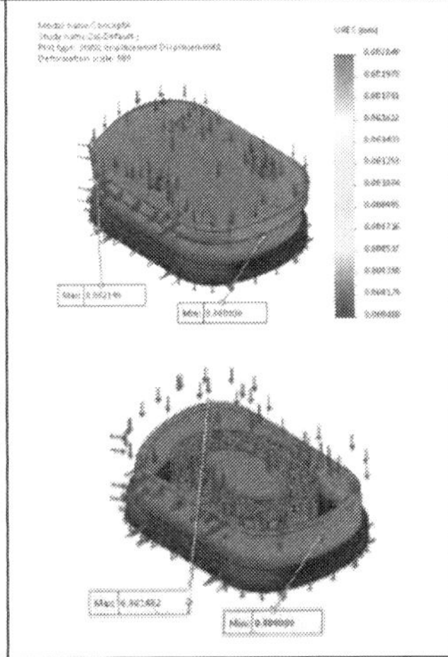	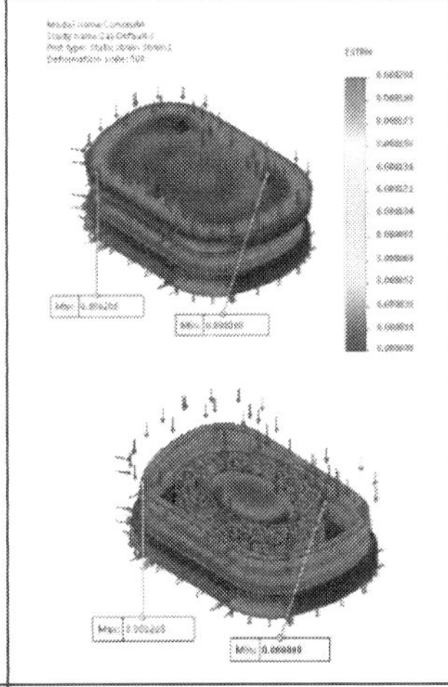
2b	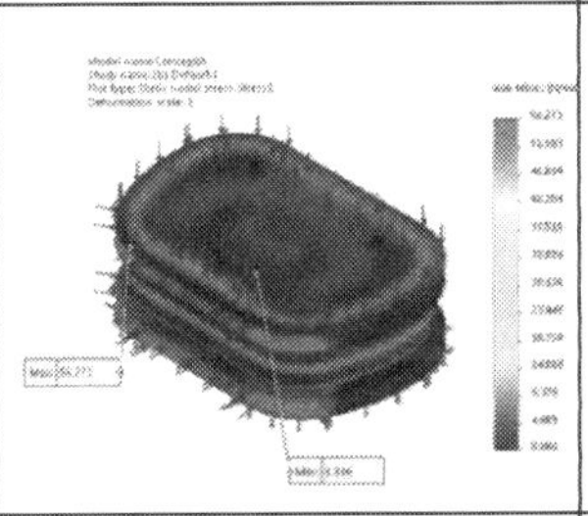	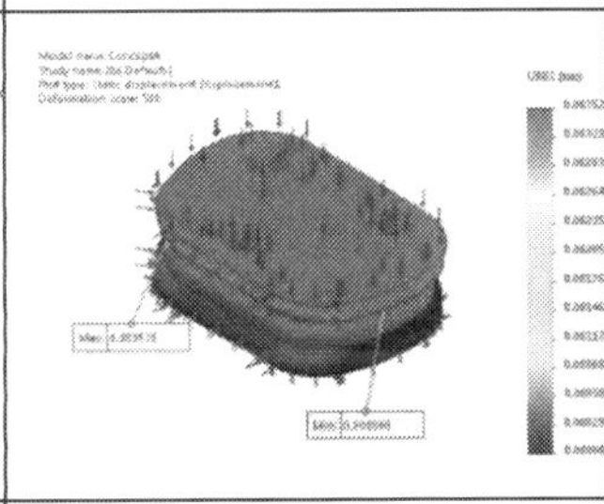	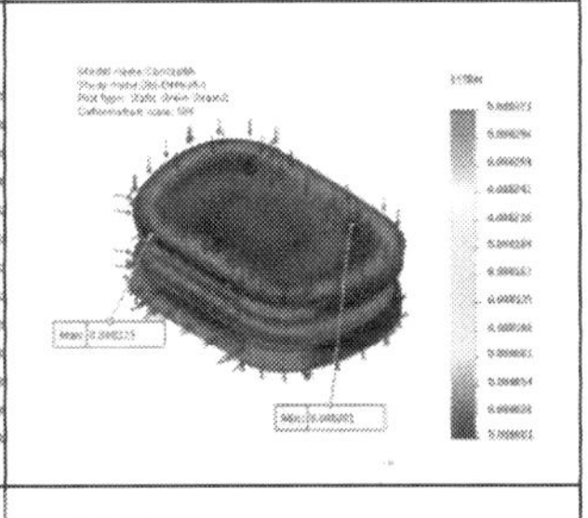
3	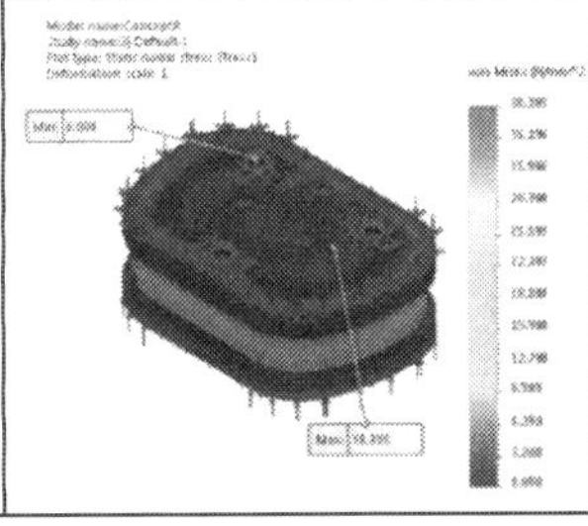	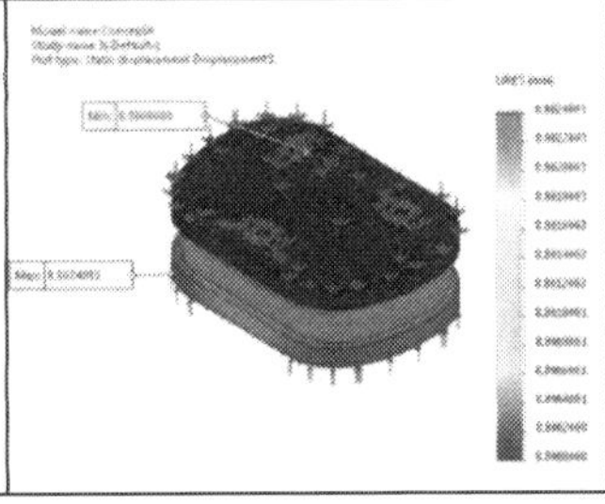	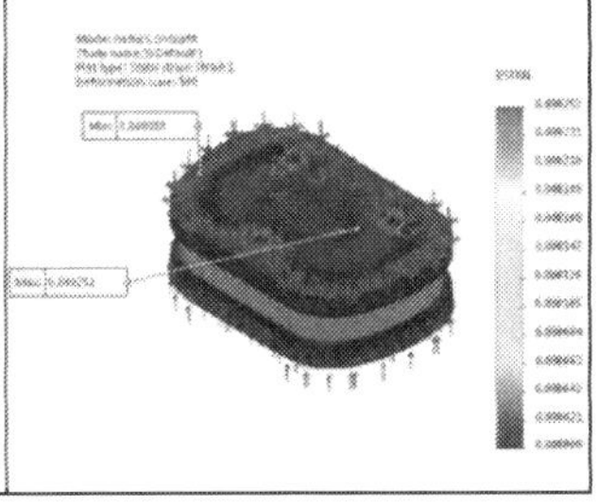

The von Misses Stress (Table 6) distribution obtained by FEA states weather the assembly will withstand the design load. All simulation maximum values were registered on the endplates components, thus the material characteristics of titanium alloy were taken into consideration for verification. Von Misses stress simulation results should be situated below the yield point value of Ti-6Al-4V which is 827.3708 *MPa*. As it can be seen in Table 6 concept 1 and concept 2 have Von Misses Stress values under the yield point of the used titanium alloy, thus the concepts will withstand the design load. Concept 3 does not withstand the design load for hypothesis *2a, 2b* and *3*, with the following maximum obtained values: for *hypothesis 2a* a maximum value of 1235.884 *MPa* was obtained in mesh node 27045; for *hypothesis 2b* a maximum value of 1629.21 *MPa* was obtained in mesh node 27045; for hypothesis 2a a maximum value of 16111.806 *MPa* was obtained in mesh node 284.

Table 6. Simulation results comparison for three lumbar intervertebral disk implant concepts: Von Misses Stress, URES and ESTRN.

CONCEPTS		CONCEPT 1	CONCEPT 2	CONCEPT 3
Hypothesis 1 – Normal up-right position				
Von Misses Stress [MPa]	Min	$3.32 \cdot 10^9$	0.037	$4.81 \cdot 10^3$
	Max	2.484	34.891	127.256
URES [mm]	Min	0	0	0
	Max	$1.83 \cdot 10^4$	$4.11 \cdot 10^4$	0.025
ESTRN	Min	$1.81 \cdot 10^8$	$2.06 \cdot 10^7$	$8.77 \cdot 10^8$
	Max	$1.52 \cdot 10^5$	$4.18 \cdot 10^5$	$6.93 \cdot 10^4$
Hypothesis 2a – Weight lifting at $\alpha = 80^0$				
Von Misses Stress [MPa]	Min	$7.54 \cdot 10^8$	0.478	0.044
	Max	34.108	344.757	1235.884
URES [mm]	Min	0	0	0
	Max	$2.15 \cdot 10^3$	$4.012 \cdot 10^3$	0.226
ESTRN	Min	$4.27 \cdot 10^7$	$1.93 \cdot 10^6$	$1.32 \cdot 10^6$
	Max	$2.08 \cdot 10^4$	$4.10 \cdot 10^4$	$6.73 \cdot 10^3$
Hypothesis 2b – Weight lifting at $\alpha = 65^0$				
Von Misses Stress [MPa]	Min	$1.15 \cdot 10^7$	0.520	0.056
	Max	56.273	388.241	1629.21
URES [mm]	Min	0	0	0
	Max	$3.525 \cdot 10^3$	$5.330 \cdot 10^3$	0.2899
ESTRN	Min	$7.44 \cdot 10^7$	$4.01 \cdot 10^6$	$1.69 \cdot 10^6$
	Max	$3.23 \cdot 10^4$	$5.19 \cdot 10^4$	$8.87 \cdot 10^3$
Hypothesis 3 – Shock in the lumbar region				
Von Misses Stress [MPa]	Min	$6.009 \cdot 10^7$	0.757	0.307
	Max	38.395	316.117	16111.806
URES [mm]	Min	0	0	0
	Max	$2.41 \cdot 10^3$	$6.06 \cdot 10^3$	0.435
ESTRN	Min	$2.63 \cdot 10^7$	$5.06 \cdot 10^6$	$4.22 \cdot 10^6$
	Max	$2.52 \cdot 10^4$	$5.83 \cdot 10^4$	0.0261

URES study is used to determine how the CAD model moves under the defined loads. The Resultant Displacement (Table 6) generates the magnitude of the resultant (compounded from the X, Y and Z directions) reaction force. According to standard ISO 18192-1:2011 [23] the maximum deflection value for a spinal artificial disk is ±*1.5 mm* displacement. According to simulation results presented in Table 6, all concept values fall within the deflection limits of the abovementioned standard. The highest displacement values are recorded for Concept 3, simulated in hypothesis 3, namely a maximum displacement computed value of 0.435 *mm* in mesh node 9164.

The equivalent strain was used to measure the geometric response and the change in shape (deformation) due to applied forces for the three defined load scenarios. ESTRN: equivalent strain is computed as the rapport between the maximum triaxial stress and the Elastic Modulus. In order for the assembly to withstand the design load, ESTRN values (Table 6) should be less than the coefficient given by the ratio between the Yield Strength and the Elastic Modulus. Ti-6Al-4V has an Elastic Modulus of 104800.31 *MPa*, thus all simulation results values were compared with a 0.0079 coefficient. Concept 1 and Concept 2 both have ESTRN values situated under 0.0079, thus withstanding the design load.

According to the obtained results, concept 3 withstands the design load only for hypotheses *1* and *2a*. For hypothesis *2b* and *3* simulated values exceed the calculated coefficient, leading to design load failure. In case of hypothesis *2b* failure was due to an $8.87 \cdot 10^3$ ESTRN value generated in mesh node 13759. For hypothesis 3, strain failure occurred due to a 0.0261 ESTRN value in mesh node 13346.

Based on simulation results, with the lowest stress, strain and displacement values, Concept 1 was selected for further research and development.

4 Conclusions

The research paper proposed a design process of a new lumbar intervertebral disk implant, which replicates the functions of a natural intervertebral disk, by limiting the anatomical movements of the trunk. Three main stages were used in the development of the new concept model: CAD modelling, FEA simulations and rapid prototyping. FEA simulations were undertaken for three concepts, using three study types: Von Misses Stress; URES: Resultant Displacement; ESTRN: Equivalent Strain. IVD Concepts 1 and 2 obtained simulation values which fell within imposed limits. Concept 3 failed design loads for two hypotheses, being eliminated from concept selection. Rapid prototyping was done for concept validation and component fitting using a 3D Kreator equipment with 1.75 mm standard PLA filament. Further research and development will be undertaken on Concept 1, which showed the optimum simulation results.

Future research includes four main stages, based on surgeons feedback: position and fit validation of the implant with custom patient anatomical data; 3D printing of an intermediary prototype with flexible components to validate the prescribed anatomical motions of the trunk, namely lateral flexion (0°/ 40°), flexion (0°/ 100°), extension (0°/ 50°) and torsion (0°/ 40°); manufacturing of material samples for in vitro and in vivo testing; develop an implantation protocol. Validation with DICOM patient data will be fairly easy as parametric CAD models were used in the development process of the intervertebral disk implant. For validation of the prescribed trunk motions, a model of the lumbar spine will be 3D printed for assembly and fitting with the implant prototype. An intermediary prototype will be 3D printed with Form 2 equipment using tough resin (TOTL04) for the endplates and gliding ellipse, and flexible resin (FLGR02) for the elastic annulus and sheath. Material samples for in vivo and in vitro tests will be manufactured from titanium alloy using SLM technology and from silicone RTV medical rubber using an open cast process in custom cast mould. Implantation protocol will be developed in collaboration with medical doctors and other industry specialists.

This work has been funded by University Politehnica of Bucharest, through the "Excellence Research Grants" Program, UPB – GEX 2017. Identifier: UPB- GEX2017, ID No. 26 /2017 (BIOSHIELD)".

References

1. A. Officer, A.Posarac (Executive editors), *World report on disability*, World Health Organization, ISBN 978-92-4-068521-5 (2011)
2. C. Thompson, J. Mutch, S. Parent, J.M. Mac-Thiong, J Spinal Cord Med **38** (2), 214–223 (2015)
3. A. Singh, L. Tetreault, S. Kalsi-Ryan, A. Nouri, M.G. Fehlings, Clin Epidemiol **6**, 309–331 (2014)
4. *Spinal Implants and Surgical Devices Market - Global Forecast to 2021*, Global Report, ID: 3978582 (2016)

5. S.M. Kurtz, A.A. Edidin, *Spine Technology Handbook*, ISBN: 978-0-12-369390-7 (2006)
6. P.J. Slosar, J. Spinal Research Foundation **9** (1), 34-41 (2014)
7. L. Harvey, *Management of Spinal Cord Injuries: A Guide for Physiotherapists*, Elsevier Health Sciences, ISBN 0-443-06858-5 (2008)
8. M.N. Melkerson, J.S. Kirkpatrick, S.L. Griffith, *ASTM International*, West Conshohocken (2003)
9. M. Nordin, V.H. Frankel, *Basic Biomechanics of the Musculoskeletal System*, 4th Edition, LWW Publisher, ISBN-13: 978-1609133351 (2012)
10. D.H. Kim, F.P. Cammisa, R.G. Fessler, *Dynamic reconstruction of the spine*, Thieme (2006)
11. C.J. Korge, F. Heider, H.M. Mayer, Oper Orthop Traumatol **9**, 480-494 (2010)
12. R. Putz, J. Waschke, *Sobotta Atlas of Human Anatomy, Vol. 1: General Anatomy and Musculoskeletal System*, Editors Paulsen F. and Waschke J., 15th Ed., Elsevier Urban & Fischer, ISBN 978-0-7234 -3733-8 (2011)
13. R.C. Reis, M.F. de Oliveira, J.M. Rotta, R.V. Botelho, Open Orthop J **9**, 20–25 (2015)
14. E.C. Benzel, *Spine Surgery: Techniques, Complication Avoidance & Management*, 3rd Edition, **1**, Elsevier, Saunders, ISBN: 978-1-4377-0587-4 (2012)
15. T. Özel, P.J. Bártolo, E. Ceretti, J. De Ciurana Gay, C.A. Rodriguez, J.V.L. Da Silva, *Biomedical Devices: Design, Prototyping, and Manufacturing*, Wiley Publishing, ISBN: 978-1-118-47892-9 (2016)
16. M. Marcolongo, S. Sarkar, N. Ganesh, **6**, 127-145 (2011)
17. T. Wholers, T. Caffrey, *Wholers Report 2014 - 3D Printing and Additive Manufacturing State of the Industry*, Wholers Associates (2014)
18. G.H. Snyder, M. Cotteleer, B. Kotek, *3D opportunity in medical technology: Additive manufacturing comes to life*, A Deloitte series on additive manufacturing, Deloite University Press, US (2014)
19. N. de Beer, A.van der Merwe, Rapid Prototyping J **19** (2), 126-139 (2013)
20. V.N. Chougule, A.V. Mulay, B.B. Ahuja, Procedia Eng **97**, 212-219 (2014)
21. A.S. Frîncu, Development of a production system project and process for a spinal implant, Graduate Project, M.E. Ulmeanu, C.V. Doicin (Coord.), Bucharest (2017)
22. C.V. Toen, M.M. Sran, S.N. Robinovitch, P.A. Cripton, Spine (Phila Pa 1976) **37** (9), E519–E527 (2012)
23. ISO 18192-1:2011, *Test Implants for surgery — Wear of total intervertebral spinal disc prostheses — Part 1: Loading and displacement parameters for wear testing and corresponding environmental conditions for test*, International Organisation for Standardization, Geneva, Switzerland (www.iso.org)

An experimental study on the dimensional accuracy of holes made by abrasive waterjet machining of Hardox steels

Alexandru Catalin Filip[1,*], *Laurentiu Aurel* Mihail[1] and *Mircea Anton* Vasiloni[1]

[1]Transilvania University of Brasov, Department of Manufacturing Engineering, Mihai Viteazul No.5, Romania

Abstract. The manufacturing process of special steels like Hardox is usually difficult. Abrasive waterjet cutting (AWJ) is one of the machining processes which can be used in this case, with good results, as proved in previous research of the authors. This paper presents an experimental study on the influence of the AWJ main process parameters to the dimensional accuracy of holes manufactured in Hardox steels. The experiments were conducted using a $L_8(2^4)$ fractional factorial design matrix considering four parameters: the traverse speed, the part thickness, the hole dimension and the material type. Based on the measurement of the dimensions of the holes, the influence of those parameters was revealed and analyzed. The manufacturing time was also registered, as this parameter directly influences the production cost. Further research aims developing a mathematical model which will be a useful tool in workshop process planning.

1 Introduction

Heavy industries like mining or the construction of roads use equipments with special requirements such as a strong wear resistance and a high tenacity. A range of materials which meet those requirements is the Hardox steel alloys family [1, 2]. Such steels can be processed by traditional mechanical cutting processes like milling or drilling [2] and also by non-traditional ones like thermal cutting or AWJ. One main advantage of AWJ is the absence of modifications of the material properties during the manufacturing process. In all other cases, there is a so-called HAZ – Heat Affected Zone, which for Hardox steels means a decrease of its initial hardness.

Abrasive waterjet machining is a flexible manufacturing process, considered to be one the most developing methods in the last decades [3, 4]. It is based [5], Figure 1a, on using a stream of water (1) at high pressures, above 3000 bar, guided through a small calibrated orifice (2) and mixed with fine abrasive particles (5) in a mixing chamber (3). The abrasive stream is then sent through a nozzle (4), towards the surface of the material. The part (6) is fixed above a matrix of slats (7) which allows the water, the abrasive and the material removed to be collected into the tank below.

* Corresponding author: filipal@unitbv.ro

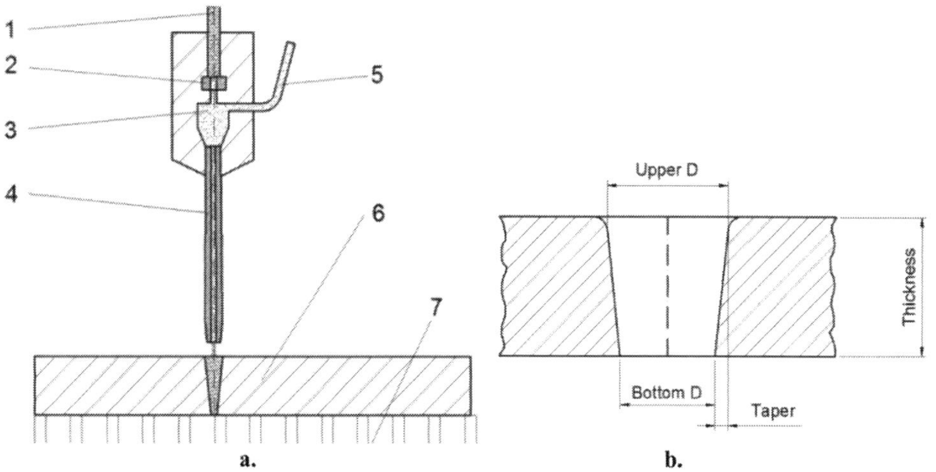

Fig. 1. a. Principle of abrasive waterjet cutting; **b.** Taper kerf formation.

The high speed of the abrasive jet creates enough force able to cut the material of the part. The cutting head is moved by a CNC controller, using dedicated software.

The main advantages of the AWJ method are: the capacity to cut almost any type of material, no thermal influence, easy and fast setup and a high flexibility. The process is also environmentally friendly as the "tools" – water and abrasive particles – are fully recyclable.

The quality of the part is defined by the geometrical and dimensional accuracy and the roughness of the surface. The typical shape of the slot is presented in Figure 1b. The transversal section of the cut is not rectangular, but a tapered one, usually called kerf. Normally, the dimension of the slot is greater on the AWJ entrance surface (the upper one) than the exit surface (the bottom one). The latest modern equipment for AWJ manages to compensate the kerf formation by using the machine dynamics. But if the machine does not have kerf compensation options, this phenomenon must be controlled by other process parameters.

The AWJ process is complex. There is a larger number of parameters which have an influence on the quality and cost of the process [5]. In the last decade there were published several research studies on AWJ, focusing mainly on the surface roughness, the kerf formation and the influence of the traverse speed on those parameters [6-13]. But each research analysed one type of material and the results have the validity for that type of material.

Since AWJ is not a common manufacturing process, today there is a lack of technological data for many materials when someone wants to process them by AWJ. Each abrasive waterjet commercial machine uses its own cutting model and it has a materials database from which the operator can select the one needed. If the material is not in that database, there will have to be selected a similar one, but the cutting model will not be accurate and the results may not be as expected.

In the case of Hardox steels family there is a very few data for the abrasive machining method. The authors of this paper have made some previous studies on the AWJ of this materials [14] aiming to reveal the machinability of Hardox steels by abrasive waterjet machining. Some other research papers [11,13] proposed regression models of the dependence between the traverse speed, the water pressure, the abrasive flow rate and the roughness of the surface which is obtained after the process.

The present research continues the previous studies, analyzing the influence of the main process parameters on the dimensional accuracy of circular holes, manufactured by AWJ on

Hardox steels parts. The kerf level was investigated and how the process parameters influence its values.

2 Setup of experiments and measurement of the parts

The experiments were realized on two types of Hardox steel, Hardox 450 and Hardox 500, which are the most commercially used grades [2]. Their mechanical properties [1-2] are according to Table 1.

Table 1. Mechanical properties of the steels type Hardox.

Material Grade	Hardness HBW Brinell	Yield Strength [N/mm^2]	Tensile Strength [N/mm^2]
Hardox 450	425…475	1200	1400
Hardox 500	470-530	1300	1550

The tests were performed in the laboratory of Advanced Manufacturing Technologies, on a waterjet machine type Maxiem, having a 20 HP pump and a maximum water pressure of 3500 bar. The dynamics of the machine does not have the possibility of automatic tilting, to compensate the effect of taper surface. The abrasive feeding system is a regular one, with constant feed rate, which was carefully measured before the experiments.

The main technological parameters of influence on the process of AWJ which are considered by the comprehensive researches until present day [3-13] are: the water pressure, the traverse speed, the abrasive flow rate and the distance between the nozzle and the surface of the material to be cut, usually called standoff distance. The water pressure directly influences the process productivity and costs and almost all researches agree that it is better to use the maximum value given by the equipment. Because the machine used for experiments has a unique flow rate of the abrasive, there was used this default rate during the tests. For this research, the standoff distance was also considered constant. All the values of the constant parameters are given in Table 2. Usually, the values of these parameters are typical for many industrial applications of AWJ [4].

Table 2. Values of the constant parameters used during the tests.

Water pressure [bar]	Abrasive flow rate [g/min]	Abrasive type	Standoff distance [mm]	Jewel orifice diameter [mm]	Nozzle orifice diameter [mm]
3500	370	garnet 80 mesh	1.5	0.279	0.838

There was designed a four parameter experiment, considering the material type, the traverse speed, the part thickness and the diameter of the hole. The levels of these parameters are presented in Table 3. The levels of the traverse speed were chosen based on previous experiments done by others [6-13] and by the authors [14] for the type of material which was tested. The levels for the other parameters were chosen based on some practical considerations about the use of Hardox steels in the real workshop of the company which is partner in this research project.

Based on the levels of the parameters from Table 3, a $L_8(2^4)$ fractional factorial plan of experiments was designed. The matrix of the experiments is according to Table 4 which contains also the cutting time (as response variable), calculated by the machine software. As seen in Table 4 the experiment contains 8 trials for the 4 factors, each one variated at 2 levels. Each test was repeated five times for obtaining a relatively enough high statistical confidence on the data.

Table 3. Levels of the variable parameters chosen for the experiment.

No	Process Parameter	Units	Level 1	Level 2
1	Material type	n.a.	Hardox 450	Hardox 500
2	Traverse speed	mm/min	30	60
3	Material thickness	mm	12	20
4	Diameter of the hole	mm	20	30

Table 4. The matrix of the experiment.

Trial Nr.	Material type	Traverse speed [mm/min]	Part thickness [mm]	Hole diameter [mm]	Cutting Time [min]
1	1	30	12	20	3.14
2	1	30	20	30	5.26
3	1	60	12	30	2.42
4	1	60	20	20	1.97
5	2	30	12	30	4.78
6	2	30	20	20	3.67
7	2	60	12	20	2.44
8	2	60	20	30	2.77

The dimensional accuracy and the taper effect were analyzed by measuring the dimension of each hole, with an optical digital profilometer, Mitutoyo Quick Image Vision measuring system (Figure 2). The dimension was measured on the both sides of the part, named the upper (entrance of the abrasive jet, "IN diameter") and the bottom (exit of the abrasive jet, "OUT diameter").

Fig. 2. The measuring setup from Mitutoyo.

In Figure 3 there are presented two sets of experiments, for a certain material type, for the two levels of thickness chosen.

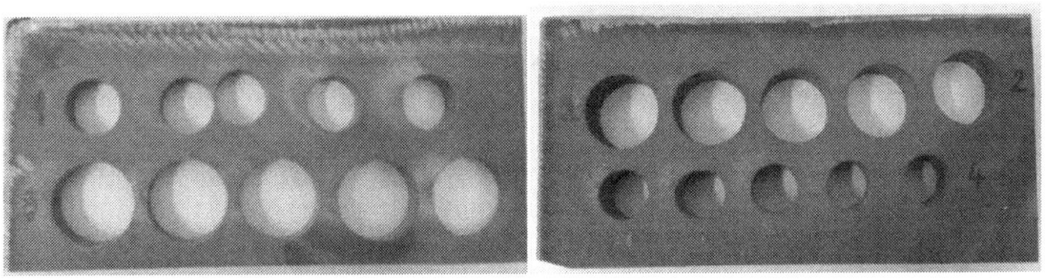

Fig. 3. Set of holes machined during the experiments, for Hardox 450 steel. Left – thickness of 12 mm. Right – thickness of 20 mm.

3 Analysis of the results

The AWJ process is a complex one. During the present research, the main aim was to point out the influence of four parameters on the dimensional accuracy of circular holes machined by AWJ in Hardox steels. While using two levels of variation for each of the four parameters (Table 3), the other process parameters were considered constant at the levels indicated in Table 2. Based on these facts, the performance diagram of the experiment is illustrated in Figure 4.

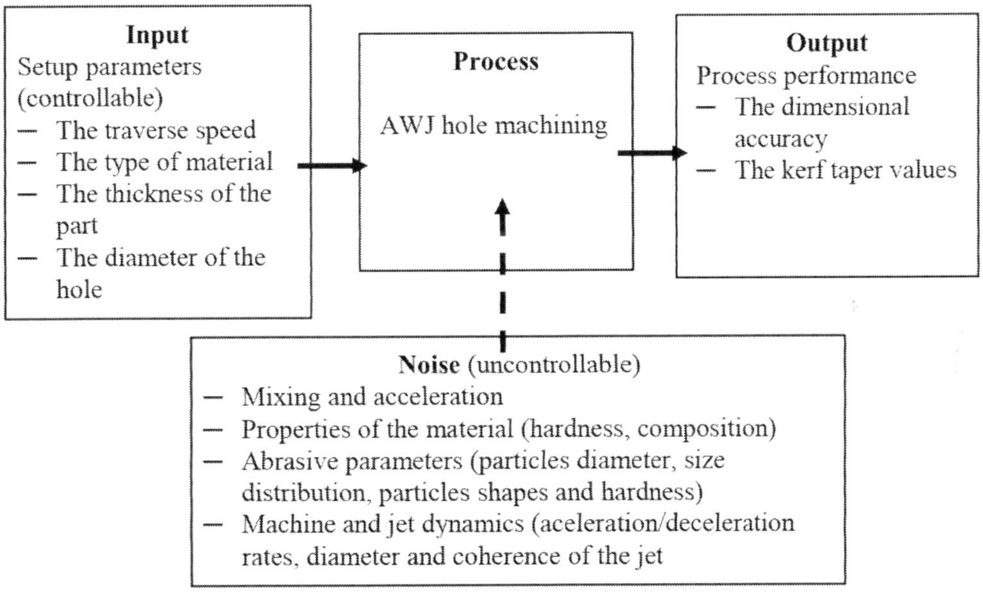

Fig. 4. The P Diagram for the studied AWJ process.

For the statistical analysis, there were calculated the values of the difference between the upper and the bottom value of the hole diameter:

$$Diff=IN\_diam- OUT\_diam. \qquad (1)$$

The values of the difference calculated with the Equation (1) are, in fact, the double of the kerf (Figure 1b). The arithmetic mean of values of the hole's diameter measured after the experiments and their deviations from the nominal value are presented in Table 5.

Table 5. Experiment mean values of the dimensional accuracy.

Trial No	IN Diameter [mm]	OUT Diameter [mm]	Difference [mm]	IN Deviation from nominal [mm]	OUT Deviation from nominal [mm]
1	20.315	20.086	0.229	0.315	0.086
2	30.375	30.114	0.261	0.375	0.114
3	30.261	29.981	0.280	0.261	-0.019
4	20.338	20.086	0.252	0.338	0.086
5	30.318	30.082	0.236	0.318	0.082
6	20.307	20.133	0.174	0.307	0.133
7	20.337	20.008	0.329	0.337	0.008
8	30.326	30.136	0.190	0.326	0.136

Generally, the Taguchi method analyses with statistical tools the performance of the process for optimizing it [15, 16]. The signal to noise ratio (S/N) is used as decision function assessing the quality characteristic (S) for robust optimisation. In this case, both of the performance parameters – the dimensional accuracy and the kerf taper values – are following the principle "the smaller is better". Thus, the objective function is defined like following:

$$S / N = -10 \cdot log\left(s^2 + \overline{y}^2\right). \tag{2}$$

Using the mean values of the difference calculated with the Equation (1), the analysis of means (ANOM) was applied for the computed Signal to Noise indexes, considering the principle "Smaller is Better" (S/N STB). The results are presented numerically in Table 6 and, graphically, in Figure 5. According with it, the highest influence on the studied quality characteristic is given by the traverse speed, followed by the type of the material and the part thickness. The lowest influence is given by the hole diameter. The values of the ranges which suggest the relative importance of the four parameters is specified in Table 6, by the row "*Delta*" and, accordingly, the order of significance is given by the row "*Rank*".

Table 6. Response table for Signal to Noise Ratios "Smaller is better".

Level	Type of material	Traverse speed	Part thickness	Hole diameter
1	11.22	12.73	11.40	11.93
2	12.62	11.11	12.44	11.91
Delta	1.40	1.62	1.04	0.02
Rank	2	1	3	4

The response table for means, on the Taguchi analysis, puts in evidence the magnitude of the studied quality characteristic, relatively at the dimensions measured. Table 7 and Figure 6 underline the impact on the studied quality deviation of the four experimental parameters. The highest influence is given by the part thickness, followed by the traverse speed. The type of material has a relatively medium influence and the hole's diameter has a very low contribution. The magnitude, expressed in millimetres, is given in Table 7 by the row "*Delta*".

Assuming a linear model for the response quality characteristic, the difference between the upper and the bottom diameter of the hole, *Diff*, Equation (1), the model will be:

$$Dd = 0,244025+0,011525*Mt-0,018875*Ts+0,024325*Pt+0,002225*Hd. \quad (3)$$

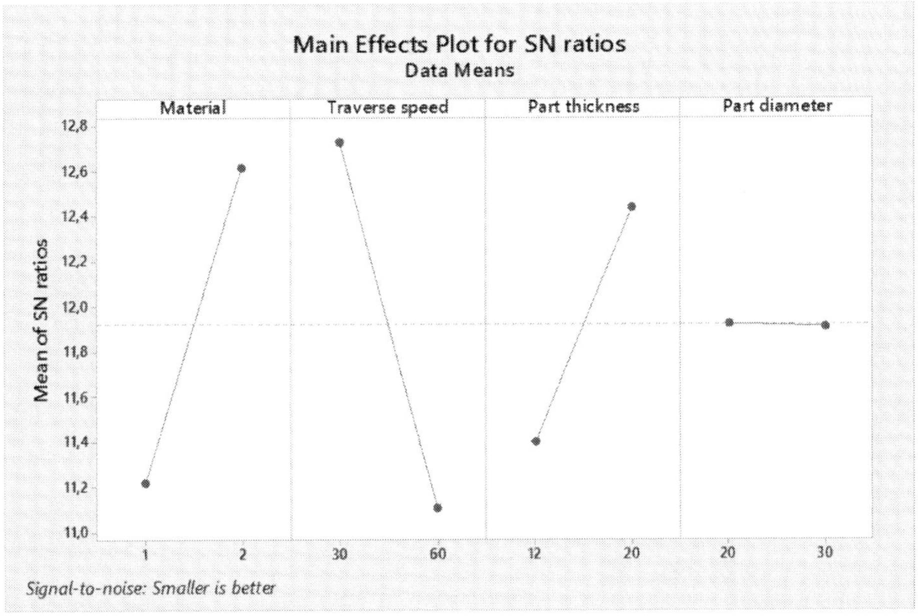

Fig. 5. Response Graph for Signal to Noise Ratios Smaller is better.

Table 7. Response table for means.

Level	Type of material	Traverse speed	Part thickness	Hole diameter
1	0.2556	0.2251	0.2683	0.2463
2	0.2325	0.2629	0.2197	0.2418
Delta	0.0230	0.0378	0.0486	0.0045
Rank	3	2	1	4

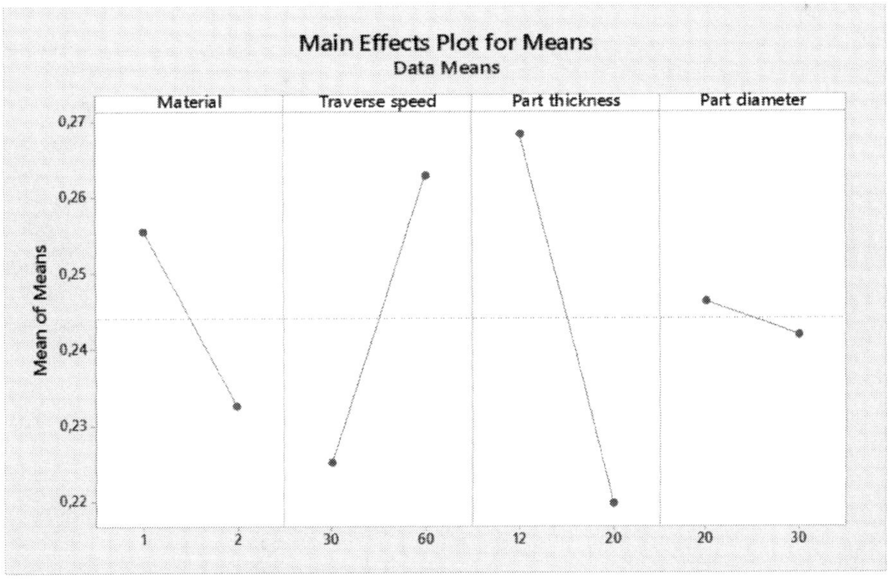

Fig. 6. Response Graph for means.

In Equation (2), *Mt* is the type of material, *Ts* is the traverse speed, *Pt* is the part thickness and *Hd* is the hole diameter.

The model estimators are:
- s = 0,05216 (the squared error);
- R-Sq = 51,6% (the determination coefficient);
- R-Sq(adj) = 0,0% (the adjusted determination coefficient).

Those statistics put in evidence the variation of the predictors relatively at the response variable. The R-Sq(adj) obtained value underlines the fact that this parameter is not relevant for the obtained mathematical model.

In Table 8 there is presented the analysis of variance (ANOVA) of means for the linear mathematical model, given by Equation (3). According with it, the statistical significance can be improved, by raising the confidence level of the data (a higher number of observations). Also, the ANOVA analysis takes into consideration the fact that for the linear model, there are made the following assumptions about the probability distribution of the responses:
- independence of observations – this is an assumption of the model that simplifies the statistical analysis;
- normality – the distributions of the residuals are normal;
- equality (or "homogeneity") of variances, called homoscedasticity — the variance of data in groups should be the same.

The ANOVA analysis presented in Table 8 shows that the error variances are very low, so the level of confidence can be considered good enough. Meanwhile, the values of F and P factors show that for a higher level of confidence, the number of tests should be higher.

Table 8. ANOVA for the linear mathematical model.

Source	Degrees of Freedom	Seq SS	Adj SS	Adj MS	F	P
Material type	1	0.001063	0.001063	0.001063	0.39	0.576
Traverse speed	1	0.002850	0.002850	0.002850	1.05	0.381
Part thickness	1	0.004734	0.004734	0.004734	1.74	0.279
Hole diameter	1	0.000040	0.000040	0.000040	0.01	0.912
Residual error	3	0.008161	0.008161	0.002720		
Total	7	0.016847				

The dispersion of the data was also analysed. The data is presented in Table 9 and Figure 7. According to the values, the variability of the data is the highest for the part thickness, followed at relatively the same level by the type of material and the traverse speed. The lowest variability level is for the part diameter.

Table 9. The response table for standard deviations.

Level	Type of material	Traverse speed	Part thickness	Hole diameter
1	0.10403	0.06230	0.04819	0.08320
2	0.05793	0.09966	0.11377	0.07876
Delta	0.04610	0.03736	0.06558	0.00445
Rank	2	3	1	4

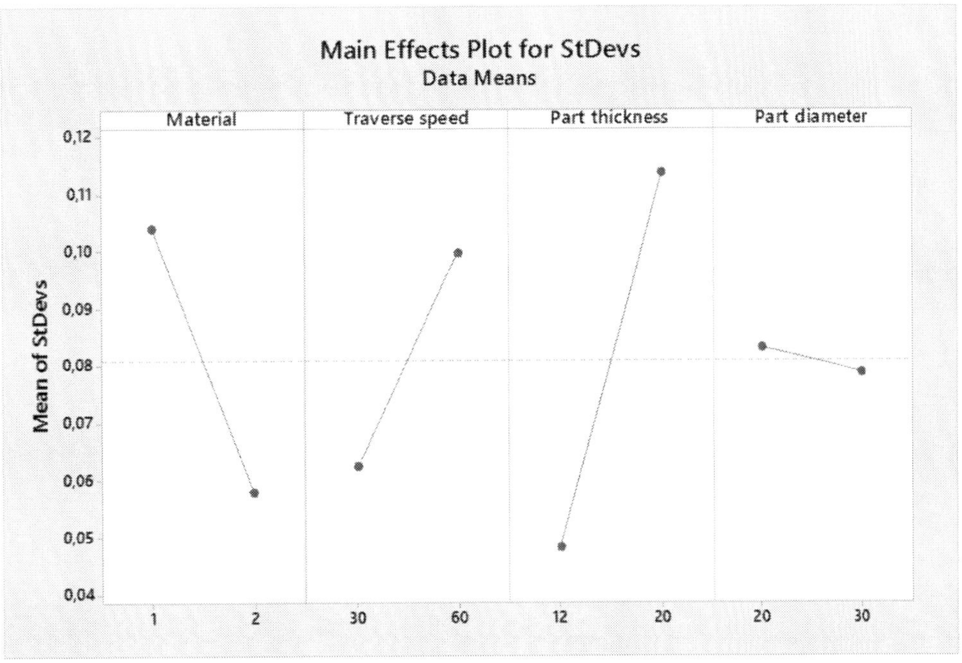

Fig. 7. Response Graph for standard deviations

4 Conclusions

This experimental research aimed to reveal the influence of the process parameters on the dimensional accuracy of holes machined by AWJ in Hardox steels.

An experimental plan was designed, with a four-parameter matrix, presented in Table 4. All the designed combinations were conducted, so the orthogonality of the experiment was assured. The dimensional accuracy was assessed by measuring the dimension of each hole on the two sides of the part, the upper and the bottom (Figure 1b). This allowed the evaluation of the kerf values, as this phenomenon directly affects the part geometry.

The measured data was analysed using the Taguchi method with statistical tools. The main conclusions of the research are:

• The S/N ratios analysis (Figure 5) revealed that the traverse speed is the most important parameter of influence on the dimensional accuracy of parts. It is followed by the material type and thickness. The diameter of the hole has a very small influence. The lower the traverse speed and the higher the part thickness are, the better is the dimensional accuracy. The diagram S/N ratios shows also that the best levels for obtaining the best performance level are level 2 for the material type (Hardox 500), a traverse speed of 30 mm/min and a thickness of 20 mm.

• The analysis of means (ANOM) shows that the part thickness is the most important parameter of influence, followed nearly by the traverse speed and the type of material.

• The ANOVA analysis shows that the error variances are very low, but the level of confidence must be raised by raising the number of tests.

• All of the statistical tests revealed that the hole's diameter has a very low influence on the dimensional accuracy and the type of the material has a medium influence together with the part thickness.

• The analysis of the experimental results are in accordance with other previous researches on Hardox and other materials, revealing that the traverse speed is the most significant parameter of influence on the part's accuracy, as well as on the part's surface quality measured by its roughness.

For a more accurate estimation of the process performance, further research will be developed, using also other types of Hardox steels to establish a mathematical model for predicting the process parameters when this type of material needs to be manufactured by abrasive waterjet cutting (AWJ).

The experiments of this research were made with the support of the project PN-III-P2-2.1-BG-2016-0206, ProWaterjetTech, financed by UEFISCDI. The measurements were done on the infrastructure of the "Metrology Lab Mitutoyo" from the "Institute High Tech Products for Sustainable Development: PRO-DD", lab created with the support of Mitutoyo Romania S.R.L.

References

1. H. Bugłacki, M. Smajdor, Advances In Materials Science, **4**, 2 (2003)
2. Hardox steels. http://www.ssab.com/products/brands/hardox (2017)
3. R. Kovacevic, M. Hashish, R. Mohan, M. Ramulu, TJ. Kim, ES. Geskin, J Manuf Sci Eng, **119**, 776-785 (1997)
4. J. Folkes, J Mater Process Tech, **209,** 6181-6189 (2009)
5. A.W.Momber, R.Kovacevic, *Principles of Abrasive Waterjet Machining*. Springer London Limited (1998)
6. YW. Seo, M. Ramulu, D. Kim, Proc Inst Mech Eng B, **217** (12) 1709 (2003)
7. A. Akkurt, M K. Kulekci, U. Seker, F. Ercan, J Mater Process Technol **147** (3) 389 (2004)
8. D. Krajcarz, Procedia Engineering, **69** 838-843 (2014)
9. J. Kechagias, G. Petropoulos, N.Vaxevanidis, Int J Adv Manuf Technol **62** 635–643 (2012)
10. L.M. Hlavac, I.M. Hlavacova, V. Geryk, S. Plancar, Int J Adv Manuf Technol **77** 1811–1818 (2015)
11. L.M. Hlavac, S.Spadlo, D. Krajcarz, I.M.Hlavacova, *Proceedings of the 24th International Conference on Metallurgy and Materials Metal*, Brno (2015)
12. H. Li, J. Wang, Int J Adv Manuf Technol, **81** 361-369 (2015)
13. M. Zohoor, S.H. Nourian, M. Salehi, Int J Advanced Design and Manufacturing Technology, **5** (2012)
14. A.C. Filip, M. Vasiloni, L.A. Mihail, MATEC Web of Conferences, **94** (2017)
15. G. Taguchi, S. Chowdhury, Y. Wu, *Taguchi's Quality Engineering Handbook*. John Wiley & Sons and ASI (2004)
16. Minitab release 17 statistical software, Minitab Inc., USA

MATEC Web of Conferences 137, 02004 (2017) DOI: 10.1051/matecconf/201713702004
MTeM - AMaTUC 2017

Investigations of Surface Topography of Hot Working Tool Steel Manufactured with the Use of 3D Print

Pawel Grobelny [1,*], *Lukasz* Furmanski[1] and *Stanislaw* Legutko[1]

[1]Faculty of Mechanical Engineering and Management, Poznan University of Technology, 3 Piotrowo St., 60-965 Poznan, Poland

Abstract. The paper presents the possibilities of 3D printing of chosen hot working tool steel for manufacturing ready made parts. Results of examination of the surface topography of material manufactured by the technology Laser CUSING®B (Laser melting with metals) on the machine, Concept Laser M1 3D printing of metal parts has the potential to revolutionize the market of manufacturing and supplying parts. It makes it possible to dissipate manufacturing and to produce parts on request at lower cost and less energy consumption. The parameters of the surface topography of the hot working tool steel directly after printing can differ depending on the distance from the base plate. The differences of surface roughness values can amount from 32% to 85% for *Ra* and from 59% to 85% for *Rz* in comparison of the sample bottom to its top.

1 Introduction

3D print is one of the most perspective technologies of manufacturing. Nowadays, 3D print develops very quickly and becomes more and more competitive to the traditional methods of prototyping, modelling and production. It finds application in many aspects of our life, from medicine and biomechanics through such branches as automotive, aircraft industry, film production to art, jewellery, furniture production and fashion [1]. Application of hot working tool steel:

• inserts with conformal cooling for injection moulds [2], [3],
• inserts with conformal cooling for pressure casting dies of zinc, magnesium and aluminium alloys,
• bodies of cutting tools with designed cooling channels,
• building of highly complicated parts for which strength and hardness are essential,
• prototyping small steel elements with elevated mechanical strength, and resistant utility parts.

2 Material used in the tests

The application of the Laser CUSING technology makes it possible to create element of metal with uniform structure, directly from a computer elaborated design [4].

* Corresponding author: pawel.grobelny@gflean.pl

DOI: 10.1051/matecconf/201713702004

The print is created thanks to the power of high quality laser which precisely sinters the subsequent layers of the selected metal powder. The process takes place in a closed chamber with the participation of nitrogen or argon. The device automatically applies a new powder layer and, depending on the requirements (speed, quality), it is able to make a layer with the thickness of 15 – 500 μm.

Once the process of sintering a layer is completed, the platform, on which the printed object is located, descends, after which the subsequent layer of powder is applied. The Laser CUSING technology makes it possible to produce parts of complex geometry of a material identical to that used in traditional production (e.g. in foundry). General diagram of the Laser CUSING technology functioning (Figure 1). The manufactured elements do not require additional machining of pre-production investments but, in order to improve the technical parameters heat treatment is applied. One of the materials used in incremental manufacturing is hot working tool steel named CL 50 WS whose composition is in accordance with the European Standard 1.2709, German Standard X3NiCoMoTi 18-9-5 and American Standard 18% Ni Maraging 300.

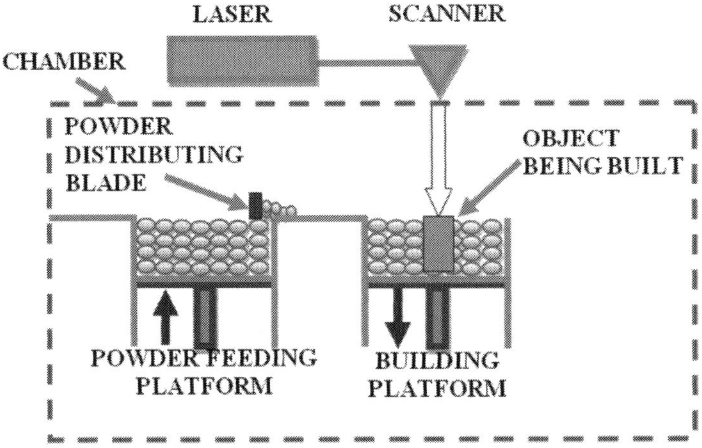

Fig. 1. General diagram of the Laser CUSING technology functioning.

3 Methodology of investigation

In the investigation, hot working tool steel with the chemical composition in accordance with the European standard 1.2709 has been used. Due to its high strength and hardness, the alloy is perfectly suitable for building elements of injection moulds [5], [6]. Thanks to the layer manufacturing by the method of Laser CUSING, it is possible to build forming inserts with cooling channels matching the shape of the part being formed, i.e. with conformal channels. The steel acquires high hardness at the level of up to 52 HRC and competitive strength as compared to the standard material used for building injection and casting moulds [7].

Samples for testing have been printed in vertical position in the shape of shafts diameter of 10 mm and length of 100 mm. They have been designated from A1 to A6 (Figure 2) and have been made on the Concept Laser M1 device.

The authors have made an attempt of characterization of the surface topography of hot working tool steel whose chemical composition is in accordance with the European Standard 1.2709. The topography examination results have been analysed, both for the samples just after 3D printing and after the process of finish machining.

DOI: 10.1051/matecconf/201713702004

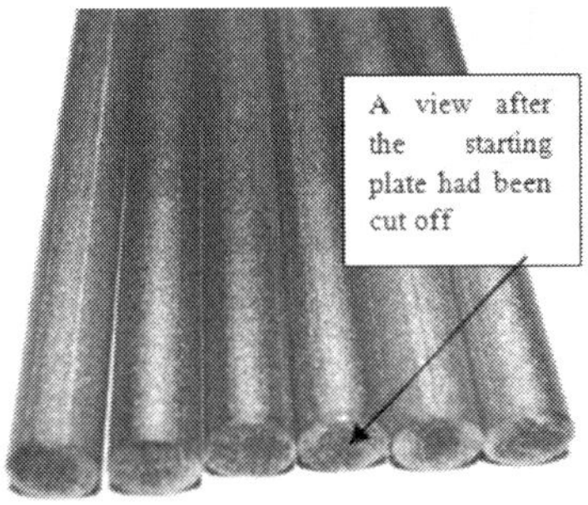

Fig. 2. Samples used in the tests.

The selection of above cutting parameters is recommended for the finish milling of hardened steels [8-10]. The machining tool was equipped with the inserts designated with the following ISO symbol: DCMT 0700204E-UR; T9325 with holder, SDHCL 2525 M11. The printed samples have been subjected to machining under production conditions on the machine MAZAK QTS 250M. The analysis of the machined material surface topography after the process of turning has been performed with the use of HOMMEL-ETAMIC T8000 R device.

4 The results of investigation

The results of the topography measurements have been subjected to statistical processing with the use of the program, digital surf Mountains® surface imaging & metrology software. The diagrams of the alterations of selected roughness parameters of the printed samples, depending on the place of measurement can be seen in Figure 3.

a) b)

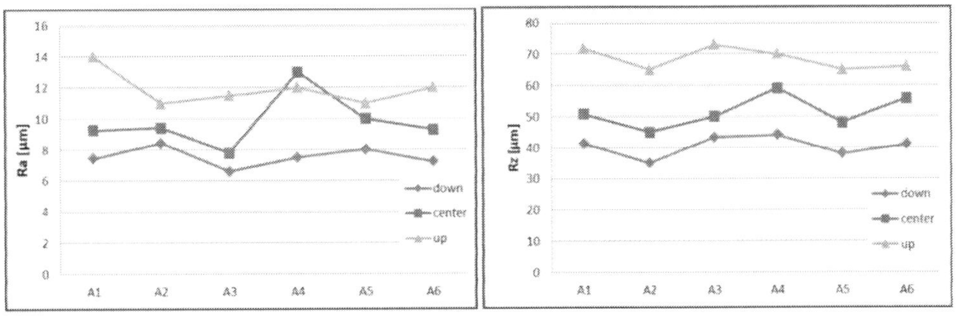

Fig. 3. Alterations of the selected roughness parameters depending on the place of measurement: a) parameter *Ra*, b) parameter *Rz*.

In the analysis of the selected roughness parameters, it has been found that the higher from the base plate, the higher are the values of the parameters under consideration, from 32% to 85% for *Ra* and from 59% to 85% for *Rz* in comparison of the plate bottom to its top. The next step in the analysis of the hot working tool steel, after determination of its roughness parameters, is presentation of the technological surface in the graphic form. Analysis of contour maps and isometric images of constructional surfaces has shown some specific features of those surfaces influencing the functional properties of the surfaces. The appearance of the surface topography has been visually analysed in order to present the characteristics of the out print and the process of turning. Figure 4 shows the out printed surface of sample A1 depending on the place of measurement. It has been found that the surface closer to the base plate is isotropic, i.e. has more evenly distributed peaks (Figure 4a) than the surface in the middle and upper part of the sample. In Figure 4c, one can see a characteristic valley, i.e. long depression of unevenness which can result from the out print kinematics. Higher average roughness values have been observed in the upper part of the sample, which has been reflected in the legend scale.

a)

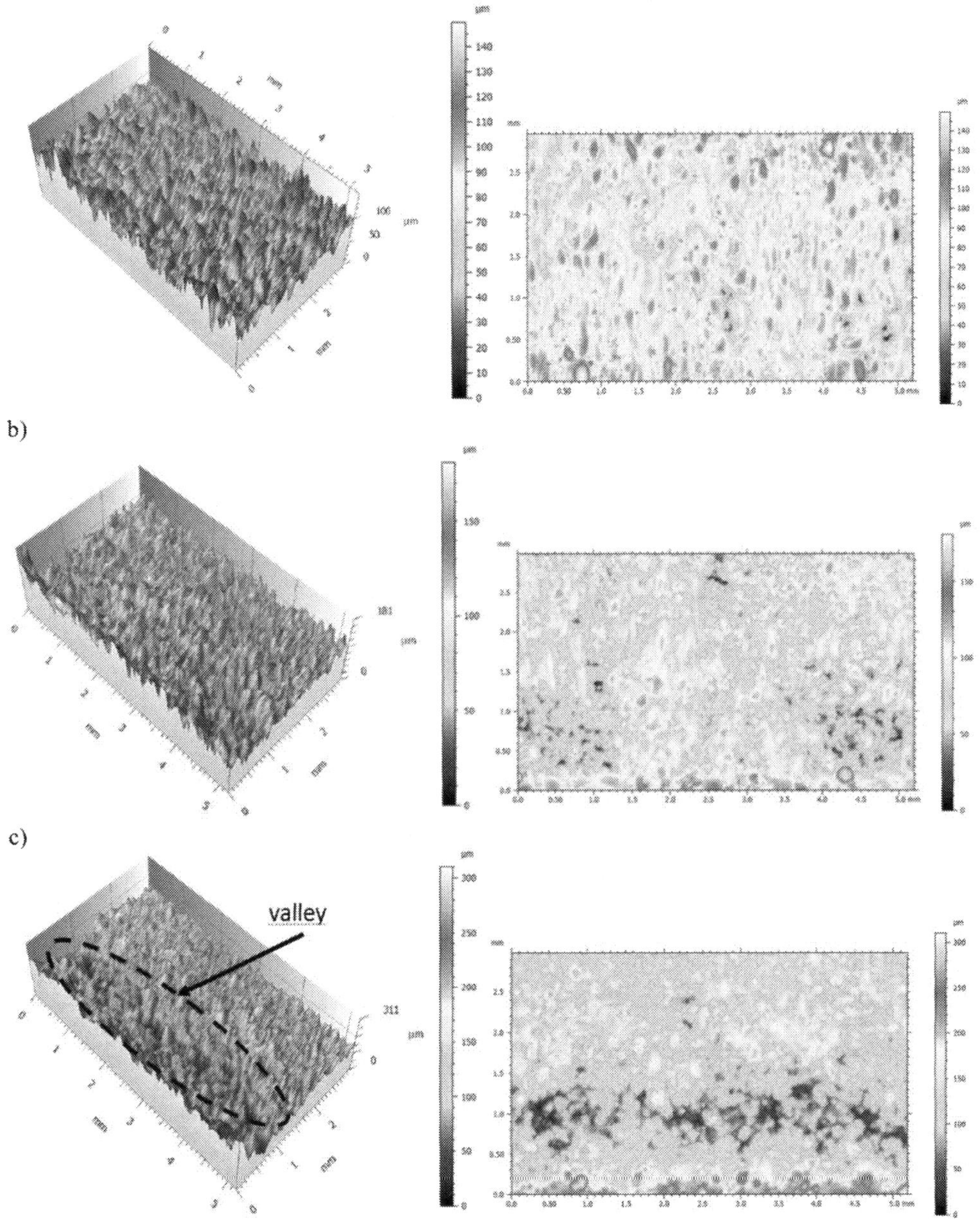

Fig. 4. Surface of sample A1 of 1.2709 after out print depending on the place of measurement, isometric view and contour map: a) bottom, b) middle, c) top.

The parameter of obliqueness *Ssk*, provides information about the surface asymmetry. The value of the *Ssk* parameter indicates prevalence of peaks on the surface for *Ssk* > 0 or prevalence of valleys for *Ssk* < 0 [11]. When the obliqueness values, *Ssk* are negative, most material is located close to the surface peaks. The *Sku* parameter (kurtosis) indicated the occurrence of unexpectedly high peaks or deep valleys for *Sku* > 3 or their absence from the surface for *Sku* < 3. Graphical presentation of the pair of roughness parameters: kurtosis (*Sku*) as compared to obliqueness (*Ssk*) can be seen in Figure 5 [12].

 DOI: 10.1051/matecconf/201713702004

In the process of wear due to, e.g. flow of liquids in inserts with conformal cooling or in tool bodies with cooling channels, the surface obliqueness *Ssk*, is one of the most important parameters due to the effect of roughness shortening [13].

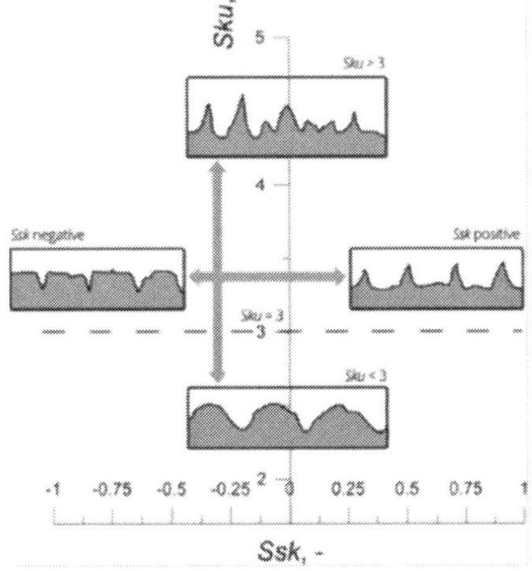

Fig. 5. Graphical representation of surface for *Sku–Ssk* maps [12].

Surface wear causes reduction of the *Ssk* value. In Figure 6, one can see the pair of amplitude roughness parameters *Ssk* and *Sku*, of the 1.2709 steel after out print depending on the place of measurement. The particularly high *Sku* for measurement A1 down proves the presence of high peaks on the surface. All the analysed surfaces, regardless of the surface asymmetry, have leptokurtosis shape.

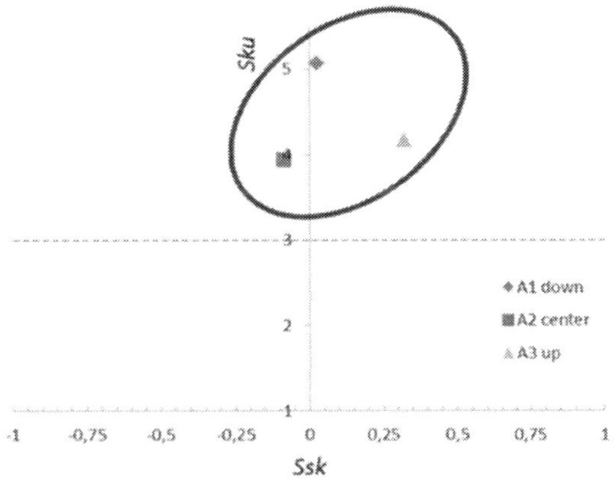

Fig. 6. A pair of amplitude surface roughness parameters, *Ssk* and *Sku*, of the steel, 1.2709, after out print depending on the place of measurement.

After printing the samples have been subjected to machining under production conditions on the machine tool MAZAK QTS 250M. The analysis of the machined material surface topography after the turning process has been performed. For samples A1, A2, A3 areas A (top), B (centre), C (bottom) were cut, which were machined (Fig. 7).

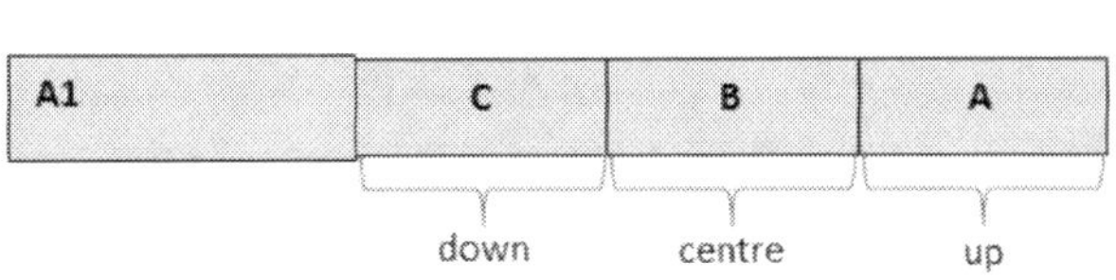

Fig. 7. The areas of sample subjected to turning.

Variable machining parameters for particular areas have been selected: cutting speed v_c = 70 ÷ 100 m/min, feed f = 0.1 ÷ 0.2 mm/ rev, cutting thickness a_p – 0.5 ÷ 0.6 mm. The detailed processing parameters and obtained results of selected surface topography parameters are presented in Table 1.

Table 1. Turning parameters and values of chosen surface topography parameters.

	Sample	A1			A2			A3		
	Area	A	B	C	A	B	C	A	B	C
	v_c - m/min	100	100	100	70	70	70	100	100	100
	f - mm/ rev	0,1	0,15	0,2	0,1	0,15	0,2	0,1	0,15	0,2
	ap - mm	0,5	0,5	0,5	0,6	0,6	0,6	0,6	0,6	0,6
Height Parameters	Sq - µm	2.04	2.89	4.14	1.97	2.15	3.48	2.00	2.40	2.46
	Ssk	-0.253	0.142	-0.0375	-0.147	-0.326	-0.0107	0.374	0.395	0.151
	Sku	2.50	2.61	2.19	3.53	4.15	2.61	3.34	2.49	2.34
	Sp - µm	6.50	9.50	11.5	16.3	21.1	17.8	27.0	9.80	13.6
	Sv - µm	11.8	8.20	10.8	21.9	16.1	10.8	8.80	10.3	12.8
	Sz - µm	18.3	17.7	22.3	38.2	37.2	28.6	35.8	20.1	26.4
	Sa - µm	1.71	2.32	3.47	1.57	1.65	2.78	1.68	1.99	2.10
Amplitude parameters - Roughness profile	Ra - µm	1.66	2.02	2.39	1.53	1.43	1.99	1.52	1.63	1.54
	Rz - µm	6.76	11.0	10.9	7.30	8.61	10.4	6.91	7.03	7.90

5 Conclusions

The development of additive manufacturing methods in relation to metal materials is an impulse to identifying the surface topography. The parameters of the surface topography of the hot working tool steel can differ depending on the distance from the base plate. The differences of surface roughness values for samples after printing can amount from 32% to 85% for *Ra* and from 59% to 85% for *Rz* in comparison of the sample bottom to its top. One can note also the differences in areas A, B, C what about the values of surface topography parameters in the case of turning.

References

1. S. Hong, C. Sanchez, H. Du, Journal of Electronic Materials, **44**, 3 (2015)
2. Laser CUSING® macht Rowenta-Bügeleisen ordentlich Dampf, Betrieb&Meister, **5**, 4-5 (2004).
3. K.M. Au, K.M. Yu, W.K. Chiu, Computer-Aided Design, **43**(4) 356 (2011)
4. E. Martinez, L.E. Murr, K.N. Amato, J. Hernandez, P.W. Shindo, S.M. Gaytan, D.A. Ramirez, F. Medina, R.B. Wicker, Proceedings of the 1st Int Conference on 3D Mat Sci, **73** (2012)
5. M.C. Sinirlioglu, US – Turkey Workshop On Rapid Technologies, **24** (2009)
6. I. Hatos, H. Hargitai, L. Solecki, Analysis of Tool Materials, **8**(2) 165 (2015)
7. F. Cajner, D. Landek, V. Leskovek, UDK 69.14.018.298:539.4, ISSN 1580-2949, MTAEC9, **44** (2), 85 (2010)
8. S. Wojciechowski, P. Twardowski, M. Pelic, Procedia CIRP **14,** 113 (2014)
9. S. Wojciechowski, P. Twardowski, Procedia CIRP **1**, 289 (2012)
10. S. Wojciechowski, R.W. Maruda, et al., Int J of Mech Sci, **119**, 360 (2016)
11. S. Kumar, Elsevier, Oxford, 93, DOI: 10.1016/B978-0-08-096532-1.01003-7 (2014)
12. G.M. Krolczyk, J.B. Krolczyk, R.W. Maruda, et al., Measurement, **88** 176 (2016)
13. A. Pereira, P. Hernández, J. Martinez, J.A. Pérez, T.G. Mathia, Scanning, **36**, 39 (2014)

Basic design rules of unit cells for additive manufactured lattice structures

Julia Kessler[1,*], *Nicolae* Balc[2], *Andreas* Gebhardt[1], and *Karim* Abbas[1]

[1]University of Applied Sciences Aachen, Faculty of Mechanical Engineering and Mechatronics, Goethestr. 1, 52064 Aachen, Germany
[2]Technical University of Cluj-Napoca, Department of Manufacturing Engineering, B-dul Muncii, 103-105, 400641 Cluj-Napoca, Romania

Abstract. Additive Manufacturing is a cutting-edge technology for the production of complex, functional, filigree and topology-optimized components. The growing demand for high-tech light-weight structures increasingly requires support in the design process of fine lattice structures. However, due to the high information density regarding lattice structures, which is based on many scientific research studies, the effective design of a specifically product-oriented lattice structure is made more difficult. The focus of this work is to provide a user-friendly information basis for the user or constructor. Therefore, this scientific work serves to show the first basic rules for the design of simple unit cells for additive lattice structures. The approach is based on the evaluation of the limiting manufacturing factors, which can be reduced both to the minimum and maximum angle and length, as well as to the cross-sectional geometry. The analysis reveals a considerable extent to the cross-sectional geometries and their welding pattern in various quality forms. The results obtained form the basis for the development of complex functional unit cells. However, the construction of a unit cell is only a part of a lattice structure, so additional factors must be added to the unit cell assembly.

1 Introduction

Additive Manufacturing (AM) is the world-wide standard for the manufacturing technology of layered processes known as Rapid Prototyping in the late 1980s. Where the conventional production technologies reach their limits, additive production shows its strengths. The key benefits of this technology are faster product development and recovery, better product characteristics and significant cost savings. In order to tap the potential, however, it is necessary to have experience, knowledge and the right tools about these technologies. The additive technologies offer various advantages, especially in the production of highly complex components with integrated functionality or lightweight structures. This applies in particular in the area of the layer-wise construction of metal components by means of Selective Laser Melting. Specifically, the higher complexity of a component through functional integration, internal structures or a topology-optimized design can be implemented without additional costs with SLM. The SLM process allows the manufacture of components with integrated hollow or lattice structures that significantly reduce the

DOI: 10.1051/matecconf/201713702005

volume to be built. But an existing problem lies in the handling and the direct integration of various grid structures into an existing component. Especially the commercial user is often overstretched, as there are many scientific knowledge but no explicit procedure for the application of lattice structures within the scope of the SLM production. Various software solutions exist on the market for the design of such lightweight structures within a component, such as Magics Structures or Autodesk Within. The commonality of these software solutions is that the structures can be multiplied by unit cells of uniform size in the component interior and can be adapted to mechanical loads, especially with Autodesk Within, by means of thickness optimization. Empirical studies and simulations using the finite element method (FEM) have shown that the size and shape of the unit cells influence the mechanical properties of the lattice structure. The only deviation from the regular unit cell is currently made to the transition area to the outer surfaces of the hollow body by rotating or distorting the unit cells there [1].

The most important element of a lattice structure is the unit cell, which may have different properties. In addition to the packing density of a cell, the single-cell cell shape and its webs are particularly decisive. The current state of the art shows that mainly the additive production of periodic lattice structures has been investigated. Extensively, cubic lattice structures were investigated by different researches, especially by Olaf Rheme. At first, under the consideration of the essential production restrictions of the SLM, cubic lattice structure types are analysed and classified by crystallography. It is clear to say that rarely other non-cubic structures, such as Honeycomb structures, Voronoi structures and topology optimization structures are analysed. At present, more researchers and users are discovering the potential of generatively fabricated lattice structures. For the SLM, periodic lattice structures are suitable, due to the targeted geometric influence. Open cell types are necessary to remove the excess powder material. Various production processes, such as, for example, the fine casting, are used for producing grid structures. In contrast to other manufacturing processes, SLM can produce virtually any desired structures arranged in space and is thus also suitable for topology optimization of components [2].

The evaluation of lattice structures requires a differentiated approach of several influencing factors. These include first and foremost the process parameters, machine specific influencing variables and the geometric properties of the to be printed parts.

All these three main focuses are in a corresponding constellation and depend on each other. The machine specific influencing variables are constant and they describe the values that are mainly defined by the plant technology. These for example include the integrated laser scanning system or rather the laser itself. Properties like the scanning speed or the laser power are based on the plant technology but finally are not constant. These parameters can be adjusted and controlled manually with an integrated machine software. Adjustable parameters like the bevor mentioned ones, belong to the process specific influencing factors. Furthermore, an important value is defined by the construction of the product. The main SLM process, the material and the functional principle of the SLM plant define and set the process boundaries as well as the construction boundaries. All adjustments, values and parameters form the basis for the product quality and thus the resulting properties of the generated product. The process specific influencing factors primarily influence the quality of the generated parts. First of all, the laser and the original material are the essential values within the SLM process. The interaction between metal powder and laser decide about the precision and the boundaries of the to be generated parts. [2, 3, 4]

2 State of the art and current research

Process based pores that appear in the struts reduce the mechanical properties and the actual aimed quality of structures. Concerning this, Chunlei Qin et *al*. investigated the relationship between the laser beam and the molten pool. The influences of the laser-power and the scanning speed were the focus of this investigation. The evaluation was performed with the help of microscopes and high-speed-imaging. The results showed that the laser-power has a decisive influence on the strut-diameter and that an interaction of the laser beam and the molten lead to a violent molten pool. This violent interaction causes the generation of pores and bad surface-roughness. [5]

One basic work in this field of research titled "Cellular Design for Laser Freeform Fabrication" was presented by Olaf Rehme [2]. His work contributes to understand the mechanical properties of lattice structures made by SLM. After a detailed description of all process constraints and boundary conditions for the manufacture of SLM lattice structures, he makes a theoretical analysis to comprise the examination of structural hypotheses for isotropic cellular materials before a generalized theory is developed for anisotropic SLM lattice structures. He evaluates different cell types such as polyhedral, cubic and rhombic cells towards their producibility. His scientific work gives information about extensive experimental evaluation of test specimens of the properties of SLM solids and lattice structures by SLM. He shows that SLM lattice structures allow a very high degree of design freedom in comparison to conventional cellular materials. The most important finding of his thesis is that power laws, which are normally used to describe mechanical properties of cellular materials, break down above a certain critical relative density. He identified one specific cell tyy (f2ccz) which excels other cell types by far and is overall best performing due to its high specific strength, high specific energy absorption and high specific stiffness when loaded under compressive stresses, especially in z-direction. Besides he gives two examples for implementation of the SLM lattice structures into applications, e.g. medical endoprostheses in order to prove how innovative products can be furnished with new functionalities, such as faster ingrowth of bone tissue.

Chunze Yan, Liang Hao, Ahmed Hussein, Philippe Young, and David Raymont describe in their paper "Advanced lightweight 316L stainless steel cellular lattice structures fabricated via selective laser melting" [6] the manufacturability and performance of advanced and lightweight stainless steel cellular lattice structures fabricated via selective laser melting. A unique cell type called gyroid is designed to construct periodic lattice structures and utilize its curved cell surface as a self-sup-ported feature which avoids the building of support structures and reduces material waste and production time.The experimental set-up is about a cell size of 5mm, with a volume fraction for massive parts of 6, 8, 10, 12 and 15% and 9 x specimens with the dimensions 25x25x15mm. The cellular lattice structures were made from a 316L stainless steel powder with average particle size of 45 ± 10μm. The results of the compression properties describe that the worst orientation lattice structure offers higher mechanical properties than the normal orientation lattice structure although both of them have the same volume fraction of 15% and unit cell size of 5 mm, indicating anisotropy of the gyroid cellular lattice structure. The Young's modulus of the worst orientation lattice structure is 302.57 MPa, which is 20.37% greater than that of the normal orientation lattice structure, 251.36 MPa, and the yield strength of the worst orientation lattice, 15.53 MPa, is 7.78% higher compared with the yield strength of the normal orientation lattice structure, 14.41 MPa.

Based on these and many other investigations, the application of lattice structures as mechanical design elements requires to take into account the final geometry of the struts, the material to be integrated into the engineering design and the material properties obtained from metal printing process. For this reason, the aim of the work is to create a basis for the Commercial User to facilitate the handling of lattice structures and their integration into special components. In addition to the various types of grid structures, their mechanical properties must also be analysed in detail in order to enable the user to select a corresponding grid structure individually and according to his selection criteria.

3 Research approach

First, a pre-study was carried out to limit the process parameters and to evaluate the material used and the associated possibilities, such as the powder grain size, which has a decisive influence on the maximum resolution and the smallest possible construction elements. Because of the fact that the fundament of every periodic lattice structure are the struts for this purpose, it was analyzed which geometries of the struts can be realized. They determine the value of the mechanical properties, the performance and finally the quality. Concerning the struts conclusions have been made in previous researches. Topics included the feasibility and the reached quality. Most of the time struts with a circular shape are set as the main shape and used for the investigation. Just a few use elliptical or different shapes. [6, 7] Thus the influence of the cross-sectional shape on the loading resistance is rarely taken into consideration. But basic researches are essential and create the necessity to take a closer look on this topic. At the beginning, it is important to know what causes a limitation on the generation of struts. The focus is on the thinnest to be reached struts and the building accuracy of the cross-sectional area. All in all, the main aim is a feasibility study. Properties of five different cross sectional shapes are examined: a circular-, an elliptical-, a square-, a triangular- and a rhombus shape. On several platforms, a matrix of struts is placed with the bevor mentioned shapes (Figure 1).

Fig. 1. Matrix of different struts and geometries.

These are designed with a constant increasing cross sectional area in x-direction and a length variation in y-direction up to 10mm. The source for each platform is a digital 3D model. For a better evaluation of the structural limits, the experimental set up was split into three different constructive starting points. At first a reference surface (RS) was set to find out what the limits are, regarding to the cross sectional are. Furthermore, the limits of the outlines of the shapes were evaluated with setting a collective greatest diagonal line or width (RO). The last specimen pool had also a predefined cross-sectional area, which was based on the area of a circular shape (RC), because most of the lattice structures consist of circular struts. Within this investigation three different initial constructive approaches have been compared. Therefore, the optical evaluation with a microscope was essential. The greatest strut width was measured and subsequently the average and the deviation were calculated. The aim was the comparison of the set point and the average value. To analyse the most important focuses of the results the quality of the struts is additionally consulted. The evaluation of the struts with a circular cross section shows some characteristics. It is noticeable that struts with a nominal diameter smaller than 0,15mm cannot be generated. The smallest reached diameter aggregates towards an average of 0,201mm, the constrained set point is 0,16mm, which almost is a deviation of 25%. Striking is that every strut within the set point range of 0,15mm up to 0,25mm, does not overwrite an average value of 0,2mm up to 0,25mm. Every reached diameter overwrites the actual to be reached design value. In all three experimental set ups it is noticeable that from a value of 0,35mm a turning point appears. This turning point describes the moment from which the deviations shift into the negative values and thus the average values are smaller than the actual set point. Comparing the conditions of the manufactured struts a characteristic is recognizable, that describes the above-mentioned phenomenon. It is obvious that struts up to a diameter of 0,25mm are just towered welding beads. They don´t appear to be cylindrical outlines und consist of irregular spherical shapes. A selection of the cross-sections to be examined could be made over the weld pattern of the struts. The determination of the smallest (0,4mm2), medium (0,7mm2), and largest (1,0mm2), cross-section area was carried out.

4 Experimental Set-Up

The above-described pre-study was necessary in order to determine the feasibility of the specimens with different geometries of the bridge and to show which sizes and geometries the process can produce. A basic research based on this aim is to create an evaluation of the boundaries for the design, based on the interplay of geometry and process properties.

In this series of experiments, the aim is to define significant values for the construction of the unit cell and its struts, both through different angular variations as well as different lengths and three selected cross-sectional areas of the struts.

For the experiment, a total of 33 specimens with different geometries circle, ellipse, square, triangle, rhombus, were tested. Each sample was build up at different angles in 10° degree steps. As can be seen in Figure 2, a sample body consists of a 5mm wide sheet and ten struts arranged thereon. The length as well as the cross-sectional area remain constant for every angular change. Beyond the five geometries, three different cross-sections 0.4 mm^2, 0.7 mm^2 and 1.0 $mm^2$are tested, which also result from the preliminary test. In addition, the lengths are varied. Generated are 5mm long struts as well as 10mm long struts. The bridge-to-base connection is a flat plane and not spherical (see picture 3 with a purple mark).

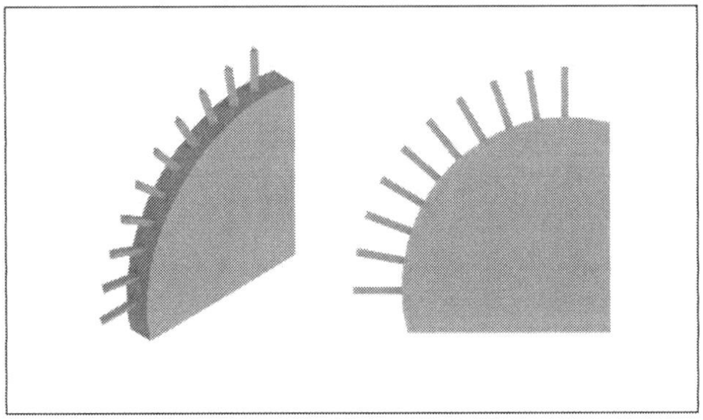

Fig. 2. Sample with different angles in 10° degree steps.

The samples are manufactured by SLM with a M1 Cusing machine (Concept Laser). All components are printed with constant process parameters, which have proven to be the optimum in the previous tests. However, this will not be discussed further. The most important parameters are defined by a laser power of 180W, a scanning speed of 1450mm/s for contours and 1000m/s for surface exposure. The layer thickness is 45µm and a grain size distribution of 30-60µm.

Fig. 3. Bridge-to-base connection as a flat plane.

The analysis of the samples is carried out optically with the aid of a microscope. The struts are microscopy and photographed from different perspectives in order to be subsequently evaluated. The analysis and evaluation of the samples gives information about the minimum and maximum angle as well as the length to be realized. In addition, the quality of the bridges and the realized geometries can be assessed.

Fig. 4. Samples on the building platform of the SLM machine.

5 Results and discussion

Since this is an optical evaluation, criteria have previously been defined which serve for a better assessment, which is used to evaluate all struts. One special focus was on the welding traces and welding beads. Here, the front and the lateral sides of the struts were considered.

The aim of the test series is to evaluate the boundaries and influences of different strut geometries on the generation of struts in different angles. As a result, struts are determined that are optimally defined according to the criteria and meet the requirements. The following table shows such an assessment and analysis.

Table 1. Protocol of the test series (example triangle, 5mm strut length, 0,4mm2 cross-sectional area).

Protocol Microscopy angle samples

Geometry: Triangle | Cross-sectional area: 0,4 mm² | Strut lenght: 5 mm

Probe	Angle (°)		Geometrie realized	Alignment geometry to the building platform	Strut realized	Surface quality	Welding beads front apparent	Welding beads coat apparent	Features
1	0		indefinable	a parallel zur BP		indefinable	indefinable	indefinable	
2	10		indefinable	a parallel to building platform		hardly adherent fragments, very many adherent fragments	indefinable	medium	Blue discolored downhill
3	20		approximately triangular, strong deformation downhill	a parallel to building platform	x	hardly adherent fragments, very many adherent fragments	strong	medium	Orange/Blue discolored downhill
4	30		approximately triangular, strong deformation downhill	a parallel to building platform	x	hardly adherent fragments, very many adherent fragments	strong	edium	Nose at the corner uphill, welding points at an edge uphill
5	40		approximately triangular, strong deformation downhill	a parallel to building platform	x	hardly adherent fragments, many adherent fragments downhill	strong	medium at the edges, light on the surface	Nose at the corner uphill, welding points at an edge uphill
6	50	x	Slight discontinuities, slight deformation downhill	a parallel to building platform	x	hardly adherent fragments, some adherent fragments	strong	medium at the edges, light on the surface	Welding points on an edge uphill
7	60	x	Slight discontinuities, less stable downhill	a parallel to building platform	x	hardly adherent fragments, less adherent fragments downhill	strong	medium at the edges, light on the surface	Nose at the corner uphill, welding points at an edge uphill, bridge slightly curved upwards
8	70	x	Slight discontinuities, less stable downhill	a parallel to building platform	x	hardly adherent fragments	strong	medium at the edges, light on the surface	strut slightly curved upwards
9	80	x	Minimal discontinuities	a parallel to building platform	x	normal	strong	light on the surface	Nose at the corner uphill
10	90	x	Minimal discontinuities	a parallel to building platform	x	normal	strong	light on the surface	On the forehead slightly discolored

The analysis shows recurring abnormalities in all geometries, which show deviations to the models constructed. For all test specimens, the horizontal ridge is not fully generated. However, only a partial segment of the strut is produced, whereas the major part is deformed and distorted. The cross-section of the partial segment increases toward the base of the whole probe. This is comparable with struts with a 10°-building direction.

Table 2. Total evaluation of test series.

Strut length	**5mm**				
cross-section area (mm2)	**Triangle**	**Ellipse**	**Circle**	**Rhombus**	**Square**
0.4	50°-90°	50°-90°	50°-90°	60°-90°	50°-90°
0.7	30°-90°	60°-90°	50°-90°	50°-90°	50°-90°
1.0	20°-90°	30°-90°	30°-90°	40°-90°	60°-90°
Strut length	**10mm**				
cross-section area (mm2)	**Triangle**	**Ellipse**	**Circle**	**Rhombus**	**Square**
0.4	50°-90°	60°-90°	50°-90°	60°-90°	50°-90°
0.7	30°-90°	60°-90°	50°-90°	50°-90°	50°-90°
1.0	30°-90°	20°-90°	30°-90°	40°-90°	60°-90°

Both for the horizontal and 10°-struts, significant oxidation is to be found on all areas which are directed towards the building platform. This discoloration only decreases at further larger angles. The unrealized subrange also decreases. Table 2 shows the results for the individual geometries and the corresponding fully realized struts with respect to the angle.

A further conspicuity shows that the actual geometry is very stretched or compressed at small angles of 20°-50° in all structures and does not correspond to the original model. This is illustrated in Figure 5. Along with the stretching and compression, an enlarged cross-sectional area of the geometries can also be seen. In this case, the enlargement of the cross-section decreases to the extent up to the vertical strut is also shown in Figure 5 (top actual state - lower target state).

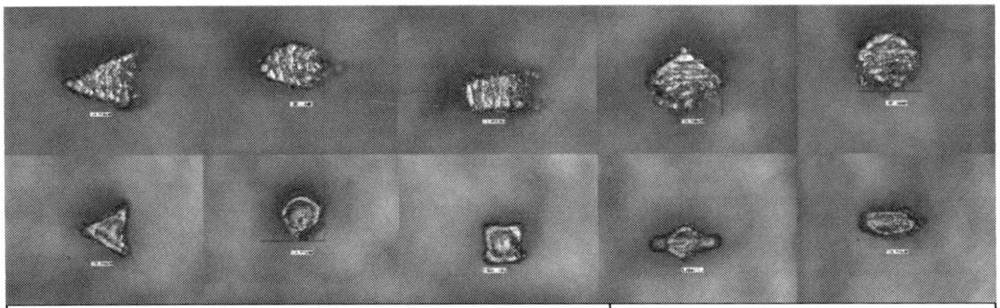

Stretching of triangle-circle-square	**Compression of rhombus-ellipse**

Fig. 5. Stretching of triangle, circle and square struts and Compression of rhombus and ellipse struts.

A conspicuousness in the welding traces is associated with the not fully generated struts. In the case of struts with visible deformation (pointing to the construction platform), two special areas are shown within the strut. These are similar for all geometries. These regions can be subdivided into welding traces which are located within the printing plane and are parallel to each other, as well as welding traces which are pointing towards the building platform and which are warped or deformed. This can be seen in Figure 6. However, this effect of bagging decreases with increasing angular size.

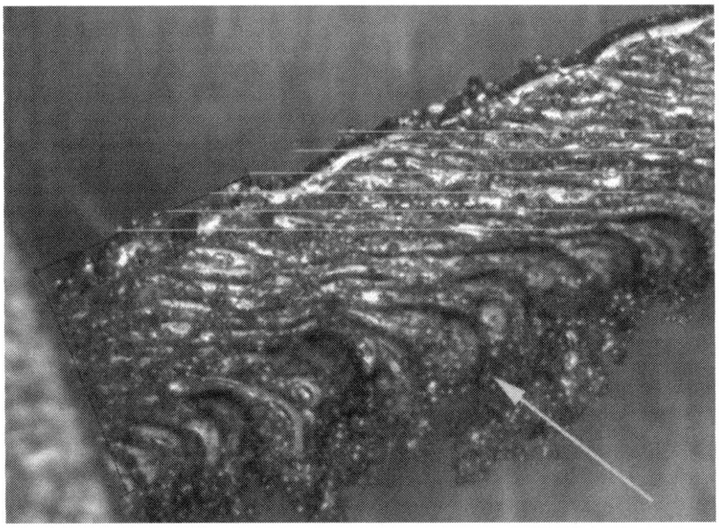

Fig. 6. Strut with welding traces.

The geometries are different basic geometries, but they are comparable in some respects, for example the triangle and square geometries. Both geometries are generated with a flat side pointing towards the building platform. There are similarities as well as differences in the comparison. Viewing the 20°-strut (Figure 7, third column) shows that the tri-

DOI: 10.1051/matecconf/201713702005

angle geometry is strongly deformed and a partial segment is missing, but a part is generated so far that the welding layers and the plane are visible. On the other hand, in the case of the square, the strut is slightly warped and a larger sub-segment is missing. As a result of the geometry itself, the cross-section of the square is constant over the planes, while the cross-section of the triangle becomes smaller and tapers. The result for the process itself is that in the case of an advanced printing process, either constant thermal energy is induced or the thermally induced energy decreases with decreasing cross-section and thus less welding traces.

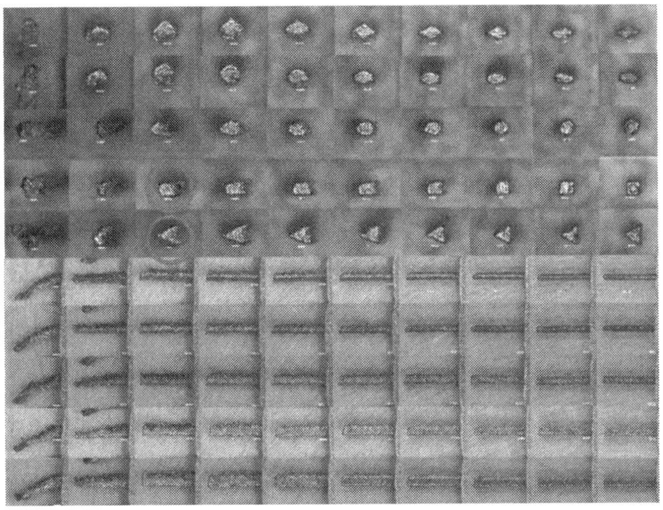

Fig. 7. Overview of all geometries and different angels (above: front side below: lateral sides of the struts).

A crucial question at the beginning of this study was if there are significant contour geometries that are significantly better than others at certain angles? But the research showed that it is only to a certain extent that struts with different geometries are better represented.

A further result of this survey is that the warping causes layers to lie in overlying planes and are exposed several times, resulting in clumping and tearing off the structure, what causes the welding process and oxidation.

The first assumption was that the strut length has a decisive effect on the geometry to be realized as a function of the cross-sectional area and the corresponding angles. The reason for this is the thermal distortion, since unsupported struts tend to deform at different angles. This assumption could not be confirmed except for minor deviations in the case of triangle and ellipse (deviation of 10°). Conclusion and outlook

The second assumption was that the generation of struts is decisively determined by the geometry and its cross-sectional areas. The assumption could be confirmed partly by means of the experimental series, however, it was found that only minor deviations could be analyzed when comparing the 0.4 series to the 0.7 series as well as the deviations of the 0.7 series to the 1.0 series. This leads to the exclusion of the 0.7 series for further experiments.

6 Conclusion and outlook

The aim of the investigations discussed in this paper was to develop a basis for a guideline for the design of simple unit cells for additive lattice structures. The main focus was on the evaluation of which geometry could be generated with which cross-section and angle. All geometries can be realized with a build-up angle of 20°, some with clear contours, the others with better contours with increasing build-up angle. Below the 20°, the struts are not very poor and have been realized with great deformations. It is striking that the trial triangle and ellipse have achieved very good results in the lower range (angles of 20°-40°) compared to the other three geometries a possible reason for the triangle geometry could be the change of the temperature gradient due to the geometry. Due to the geometry itself, e.g. While the cross section of the triangle becomes smaller and tapers as the square crosses across the planes.

Further mechanical tests, such as the tensile test, should now provide information about the mechanical properties of these five different web geometries. After evaluating the tensile tests, a selection of the geometries which provide the best mechanical characteristic values is to be carried out. With these geometries, the newly developed unit cells are then constructed for the construction of individual lattice structures. These different lattice structure types are deposited with mechanical characteristics and will be main part of the guideline. The feasibility of geometries is not sufficient for the construction of a single cell, so the mechanical properties have to be evaluated. The mechanical tests include, but are not limited to, tensile/compression test, fatigue test, three-point bending test and torsion. These results will determine the decision criteria for the commercial user.

Most of the scientific work has concerned itself with cubic unit cells, and with the geometries of the individual bridges rather than with the geometry of the entire unit cell. This survey ensures that both complex polygon structures and cubic cells are possible. The aim of this research work is to generate a matrix with 12 different lattice structure types with both cubic and polygonal unit cells for the possibility of user-friendly selection of grid structures to allow the commercial user the possibility of user-friendly selection of grid structures for integration into certain components.

The guideline for the implementation of grid structures in additive manufactured components should include the essential steps: Analysis of the component, possible areas for integrating the grid structures, define the stress of the areas (stiffness/tension/pressure/bending...), lattice structure type selection by using the matrix, lattice structure implementation in the defined areas of the part, additive manufacturing of the part. The core function of the guide is the selection of a suitable grid structure as a function of the desired mechanical properties or stresses of the component.

This research was supported by the OpTi-DeP Project (no. BG101/2016) financed from the UEFISCDI by the Romanian Government and also by the HORIZON 2020 AMaTUC project (GA 691787) financed from EU.

References

1. R. Naguschewski, *Stress-Based Construction of Inhomogeneous Lattice Structures and their Thermomechanical, Evaluation for Selective Laser Melting*, Master Thesis (2016)
2. O. Rheme, *Cellular Design for Laser Freeform Fabrication*, Institute of Laser and System, (Dissertation, 2009)
3. S. J. Merkt, *Qualification of generatively manufactured lattice structures for tailormade component functions*, (Dissertation, 2015)
4. D. Ziebura, *Development of a system for the targeted parameter determination in the SLM process using the example IN 625*, (Master thesis, 2015)
5. Ch. Qiu et al., MSE, A, **638**, 228-231. (2015), doi.org/10.1016/j.msea.2015.01.031
6. Ch. Yan, L. Hao, A. Hussein, P. Young and D. Raymont, Materials and Design, **55**, 533-541 (2014)
7. S. McKown, Y. Shen, W.K. Brookes, C.J. Sutcliffe, W.J. Cantwell, G.S. Langdon, G.N. Nurick, M.D., Journal of Sandwich Structures & Materials, **13**(4) (2011)

Influence of the layer thickness in the Fused Deposition Modeling process on the dimensional and shape accuracy of the upper teeth model

Ján Milde[1*], *Ladislav* Morovič[1], and *Jakub* Blaha[1]

[1]Slovak University of Technology in Bratislava, Faculty of Materials Science and Technology in Trnava, Institute of Production Technologies, Department of Machining and Forming, Jána Bottu 25, 917 24 Trnava, Slovakia

Abstract. The article is focused on research of the influence of the layer thickness in the Fused Deposition Modeling (FDM) additive manufacturing method on shape geometrical and dimensional accuracy of parts. The digital model of the object (upper teeths) was obtained by intraoral 3D scanner 3Shape TRIOS used in dentistry. Based on the 3D (Three Dimensional) scanned digital model, the manufacturing of the upper teeths was realized on the FDM 3D printer Zortrax M200. The parts were manufactured in three different layer thicknesses. The manufactured parts was digitized by optical 3D scanner GOM ATOS Triple Scan. The dimensional and shape accuracy of the parts was evaluated in the software GOM Inspect.

1 Introduction

According to Békés [1], parts may be produced by adding materials (additive manufacturing methods), moving materials (casting technology, forming technology) and removing materials (machining technology). Additive methods of parts production include so called Rapid Prototyping (RP) that produces components a layer by layer [2]. The layer thickness in the RP process significantly determines the printed part's quality. Lower layer thickness typically results in a part being printed with smoother surfaces and also better quality [3]. All 3D printing parameters are critical. The most solved parameters in the other publications are: layer thickness, part orientation, internal structure and used material.

The aim of this article is to publish the results of the experiment *"Influence of the layer thickness in the process of FDM on dimensional and shape accuracy of the upper teeth"*. It is known, that in the process of injection molding and additive manufacturing of plastic materials, some plastic materials, especially ABS (Acrylonitrile - Butadiene - Styrene) are susceptible to shrinkage. The model used for the experiment was a human upper teeth, obtained by 3Shape TRIOS optical intraoral 3D scanner. In second phase, the upper teeth were manufactured and subsequently 3D digitized by optical 3D scanner GOM ATOS II

* Corresponding author: jan.milde@stuba.sk

Triple Scan. The final step was the comparison of the printed models with the reference CAD model [4, 5, 6, 2] in the GOM Inspect software [7, 8].

2 3D model of upper teeth used for the experiment

The model used in the experiment is focused on dentistry. In dentistry, models of this type are used for a design and manufacturing of invisible teeth braces. The digital model, which was used in the experiment was a model of upper teeth. This model was chosen due to its importance for planning purposes in invisible teeth braces industry, where it is very important to preserve the best dimensional and geometrical accuracy [9, 10].

Acording to the collaboration between Slovak University of Technology, Faculty of Materials Science and Technology and Comenius University, Faculty of Medicine, we made the experiment, which is based on production technologies, materials and it is also connected with medicine.

On the 3D scanned model, there were also some significant areas (asymmetrical volumes) and also was necessary to create a cube for alignment in a fixture for a 3D scanning. Due to the facts, the editing of the upper teeth model was necessary. For model editing, the software Autodesk PowerSHAPE 2017 was used (Fig. 1.a).

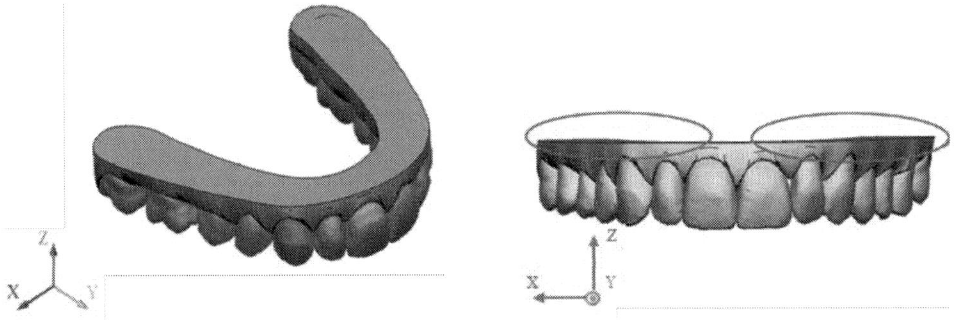

a) b)

Fig. 1. Original STL model captured by intraoral 3D scanner 3Shape TRIOS: a) view ISO 3, b) view +Y (the asymmetrical areas are marked by ellipses).

2.1 Editing of the upper teeth digital model

At the beginning of editing, in the software Autodesk PowerSHAPE 2017, on the polygon mesh of the upper teeth, coordinate system was created. On the upper area of the upper teeth model, a surface was created (Fig. 2.a) due to separation for two single volumes. The asymmetrical areas divided by surfaces are shown on Fig. 2.b. This separation was made for creating of flat and symmetrical surface, which is situated on the bottom area of the teeth.

2.2 Creating of new elements on the upper teeth model

The surface used for dividing of models was created using the function *"Create primitive surface"*. After this step the function *"Split active solid"* was used for dividing of the two models. The upper volume was then removed and the original was closed and converted to the solid (Fig. 3.b) for a better alignment on 3D printer building platform. In the front of the upper teeth model, there was also created a cube (Fig 3.a), which was later used for alignment in a fixture for a 3D scanning.

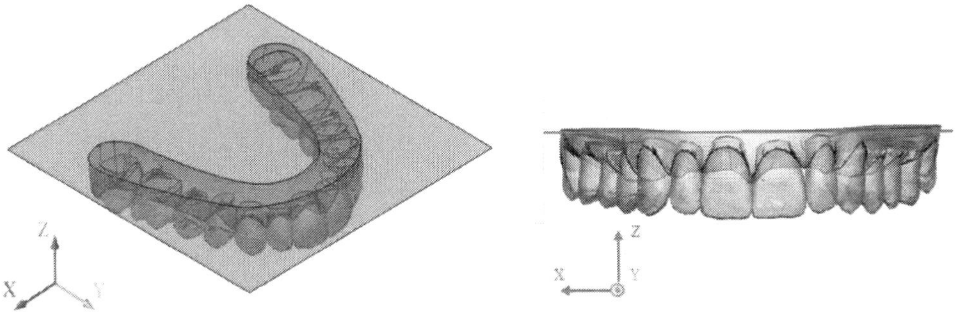

a) b)

Fig. 2. Editing of the upper teeth model in software Autodesk PowerSHAPE 2017:
a) surface created in the Z axis, b) asymmetrical areas divided by surface.

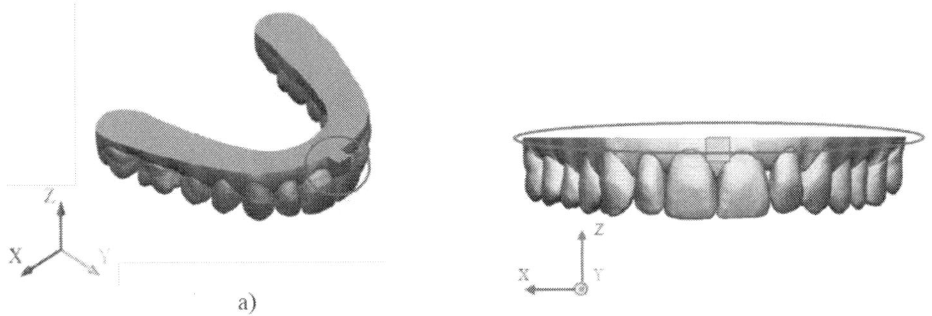

a) b)

Fig. 3. Upper teeth model used for manufacturing:
a) cube for alignment in the front of the model, b) final symmetrical surface.

3 Fused Deposition Modeling method

FDM is a method of additive manufacturing in which a part is produced using layer by layer deposition of material. It is an important technology as it has potential to reduce the manufacturing lead time of the product. A lot of steps of FDM process are automatic. For example, layer thickness is selected manually among few options shown by the software. Layer thickness in FDM process has significant effect on many key characteristics which have influence on final quality and cost of the product [11, 12].

In the FDM technology, the used material is in a plastic wire form of diameter 1.75 or 2.85 mm, unwound from a coil. This fiber material is fed to the liquefier head where it is heated, fused and extruded through the nozzle in the form of fine fiber. There are several kinds of kinematic structures used in FDM. The movements of the liquefier head and 3D printer can be divided into three most used groups:

- head is moved in axis X, Y, Z,
- head is moved in axis X, Y and build platform is moved in Z axis.
- head is moved in axis X, Z and build platform is moved in Y axis [13].

The scheme of the FDM method (head is moved in axis X, Y and build platform is moved in Z axis) is shown in Fig. 4.

The fiber material used in the experiment was ABS with a diameter of 1.75 mm. ABS material was used in the experiment due to his minimal use in the production of the teeth models in dental applications for production of invisible teeth bracers.

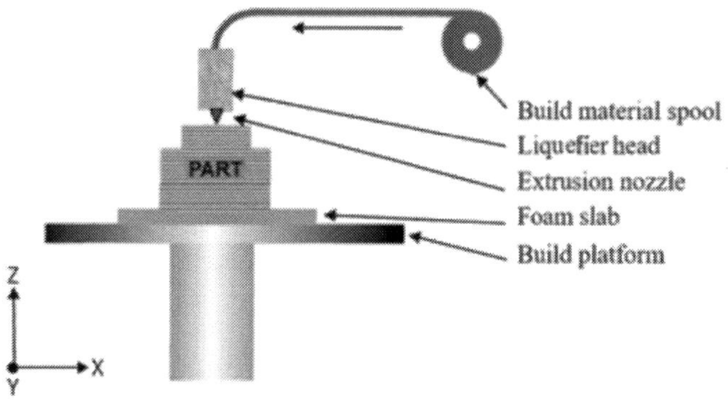

Fig. 4. Scheme of Fused Deposition Modeling [13].

ABS is an engineering plastic common used in FDM technology, that has butadiene part uniformly distributed over the acrylonitrile-styrene matrix. It possesses excellent toughness, good dimensional stability, easy processing ability, chemical resistance and cheapness. It is a low cost engineering plastic that is easy to machine [14, 15, 16]. ABS is an ideal material for structural applications when impact of resistance, strength and stiffness are required. It is widely used in FDM methods since it has excellent mechanical properties and is easy to paint and glue [17].

Parameters, which have influence on the dimensional and shape accuracy in the process of FDM are the following.

• infill,
• layer thickness,
• model orientation,
• material,
• nozzle temperature.

Manufacturing of the upper teeth was performed by FDM 3D printer Zortrax M200.

4 Layer thickness setting of the upper teeth digital model

For other additive manufacturing methods as Selective Laser Sintering (SLS), Selective Laser Melting (SLM) [18, 19, 20, 21], Poly-Jet Modeling (PJM) or Stereolitography (SLA) selection of layer thickness is also critical as for Fused Deposition Modeling (FDM). Layer thickness in all additive manufacturing methods have impacts on the printing process (time and cost) and the visual and physical properties of a printed part. Often the difference between an identical part printed at 100 μm and 200 μm is very hard to distinguish (Fig. 6.) however the 100 μm part will take twice as long to print and cost more [3, 10].

Understanding the final application of a part is always important when selecting the layer thickness. If a part has sharp edges that are required for the functionality of the part or if a part fits into another part then a smaller layer height is ideal [3, 22].

Fifteen prints were made in total. Five prints were manufactured for each type of layer thickness. The layer thickness which is recognized as the height of the deposited slice from the FDM nozzle as shown in Fig. 5. The layer thickness parameter is used to examine the influence of building thicker or thinner layers on the outcome quality [23].

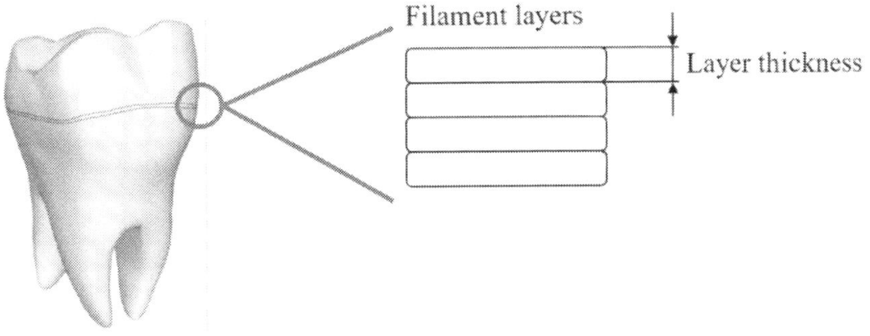

Fig. 5. Scheme of layer thickness in Fused Deposition Modeling process [22].

The layer thicknesses, which were used in the experiment:
- 0.09 mm,
- 0.19 mm,
- 0.29 mm.

a) b) c)

Fig. 6. A macro view of the FDM prints shown at the same scale for comparison: a) 0.09 mm, b) 0.19 mm, c) 0.29 mm [3].

5 Evaluation of the accuracy of shape and dimensions of the upper teeth

The printed models of upper teeth were digitized (Fig. 7.) using optical 3D scanner GOM ATOS II TripleScan. The measuring volume (MV), used for the digitization was MV 170 (170 x 130 x 130 mm). The resolution of the cameras was 5 Mpix (the measuring point distance was 0.07059 mm). The printed upper teeth models was matte coated [24].

Process of the 3D scanning is shown on Fig. 7. The comparison of each scanned model with the reference CAD model and evaluation of the dimensions was made in the software GOM Inspect V8 SR1. Parameters, which were measured are the distances between two points in the 5 single sections in total, from tooths marked as *Second molars (17 – 27)* up to tooths marked as *Canines (13 – 23)* (Fig. 8.). In the GOM Inspect software, the distance between points was measured by the function *"Construct – Distance – 2-Point Distance"*. These points were created by the function *"Fitting Point"*. All fitting points were created by the Chebyshev principle [25]. The scheme of the measured distances between point pairs is on the Fig. 9. and Fig. 10.

The shape deviation of the upper teeth model is visualised by the colour deviation map, which is created by the comparison of the reference CAD model and the 3D digitized parts.

DOI: 10.1051/matecconf/201713702006

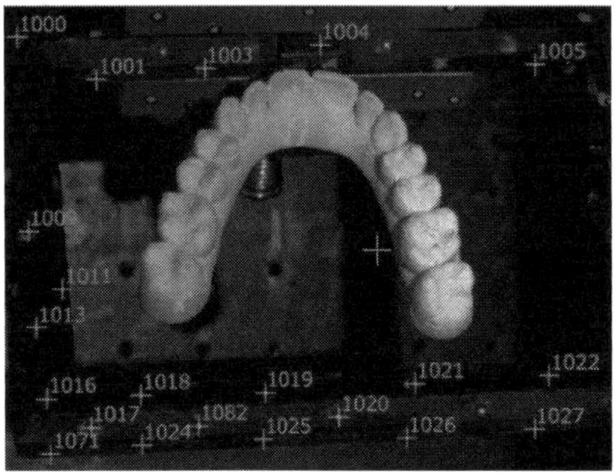

Fig. 7. Printed upper teeth model in the process of 3D digitization with recorded reference points.

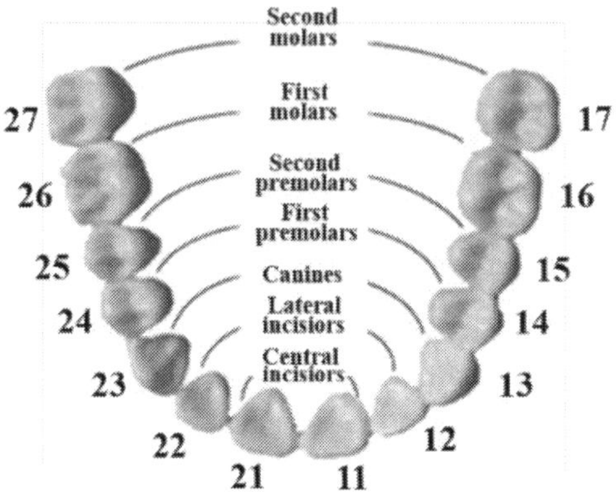

Fig. 8. Tooth marking (upper teeth) [9].

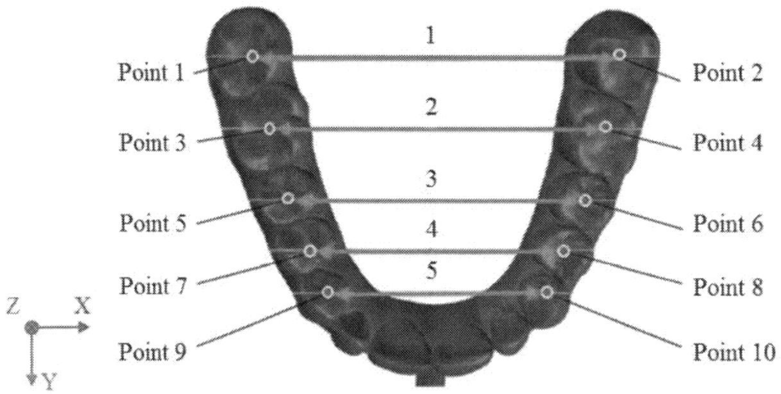

Fig. 9. The distance between point pairs in sections [7].

Point 1 ↔ Point 2				
	Nominal	Actual	Dev.	Check
L	+52.67	+52.34	-0.33	

Point 3 ↔ Point 4				
	Nominal	Actual	Dev.	Check
L	+48.29	+48.00	-0.29	

Point 5 ↔ Point 6				
	Nominal	Actual	Dev.	Check
L	+41.88	+41.68	-0.20	

Point 7 ↔ Point 8				
	Nominal	Actual	Dev.	Check
L	+35.76	+35.68	-0.09	

Point 9 ↔ Point 10				
	Nominal	Actual	Dev.	Check
L	+26.94	+27.15	+0.20	

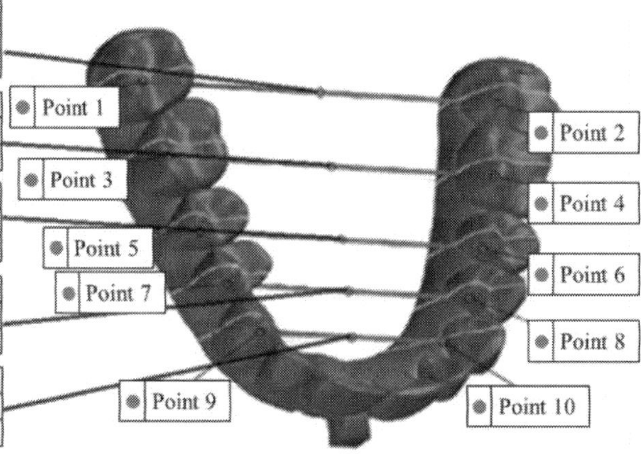

Fig. 10. The distance between two points on the upper teeth (layer thickness: 0.29 mm, print 4) [7].

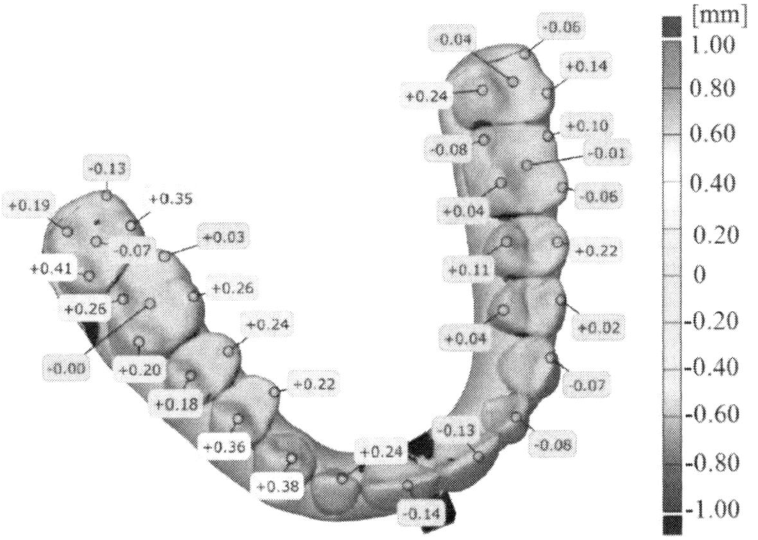

Fig. 11. Colour deviation map of the upper teeth model (layer thickness 0.09 mm) [7].

Table 1. Nominal distances, actual distances, deviations and arithmetic means of deviations [7].

Layer thickness: 0.09 mm							
Sections (points)	Nominal distances [mm]	Deviations [mm]					Arithmetical means [mm]
		Print 1	Print 2	Print 3	Print 4	Print 5	
1 (1 – 2)	+52.67	-0.27	-0.23	-0.21	-0.26	-0.27	**- 0.25**
2 (3 – 4)	+48.29	-0.25	-0.24	-0.19	-0.23	-0.24	**- 0.23**
3 (5 – 6)	+41.88	-0.19	-0.18	-0.16	-0.14	-0.20	**- 0.17**
4 (7 – 8)	+35.76	-0.10	-0.09	-0.11	-0.07	-0.14	**- 0.10**
5 (9 – 10)	+26.94	-0.02	+0.04	+0.04	+0.06	-0.01	**+ 0.02**

Table 2. Nominal distances, actual distances, deviations and arithmetic means of deviations [7].

Layer thickness: 0.19 mm							
Sections (points)	Nominal distances [mm]	Deviations [mm]					Arithmetical means [mm]
		Print 1	Print 2	Print 3	Print 4	Print 5	
1 (1 – 2)	+52.67	-0.30	-0.31	-0.33	-0.30	-0.34	**- 0.31**
2 (3 – 4)	+48.29	-0.25	-0.28	-0.30	-0.25	-0.25	**- 0.27**
3 (5 – 6)	+41.88	-0.24	-0.22	-0.21	-0.20	-0.23	**- 0.22**
4 (7 – 8)	+35.76	-0.11	-0.12	-0.14	-0.11	-0.09	**- 0.11**
5 (9 – 10)	+26.94	+0.11	+0.08	+0.10	+0.13	+0.09	**+ 0.11**

Table 3. Nominal distances, actual distances, deviations and arithmetic means of deviations [7].

Layer thickness: 0.29 mm							
Sections (points)	Nominal distances [mm]	Deviations [mm]					Arithmetical means [mm]
		Print 1	Print 2	Print 3	Print 4	Print 5	
1 (1 – 2)	+52.67	-0.30	-0.33	-0.38	-0.33	-0.44	**- 0.36**
2 (3 – 4)	+48.29	-0.27	-0.31	-0.29	-0.29	-0.29	**- 0.29**
3 (5 – 6)	+41.88	-0.19	-0.18	-0.22	-0.20	-0.25	**- 0.21**
4 (7 – 8)	+35.76	-0.07	-0.05	-0.12	-0.09	-0.11	**- 0.09**
5 (9 – 10)	+26.94	+0.21	+0.23	+0.16	+0.20	+0.18	**+ 0.19**

6 Conclusion

According to Table 1, 2 and 3 the most suitable layer thickness from the aspect of dimensional deviations was found to be layer thickness 0.09 mm with the smallest deviation. In the case of dimensional deviations, the most suitable layer thickness, with respect to the dimensional accuracy was found to be layer thickness with the smallest arithmetical mean deviation of -0.25 mm.

The second lowest dimensional deviations were found in layer thickness 0.19 mm with the arithmetical mean of deviation -0.31 mm.

The worst layer thickness in terms of dimensional accuracy was 0.29 mm, which created the greatest inaccuracies and distortions from required arithmetical mean of dimension -0.36 mm on the model of the upper teeth.

It was found that the layer thickness has a significant influence on dimensional and geometrical accuracy. The lowest shrinkage effect was found on the parts, where the amount of the single layers was the highest. In other words, the biggest shrinkage effect was found on parts where the amount of single layers was the lowest. Research of the layer thickness in the process of manufacturing by FDM method of upper teeth model was important due to the accuracy of the invisible teeth bracers. From the previous research is known, that layer thickness have a critical influence on the dimensional and shape accuracy.

Printed teeth models with the accuracy lower than ±0.25 mm are suitable as a model for the manufacturing of invisible teeth bracers. In the next step, the invisible teeth bracers are fabricated from thin transparent plastic plates using the vacuum press machine. The parameter *"layer thickness"* was chosen due to importance of the time build. Result from this research can be used in the optimal settings of *"layer thickness"* parameter in the process of manufacturing of teeth models with the assistance of FDM method.

The article was written with the support of the project VEGA 1/0477/14.

References

1. J. Békés, 6th International Scientific Conference CO-MAT-TECH, 283, (1998)
2. L. Morovič, Non-Contact Measurement of Free-Form Surfaces, 89, (2016)
3. P. Cain, 3D Hubs, [online]. [cit. 01. 07. 2017], Available at: https://www.3dhubs.com/knowledge-base/impact-layer-height-3d-print, (2017)
4. P. Bere, C. Neamtu, Open Engineering, **4**, 148, DOI: 10.2478/s13531-013-0158-x, (2014)
5. L. Morovič, *Návrh, obrábanie, meranie a skenovanie tvarovo zložitých plôch (Design, machining, measuring and scanning of free form surfaces)* (STU MTF, Trnava, 2013)
6. L. Morovič, P. Pokorný, Applied Mechanics and Materials, Advanced Materials Researc, **2269**, 468-471, (2012)
7. J. Blaha, *Vplyv hrúbky vrstvy objektu pri metóde FDM na rozmerovú a tvarovú presnosť dielu (Influence of the layer thickness of the object on the FDM printer on the dimensional and shape accuracy of the part)* (STU MTF, Trnava, 2017)
8. R. Beláni, *Vplyv vybraných parametrov Fused Deposition Modeling na kvalitatívne vlastnosti súčiastky (The influence of selected 3D printing parameters on quality characteristics of printed parts)* (STU MTF, Trnava, 2016)
9. R. K. Jaiswal, J. Bhagchandani, S. Agarwal, R. Mehrotra, A. Chaudhari, Journal of Dentofacial Sciences, **3**, 47, (2014)

10. S. Vijay, K. Jattib, N. Khedkarc, T. Singhd, 3rd International Conference on Materials Processing and Characterisation, **1625** (2014)
11. K. Thrimurthulu, P. Pandey, N. Reddy, International Journal of Machine Tools and Manufacture, **44**, 585 (2004)
12. Daekeon, K. Jin-Hwe, K. Soonman, S. Jungil, L. Seokhee, Journal of Materials Processing Technology, **209**, 5593 (2009)
13. J. Milde, L. Morovič, Research papers of the Faculty of Materials Science and Technology in Trnava, Slovak University of Technology in Bratislava, **24**, 73 (2016)
14. I. A. Popan, N. Bâlc, B. Luca, A. Popan, A. Carean, Applied Mechanics and Materials, **808**, 339, DOI: www.scientific.net/AMM.808.339, (2015)
15. J. Líška, J. Kodácsy, Advanced Materials Research, **325**, 381 (2011)
16. K. Líška, 3rd International Scientific and Expert Conference: *TEAM 2011*. **191** (2011)
17. S. Olivera, H. B. Muralidhara, K. Venkatesh, K. Gopalakrishna, Ch. S. Vivek, Journal of Materials Science, volume, číslo strany, DOI: 10.1007/s10853-015-9668, (2016)
18. N. Bâlc , C. S. Cosma, J. Kessler, V. Mager, *Applied Mechanics and Materials*, **808**, 199, DOI: http://www.scientific.net/AMM.808.199, (2015)
19. C. Ş. Borzan, P. Berce, H. Chezan, E. Sabău, S. A. Radu, M. Ridzoň, Academic Journal of Manufacturing Engineering, **11**, 108 (2013)
20. R. Pâcurar, A. Pâcurar, *16. chapter "Applications of the Selective Laser Melting Technology in the Industrial and Medical Fields",in the book entitled "New Trends in 3D Printing*" (IN TECH, Rijeka, 2016)
21. C.Ş. Miron-Borzan, A. Popan, V. Ceclan, A. Popescu, P. Berce, Applied Mechanics and Materials, **808**, 169, DOI: 10.4028/www.scientific.net/AMM.808.169, (2015)
22. D. Chakraborty, A. B. Reddy, A. R. Choudhury, *Computer-Aided Design,* **40**, 235 (2008)
23. F. Rayegani, G. C. Onwubolu, The International Journal of Advanced Manufacturing Technology, **73**, 509 (2014)
24. D. Paloušek, M. Omasta, D. Koutný, J. Bednář, T. Koutecký, F. Dokoupil, Optical Materials, **40**, 1, DOI: 10.1016/j.optmat.2014.11.020, (2015)
25. M. Vittek, Vplyv orientácie CAD modelu pri metóde FDM na rozmerovú a tvarovú presnosť súčiastky (Influence of CAD orientation in the process of Rapid Prototyping on geometrical and dimensional accuracy of components) (STU MTF, Trnava, 2016)

 DOI: 10.1051/matecconf/201713702007

Implementation of additive technologies in elementary education

Melita Milić[1,*] *Sven* Maričić[1], and *Donald* Radolović[1]

[1]University of Juraj Dobrila, Zagrebačka 30, 52100 Pula, Croatia

Abstract. The use of additive manufacturing (AM) in the wider sense at the level of elementary education is not sufficiently analysed in Republic of Croatia at the moment. Partially, CAD/CAM technologies requires far more specialized knowledge than what the Croatian curriculum provides. The application of different technological devices in the educational process, accelerates the teaching process, and makes it more interesting and more acceptable. The knowledge that pupil need to have at the end of schooling is rapidly changing. It is quite possible scenario that most of the knowledge gained during the elementary and secondary education will be obsolete before future student applies to the university. In order to achieve further progress, it is necessary to systematically explain all the possibilities of using these technologies from the elementary level. By developing different strategies, the young generation can focus on the knowledge they will need in their future jobs.

1 Introduction

Technology development has always led to various advances that have contributed to the improvement of living conditions largely. At present, new technological advances have been applied, some even call it the 4th technological revolution. Any change in this direction, besides improving the conditions of life and work, required the users to adapt to new conditions and new knowledge. Robots have already taken up their places in the households as technological devices [1, 2, 3]. It is no longer enough to turn it on or off, but to program it and adapt it to the user habits and needs.

2 Curriculum with iPad and STEM topics included

Taking into the consideration the fact that iPad started to be used in Croatian education for less than two years in the first class, there are many questions about the benefits and disadvantages that students will have when they will complete the elementary school. What it is wanted to be achieved? In consideration that students have not completed their

* Corresponding author: melita.milic@skole.hr

education cycle (Figure 1), and in all schools, the students do not use the new technology in a sufficient extent [4], the question is rightly whether students will take advantage of, or the use of early learning technology will make it more difficult for them to continue their education. It is considered that through this step, the school develops and strengthens the natural and technological interdisciplinary area and by doing that, it brand itself as well. The learning solution proposed by the authors is reinforcing students in acquiring the ability to learn and apply knowledge in later application.

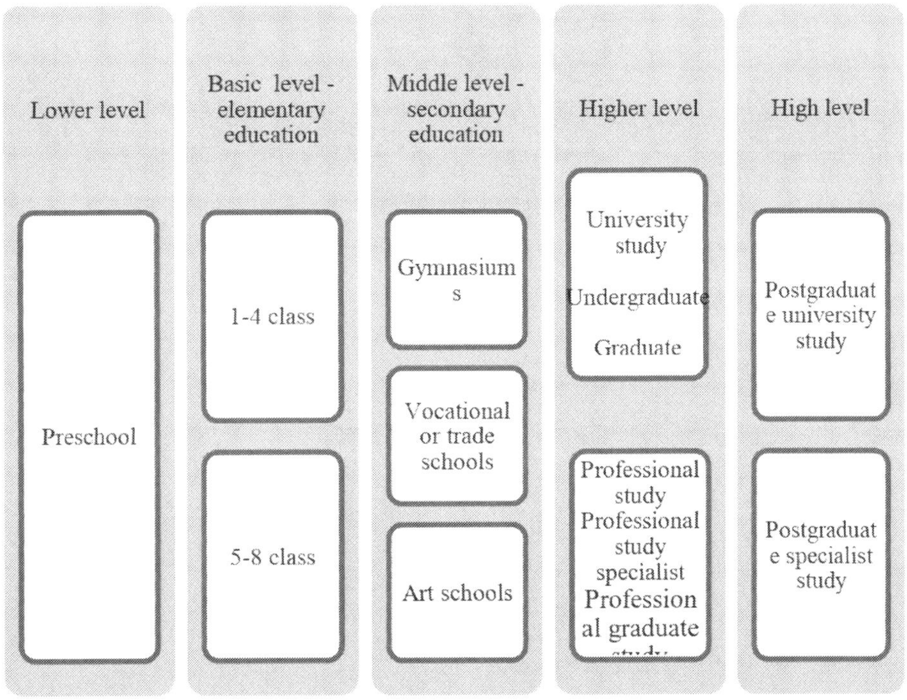

Fig. 1. Schematic overview of the educational system in Croatia. [5].

In the world, teaching with the everyday use of modern technology (computers, tablets, mobile phones) has become the standard. Different strategies have been developed, adapting content and ways of access to self-instruction. In Croatia, this trend is weaker, more sporadic and is left to directors, sometimes to teachers.

And while the usage of an iPad is popular topic and has enough material on using them in the class, the active use of additive technologies in teaching is somehow weaker. Therefore, it is even more important to design and prepare activities and areas of application. Teaching the STEM area is of particular importance in elementary schools. Various initiatives are making different programs for education as a supplement for official curriculum.

2.1 Questions for research

The Croatian Makers Project called on all educational institutions to respond to the equipment of their school with BBC micros bits in 2017. Significant number of universities joined and gained micro:bit as well as the vast majority of elementary schools. (Figure 2).

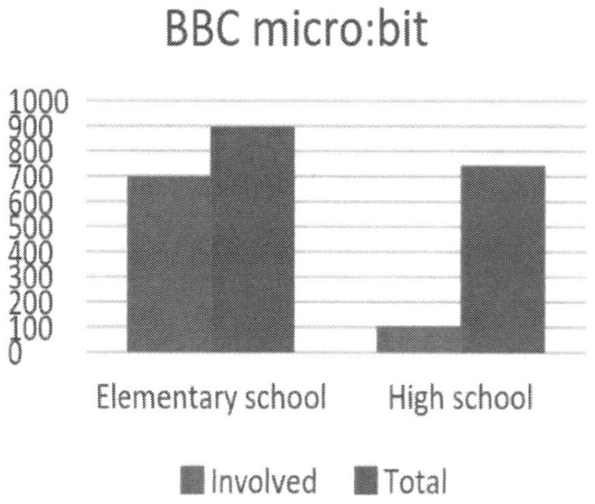

Fig. 2. Involved schools in BBC micro:bit [6].

In high school system the response was somehow weaker. The great response from elementary schools and faculties was the reason for reflection on the next change - the use of additive technology in teaching.

Reflecting on the use of additive technologies in Croatia, besides the current situation of school equipment, one should consider educating teachers to apply them. In this article, it was assumed the fact that all schools are well equipped. In order to make a good teacher education, it is necessary first to devise the area of application and curriculum. The proposed matter will be the use of additive technologies in teaching. In this connection, the following questions arise:

- In which other items could be applied the additive technology;
- What are the ways in which it could be applied?
- What could be positive side aspects related to:
    - Teachers
    - Student
- What could be side-by-side aspects related to:
    - Teachers
    - Student
- What conditions would have to be achieved in order to apply the additive technology in regular teaching (apart from equipping the school).

2.1.1 Subjects and how to do it?

The interdisciplinary construction of robots, which involves motors, sensors, and programming, makes it useful pedagogical tool for all STEM areas [2, 7]. Beside STEM subjects, robotics is usable in every subjects. Darling-Hammond [8] said that effective teachers encourage learning by:

- Creating ambitious and meaningful tasks that reflect how knowledge is used in the field;
- Engaging students in active learning, so that they apply and test what they know;
- Drawing connections to students' prior knowledge and experiences;
- Diagnosing student understanding in order to scaffold the learning process step by step;
- Assessing student learning continuously and adapting teaching to student needs;

• Providing clear standards, constant feedback, and opportunities for work;
• Encouraging strategic and metacognitive thinking, so that students can learn to evaluate and guide their own teaching.

2.1.2 Simple robotic arm design

To illustrate the practical example of what the content proposal and the design looks like, it is possible to describe the following case: CAD design consisting of several phases. In the first phase, the choice of software solution follows. The second phase consists of the design of basic elements, keeping in mind the basic rules for 3D printing. In practice, this means avoiding parts for which a larger amount of support is required in order to speed up the production. In the third stage, parts are analysed by means of a slicer program and made on one of the additive machines. Finally, in the fourth phase, a robotic arm is assembled and demonstration of the basic robotic principles are done. Figure 3 shows an example of a robust robotic arm used for educational purposes.

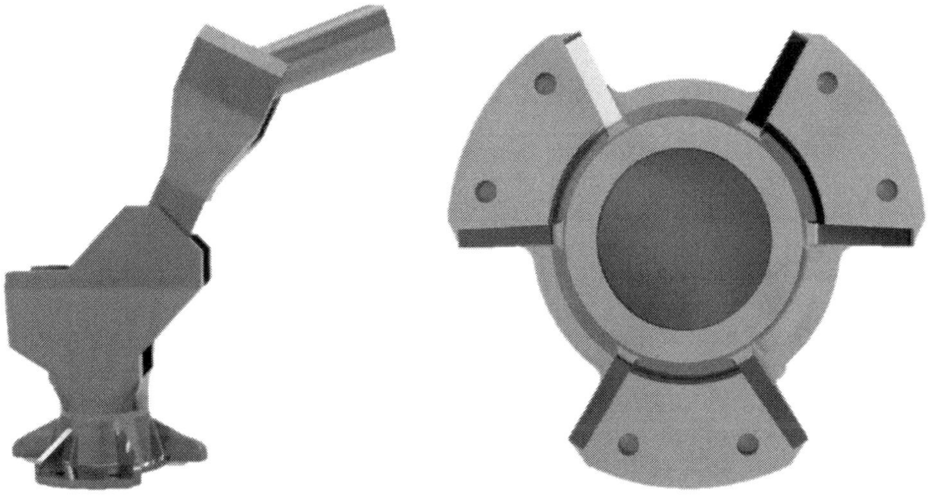

Fig. 3. Sample design of robotic arm modules and base housing.

2.1.3 Additive technology

Although the 3D prints are well known, the creation and application of a robotic hand, especially in education, is area with significant potential (Figure 4).

According to present trends [9], robotic education should be viewed as a means of stimulating important life skills. Educational robots can be divided into several roles: mentor, guardian, colleague, friend, and applied in different topics for learning: foreign language, science, math, physics, geometry and technology, meanwhile the use of the robotic hands for educational purposes needs to be implemented. It was assumed that the robot hand can be included in the curriculum of elementary school as an integral part of all subjects in school in some extent. For the purpose of this research, it was selected one obligatory subject in elementary school.

 DOI: 10.1051/matecconf/201713702007

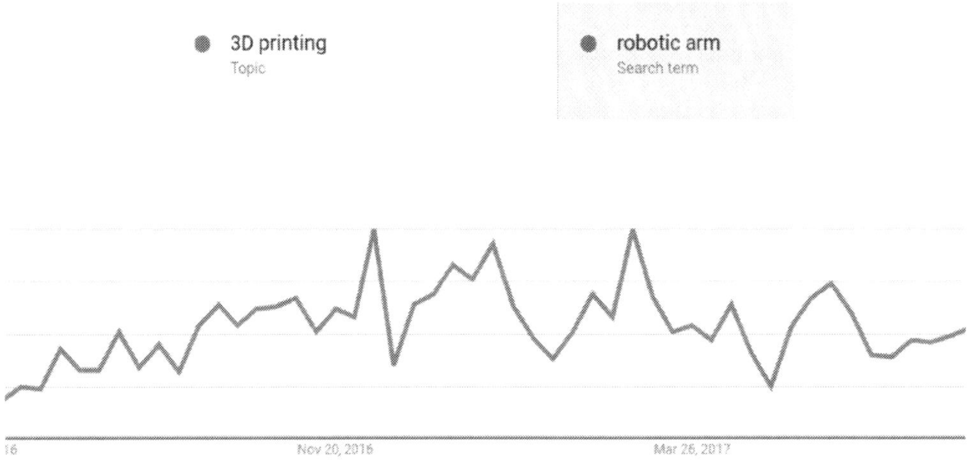

Fig. 4. Google Search Trend Lines.

Table 1. Curriculum options for applying a robotic system to elementary school program.

Class	**STEM domain**				
	numbers	algebra	shape and space	measures	data handling
1	+		+		
2	+		+		
3	+		+		
4	+		+	+	
5	+		+	+	
6	+		+	+	
7	+	+	+	+	+
8	+	+	+		+

As it is presented in the Table 1, there are many possibilities of application in all mathematical areas. In some mathematical areas, such as numbers and shape and space, it was considered that a robot arm can be applied in all classes. For some domain, such as algebra and data, it was considered it to have fewer options for application. Table 1 shows that a robot hand can be applied in all grades of elementary school in the subject of mathematics. If a robotic hand with AR or VR it is used [10], the possibilities are unlimited.

3 Conclusion

CAD/CAM design in additive manufacturing education system is of great importance. As an obious trend, a robotic systems design for 3D printing represents a challenge for the education system. Johnson [11] said that robotics is timeless. Impact of robotic arm to educational process will be shown in time. Like robots in education, the robotic arm and their application will largely depend on the support of key factors. If quality education is not provided to the teachers, all options possibly will not be utilized. In Croatia, educational robotics is extracurricular and optional for most of the children. By high positive response of all surveyed schools and faculties, education in area of robotics is one of the most popular topics to enter in each school in the curriculum. A logical upgrade is the use of a robotic arm in at least some subjects in schools in the near future.

References

1. H. Suto, IJCSE, **6**(2), 139-144 (2013)
2. A. Sergeyev, N.Alaraje, S. Kuhl, M. Meyer, ASEE, **2** (2015)
3. A. Sergeyev , TTIJ, **10**(3) (2010)
4. M. Pérez-Sanagustín, M. Nussbaum, I. Hilliger, C. Alario-Hoyos, R.S. Heller, P. Twining, C.-C. Tsai., CE, (2016)
5. *** https://mzo.hr/hr/rubrike/obrazovanje, (accessed 09.07.2017.)
6. *** http://croatianmakers.hr/hr/ustanove-stem-revolucije/, (accessed 09.07.2017.)
7. C. Kim, D. Kim, J. Yuan, R.B. Hill, P. Doshi, C.N. Thai, CE, (2015)
8. I. Darling-Hammond, CA, (2008)
9. G. Nikolić, AG, **20**(1-2), 36 (2016)
10. Z. Tang, J.WU, EMASS, (2016)
11. J. Johnson, AROB, **7**(1-2), 16-21 (2002)

Research Regarding the Manufacturing through AM Technologies of an Implant for Cervical Disc Replacement

Cristina Stefana Miron-Borzan[1,*], *Emilia* Sabau[1], *Mircea* Mera[1] and *Petru* Berce[1]

[1]Technical University of Cluj-Napoca, Manufacturing Engineering Department, 103-105 Muncii Bvd. Cluj-Napoca, Romania

Abstract. Worldwide, accidents are one of the main causes of illness in developed and undeveloped socio-economic countries. Additive Manufacturing (AM) technologies bring extremely useful and advantageous applications for the new neurosurgical procedures. Because the surgical insertion of devices for cervical disc replacement is very difficult, the development of new devices that can minimize these disadvantages, are needed. The aim of this paper was to improve the characteristics of an implant for cervical intervertebral disc replacement, a model based on an existing implant. A cervical cage designed for stabilization and arthrodesis between the cervical vertebrae was analyzed. A new design of a cage that have some improvements, useful for patient safety, as well as for facilitating the surgery was developed. The new proposed design was verified through Fine Element Analysis.

1 Introduction

The human body is a complex system composed of structured devices coordinated by the central nervous system. Each part of the human body must function at optimal parameters. The components of the human body are part of some complex subsystems and their optimal function makes the proper functioning of the other components. In the case of an accident or disease the intervention of a surgeon is required and a prosthesis can resume the activity of an organ, or part of the body [1].

The cervical cage is an intersomatic cage whose main function is to maintain the disc height and the radicular decompression by intersomatic distraction. Its design is favoring osteogenesis. It have a convex shape on its supper face and has openings on the upper and bottom faces. It comes in several heights so it is able to adapt to the variable morphology of the intervertebral spaces [2]. The criteria required to an ideal cage for cervical interbody fusion are the following: providing immediate stability, maintaining spinal alignment and foraminal height, achieving higher or at least equal fusion success rate, and obviating complications by using autograft [2].

Cervical total disc replacement has been accepted as a viable option for surgical management of cervical spondylosis or degenerative disc disease [3].

*Corresponding author: borzan_cristina@ymail.com

Performing a cervical discectomy by an interbody fusion is one of the surgical procedures for the treatment of cervical degenerative disc disease. Although high fusion rates have been reported with autogenous bone grafts, the immediate and long-term morbidity associated with an iliac crestharvestis well recognized [4].

The requirements for an appropriate cage height and shape must be estimated prior to surgery. Must been compared the X-ray template for Cage with the adjacent intervertebral discs on a lateral radiograph. With the segment fully distracted, the implant must fit tightly and accurately between the end plates. The final choice of height and shape will be made with the help of a trial implant during surgery. To achieve maximum segment stability, it is essential to implant the largest possible cage [5]. The sizes of an existing Cage are presented in Figure 1.

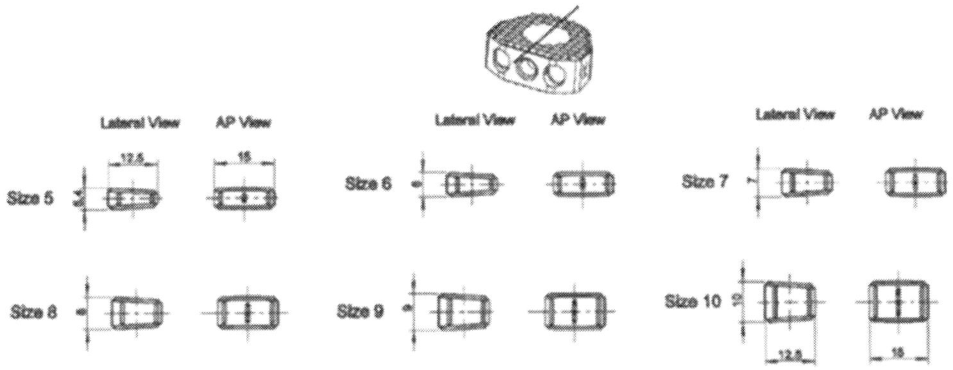

Fig. 1. The sizes for a Cage [5].

Currently, worldwide there are many companies that produce medical devices widely, used in new techniques of spinal neurosurgery.

Polyetheretherketone (PEEK) and Titanium (Ti) are two biocompatible materials that are used in manufacture of Cages through AM technologies. In the scientific literature there are many studies focused on PEEK and Ti cages. Initially, the threaded cylindrical titanium cages that became popular in the late 1990's helped the success rate of the procedure by providing more firm fixation of the disc space. Petter Ullrich [6] presents anterior interbody cages from titanium cylinders that are placed in the disc space. The cages are porous and allow the bone graft to grow from the vertebral body through the cage and into the next vertebral body. The cages offer excellent fixation, so most patients do not need additional instrumentation (e.g. pedicle screws) or post-operative back braces for support. Other authors [7-9] showed that Titanium or carbon fiber cages were widely used for cervical interbody fusion, but in some cases subsidence, migration, or structure failure had occurred.

PEEK cages have been widely used during the past decade in patients with degenerative disorders of the cervical spine [10]. The PEEK cages are biocompatible, radiolucent, and have modulus of elasticity similar to the bone [7].

Studies focused on clinical cases also showed that the actual used PEEK cages have a better behavior than pure Ti cages after implantation. A direct comparison of cervical Ti and PEEK cages in a clinical setting is very rarely found in the literature [11, 12, 13, 14], and even less studies consequently compare the radiological results [16, 18]. The latter studies showed the PEEK-implants being superior in maintaining cervical interspace height and achieving radiographic fusion [11, 13], even suggesting to cease the application of Ti-cages in cervical spine surgery [11, 15].

The Cages can be manufactured through AM Technologies. If the Ti powder is used, the recommended method is Selective Laser Melting process (SLM) and in the case of PEEK, the Selective Laser Sintering process (SLS). Selective laser sintering, 3-dimensional printing is currently used for industrial manufacturing of plastic, metallic and ceramic objects [16]. The SLS process is based on the realization of a CAD product by adding successive layers. The laser (carbon dioxide type) covers the entire surface of the section (point by point), sintering fine layer of material deposited on the working platform [17].

Selective Laser Melting (SLM) is an additive metals manufacturing technology with a presence in medical orthopedics, dental, aerospace, high technology engineering and electronics sectors. The process uses a high powered laser to fuse fine metal powders together, layer by layer direct from CAD data, to create functional metal parts [17].

The aim of this paper was to improve the characteristics of an implant for cervical intervertebral disc replacement in order to prevent its migration, to facilitate the surgery and to evaluate his behavior during extension of the head through Fine Element Analysis.

2 Materials and Method

Based on the analysis of the existing Cage implant was developed a new Cage, that have some improvements, useful for the patient's safety and to facilitate surgery. Initially a standard-type Cage, which kept the dimensions of an existing Cage was made (Figure 2). The design was done in Solid Works software with the maximum dimensions required for a Cage- size 10.

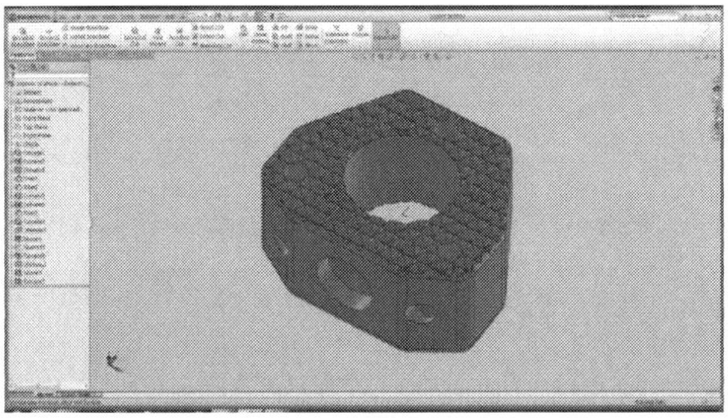

Fig. 2. The proposed model for analyze.

To prevent the migration after implantation, the new proposed implant has two "ears" for fixing the Cage on neighboring vertebrae. The implant can be fixed with the help of two self-tapping screws, from Titanium Alloys, biocompatible. The new model presents holes in correlation with the dimensions of the screws' head. In order to verify the new model, the implant was manufactured from pure Ti using SLM process (120 W laser power and 500mm/s scan speed).

In the scientific literature, there are studies made in the field of stresses from the cervical spine. One of the authors [18], showed that the weight on the spine dramatically increases when flexing the head forward at varying degrees. An adult head weighs 10 to 12 pounds in the neutral position. As the head tilts forward the forces seen by the neck surges to 27 pounds at 15 degrees, 40 pounds at 30 degrees, 49 pounds at 45 degrees and 60 pounds at 60 degrees. At 90 degrees the model prediction was not reliable.

For analyzing the forces in the cervical spine, taking into account flexing the head, the pounds were transformed into Kilograms [Kg] and after in Newtons [N] (table 1).

Table 1. The forces in the cervical spine taking into account flexing the head forward

Position	Neutral	15	30	45	60
Force applied to the Cervical Column [N]	45-54	122	181	222	271

 DOI: 10.1051/matecconf/201713702008

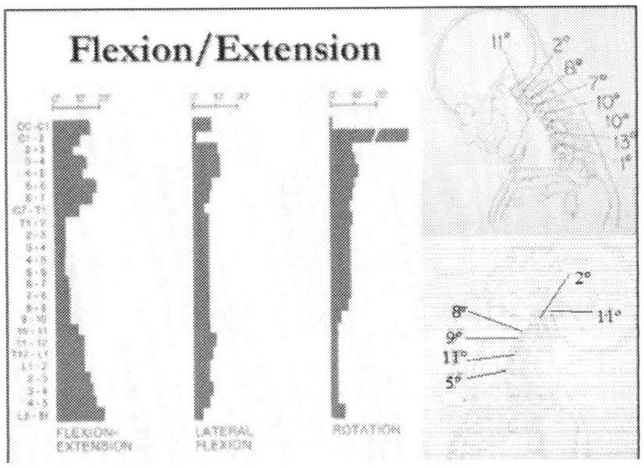

Fig. 3. The flexing positions and angles for the human neck [19].

We considered the theoretical case, when the flexing and extension forces are equal, figure 3. In order to analyze the forces and the stresses in the fixing area of the implant, a FEA analyze was made in the ANSYS Software. The analysis was done in the case of the maximum load (271 N), for pure Ti and PEEK, materials that can be used in the Cage manufacturing. The mechanical properties for PEEK parts manufactured through SLS are: Density 1310 kg/m^3, Tensile Modulus 4250MPa, Tensile Strength 90MPa, Strain at break 2,8 % [20] and for pure Ti, manufactured at 120 W, the mechanical properties are: Density 4510 kg/m^3, Tensile Strength 440MPa, Young Modulus 100GPa, Poisson's Ratio 0.35 [21].

3 Results and Discussions

Nowadays, numerical simulation represents a very effective method for simulating various processes including all common engineering materials [22]. The Finite Element Analysis (FEA) is a numerical solution of the problems that can be expressed in mathematical language in the form of partial derivative equation systems. The meshing concept is found at FEA base. In very intuitive terms, the meshing operation is a division of the spatial domain occupied by the analyzed body into regions of limited sizes known as finite elements. At present, there are a large number of commercial programs for finite element analysis. Among the best known are ANSYS, ASTRAN, ALGOR, ABAQUS etc.

For this research the ANSYS 16.0 Software was used. The first proposed objective was to analyze and to determine the stress state and the deformation of the implant. The following kinematic restrictions were imposed on the model, namely: applying forces on the right and left sides of the implant (where will be the contact area with the screws), like in the Figure 4. The left side and the right side of the piece takes up a force of 135,5 N (traction). We considered that the value strength of 271 N is equal taken by the both sides. Finite elements were used in order to meshing the implant, the thickness of the elements were of 0.50 mm (the same size on the entire surface of the model), Figure 5.

Figure 6 shows the distribution of the von Mises equivalent stress in the implant model. It can be seen that the maximum value for the pure Ti material was 35,182 MPa (Figure 6, a) and the maximum value for the PEEK: 34,889 MPa (Figure 6, b).

The difference between these two materials is very small, about 0,293 MPa. We can conclude, that from the point of view of stress trials, this two used materials have a good behavior.

In the case of the total deformation for the pure Ti material the maximum value was 0,009 mm (Figure 7, a) and the maximum value for the PEEK was 0,233 mm (Figure 7, b).

Fig. 4. Kinematic restrictions

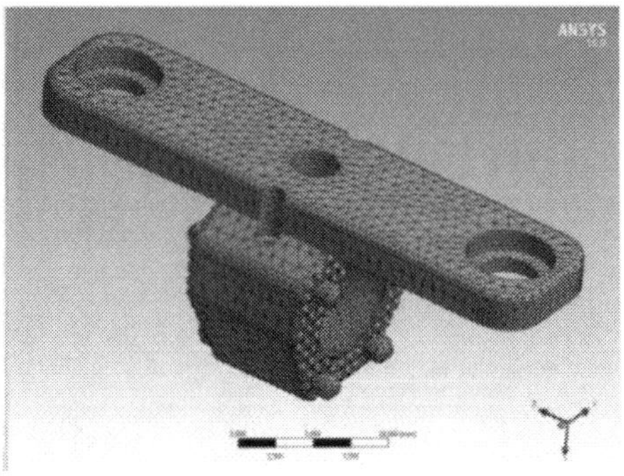

Fig. 5. Finite element network

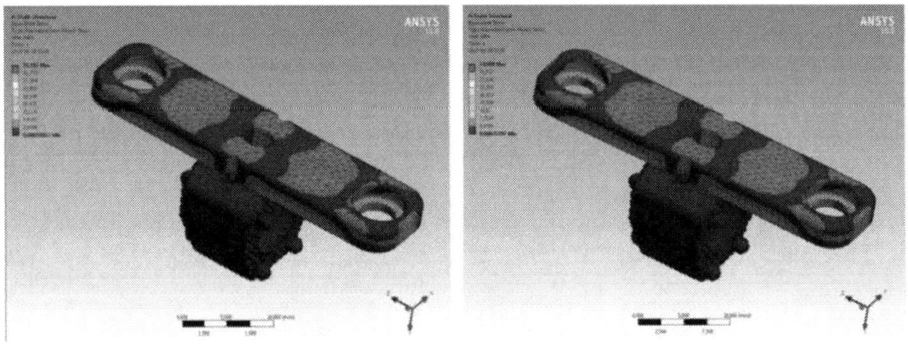

a. Pure Ti material b. PEEK material

Fig. 6. Distribution of the von Mises stress

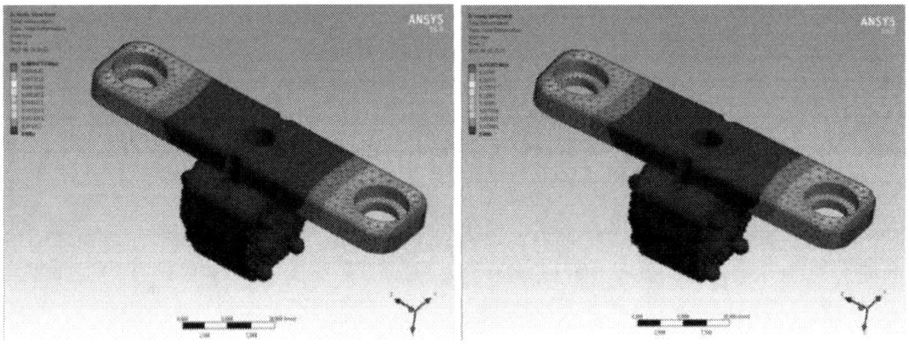

a. Pure Ti material b. PEEK material

Fig. 7. Total deformation.

The difference between these two materials is about 0,224 mm. In this case the deformation for the PEEK material is larger than in the case of pure Ti material. We can conclude that for this model, the pure Ti is more rigid than the PEEK material which has a higher elasticity. PEEK has a better behavior for this kind of application and for the anatomical structure taking into account his modulus of elasticity similar to the bone. For the proposed design of the cage even if pure Ti cages have a good resistance and behavior from the point of view of stresses, PEEK cages seems to be more close to the required anatomical structure in cervical area. The presented example demonstrates the possibility of implementing the proposed model in the structure of finite element analysis programs. If the material resistance is undersize or oversize, we will also optimize the shape. The decision of optimization will be taken by reporting to a safety factor of the von Mises stress C = 1,50.

4. Conclusions

A new implant for cervical disc replacement (cage) was developed. The new proposed model was manufactured through SLM process and has some improvements, helpful for the patient's safety and for facilitating the surgery. This paper analyzed implants from pure Ti and PEEK using the ANSYS 16.0 Software. The analysis proposed to determine the stress state during extension of the head and the deformation of the implant. We considered that the value strength of 271 N is equal took by the both sides of the implant. The left and the right side (ears) of the piece takes up a force of 135,5 N (traction). From the point of view of stress trials these two used materials have a good behavior. The pure Ti is more rigid than the PEEK material, but PEEK has a higher elasticity. For the proposed design of the cage even if pure Ti cages have a good resistance and behavior from the point of view of stresses, PEEK cages seems to be more close to the required anatomical structure in cervical area. Also, the presented example demonstrates the possibility of implementing the proposed model in the structure of finite element analysis programs.

This work was supported by a grant of the Romanian National Authority for Scientific Research and Innovation, CNCS/CCCDI–UEFISCDI, project number PN-III-P2-2.1-BG-2016-0216, within PNCDI III.

References

1. C. Borzan, *Manufacture of Customized Implants from Biocompatible Materials* (Master Thesis, UTCN, Cluj-Napoca, 2011)
2. Corbel Instructions, Établissement SpineNetwork, Surgiview, France
3. T.-H.Tu, J.C.Wu, L.-Y.Fay, C.-C.Ko, et all, J Neurosurg Spine, **16**, 211 (2011)
4. M. Y. Chang, M. H. Chen, C. J. Chang, J. S. Huang, FJS, **46**, 4 (2013)
5. Synthes, Cervios and CerviosChronOS, Technique Guide
6. P. F. Ullrich, Spine-Health, (2004)
7. J.C. Liao, C. C.Niu, W.-J. Chen, L.H. Chen, IntOrthop, **32**, 5 (2008)
8. C.C. Niu, L.H. Chen, P.L. Lai, T.S. Fu, W.J. Chen, Chang Gung Med J.,**28**, 4 (2005)
9. I Haven, PJ Loon, RH Bartels, JL Susante ActaOrthop Belg., **71**, 5 (2005)
10. F. Roel, M.R.Kersten, S.M.van Gaalen, et all, The Spine J., **15**, 6, (2015)
11. Y.C. Chou, D.C. Chen, W.A. Hsieh, W.F. Chen, et all, J Clin Neurosci. **15**, 11 (2008)
12. U. Meier, D. Kemmesies, Orthopade, **33,** 11 (2004)
13. C. Niu, J.C. Liao, W.J. Chen, L.H. Chen, J Spinal Disord Tech., **23**, 5 (2012)
14. G. Matge, Acta Neurochir, **144**, 6 (2002)
15. M. Cabraja, S. Oezdemir, D. Koeppen, S. Kroppenstedt, BMC Musculoskelet Disord, **13**, 172 (2012)
16. F. Fina, A. Goyanes, S.Gaisford. A. W.Basit, Int. J. of Pharmaceutics (2017)
17. P.Berce, N. Bâlc, ş.a. *Fabricarea Rapidă a Prototipurilor* (EDP Tehnică, Buc., 2000)
18. K. K. Hansraj, Neuro and Spine Surgery, Surgical Techn. Int. **XXV** (2014)
19. *** http://web.iitd.ac.in/~achawla/public_html/736/12-biomech-I-head-neck-v2.pdf
20. EOS GmbH - Electro Optical Systems, EOS PEEK HP3, Material Data Sheet
21. A. Ispas, C. Cosma, et all, J Optoelectron Adv M, **18**, 9-10 (2016)
22. P. Bella, P.Buček, M.Ridzoň, M. Mojžiš, L.Parilák, Key Engineering Materials, **716** 708 (2016).

Influence of the volume ratio of solid phase on carrying capacity of regular porous structure

Katarina Monkova[1,*], *Peter* Monka[1,2], *Jozef* Tkac[1]

[1]Technical University of Kosice, Faculty of Manufacturing Technologies with the seat in Presov, Sturova 31, 080 01 Presov, Slovakia

[2]Regional Technological Institute, University of West Bohemia, Univerzitni 8, 306 14 Pilsen, Czech Republic

Abstract. Direct metal laser sintering is spread technology today. The main advantage of this method is the ability to produce parts which have a very complex geometry and which can be produced only in very complicated way by classical conventional methods. Special category of such components are parts with porous structure, which can give to the product extraordinary combination of properties. The article deals with some aspects that influence the manufacturing of regular porous structures in spite of the fact that input technological parameters at various samples were the same. The main goal of presented research has been to investigate the influence of the volume ratio of solid phase on carrying capacity of regular porous structure. Realized tests have indicated that the unit of regular porous structure with lower volume ratio is able to carry a greater load to failure than the unit with higher volume ratio.

1 Introduction

Since the beginning of the 1980s additive manufacturing evolved from the first processes for the rapid production of prototypes into a number of different processes of which some are capable of direct part production. Today processes like selective laser melting (SLM), selective laser sintering (SLS) and with some limitations fused deposition modelling (FDM) are capable to produce direct parts in end-user quality out of metal or thermoplastics. Additive manufacturing processes are technologically specified for industrial production and due to a rising competition between service providers additive manufacturing becomes economically feasible for a growing number of industrial and end-user applications. [1]

Metal AM is an 'additive' technology that works by fusing together very fine layers of metal powder using a focused laser beam. This powder fusion process can produce complex geometries which might not have been possible using traditional manufacturing techniques. The process begins by slicing the CAD model to obtain a 2D contour at each level of the build axis, which is fixed as the z-axis. Starting from the base 2D contour, a user defined slice thickness is added cumulatively at successive slicing planes. This layer by layer

* Corresponding author: katarina.monkova@tuke.sk

stacking gives rise to an error called the staircase effect which diminishes the surface finish of the part. It is the preferred method of 3D rapid prototyping for complex geometries that allows users to reduce the time and cost by completing multiple prototypes directly from CAD data. From a design perspective, the challenge of additive manufacturing is to understanding the limitations and opportunities of these new processes and to use them in the right applications. [2-4]

Scientists at the Massachusetts Institute of Technology have discovered new ways to use complex geometry to make strong, lightweight shapes that might be useful as structural materials for airplanes, automobiles and buildings. Authors´ research deals with producing one type of regular porous structures – so called gyroid structure. [5]

2 The samples of regular porous structure - gyroid

Generally, porous structures are divided into two categories: regular and unregular - also called foams. The regular structures are better controlled in of mechanical engineering industry. Basic building units of geometrically defined porous structures are cells that are created by simple bodies, e.g. cube, cuboid, cylinder or sphere. These geometries are usually well modelled in CAD applications. More complex structures have their geometries created by curved surfaces. The scientists have found an inspiration related to the regular porous structure in nature. Many natural structures such as bone and some corals are built from strong microstructure building blocks. Using microscopes with three-dimensional nanoscale resolution, Yale University researchers found that shades of green in the wings of five butterfly species are produced by crystalline structures called gyroids. In the Fig. 1 is presented wing scale photonic nanostructure that was base of three steps towards to the virtual 3D model creation. [6, 7]

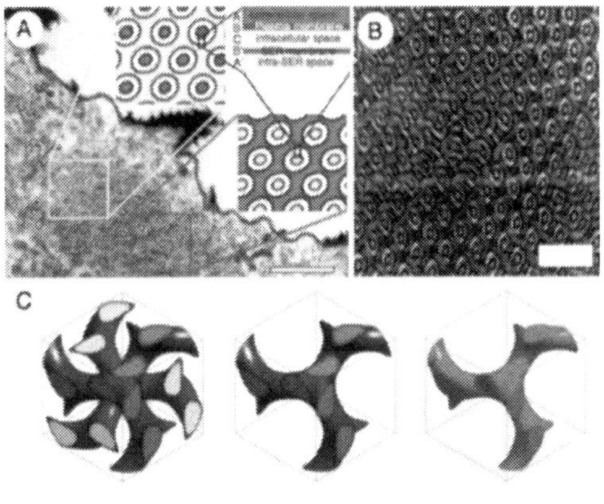

Fig. 1. Wing scale photonic nanostructure, from electron microscope to model. [7]

The shape of Shoen gyroid structure are described by mathematical formula (1) [8]

$$sin\left(\frac{2\pi}{L}x\right)cos\left(\frac{2\pi}{L}y\right) + sin\left(\frac{2\pi}{L}y\right)cos\left(\frac{2\pi}{L}z\right) + sin\left(\frac{2\pi}{L}z\right)cos\left(\frac{2\pi}{L}x\right) - t = 0, \quad (1)$$

where factor L influences the size of basic cell and parameter t decides about cross section of the strut, what influences the relative density of structure. Next parameter - volume ratio Vr of solid phase expresses how many percent of the cell space is filled by material. It is given by equation (2): [9]

$$V_r = \frac{Volume\ of\ solid\ phase}{Total\ volume} 100 \quad [\%]\,. \quad (2)$$

The samples for testing were produced by DMLS technology using EOS M290 machine with default technological parameters (laser power 400 W, diameter of laser beam 100 μm). Material of samples was the Maraginsteel MS1. The particles of the powder were spherical with the diameters in the range of 15-50 μm. (Fig. 2) The difference in grain size is welcome, because smaller particles can fill in the space between bigger grains. [10, 11]

Fig. 2. Metal powder of Maraginsteel MS1 [10]

Conventional steel 1.2709 according to the European Standard has excellent mechanical properties, high yield strength and tensile strength, is stretchable and resistant to dynamic stress. In the case of additively sintered material, it has reduced ductility and resistance to dynamic stress. It can be cured with a simple heat treatment to an approximate hardness of about 54 HRC. Table 1 provides the chemical composition of the starting material of the EOS Maragingsteel MS1 powder. [12]

Table 1. Chemical composition of EOS Maragingsteel MS1 [12]

Component	Fe	Ni	Co	Mo	Ti	Al	Cr, Cu	C	Mn, Si	P, S
[%]	remnant	17-19	8.5-9.5	4.5-5.2	0.6-0.8	0.05-0.15	≤0.5	≤0.03	≤0.1	≤0.01

Preliminary experiments have been aimed at assessing the ability of additive technology to print gyroid structures without supporting structures. All the samples were designed by means of software Siemens NX 10 CAD, where they were edited and adapted to additive manufacturing.

The specimens with basic cell, sizes ranging from 3 mm to 8 mm with a volume of 15 %, were printed. All these samples (with constant volume ratio of 15 %) were printed without defects, thereby confirming the self-supporting property of Gyroid within a given range of sizes. Similar results were obtained by [13], whose production equipment was able to print the smallest aliquots of the tested series, having a cell size of 2 mm and a volume of 15 %.

In this phase of preliminary experiments, next set of samples with constant cubic cell size of 6.5 mm and volume ratios 13.5 %, 9.5 %, 7.3 %, 5.8 %, 5.0 % and 4.2% were printed. They are shown in Fig. 3. [14]

Fig. 3. Set of test specimens and the sample with cubic cell size of 6.5 mm, volume ratio 9.5 % [14]

However, this preliminary experiment had some specifications. There have been some problems with deformation of samples during their production or during preliminary tests, which were necessary to solve.

3 Some problems of samples production

The first problem was related to the fact that some porous structures have deformed near the surface, where the sample was separated from a platform, see Fig. 4. [14]

Fig. 4. Deformation of samples during preliminary pressure tests [14]

It was the reason why the basic structure (a deformation core) of sample has been equipped with two 2.5 mm high contact plates on both joining surfaces. The insertion of the end struts should support deformation throughout the porous structure during pressure tests to avoid local deformity at the point of contact with the test device head. Parallelism of the bearing surfaces should be ensured. A sample model of the test body from perspective is shown in Figure 5.

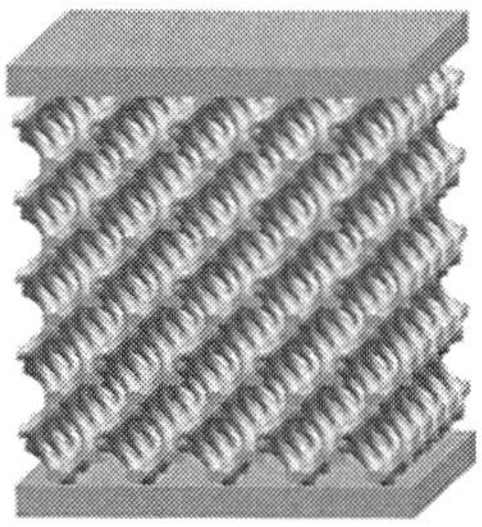

Fig. 5. The model of sample with bottom and top plate of continuous solid material

Figure 6a shows a model of a test body whose self-supporting function at the edges was disturbed. The deformation near the area of separation could be caused by stiffness reduction due to bending of strut structure during the cutting of sample with a band saw. To solve this problem, in relevant direction next one cell to the basic matrix of the sample was added and the sample was cut from support so to be its self-supporting function was kept. The sample with fully self-supporting function is shown in (Fig. 6b). [15]

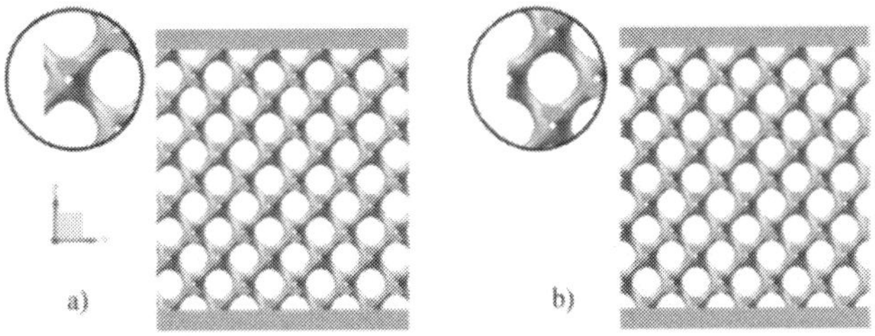

Fig. 6. Self-supporting function of samples, a) disturbed, b) kept [15]

 DOI: 10.1051/matecconf/201713702009

In next phase of preliminary experiments the effect of orientation of gyroid structure on the machinability was investigated.

The lack of practical knowledge gave rise to another experiment in which the various orientations of the pressure samples during the additive production were evaluated. Figure 7 shows different position variations. The sample in Fig. 7a was printed in the vertical position. Subsequent rotation around the *y* axis by 5° was given a position for another sample shown in Fig. 7b. Larger rotation of the sample would considerably complicate the design of the support structures, as some portions of the porous core struts would need to be deposited with the support elements so that these critical parts did not begin to build in the bulk base of the production powder, resulting in defects in the porous core construction.

The last variant was produced in a horizontal position, see Fig. 7c. The specimen was extended by one row of cells at the point of contact with the platform to provide sufficient space for the cut during separation of the samples from the platform by the band saw. After cutting, the excess struts were manually cut off so that samples with a cubic porous core could originated. [14]

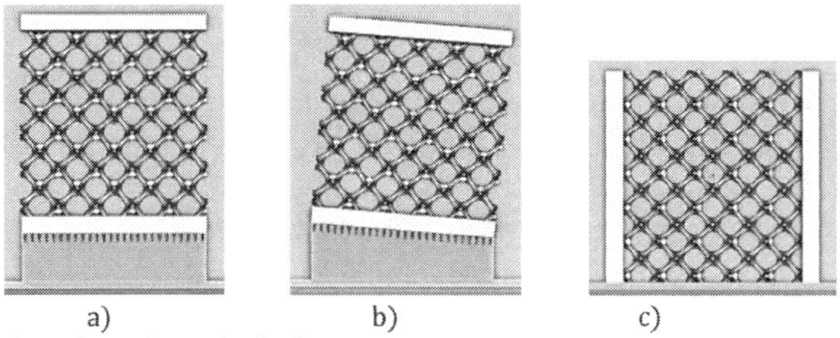

Fig. 7. Orientation of tested samples [14]

Due to the fact that the structures positioned in version c), at which the continuous layers of material were on sides of structure (on left and right sides), had no imperfections, they were smooth and reliable in every manufacturing repetition, the horizontal position of sintering was evaluated as the most suitable for samples production, as it is displayed in Fig. 8.

Fig. 8. Horizontal position as the best position for structure production

4 Experiments

The goal of test was the investigation of influence volume ratio on carrying capacity. Based on preliminary experiments, the sample size with five cells in one direction was set for experiments, because of the balance between the accuracy of the measured values and the economic indicator (the production cost per unit of sample). In the Fig. 9 is presented the gyroid structure at the same cell size, but with various volume ratios.

 DOI: 10.1051/matecconf/201713702009

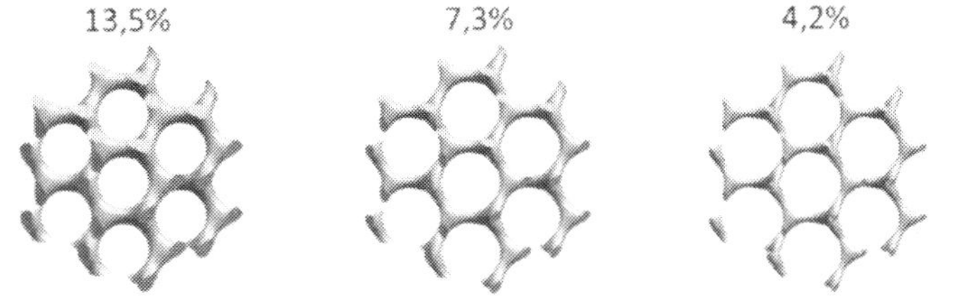

Fig. 9. The gyroid structure with various volume ratio

The samples were loaded up to the pressure that after exceeding the yield point of the material causes the irreversible plastic deformation. The test is finished if the set compression is achieved, or if the sample appeared significantly destructive conditions. The deformation characteristics, such are yield strength and maximum power load or load capacity have been evaluated from pressure diagram. Equipment Zwick/Roell Z250 with force capacity 250 kN was selected for the test. The test specimens must not be subjected to a pressure deformation prior to the test. However, overloading may be necessary to eliminate clearance between the test and the test body, so the test samples were preloaded with a force of 10 N. The compression speed of samples was 0.1 mms^{-1} at the environment temperature of 20 °C. The results of tests are presented in Fig. 10. [14]

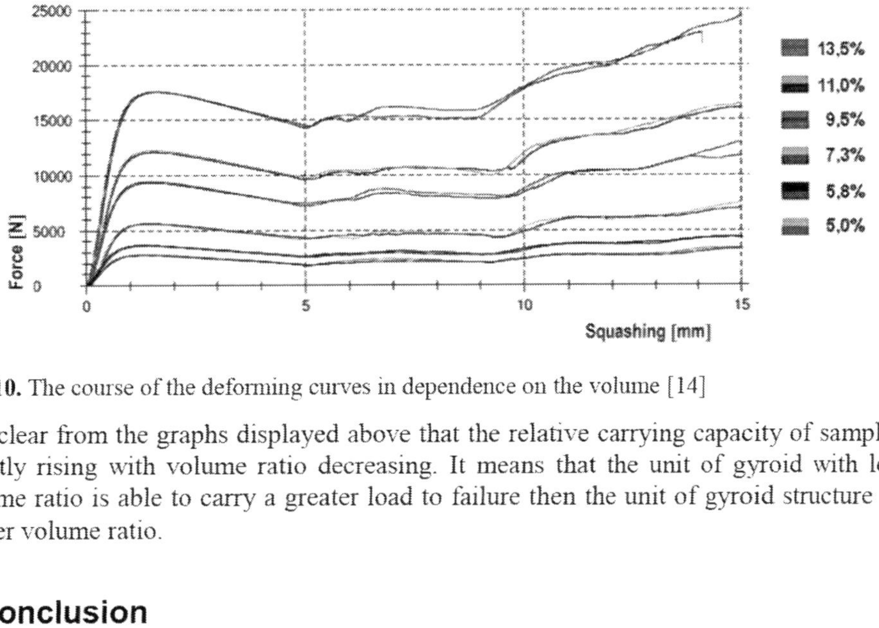

Fig. 10. The course of the deforming curves in dependence on the volume [14]

It is clear from the graphs displayed above that the relative carrying capacity of samples is slightly rising with volume ratio decreasing. It means that the unit of gyroid with lower volume ratio is able to carry a greater load to failure then the unit of gyroid structure with higher volume ratio.

5 Conclusion

Direct metal laser sintering is the preferred method of 3D rapid prototyping for complex geometries that allows users to reduce the time and cost by completing multiple prototypes directly from CAD data. From a design perspective, the challenge of additive manufacturing is to understanding the limitations and opportunities of these new processes and to use them in the right applications. [16]

Authors in the article have tried to point on some effects that influence the manufacturing of regular porous structures in spite of the fact that input technological parameters were set the same as the default parameters of the machine. These parameters are: geometry, slope of the sample and its positioning during laser sintering. Realized pressure tests also have indicated that the unit of gyroid with lower volume ratio is able to carry a greater load to failure than the unit of gyroid structure with higher volume ratio.

The present contribution has been prepared under projects VEGA 1/0614/15, KEGA 087TUKE-4/2015, KEGA 007TUKE-4/2018 and LO1502 'Development of the Regional Technological Institute' under the auspices of the National Sustainability Programme I of the Ministry of Education of the Czech Republic aimed at supporting research, experimental development and innovation.

References

1. D. Paramita et al.:, Procedia Manufacturing, **1**, 343 - 354 (2015)
2. J. Dobransky, et al.:, Key Engineering Materials, **669**, 532 - 540 (2016)
3. M, Cloot et al.: Assessing new support minimizing strategies for the additive manufacturing technology SLM. Solid Freeform Fabrication Symposium (2013)
4. S. M. Hur et al.: Determination of fabricating orientation and packing in SLS process. Journal of Material Processing Technology 236 - 243 (2001)
5. Beno, P., Kozak, D., Konjatic, P.:, Tehnicki Vjesnik, **5**(21), 1051 - 1055 (2014)
6. F. Ferroudji et al.,Inter. J.of Mechanical & Mechatronics Engineering IJMME/ IJENS, **04**, 92 - 97 (2014)
7. S. H. Masood, W. Rattanawong, Int. J. of Advance Manuf. Technol., **19**, 209 - 216 (2002)
8. V. Majstorovic et al., Procedia CIRP, **25**, 27 - 32(2014)
9. D. T. Pham et al., Int. J.of Advance Manufacturing Technology, **15**, 674 - 682 (1999)
10. P. Baron, et al., J. of the Inter. Measurement Confederation, **94**, pp. 245 - 253 (2016)
11. S. Allen, D. Dutta, Solid Freeform Fabrication Symposium, Austin, 259-269 (1994)
12. A. Panda, et al.: *Monitoring and Evaluation of Production Processes: An Analysis of the Automotive Industry*, Springer International Publishing (2016)
13. K. Monkova, et al., Advances in Mechanical Engineering, 1-11 (2014)
14. P. Hanzl, M. Zetek, I. Zetkova:, Proceedings of the 26th DAAAM, 748 - 752 (2015)
15. P. Hanzl, I. Zetkova, J. Mach, Manufacturing Technology, **1**, 34 - 38 (2017)
16. M. Ungureanu et al., Procedia Engineering, **149**, 495 – 500 (2016)

Finite Element Analysis to determine the optimum contact pressure between the components of a hip implant made by using the Selective Laser Sintering and the Selective Laser Melting Technologies

Răzvan Păcurar[1,*], *Ancuţa* Păcurar[1], and *Anna* Petrilak[1]

[1]Technical University of Cluj-Napoca, Faculty of Machine Building, Department of Manufacturing Engineering, B-dul Muncii no. 103-105, 400641, Cluj-Napoca, Romania

Abstract. The article presents a case study that was realized at the Technical University of Cluj-Napoca (TUC-N) in the field of customized medical implants made by using different additive manufacturing (AM) technologies, such as the Selective Laser Sintering (SLS) and the Selective Laser Melting (SLM). Finite element analysis was successfully used to determine the mechanical behavior of an acetabular liner that was made from PA 2200 powder material by SLS and to determine the optimum technological parameters required to be used in the manufacturing process of the acetabular liner by SLS, so as the contact pressure between the acetabular liner made from PA 2200 material by SLS and the femoral head made from TiAl6V4 material by SLM will be optimum at the end and the displacement between the components will remain in the admissible limits.

1 Introduction

Selective Laser Sintering (SLS) and Selective Laser Melting (SLM) technologies are two of the most developed Additive Manufacturing (AM) technologies which nowadays are widely used for producing medical implants (e.g. hip implants) [1,2]. Total knee and hip replacement arthroplasty are commonly used procedures in which the knee or hip joint is replaced by an artificial joint [3]. The success of the engineered artificial joint depends on the biomechanical and material properties [4-6]. A total hip replacement is shown in Fig. 1 and is composed of an acetabular component, which usually is made of metal (titanium alloys etc), a Plastic Liner (made of polymers), a femoral head and a femoral stem that are usually made of the same material as the acetabular cup [7]. One of the most important challenges regarding AM technologies is to know how to control the process parameters to obtain the adequate characteristics of a part from mechanical, thermal and biomedical point of view [8,9]. The orientation of the parts during the manufacturing process has also a high influence on the accuracy and characteristics of the realized medical implants, as well [10].

*Corresponding author: razvan.pacurar@tcm.utcluj.ro

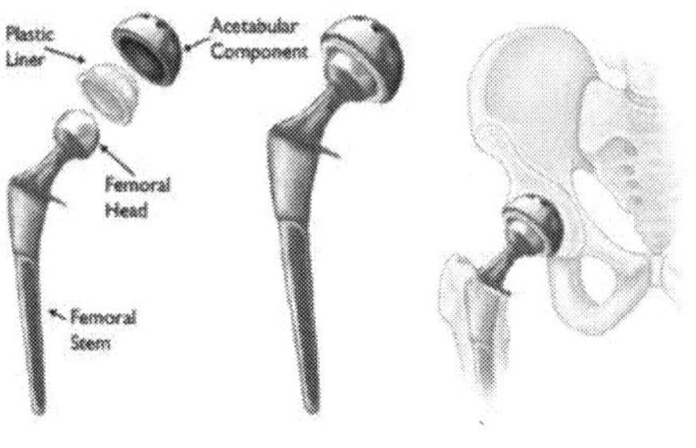

Fig. 1. The individual components of a total hip replacement [7].

Finite element analyses were performed using dedicated FEA software programs, such as ABAQUS, ANSYS, etc. to simulate the positioning between the femoral head and polyethylene liner, which is shown in Fig. 1. All the material were modeled as homogenous, isotropic âand linear elastic, excepting the plastic liner which was considered as being a non-linear elasto-plastic [11]. The contact areas on the articulating surface were located about the superior region of the liner. The conclusion of the made analyses was that in the case of a microseparation equal with 2000 μm, the contact pressure is centralized at the rim of the liner, leading to a high contact area at the rim liner and elevated stress which can cause severe plastic deformation of the liner. The microseparation distance and the angle and positioning orientation during the manufacturing process on the machines have a significant influence on the von Misse stress and contact pressure that results at the end [12]. This paper presents results that were obtained at the Technical University of Cluj-Napoca (TUC-N) related to the Finite element analysis that was made in the case of a total hip replacement by using Solidworks Simulation FEA program in order to determine the stresses and contact pressure that occur in the case of an acetabular liner made of PA 2200 material by SLS and an acetabular cup made of TiAl6V4 material by SLM, components that were manufactured at the end at TUCN using dedicated AM equipment.

2 Steps of the Finite Element Analysis

The contact pressure analysis of total hip replacements was realized using the 3D models that were designed by using the SOLIDWORKS program as presented in Fig. 2. The dimensions of the Ti6AlV4 alloyed acetabular cup that was made at TUCN by SLM were measured and transposed to the 3D model. The outside diameter of the acetabular cup (marked with 1 in Fig. 2) was 56 mm, the inside diameter being 46 mm. The acetabular liner dimension (marked with 2 in Fig.2) made from PA 2200 material by SLS and the femoral head (marked with 3 in Fig. 2) made from TiAl6V4 by SLM were chosen according to the specifications of the Waldermark Link Company from Germany [13]. The plastic acetabular liner has been designed in consequence having the outside diameter of 46 mm, the inside diameter being designed as having a dimension of 32 mm. The femoral cup is a 31 mm spherical component. In order to simplify the FEA model, the 3D models have been designed as simple hemispheres as shown in Fig. 2 (without lattice structures on the surfaces of the acetabular cup and without reinforcement on the acetabular liner outer surface. In order to determine the mechanical behavior of the acetabular liner made of PA 2200 material, a static study was performed using SOLIDWORKS Simulation program.

After the character of the study was defined, the materials of the individual components were set, with Linear Elastic Isotropic characteristics, specifying as following: for both, the acetabular cup and femoral head made by SLM from Ti6Al4V material, the mechanical and thermal characteristics (e.g. Young Elastic modulus E=104 GPa, the coefficient of Poisson ξ=0.31) were selected as they were found within SOLIDWORKS Simulation library.

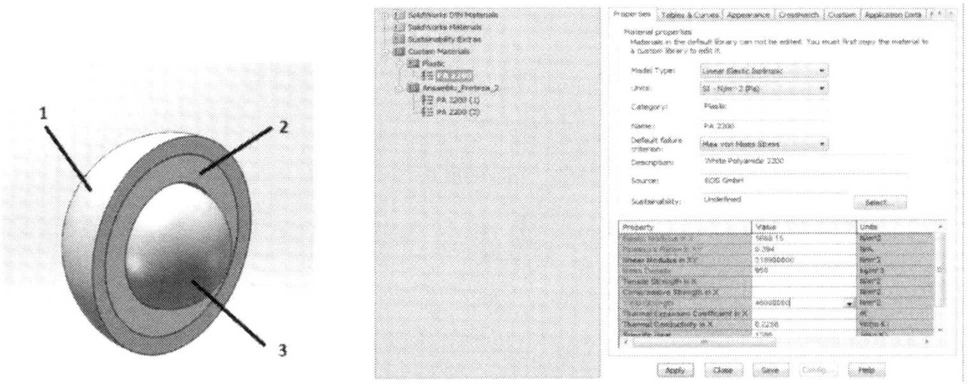

Fig. 2. FEA model of hip replacement. Definition of PA2200 characteristics in SolidWorks.

Further on, due to the fact that the type of material the plastic liner is made from (PA 2200) was not found within the Solidworks Material Library feature, the definition of a new type of material was required to be realized, with the mechanical and thermal characteristics determined based on experimental tests that were made at TUCN by using different samples, required for tensile or compressive test experiments (see Fig.2).

The next step in FEA analyses was the definition of the boundary and displacement constrains (see Fig.3).

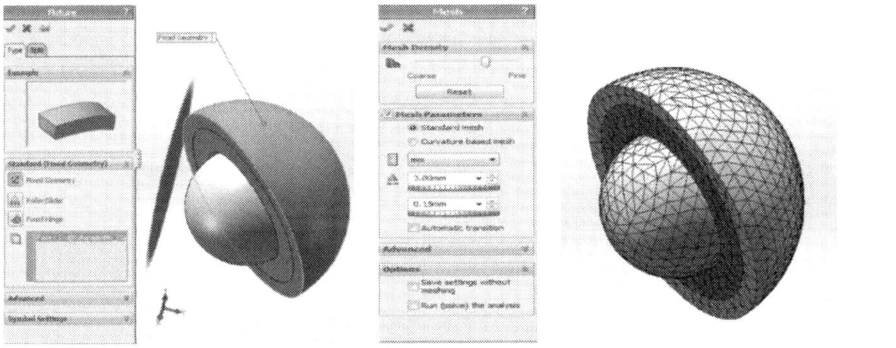

Fig. 3. Boundary and displacement constrains imposed within the FEA analysis. Mesh characteristics.

First, the Component Contact Type was set as Global Contact with No Penetration between the femoral head, the acetabular liner and the cup subassembly and with a friction coefficient equal to 0.6 μm according to the values found in other research that was made regarding the contact pressure values between the hip replacement components [14]. From the designing stage a boundary condition was given to the acetabular cup and liner assembly, allowing a rotation of 20 degree around x axis (anteversion of the cup) and 45 degree around y axis (cup abduction) .The main boundary condition was given as fixture of the acetabular cup as shown in fig.3 in order to keep fixed the model in the same position as it was defined by the inclination in the 3D designing stage of the models.The application of boundary conditions in this way for the realized biomechanical Finite Element Analysis was based on assumptions including forces and pressures acting in the human body as well as displacements and symmetry boundary conditions based on simplifications in the model.

Within this analysis an external load was applied as displacement of the femoral head equal with 0.1 m. Several other research that were made demonstrated that a possible displacement between the acetabular cup and femoral head is between 0.03 and 0.35 mm (depending on the bone structure the implant is implanted in and the age of the patient)

DOI: 10.1051/matecconf/201713702010

[11]. The next step required to be realized before running the FEA analysis was the mesh generating procedure which was performed as shown in fig.3. The mesh parameter (standard type) was selected so as 3 mm dimension were set for the tetrahedral elements.

3 Results of the Finite Element Analysis

After following all the steps of the analysis, as decribed above, it was possible to run the analysis to determine the displacement, stress and contact pressure between the components (see fig. 4).

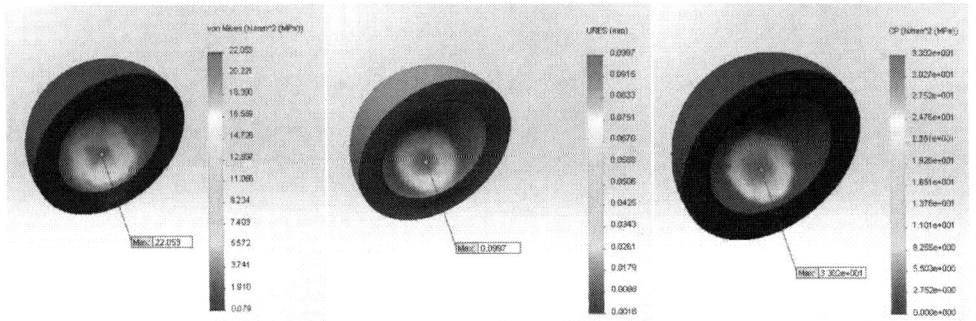

Fig. 4. Distribution of the von Mises stress, displacement and contact pressure between components.

The finite element analysis was performed by taking into consideration different values of the mechanical characteristics (the average values of the Young modulus E, the compression limit strength Rdx and the mechanical resistance at the failure Rm), as these characteristics were determined in the case of different batches of samples that were made from PA 2200 powder by Selective Laser Sintering (SLS), using different settings of laser power (LP) and orientation in the working chamber of the SLS machine (see Fig. 5 and Table 1).

Table 1. Technological parameters, compression test values and finite element results for different settings of PA2200 material characteristics.

PA2200 Samples	LP [W]	Orientation (°)	E [N/mm²]	Rdx [N/mm²]	Rm [N/mm²]	σmax [N/mm2]	Cp [N/mm2]
I (1-5)	4.5	y=0°	1,413,735	33,33	99,08	22.53	33.02
II (11-15)	4.5	y=90°	1,174,300	32,95	90,48	18.33	27.45
III (21-25)	4	y=0°	678,487	15,23	40,34	10.6	15.88
IV (31-35)	4	y=90°	591,062	12,08	31,51	9.24	13.84
V (41-45)	4.5	z=-45°	419,724	10,63	37,41	6.57	9.83
VI (46-50)	4.5	z=+45°	452,354	10,81	40,39	7.08	10.59

The maximum value of the von Mises stress was obtained in the case of first set of samples made by SLS (batch I) as shown in Fig. 4 (the value of σ_{max} was 22.053 MPa in this case).

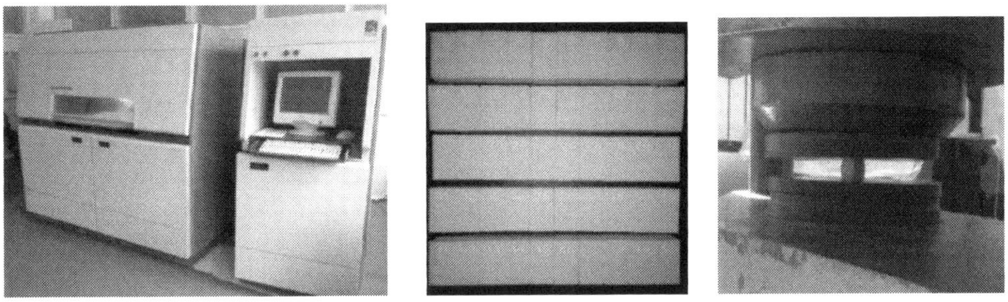

Fig. 5. DTM Sinterstation 2000 equipment, samples and compressive tests performed using the ZD 40 Tension-Compressive Testing Machine from TUCN.

One of the most important condition for a part to resist is to respect the relations:

$$\sigma_{max} \leq \sigma_a \quad (1)$$

$$\sigma_a = \frac{Rdx}{c} \quad (2)$$

in which: σ_{max} is the maximum von Misses stress obtained by the FEA $[N/mm^2]$, σ_a is the admissible stress$[N/mm^2]$, Rdx is the compression limit strength $[N/mm^2]$ and c is the safety coefficient (c=1.5).As it was possible to notice from the results obtained as presented in table 1, in the case of batch I and II, condition (1) was respected (22,53 MPa ≤ 24.22 MPa and 18.33 MPa ≤ 24.22 MPa).In the case of samples manufactured with a laser power equal to 4W under the same loading condition the relation (1) and with a laser power equal to 4.5 W oriented as compared to the Z-axis at ±45° this condition has not been respected, so in these cases it was concluded that the manufactured parts will not resist.

As regarding the displacement distribution, the maximum value which was obtained in the case of made analyses was 0.0977 mm as shown in Fig. 4. The obtained value is in the admissible range of displacement between the acetabular cup and femoral head (0.03 and 0.35 mm), values that are acceptable and reported by the researchers in this field [11]. Last but not least, regarding the contact pressure (Cp) that has been obtained after the made analyses (see Fig.4), it is important to state the fact that the results (Cp = 9.83 – 33.02 MPa) are comparable with those that were obtained by other researchers that have determined the contact pressure between the acetabular cup made from TiAl6V4 material and the acetabular liner made by polyethylene materials with different dimensions (36 mm and 40 mm). The contact pressure values that was determined by this group of researchers have varied as reported between 8.2 to 60.4 MPa, in concordance with the part orientation and technological parameters that was used on the SLS machine [14]. The contact pressure value is influenced not only by the technological parameters and part orientation point of view, but also by the Young modulus, which is variable in concordance with the age of the patient and the bone characteristics (bone quality, coefficient of friction, acetabular cup – acetabular liner – femoral implant - bone interaction, etc). Based on the results obtained after the finite element analyses regarding the values of the von Mises maximum stress and the contact pressure between the realized components, it was decided that the acetabular liner will be made from PA2200 material by SLS with a laser power of 4.5 W and an orientation of 0° along the Y-axis.

4 Conclusions

Taking into consideration the results obtained regarding the mechanical characteristics of the PA 2200 samples that were manufactured by selective laser sintering (SLS) technology at the Technical University of Cluj-Napoca (TUCN) and considering the obtained results of the finite element analyses that were made for estimating the maximum stress and contact pressure between the implant components, it was decided to manufacture the acetabular liner presented in Fig. 6 with a laser power of 4.5 W and an orientation of 0° along Y-axis, at the end. The maximum von Misses stress obtained in this case (22,53 MPa) was in the admissible limits, as well as the contact pressure (33.02 MPa), which was in the reported limits that were determined and presented by other researchers that have analyzed the variation of the contact pressure of the acetabular liner made from PA 2200 material by SLS, with different part orientations and different settings of the technological parameters.

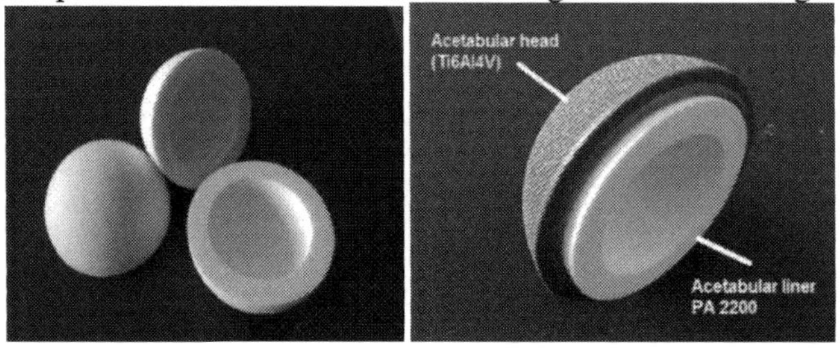

Fig. 6. Acetabular head and acetabular liner made by AM technologies (SLS and SLM).

This research was supported by the OpTi-DeP Project (no. BG101/2016) financed from the UEFISCDI by the Romanian Government and also by the HORIZON 2020 AMaTUC project (GA 691787) financed from EU.

References

1. F. Bartolomeu, M. Buciumeanu, E. Pinto, N. Alves, F.S. Silva, O. Carvalho, G. Miranda, Transactions of Nonferrous Metals Society of China, **27** (2017)
2. D. I. Băilă, Procedia Engineering, **69** (2014)
3. C. Song, A. Wang, Z. Wu, Z. Chen, Y. Yang, D. Wang, Materials & Design, **117** (2017)
4. R. E. Huxtable, T. R. Ackland, G. C. Janes, J. R. Ebert, Clinical Biomechanics, **46** (2017)
5. J. Bai, B. Zhang, J. Song, G. Bi, P. Wang, J. Wei, Polymer Testing, **52** (2016)
6. P.-I. Tsai, C.-C. Hsu, S.-Y. Chen, T.-H. Wu, C.-C. Huang, Computers in Biology and Medicine, **76** (2016)
7. Ortho Info, Total hip replacement, http://orthoinfo.aaos.org/topic.cfm?topic=a00377
8. D. Drummer, D. Rietzel, F. Kühnlein, Physics Procedia, **5,** Part B (2010)
9. X. Wang, S. Xu, S. Zhou, W.Xu, M. Leary, P. Choong, M. Qian, M. Brandt, Y.M. Xie, Biomaterials, **83** (2016)
10. K. Senthilkumaran, P. M. Pandey, P.V.M. Rao, Materials & Design, **30**, (2009)
11. X. Hua, J. Li, L. Wang, Z. Jin, R. Wilcox, J. Fisher, Journal of Biomechanics, **47** (2014)
12. C.S. Miron Borzan, M.C. Dudescu, V.Ceclan, A. Trif, M. Ridzon, P. Berce, Materiale Plastice, **53** (2016)
13. Waldermark LINK Company of Germany, McMinn Acetabular Reconstruction Cup, https://www.linkorthopaedics.com/en/for-the-physician/products/huefte/primary/cups-cementless/mcminn/
14. A. Goldsmith, D. Dowson, G. H. Isaac, J. G. Lancaster, Proceedings of the Institution of Mechanical Engineers, Part H: Journal Engineering in Medicine, **214** (2000)

Optimized anodization setup for the growth of TiO_2 nanotubes on flat surfaces of titanium based materials

Gabriela Strnad [1,*], *Razvan* Cazacu [1], *Paul* Chetan [1], *Zoltan* German-Sallo [2], *Laszlo* Jakab-Farkas [3]

[1] "Petru Maior" University of Tirgu Mures, Faculty of Engineering, Department of Industrial Engineering and Management, Nicolae Iorga 1, Tirgu Mures, Romania
[2] "Petru Maior" University of Tirgu Mures, Faculty of Engineering, Department of Electrical Engineering and Computers, Nicolae Iorga 1, Tirgu Mures, Romania
[3] Sapientia University of Cluj Napoca, Faculty of Technical and Human Sciences, Tirgu Mures, Romania

Abstract. An extensive research work on development of nanostructured TiO_2 layers on the surface of titanium based materials for biomedical implants led the authors to the optimization of process parameters of electrochemical anodization in phosphate/fluoride based electrolytes. Based on those parameters, a dedicated optimized electrochemical anodization setup was originally designed and realized. The anodization bath was designed in order to provide a proper circulation of electrolyte and the possibility of distance anode-cathode modification, the DC power supply was designed accordingly to the electrical parameters requested by the nanotubes development, and a dedicated software (Nanosource) was developed for process control and ease and flexibility of process parameters acquisition, storage and processing.

1 Introduction

Metallic biomedical implants made of pure titanium or titanium alloys, especially Ti6Al4V, are nowadays widely used in dental implantology, cranio-facial surgery and orthopedic surgery. They present good mechanical properties, but more important, due to the spontaneous formation of a thin oxide layer onto the surface, they are highly resistant to corrosion and biocompatible. Their osseointegration is the clinical goal of any surgical intervention, and today is well acknowledged that the surface modification of implants enhances the osseointegration.

In the recent years, the surface modification at nanoscale level is a research direction that is followed by many teams. Nanostructured TiO_2 layers with nanoporous or nanotubular morphologies are under intense research focus, due to the fact that studies show their enhanced osseointegration [1-8]. Cells sense and interact with titania nanotubes (TNT) through changes in morphology, proliferation, differentiation, and absorption. TNT

* Corresponding author: gabriela.strnad@ing.upm.ro

are capable of promoting bone cell function mainly due to: increase surface roughness/energy, the incorporating of fluoride ions (in the case of TNT developed in fluoride based solutions), and the ability to mechanically stimulate cells (this means the geometrical fitting between nanotubes opening of 15…100 nm and the diameter of surface proteins, ~ 10 nm, which are absorbed onto the nano modified surface; these proteins are sensed by integrins, the surface receptors of bone cell membrane, and to them bone cells are attaching to).

Electrochemical anodization is a method which allows the modification of titanium based surfaces at nano scale level if process parameters are optimized [9-13]. Depending on electrolyte properties (composition, temperature, pH, conductivity), current parameters (anodization potential, sweep rate, anodization duration), initial surface shape and preparation, other specific features in anodization cell (distance anode-cathode), titania layers that are developing can be either compact, either nanostructured.

The electrochemical anodization process was optimized by the authors allowing the development, in a reproducible manner, of uniform, self-organized TiO_2 layers on the flat, polished or micro rough surfaces of pure titanium and dual phase ($\alpha+\beta$) Ti6Al4V alloy, in phosphate/fluoride based electrolyte. The optimization of the process was made by using a custom-build anodization setup (that will be denoted as *initial electrochemical anodization setup - IEAS*), by several sets of experiments whose results are presented elsewhere [14-18], and which allowed the tuning of the process parameters. Based on these results, an *optimized electrochemical anodization setup (OEAS)* was originally conceived and realized by the authors, being presented in current paper. The DC power supply was designed accordingly to the electrical parameters requested by the nanotubes development. The anodization bath was designed in order to provide a proper circulation of electrolyte. The holding system for the samples allows the modification of their position and distance anode-cathode. The setup is completed by all necessary arrangements for process monitoring instrumentation (thermometer, ph-meter, conductivity meter) installation.

2 Methodology

The anodization experiments presented in present study were performed in new optimized electrochemical anodization setup (OEAS), using the same type of samples (shape, dimensions, surface preparation) and anodization process parameters, as in the old initial electrochemical anodization setup (IEAS), in order to be possible to realize a comparison of resulting titania nanotubular layers.

Samples, made from pure titanium or Ti6Al4V alloy, are in form of disc shape of 16 mm in diameter and 3 mm in height. Before the anodization process the samples were subjected to a thorough cleaning process. The samples were connected at the anode of the cell; cathode: ∅16 x 3 mm disc, made of high purity copper (99.99%); distance anode-cathode: 15 mm; electrolyte: 1M H_3PO_4 + 0.5 wt% HF; room temperature; anodization potential: 20 V, applied with a sweep rate of 0.1 V/s; anodization time: 30 min. The surface morphology was evaluated by using scanning electron microscopy (SEM), performed in a JSM 5200 (JEOL) scanning electron microscope, operated at 25 kV. The dimensions of nanostructures developed on the modified surfaces were measured on SEM micrographs by using an open source graphical image processing software (Gimp).

3 Results and discussion

Figure 1 presents the *block scheme of the optimized electrochemical anodization setup* that has been conceived. The main components of the OEAS are: anodization bath (tank), DC power supply, peristaltic pump (with its power supply), instrumentation for electrolyte parameters monitoring (thermometer, pH – meter, conductivity meter). The electrical anodization process parameters are computer controlled and monitored by dedicated software originally developed – Nanosource.

Each component of the OAS was carefully designed: the anodization bath was designed in order to provide a proper circulation of electrolyte, and to have a holding system for the samples that allows the modification of their position and distance anode-cathode; the DC power supply was designed accordingly to the electrical parameters requested by the nanotubes development.

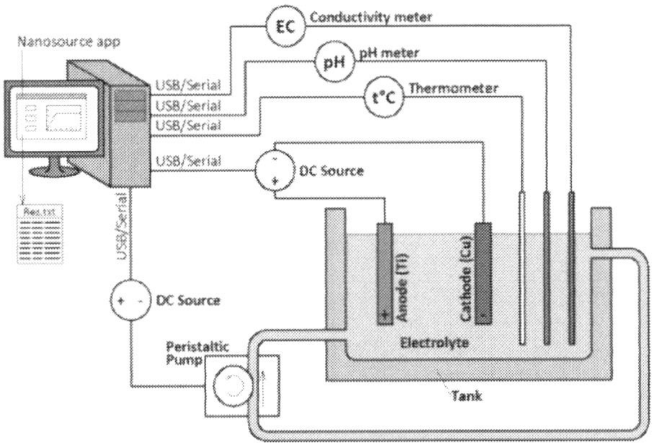

Fig. 1. Block scheme of optimized electrochemical anodization setup (OEAS).

Fig. 2 presents the design of the *anodization bath* complex. Its main feature is the tank (1), perforated with two holes (2) for the recirculation of the electrolyte by means of a peristaltic pump (not represented here). At the top, the tank is capped by a cover piece (3). The mobile support (4) is used to fix the copper cathode (5) with the help of a clamping bolt (6) and two gaskets (7 and 8) to prevent the infiltration of the electrolyte inside the channel of the electric wire (9) coming from the DC source. The titanium-based anode (10) is fixed on the opposite side using a second clamping bolt (11) and its corresponding gasket (12). The anode is pressed into position with a rod and a spring (13) which also have the role of passing the current from the DC source.

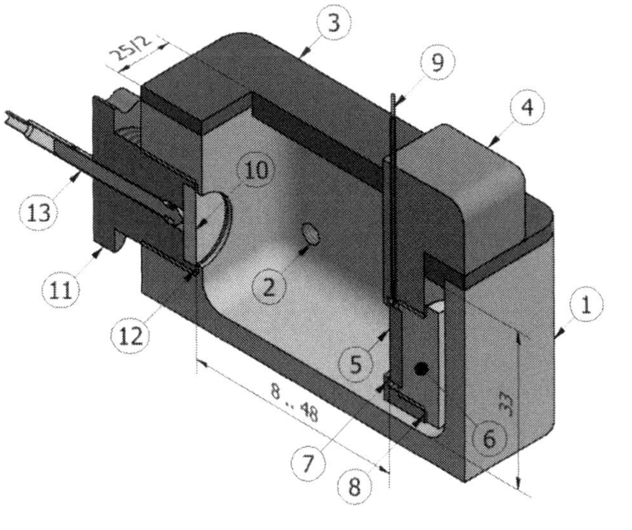

Fig. 2. Design of anodization bath.

The mobility of the cathode support makes possible the precise adjustment of the distance between the anode and the cathode (using a scale on the cover of the tank) in the range of 8 mm to 48 mm. This allows the testing of the influence of this distance on the formation and quality of the nanotubes.

A *dedicated DC power supply* was designed in order to obtain the required supply voltage, applied with a specific sweep rate, using discrete non-programmable electronic components, such as counter, digital to analog converter, linear amplifiers using rail to rail operational amplifier (Fig. 3 and Fig. 4). The ramp supply voltage generator has two major parts, the ramp generator and the analog output circuit. The first part assures the required ramp (in V/s), while the second part interfaces the generators output with the load.

The ramp generator provides a ramp with n bit resolution, where the value is increased by each clock cycle and the start and stop value can be defined by the user. The slope is defined by the clock input.

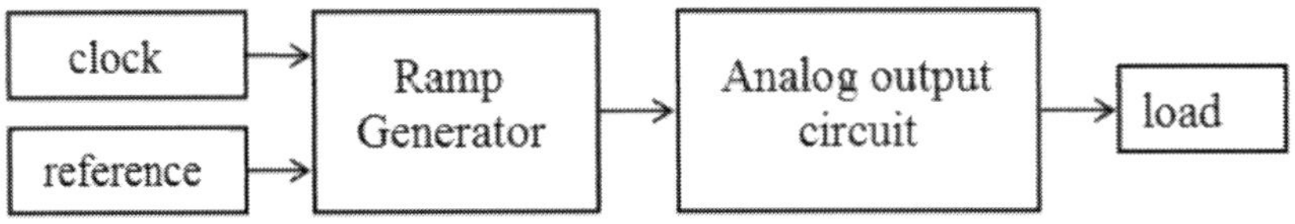

Fig. 3. Block scheme of DC power supply.

The ramp is generated by a n digital-analog converter (DAC) which has at the input a modulo N (with $N \leq 2^n$) counter. After reaching the value N, the counters outputs remain in this state until the process ends. The reference voltage of the DAC controls the ramps step.

The counters clock can be set at different values to be adapted for the required values as the output of the ramp generator will have a step of value $(N/2^n) \cdot U_{REF}$ in every period of the clock signal. The ramp duration is $T_{RAMP} = N \cdot T_{clock}$, the duration until the final time can be set manually, depending on the controlled process. The maximum value of the output voltage is $U_{OUTmax} = A \cdot N \cdot q$, where A is the gain of the non-inverting amplifier and $q = U_{REF}/N$ is the quantum corresponding to one step.

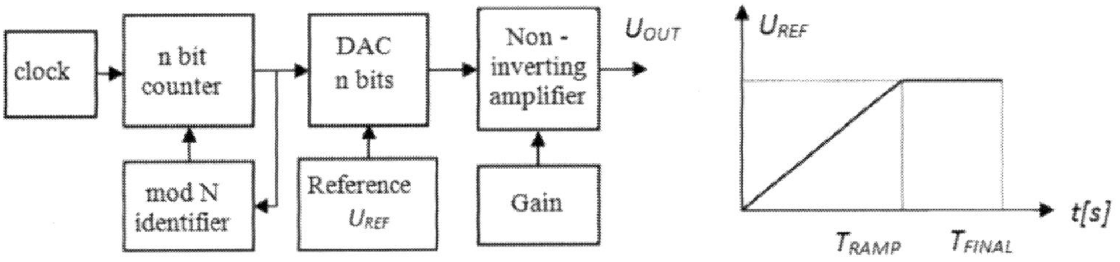

Fig. 4. The ramp supply voltage generator.

For a supply voltage domain $U_{OUT} = 1 \div 20\ V$ with a ramp of $0.1\ V/s$ results that T_{RAMP} is $200\ s$. By using a reference $U_{REF} = 5\ V$, the quantum is $q = 5/256 = 0.01953125\ V$, after 200 steps the output of DAC is $3.90625\ V$. By setting the clock to $1\ Hz$ and the gain of the analog output circuit at $A = 20/3.90625 = 5.12$ the proposed values are obtained. The used DAC is a CMOS 8-bit successive approximation converter (DAC0804) having a resolution $\pm(1/4)LSB$ with a conversion time of 100 μs. The non-inverting amplifier is realized with an extended supply voltage operating range and rail-to-rail output operational amplifier. The current consumption is usually under $50 \div 100\ mA$, there is no need for current amplification.

In order to conduct repeatable experiments, the parameters of the anodization system need to be monitored and controlled. Solving of this problem was realized by creating a *dedicated control and monitoring software - Nanosource*, developed in Labwindows CVI. This software is able to control the anodization source voltage and current limit and also to log the temperature of the anodization bath together with the measured voltage and current from the source. The equipment items are connected with the process control computer through electrically isolated USB ports which at device level are accessible as virtual serial ports. The communication with the devices is controlled by a predefined timer with a period of 0.1s. The developed software has two modes of operation: a *manual* one, in which the prescribed value is entered in the *Uset* or *Ilim* box on the graphical user interface; the second operation mode is an *automated* mode, with a starting ramp with specified slew rate for starting the anodization process, then a potential hold mode for the continuous growth mode. In this mode, the user can set the starting voltage *U_start*, the slew rate *V/S*, the hold voltage *U_hold* and the time for the continuous growth *Hold time (s)*. The total time of the process *Total (h:m:s)* is displayed on the user interface by following the setting of the process parameters. The output of the anodization source is switched automatically on at the start, and off at the end of the process. Data is logged at every measuring interval and files for every process conducted are recorded separately.

By using this OEAS there were developed TiO_2 layers on flat surfaces of pure titanium and Ti6Al4V alloy in phosphate/fluoride based electrolytes, at an anodization potential of 20 V. The obtained nanotubes were uniform, self-arranged and presented diameters in 50-110 nm range and are fully comparable with those obtained in IEAS.

Figure 5.a presents the SEM image (top view, collected at 35000X magnification), of the TiO_2 layer developed using OEAS, showing well developed nanotubular morphology, with nanotubes diameter in 50-110 nm range. The nanotubes walls are well separated and the nanotubes are uniformly distributed and self-arranged. This proves that the OEAS provides reliable results on proper nanostructures development.

Fig. 5.b presents the registration of the anodization potential U [V] and the anodization current I [A] taken by the Nanosource software during the anodization process, and the graph of potential and current evolution, processed in Excel by importing data from Nanosource. This shows the advantages of OAES over IAES, which are the ease of process control and monitoring, and the ease and flexibility of data collecting, storage and processing.

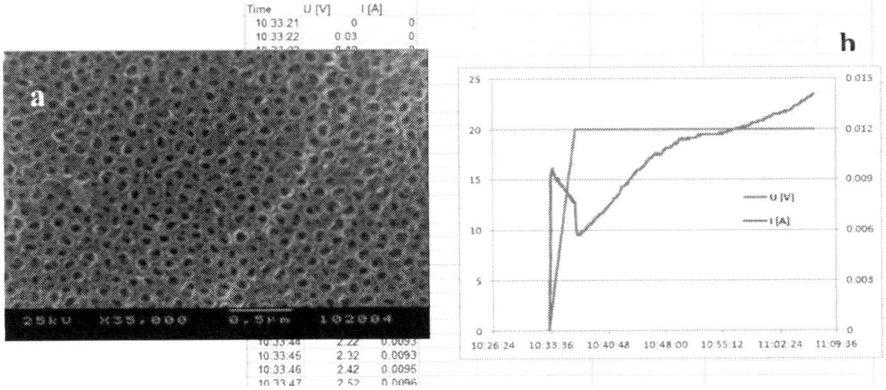

Fig. 5. SEM micrograph of TiO_2 nanotubes developed in OEAS (a). The evolution of anodization potential and current during the electrochemical anodization; data were collected and stored using dedicated Nanosource software (b).

All the further work that will be carried on nanostructured TiO_2 layers development on flat surfaces of pure titanium and titanium based alloys by electrochemical anodization will be performed from now by using the OEAS.

4 Conclusions

The obtained results have proven that it was possible to develop nanotubular oxide layers on the surfaces of titanium based materials for biomedical implants by using an optimized electrochemical anodization setup (OEAS) originally designed and developed by the authors.

The carried-out work on setup optimization led to the realization of: a dedicated electrochemical bath (tank), specially designed in order to allow a proper circulation of chemical electrolyte and the modification of anode-cathode distance; a dedicated DC power supply, which was designed and realized in concordance with the electrical parameters requested by the nanotubes development; and a dedicated software Nanosource which was designed and realized for process control and monitoring.

This work was supported by a grant of the Romanian National Authority for Scientific Research and Innovation, CNCS/CCCDI-UEFISCDI, project number PN-III-P2-2.1-PED-2016-0142, within PNCDI III.

References

1. R. R. Baena, S. Rizzo, L. Manzo, S. M. Lupi, Journal of Nanomaterials, article ID 6092895 (2017)
2. T. Monetta, A. Acquesta, A. Carangelo, F. Bellucci, Metals, **7**(6) 220 (2017)
3. K. Gulati, S. Maher, D. M. Findlay, D. Losic, Nanomedicine, **11**(14) 1847-1864 (2016)
4. D. Karazisis, A. M. Ballo, S. Petronis, H. Agheli, L. Emanuelsson, P. Thomsen, O. Omar, Int J of Nanomed, **11**, 1367-1381 (2016)
5. A. Mazare, Corrosion Science, **103** 215–222 (2016)
6. M. Kulkarni, A. Mazare, E. Gongadze, S. Perutkova, V. Kralj-Iglič, I. Milosev, P. Schmuki, A. Iglic, M. Mozetic, Nanotechnology, **26** 062002 (2015)
7. K. Narendrakumar, M. Kulkarni, O. Addison, A. Mazare, I. Junkar, P. Schmuki, R. Sammons, A. Iglic, Dental Materials, **31** 1460-1468 (2015)
8. E. Beltrán-Partida, A. Moreno-Ulloa, B. Valdez-Salas, C. Velasquillo., M. Carrillo, A. Escamilla, E. Valdez, F. Villarreal, Materials, **8** 867-88 (2015)
9. T. Monetta, A. Acquesta, A. Carangelo, F. Bellucci, Metals, **7**(5) 167 (2017)
10. M. Kulkarni, et al., Advanced Material Letters, **7**(1) 23-28 (2016)
11. K. Gulati, A. Santos, D. Findlay, D. Losic, Journal of Physical Chemistry C, **119** (2015)
12. A. Mazare, et al, Surf. Interface **46** Anal., 186-192 (2014)
13. K. Lee, et al., Chemical Reviews, **114** 9385–9454 (2014)
14. G. Strnad, C. Petrovan, O. Russu, L. Jakab-Farkas, IoP Conf. Series: Materials Science and Engineering, **161** doi: 10.1088/1757-899X/161/1/012051 (2016)
15. G. Strnad, L. Jakab-Farkas, D. Portan, Academic Journal of Manufacturing Technology, **14(4)** 112-118 (2016)
16. G. Strnad, L. Jakab-Farkas, C. Petrovan, O. Russu, Procedia Engineering, **181** 242-248 (2017)
17. G. Strnad, Z. German-Sallo, L. Jakab-Farkas, C. Petrovan, D. Portan, MATEC Web of Conferences, **112** 04021 (2017)
18. G. Strnad, D. Portan, L. Jakab-Farkas, C. Petrovan, O. Russu, Materials Science Forum, **907** 91-98 (2017)

Effect of phosphate/fluoride electrolytes on mass and dimensional stability of anodization bath manufactured by FDM

Gabriela Strnad [1,*], *Razvan* Cazacu [1], *Paul* Chetan [1], *Andrei Serban* Gaz Florea [2], and *Ferencz* Peti [3]

[1] "Petru Maior" University of Tirgu Mures, Faculty of Engineering, Department of Industrial Engineering and Management, Nicolae Iorga 1, Tirgu Mures, Romania
[2] University of Medicine and Pharmacy of Tîrgu Mureş, Faculty of Pharmacy, Department of Fundamental Pharmaceutical Science, Organic Chemistry, Gh. Marinescu 38, Tirgu Mures, Romania
[3] CIE Matricon, Department of Quality, Gh. Doja 155, Tirgu Mures, Romania

Abstract. Present paper is an experimental study on mass and dimensional stability of components manufactured by additive technology of Fused Deposition Modeling (FDM) from PLA and ABS filaments, components to be subjected to the action of aqueous phosphate/fluoride solutions during the process of surface modification and TiO_2 nanotubes development on the surface of titanium based materials by electrochemical anodization. Several specimens were printed with 30% and 100% fill density; we used control samples of PP, PLA and ABS in order to compare the results. The specimens and control samples were in contact with 1M H_3PO_4 + 0.5 wt% HF electrolyte, for 2 hours and 48 hours. Regarding mass stability we found that the specimens' mass is increasing after exposure to electrolyte, showing absorption on to the material, the mass gain being up to 0.2% from initial mass. Dimensional stability is also questionable; there are modifications of up to 0.05 mm after 48 hours exposure to electrolyte. All of our results lead to the conclusion that, even if FDM has certain advantages in terms of flexibility of design and short design to product time, drawbacks appear in terms of mass and dimensional stability when the printed components work in aqueous acid solutions, raising questions regarding their safe utilization over time.

1 Introduction

Additive manufacturing is becoming nowadays more and more relevant for products where customization is of most importance. The advantages that additive manufacturing brings in terms of: flexibility, complexity and variety of the parts, little lead time, little skill manufacturing, few constrains, less waste, no assembly operations required, boost its applications on almost every market (automotive, aerospace, medical, robotics).

* Corresponding author: gabriela.strnad@ing.upm.ro

Fused Deposition Modeling (FDM) works with thermoplastic resins, in form of filaments; the material is heated and extruded out of a small nozzle and the model/part is built-up layer by layer by the deposition of the extruded material which rapidly cools down and solidifies. FDM is one of the quickest additive manufacturing technologies and a large number of materials for plastic filaments are available on the market: polylactic acid (PLA), acrilonitrile butadiene styrene (ABS), polycarbonate (PC), polyamide (PA), polystyrene(PS), polyether ether ketone (PEEK). PLA filament is made from organic and renewable sources; it is safe and easy to print. The material presents good tensile strength, allows 3D printing at high print speeds with good surface quality of printed parts. ABS provides excellent mechanical properties, enhanced interlayer adhesion, minimal warping and reliable bed adhesion.

Taking into account the FDM advantages we planned to use it for the manufacturing of the bath (tank) that we use in our custom-built electrochemical anodization cell. In this laboratory installation, we perform the surface modification of titanium based alloys in order to develop nanostructured oxide layers. Self-ordered nanotubular TiO_2 layers attracted lately tremendous research efforts due to their outstanding applications in medicine (implants, stents, valves, and drug delivery systems), solar cells, batteries for electric vehicles, catalyst for fuel cells, and oil-water filtration.

Currently we are able to develop uniform, self-arranged nanotubular TiO_2 layers on the surface of pure titanium and Ti6Al4V alloy, by using electrochemical anodization in phosphate/fluoride solutions (Fig. 1). Depending on the process parameters we can develop nanotubes of diameter in 30-120 nm range and thickness in 100-500 nm range.

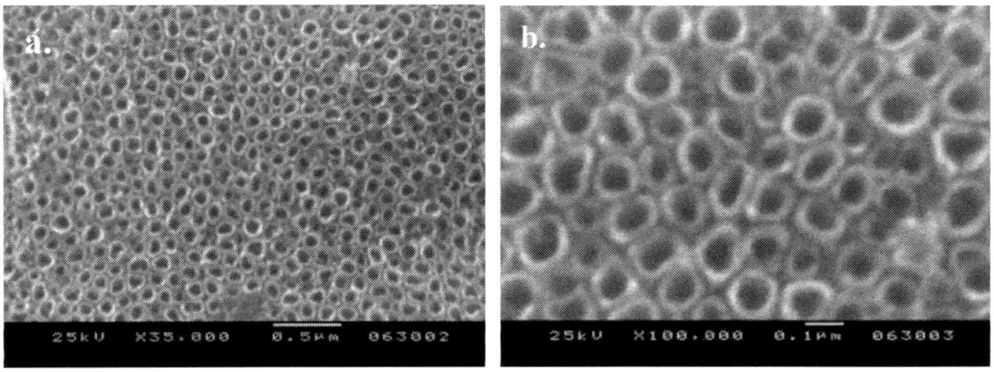

Fig. 1. SEM micrographs of Ti6Al4V surface, anodized in 1M H_3PO_4 + 0.5 wt% HF, at U=20 V, 2 hours, showing the development of well defined, highly ordered nanostructured oxide layer (a – 35000X magnification; b – 100000X magnification) [1].

Our research is focused on applications of these layers in biomedical implants. This is due to the fact that recent results demonstrate that surface modification at nano scale level, by producing an ordered nanostructured surface, promotes enhanced osseointegration process [2-7]. Electrochemical anodization technology can control the morphology, structure and chemical composition of TiO_2 layers, leading to enhancement of the surface properties. By optimizing the process parameters of anodization, highly order and self arranged nanostructured surfaces can be developed [8-11]. Our recent results [1, 12-14] are in line with state of the art in this research field, and we bring also new contributions by being able to superimpose nanotubular layers over micro rough topographies [15].

In this context, the aim of present experimental research is to find if the manufacturing of our electrochemical anodization baths by FDM, in order to benefit from this technology advantages, is a good option when the application is related to the usage of 3D printed components in contact with aqueous acid solutions. There is no available data on filament materials mass and dimensional stability. Moreover, there is no data on stability of FDM printed components as their properties depend on the many process parameters (nozzle diameter, nozzle temperature, layer height, shell thickness, print speed, fill density, outer shell speed, inner shell speed, infill speed, etc.) that can be set at different values.

2 Materials and methods

The specimens designed for the experimental study performed by us were manufactured by additive technology of Fused Deposition Modeling (FDM) using an Ultimaker 2 Extended + 3D printing machine.

We printed several baths (specimens 1-4) using PLA (polylactic acid) and ABS (acrylonitrile butadiene styrene) filaments, materials whose main properties are listed in Table 1. (filaments are of original Ultimaker brand).

Table 1. Main properties of filaments used for specimens manufacturing by FDM.

	PLA filament	**ABS filament**
Diameter [mm]	2.85 ± 0.1	2.85 ± 0.1
Max. roundness deviation [mm]	0.1	0.1
Density [g/cm^3]	1.24	1.10
Melting temperature [°C]	145-160	225-245
Glass transition [°C]	~ 60	-
Melt mass flow rate [g/min]	0.609	0.410
Tensile modulus [MPa]	2346.5	1681.5
Tensile stress at yield [MPa]	49.5	39
Tensile stress at break [MPa]	45.6	33.9
Elongation at yield [%]	3.3	3.5
Elongation at break [%]	3.2	4.8
Flexural strength [MPa]	103	70.5
Flexural modulus [MPa]	3150	2070
Hardness [Shore D]	83	76
Water solubility	insoluble	insoluble
Incompatible materials	oxidizing agents, strong bases	strong oxidizing agents
Chemical stability	chemically stable	chemically stable
Stability (reactivity)	no data available	no data available

The specimen's design was made in CATIA V5, the .stl file was saved and imported in Cura 15.01 software. The parameters used for .gcode generation of our specimens are listed in Table 2. Basically, we used the same parameters, except fill density; from each material we print two specimens: one with hollow interior (infill 30%) and one with massive interior (infill 100%).

Chemical solution of 1M H_3PO_4 + 0.5 wt% HF was prepared from reagent grade chemicals (Chemical Company) and deionized water. Experiments were carried out at room temperature.

After the specimens were manufactured we subjected them to the following experimental procedure:

1. the *initial mass* and *initial specific dimensions* of the specimens were measured;

2. the chemical solution was put in the interior space of the specimens and maintained there for a duration of *2 hours*;
3. after 2 hours the chemical solution was evacuated, the specimens were cleaned by washing in distilled water, carefully wiped, dried in air, and their mass was measured at 1, and 16 hours after extraction from solution;
4. after 16 hours the selected dimensions of the specimens were measured again;
5. the above-mentioned procedure was repeated in the same way, but with a maintaining duration of *48 hours* of chemical solution inside the specimens.

Table 2. Main parameters used for .gcode generation in Cura 15.01.

	Specimen 1 **S1 – PLA 30**	**Specimen 2** **S2 – PLA 100**	**Specimen 3** **S3 – ABS 30**	**Specimen 4** **S4 – ABS 100**
Nozzle size [mm]	0.4	0.4	0.4	0.4
Layer height [mm]	0.15	0.15	0.15	0.15
Shell thickness [mm]	0.8	0.8	0.8	0.8
Print speed [mm/s]	50	50	50	50
Fill density [%]	30	100	30	100
Outer shell speed [mm/s]	40	40	40	40
Inner shell speed [mm/s]	80	80	80	80
Infill speed [mm/s]	100	100	100	100

The weighting of the specimens was measured using Adam AEA-100G analytical balance with an accuracy of ± 0.1 mg.

Dimensional control was performed on a Zeiss Contura G2 coordinate measuring machine (CMM) equipped with VAST XXT scanning sensor, a Renishaw stylus with a ball radius of 2 mm, scanning error of ± 2.5 μm. A measuring program was designed accordingly to the specific dimensions that were planned to be inspected and whose variations we aim to evaluate.

In order to be able to analyze our results on mass stability we subjected *3 control specimens* to the same experimental procedure. The control specimens were: CS – PP (a bath made from polypropylene by injection), CS - PLA (a specimen cut from PLA filament used for FDM), and CS - ABS (a specimen cut from ABS filament used for FDM). CS – PLA and CS – ABS were subjected to the electrolyte action by immersing them in the chemical solution.

3 Results and discussion

The specimens were designed in order to allow us to conclude on *mass stability* of materials after exposure in aqueous chemical solutions, and also on the *stability of specific features' dimensions*. These features are in direct link with the final design of our electrochemical anodization bath where we perform our experiments in order to modify, at nanoscale level, the surface of titanium based alloys for biomedical implants.

Fig. 2 presents the isometric drawing of the specimens and the specific dimensions of interest for present research work.

In Fig. 3.a the specimen's model loaded in Cura software for the generation of the .gcodes for 3D printing is presented. We generate four different .gcodes using different settings for: material (PLA, ABS) and fill density (30%, 100%). Printing time was ~1 h 10 min for the specimens with 30% infill and ~3 h 20 min for the specimens with 100% infill.

Fig. 3.b presents an image taken during one specimen's measuring process on Zeiss Contura G2 CMM.

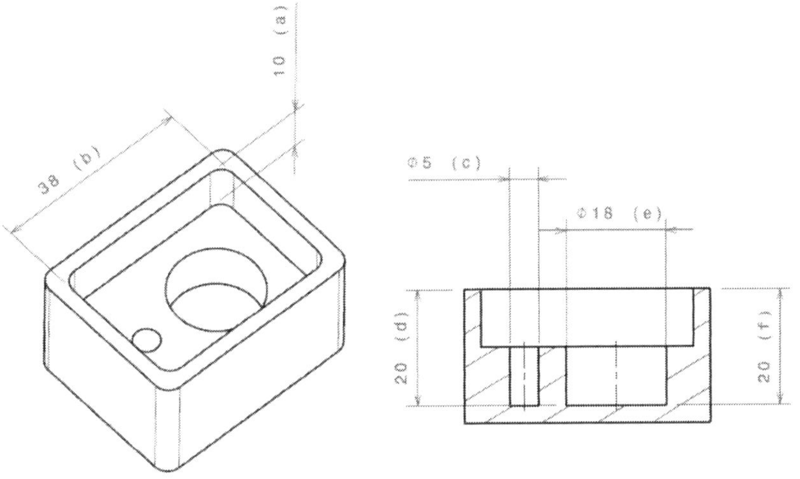

Fig. 2. The specimen's drawing and indication of nominal dimensions that were measured.

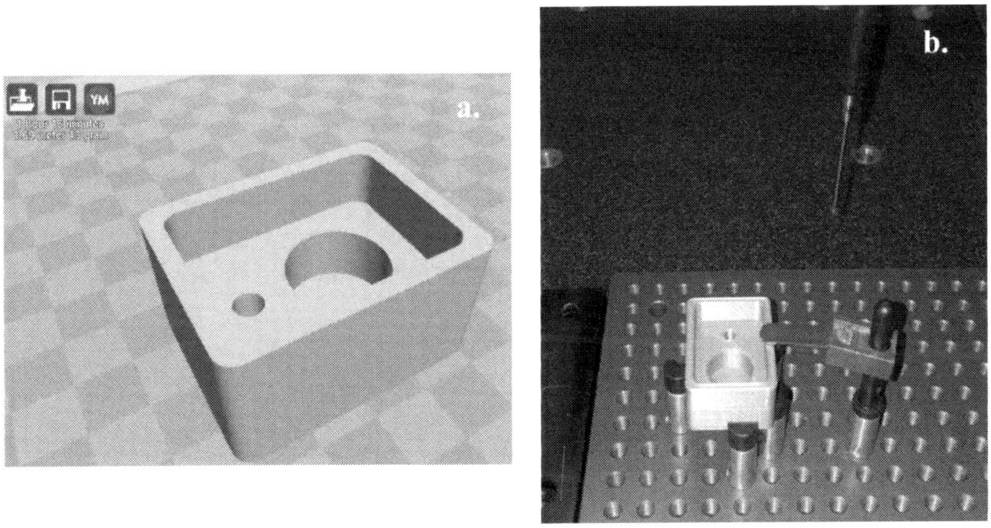

Fig. 3. The specimen's model loaded in Cura 15.01 software (a); Dimensional control on CMM (b).

Tables 3 and 4 summarize the results of our experiments on the influence of the electrolyte on *mass stability*. By analyzing the results, we can affirm that there is a non negligible influence of chemical solution on specimens mass.

After 2 hours of exposure (Table 3) to the aqueous acid based solution the mass of control samples does not suffer any changes (after complete drying, which occurs in a time shorter than 1 hour for PP and ABS filament), but the mass of FDM manufactured specimens is changing, demonstrating that the surfaces in contact with chemical electrolyte absorb it (Table 3). This is due to the inherent porosity of the surfaces resulting from layer by layer deposition during FDM process. It is interesting to observe that in the case of PLA the difference in mass is bigger in the case of specimen printed with 100% infill compared with the one printed with 30% infill (0.12% compared with 0.08%). In the case of ABS the situation is opposite, the specimen printed with 100% fill density exhibit a lower mass difference, this sample having the lowest electrolyte action influence (0.02% from its initial mass).

Table 3. The results on the influence of 2 hours exposure to the action of phosphate/fluoride electrolyte on specimens and control samples mass.

Specimen	Initial mass M_1 [g]	Mass after *2 h in electrolyte* M_2 [g]		Difference in mass after *2 h in electrolyte* $\Delta M = M_2 - M_1$			
		1 h of drying	16 h of drying	1 h of drying		16 h of drying	
				[mg]	%	[mg]	%
S1 - PLA 30	14.1244	14.1431	14.1357	18.7	0.13	11.3	0.08
S2 - PLA 100	24.4316	24.4748	24.4605	43.2	0.18	28.9	0.12
S3 - ABS 30	14.3764	14.3988	14.3970	22.4	0.16	20.6	0.14
S4 - ABS 100	23.2452	23.2476	23.2495	2.4	0.01	4.3	0.02
CS - PP	6.6384	6.6384	6.6384	0.0	0.00	0.0	0.00
CS - PLA	0.3424	0.3428	0.3424	0.4	0.12	0.0	0.00
CS - ABS	0.3336	0.3336	0.3336	0.0	0.00	0.0	0.00

After 48 hours of exposure (Table 4) to the aqueous acid based solution the mass of ABS control sample (measured after 16 hours of drying) is also influenced by the contact with electrolyte. The PLA control sample mass remain unchanged, and the mass of ABS control sample is changing, showing that even extruded filament of ABS absorbs the solution in which it is put. The mass of PP control sample appears to be less the initial one, that can be the results of a slight dissolution of the material (0.6 mg mass loss); but the value is very low, is the range of analytical balance accuracy, so we cannot affirm that we are certain that the material dissolution occurred. The mass differences in the case of PLA specimens are bigger than those measured after two hours of exposure (0.15% vs. 0.08% and 0.2% vs. 0.12%), and also the difference is higher in the case of S2 - PLA 100 compared with S1 – PLA 30 (0.2% vs. 0.15%). In the case of ABS the mass differences are of 0.1% from initial mass, being lower compared to PLA.

Table 4. The results on the influence of 48 hours exposure to the action of phosphate/fluoride electrolyte on specimens and control samples mass.

Specimen	Initial mass M_1 [g]	Mass after *48 h in electrolyte* M_2 [g]		Difference in mass after *48 h in electrolyte* $\Delta M = M_2 - M_1$			
		1 h of drying	16 h of drying	1 h of drying		16 h of drying	
				[mg]	%	[mg]	%
S1 - PLA 30	14.1357	14.1852	14.1567	49.5	0.35	21.0	0.15
S2 - PLA 100	24.4605	24.5801	24.5083	119.6	0.49	47.8	0.20
S3 - ABS 30	14.3970	14.4161	14.4116	19.1	0.13	14.6	0.10
S4 - ABS 100	23.2495	23.2735	23.2727	24.0	0.10	23.2	0.10
CS - PP	6.6384	6.6380	6.6378	-0.4	-0.01	-0.6	-0.01
CS - PLA	0.3428	0.3438	0.3428	1.0	0.29	0	0.00
CS - ABS	0.3336	0.3344	0.3340	0.8	0.24	0.4	0.12

As a general observation, we can affirm that there is a certain influence of aqueous 1M H_3PO_4 + 0.5 wt% HF electrolyte on mass stability of FDM manufactured parts. The chemical solution is absorbed onto the surface of the specimens, these raising questions of the surface contamination and further possibility of cleaning. This detrimental influence does not exist in the case of control sample manufactured by injection.

As regards *dimensional stability under the influence of electrolyte*, Table 5 presents the results of our CMM investigation of specific dimensions on each specimen.

Table 5. The results on the influence of exposure to the action of phosphate/fluoride electrolyte on specimens' dimensional stability.

	Nominal dimension d_N [mm]	Initial dimension d_1 [mm]	Dimension after *2 h in electrolyte* d_2 [mm]	Dimension after *48 h in electrolyte* d_3 [mm]	Difference after *2 h in electrolyte* $\Delta d = d_2 - d_1$ [μm]	Difference after *48 h in electrolyte* $\Delta d = d_3 - d_1$ [μm]
S1 – PLA 30	**10** (a)	9.931	9.943	9.948	12	17
	38 (b)	37.885	37.893	37.909	8	24
	Ø 5 (c)	4.798	4.799	4.800	1	2
	20 (d)	19.880	19.885	19.894	5	14
	Ø 18 (e)	17.725	17.728	17.734	3	9
	20 (f)	19.901	19.916	19.926	15	25
S2 – PLA 100	**10** (a)	9.891	9.889	9.896	-2	5
	38 (b)	37.953	37.967	37.983	14	30
	Ø 5 (c)	4.892	4.893	4.894	1	2
	20 (d)	19.895	19.906	19.917	11	22
	Ø 18 (e)	17.818	17.823	17.829	5	11
	20 (f)	19.898	19.905	19.921	7	23
S3 – ABS 30	**10** (a)	9.781	9.786	9.829	5	48
	38 (b)	37.680	37.671	37.677	-9	-3
	Ø 5 (c)	4.604	4.603	4.604	-1	0
	20 (d)	19.859	19.866	19.863	7	4
	Ø 18 (e)	17.484	17.483	17.487	-1	3
	20 (f)	19.797	19.794	19.825	-3	28
S4 – ABS 100	**10** (a)	9.787	9.781	9.770	-6	-17
	38 (b)	37.800	37.795	37.795	-5	-5
	Ø 5 (c)	4.534	4.533	4.532	-1	-2
	20 (d)	19.628	19.595	19.576	-33	-52
	Ø 18 (e)	17.431	17.429	17.428	-2	-3
	20 (f)	19.497	19.487	19.469	-10	-28

Analyzing the results, a first discussion can be made on the dimensional accuracy of printed specimens. All dimensions on the manufactured specimens are smaller than the prescribed ones, the deviations from nominal dimensions being in the range: 0.069 – 0.275 mm for S1- PLA 30, 0.047 – 0.182 mm for S2 – PLA 100, 0.141 – 0.516 mm for S3- ABS 30, and 0.200-0.569 for S4 – ABS 100. Deviations of such values show a poor dimensional accuracy of FDM process.

Regarding the influence of electrolyte on specific dimensions of the specimens our results show that there are notable changes even after a short exposure of 2 hours. The PLA specimens show a slight increase in dimensions, while ABS specimens show mainly a decrease in dimensions. This shows that ABS is swelling in aqueous acid solution. The measured difference from initial dimensions can be as big as 0.033 mm. At long exposure duration of 48 hours, PLA show also high modifications of initial dimensions of up to 0.030 mm, which can suggest a dissolution process. It is interesting that S3 – ABS 30 show an increase in dimensions, which compared with the decrease of them at 2 hours of exposure, seems to indicate that in the same time the swelling and dissolution occurs, and for long exposure time the dissolution effect is higher than the swelling one. For S4 –ABS 100 all dimensions are lower than initial ones, showing a predominant swelling effect, the differences are up to 0.052 mm.

This dimensional instability in chemical electrolyte, together with the initial poor manufacturing accuracy leads us to the conclusion that FDM is not a suitable solution for manufacturing of parts that will be in contact with acid based phosphate/fluoride electrolytes. Furthermore, our results show an increase in all specimens mass, showing absorption of chemical solutions, raising questions regarding their safe utilization.

4 Conclusions

As the manufacturing of the mechanical components by using additive technology brings important advantages in terms of flexibility of design and short design to product time, present research aimed to study the opportunity of manufacture, by FDM process, of specific component of an electrochemical anodization set-up, which is anodization bath.

Using PLA and ABS we printed 4 specimens (30% and 100% fill density) and subjected them to the action of 1M H_3PO_4 + 0.5 wt% HF electrolyte (for duration of 2 hours and 48 hours). This specific acid mixture is the aqueous solution that allows the development of titania nanotubes on titanium based materials, by using electrochemical anodization. We aimed to study the mass and dimensional stability of these specimens, in the absence of detailed information on this issue, and in conditions in which the filaments technical specifications are that the materials are chemical stable, but the specifications on stability are of "no data available".

Our results on mass stability show an increase of it after exposure to chemical solution, in the range of 0.01% from initial mass to 0.5% from initial mass. After 48 hours of exposure and 16 hours of drying the specimens still present a mass gain of 0.1%-0.2%, raising serious questions regarding the surface contamination and further possibility of cleaning. The control sample made from PP by injection does not show this absorption effect.

Our results on dimensional stability show a poor initial manufacturing accuracy of FDM process, followed by changes in specimens' dimensions as a result of exposure to chemical solution. The modification of dimensions depends on material and infill density. On PLA (regardless of infill) the dimensions are increasing at short and long exposure time, showing a slight dissolution effect, being up to 0.030 mm. ABS exhibits a decrease in dimensions, showing a swelling effect, except the case of the specimen with 30% infill exposed for 48 hours to electrolyte, which have bigger dimensions that initial ones, showing that dissolution effect is stronger than swelling one. The modification of ABS specimens' dimensions is up to 0.052 mm.

To sum up, our results lead us to the conclusion that FDM is not a suitable solution for manufacturing of parts that will be in contact with acid based phosphate/fluoride electrolytes. Even it is a flexible and rapid method to deliver finished product, in the specific application studied by us, which can be extended to other applications that deal with aqueous solutions or chemical mixtures, the drawbacks that appear in terms of mass and dimensions stability have to be taken in account.

This work was supported by a grant of the Romanian National Authority for Scientific Research and Innovation, CNCS/CCCDI-UEFISCDI, project number PN-III-P2-2.1-PED-2016-0142, within PNCDI III.

References

1. G. Strnad, C. Petrovan, O. Russu, L. Jakab-Farkas, IoP Conf. Series: Materials Science and Engineering, **161** doi: 10.1088/1757-899X/161/1/012051 (2016)
2. K. Gulati, S. Maher, D. M. Findlay, D. Losic, Nanomedicine, **11** (14), 1847-1864 (2016)
3. D. Karazisis, A. M. Ballo, S. Petronis, H. Agheli, L. Emanuelsson, P. Thomsen, O. Omar, Int J of Nanomed, **11**, 1367-1381 (2016)
4. A. Mazare, Corrosion Science, **103**, 215–222 (2016)
5. M. Kulkarni, A. Mazare, E. Gongadze, S. Perutkova, V. Kralj-Iglič, I. Milosev, P. Schmuki, A. Iglic, M. Mozetic, Nanotechnology, **26**, 062002 (2015)
6. K. Narendrakumar, M. Kulkarni, O. Addison, A. Mazare, I. Junkar, P. Schmuki, R. Sammons, A. Iglic, Dental Materials, **31**, 1460-1468 (2015)
7. E. Beltrán-Partida, A. Moreno-Ulloa, B. Valdez-Salas, C. Velasquillo., M. Carrillo, A. Escamilla, E. Valdez, F. Villarreal, Materials, **8**, 867-88 (2015)
8. M. Kulkarni, et al., Advanced Material Letters, **7 (1)**, 23-28 (2016)
9. K. Gulati, A. Santos, D. Findlay, D. Losic, Journal of Physical Chemistry C, **119**, 16033-16045 (2015)
10. A. Mazare, et al, Surf. Interface Anal., **46**, 186-192 (2014)
11. K. Lee, et al., Chemical Reviews, **114**, 9385–9454 (2014)
12. G. Strnad, L. Jakab-Farkas, D. Portan, Academic Journal of Manufacturing Technology, **14** (4), 112-118 (2016)
13. G. Strnad, L. Jakab-Farkas, C. Petrovan, O. Russu, Procedia Engineering, **181**, 242-248 (2017)
14. G. Strnad, Z. German-Sallo, L. Jakab-Farkas, C. Petrovan, D. Portan, MATEC Web of Conferences **112**, 04021 (2017)
15. G. Strnad, D. Portan, L. Jakab-Farkas, C. Petrovan, O. Russu, Materials Science Forum, **907**, 91-98 (2017)

MATEC Web of Conferences 137, 02013 (2017) DOI: 10.1051/matecconf/201713702013
MTeM - AMaTUC 2017

Education packed in technology to promote innovations: Teaching Additive Manufacturing based on a rolling Lab

Laura Katharina Thurn[1,*], *Nicolae* Bâlc[2], *Andreas* Gebhardt[1] and *Julia* Kessler[1]

[1]Aachen University of Applied Sciences, Faculty of Mechanical Engineering and Mechatronics, Goethestr. 1, 52064 Aachen, Germany
[2]Technical University of Cluj-Napoca, Dept. of Manufacturing Engineering, B-dul Muncii. 103 – 105, 400641 Cluj-Napoca, Romania

Abstract. 3D Printing is about to revolutionize the way we design and produce future products. This causes not just immense research and development activities worldwide but requires intensive education and training efforts. As 3D Printing is an ideal tool to set up a decentralized production, it is one key to integrate even rural and under-industrialized parts of the country into high tech education and professional training needed. But how to reach the relevant groups - pupils from schools as well as professional staff of preferably companies? Remembered the old proverb: "If the mountain won't come to the prophet, the prophet must go to the mountain" the plan of a mobile 3D printing lab was born. Based on our experiences, we designed the technical infrastructure and the educational approach to adopt Additive Manufacturing courses to a short term comprehensive teaching. To create a fully equipped workplace for every participant, we redesigned a double-decker bus. The paper addresses the strategy behind the bus, the infrastructure in detail, the developed courses taught and the resonance on this project. Looking forward, the paper gives a sneak preview on the further development in terms of specialized courses to be given in the bus as well as in the frame of accompanying activities such as engineering design of parts, application of new materials for Additive Manufacturing such as metals and follow-up production like plastic injection molding.

1 Introduction

Additive Manufacturing (AM) is the world-wide term for the layer-based technology, which became known as Rapid Prototyping in the late 1980s. Additive Manufacturing, 3D Printing and Desktop Manufacturing are today identical names for those new production technologies that, based on a geometrical representation, creates physical objects by successive addition of material [1] and which are said to be based on a further industrial revolution.

Additive Manufacturing is capable to accelerate the product development by a quickly making complex prototypes with improves quality. AM-processes offer the unique possibility

* Corresponding author: thurn@fh-aachen.de

to build complex devices with filigree structures and integrated functionalities and to do this in a resource-efficient way. It is not only an indispensable tool for the direct digital production of models and prototypes, but also a manufacturing process for the production of end products of plastic, ceramics and metal as well as for the production of tools, molds and dies [2].

Everyone involved in design and production, but also in strategic planning, should know about the AM technology to be able to do a qualified assessment and selection.

The application fields of the technology are broad: aerospace, automotive, tool making, medical and electronic engineering, furniture, architecture, design and many more [2].

According to Wohlers Report 2017, the Additive Manufacturing industry, including all AM products and services worldwide, grew 17.4 % (Compound Annual Growth Rate - CAGR) to $6.063 billion in 2016. This includes revenues from the primary AM market that consists of all products and services directly associated with AM worldwide [3].

2 Education in the Field of Additive Manufacturing

Additive Manufacturing, 3D Printing is about to revolutionize the way we think about, design and produce future products. This causes not just immense research and development activities worldwide but requires intensive education and training efforts.

All around the world higher education institutions are developing new curricula and acquiring equipment, software, and materials to support research and instruction. The relatively low prices of desktop 3D printers make them more attractive for training in Today STEM (science, technology, engineering and mathematics) programs at different levels [3].

After initially isolated, informal qualification measures in the enterprises and through trainings from machine manufacturers, today providers of formal advanced training also take up this new technology and offer further training on Additive Manufacturing. The continuing education market as such is made up of offers from the state education system, from institutions close to the economy, such as chambers and associations, from social groups and institutions such as churches, trade unions or foundations, which also provide establishments offering continuing vocational training. In addition, private-sector enterprises with a commercial or non-profit background offer continuing education. Today, In various training programs exist for AM, such as e.g. day seminars, multi-day courses or in-house trainings. In addition, Additive Manufacturing is also integrated into technical advanced training courses, which are not specifically oriented to AM [4].

There is a huge gap between the technical state of the AM technology as a cutting edge technology and present day education that still mostly is done as always. To exploit the whole potential of this new technology that is oriented to practically make things and to remain competitive as an economic (business) location we need to search new ways of teaching.

3 "Upside down" Teaching Approach

Additive Manufacturing is a typical cross section technology. Education in this field has to include a lot of different disciplines e.g. engineering design, mechanics, electrical engineering, IT, control engineering, programming and materials sciences. In addition, education in the field of AM has to impart a high amount of practical skills.

Teaching new and emerging technologies is not just a multi-disciplinary but also a multi-level task in terms of technical knowledge. The task is not limited to teach how to operate the printer but to teach the necessary computer aided design (CAD) capabilities. Because both design and printing is necessary but time consuming, it is mandatory to keep the target group interested and motivated. For the long-term success, it is inevitable to teach not just operating skills, but a deeper involvement in the topic by analysing and feed back to improve the results.

The "GoetheLab for Additive Manufacturing", a research group of the University of Applied Sciences Aachen works in the field of Training on Additive Manufacturing (AM) and has developed, tested and evaluated a wide range of education formats for differentiated target groups.

For the training, we developed an upside down approach, better known as the "v" procedure used for the development of integrated mechanical and electrical systems [5], in our case for both: The CAD, and 3D Printing part. We started with the most complex situation (which mostly is the original), then stripping it down to maximum simplicity and finally working it out to an optimum, compare figure 1.

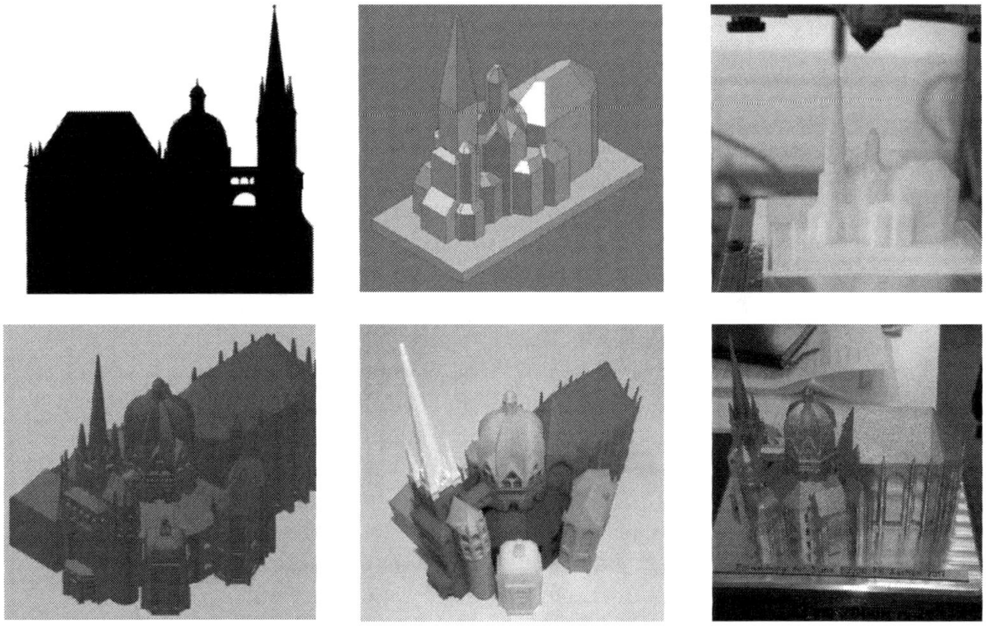

Fig. 1. From 2D drawing to a 3D print; top: left – 2D silhouette; middle – 3D CAD model; right – printed model (personal printer); down: left – complex 3D CAD model; middle – printed model plastic; right – printed model metal.

3D Printing can be taught very effectively if a brought theoretical approach is skipped and the trainees start into the real complex task right from the beginning. This leads to very fast results even accompanied by "brave trail and minor error". As these results are quite rough, the trainees themselves want to improve it and consequently become the driving force.

It turned out that the combination of a direct hands-on course with a comprehensive, theoretical approach to 3D technology meets the demands to teach not just 3D Printing but a wide range of new technologies and is necessary to transfer the knowledge.

This concept can be regarded as an example of how to stimulate social competences by means of technical, interdisciplinary knowledge with Additive Manufacturing: participants gain self-confidence by acting, they experience, give appreciation and learn from/with each other. Simultaneously the enthusiasm for STEM is awakened by an exploratory and playful approach to new technologies [6].

4 Conceptual Design and Development of a mobile Training Lab

As Additive Manufacturing is an ideal tool to set up a decentralized production, it is one key to integrate even rural and under-industrialized parts of the country into high tech education and professional training needed. But how to reach the relevant groups - pupils from schools as well as professional staff of preferably SME´s (Small and medium-sized enterprises)? Extra training, like design and operation of 3D Printers, does not fit in the regular timetable of schools nor can a company accept off-hours of their employees just to reach the training site. Consequently, rural parts of the country are again on the lose end. Small and medium-sized enterprises (SMEs) form the backbone of the economy and especially the middle class outside the metropolitan areas has problems to keep up with the latest findings of science and research. At the same time, SMEs are increasingly lacking technologically advanced specialists to implement new technologies in companies.

Remembered the old proverb: "If the mountain won't come to the prophet, the prophet must go to the mountain" the plan of a mobile 3D printing lab was born. Based on our experiences, we designed the technical infrastructure and the educational approach to adopt Additive Manufacturing courses to a short-term comprehensive Teaching.

The aim is to develop a mobile laboratory facility for flexible use in schools, educational institutions and companies. The lab should be self-sufficient and able to operate a large surface area. To create a fully equipped workplace for every participant, we redesigned a double-decker bus, see figure 2. The first floor is equipped with a teacher seat and a filament printer on each of the eight pupil seats (see figure 3 - left). The seats consist of eight computer spaces with all required software (CAD, AM, simulation and printer software). Each floor provides a conference area with enough space to hold meetings or presentations with up to twenty people. The ground floor (figure 3 - right) serves as showroom with industry-level machines and demonstration parts including printed metal parts. The round-up are chill-out-space with literature boards and a coffee bar.

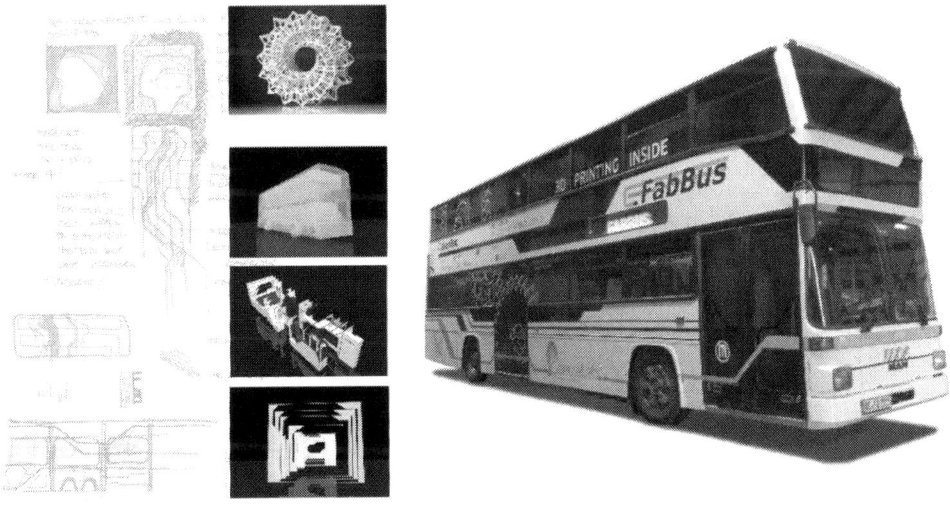

Fig. 2. Left: Design concept (Source: smd partner, bültmann-nießen gbr), right: redesigned double decker bus – 3D printing inside.

The infrastructure of the bus is designed in such a way that adapted and individualized courses can be offered to address all target groups. The bus bears the name "FabBus - 3D printing inside": the verb "to fab" (from fabricate) has prevailed in the AM scene and stands for printing on small desktop-printers.

Fig. 3. Left: First floor – Teaching area; right: ground floor – show room and "industry-area".

Especially the local industry will benefit from the mobile lab, in order to present the technology of Additive Manufacturing to their employees and technicians and to enable a "hands-on-training".

5 Experimental Research on using the FabBus in AM Education

In May 2015, the bus was inaugurated and has been touring since the borders between Germany and neighbouring countries such as the Netherlands and Belgium. The experience has shown, that completely different target groups express interest and offers, the resonance is huge.

In the period from May 2015 to the end of 2016 more than 50 missions were done using the bus. An internal evaluation showed the following main target groups:

- Schools, universities and other training facilities,
- Small and medium-sized enterprises / corporations / group of companies,
- Trade fairs, congresses and public events.

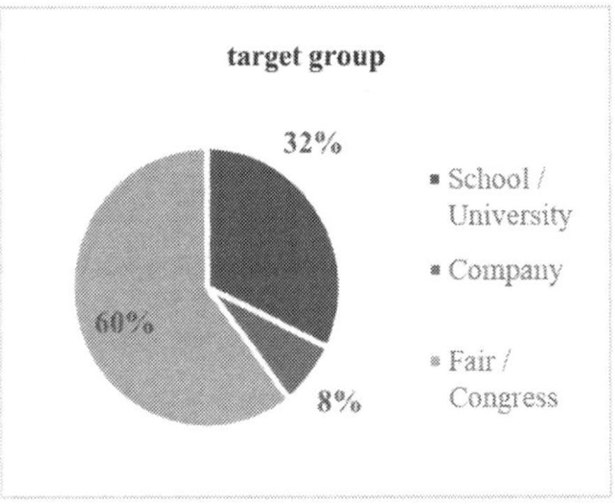

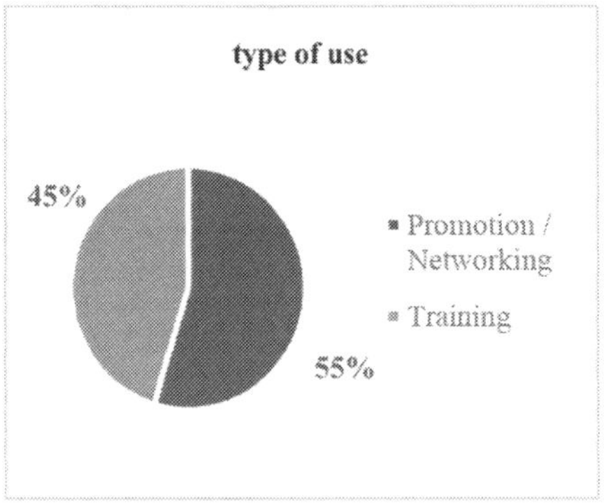

Fig. 4. Percentage distribution of bus operations (n = 53) in the period: 05/2015 to 12/2016; Left: Target group distribution, right: Distribution by application type.

There are two main applications for the mobile training lab and a distinction is made between training and promotion/networking. Training includes all events where the bus has been booked to provide training for up to 16 people at a company or at a school / university. Promotion/Networking covers those events where the bus was used as a promoter of innovation, promoter of the AM Technology. Figure 4 gives an overview of the two distributions.

It is striking that 60% of the events took place at trade fairs and congresses and only 8% at companies. It should be noted that corporate demand is increasing at the end of the review period, which is explained by the growing awareness of AM technology in general, but also by the hypothesis that training in the field of AM is becoming more and more important.

The application areas of AM technology are broad and many different industries are interested in training and further education. Because of this, it is important to offer different formats. To meet the different needs of the various target groups, new (teaching) concepts have been developed. Depending on the experiences of the participants and the desired learning objectives, the following workshops were designed and implemented inter alia with the application of the upside-down approach (compare chapter 3):

- 3D printing and its application (3 hrs.)
- How does a 3D printer work? (3 hrs.)
- From scan to a 3D print (6 hrs.)
- Simple way of 3D construction (6 hrs.)
- Understanding AM in 10 easy lessons (5 days)
- Consulting and Training for the integration of AM in businesses (individual)
- Design for Selective Laser Melting
- Demonstration workshops for beginners and advanced
- Customized Seminars and Workshops

6 Conclusion

The rolling laboratory is the core component of a sustainable and future-oriented promotion of young talent and brings Additive Manufacturing (3D Printing) from the laboratories of science to interested people in order to inform them at an early stage and to promote their ideas. With specialist knowledge, the know-how is to be applied to schools and enterprises, so that pupils, employees and technology-oriented people can experience and take advantage of the latest achievements in science and technology. As 3D Printing is an ideal tool to set up a decentralized production, it is one key to integrate even rural and under-industrialized parts of the country into high tech education and professional training needed.

Through a mobile training laboratory, the various needs and the changing (requirements) of training and further education in the field of AM can be addressed. "From lab to fab" is the task of today and from a fixed classroom to a flexible rolling lab is the way how it works.

This research was supported by the OpTi-DeP Project (no. BG101/2016) financed from the UEFISCDI by the Romanian Government and by the HORIZON 2020 AMaTUC project (GA 691787) financed from EU.

References

1. ISO/ASTM 52900 Standards: *Additive manufacturing - General principles - Terminology* (2015)
2. A. Gebhardt, J. S. Hötter, *Additive Manufacturing: 3D printing for prototyping and manufacturing*, Carl Hanser Verlag GmbH Co KG, (2016)
3. Wohlers (Wohler's Associates), *3D Printing and additive manufacturing state of the industry: Annual worldwide progress report*, (2017)
4. M. Herbert, *Personal für die additive Fertigung – Staff for Additive Manufacturing*, Springer Verlag, ISBN: 978-3-658-13307-8, (2016)
5. Gerrad Consulting, "*The W–Model*", https://gerrardconsulting.com/blog/the-w-model/, last visit 5th July (2017)
6. L. Thurn, A. Gebhardt, *Arousing Enthusiasm for STEM: Teaching 3D Printing Technology*, libreriauniversitaria.it Edizioni (2017)

Comparison of the Accuracy of 3D Printed Prototypes Using the Stereolithography (SLA) Method with the Digital CAD Models

Emil Yankov[1,*] and *Maria P.* Nikolova[1]

[1]University of Ruse, Dept. Material Science and Technology, Ruse, 8 Studentska Str., Bulgaria

Abstract. Each printing method and the materials used for prototype construction have distinct advantages and drawbacks. A large part of the 3D printer producers indicates high printing accuracy (positioning accuracies from 100 μm up to 1 μm and layer thickness from 100 μm to 5 μm). In the process of layers deposition, the material is softened, melted or irradiated with a certain light source to polymerize. Thus, during the printing process, the volume of the materials is changing causing shrinking or expansion which often leads to the occurrence of precision errors in the prototype as opposed to the digital model. This phenomenon leads to distortions of the printed object known as a curling effect. Certain chemical compounds are added by the producers to multiple materials in order to enhance their mechanical and technological properties. However, this often results in changes of some printing parameters influencing the precision and quality of the manufactured object. The aim of the particular study is to establish the allowances with regard to the accuracy during manufacturing precise objects used in micro technologies as well as the necessary adjustments in dimensions because of the shrinkage and the positioning of the SLA 3D printed parts from photopolymer material.

1 Introduction

In recent years the development of electronics, machine building and micro-technologies has imposed extreme progress of different kinds of 3D printers (FDM, SLS, SLA, SLM, DLP, DMLS, EBM, etc.) working with wide variation of materials (PLA, ABS, PETG, PVA, photopolymers, ceramics, metals, etc.). The stereolithography is an additive process of 3D printing with liquid photopolymer that polymerizes under the laser irradiation. The work table is moving up- and downwards and after immersing in the resin the photopolymer hardens by selective irradiation of the laser beam. In that way, the printed object is formed as a bottom-up physical model with clearly discernible details.

In recent years 3D models of various anatomical regions have been presented in the international "3D Printing for professional - London 2016". The scientists from Washington State University are printing bone-like structures („Inspiring ingenuity - Printing new bone") that could be used in orthopedic and dental procedures [1, 2]. This

* Corresponding author: eyankov@uni-ruse.bg

bone-like material is acting as a scaffold for the new bone growth and after that, it dissolves with no side effects or needs for surgical interventions. Additionally, the complex anatomical shapes of the craniofacial or ear prosthesis, for example, are hard to be manufactured by conventional methods. The 3D printing can cope with this challenges of today more quickly and at less cost. After scanning the object with 3D scanners or cameras "https://digitalscan3d.com/" and thus generating the high-quality mesh, the digital model for the 3D printing is acquired. Simultaneously, the 3D solid model accuracy depends on the physical properties of the material during heating and cooling that results in changes of the object dimensions. With the increase of the model size, the errors accumulate for each single layer. These errors show positive or negative absolute value that could affect the durability and exploitation of the object afterward. The ultraviolet oven treatment changed also the solid model dimensions because of the residual stresses locked in between the layered structure. That's why the aim of the particular study is to establish the allowances with regard to the accuracy during SLA 3D printing as well as the necessary adjustments in dimensions because of the shrinkage and the positioning of the printed parts from photopolymer material.

2 Experimental procedures

The digital model was constructed by CAD system SolidWorks. The size of the model was 30×30 mm and 3 mm thickness. The micro-squares within the model had dimensions of 200×200 μm and depth of 500 μm with a separating interval of 400 μm in X and Y directions (Fig. 1 a). The printed model was received by means of SLA printer Formlabs 2.0 (Fig. 1 b) using standard photopolymer "Clear" and 25 μm prescribed values of printing precision. The model was 45° tilted with respect to X, Y and Z axis in the work space and the supporting props were fitted at the edges of the object in order not to affect the examined area. For ensuring the fixation stability the distance between the props was selected to be 600 μm and the size of fastening point between the model and the supporting prop was 400 μm. The supporting props were reinforced by a belt with 3.5 mm thickness positioned in bedplate of the model. The depth of the base sticking to the work table was equal to 1 mm.

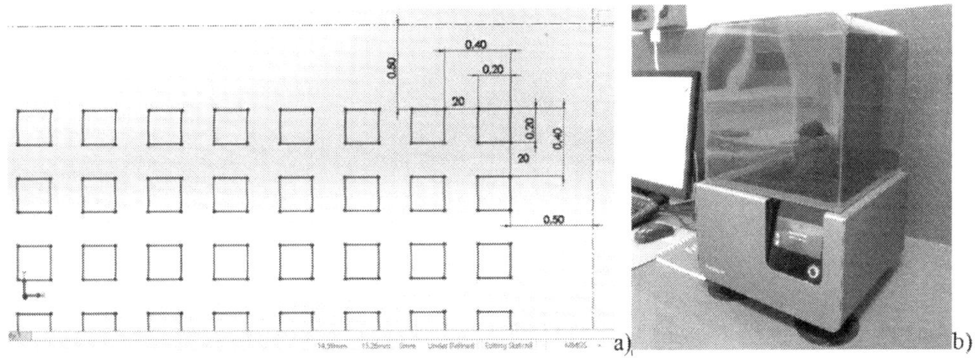

Fig. 1. Manufacturing of physical 3D model: a) providing of the net of micro-squares; b) Formlabs 2 3D printer.

The measurement of the coordinates of the micro-squares after the 3D printing was carried out by a "Carl Zeiss" optical microscope at 12.5× magnification. The digital micrograph acquired by 5 Mpix camera and S-EYE software (Fig. 2) was used to measure the micro-grid deviations of the printed model.

DOI: 10.1051/matecconf/201713702014

As shown in Fig. 3, the coordinate grid method [3-5] that consists of measuring the relative "distortions" in different directions (ε_x and ε_y) was used to calculate the deviations during the photopolymerization of the printed model. The following equations were used:

$$\varepsilon_X = \frac{\Delta X}{X_i}, \quad \varepsilon_Y = \frac{\Delta Y}{Y_i} \tag{1}$$

Where ΔX was the drift value of the X axis and ΔY was the drift value of the Y axis. The X_i and Y_i were the original (initial) dimensions in the model and $X_{n,i}$ and $Y_{n,i}$ was the changed values in dimension (Fig. 3).

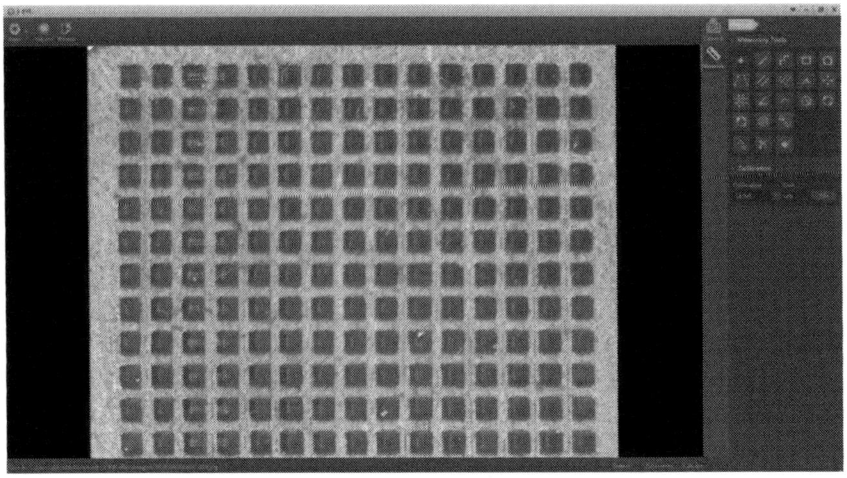

Fig. 2. Digital image and measured values of the 3D printed model.

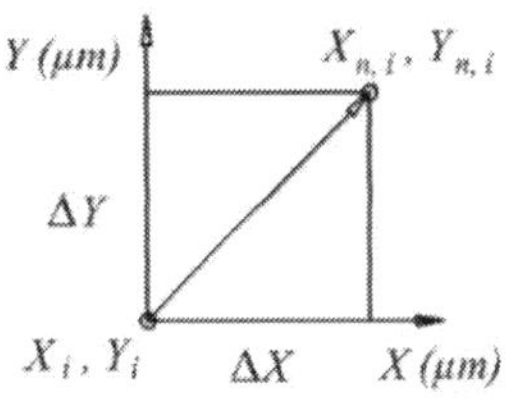

Fig. 3. The setting of the deviations by the coordinate grid method.

The coordinates of one of the corner point were used as reference fiducial marks. These values were entered into Excel tables in order to determine the differences between the coordinates of the specified digital model and those of the 3D printed prototype. The measured X and Y values were split into two separate tables for the purpose of defining appropriate tolerances between the prototype and the digital model of both axes.

3 Experimental results

After the mathematical processing of the results obtained (Fig. 4), the data are presented in graphical form (Fig. 5) showing the relative deviations of each cell against the reference digital model. Fig. 4 shows the tabular deviation in the values by color code in X and Y directions. A positive displacement could be noted in the upper left part of the model. That is probably a consequence of the chosen option of positioning and the selected fixation scheme. A further reason for this effect could be the model hanging and the near located supporting props.

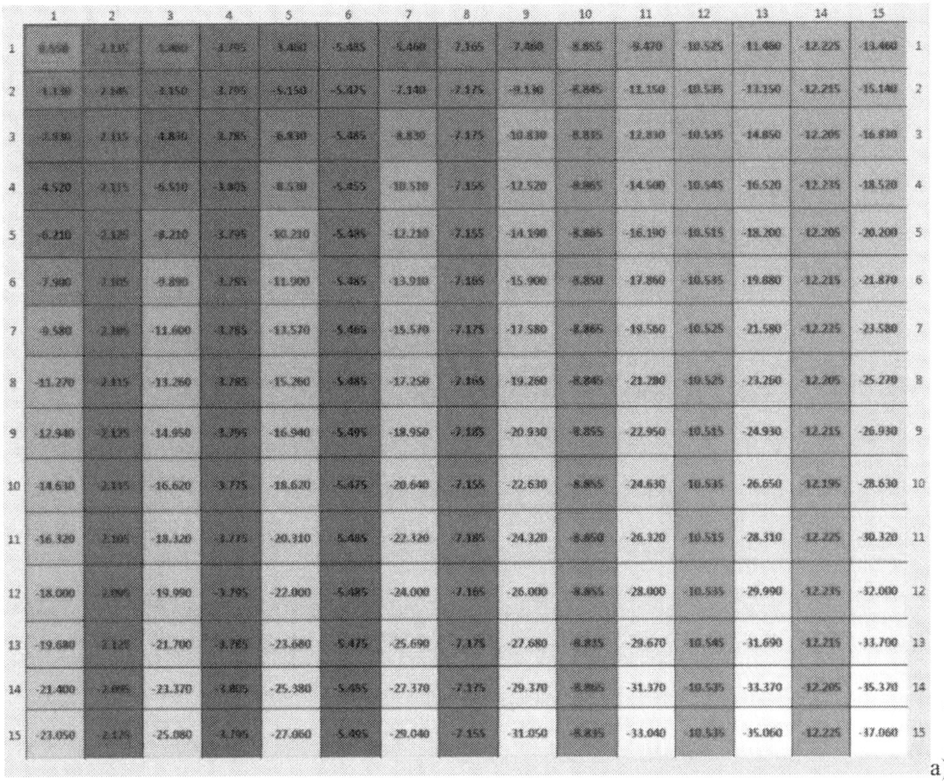

a)

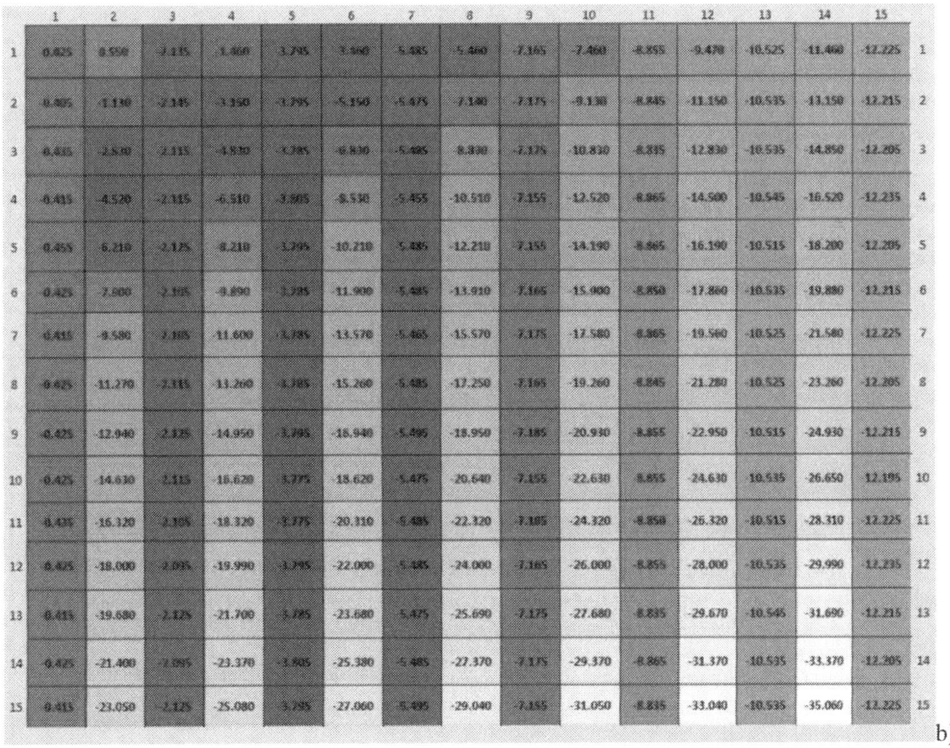

b)

Fig. 4. Colour relative deviation of the coordinates of: a) X axis; b) Y axis.

DOI: 10.1051/matecconf/201713702014

In contrast, in the lower right corner, there is a size reduction of the model. The reason for this could be the model shrinking after exposure to the UV lamp and the additional heat setting.

The calculated relative values are presented in 3D graphical form (Fig. 5) with the aim of obtaining a dimensional picture for the tolerances in X and Y directions after the 3D printing.

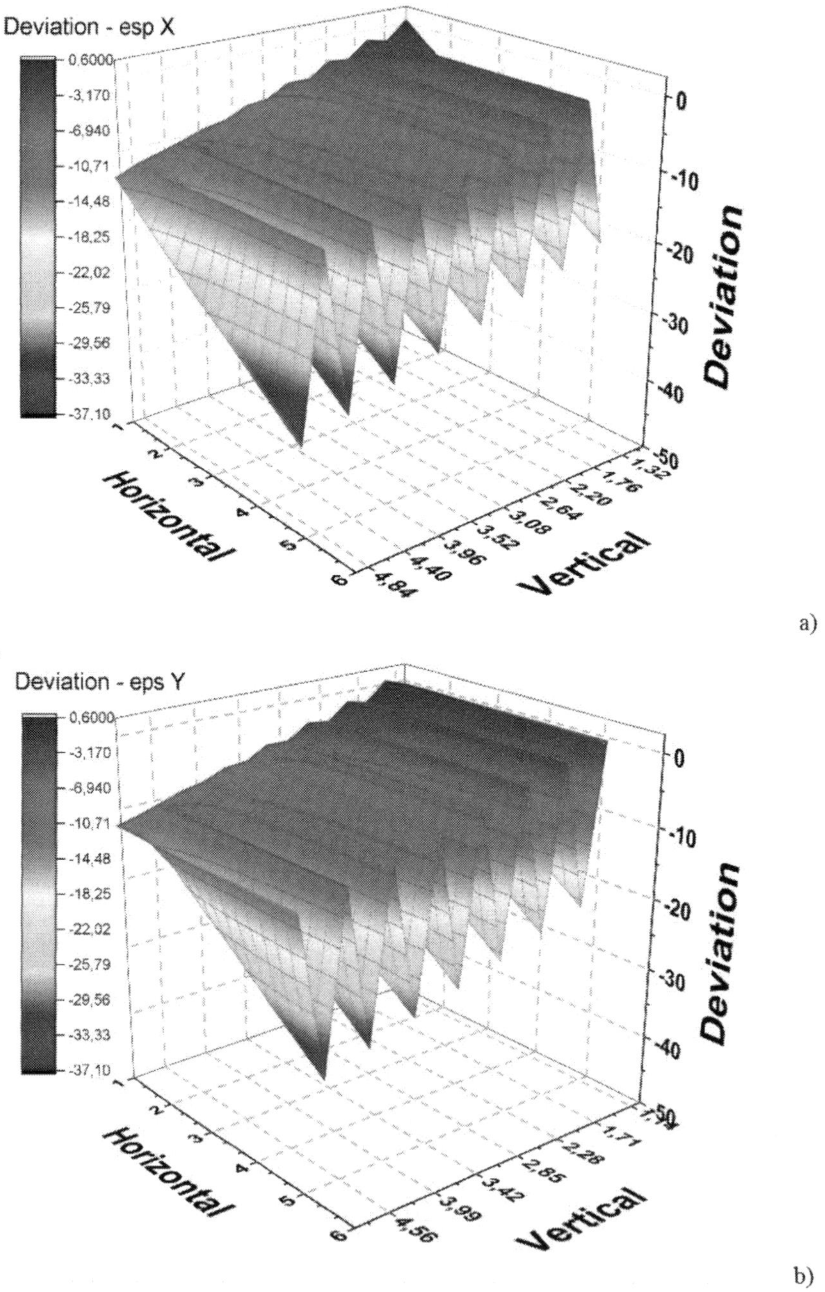

Fig. 5. 3D graph of the size deviation of: a) X axis; b) Y axis.

The deviations in both X and Y axis increase with the rise in the number of layers (Fig. 5). These errors accumulate and show permanent character shifting towards the negative values during the building up of the solid model.

4 Conclusions

The main conclusions drawn on the basis of the particular research are:
• The coordinate grid method allows rapid and efficient measurement of the deviations in the nominal size values of the printed (micro) parts. Additionally, the changes in size could also be determined in different directions (for example X and Y).
• The lowest divergences in the nominal size values are measured in the upper part of the printed model where the first layer is built and the model is near to the moving table. This solid part of the object is connected to the table with shorter props. This means that the random and systematic deviations in the X direction (parallel to the table surface) are less. •
• An additional factor to consider is the immersion and removal of the part. The more the built layers are moving away from the table, the more their "mobility" will grow and the deviations increase. The highest is the size deviations in the last polymerized layers.
• With regard to the Y axis deviations - ε_y (eps Y), the same trend is seen: the greater the distance from the table, the higher and more irregular the deviations with respect to the absolute values of the digital model.
• In the upper part of the object near to the table, the dispersion of the values is similar with respect to the end of the solid model but the latter are shifted towards negative values at both X and Y axis.
• The results obtained will be valuable for introducing of size corrections in the digital models in order to print more accurate 3D physical model.

In order to determine the reasons and confirm the results, additional examinations will be needed.

References

1. H. Y. Yu, D. D. Ma, B. L. Wu, Nan Fang Yi Ke Da Xue Xue Bao (Journal of Southern Medical University), **37** (5) 668-672 (2017)
2. M. A. Kuss, R. Harms, S. Wu, Y. Wang, J. B. Untrauer, M. A. Carlson, B. Duan, RSC Advances **7**, 29312-29320 (2017)
3. N. Feschiev, V. Gagov, E. Minev, Conf. Proc. Avangard technologies and machine building materials, VMEI Gabrovo, 187-191 (1992)
4. B. Tomov, V. Gagov, N. Feschiev, E. Minev, Proc. of 9th Machine Tool Conference, University Politechnica of Bucharest, **9** 264-269 (1994)
5. V. Gagov, Testing and Processing of Contemporary Materials, Ljubljana-Ruse-Gliwice, 125-136 (1997)

 DOI: 10.1051/matecconf/201713703001

The influence of hole finishing strategies on quality

Vlad Bocanet[1] and *Nicolae* Panc[1,*]

[1]Technical University of Cluj-Napoca, Department of Manufacturing Engineering, no.103-105, Bvd. Muncii, Cluj-Napoca, Romania

Abstract. There are many strategies that the operator can use for hole finishing, depending on the equipment available to him. The issue at hand is choosing the optimal strategy when equipment is available for more than one possible manufacturing method. The present work analyses four types of finishing strategies for holes and determines the advantages and disadvantages of each, from a quality perspective.

1 Introduction

In our days, there are many technologies for manufacturing a part. Each technology has its advantages and disadvantages. Depending on the requirements that the part must satisfy, the most appropriate technology is chosen. If theoretically this is easily done, in practice it is limited by existing possibilities. The usage of optimal technologies is not always possible and, thus, it is necessary to use the optimal existing technologies [1].

In the case of hole finishing there are multiple ways of obtaining the same hole through different strategies or by using different tools. The same diameter hole can be obtained by using a mill, reamer, boring bar, turning tool or abrasive body [2]. Choosing the optimal strategy depends on the existing equipment, the machine tools used, the shape of the part being machined and many other factors that influence the machining strategy. A precise hole is an interior revolution surface that is different from a hole through its superior accuracy, low shape errors (which are standardized) and high surface quality [2-4].

The aim of this article is to study the way in which the quality of a precision hole is influenced by the chosen manufacturing strategy. The quality of a precision hole considers four aspects deemed relevant: dimensional, shape and actual position accuracy, as well as the resulting surface. All these aspects are important in precision holes with a functional role (sliding or ball bearings, guide bushings, centering elements etc.). Thus, the surface is important for precision holes that have shafts performing rotational or reciprocating movement because of the need of adequate lubrication. In such situations, the surface can have an influence on the maintenance of the oil film [5]. The actual position accuracy is important in the case when the precision hole is used for guiding an element passing through it. The form accuracy is also important when the precision hole needs to ensure a certain load capacity for its corresponding element. The lack of an appropriate form

* Corresponding author: nicolae.panc@tcm.utcluj.ro

accuracy can lead to an uneven distribution of the load on the hole surface leading to uneven wear and, as a result, a low reliability [6, 7].

Next, we will compare the quality of the precision holes obtained though the most common manufacturing techniques and identify the optimal technology but also determine the influence of certain factors on hole quality.

2 Research methodology

To reach the proposed aim, a certain methodology was used, which is presented briefly in figure 1. In total four different techniques were used on three different material. After the rough boring, the four methods were applied for the finishing part. The characteristics were measured and the data was analyzed. The obtained results were analyzed quantitatively.

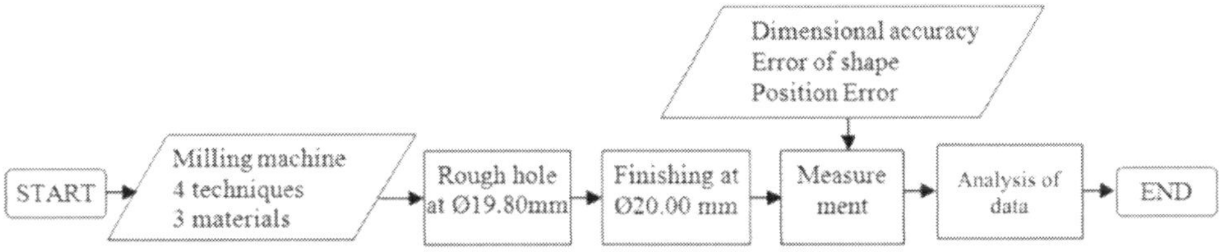

Fig. 1. Flowchart of the proposed methodology.

2.1 Experimental design

The experiment consisted of machining three different types of materials from the most commonly used material groups in industry: steel and aluminum alloys and plastics. For each group, a material was chosen that is frequently used:
• Steel alloy – S395;
• Aluminum alloy – AW7075;
• Plastics – copolymer POM-C.

The two parts that were machined from each material are presented in Figure 2 and 3. Figure 2 shows the part drawing while Figure 3 shows the actual parts that were machined.

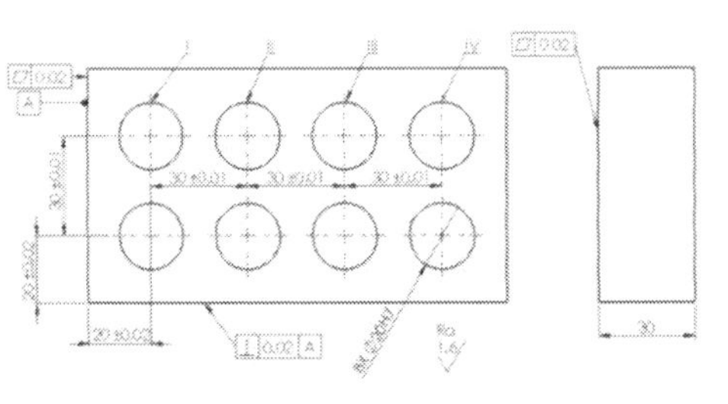

Fig. 2. Drawing test part.

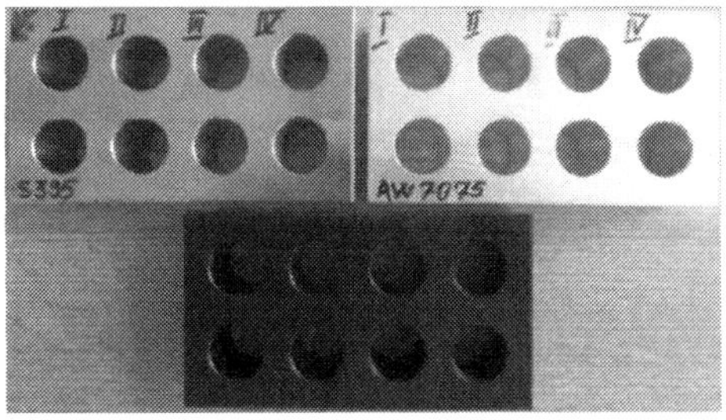

Fig. 3. Manufactured parts.

Two holes were executed in each part, for each of the four strategies, a total of 24 precision holes resulting. The four strategies used are:
I – helical milling;
II – profile milling;
III – head boring;
IV – reaming.

The holes were first drilled to Ø19.80 mm and then, the Ø20H7 diameter was obtained through a single pass of the tool.

2.2 Equipment and measurements

The following equipment was used for machining and measuring the precision holes:
• Vertical machining center (VMC);
• TRM50/50 Boring head Testarossa D'Andrea;
• HSS Reamer for Ø20H7;
• Drill Ø19.80 mm;
• Solid carbide end mill, Ø16, z=4 mm;

The machining parameters that were used are presented in Table 1. The same work conditions were used for all material types in order to do a comparative quantitative study.

Table 1. Parameters of cutting bores.

Method	*Feed*	*Spin*	*Tooling allowance*
	[mm/min]	[rev/min]	[mm]
Helical milling	1000	3500	0.02
Contour milling	400	3500	0.02
Head boring	120	2800	0.02
Reaming	160	80	0.02

The measurement was done on a Zeiss Prismo Navigator Coordinate Measuring Machine (CMM) equipped with a VAST sensor and 5 mm ruby tactile probe. According to the manufacturer the CMM has a length measurement error of 0.5 + L/500 µm.

3 Results

The quality of the precision holes was determined through the measuring of the dimensional and surface quality characteristics. In the case of the dimensional measurements the dimensional accuracy, error of shape and position error were determined.

After the measurement on the CMM, the data was stored in an Excel worksheet. When processing the data, two main variables were taken into account: the working method used and the material of the part. The averages were calculated for the diameter, the circularity and the cylindricity.

3.1. Dimensional accuracy

If we compare the methods numerically, we get the data in Table 2. From the table, it can be seen that although head boring and reaming give on average the least deviation from the nominal value, their spread is much greater than that of helical and profile milling.

Table 2. The average diameter and the standard deviation for the values of each working method, in millimeters.

	Helical milling	Profile milling	Head Boring	Reaming
Average	20.007	19.993	19.995	19.995
Std. Dev.	0.008	0.010	0.018	0.054

Graphically we can compare either the nominal diameter obtained (Figure 4.a) or the deviation from the nominal value (Figure 4.b). If we also add the "material" dimension into the analysis, it can be seen from Figure 4 that reaming and head boring have the greatest variation while the first two methods have comparable spreads across the material variable.

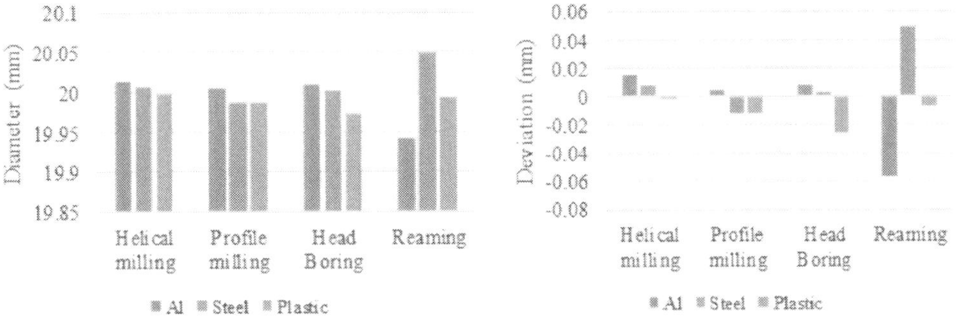

a) b)

Fig. 4. Influence of the working method on the diameter, a) absolute values, b) relative to the nominal value.

If we consider the influence of the material (Figure 5) we notice that plastic has the least variation across working methods in comparison with metals. We also see that, in the case of metals, reaming skews the distribution of values quite a lot.

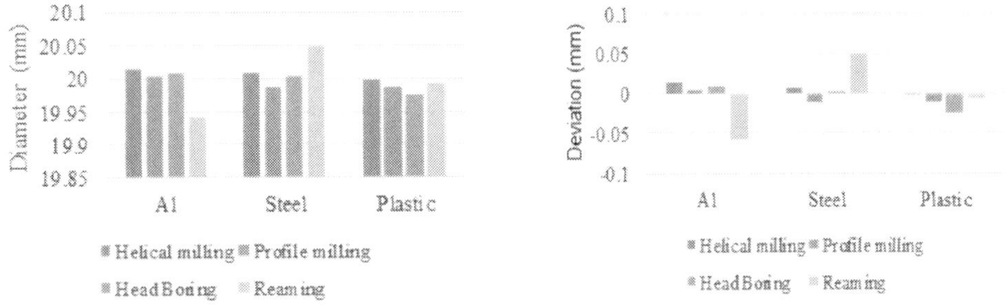

a) b)

Fig. 5. Influence of the material on the diameter, a) absolute values, b) relative to the nominal value.

3.2 Shape deviation

Shape deviations were measured by using circularity at three different levels and the cylindricity. The three considered levels were: upper section (3 mm from the top), middle section (15 mm from the top) and lower section (27 mm from the top). The profile of the precision hole can be determined by observing the variation in diameter from one section to the other (Figure 6). It can be seen that deviations from the nominal value tend to be higher in the upper section and lower in the lower section. For aluminum and plastic, the initial diameter tends to be smaller and then approaches the nominal value while for steel the upper diameter is larger (Figure 6.a).

If we compare the average circularity (computed by averaging the circularity of the three sections) for the precision holes, taking into account working method and material, we get the graphs in Figure 7. It can be seen that the highest deviation is for plastic and for reaming.

We get similar results for cylindricity, as shown in Figure 8 although the other values seem to be close together.

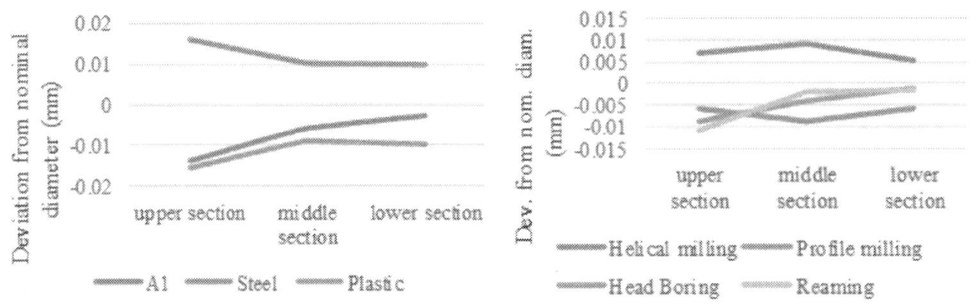

a) b)

Fig. 6. Influence of the precision hole shape, a) material, b) working method.

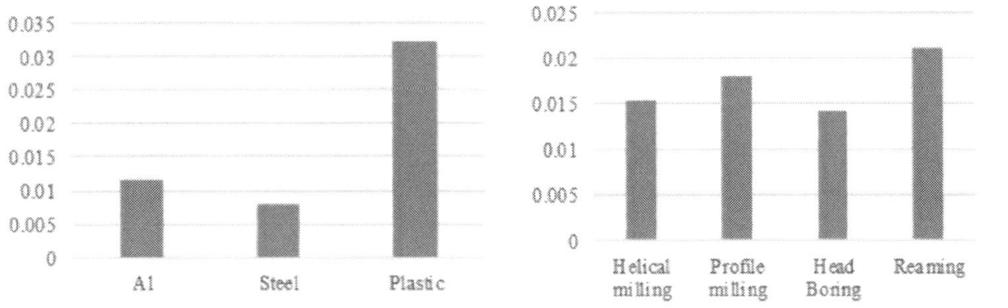

a) b)

Fig. 7. Influence on the average circularity of the precision hole, a) of the material, b) of the working method.

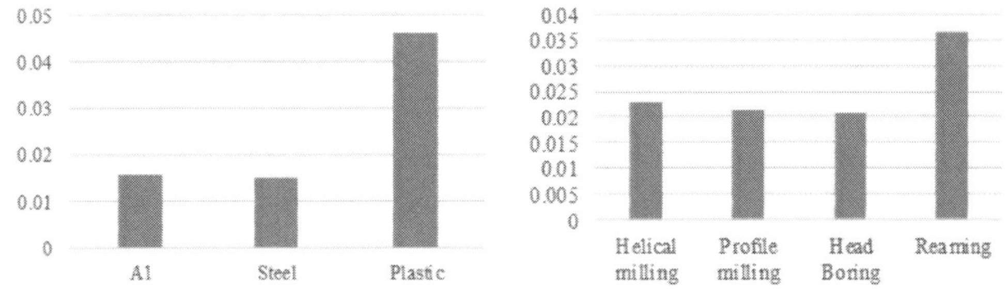

a) b)

Fig. 8. Influence on the average cylindricity of the precision hole, a) of the material, b) of the working method.

4 Conclusions

The present paper analyzed the influence of machining strategy on hole quality. During the research both machining strategy and material were taken into consideration.

After interpreting the results of this study, the following conclusions were reached:

• The highest quality can be obtained by helical milling, it being the most stable strategy, with a high repeatability rate resulting from the low standard deviation (<0.01 mm);

• Reaming is the most unstable strategy, being highly influenceable by tool wear;

• Plastics (POM) seem to be fairly indifferent to machining strategy with a constant dimensional accuracy;

• Helical milling is recommended for obtaining a high form precision for holes;

• Head boring followed by helical milling is recommended for obtaining a high circularity;

• To obtain a good cylindricity, one could use either head boring, helical milling or profile milling.

The conclusions drawn in this paper are based on the results of this study. However, they need to be confirmed through a more extensive study, containing a higher number of parts in order to get more statistically relevant results.

The results presented in this paper were obtained with the support of the Technical University of Cluj-Napoca through the research Contract no. 2002/12.07.2017, Internal Competition CICDI-2017.

References

1. I.Vuscan, N.Panc, *Bazele prelucrarilor mecanice*, (Ed. Scoala Ardeleana, Cluj-Napoca (2015)
2. N. Panc, V. Bocanet, M. Bulgaru, C. Beldean, Acta Technica Napocensis, Series Applied Mathematics and Mechanics, **57** (1) (2014)
3. I.A. Popan, A. Popan, S.C. Cosma, A. Carean, International Conference on Computing and Solutions in Manufacturing Engineering - CoSME'16, published in MATEC Web of Conferences **94** (2017)
4. I.A. Popan, N. Balc, A. Carean, I.A. Popan, Proceeding of 16th International Scientific Conference - Automation in Production, Planning and Manufacturing, Zilina-Oscadnica, Slovakia, 117-122 (2015)
5. J. Paulo Davim Editor, Surface Integrity in Machining, (Springer-Verlag London Limited) (2010)
6. S.A. Radu, A. C. Popescu, N. Balc, N. Panc, Proceedings in ARSA-Advanced Research in Scientific Areas (2014)
7. I.A. Popan, N. Balc, A. Popan, D. Fratila, A. Trif, Applied Mechanics and Materials, **808**, pp 54-59 (2015)

Studies of the laser etching on painted plastic parts to prevent the risks of engraving failures at mechatronic devices

Ion Cristian Braga[1,*], *Anisor* Nedelcu[1] , and *Razvan* Udroiu[1]

[1] Transilvania University of Braşov, Department of Manufacturing Engineering, Mihai Viteazu No.5, Braşov, Romania

Abstract. Nowadays, the automotive mechatronic devices are developed with many symbols engraved over the painted plastic parts. Many risks are coming due to painting layer, which could be dry or not, the paint thickness variability it is also one important aspect. The paper's aim is to present the studies of the laser etching process and the risk reduction of the possible failure by using the 2D optical measurement of different parameters for the laser beam, also using the 3D scanning in order to see the deep etching of the paint. The experiments will let the team know what should be the parameter setting for the painted plastic materials delivered in production in different time after the painting, therefore to reduce the over-adjusting and first part approval process duration.

1 Introduction

The word laser is coming from an acronym, LASER, standing for "Light Amplification by Stimulated Emission of Radiation". History of the laser began in 1916 with the first concept of the stimulated emission published by A. Einstein in the paper "On The Quantum Theory of Radiation" [1]. But the first working LASER was invented by Theodore Maiman who built this in nine months based on the ruby rod from his MASER (predecessor of the LASER, but using microwave instead of light) research, so in 16th of May, 1960 "a trace of red light pulse was registered on a Hughes Memoscope" [2].

There are different types of the laser according to the physical state of view: solid state laser, liquid laser, gas laser, X-Ray laser, chemical laser, excimer laser, fiber hosted laser. The solid state laser could be classified based on an insulator as Ruby Laser, Nd: YAG and Nd: Glass Lasers, Yt: YAG laser, Color Center Laser, Alexandrite Laser, Ti-Saphire Laser, also based on a semiconductor as Gallium Arsenide Laser or Gallium Nitride Laser. The gas laser could be classified based on the atom laser as Helium-Neon or Helium-Cadmium, molecule as CO_2 (Carbon Dioxide) or N_2 (Nitrogen), ion laser as Argon ion or Krypton ion, and metal vapor laser as Copper Vapor or Gold Vapor. The fiber hosted lasers are based on Erbium and Yttrium ions in double clad forms [3].

* Corresponding author: jbc69mc@gmail.com

In automotive industry are often met CO_2 gas laser where the main emission wavelengths are 10.6μm and 9.4μm, and the Nd: YAG laser where the main emission wavelength is at 1.064μm.

The Nd: YAG is an acronym stands for Neodymium: Yttrium-Aluminium-Garnet and is a rod based on the synthetic crystalline material of garnet with formula $Y_3Al_5O_{12}$ doped with Neodymium, a rare earth element, and where some Y^{3+} ions are replaced by the other ions of Nd^{3+}. The primary emission of Nd: YAG wavelength is 1064 nm [4].

The Nd: YAG laser is optically pumped by a flash tube or laser diodes. In the laser cavity is a Q-switch which control the maximum population inversion of the neodymium ions to generate modulated pulses. The optical resonator consists also in a highly reflective mirror (HR) which amplifies the light and an output mirror (OM) a partially reflective mirror with small hole which reflect the most of all light inside of cavity and let some photons to escape – the beam of the laser (see Figure 1) [5,6].

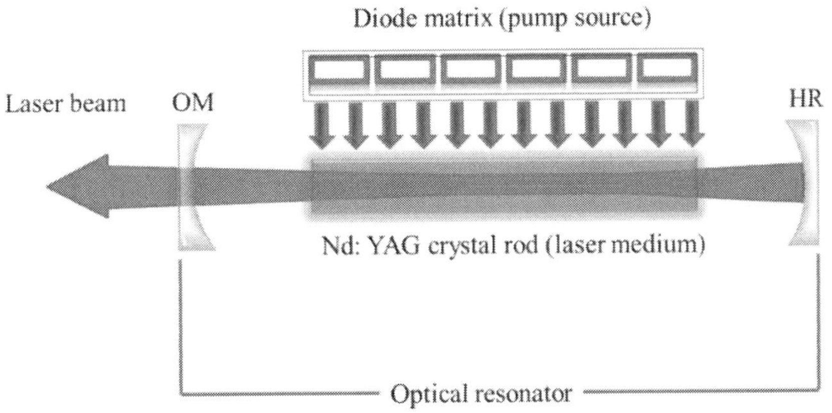

Fig. 1. Schematic diagram of the Nd:YAG laser diode pumped.

The Nd: YAG lasers are often met in the automotive industry and are used for cutting, welding, engraving, marking, etching or ablation. As similar researches in the field are related to the generalities of the using of the Nd:YAG laser for etching, but not for the influence of the parameters over the final result, the aim of this paper is to make the analysis of the Nd: YAG laser parameter variation at etching painted plastic parts where the drying of the painted parts is variable too.

2 Case study

The case study is made in an automotive manufacturer which produces mechatronic devices. As the manufacturing processes, these devices consist of assemblies of electronic, metallic and plastic components. Many of them are used in the interior of the cars and are in the visible area of the driver or the other passengers.

The plastic components from the visible area with different symbols are illuminated by the light behind coming from the electronic components. Therefore, the plastic molded injected parts are mainly white or transparent and painted over the surface. The painting systems are water based, as solvent based too. At the end, the symbols are made by removing the paint by laser etching as is presented in Figure 2, a) and b).

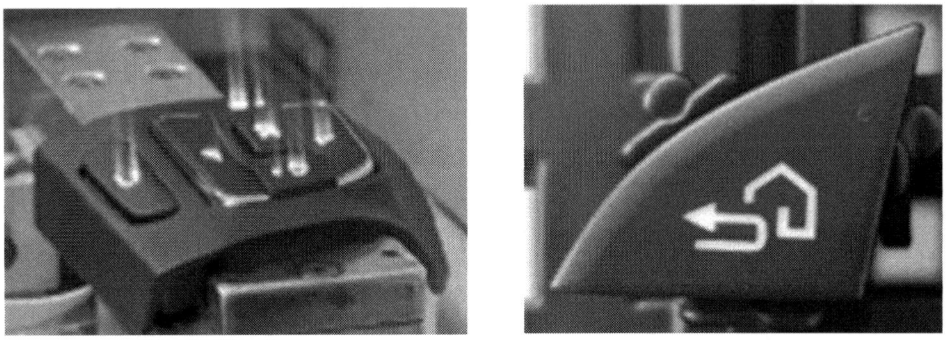

Fig. 2. a) Laser etching of painted plastic parts. b) Symbol after laser etching.

The aim of the study is to find a correlation between the laser setting parameters and the drying stage of the painted parts, as was known that sometimes the painted parts are delivered to laser etching almost immediately after the end of painting line or even after more than 8 hours. The risk to deliver parts with not good engraved symbols is present, the non-conform parts could be with a symbol not clean (paint residues in the white or transparent areas) or burn material due to the intensity of the laser beam. An adjustment is needed at the change of the batch as every program for each dedicated project part number is uploaded by the operator.

The laser device used in the experiments is a Rofin-Sinar Laser, model Power Line E-20, maximum power P_{max} = 1.77 kVA and maximum current I_{max} = 7.7 A, Diode pumped Nd: YAG laser, pulsed (Q-switch), with wavelength: 1064 nm. The device is integrated into the end of line testing equipment and it is used only for specific families of a product.

The plastic components used in the mechatronic device are molded injected from a polycarbonate (PC), Makrolon 2405, a translucent thermoplastic coated by a paint system from Alexit, hardener mixed with the laser coating, one layer sprayed in a Venjakob painting line [7,8].

The experiments were developed based on the etching of the parts at different timing direct after the painted batch was out of the line. According to the process, the plastic painted parts are oven-dried around 40 minutes at almost 80 degrees Celsius, as is presented in the curve from Figure 3, measured with the Voltcraft - termologger 309, digital, with sensors Tc type K.

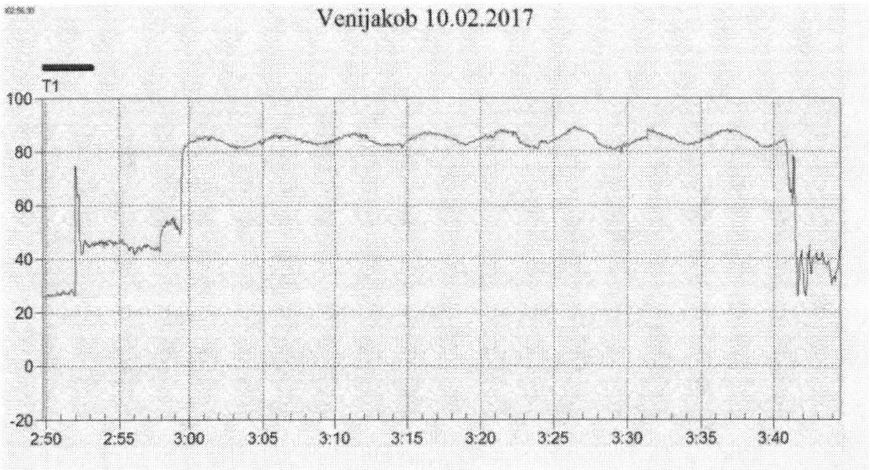

Fig. 3. Drying curve oven Venjakob.

The number of selected parts for testing was 18 button sets, for each of one experiment out of six have been used 3 button sets, first two sets for testing of the adjustment, and only the last one evaluated and measured.

The time for laser etching was established based on the issues coming from manufacturing in the last half year, and the investigation shows that the most of all parameter changing have been made when the parts arrive in laser between 45 minutes to 3 hours after end of painting line (the parts need to be visually checked after the conveyor and the duration of this operation is around 30-40 minutes). Therefore, the test samples have been sent to laser etching as is presented in Table 1. Must be mentioned that the storage environment is controlled by the sensors and assured by an installation that provides a certain level of humidity and the temperature (according to the requirements from the electronic production in automotive). As can be seen in the table, the humidity as well the temperature has almost no variation.

Table 1. Sampling for testing.

Sampling		**The storage environment**	
Test No.	**Duration (minutes)**	**Humidity (%)**	**Temperature (Celsius)**
T1	10	43.20%	23.8
T2	60	43.80%	23.9
T3	120	43.50%	24.0
T4	240	43.50%	24.0
T5	640	43.60%	24.1
T6	1440	43.50%	23.9

The chosen parameters in analysis from laser equipment to work with are the following: the laser current, the speed, and the pulse width.

Nowadays, the manufacturer is challenged by the large number of parts to be produced, it means that is very important which is the efficiency of each step from the processes, then the reducing of the speed and the pulse width is not a good option because will increase the tact time, reducing number of parts per hour. As a consequence, the speed remains at 1000 mm/s and the pulse width at 100 us (microseconds). The study was based visual validation of the part by the quality inspector, and by the laboratory measurements in the 3D scan and 2D optical equipment. The requirements are established for a clean area after removing the paint, but also the dimension of the symbol width, as is presented in Figure 4, a) and b).

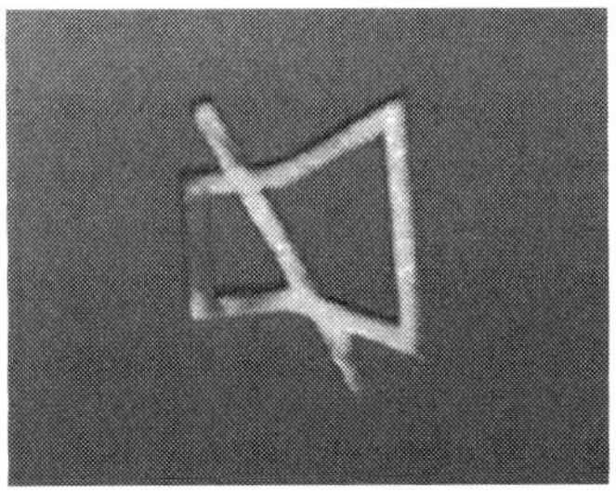

Fig. 4. a) Incomplete removal of the paint. b) Symbol width too big.

The adjustment of the current was performed in three steps: the first step, based on the experience, the current value was set at one estimated value, and then the part was visually inspected. Based on the result of the inspection, the second step was made an increase between 0.2A to 0.6 A, and in the last step usual only fine adjustment, between 0.1A to 0.2A. As the most of all situations, the parts are dry, for the wet parts (direct from the painting line) the current was decreased because of the wrong estimation at the first adjustment (see Table 2).

Table 2. Adjustment of the parameters of laser.

Sampling		Trial 1	Trial 2	Trial 3
Test No.	Duration (minutes)	Current (A)	Current (A)	Current (A)
T1	10	21.0	20.5	20.1
T2	60	20.0	20.6	20.7
T3	120	20.6	20.8	21.0
T4	240	20.8	21.0	21.1
T5	640	21.5	21.5	21.5
T6	1440	21.5	21.5	21.5

As the result of the measurements can be observed that the current adjustment for the last tests, T5 and T6, was the same and not needed to be changed from the steps. Also can be notified that risk to produce a part with laser etching non-conform is very big if the parts are wet and delivered immediately after the end of painting line, and still big is the parts arrive in the assembly in less than an hour. And this, because usually the laser is adjusted at the bigger current value which will burn the plastic material and the symbol width will be out of the tolerance (Figure 5).

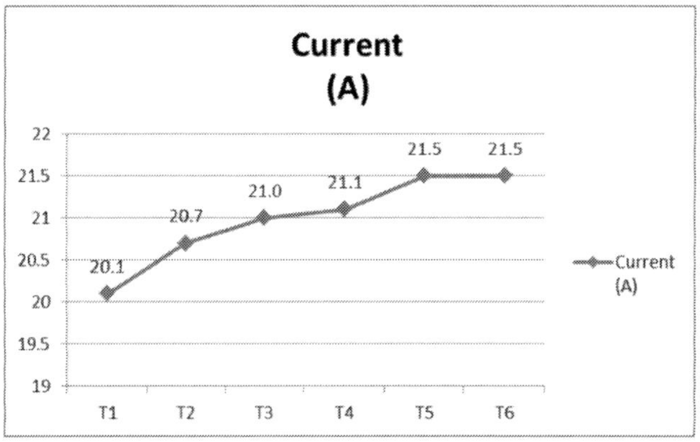

Fig. 5. Graph of laser current variation.

All the parts from the last adjustment were measured in the laboratory in 2D O-Inspect Zeiss equipment. The microscope was programmed to measure the distance between the line 1 and line 2, in fact, the border from the removed paint and the plastic material, as can be seen in Figure 6, a) and b).

Fig. 6. a) Symbol width programming – line 1. b)Symbol width programming – line 2.

For instance, in Figure 7, a) and b), can be easily observed that the test T2 had the NOK result of first step adjustment at 20.0 A, and the laser etching was not completely, then in third step the dimension was according with the requirements (0.4 MM +/ 0.1 MM).

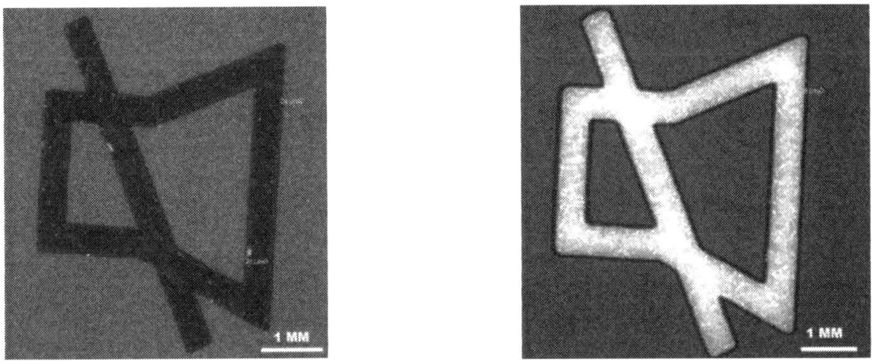

Fig. 7. a) Symbol width T2 step 1. b) Symbol width T2 step 3.

In order to prove the hypothesis that the current adjustment is enough to be adjusted and as wet the painted parts are, so small the current must be, the depth of laser etching was measured in 3D measuring techniques with ATOS ScanBox - Optical 3D Coordinate Measuring Machine, Series 4, ATOS CORE MV 135. Based on the evaluation of all tests, was observed a variation in an average of the depth around 5 microns if compared the test T1 and test T5 - see Figure 8, a) and b).

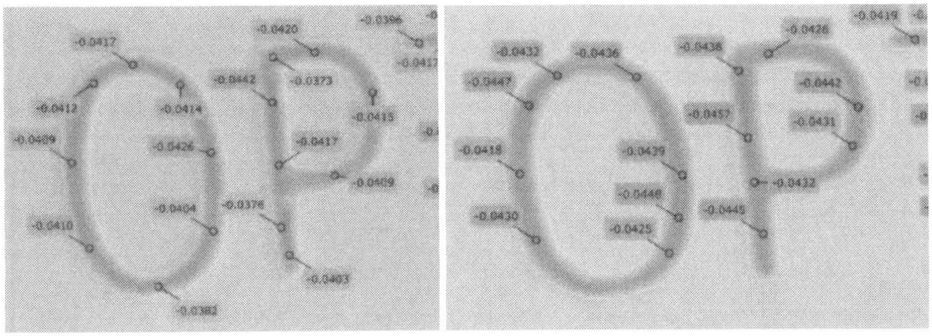

Fig. 8. a) Symbol depth T1 step 3. b) Symbol depth T5 step 3.

Conclusion and next steps

The study shows the influence and adjustment of the current at laser etching of painted parts for the situation when the parts are wet or dry. In that way, the process can be developed in such a way to give the signal to the next operation when the parts are delivered directly from the painting area or they are stored in the warehouse. The experiments used the plastic molded injected parts from polycarbonate (PC) coated with a paint system water based and only one layer on the parts.

Similar studies were made of the other pieces of the same plastic material but using the paint system in two layers (from the same provider), the results are not relevant and further tests with more samples will be done.

As a lesson learned, in the time of the studies were observed differences between the lasers etching equipment in regard to the type used, the presented study has been made on Rofin equipment, and the second study on the Foba equipment. More, the life duration of the laser diode for the Rofin equipment was at the beginning, and the life duration of the laser diode from Foba equipment was almost half. Therefore, a study to understand the variation due to changing of the current and the life time of the laser diode is to be made. Also is good to know if an optimal configuration of the parameters such as current, pulse width, speed and line width or other beam parameters could be found by using the design of experiment Taguchi method in order to obtain a higher level of generalization of the research.

References

1. J. Hecht, *Short history of laser development, Optical engineering*, **49**(9), 091002 (2010)
2. A. H. Rawicz, *Theodore Harold Maiman and the invention of Laser*, Proc. of SPIE 713802-1 (2008)
3. C. Nitish, B. Sudeep, IJRAP, **2**(1), 132-141 (2011)
4. M.C. Rao, Research Journal of Material Sciences, **1**(2), 20-24 (2013)
5. J. Dong, X.-S. Liu, C. Peng, Y.-Q. Liu, Z.-Y. Wang, Applied Sciences, **5**, 1837-1845 (2015)
6. ***, LASER Light Amplification by the Stimulated Emission of Radiation, https://lasersdbw.larc.nasa.gov/tutorials/laser/ (2017)
7. ***, Venjakob, Ven Spray Perfect, https://www.venjakob.de/en/our-plants/detail/ven-spray-perfect-17/index.nc (2017)
8. ***, Mankiewicz, Product and markets, Automotive, http://www.mankiewicz.com/english/products-markets/automotive/ (2017)

DOI: 10.1051/matecconf/201713703003

Tracking of the development wear of cutting tools with different geometry during parting-off

Robert Čep[1,*], *Šárka* Malotová[1], *Tomáš* Zlámal[1], *Vladimír* Vrba[1] *and Marian* Borzan[2]

[1]Faculty of Mechanical Engineering, VŠB-TU Ostrava, 17. listopadu 15/2172, Ostrava-Poruba, 708 33, Czech Republic
[2]Faculty of Machine Building, Technical University of Cluj-Napoca, Memorandumului street, nr. 28, Cluj-Napoca, 400114, Romania

Abstract. The article deals with the tracking of the wear development of cutting tools made from sintered carbide during parting-off of steel C45 (ČSN 12 050.1). For experiment were selected inserts from two producers of cutting tools with different geometry. The parting-off was realized in the workpiece axis and the whole experiment was divided into two parts, parting-off to zero and parting-off to pre-drilled hole into material. The experiment was evaluated on the principle direct microscopic method and was observed the wear on the flank of cutting tool VBB. Tests were realized in cooperation with Pramet Tools Šumperk, Czech Republic.

1 Introduction

Parting-off is one of the special turning methods, where is required special cutting tools. The aim of every process of machining is to achieve the best results in terms of the quality of surface and the durability of the cutting tools. During parting-off is payed attention especially to toughness and strength of tool and formation and leaving of chip. Parting-off of bigger diameter of workpieces or production of deeper grooves can cause very strong vibrations, which influence surface and lifetime of cutting tool. This is associated with chip formation, when unsuitable formation the chip winds up on cutting tool and damages it. During parting off is very important to become conscious of the cutting tool moves from maximal diameter to its axis, where the cutting speed is zero and after separation of the component, a small pin was created (figure 1). [1, 2]

Grooving or parting-off cutting tools are designed in different shapes and geometry with respect to build up edge, which is typical mechanism of wear for these methods. Creation of build up edge is confirmed for machining of softer materials, like as steel C45. The build up edge causes pulling up of the parts of cutting tool. The wear on the flank and creation of the groove on the rake are caused by high cutting speed. For parting-off is important to set up suitable entry conditions (e.g. width of cutting tool, clamping, type of inserts or cooling), which ensures durability and quality of surface. [2, 3]

* Corresponding author: robert.cep@vsb.cz

DOI: 10.1051/matecconf/201713703003

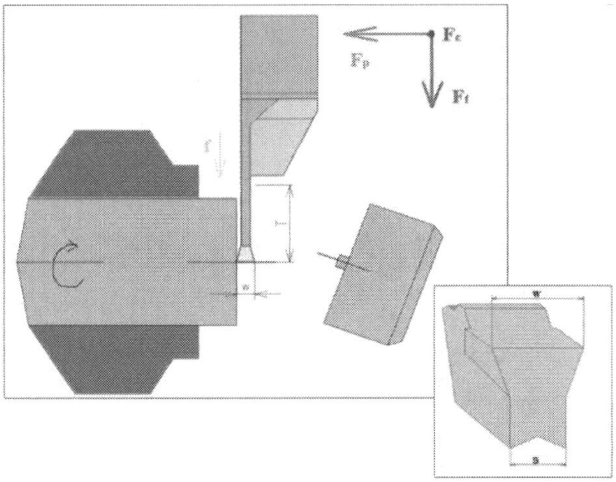

Fig. 1. Parting-off [2].

2 Design of experiment

Experimental part was focused on the tracking of behaviour wear of cutting tool with different geometry during parting-off of material C45 (ČSN 12 050.1) about diameter 80 mm. The inserts were made from sintered carbide, width w = 3 mm, from producer Pramet Tools and Iscar (table 1). The insert marked G362-AA-U02-P14 is designed with bevel and is suitable for production grooves with mechanical stress and parting-off during high cutting speeds, where is not the creation of build up edge. Tested ISCAR inserts are designed with bevel and without it and with different geometry. For all type of inserts were tested two samples due to statistical evaluation. All cutting tools were clamped into blade (ISO X). The clamping was ensured by cutting force into self-locking bed. During parting – off was used cooling for smoother machining.

Table 1. Tested inserts [4].

Producer	Insert	Geometry	Grade	Substrate	Coating	Sample	Figure
Pramet	LCMR 031602	CM	T8330	087	P523	A1, A2	
	G362-AA-U02-P14	P14	T8330	087	P523	B1, B2/ E1, E2	development
ISCAR	DGN 3102J	J	IC808	-	-	C1, C2	
	DGN 3102C	C	IC808	-	-	D1, D2 / F1, F2	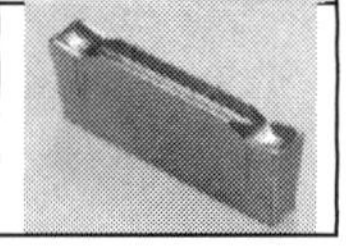

Due to previous tests, when there was not significant wear for cutting tools above or below workpiece axis, cutting edge was set up into the workpiece axis. The experiment was divided into two parts – parting-off to zero diameter of workpiece and for parting-off with pre-drilled hole.

2.1 Parting-off to zero (ø 80 mm – 0 mm)

For the economic point of view, for the both cases were designed procedure of parting-off for one cycle (figure 2a parting-off to zero and figure 2b parting-off with pre-drilled hole). In table 2 there are cutting conditions. Feed rate and width of cutting tool was constant. Due to designed cycle of parting-off was wear measured in interval t_{As} = 2,5 min

Table 2. Cutting conditions.

Cutting conditions – parting-off to zero			
Ø 80 – Ø 25 mm		**Ø 25 – Ø 0 mm**	
v_c [m · min^{-1}] (const.)	180	n [min^{-1}] (const.)	2 300
f [mm]	0,14	f [mm]	0,14
w [mm]	3	w [mm]	3

2.2 Parting-off with pre-drilled hole

To exclude the impact of the build up edge testing was with modified semi-finished product. In axis of workpiece was pre-drilled hole of diameter 20 mm, so that cutting speed was constant. Due to save material and time consuming, the tests were tested with using cutting tools marked G362-AA-U02 a DGN 3102 C. Process diagram is shown in figure 2b. The cutting conditions were set up the same like as previous case, when cutting speed was 180 m · min^{-1}, feed rate 0,14 mm and width of cutting tools 3 mm. The wear was measured in interval t_{As} = 3,5 min after twenty grooves.

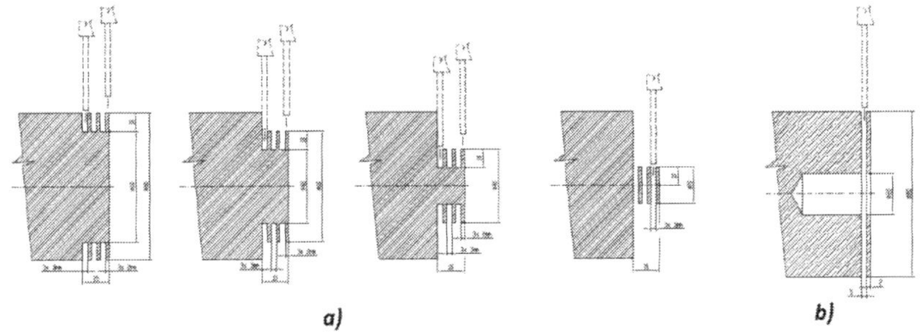

Fig. 2. a) Diagram of parting-off to zero **b)** Diagram of parting-off with pre-drilled hole.

3 Results and Evaluation

For evaluation of wear was used direct microscopic method. The wear size was determined by direct measuring of linear dimensions in different time intervals for parting-off to zero and pre-drilled hole. All measuring were realized by same ways. Critical value for flank wear was determined VBcrit = 0,2 mm. When the critical wear or interval of machining t_{As} = 30 min were reached, the tests were stopped and ended. In table 3, there are shown results for individual testing cutting tools. There is always shown the wear in final phase of parting-off to zero.

Table 3. Parting-off to zero.

PRAMET			
LCMR 031602 – geometry CM			
Sample A1	**Rake**	**Flank**	**Secondary flank**
t_{AS}=12,5 min **VB_B=0,088 mm**		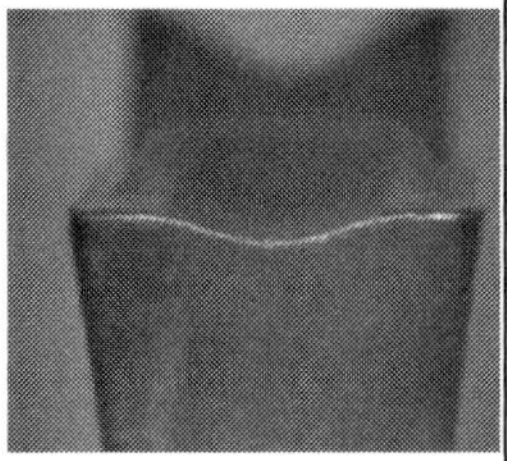	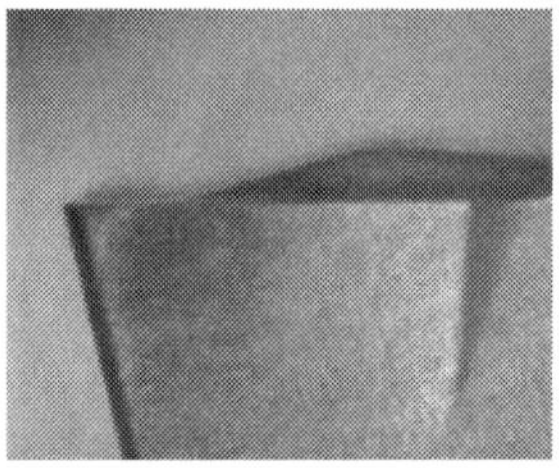
The samples A1 and A2 were poorly formed the chips. Visible wear of insert has emerged after 10 min in the cut. It was breaking off of cutting edge and damage of secondary flank. Sample A2 was damaged due to the created build up edge on the cutting edge and the test was stopped.			
G362-AA-U02 – geometry P14			
Sample B1	**Rake**	**Flank**	**Secondary flank**
t_{AS}=5,1 min **VB_B=0,076 mm**		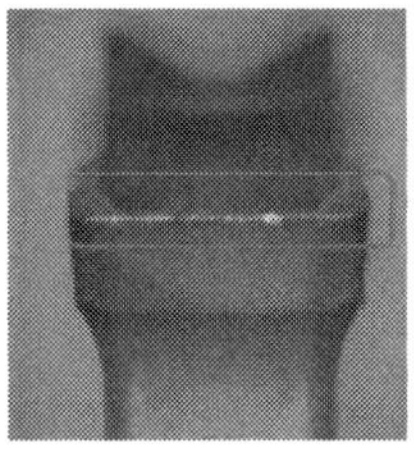	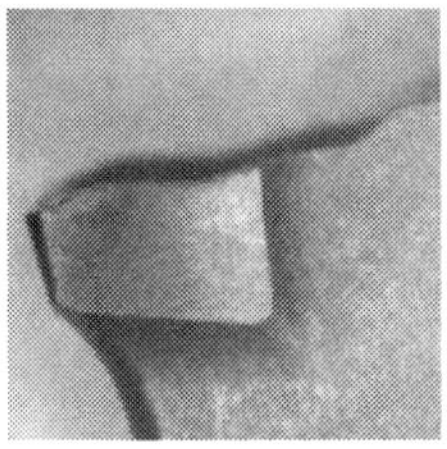
ISCAR			
DGN 3102J – geometry J			
Sample C1	**Rake**	**Flank**	**Secondary flank**
t_{AS}=12,5 min **VB_B=12,5 mm**	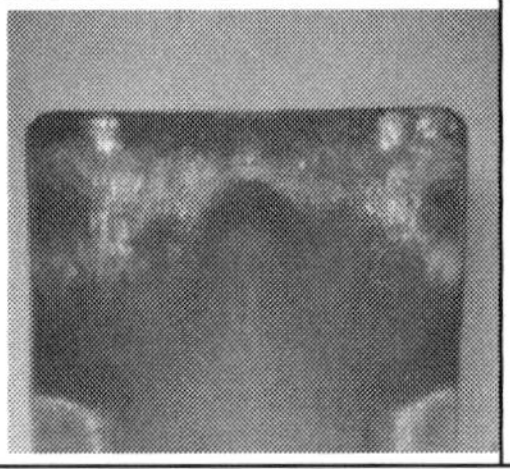		
Competitive inserts formed chip very good and the whole process was quiet. There was abrasion wear and a plastic deformation at 9 minutes in cut. The test could continue with this insert. Due to time difficulty the test was stopped.			
DGN 3102C – geometry C			
Sample D1	**Rake**	**Flank**	**Secondary flank**
t_{AS}=7,5 min **VB_B=0,054 mm**	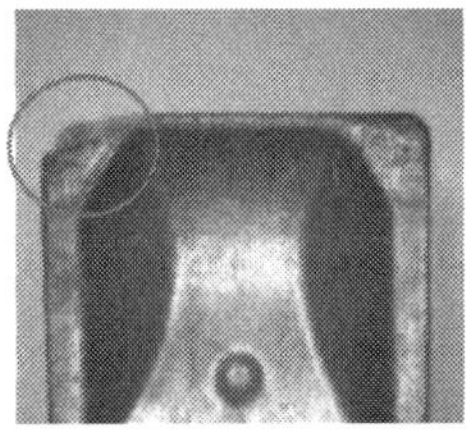		
Again, the chip formed very good with using competitive inserts. It was same cutting tool, but with different geometry. The test was for the both samples (C1, C2) prematurely stopped, because after 7 minutes the tip was damaged.			

DOI: 10.1051/matecconf/201713703003

Sample B1 poorly formed the chip and it was winded up to tool (figure 3). Test was prematurely stopped, because on the insert was mechanical damage after six minutes in cut. Development of wear for sample B2 was similar to previous inserts. There were cutting edge break off and test was prematurely stopped.

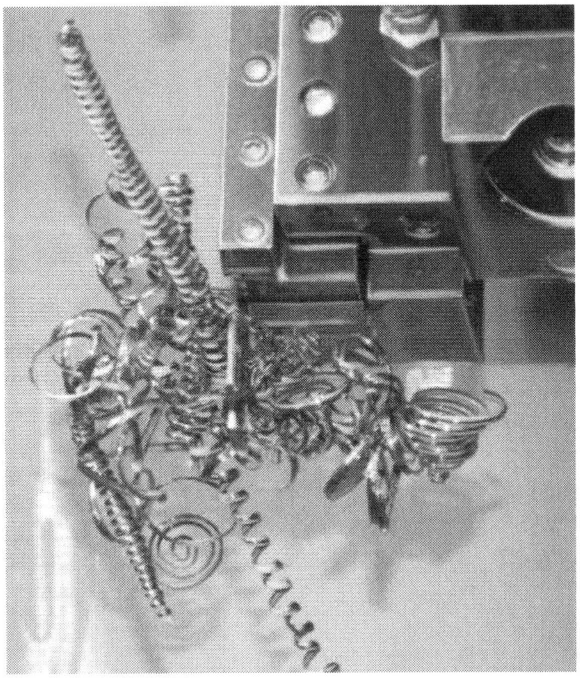

Fig. 3. Forming of the chip – sample B.

Figure 4 is shown graphical development of wear during parting-off to zero.

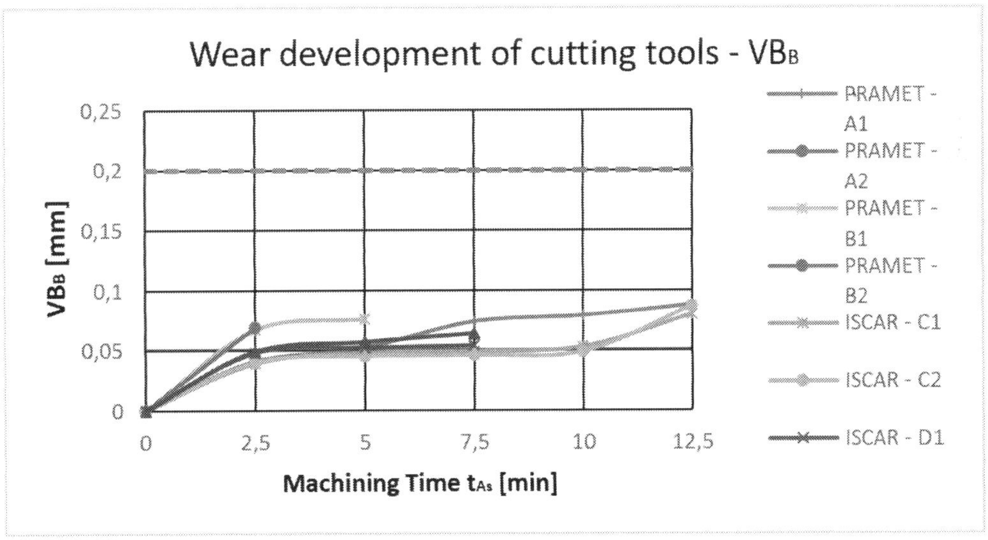

Fig. 4. Development of wear during parting-off to zero.

In table 4, there are results for testing cutting tools during parting-off to pre-drilled hole. Due to time difficulty were chosen two samples for testing (Pramet G362-AA-U02, Iscar DGN 3102C).

DOI: 10.1051/matecconf/201713703003

Table 4. Parting-off to pre-drilled hole.

PRAMET TOOLS			
G362-AA-U02 – geometry P14			
Sample E1	**Rake**	**Flank**	**Secondary flank**
t_{AS}=31,5 min **VB_B=0,069 mm**	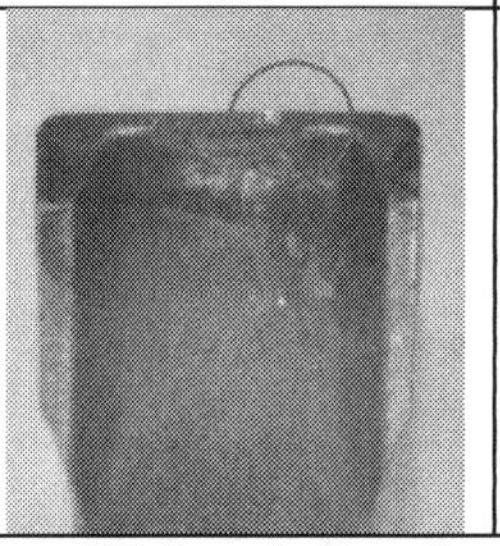	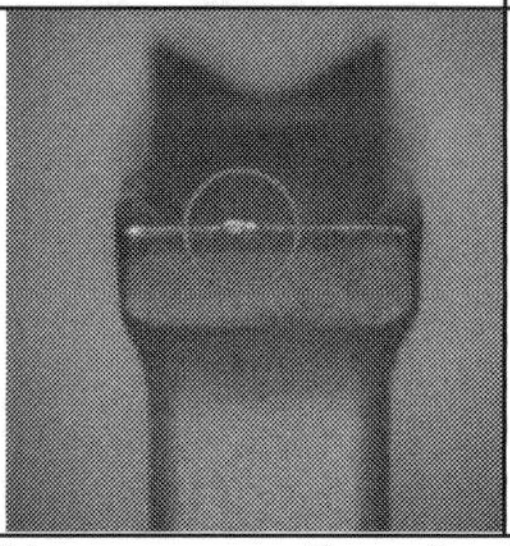	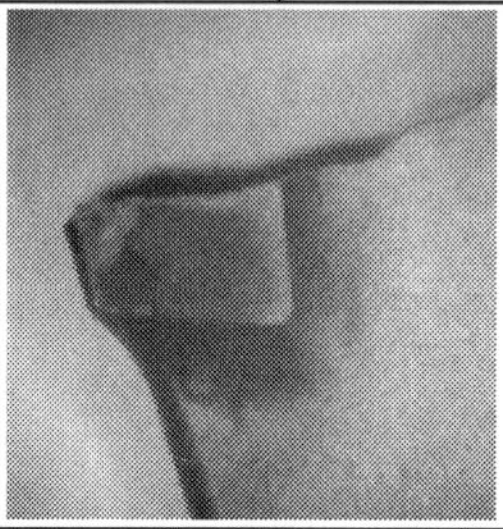
In the case of a modified workpiece, the chip was wrong formed, wind up to cutting tool, created tangled mass of machined material. Better chip forming was measured when cutting tool was deeper into workpiece.The test was stopped after half hour of machining, when was recorded small break off			
ISCAR			
DGN 3102C – geometry C			
Sample F1	**Rake**	**Flank**	**Secondary flank**
t_{AS}=31,5 min **VB_B=0,093 mm**		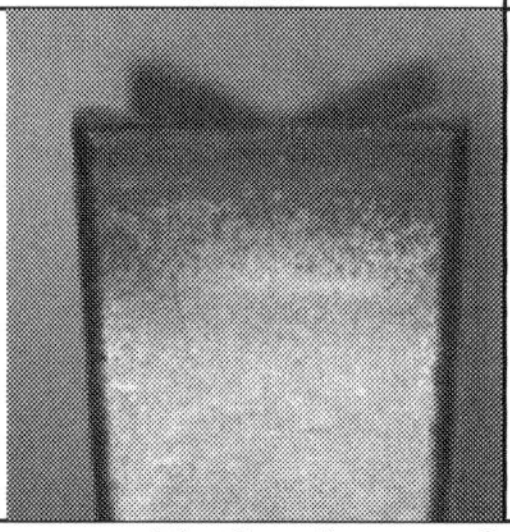	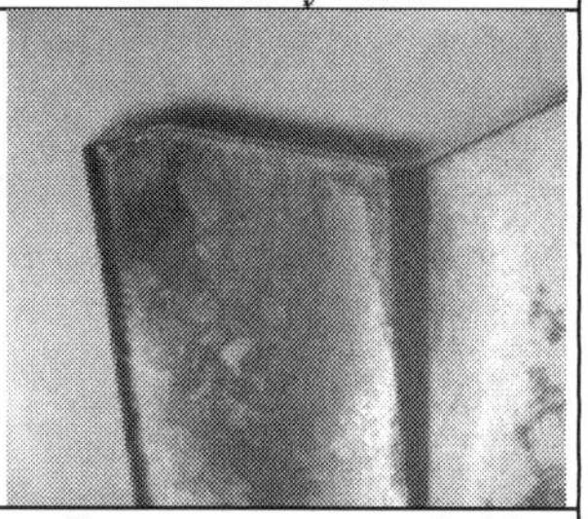
In the first third of the diameter, there was wrong chip forming, parting-off process was attended by excessive noise. Still, the process could continue, cutting edge was not damaged, but due to time difficulty the test was stopped.			

Figure 5 is shown graphical development of wear during parting-off to pre-drilled hole.

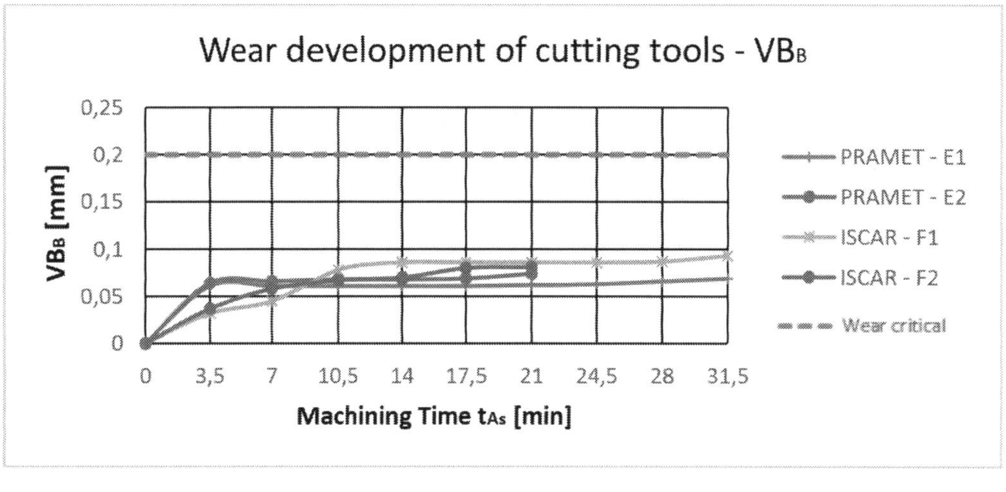

Fig. 5. Development of wear during parting-off to pre-drilled hole.

4 Conclusion

The aim of experiment was found out behaviour of removable cutting inserts made from sintered carbide during parting-off into axis workpiece. The experiment was divided into two parts, first was the evaluation the wear during parting-off to zero and parting-off to pre-drilled hole of diameter 20 mm. With decreasing diameter, decrease cutting speed and at this moment is danger of build up edge. The cutting tools were selected from offer firm Pramet Tools and Iscar.

The Pramet cutting tools with geometry CM stayed in the cut approximately 13 minutes. After ending test was measured visible break off of cutting edge, which was caused by build up edge. It is typical mechanism for parting-off during small cutting speed. The insert with geometry P14 formed long chip, which winded up to cutting tool and holder and caused damage cutting edge and the tests were stopped. Competitive inserts made by ISCAR with geometry J formed chip very good and had quiet process. The both samples (C1, C2) stayed in the cut 13 minutes and wear was determined like as abrasive on the flank. Due to time difficulty teste were stopped, but these inserts could continue in parting-off. The inserts wit geometry C had wrong wear development. On the first sample D1 with this geometry had break off on the cutting edge due to build up edge. The test with sample D2 was prematurely stopped, the insert ceased form. The samples G362-AA-U02-P14 (PRAMET) a DGN 3102-C (ISCAR) has bevel, which has a role to improve of inserts properties. In conditions of parting-off to zero (low cutting speed) vigorously decrease their durability. Machined material sticks on the insert and is created build up edge, which caused damage of cutting edge and prematurely stopped tests. Due to this reason these inserts were chosen for parting-off to pre-drilled hole, where cutting speed is still constant. The both cutting tools had very similar wear development. From the start of parting-off the chip wrong formed, this problem was removed after cut into workpiece and the whole process was quiet. The both type of inserts stayed in cut over the 30 minutes, then tests were stopped due to time and economic difficulty. The Pramet insert had visible break off of cutting edge, but it didn't influence its functionality. Competitive insert had very quiet process and better chip forming. The wear was only in abrasive form on the flank of cutting tool.

The experiment follows the recommendations, the grooves, where are shocks, dynamic stress and for parting-off outside the build up edge area (high cutting speed) is suitable used inserts with bevel.

Article has been done in connection with projects Education system for personal resource of development and research in field of modern trend of surface engineering - surface integrity, reg. no. CZ.1.07/2.3.00/20.0037 financed by Structural Founds of Europe Union and from the means of state budget of the Czech Republic and by project Students Grant Competition SP2016/172 and SP2016/174 financed by the Ministry of Education, Youth and Sports and Faculty of Mechanical Engineering VŠB-TUO.

References

1. T. Zlámal, J. Petrů, O. Vortel, M. Pagáč, P. Krajkovič, Adv. in Sci. and Tech. – Research J., Soc. Polish Mech. Eng. & Techn., **10**, 144-150, ISSN 2299-8624 (2016)
2. Š. Malotová, *Contribution to the evaluation of the durability of cutting tools for parting-off of stainless steel*, Thesis of Dissertation, Ostrava, VŠB – Technical university of Ostrava, 26 (2017)
3. *** http://www.productionmachining.com/articles/five-process-security-tips-when-parting-off
4. J. Veselovský, *Tracking of the Type and Development of the Wear Insert During Grooving*, Diploma Thesis, Ostrava, VŠB – Technical University of Ostrava, 79 (2016)

 DOI: 10.1051/matecconf/201713703004

A case study about acquisition of mechanically fixed cutting inserts

Raluca Daicu[1,*], *Valentin* Dițu, and *Gheorghe* Oancea[1]

[1] Transilvania University of Brașov, Department of Manufacturing Engineering, Mihai Viteazu No. 5, Brașov, Romania

Abstract. During the process of turning metals there are used cutting inserts that can be acquisitioned from different producers. Because the offer is varied the acquisition must be made after an economical analysis that includes experimental tries. The paper presents an acquisition methodology for metallic carbide cutting inserts and presents a case study. This research was developed using three batches of cutting inserts acquisitioned from different producers. The specimen material was 1C45 and using the electrical current at cutting, the batches were analysed from qualitative and economical point of view.

1 Introduction

During metal cutting process with good electricity conductor edges it appears an electrical current due mainly to the difference of temperature between the cutting zone and the cold elements of the part-tool thermocouple. Previous research [1-8] showed that it exists a direct dependency between the voltage of the electrical current at cutting and the temperature in the cutting zone: the higher the temperature is, the higher the voltage of the electrical current is.

The authors demonstrated, in a previous research [9], at drilling process, a connection between the electrical current at cutting and the quality of the cutting edges, by microscopically examination of the drills edges and identifying causes of the high electrical current voltage, and led to the conclusion that the electrical current is sensible to any imperfection of the drill edge. Based on this research, the authors developed, in research [10], a methodology for drills acquisition, where two drill batches from two different suppliers were compared from qualitative and economical point of view.

Regarding turning process, the previous research [11] uses the influence of cutting speed on the electrical current at cutting to highlight quality differences between cutting inserts produced in different countries, and research [12] shows that the voltage value of the electrical current at cutting can be used for quality appreciation of the metallic carbide cutting inserts.

Based on the presented facts, in this paper is used the electrical current at cutting to develop an acquisition methodology for metallic carbide cutting inserts and presents a case study, comparing three batches of cutting inserts.

*Corresponding author: raluca.daicu@unitbv.ro

2 Analysis of the quality of metallic carbide cutting inserts coated with titanium

There were acquisitioned three batches of metallic carbide cutting inserts, from three different producers, at different prices, each batch having 10 inserts. The producers are named A, B and C. The type of the metallic carbide cutting inserts is WNMG080408 and it was used a MWLNR 2020 K 08N type support. The acquisitioned cutting inserts are similar from the material point of view (type P for steel), coated with titanium, and with a similar shape of the rake surface.

The experimental stand is presented in Figure 1 and its elements are: an universal lathe (1), part (2), cutting tool (3), electrical current collector (4), a digital multi-meter with a data acquisition function (5) and a PC (6). The multi-meter has a software that allows data acquisition from second to second and exports the measured values in an Excel file (Figure 2).

Fig. 1. Experimental stand used in measuring process of the electrical current at turning.

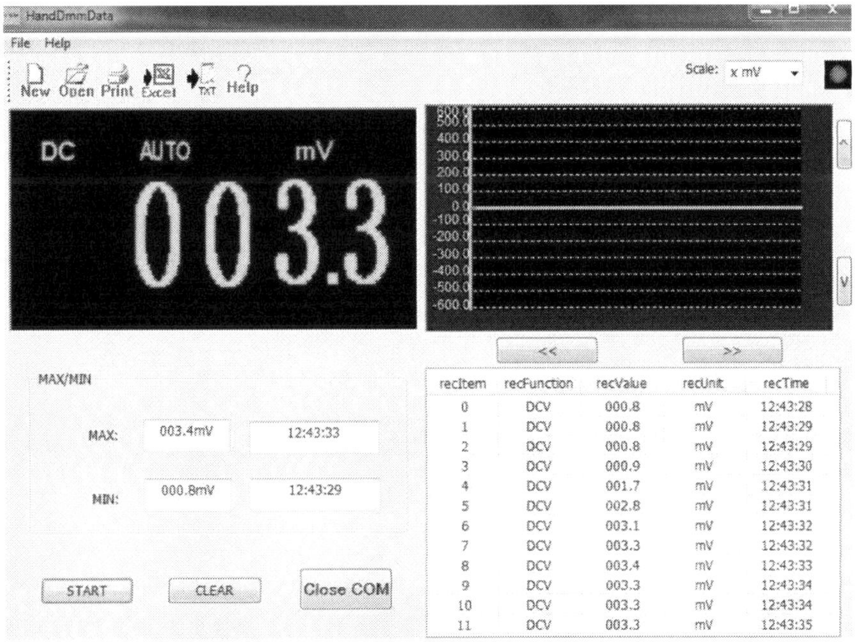

Fig. 2. Data acquisition using the presented experimental stand.

The experimental steps are:

• voltage measurement of the electrical current at cutting, for 18 cutting edges from each batch, with the installation presented in Figure 1;

• analysis of the voltage values of the resulted electrical current;

• drawing a set of conclusions about the quality of the cutting inserts batches.

The experiments were done on 1C45 material which has the following chemical composition: C = 0.45%, Si = 0.22%, Mn = 0.56%, P = 0.025%, S = 0.034. The tensile strength of 1C45 steel is 660 MPa (66 daN/mm^2) and the hardness is 224 HB.

The tests results for 18 cutting edges from each batch (producer A, B and C) are centralized in Table 1.

Table 1. Obtained values for cutting electrical current voltage at turning 1C45 steel with cutting inserts from producer A, B and C.

No. of edge	Cutting depth [mm]	Cutting speed [m/min]	Feed [mm/rev]	Voltage "U_A" [mV]	Voltage "U_B" [mV]	Voltage "U_C" [mV]
1	1.5	107	0.208	3.9	3.2	12.6
2				4.8	3.5	10.3
3				4.3	3.1	9.7
4				4.1	3.1	11.7
5				4.3	3.1	11.6
6				5	3.6	11
7				4.1	3.2	10.1
8				4.2	3.3	12.3
9				4	3.2	11.1
10				4.2	3.2	12
11				4.2	3.3	12.9
12				4.7	3.6	12.3
13				4.2	3.3	7.5
14				4.6	3.2	11.4
15				5.6	3.3	12.1
16				4.5	3.2	10.1
17				6.1	3.1	12.6
18				4.9	3.3	8.7

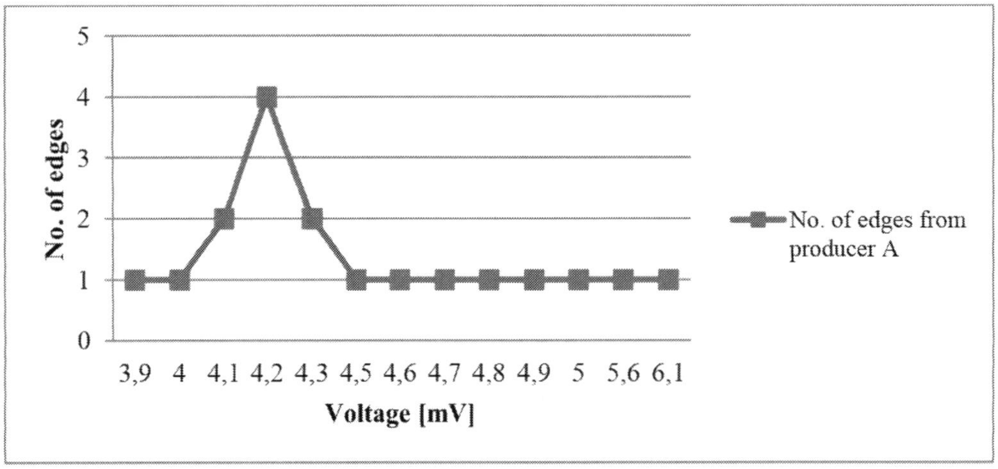

Fig. 3. Number of cutting edges from producer A that have the same value of the electrical current at cutting.

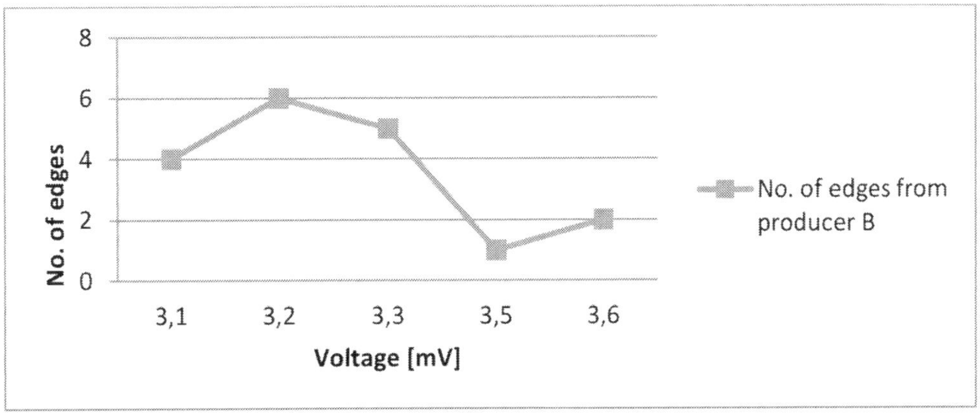

Fig. 4. Number of cutting edges from producer B that have the same value of the electrical current at cutting.

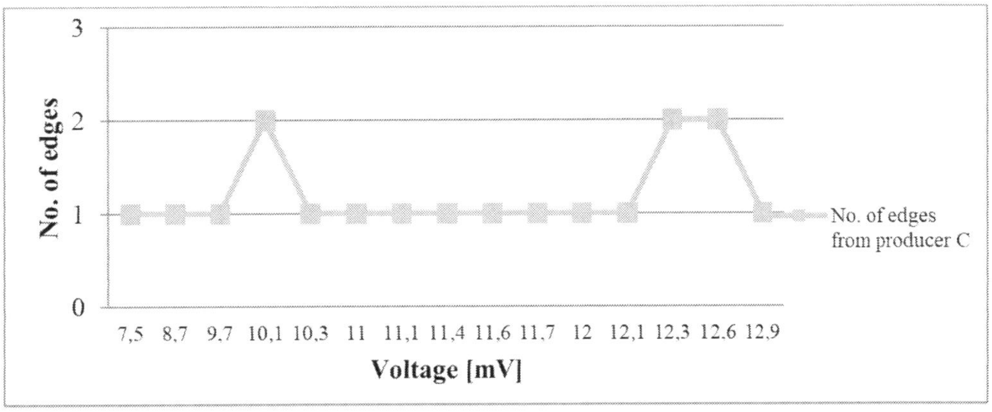

Fig. 5. Number of cutting edges from producer C that have the same value of the electrical current at cutting.

The graphical representations from Figures 3, 4 and 5 are based on the data from Table 1, and present, for each producer, the number of cutting edges that had the same voltage value of the electrical current at cutting.

Fig. 6. Surface roughness obtained with cutting edges that had different voltages of the electrical current.

During the experimental tries it was observed that the surface roughness is different when the value of the electrical current is different (Figure 6). For a higher value of the voltage, the surface roughness is greater.

Fig. 7. Surface roughness obtained with cutting edges from producer C.

It was also observed that even though the voltage of the electrical current for the edges from the producer C is higher, the surface quality is good (Figure 7), fact that leads to the conclusion that the titan coating of the plate is improperly done, the titan layer being very thin and it is easily exfoliated.

During the experimental tries it was also observed that the shape of the rake surface for producer A, presented in Figure 8, leads to chip fragmentation only when the cutting depth is greater than the nose radius of the cutting insert.

Fig. 8. Rake surfaces for cutting inserts from producers A, B and C.

Analysing the data from Table 1 and the graphics from Figures 3, 4 and 5 it can be observed that the qualitative order of the producers is: the cutting inserts from producer B present the best quality, followed by cutting inserts from producer A, and producer C presents the lowest quality.

3 Economical analysis of three metallic carbide batches from 3 different producers

The metallic carbide cutting inserts coated with titanium were acquisitioned with the next prices: inserts from producer A with 8.2 EUR, from producer B with 10 EUR, and from producer C with 8.5 euro.

It was calculated the average value of the voltage of electrical current at cutting obtained for the analysed edges from each batch and the results are: for producer A the average voltage is 4.54 mV, for producer B is 3.24mV and for producer C is 11.11 mV.

Considering the producer B as reference, the tool life for cutting inserts from producer A is with 40% lower, and for the cutting inserts from producer C is with 242% lower. The tool life proportions were calculated inversely proportional with the differences between the average voltage values obtained for the cutting inserts from the three producers.

The cost difference at the acquisition of a cutting insert from producer B and from producer A, taking producer B as reference, is 18%, and the cost difference at the acquisition of a cutting insert from producer C is 15%. Considering the tool life differences, results that for an acquisitioned cutting insert from producer B, rather than from producer A, it is made an economy of 2,2 EUR/ insert. The comparison with cutting inserts from producer C cannot be made because these are improperly coated with titanium.

4 Conclusions

Due to the experiments, the next conclusions are drawn:

• the voltage of electrical current at cutting can easily determinate the tool life differences between the same type of cutting inserts;

• in this research it was observed that the cutting inserts from producer B lead to a greater economy if we consider also their tool life;

• it was observed a connection between surface roughness and electrical current at cutting when turning with cutting edges from the same producer;

• the experimental tries conducted to the conclusion that the titanium coating for the cutting inserts from producer C was improperly done.

References

1. R. Daicu, G. Oancea, AMM **808**, 40 (2015)
2. H. Bagchi, S.K. Basu, Wear **26**, 39 (1973)
3. A.L. Plotnikov, E.G. Krylov, E.M. Frolov STIN **1**, 2 (2009)
4. M. Murata, S. Kurokawa, O. Ohnishi, CSME **79**, 2546 (2013)
5. M. Murata, S. Kurokawa, O. Ohnishi, M. Uneda, T. Doi, CSME **6**, 958 (2012)
6. H. Hirota, M. Murata, A. Kojima, K. Yanagihara, JSPE **64**, 1191 (1998)
7. V. Dițu, AJME **8**, 18 (2010)
8. R. Tanaka, A. Hosokawa, K. Yamada, T. Ueda, *Proceedings of the Seventh International Conference on Progress of Machining Technology*, 786 (2004)
9. V. Dițu, G. Oancea, R. Daicu, MATEC Web Conf. **94**, 02005 (2017)
10. R. Daicu, V. Dițu, G. Oancea, MATEC Web Conf. **112**, 01023 (2017)
11. A.K.M. Nurul Amin, M. R. A. Sarker, M. Ahmed, A.N. M. Karim, J. Mater. Process. Technol **77**, (1998)
12. V. Dițu, B. Lepădătescu, WSEAS **12**, (2013)

Ergonomic requirements for the operation of machines and technical equipment

Adam Górny[1]*

[1]Poznań University of Technology, Faculty of Management Engineering, 11 Strzelecka St., room 318/2, PL-60-965 Poznań, Poland

Abstract. In order to operate machinery and equipment safely, it is critical for the solutions in place to conform to design-related and operating requirements. Design-related requirements are primarily the responsibility of machine designers/developers and manufacturers. Operating requirements should be satisfied by employers, who are responsible for ensuring safe working conditions for their employees. Under applicable laws, machinery and equipment should be designed, produced and then operated without placing excessive loads on workers and in keeping with machine functionality and intended use. One should also ensure that machinery and equipment can be maintained and adjusted without exposing their operators to hazards. Ergonomic criteria are an integral part of such requirements. They ensure that human users and operators of technical equipment are enabled to function properly. Design-related requirements are viewed as a priority safety consideration. While they facilitate the use of technical tools, the actual safety of employees ultimately depends on the satisfaction of specific requirements during operation.

1 Introduction

Employers have the responsibility to ensure safe and healthy working conditions for their employees. This obligation includes fitting workstations with machinery and equipment that meets safety requirements [1, 2]. The requirements themselves are not new. However, recent years have seen significant changes in the approach to their implementation [3-6] prompted by the requirement to account for the needs of specific employee groups [7, 8] and doing work without hazards [9-11].

The key to securing safe operation of machinery and equipment are essential requirements, which make their proper use possible. To ensure safe operation, it is crucial to fulfill the minimum criteria guaranteeing minimal safety levels, as required by the European Union [4, 12, 13].

An integral part of the minimum requirements are ergonomic criteria designed to create an environment for the efficient operation of equipment without hazards and excessive strains.

* Corresponding author: adam.gorny@put.poznan.pl

DOI: 10.1051/matecconf/201713703005

2 Operating safety guidelines

Work requires resources which make possible its performance without excessive strenuousness and impediments. Such resources include tools and equipment in the workplace [14, 15]. One significant requirement for machinery and equipment design is to account for knowledge on people's psychological and physical capabilities. Once fulfilled, the requirements guarantee safe operation. A central factor for the desired level of safety is to achieve compliance with mandatory laws and regulations.

Such compliance should be viewed as the most fundamental prerequisite for the safe operation of technical equipment. It applies to proper design and development as well as rules governing equipment operation. The essence of applying such requirements in conjunction with the principles of ergonomics has been captured in Figure 1.

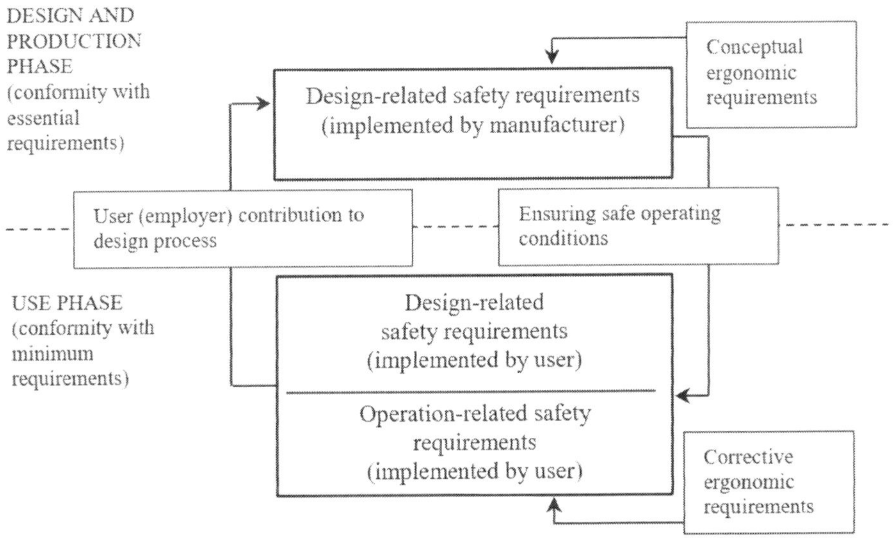

Fig. 1. Implementation of safety requirements (in conjunction with the principles of ergonomics) during the design and operation of machinery and equipment.

Adherence to ergonomic requirements helps reduce work strenuousness, achieve comfort and boost workers' ability to operate machinery and equipment efficiently [16]. To ensure the safe operation of technical equipment, it is critical to comply with:

- Anthropometric requirements applying to the adjustment of technical equipment to match the size and weight of the human body,
- Physiological requirements applying to the adjustment of technical equipment to match physiological characteristics, such as muscular loads,
- Psychophysical requirements, associated with human senses,
- Hygienic requirements, designed to promote working comfort.

To deploy the above measures effectively, organizations need to put in place a proper surveillance system to monitor operation and apply improvement measures where necessary.

3 Role of ergonomic criteria

3.1 Ergonomic criteria within the essential requirements

Essential requirements should be seen as common (EU-wide) and critical for the safe operation of machinery and equipment. Their responsibility for their incorporation into machine design and development rests with the manufacturer.

Essential requirements refer, inter alia, to materials, control and supervision, operator postures, visibility, lighting, etc. The tools and equipment used at work should accommodate user needs. It is therefore vital to be aware of human characteristics, people's physical and mental performance levels, acceptable physical and psychological loads, body sizes and build types as well as capabilities and skills [1, 5, 4, 17-20]. Significant design issues include machine-to-man information transfers, human handling of machines by means of controls and conformity with anthropometric requirements applicable to technical equipment [21].

In the course of design, whose outcomes need to comply with mandatory laws, organizations should identify the safety requirements whose satisfaction will help achieve the desired operating parameters and define the scope of improvement measures [2].

In order to satisfy ergonomic criteria, proper reference needs to be made to the conditions in which humans operate in the work environment. By accounting for ergonomic requirements, organizations can define the extent to which the level of conformity they achieve can be expected to reduce waste and improve performance capacities [1]. Ergonomic conformity should minimize adverse effects of work such as injuries, diseases and errors. This is the result of reducing strains brought about by failures to create proper conditions for the performance of tasks and processes [3, 12]. It is equally crucial to create an esthetically pleasing environment that motivates workers to work effectively and safely [21].

3.2 Ergonomic criteria in minimum requirements

Minimum requirements are defined as mandatory conditions for the operation of technical equipment. They may be viewed as fundamental for ensuring worker safety and proper technical condition and serviceability at all work stages [2, 5, 13, 17]. Such requirements do not go as far as more stringent safety requirements (such as industry requirements), which employers may find necessary to e.g. minimize accident-related losses [12, 13].

Minimum requirements apply, among others, to shape, color, touch, size, and current situation assessment. Through their satisfaction, organizations lower risks to acceptable levels. The key aim is to comply with statutory/regulatory requirements and optimize loads.

A significant part of minimum requirements are ergonomic criteria, which help meet user expectations and consequently boost work efficiency. They are stipulated in legal requirements and normative technical criteria. Ergonomic requirements point to the need to accommodate user needs and other factors contributing to working comfort. Their nature makes such requirements particularly critical for ensuring efficient information flows between technical equipment (machines) and man (operator) [11, 20, 22]. The dependencies that arise can be identified through criteria determining the ability to perform work. These may be seen as hazard mitigators and a necessary condition for having technical equipment approved for use.

4 Practical aspects of conformity with ergonomic requirements

The job of an employer is to launch measures aimed at ensuring that the tools and equipment provided to workers are fit for use and properly adjusted to meet specific performance demands [12]. In selecting equipment, it is crucial to recognize the conditions and nature of work as well as any work safety hazards [2, 13].

Organizations seeking to comply with ergonomic requirements must account for the conditions in which technical equipment is used as well as other significant factors affecting users' ability to operate in the work environment [1, 2]. To that end, they should [20]:

- Accommodate the physical features of operators, including their strength and endurance,
- Provide sufficient room to allow operators to move freely,
- Avoid forcing operators to perform work at a pace imposed by machines,
- Avoid work in unnatural or forced postures or postures maintained over extended periods,
- Avoid forcing concentration over prolonged periods to monitor operation,
- Adjust the man-machine interface to predictable operator characteristics,
- Allow freedom in the performance of work without forcing workers to behave unnaturally due to organizational constraints.

Regardless of the solution-selection methodology that organizations adopt, they must verify their ability to ensure safe operation. Such safety audits identify discrepancies between the existing and desired states and, with respect to ergonomic requirements, the effectiveness achieved in preventing excessive loads. Ultimately, the audits show what kinds of solutions have been adopted and whether they are sufficient to ensure compliance with requirements relating to the operation of machinery and equipment. In conducting such audits, one should bear in mind that operating safety rests also on machine and device reliability [2]. In this sense, one can speak not only about designing appropriate solutions but also about ensuring proper conditions for their deployment.

Guidelines for the scope of audits of conformity with ergonomic and operating safety requirements are summarized in Table 1.

Table 1. Scope of audits verifying the safety of machinery and equipment operation.

Specification	Description
Exposed individuals	Persons who can reasonably be described as exposed to the impacts of irregularities associated with the operation of machinery and equipment,
Exposure type, frequency and duration	Factors associated with access to hazard zones during entry to adjust settings, program, modify and correct process parameters, maintain and clean machinery and equipment,
Relationship between exposure and actual effects	Requires the organization to identify adverse work-related factors and the effect of accumulated exposures to risk factors and any related synergies,
Human factor	Associated with the direct application of ergonomic requirements in machine design, building and equipping; recognition of worker inclinations to deviate from safety procedures,
Reliability of safety devices and systems	Requires the identification of circumstances potentially leading to waste (such as damage to items of critical importance for proper performance, electric power failures and power supply disruption),
Danger of safety measure failure	Associated with the application of safety measures that reduce work efficiency and effectiveness,
Ability to keep means of ensuring safety in proper condition	Requires an indication of whether the existing safety precautions can be maintained to an adequate safety standard
Information regarding use	Associated with the availability of information on machine use, especially when it is critical for the effectiveness of the safety measures in place

The results of the ergonomic and safety evaluation of the controls of the eccentric press referred to in Figure 2 are shown in Table 2.

Fig. 2. The eccentric press whose controls were assessment.

Table 2. Assessment of solutions in place.

Sample identified non-conformities
- During work performance (and specifically when handling machine controls), additional personnel enter the hazard zone and become exposed to hazards, - The way the controls are operated does not eliminate having to access the hazard zone, - To access controls, workers are forced to work in awkward bending postures, - Controls descriptions and labels do not ensure an unambiguous identification of controls, - Controls descriptions and labels fail to meet requirements concerning e.g. font size and operator's native language, - The available work manual does not clarify the meaning of symbols on press controls.

5 Conclusion

For the safety of persons exposed to risks during the operation of machinery and equipment, it is essential to ensure that ergonomic requirements are adequately satisfied. Such satisfaction can be achieved by employing specific design solutions or, should that be impossible, by incorporating proper safeguards into machinery and equipment (collective protective equipment) [13].

If the existing improvement measures are aimed at mitigating the risks that result from the machines failing to meet worker needs, an ergonomics audit is in order to verify the choice of such measures. Any follow-up steps should be proportionate to the extent of the problem while ergonomic requirements should constitute improvement guidelines specifying the need for change.

To achieve conformity with requirements that account for ergonomic criteria at the operational stage, organizations need to employ corrective ergonomics to make sure their existing solutions offer a sufficient response to any non-conformities with operational requirements without contributing to adverse impact on workers [1, 12, 13]. In such circumstances, it is essential to recognize employee needs, whose satisfaction is central for efficient work performance.

References

1. A. Górny, Procedia Manufacturing **3**, 4700-4707 (2015)
2. A. Górny, B. Mrugalska, CCIS **374**, 494-498 (2013)
3. A. Mazur, H. Gołaś, A. Rosińska, M. Drzewiecka, Quality improvement of products,. 117-130 (2011)
4. B. Mrugalska, P. Arezes, Work **51(4)**, 747-755 (2015)

5. Directive 2006/42/EC of the European Parliament and of the Council of 17 May 2006 on machinery, and amending Directive 95/16/EC (recast); OJ L 157, 9.6.2006, 24-86 (2006)
6. Directive 2009/104/EC of the European Parliament and of the Council of 16 September 2009 concerning the minimum safety and health requirements for the use of work equipment by workers at work (second individual Directive within the meaning of Article 16(1) of Directive 89/391/EEC); OJ L 260, 3.10.2009, 5-19 (2009)
7. M. Butlewski, M., Applied Mechanics and Materials, **657**, 1061-1065 (2014)
8. M. Rembiasz, MATEC Web of Conferences, **94**, 07009 (2017)
9. A.M. Lasota, K. Hankiewicz, Occupational Safety and Hygiene, SHO 2016, 261-264 (2016)
10. A.M. Lasota, K. Hankiewicz, Occupational Safety and Hygiene, SHO 2016, 138-140 (2016)
11. A. Misztal, M. Butlewski, A. Jasiak, S. Janik, Metalurgija **54**(2), 429-432 (2015)
12. A. Górny, Occupational Safety and Hygiene, SHO 2013, 164-165 (2013)
13. A. Górny, Occupational Safety and Hygiene, 497-501 (2013)
14. A. Górny, MATEC Web of Conferences **94**, 04011 (2017)
15. R. Chaib, I. Verzea, I. Cozminica, M. Bendir, Safety Science **76**, 145-150 (2015)
16. P.M. Arezes, A.S. Miguel, Ergonomics in Contemporary Enterprise, 47-56 (2007)
17. K.P. Hankiewicz, J.S. Marcinkowski, Advances in Industrial Ergonomics and Safety, part III, 701-708 (1991)
18. G.H. Choi, Safety and Health at Work **8**(1), 59-66 (2017)
19. K.K. Milovanovic, A.K. Arsic, I. Savovic, 6th International Quality Conference, 927-938 (2012)
20. EN ISO 12100:2010 (PN-EN ISO 12100:2012), *Safety of machinery. General principles for design. Risk assessment and risk reduction*, PKN, Warsaw (2012)
21. J.R. Wilson, H.M. Hains, *Participatory ergonomics*, In: G. Salvendy (ed.), Hendbook of the Human factors and Ergonomics (Wiley, Chichester, 1977)
22. A. Górny, Zeszyty Naukowe Politechniki Poznańskie, Seria: Organizacja i Zarządzanie **63**, 51-66 (2014)

SPH Simulation of single grain action in grinding

Dănuț Julean [1,*]

[1]The Technical University of Cluj-Napoca, Department of Engineering Design and Robotics, B-dul Muncii, 103-105, 400641, Cluj-Napoca, Romania

Abstract. The paper presents a study of chip formation in single grain grinding using a convenient FEM formulation, the Smooth Particle Hydrodynamics (SPH) method. The chip formation process was geometrically idealized. It was simulated as an inclined linear scratching of a cuboid part. After a systematic study, the results prove that this approach is useful to study the influence of grinding speed, friction conditions and uncut chip thickness on grinding forces, stresses and strains occurring during the grain action.

1 Introduction

Grinding is a cutting surface generation procedure using a multi-cutting edges tool with undefined geometry. These cutting edges belong to a large number of abrasive grains connected by a special property bond and remove the material from the workpiece surface with a high cutting speed. The goal of grinding is the acquiring a new surface with high accuracy (dimensional, of form and of position), high surface finish and texture, special behaviour in functioning (endurance, corrosion resistance, interchanging capability) and even aesthetic qualities.

In the last decades, research efforts were made to improve the results of grinding, envisaging the increasing removal rate, increasing the cutting speed, optimizing the grinding wheel geometry and topography, reducing the thermal influence on machined surface, protecting the industrial environment by minimizing the grinding fluid and raising its efficiency, looking for new abrasive materials with high wear resistance, finding efficient dressing methods for grinding wheels, and manufacturing grinding machines with high quality dynamics.

All these achievements were based on careful studies on the single grain cutting mechanism. The particularities of chip formation were studied experimentally and theoretically to achieve satisfactory models with the ability to prognosticate the influence of grinding factors on grinding results.

The experimental studies on chip formation are difficult because of microscopic size, small forces, high cutting speeds, and very quick phenomena, so a useful tool in studying abrasive grain cutting is the simulation using finite element method (FEM).

[*] Corresponding author: danut.julean@staff.utcluj.ro

2 SPH simulation and grinding

A literature review revealed that grinding and especially single grain grinding were in the attention of several researchers with the aim of finding a detailed description of grain action [1-6]. A useful tool in such studies, besides of experimental research, proved to be Smooth Particle Hydrodynamics (SPH), [7-10]. This is a meshless method implemented in LS-Dyna software [11].

2.1 Single grain process modelling

2.1.1 Model geometry

The general geometry of the SPH model is presented in figure 1. With the help of Ls-PrePost [12], the single grain grinding has been reproduced. The abrasive grain has been designed as a conical frustum with dimensions convenient to the scale of the process (Dg = 0.05 mm, dg = 10% Dg, and Hg = 40% Dg) (fig 1). The grain was meshed via the automatic mesher into solid elements. The workpiece has been designed as a box filled with particles evenly distributed in the 3-D directions. To simplify the simulation and to keep it in convenient computer time, the adopted dimensions were L= 0.05 mm, B= 0.04 mm and H= 0.01 mm. The number of considered particles within the workpiece volume was N=160,000.

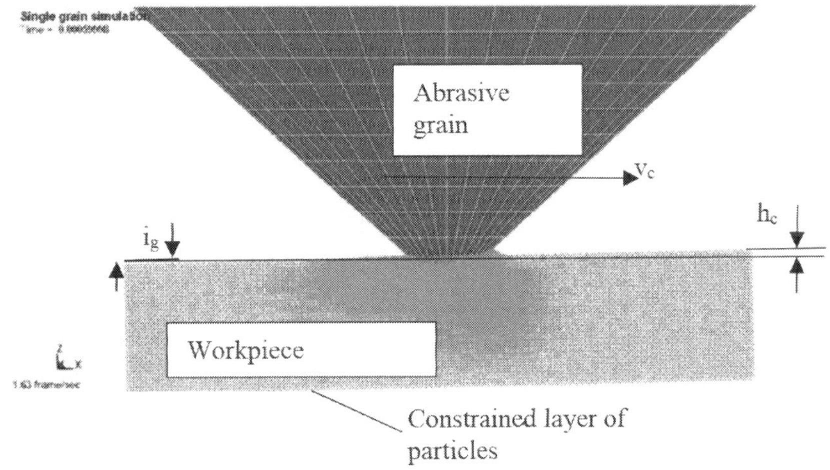

Fig. 1 Single grain grinding model in Ls-Prepost

2.1.2 Boundary conditions of the simulations

The grain movement is constrained to one degree of freedom translation along the X axis and is achieved as the movement of the rigid part (abrasive grain) with constant velocity, namely the grinding speed v_c. The different uncut chip thicknesses h_c were obtained by rotating the workpiece around the Y axis with different i_g angles. All particles from the base of the workpiece were constrained to be immobilized by constraining all the degrees of freedom for them. An automatic surface to nodes contact algorithm with different friction coefficient was consequently used for modelling grain-workpiece contact.

2.1.3 Materials modelling

The grinding grain was considered made from a conventional abrasive material, Al_2O_3 and was modelled as a perfectly rigid material (*MAT_RIGID), with mechanical properties listed in Table 1.

Table 1. Aluminium oxide grain properties [13]

Mechanical Property	Units of Measure	Value
Density	g/mm^3	0.00389
Flexural Strength	MPa	379
Elastic Modulus	GPa	375
Shear Modulus	GPa	152
Bulk Modulus	GPa	228
Poisson's Ratio	—	0.22
Compressive Strength	MPa	2600
Hardness	Kg/mm^2	1440
Fracture Toughness K_IC	MPa $m^{1/2}$	4
Thermal Property		
Thermal Conductivity	W/m °K	35

The workpiece was taken as hardened VSIMoCr52H13 chromium hot-work steel SR NE ISO 4957:2002, equivalent to AISI H13 steel, which is widely used in hot and cold work tooling applications due to its excellent combination of high toughness and fatigue resistance. Its composition is displayed in Table 2 and its thermo-mechanical properties in Table 3.

Table 2. Steel chemical composition [min-max%] [14]

C	Si	Mn	P	S	Cr	Ni	Mo	W	Co
0.37-0.43	0.9-1.20	0.30-0.50	0.030	0.030	1.20-1.50	1.90-1.10	0.35	0.3	4.8-5.50

Table 3. Workpiece material properties [14]

Mechanical Property	Units of Measure	Value
Density	g/mm^3	0.00783
Young's Modulus	GPa	206.84
Shear Modulus	GPa	8.0.172
Bulk Modulus	GPa	164.161
Poisson's ratio		0.2900
Yield stress at offset	GPa	0.85495
Engineering ultimate stress	GPa	1.0342
Yield offset	%	0.2000
Thermal Properties		
Specific Heat Capacity	J/g°C	0.460
Thermal conductivity	W/m °K	37

The constitutive Johnson-Cook material model was involved in modelling the workpiece material. This plasticity model is widely employed in simulations of high-speed metal cutting or hardened materials cutting because it can accommodate high strain rate and temperature effects which are definitory features for chip formation in grinding. This material model is implemented in Ls-Dyna as *MAT_JOHNSON_COOK (*MAT_15) and has the general form:

$$\bar{\sigma} = \left[A + B\bar{\varepsilon}_p^{\ n}\right]\left[1 + C\ln\frac{\dot{\bar{\varepsilon}}_p}{\dot{\varepsilon}_0}\right]\left[1 - \left(\frac{T - T_r}{T_m - T_r}\right)^m\right] \quad (1)$$

where $\bar{\sigma}$ is the effective stress, $\bar{\varepsilon}_p$ is the effective plastic strain, $\dot{\bar{\varepsilon}}_p$ and $\dot{\varepsilon}_0$ are the effective and reference strain rates, and$T$, T_r, and T_m are the current, room and melting temperatures.

The separation of the chip was simulated by using the damage law of the Johnson-Cook model. This considers strain, strain rates and temperatures as in equation (2):

$$\varepsilon_f = [D_1 + D_2\exp(D_3\sigma^*)]\left(1 + D_4\ln\frac{\dot{\bar{\varepsilon}}_p}{\dot{\varepsilon}_0}\right)\left[1 - D_5\left(\frac{T - T_r}{T_m - T_r}\right)^m\right] \quad (2)$$

where: σ^* is the ratio of pressure divided by effective stress. Fracture is then allowed to occur when damage parameter D defined by equation (3) reaches unity (D=1.0) and so all the concerned particles are removed from the computation.

$$D = \sum \frac{\Delta\bar{\varepsilon}_p}{\varepsilon_f} \quad (3)$$

The Johnson-Cook model strength and damage constants considered in achieved simulations are displayed in Table 4

Table 4. J-C Model input constants [15]

A	B	n	C	m
715 MPa	329 MPa	0.28	0.03	1.5
D1	D2	D3	D4	D5
-0.8	2.1	-0.5	0.0002	2.57

2.2 Computer simulations

The software used to achieve the numerical simulation was the Ansys LS-Dyna solver R9.00 on a win64 machine and LSPrePost v 4.3 as specialized pre and post-processing interface. As hardware, a conventional system using Intel ® Core™ i7-4702-MQ @ 2.20GHz CPU with 16 Gb RAM was used. All references and keywords used by LS-Dyna solver are detailed in [17].

Due to the fact that SPH simulation in metal cutting was used especially when high cutting speeds were involved the screening simulations covered the usual range of grinding speeds. The values of uncut chip thickness were considered as the maximum values that can be produced during conventional grinding. Different friction conditions between SPH particles and the grain (rigid body) had been considered too. Figure 2 displays the results of such simulations, emphasizing the outputs as stress and strain distributions and grinding forces. The groove produced by the grain action on the surface of the work piece is visible and the pile up of material during the ploughing stage also.

2.3 Factors analysis

A full factorial DOE program with three factors and two levels (Tab. 5), was designed to analyse the influence of friction coefficient μ, cutting speed v_c and uncut chip thickness h_c on the levels of forces F_x and F_z, produced during the virtual chip formation due to the grain cutting movement on the workpiece.

Table 5. Parameters for single grain grinding simulations

Simulation parameters	Units	Low level	High level
Cutting speed v_c	m/s	50	100
Uncut chip thickness h_c	mm	0.001	0.002
Friction coefficient μ		0.1	0.6

 DOI: 10.1051/matecconf/201713703006

Material model grain/work		*MAT_20 / *MAT_15

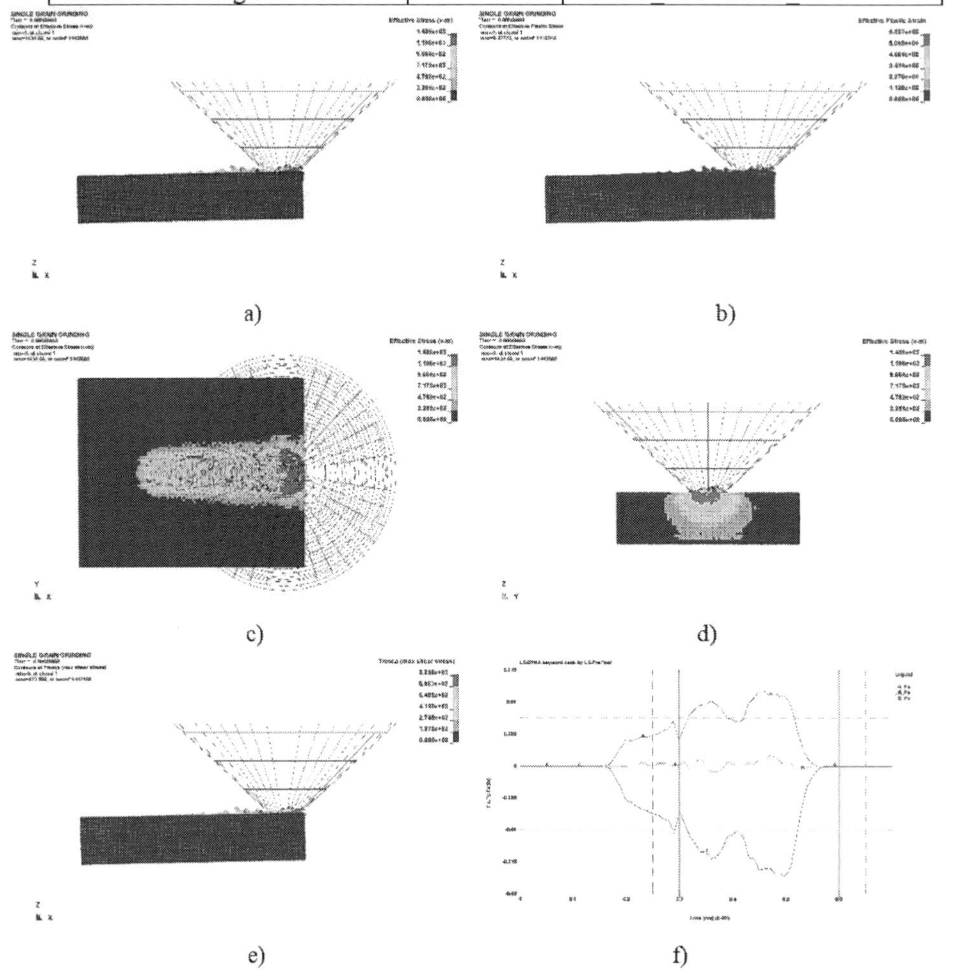

Fig. 2 SPH simulations results for μ=0.1, v_c=100 m/s and h_c=1 μm: a) effective stress; b) effective strain, c) effective stress - top view; d) effective stress - front view; e) maximum shear stress (Tresca); f) grinding forces F_X, F_Y and F_Z versus grinding time

After running the planned simulations and the response variables Fx and Fy were obtained, ANOVA analysis was done. The significant factors were then identified, the modified factorial model was finally used for both grinding forces and the following regression equations were obtained in terms of actual factors:

$$Fx = -0.0184+2.97\,E\text{-}004\, v_c + 0.0335\, h_c - 3.19\,E\text{-}004\, v_c h_c \quad (4)$$

$$Fz = -0.01237 - 0.0148\,\mu + 3.02\,E\text{-}004\, v_c + 0.0362\, h_c - 3.34\,E\text{-}004\, v_c h_c \quad (5)$$

Analysing the achieved equations (4) and (5), one may observe that the "tangential" force (F_X) is mainly influenced by grinding speed v_c and uncut chip thickness h_c, while the "normal" force (F_Z) is influenced also by the friction coefficient μ. Both forces are influenced by the combination of v_c and h_c factors.

Figure 3 displays the grinding forces per grain F_X, and F_Z, versus the main factors, uncut chip thickness h_c and grinding speed v_c, considering an average friction coefficient μ = 0.35. One may observe that increasing in grinding speed produces a slight decrease in grinding forces while the increase in uncut chip thickness produces a significant increase of forces.

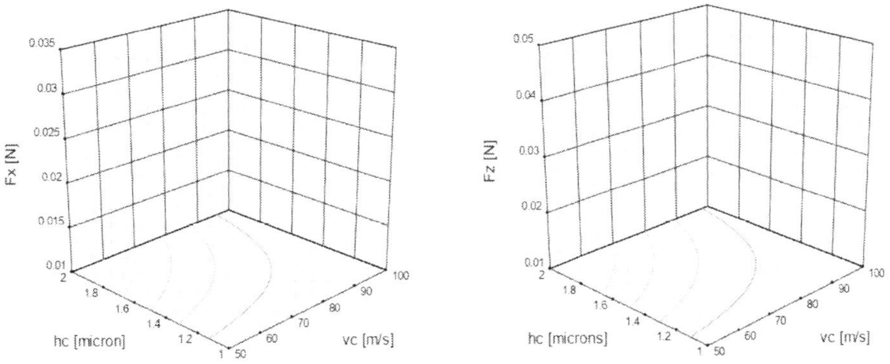

Fig. 3 Variations of simulated grinding forces

3 Conclusions

The present approach to simulating the chip formation in single grain grinding, using a numerical engineering method, FEM with a special meshless formulation, Smooth Particle Hydrodynamics (SPH), proved to be a useful instrument in studying the chip formation mechanics. It was demonstrated that the method is sensitive to the kinematic parameters of the process, grinding speed v_c and uncut chip thickness h_c, and in a smaller account to friction (μ). The method proved to be useful for getting the 3-D, stress and strain distributions in this complex deformation process. The method may be directed to study the geometry of the formed chip, because it permits to identify the ratios between rubbing, ploughing and cutting action of the grinding grain, also. Due to the sensitivity to the material formulation, it may be directed to grindability studies too.

References

1. D. Anderson, et al., Int. J. of Mach. Tools and Manuf., **51** (12) 898-910, (2011)
2. C. W. Dai, J. H. Xu, W. F. Ding, et al. Adv. Mat. Res., **1017**, 598-603, (2014)
3. Guan, Peng, Jiqiang Li, Shuang Zhu, et al. App. Mech. and Mat., **121-126,** 1879-1885, (2012)
4. X. K. Li, L. Yan, W. B. Rashid, Y. M. Rong, Advanced Materials Research, **76-78**, pp. 9-14, (2009)
5. W. S. Wang, C. Su, Z. R. Pang, J. M. Hou, Key Eng. Mat, **416**, 210-215, (2009)
6. L. Yan, Z. X. Zhou, F. Jiang, et al., Key Eng. Mat. Vols. **431-432**, 269-272, (2010)
7. N. Rüttimann, M. Roethlin, S. Buhl, et al., Procedia CIRP, **8**, 322-327, (2013)
8. R. D. Shen, X. M. Wang, C. H. Yang, App. Mech. and Mat., **483**, 3-8, (2014)
9. C. Su, J. M. Ding, L. D. Zhu, Adv. Mat. Research, **186**, 353-357, (2011)
10. C. Su, L. D. Zhu, W. S. Wang, Adv. Mat. Res, **239-242**, 3123-3126, (2011)
11. J. L. Lacome, SPH: A new feature in LS-DYNA, Tech. rep., DYNALIS, (2000)
12. *** www.lstc.com/products/ls-prepost
13. *** www.accuratus.com
14. *** www.matweb.com
15. L. Chen, T.I. El-Wardany, M. Nasr, M.A. Elbestawi, CIRP Annals, **55**, Issue 1, 89-92, (2006)
16. www.lstc.com: LS-DYNA R 9.0 Keyword Manual Vol I, II, III (2016),

DOI: 10.1051/matecconf/201713703007

The conceptual framework for physical risk assessment in multi-purpose workplaces

Andrzej Marek Lasota[1,*] and *Krzysztof* Hankiewicz[2]

[1] University of Zielona Gora, Institute of Safety Engineering & Work Science, 4 Prof. Z.Szafrana Street, 65-516 Zielona Gora, Poland
[2] Poznan University of Technology, Chair of Management and Computing Systems, 11 Strzelecka Street, 61-845 Poznan, Poland

Abstract. In industry work related musculoskeletal disorders (WRMSDs) are still a common problem which frequent cause of health problems, sick leave and it can result in decreased productivity, quality of work and increased absenteeism. Though, traditional manufacturing work practices do not enough take into account task variability issues during the work design – assessment process. Variations in task content and organizational work performance are due to effective use of equipments and reaching high level of productivity. But this variation has impact on work demand and risk to WRMSDs and makes some difficulties with assessment of risk. Therefore, this study aims develop a framework for assessment multi-purpose workplaces. The proposed method integrates complementary concepts of widely known techniques used for evaluation of physical risk factors to WRMSDs on workplaces. Additionally the research framework is highlighting major differences in worker's exposure to WRMSDs risk, potential impact on awkward postures, and how these findings can be used for finding solutions in a future ergonomic intervention.

1 Introduction

For many years, the importance of ergonomics and human factors has grown in the industry. Ergonomic work environment is directly dependent on design and ergonomic evaluation process [1]. Ergonomic evaluation is inextricably linked to work safety and workplace risk. Taking into account safety aspects the machine design and construction phases significantly determines reduction of risks on work stands [2]. It is a well-known fact that stress also determines risk. However, the results showed that workers experience workplace stress associated with the functions of the work environment in terms of work organization more often than job content stressors related to the type of tasks [3]. Additionally quality management also define work environment [4].

Employee productivity depends, among other resulting from the Work-related Musculoskeletal Disorders (WRMSDs), which cause health problems, sick leave and may reduce the quality of work, increasing cost and absenteeism in many occupations [5].

* Corresponding author: am.lasota@o2.pl

Moreover, the growing number of older workers can change statistic values and causes the need to design workplaces for these employees as well [6].

To avoid WRMSDs, musculoskeletal load must be on appropriate level and researchers have developed a wide range of methods to estimate the risk to WRMSDs: Ovako Working Posture Analysing System (OWAS) [7], Rapid Upper Limb Assessment [8], Rapid Entire Body Assessment [9], Occupational Repetitive Action [10] and Quick Exposure Check [11]. These methods are widely used in various industries [12-15].

In the scientific literature, there are plenty of studies reported health outcomes and assessment of risk to WRMSDs of occupational groups, at production and assembly process [12] used mentioned risk assessment method to evaluate jobs, but many industrial jobs are multi-task in nature [16] and there are multi-purpose workplaces. Moreover, the growing level of customization and shorter product life manifests itself in the form of smaller batch sizes. Such production system results in that on the same machines, workplaces variety of components are manufactured after retooling/rearrangement. This diversity results in variability of tasks performed by operators which could have an influence on the workload and posture of an operator during work, repetition, excessive physical load, duration of movement and finally the risk of WRMSDs. Variability of tasks on the multi-purpose workplace increase the difficulty associated with estimating the ergonomic risk and with identifying reasons of risk which in turn makes it difficult to formulate recommendations and take corrective action. Therefore, the objective of this study was to develop a framework of physical risk assessment on multi-purpose workplaces.

2 Research method

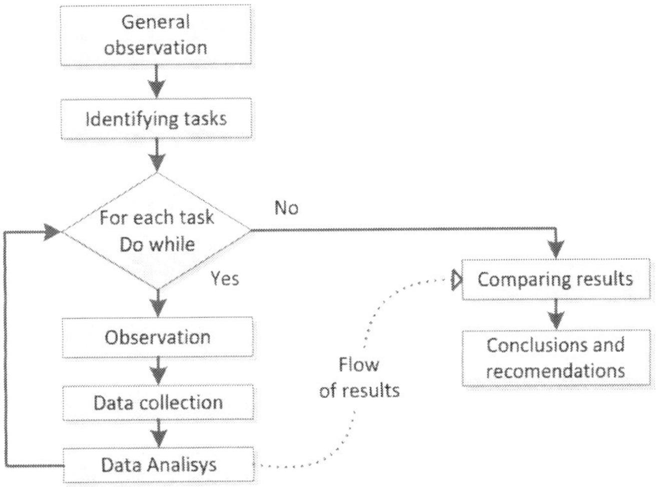

Fig. 1. Flowchart of the proposed framework

The proposed research method is a framework (Figure 1). To reach the objectives of this study, the research method began with general observation. This initial step supplies insights about job performed on a chosen workplace. After that, the identification of tasks on entire workplace is possible and evaluation of each task in based on three main steps: 1) observation of single task, 2) data collection, and 3) data analyzing.

On the next stage the results from all assessed tasks are correlated for comparison and analyses. This makes it possible to formulate conclusions and recommendations for identified reasons of abnormalities and risk, and improvement opportunities.

2.1 General observation

The purpose of this step is general observation of multi-purpose workplace for assessment in the field were multi-task is done on a single workplace. The general visual observation and interview with managers, supervisors and technologists could be helpful to understand of multi-purpose workplace. The collected information from supervisors should help to understand job and work activities performed by operator(s).

2.2 Identifying tasks

Observation employee(s) during their daily routine work and in the current and real working environment, group discussion and interview should be used to identify tasks performed on the chosen workplace. It could help establish the priority and sequence of assessment, data collection method(s), assessment strategy.

2.3 For each task

For each identified task performed on the selected workplace it should be done observation, data collection, and data analysis.

2.3.1 Observation

Observation employee during their daily routine work in the current task, real working environment, and interview with him/her should be used to explore any problems and difficulties in performing task. It also should helps in creating a friendly atmosphere for carrying research.

2.3.2 Data collection

Several methods of data collection are possible: observation, interview, task analysis. Visual observation is very simple and useful, but video recording has many advantages. It allows carrying out analysis after the event and it could be repeated many times and does not interference with work patterns. Tasks analyses and interview will depend on the objectives of the study and method(s) of evaluation.

2.3.3 Data analysis

This step is an in-depth analysis of all data collected. Analysis of risk factors and the level of risk to WRMSDs can be made through defined ergonomic evaluation criteria and WRMSDs evaluation method(s).

2.4 Comparison of results

The results from the observational study for each task indicate the levels of risk involved with the adopted working postures and additional physical factors e.g. repetitions, load/force and the final action categories. They provide additional information about workload, repetitiveness and awkward body segment posture for each task. The level of action category gives key information and guidelines to the observer which adopted working postures by workers are harmful, and what level of urgency of ergonomic intervention is required. Summary of the results of all assessed tasks on evaluated workplace provides to the observer extra information about the awkward postures, action categories and activities to which they relate. A comparison of results allows determining during which activities the employee adopting incorrect postures and in what part of these activities are common to all assessed tasks. This information is necessary to propose solutions; especially the solution focuses on changes in the workplace.

2.5 Conclusions and recommendations

In this step after identifying abnormalities e.g. vibration, awkward working postures, it is straightforward to conclude about WRMSDs risk in each task and about the reasons of harmful action categories and non neutral body segment postures. Furthermore, these conclusions have to be taken as a recommendation for improving working condition. It should further be used with fundamental ergonomics principles to develop of corrective actions and ergonomic solutions which will take into consideration demands flowing from all tasks performed on assessed workplace. Additionally it should help in formulation solution fulfilling all demands flowing from all assessed tasks.

3 The case study in a fittings factory

The study was conducted at a workplace of manual processing of rubber components. The operator aged 51 years, 3 years experience manually treated: rubber feet, rubber elbows, and compensators, with the use of: scissors, scalpel or knife. She manually cut out rubber surpluses that flowed from the mold under pressure from a hydraulic press.

3.1 General observation

The general observation of multi-purpose workplace in the field, face-to-face interview with supervisor and collected information from worker and supervisor helped to understand job and work activities performed by operator.

3.2 Identifying tasks

Observation employee in the field, discussion and previous information was used. A strategy of assessment was developed. Authors' used pen and paper observation method with sampling (interval - 25 seconds). The OWAS method as a one of the simpler technique which allows on quick assessment was used. It was identified three tasks performed on the workplace: Task#1: processing of rubber feet. The employee reaching the container from the cart, then empty out parts on the table top, empty container put off on the side, cut off rubber surpluses from parts, and then ready parts put off into a container. After graduating from the party put off a container on a cart. Task#2: processing of rubber compensators. She was reaching for two parts from the container located on the cart (one piece in left hand, second piece in right hand), and then put on the table, cut off surplus of rubber, and then put down the finished parts on the second cart. Task#3: processing of rubber elbows. A worker reached a one piece with both hands from the palette, has put on the table top, cut off surplus of rubber, and then put down the finished parts on the pallet.

3.3 For each task

For each of three identified task performed on the selected workplace it was done observation, data collection, data analysis.

3.3.1 Observation

Observation employee was done during their daily routine work in the current task and real working environment, and interview with her. She was asked about task performed at the workplace. It allowed get more information about tasks; explore problems and difficulties on assessed task, job and workplace.

3.3.2 Data collection

Data collection consisted of recording the postures through visual observation. The evaluations were recorded for each task separately. In total 212 work postures were evaluated: 48 for task#1, 75 - task#2 and 89 - task#3. The OWAS method was used. It is based on a classification of different positions for the back, arms, legs and the force/load used in the work. It uses a four-digit code to describe various postures and force/load combinations. These four-digit codes from different body parts in a specific position combined with the estimated load provide information about each postural load. The combination creates categories describing the risk of exposure to WRMSDs and action categories (AC) necessary to improve the working conditions: AC 1 – no risk, intervention is not required; AC 2 – low risk, immediate intervention is not required, but the ergonomic adjustment should be taken into account in future actions; AC 3 – medium risk, ergonomic intervention should be carried out as soon as possible; AC 4 – high risk, ergonomic intervention is required immediately.

3.3.3 Data analysis

The recorded data were separately studied for each task. Four digit code for each recorded postures and action categories included harmless (AC1) and harmful (AC 2, AC 3, AC4). For task#1, authors' findings show that the most common four-digit posture code was: 1111 – 35.4%, classified to AC1; 2111 (33.4%, AC2) and 3211 (25.0%, AC2). The 65% of postures were classified to harmful action categories. In the case of task#2 the most common four-digit posture code: 2111, 64.0% and classified to AC2, and 1211, 16.0%, AC1. The 76% of postures were classified to harmful action categories. For task#3 2131 - 36%, AC2; 2231 - 23.6%, AC2 and 1121, 13.5%, AC1 were the most common four-digit postures. The 72% of postures were classified to harmful action categories.

3.4 Comparison of results

This step shows that the most common four digit code posture were 2111, 30.2%, AC2 and 2131, 15.1%, AC2. The 72% of postures were classified to harmful action categories. In total, two postures classified to AC4: one 4141 (reaching for two parts from the container located on the cart) at task#2 and one 4151 (putting the container with ready parts on the cart) at task#1. Six postures classified to AC3: one posture coded 2151, which occurred at task#1 (reaching of container with parts from the cart); two postures: 2152 at task#3 (reaching a part from the pallet located on the floor); two postures: 2172 occurred at task#3 (transferring a part from the palette to the workstation); and one postures 2251 at task#2 (putting the part in the container positioned on the cart).

The comparison in this study shows that posture classified to AC2 and coded as: 2111, 2121, 2131, 2211 and 3211 identified on two different tasks.

3.5 Conclusions and recommendations

The results indicate differences in the AC for each evaluated task and incorrect working postures. The common problem for all evaluated tasks for this workplace is reaching in and laying on the cart or the pallet. This fact requires common solution.

4 Conclusions

The proposed framework allows identifying the physical risk factors of the multi-purpose workplace created with the each task and whole job performed on the workplace supporting improvement actions. It is based on traditional assessment method included pen and papers.

Further researches are necessary to improve efficiency of using framework. It will be focused to confirm validity and reliability of proposed method and it can be two-faceted. Firstly, in the design process – e.g. in stage of simulation in virtual environment will use digital human modeling for ergonomic assessments proposed solutions. Secondly, in the field will use on different type of multi-purpose workplaces.

References

1. A. Górny, Procedia Manufacturing, **3**, 4700 (2015)
2. B. Mrugalska, P. M. *Arezes,* Occupational Safety and Hygiene, **97** (2013)
3. A. Mościcka-Teske, J. Sadłowska-Wrzesińska, M. Butlewski, A. Misztal, A. Jacukowicz, Int J Occup Saf Ergo, **1** (2016)
4. A. Górny, The Shaping of Work Environment, C. Stephanidis (ed.), Posters Extended Abstracts: International Conference, HCI International 2015 (part II), Communications in Computer and Information Science, **529**, 136 (2015)
5. BLS, Bureau of Labor Statistics 2016, *Nonfatal Occupational Injuries and Illnesses Requiring Days Away from Work*, http://www.bls.gov/news.release/pdf/osh2.pdf (2016)
6. M. Butlewski, Applied Mechanics and Materials **657**, 1061 (2014)
7. O. Karhu, P. Kansi, I. Kuorinka, Appl Ergon, **8** (4), 199 (1977)
8. L. McAtamney, E. N. Corlett, Appl Ergon, **24** (2), 91 (1993)
9. S. Hignett, L. McAtamney, Appl Ergon, **31** (2), 201 (2000)
10. E. Occhipinti, D. Colombini, Med Lav, **87** (6), 491 (1996)
11. G. David, V. Woods, P. Buckle, Appl Ergon, **39** (1), 57 (2008)
12. A. M. Lasota, LogForum, **10** (4), 383 (2014)
13. A. M. Lasota, K. Hankiewicz, *Evaluation of ergonomic risk in the production line of frozen food products*, in: Yue, X.G., Duarte, N.J.R. (Eds.), International Conference on Economics and Management Innovations – CEMI, ACSR-Advances in Computer Science Research. Atlantis Press, Paris, France, 272 (2016)
14. A. M. Lasota, K. Hankiewicz, International Symposium on Occupational Safety and Hygiene - SHO 2016; Proceedings book. Guimaraes, Portugal, 138-140 (2016)
15. A. M. Lasota, K. Hankiewicz, Occupational safety and hygiene IV, Taylor & Francis Group, London, 261 (2016)
16. P. Drinkaus, R. Sesek, D. S. Bloswick, C. Mann, T. Bernard, Int J Occup Saf Ergo, **11(3)**, 263-281 (2005)

Surface roughness evaluation after machining wear resistant hard coats

Katarina Monkova[1,*], *Peter* Monka[1,2], *Jiri* Cesanek[2], *Jan* Matejka[2] and *Vladimir* Duchek[2]

[1]Technical University of Kosice, Faculty of Manufacturing Technologies with the seat in Presov, Sturova 31, 080 01 Presov, Slovakia
[2]Faculty of Mechanical Engineering, University of West Bohemia, Univerzitni 8, 306 14 Pilsen, Czech Republic

Abstract. The technology HVOF – High Velocity Oxygen Fuel was used to spray the alloy Stellite 6 for the purpose of this research. This method is based on using high velocity spraying flame. The article deals with the evaluation of surface quality after machining of wear resistant hard coats. The goal of the authors was to understand the main factors that influence final surface quality at machining of the coatings from Stellite 6. The findings of experiment have confirmed that there have been obtained very important information about machining of this alloy at the specific type of workpiece. Authors have evaluated the quality of the surface not only by means of Abbott curve, but they have also observed the dependency of arithmetical mean roughness value on both cutting edge radius and feed. The analyses of cutting tips wear at changeable inserts of tools have shown that the typical wear appears in the form of a notch.

1 Introduction

The machinability of a material is generally defined by the surface roughness quality and the surface integrity obtained, the tool life, heat generation in the cutting zone, chip formation difficulties, material removal rate and power consumption, machine tool dynamics, and other parameters involved in metal machining. Since the machinability of a material largely depends on its physical, chemical and metallurgical characteristics, the selection of cutting regimes, the right cutting tools and adequate machine tools is the appropriate solution to the most economically effective and sustainable manufacturing. [1]

The major quality assurance of machined parts checks geometric accuracy, surface roughness and residual stress levels of the machined parts. All these parameters and the factors related to surface integrity are largely dependent on the selected cutting regimes for machining a given material using a given cutting tool on a certain machine tool. Therefore, researching for optimal cutting regimes/parameters for machining a certain material using specific cutting inserts is the topmost priority in machining research. All other research

* Corresponding author: katarina.monkova@tuke.sk

relevant to machining would be accurate and effective if the optimal cutting regimes have been used for machining the samples under investigation. [2, 3]

Special category of the machined components creates the parts, surfaces of which are deposited. Deposits can be made by various technologies. They are organized in Fig. 1 according to the dependence of jet temperature on the particle impact velocity. [4]

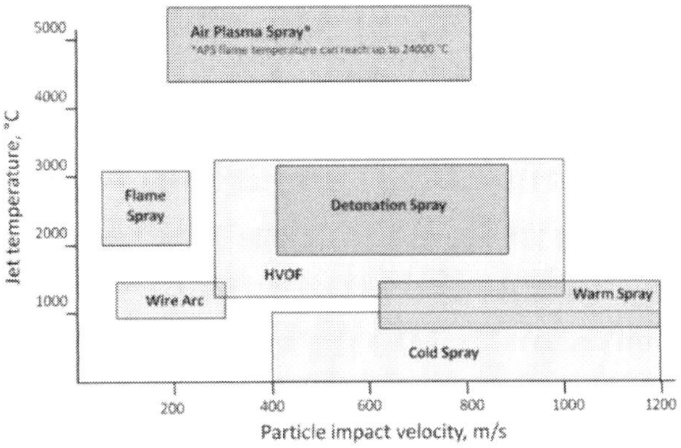

Fig. 1. The technologies used for hard coatings deposition [4].

One of the technologies for deposition and for surface quality improvement is High Velocity Oxygen Fuel (HVOF) that was used for the purpose of this research. HVOF coating is a thermal spray process used also to improve or restore wear resistance and corrosion protection. Molten or semi-molten materials are sprayed onto the surface by means of the high temperature, high velocity gas stream, producing a dense spray coating which can be ground to a very high surface finish. [5]

The principle of HVOF spraying technology is shown in Figure 2 [4].

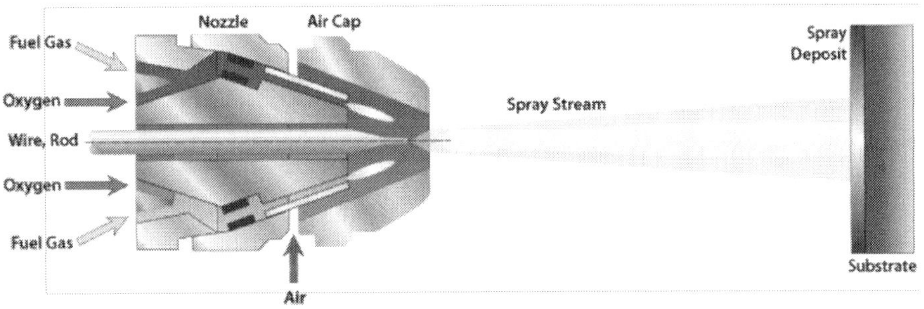

Fig. 2. The principle of HVOF technology [6].

One of the key benefits of this system's high velocity is the extremely high coating density and low oxide content. The low oxides are due partly to the speed of the particles spending less time within the heat source and partly due to the lower flame temperature (around 3,000 °C) of the heat source compared with alternative processes. Generally, the parts deposited by coats are often characterized as difficult-to-machine. They are materials which produce excessive tool wear, excessive cutting forces causing high heat generation, difficulties in chip formation resulting in poor surface finish during machining operations. As well as producing excellent bond strength, some HVOF coatings can be sprayed very thick due to the exceptionally high velocities producing coatings in compression instead of tension. This enables materials such as carbide to be applied very quickly in excess of 6mm. [6,7]

Coatings produced using HVOF have outstanding characteristics, even above other thermal spray processes, that include: high density, optimum hardness, higher coating thickness (coatings exhibit greater coating thickness limits than plasma, combustion or wire coatings of the same coating material, beneficial residual stress, excellent wear resistance, superb corrosion resistance and exceptional metallurgical properties, fine surface finishes – (coatings can be machined, ground, lapped, honed or super finished to produce very high surface finishes to precise tolerances) [8].

2. Deposited material – Stellite 6

Nowadays, many Stellite alloys are produced with combinations of various amounts/proportions of cobalt, chromium, molybdenum, tungsten, titanium, nickel, iron, aluminium, carbon, boron, manganese, phosphorus, silicon and sulphur. By virtue of their low magnetic permeability and excellent anticorrosive properties some of the stellites are extremely suitable for medical surgery, dental implants, bone replacement, artificial heart valves, and heart pacemakers [9].

The distribution of chemical composition with spectral analysis of deposited material Stellite on the selected area of sample is presented in Fig. 3 [10].

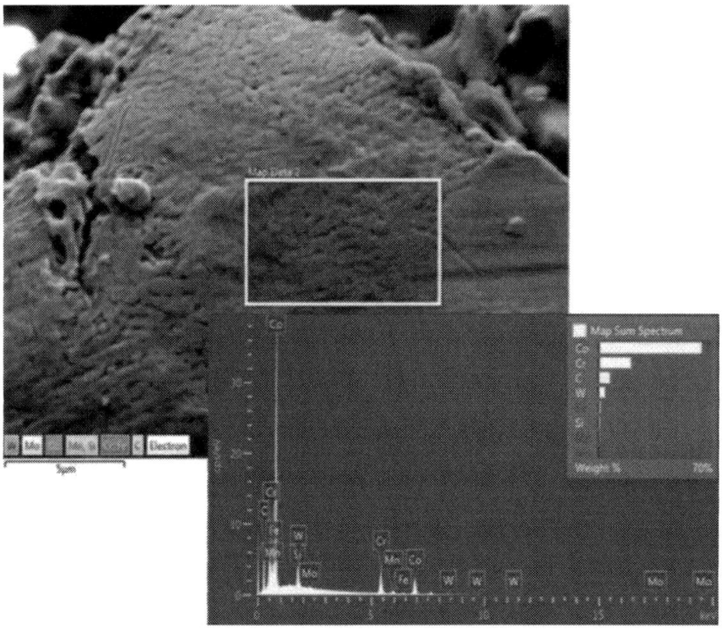

Fig. 3. Chemical composition and spectral analysis of deposited material Stellite [10].

Due to their high hardness and good ductility, highly dense molecular structure, harder and more regularly organised carbides, lower thermal conductivity, tendency to resist plastic deformation having lower Young's modulus, stellites are categorized as difficult-to-machine materials due to their poor machinability. [11,12]

There is a good prospect to improve the machining technique and the technology of stellites to overcome the machining difficulties. As parts made of stellites are widely used in various industries and very popular in specific applications, it is a big drawback that adequate machining processes for stellites have not yet been developed other than the costly and time-consuming technique of grinding. However, an attempt has been made to reveal the optimal machining parameters (cutting speed, feed rate and depth of cut) for a lower value of surface roughness for the purpose of successful machining of Stellite 6 using new grades of coated and uncoated carbide inserts with various nose radii [13-15].

3 Experiments

3.1 Conditions of coating process

As it was mentioned above, powdered Stellite 6 was used as the coating material. As an energy carrier was selected Nitrogen with flow rate 6.5 l/h. The spraying distance was 360 mm and the layer of deposited coating was 0.4 mm.

The porosity of deposited coating was measured at randomly selected three different locations and the average was used for further analysis. The results revealed that the majority of pores present in the HVOF coatings were type A pores (0–10 μm in size). These pores were formed because of interactive effects of particles and the gaseous medium. The porosity was evaluated lower than 1% of a volume.

In the next step, mechanical properties and micro-hardness of deposited coating were measured. HR15N hardness was measured on the top surface of the coating, while average of 8 measurements were done. The measured value of 82.8 ± 3.1 corresponds to approximately 45 HRC. Micro-hardness HV0,3 was measured in the cross section, at half the thickness of the coating. Average of 15 measurements were performed with result value of 650 ±75.

3.2 Conditions of experiments at the machining

Based on the research of available literature and based on previous experience of experts at Regional Laboratory Institute of Technology, the following processing factors were selected: [10, 16]

1) Tools
   a) Inserts (changeable) - with circle shapes made from cutting material CERMET with geometry as it is shown in Fig. 4. Next parameters of inserts - diameter of ø 12 mm, radius of the cutting edge with values of 10/14/18 microns.
   b) Tool holder - Capto modular system with HSK cone.

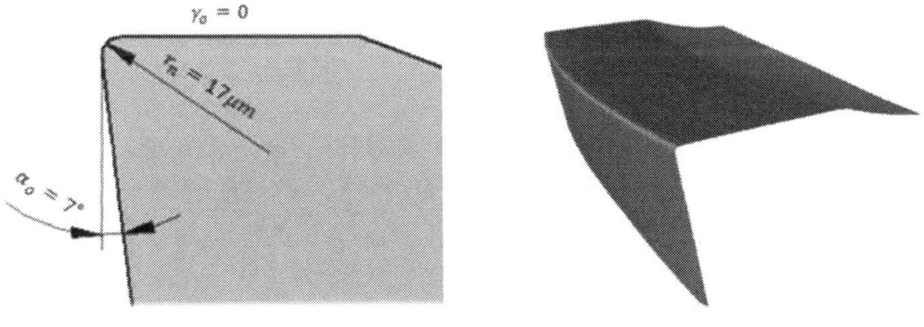

Fig. 4. Macro-geometry of insert in orthogonal plane and 3D scan of real insert.

2) Workpiece - steel cylinder from material 12 050.1 with diameter 60 mm deposited with Stellite 6by means of HVOF spraying technology (sprayed in the average thickness of 0.45 mm).
3) The machine - DMG MORI CTX beta 1250.
4) Technological factors of experiments
   a) cutting speed of 25 meters per minute

b) Depth of cut 0.2 mm

c) Feeds: 0.15 / 0.30 / 0.45 mm

d) Cooling - two variants: with cooling pressure of 100 bars / without cooling

Observed parameters were arithmetical average deviation from a mean line *Ra* and ten-point height of irregularities *Rz*.

Process of machining was realized at following cutting conditions: cutting speed $v_c = 25$ mmin^{-1}, feed per revolution $f = 0.15$ mm, cutting depth $a_p = 0.2$ mm and cooling at the pressure 100 bars.

3.3 Results of experiments

Experiments were carried out in a combination of factors described above. The selected type of workpiece (represented by a combination of structural steel and thermal spray of alloy Stellite 6 by technology HVOF) has been characterized by markedly different behaviour in comparison with the machining of this type of material in casted or moulded state. At many samples, it has occurred to the separating of material at borders of joined structural units of the coating layer and not at the path that was specified by kinematic join between the tool tip and rotating workpiece. Characteristic surface after machining of coated samples no. 13 is shown in Figure 5.

Fig. 5. Machined surface of coated sample.

At the machining with unsuitable conditions, it occurs to the oscillation of machining set, what influence stability of the machining process and surface roughness, of course. The traces of cutting tool on top surface of the workpiece are visible in Fig. 6.

Fig. 6. Traces on the top surface of the workpiece.

Achieved results were statistically processed, while remote values have been excluded and they are represented by dependencies of surface roughness on cutting tip radius and on the feed. The obtained graphs are shown in Figures 7 and 8.

The plots point to the following statements:

• Supposed significant influence of observed characteristics of surface roughness on increasing of feed per revolution.
• The surface roughness is slightly increasing with the increasing radius of cutting edge.

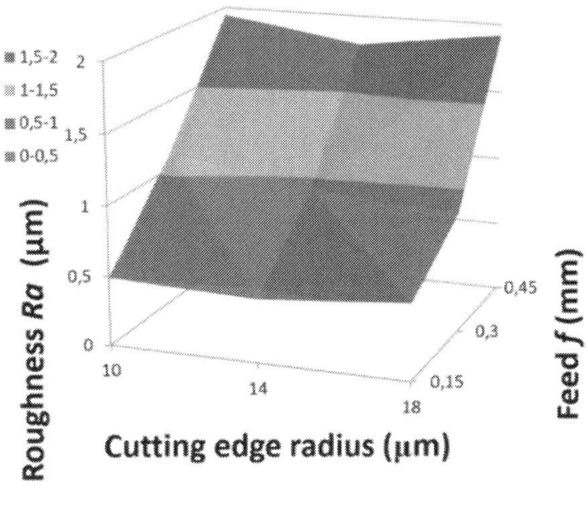

Fig. 7. Dependence of arithmetical average deviation from a mean line *Ra* on cutting edge radius and feed.

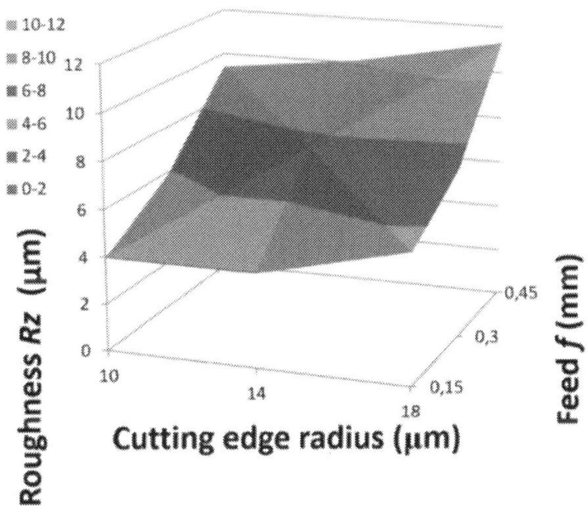

Fig. 8. The dependence of ten-point height of irregularities *Rz* on cutting edge radius and feed.

The Abbott curve was also used for evaluation of observed surface. It simulates the effects of wear or running-in, giving information on the material and void volumes of the surface topography, on damaging rate of the superficial layer, on bearing capacity and fluid retention capacity of the studied materials. Utility of Abbott curve in tribological evaluation of a surface becomes evident when studying surfaces of same material but obtained with different technologies: these surfaces may have the same *Ra* but different Abbott curves. The reasons of these differences may be: different shapes of the asperities, their distribution on height but also their spacing ranges. Studying the Abbott curves after functioning on laboratory models reveals some specific tendencies that may be correlated with other tribological parameters as wear and friction

 DOI: 10.1051/matecconf/201713703008

coefficient, and they may recommend range of load, speed or combination of them for actual application [16, 17].

The obtained Abbott curve together with the detail of coated surface after machining sample n. 13 is shown in Fig. 9.

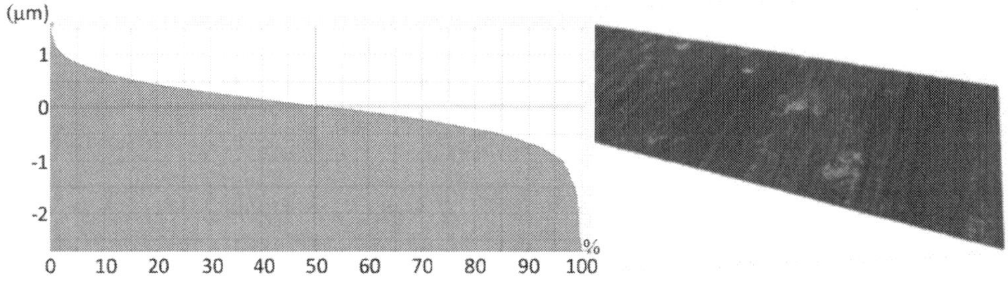

Fig. 9. Abbott curve together with the detail of coated surface after machining sample num. 13.

4 Conclusions

Thermal spray coating is an attractive technique as it offers a wide choice of materials and processes that have a reduced impact on the environment when compared to conventional plating processes. When thermally-sprayed coatings are porous, especially with connected porosities, they are not very suitable for corrosion protection. This situation can be drastically improved by curing their inter-connected porosities through sealing or by post-treatments that are generally thermal. The sealing or post-treatment process depends on the coating material and its service conditions and, also, on the cost that the customer can accept [18, 19].

During experiments, it has not been proved the fact that in the case of intensive cooling (with pressure 100 bars) would be achieved markedly better quality of surfaces than in the case of dry machining.

Based on the experiment results it can be stated that in some cases very different values of surface roughness have achieved, which may also be caused by different structure of thermal spray, e.g. due to a various number of pores or due to the smaller size of cohesive forces, leading to an easier extraction of particles in the course of machining.

The analyses of cutting tips wear at changeable inserts of tools have shown that the typical wear was VB_B type in combination with VB_N wear in the form of a notch (according to the standard ISO 3685) and there also occurred slight imprint of the face near to the cutting edge. The findings of experiment have confirmed that there have been obtained very important information about machining of alloy Stellite 6 at this specific type of workpiece, where following characteristics were required: very low values of the surface roughness, the mean depth of cut 0.2 mm, macrostructure of joined clusters with variable consistency on the border.

The effort of authors´ future research will be focused on the quality of machining surface improvement. Another type of cutting material and different geometry of cutting tips will be tested along with specification of the most suitable parameters, at which the best surface quality will be achieved.

The paper originates with the direct support of Ministry of Education of Slovak republic by grants VEGA 1/0614/15, KEGA 087TUKE-4/2015, KEGA 007TUKE-4/2018 and thanks to project ED2.1.00/03.0093 - Regional Technological Institute covered by the European Regional Development Fund and the state budget of the Czech Republic.

References

1. B. Matesa, D. Kozak, A. Stoić, I. Samardžić, Metalurgija **50** (4), 227 (2011)
2. G. Krolczyk et al., Metalurgija **54** (1), 91 (2015)
3. P. Twardowski et al., Advances in Mechanical Engineering **7** (6), 1 (2015)
4. Grigoriev, S. et al., Surface and Coatings Technology **268**, 77 (2015)
5. A. Panda et al., Applied Mechanics and Materials **616**, 183 (2014)
6. E. Gawronska, N. Sczygiol, Transactions on Engineering Technologies 17 (2015)
7. T. Temucin et al., Technical Gazette **20** (5), 787 (2013)
8. J. Petru et al., Tehnicki Vjesnik **20** (2), 225 (2013)
9. J. Jurko, Adv. Comp. Sci. Edu. Appl. **202** (2), 338 (2011)
10. P. Monka, et al., Proceedings of Manufacturing Technology conference, Pilsen, 1 (2017)
11. H. Shahanur et al., Int. J. of Engineering Materials and Manufacture **1** (2), 35 (2016)
12. S. M. Stojadinovic, V. D. Majstorovic, FME Transactions **42** (3), 249 (2014)
13. P. Fauchais, A. Vardelle, Advanced Plasma Spray Applications, InTech, 250 (2012)
14. R. Cep et al., Key Engineering Materials **581**, 341 (2014)
15. P. Monka, P., Int J Adv Manuf Technology **85**, 2325 (2016)
16. M. Moravek, *Influence of input characteristics micro and macro structure of coat with selected type on its machinability by the tool with defined tip geometry*, Diploma Thesis, Department of Machining Technology, University of West Bohemia in Pilsen (2016)
17. M. Ungureanu et al., Proceedings of the 5th Central European Conference in Regional Science (CERS), 1112 (2015)
18. D. Kinik et al., Tehnicki Vjesnik **22** (2), 351 (2015)
19. S. Malotova, Manufacturing Technology **16** (1), 168 (2016)

MATEC Web of Conferences **137**, 03009 (2017) DOI: 10.1051/matecconf/201713703009
MTeM - AMaTUC 2017

Aspects of design of experiments on machining technology by forming the internal threads

Sandor Ravai-Nagy[1,*] , *Nicolae* Medan[1], and *Ferencz* Szigeti[2]

[1]Technical University of Cluj-Napoca, North University Center Baia Mare, Dr. V. Babeş str., 62A, 430083, Baia Mare, Romania
[2]University of Nyíregyháza, Institute of Engineering and Agricultural Sciences, Sóstói út 31/B, 4400, Hungary

Abstract. The purpose of this paper is to present the necessary steps to make a design of experiment with application to study the machining of internal thread by cold deformation with roll-form taps for aluminium alloys. In the first part of the paper is presented an exploratory research. The performed experimental studies indicate influences of processing speed on thread resistance. Through the proposed study, we wish to determine the machining regimes and influence of machining regimes on the precision and resistance of the thread.

1 Introduction

Inside of processing technologies of inner thread beside the cutting method using taps the cold forming methods are used also.

It is known that the cold formed process of aluminium has an influence on the precision of the workpiece [1].

The cold formed thread is more resistant to mechanical stresses than the thread obtained by cutting [2, 3]. Other advantages of plastic deformation of threads are: high productivity, high precision, surface quality and durability of tools [4].

Because the cold forming taps manufacturing companies recommend working regimes for aluminium alloys in wide range, process speed V_c=5-30 m/min, for proper use it is necessary to perform further studies and tests. To optimize the working regime, it is necessary to carry out studies for each material and technological processing system.

To carry out an experimental research, it is recommended to follow a series of steps [5, 6]:

1. Recognition of and statement of the problem;
2. Selection of the response variable;
3. Choice of factors (process parameters), levels, and ranges;
4. Choice of experimental design;
5. Performing the experiment;
6. Statistical analysis of the data;
7. Conclusions and recommendations.

* Corresponding author: Ravai.Nagy.Sandor@cunbm.utcluj.ro

2 Exploratory research

In a first step, threads have been machined to familiarize with thread cold forming technology [7].

In the experiments, M8 threaded holes are machined using two different types of cold forming taps and a classical cutting tap to compare the manufacturing technologies. The processing speeds have been selected at 50% and 75% of the maximum speed recommended by the manufacturers. The processed material is Aluminium alloy 6005. For lubrication and cooling, 5% emulsion was used. In the experiments was used the specific (normal) emulsion for metal cutting, because in the process of manufacturing on CNC machine tools it is very difficult to make use of a different, special, lubrication liquid for the cold forming taps.

In table 1 are presented the results of exploratory experiments.

Table 1. Centralizing exploratory experimental results.

The used tool	Exploratory experiment results	
	Processing speed	
	12.5 m/min (50%)	18.75 m/min (75%)
Cold forming tap without cooling/lubrification channel: Walter D2061805-M8 M8-6Hx 86378800 HSS-E Ø7,4 17/12	Thread was formed	Thread was NOT formed. The tap was covered with material (fig. 1.a)
Cold forming tap with cooling/lubrification channel: Emuge M8-6Hx Inno Form 1-Z-1KZ HSSE TIN-T1 52/1 Ø7,45	Thread was formed	Thread was NOT formed. The tap was covered with material (fig. 1.b.)
Taps machine with helical grooves for chip release, M8	Thread was formed	Thread was formed

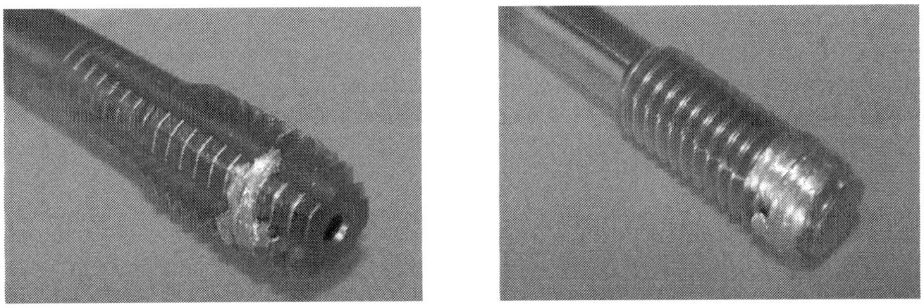

a) Cold forming tap with cooling/lubrification channel

b) Cold forming tap without cooling/lubrification channel

Fig. 1. Cold form taps with grooves covered by material.

The processed threads have been tested to tensile test to establish the shear strength. The obtained results we show in Table 2.

Following the results of the experiments it was ascertain:

• From the point of view of the deformation speed, for value of 18.75 m/min, the internal thread M8 can not be processed by forming method using cold-forming tap.

• From the point of view of shear strength of the obtained thread, differences were found depending on the used tool. Processing with cold-forming tap without cooling/lubrification channel it was ascertain a 14% increase in shear strength compared whit cutting tap and 7.25% when processing with Cold-forming tap with cooling/lubrification channel.

Table 2. The measured tensile force.

The used tool	**The measured tensile force**					**Average of the tensile force**
	[kN]					[kN]
Cold forming tap without cooling/lubrification channel Vc=12.5 m/min (500 rmp) Walter D2061805-M8 M8-6Hx	20.979	21.243	22.302	22.105	19.248	21.175
Cold forming tap with cooling/lubrification channel Vc=12.5 m/min (500 rmp) Emuge M8-6Hx Inno Form	19.035	19.188	19.447	20.785	20.257	19.742
Taps machine with helical grooves for chip release, M8 Vc=12.5 m/min (500 rmp)	18.418	18.707	18.378	18.727	18.585	18.563

3 Design of experiments

Considering the results of the performed experimental exploratory research we propose an experimental research plan to determine the influences of the technological parameters on the thread cold-forming methods.

3.1 Recognition of and statement of the problem

After completing field of bibliography and exploratory research, can be formulated the research questions: the influence of processing regimes on the shear resistance of the screw and on thread deformation.

3.2 Selection of the response variable

Given the phenomena occurring under thread cold-forming process, response parameters studied are: 1) Thread precision, 2) Thread tensile force.

3.3 Choice of factors (process parameters), levels, and ranges

To study the thread cold-forming process the following parameters were identified:
• Process speed. The range of processing speed will have the values: 12.5 m/min, 15 m/min, 17.5 m/min and 18.75 m/min.
• Tool type used. These are cold-forming tap without cooling/lubrification channel and cold-forming tap with cooling/lubrification channel;
• The size of the processed thread. The range of threads for which there are cold-forming taps is between the values: M1 - M12;

• Material processed. It is planned to perform experiments on aluminium alloys such as: 5019, 6005, 6082, 7050, 7075.
• Lubricating fluid: 5% emulsion, 10% emulsion, special liquid for cold-forming taps.
In Table 3, the experimental field is presented.

Table 3. Experimental domain.

Parameter	Values
Process speed	12.5 m/min, 15 m/min, 17.5 m/min and 18.75 m/min
Tool	Cold-forming tap without cooling/lubrification channel B519Y700. Cold-forming tap with cooling/lubrification channel B521Y700.
Tread dimension	M4, M5, M6, M8, M10, M12
Material	5019, 6005, 6082, 7050, 7075

3.4 Choice of experimental design

In the first stage, we propose an experimental research using the Taguchi method to determine the percentage of influence of the chosen parameters and the interactions between them on the studied phenomenon. Depending on the obtained results in the next step, will be chose the parameters whose influence is significant.
In the second stage, experimental research will be used using the full factorial method.

3.5 Performing the experiment

For the processing of the thread will use the technological assembly shown in Figure 2. It consists of:
• Machine tool with CNC control "NTC104N type EML510BR", n_{max} = 8000rmp, s_{max} = 30 m
• Clamping device for test piece with force and torque measuring sensor Kistler 9345B or equivalent.

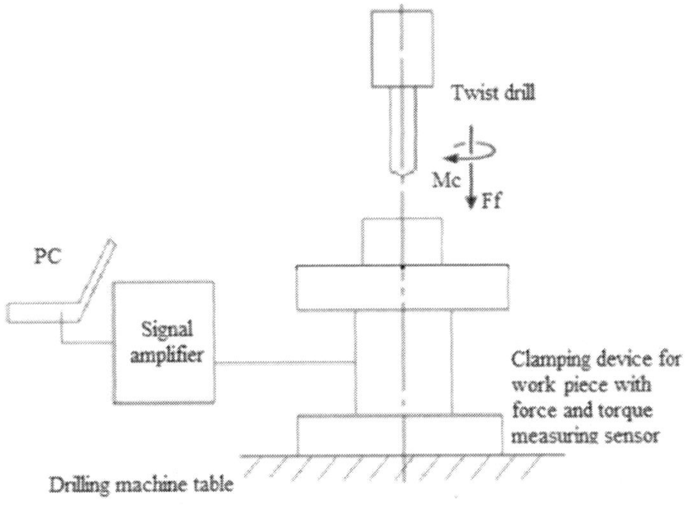

Fig. 2. Stand for manufacturing threads and measuring parameters of the processing regime.

The processed threads will try to tensile using the technological assembling set out in figure 3, in order to determine the shear resistance of the formed screw of the thread.

DOI: 10.1051/matecconf/201713703009

a) Universal testing machine whit fixture device

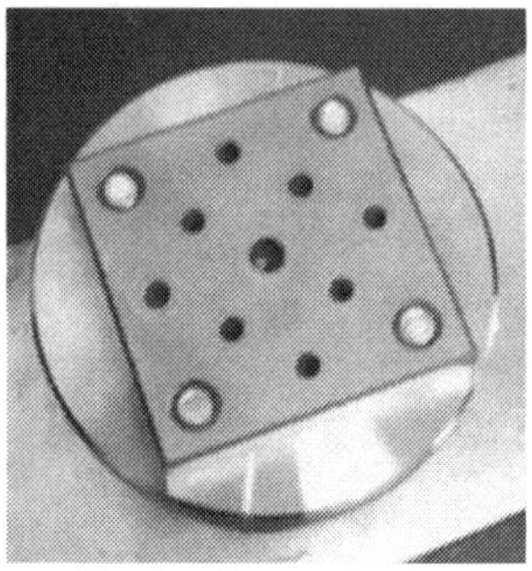

b) The fixture device

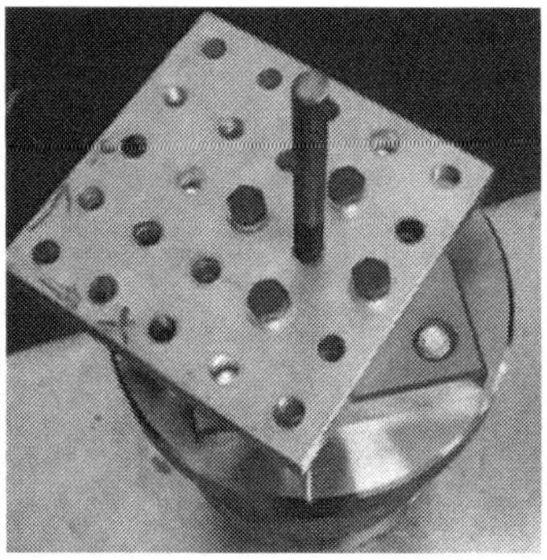

c) The specimen on fixture device and the clamping method

Fig. 3. Technological assembly used in the thread mechanical tensile test.

3.6 Statistical analysis of the data

An important role in an experimental research has statistical analysis of the data [8]. Statistical analysis of experimental data obtained certifying that the values are real values of the process studied they are not affected by the system errors or measurement errors.

Statistical analysis of the experimental data consists of:

1) verifying the aleatory character of data. This will be performed using the Young test;

2) To verify the normality of the experimental data distribution will be used Shapiro-Wilk normality test. This test is used for data sets that do not exceed 50 values. In the present case, each data set has a number of seven values;

3) Identifying data affected by aberrant errors can be accomplished by applying the Romanowski test. This test for identifying data affected by aberrant errors having applies for a number of up to 20 data.

After performing statistical analysis of the data can proceed to modelling the experimental data in order to obtain those mathematical models which best describe the process.

4 Conclusions

Performing an experimental research requires some steps to obtain conclusive results.

1) By conducting the experimental researches, we aim at obtaining results for the improvement of the cold-forming process of the threads.

2) In this paper, we present a research methodology in the field of thread cold-forming and the used technological ensemble.
3) The design of experimental research is based on exploratory results already made.
4) In a first step, the exploratory research will be expanded using the Taguchi method to determine the influence of the studied process parameters. This method has the advantage that by performing a small number of experiments it is possible to obtain concrete data about the influence of the process parameters.
5) In the second step, using the data obtained from the application of the Taguchi method, an experimental research will be carried out using the full factorial plan method for the process parameters with major influence.

We want to have data at the end of the experiments with the help of any time, according to the identified parameters we can optimize the thread deforming processing in case of cold-forming taps.

References

1. S.A. Moldovan, M. Banica, V. Nasui, MATEC Web of Conferences, **112**, (2017).
2. Emuge GmbH, Katalog 150. Thread Cutting Technology. Clamping Technology. (2016)
3. Walter AG, Product handbook. Drilling & Threading (2009)
4. O. Pruteanu, Al. Epureanu, C. Bohosievici, Cs. Gyenge, *Tehnologia fabricării mașinilor*, EDP, București (1981)
5. D.C. Montgomery, *Design and analysis of experiments*, 8th edition, Wiley, (2013).
6. M. Țîțu, C. Oprean, A. Boroiu, *Cercetarea experimentală aplicată în creșterea calității produselor și serviciilor*, Ed. Agir, București (2011)
7. S. Ravai-Nagy, D. I. Vlad, D. C. Tămâian, A. Százvai, Experimental studies on tapered hole manufacturing using roll-form taps. International Multidisciplinary Conference, 12th Edition 24-26 May, Baia Mare – Nyiregyhaza (2017)
8. N. Medan, *Hidraulica*, **4**, 41-47, ISSN 1453 – 7303 (2016)

MATEC Web of Conferences 137, 03010 (2017) DOI: 10.1051/matecconf/201713703010
MTeM - AMaTUC 2017

Research on influence of gear parameters on noise, vibrations and harshness conditions for automatic transmissions run-off cycle

Nelu Pascalau[1], *Ioan* Vuscan, *Nicolae* Panc[1,*]

[1]Technical University of Cluj-Napoca, Department of Manufacturing Engineering, 103-105 no. Bvd. Muncii, Cluj-Napoca, Romania

Abstract. Noise vibration harshness (NVH) defines, as a whole, that specific field within automotive industry, that studies mostly the noise and vibrations for different assemblies (such as chassis or drivetrain – gearbox) or complete vehicles, particularly cars and trucks. Gear quality parameters have been studied and it has been experienced that these parameters have an important relevance for NVH topic. Therefore, this paper introduces a case-study, as to highlight the influence of two of these parameters, profile angle deviation ($fH\alpha$) and tooth trace angle deviation ($fH\beta$), on run-off cycle on test benches, for high-performance automatic transmission, designed for passenger vehicles. The demand for high accuracy is mandatory, so fine adjustments are required, as could be further observed, in order to accomplish the requirements for a lower NVH run-off rate, while the whole life-time.

1 Introduction

Quality of a gear, highlighted by its elements, is defined by the accuracy of applied manufacturing processes, as to achieve the geometrical and dimensional demands, required by functional conditions and using cycles, for each component of an assembly or subassembly [1]. The influence of machining accuracy of gear parameters on noise vibration harshness conditions has been tested and experienced, therefore, it can be observed that, in order to reduce the level of noise and vibrations, it is important not only the machining accuracy, but also the fine adjustment of some parameters, as to ensure a smooth gearing process, concerning the complexity of brand-new automatic transmissions [2].

Decreasing the level of noise and vibrations while run-off cycle is a very important aspect within the automotive industry, therefore huge amounts of money are invested yearly, in order to develop new products and to optimize those in using cycle, as a result of the problems noted after a time range in use [3-6].

The sources of noise in a vehicle may vary, including the engine, drivetrain, contact between tires & rolling surface and air friction, while moving [7]. Decreasing the NVH values has been and still is further studied, as a field apart within automotive industry.

* Corresponding author: nicolae.panc@tcm.utcluj.ro

A variety of specific solutions, combined with innovative simulation and competitive manufacturing technologies, ensures the wide range of solutions for products and systems. Sound insulation, as well as lowering the level of noise and vibrations, using special materials and combining some frequency ranges between 20 and 20.000 Hz is possible nowadays [8]. Speaking about assemblies and subassemblies, the problem is being specifically applied, related to final values required by the assembly itself, as not to exceed the level of noise & vibrations [1]. This paper presents a case study, where the maximum limit for noise is 40 dB, value registered on final assembly test bench, while run-off test cycle.

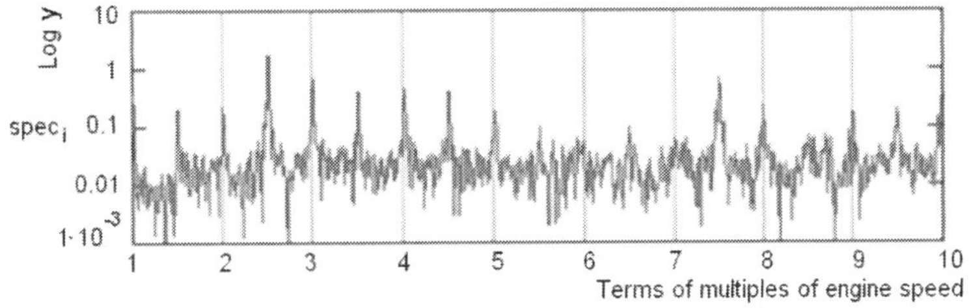

Fig. 1. The noise spectrum of Michael Schumacher's Ferrari at 16680 rpm, showing the various harmonics [9].

In order to measure the level of noise and vibrations for automatic transmissions (figure 1), there are used high performance test benches, similar to that one from figure 2, allowing functional simulation in real conditions, as well as measurement and registering of achieved values and identifying those components that produce disturbances while running.

Fig. 2. Test bench for transmission run-off test [10].

Further on, there is described how exactly this research has been performed, through fine adjustment within manufacturing process of two parameters, profile angle deviation-fHα and tooth trace angle deviation-fHβ, as to get the desired results for noise level at the end of assembly process, in different cases, meaning different values set while gear hobbing process for single parts, in order to reach the demands for the whole assembly, during complex gearing process, inside the automatic transmission.

DOI: 10.1051/matecconf/201713703010

2 Research methodology

The purpose of this case study is to highlight the influence of fine adjustment of these 2 parameters, fHα and fHβ, on transmission run-off cycle, through gearing process with matting gears, ring gears and sun gears. In order to respect the maximum allowed limit for noise of 40 dB of the whole assembly, the quality requirements for gear manufacturing process are according to Q7, DIN 3962 [11].

Table 1 contents the parameters of gear hobbing process, performed on Gleason Pfauter P90 machines.

Table 1. Gear hobbing process parameters.

Equipment	*Process parameters*				
	Roughing		*Finishing*		*Cutting tool*
	Axial feed	*Cutting speed*	*Axial feed*	*Cutting speed*	
Gleason Pfauter P90	2.0 mm/rev	320 m/min	0.4 mm/rot	850 m/min	HM, accuracy AAA+, 2 starts, DIN 3960

As can be noticed in table 1, while gear hobbing process, the work-piece is also finished, according to required accuracy class, so the two parameters fHα and fHβ will be set as to avoid as much as possible the distortions from hardening process (figure 3).

Quality after DIN 3962	= 7
Deviation	
Profile form -	0.009
Pressure angle - fHalfa	0.007
Total pitch error -	0.028
Pitch error	0.011
Normal base pitch -	0.009
Tooth-trace deviation- fHbeta	0.011

Fig. 3. Quality requirements for planetary gear.

The tests have been performed at the end of assembly process and have been identified 3 different cases, concerning the two parameters, fHα & fHβ, as can be seen in tables 2, 3 and 4. Basically, these tables content the individual values and average for these two parameters, for gear hobbing and finished part, as well.

Case 1, according to table 2:

- fHα not optimized
- fHβ not optimized

Table 2. Case 1 – fHα and fHβ not optimized.

Gear hobbing				*Finished*			
fHα	fHαm	fHβ	fHβm	fHα	fHαm	fHβ	fHβm
-3/11 µm	6/8 µm	6/18 µm	10/12 µm	-7/7 µm	-1/3 µm	-11/11 µm	-2/2 µm

Further on, there is presented the case when fHβ has been optimized, meaning asymmetrical adjustment on the right and left flank, fHα not optimized.

Case 2, according to table 3:

- fHα not optimized
- fHβ optimized

Table 3. Case 2 – fHβ is opzimized.

Gear hobbing				*Finished*			
fHα	fHαm	fHβ	fHβm	fHα	fHαm	fHβ	fHβm
-3/11 μm	6/8 μm	8/18 μm left 14/24 μm right	11/13 μm left 18/20 μm right	-7/7 μm	-1/3 μm	-11/11 μm	5/7 μm left 8/10 μm right

The last case, number 3, it is about both parameters optimized, as it can be noticed in table number 4.

Case 3, according to table 4:
• fHα is optimized
• fHβ is optimized

Table 4. Case 3 – fHα and fHβ are optimized.

Gear hobbing				*Finished*			
fHα	fHαm	fHβ	fHβm	fHα	fHαm	fHβ	fHβm
0/4 μm	1/3 μm	8/18 μm left 14/24 μm right	11/13 μm left 18/20 μm right	-7/7 μm	-6/-1 μm	-11/11 μm	5/7 μm left 8/10 μm right

These values have been established after testing several charges of planetary gears, as to reach the capability of manufacturing and assembly processes. These fine adjustments have been performed based on results registered on transmission test bench at the end of assembly process.

3 Results

Within this chapter are presented the results achieved for each case, from the three presented above. These planetary gears are assembled in the automatic transmission, presented in figure 4.

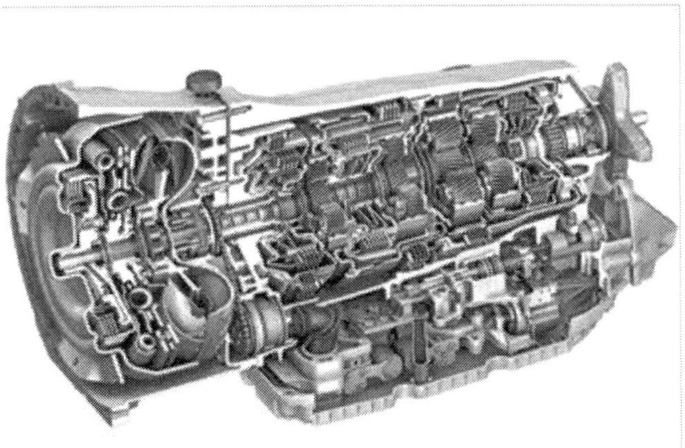

Fig. 4. 9-speed automatic transmission, with planetary gears [12].

Further on, there will be presented the results regarding the noise values, achieved on assembly test benches, for each case.

DOI: 10.1051/matecconf/201713703010

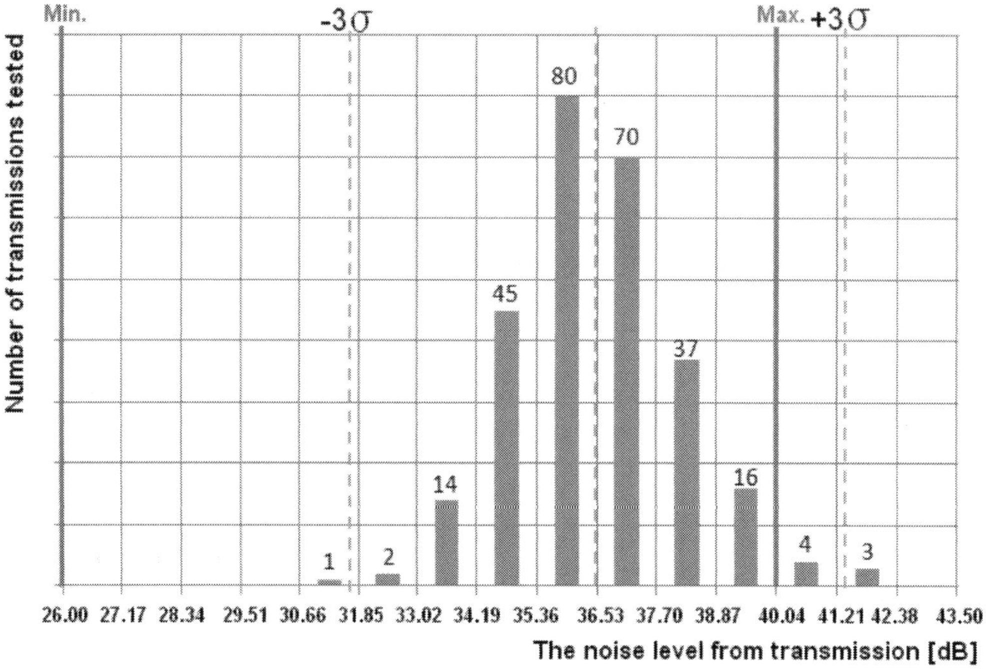

Fig. 5. Case 1 – noise results achieved with fHα and fHβ not optimized, in dB.

As it can be seen in figure 5, there have been tested 272 transmissions; 7 pieces have exceeded the limit of 40 dB, with an average of 36.53 dB, minimum value is 31.54 dB, maximum value is 41.54 dB; scrap rate in this case is 2.57%>1% maximum percentage allowed.

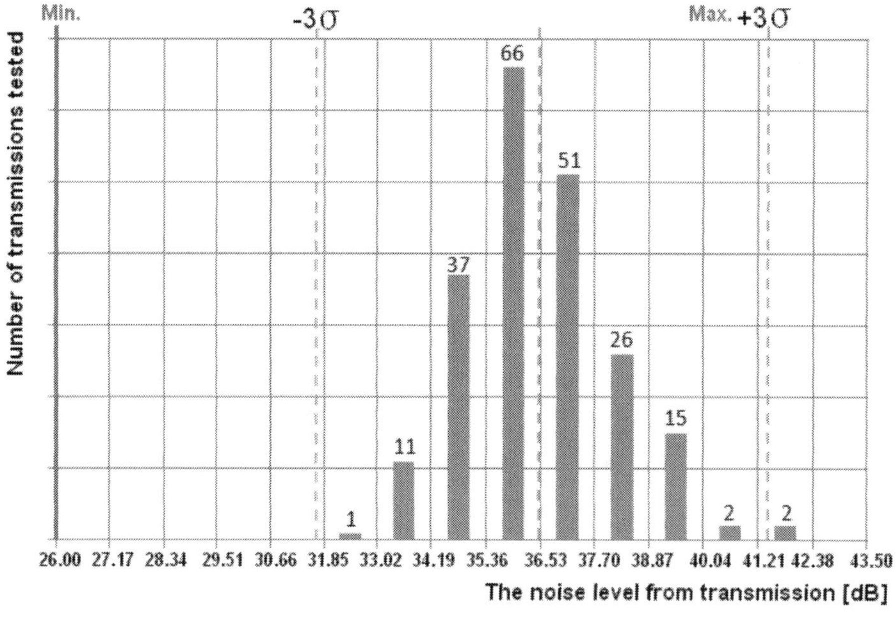

Fig. 6. Case 2 – noise results achieved only with fHβ optimized, in dB.

Applying the case number 2, there have been tested 211 transmissions, 4 pieces have exceeded the limit of 40 dB, the average is 36.53 dB and minimum value is 31.62 dB, maximum value is 41.62 dB; scrap rate in this case is 1.89%. All these can be observed in figure 6.

Results achieved when applying the case number 3, when both parameters, fHα & fHβ have been optimized, are much better, as can be noticed in figure 7. Practically, there have been tested 593 transmissions, 1 piece has exceeded the limit, the average of registered values is 35.58 dB, minimum registered value is 30.88 dB and maximum is 40.28 dB. Considering these values, the scrap rate is 0,17%<1% maximum allowed.

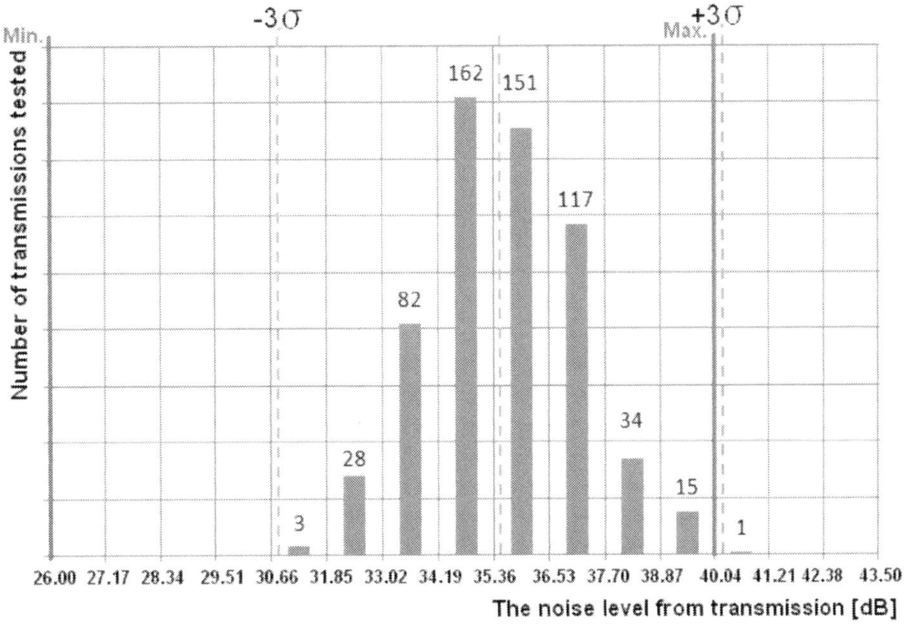

Fig. 7. Case 3 – noise results achieved with fHα and fHβ optimized, in dB.

Further on, there are presented also the effective values of these two parameters, fHα according to figure 8 and fHβ according to figure 9, measured on a Klingelnberg P26 gear measuring equipment. Klingelnberg P26 is measuring equipment used to inspect cylindrical gears, pinion type cutters and shaving cutters, worms and worm wheels, hobs, bevel gears, general dimension, shape, and positional deviations of axially symmetrical workpieces, cam and camshaft measurement and rotor measurement.

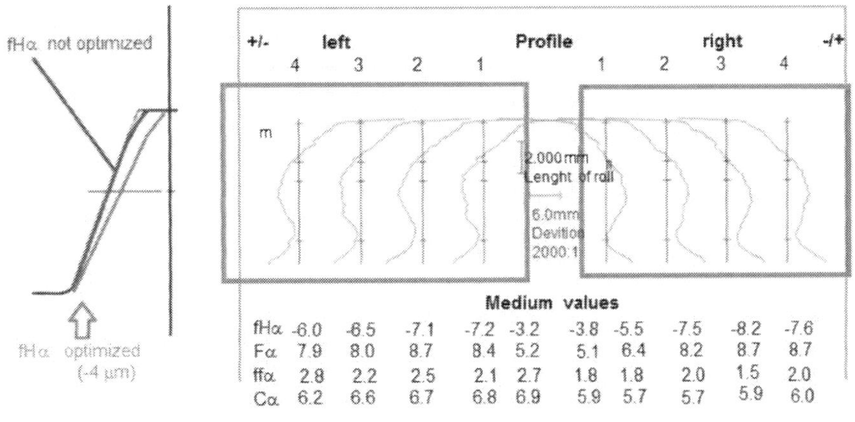

Fig. 8. Values measured for optimized fHα.

 DOI: 10.1051/matecconf/201713703010

Figure number 8 shows some values achieved for profile angle deviation, when measuring on Klingelnberg P26 equipment, after an adjustment of 4 microns in minus has been applied, as to get a smoother gearing process with matting gears, especially with sun gears. In figure number 8, there are presented some values for tooth trace angle deviation – fHβ, so it can be seen that the medium values are positive, according to the limits specified in table 4, within chapter 3.

Basically, in this case, it has been set the asymmetrical machining principle, on the left and right flank, while manufacturing the gear on Gleason P90 machines. For each case apart, the scrap rate has been calculated, only when applying the fine adjustment for both parameters, this value could be reduced under 1%, the maximum limit allowed.

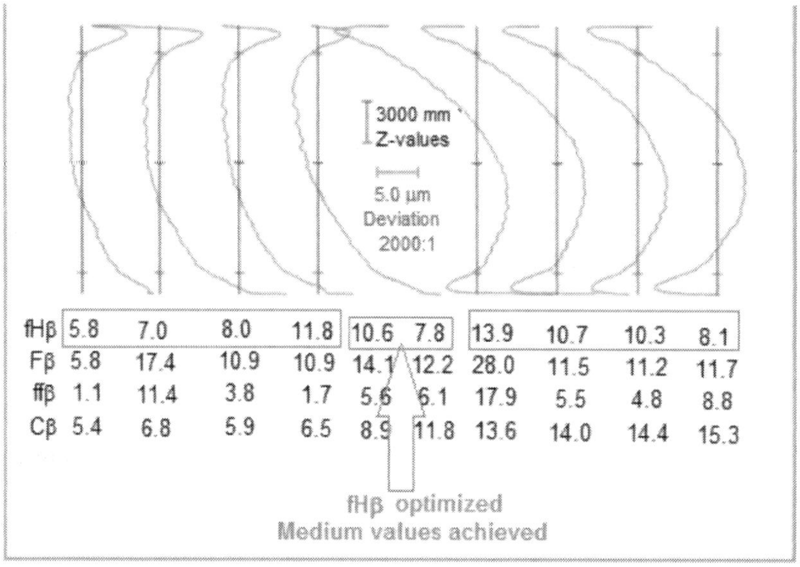

Fig. 9. Values measured for optimized fHβ.

4 Conclusions

This paper highlighted how the fine adjustment of these 2 parameters, fHα and fHβ, influences the noise level through gearing process, of an automatic transmission tested on a test bench. There have been identified 3 different cases, as specified above, within chapter 3, the results achieved have been presented within chapter 4, for each case apart. Concluding on those facts presented above, it can be noticed, such as follows:

• The noise level depends on quality not only of planetary gears, but of the other matting gears, as well, precisely ring gears and sun gears;

• Even if the quality of planetary gears as individual parts is good, the most important is the result from the end of assembly process;

• When applying the first case, the scrap rate was 2.57%, much more than 1%, the maximum value allowed;

• Considering the second case, the value for scrap rate was 1.89%, better in this case, but still needed to be further optimized;

• When applying the third case, the target has been reached, less than 1%, 0.17% precisely;

• This research is a base for further trials that could be further performed, if such problems may occur.

As presented above, these main conclusions have been stated as being the key features when discussing the results achieved, presented above. These results have been achieved applying this experimental methodology and can be further used and extended if needed, if the problem appears while testing the run-off cycle. This study might be considered as a smaller technical source of know-how for similar challenges that have to be passed when speaking about NVH topics, especially for automatic transmissions.

References

1. J. A. Speck, *Mechanical Fastening, Joining, and Assembly, 2nd Edition*, (CRC Press, Taylor by Francisc Group, 2015)
2. Ih. Jeong-Guon, C.-Y. CHOI, T.-K. KIM, S.-H. JANG, H.-J. KIM, International Journal of Automotive Technology **12**(4), 617–630 (2011)
3. P.M. Suresh, *Estimation of noise and vibration (nvh) of the sandwich constructed component of the automotive vehicle,* (Ph.D Thesis at Kuvempu University, India, 2009)
4. T. Eritenel, R. G. Parker , Journal of Sound and Vibration **331**(15), 3628-3648 (2012)
5. Y. Guo, T. Eritenel, T. M. Ericson, R. G. Parker, Journal of Sound and Vibration **333** (22), 5762-5785 (2014)
6. G. D'Elia, E. Mucchi, M. Cocconcelli, Mechanical Systems and Signal Processing **83**, 305-320 (2017)
7. D. Ibarra, R. Ramírez-Mendoza, E. López, Applied Acoustics **118**, 58-65 (2017)
8. G. Sheng, *Vehicle Noise, Vibration, and Sound Quality*, (SAE International, 2012)
9. *** https://en.wikipedia.org/wiki/Noise,_vibration,_and_harshness
10. *** http://www.powerlinkpt.com/wp-content/uploads/2017/02/powerlink-slide-en-05.png
11. DIN 3962, *Toleranzen für Stirnradverzahnungen.*
12. *** http://www.car-engineer.com/9g-tronic-automatic-transmission-mercedes/

Optimization of the Surface Roughness Equation obtained by Al7136 End-Milling

Alina Bianca Pop[1,*] and *Mihail Aurel* Ţîţu[2]

[1]SC TechnoCAD SA, Vasile Alecsandri no. 72, code 430351, Baia Mare, România
[2]Lucian Blaga University of Sibiu, Victoriei Street no. 10, code 550024, Sibiu, România

Abstract. The aim of this paper is to optimize the regression equation of the surface roughness obtained by 7136 aluminium alloy machined by end-milling process. The surface roughness is dependent on certain process parameters, which can vary, causing in this way variations of the surface quality. The research method used in this paper is the experiment and the Taguchi design of experiment. The experiment was performed using an experimental stand, in which every step to get the purpose, is presented. The measurements were made using a portable surface roughness tester. In the first part of the paper the influence percentage of the involved parameters in the machining process, was determined. Then, a multiple linear regression model, in three different ways, was realised, in order to optimise the predicted regression equation that was initially proposed.

1 Introduction

Surface roughness greatly influences the performance of mechanical parts, as well as the production cost, therefore, it is a very important measurement of the product quality. Also, the surface roughness has a great impact on the mechanical properties, like: fatigue behaviour or corrosion resistance, and functional attributes like: wear, friction, light reflection, heat transmission, and electrical conductivity. Currently, there have been many research developments in surface roughness modelling and optimization of the metal cutting process parameters, to obtain the desired surface finish, since only the proper selection of these parameters can produce a better surface finish [1]. About the previous researches, the surface roughness was studied in different research papers, in which the experiment and the mathematical and statistical methods were adopted. For example, in [1] and [2] the 7136 aluminium alloy used in the aircraft industry was studied by end milling process to analyse the cutting feed effect on the surface quality and the manufacturing time reduction, using standard tools for aluminium machining. A mathematical modelling of the surface roughness resulted by the end milling process of Al7136, depending on the cutting process parameters studied in a settled experimental field, was determined in [3]. In this paper, using the variance analysis ANOVA, the degree of confidence for the achieved results by the regression equation was determined, and the suitability of this equation on

* Corresponding author: bianca.bontiu@gmail.com,

every point of the experimental field. In [4] the author purpose was to highlight the importance of Taguchi's method use to analyse the surface roughness of 7136 aluminium alloy in end milling process. In this paper, the attention will be focused on the Taguchi design of experiments – method steps presented in [5] and [6]. Then, to determine the equation who describes the surface roughness values according to the process parameters it is necessary to realised some practical experiments to determine the values of surface roughness for different set-up values of cutting process parameters. Using the obtained data, a linear regression model will be determined to describe the studied process. A similar research made on this direction is [7]. Finally the regression model will be optimised to increase the degree of prediction for this model.

2 The experiment

First, to get this research paper purpose, it is necessary to obtain the data that has to be processed – the surface roughness measurements. For this, the experimental stand used to perform the experiment, is presented in figure 1.

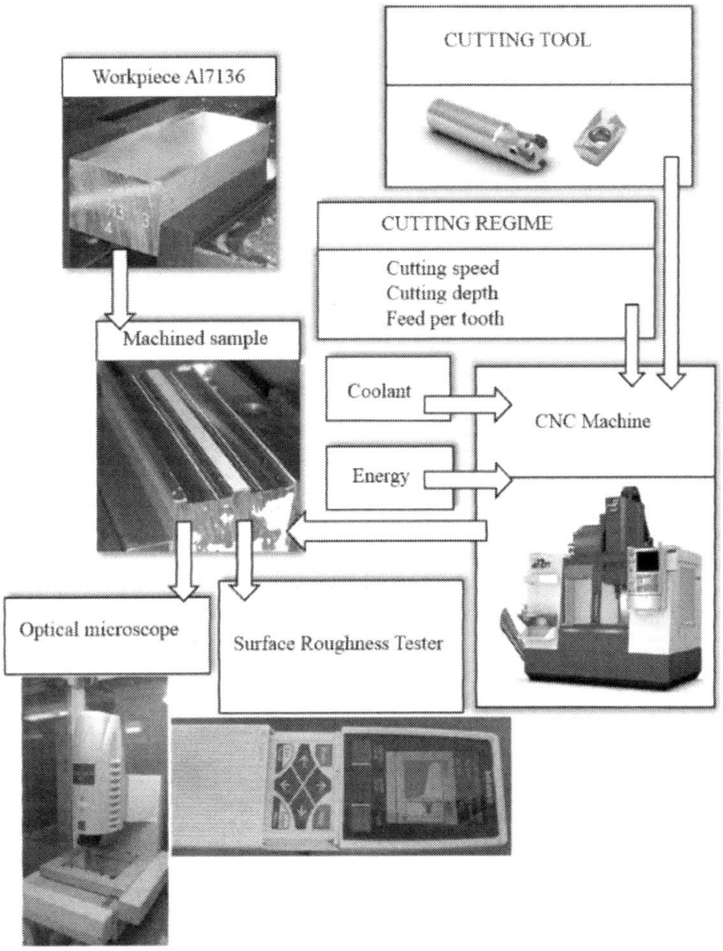

Fig. 1. The experimental stand [8].

As it can be seen here, first the workpiece material is chosen - the Al7136 aluminium alloy. The behaviour of this material will be studied in accordance with the proposed objectives taking into account the variation of the cutting regime parameters. The

DOI: 10.1051/matecconf/201713703011

experiment will be performed using a standard tools set of aluminium machining - 16 mm End milling cutter with 100% tool engagement - SECO R217.69-1616.0-09-2AN, holding two indexable cutting inserts XOEX090308FR-E05, H15. The machine used on milling tests is HAAS VF2 CNC. To make an optical determination and evaluation of the finished surface, the optical microscope Micro-Vu VERTEX 310 was used. To analyse the surface roughness for different machining conditions it was used the portable surface roughness tester - Mitutoyo SURFTEST SJ-210. After the experimental stand presentation the research methods used in this paper may be approached. For the beginning, the Taguchi's Method will be used to determine the parameters percentages and their interactions influence exerted on the surface roughness.

3 The research methodology

The research method of this study is the experiment. Then to determine the Ra values of the surface roughness obtained by Al7136 end-milling, the Taguchi design of experiment is used. Well, to determine the surface roughness, it is necessary to set up the experimental domain. The cutting process involves a series of cutting parameters. By all of these parameters, in order to achieve a good surface roughness in this paper the attention will be focused especially on the cutting speed [m/min], cutting depth [mm] and feed per tooth [mm/tooth]. The values of the cutting regime parameters are chosen in accordance to the SECO Tools manufacturer indications and also according to the technological possibilities of the cutting tool, cutting inserts and CNC machine.

Table 1. The cutting regime according to the tools manufacturer indications.

Cutting speed [m/min]	Cutting depth [mm]	Feed per tooth [mm/tooth]
495	2	0.04
530	2.5	0.06
570	3	0.08
610	3.5	0.11
660	4	0.14

But, according with the Taguchi's method, the cutting process parameters gets the levels of minimum and maximum values. These levels are presented in table 2.

Table 2. The cutting process parameters and their values according to the Taguchi's Method.

Abbreviation	Parameter	Level 1 (minimum)	Level 2 (maximum)
A	Cutting speed	495	660
B	Cutting depth	2	4
C	Feed per tooth	0.04	0.14

Corresponding to these 3 established parameters and their third degree interactions, after the Taguchi method applying, it result the L8 (2^3) orthogonal array. Well, it is necessary to conduct 8 experiments according to the Taguchi design of experiments. For each of these 8 experiments, three sets of the surface roughness measurements were made, which are presented in table 3.

Table 3. The measured surface roughness according with the parameters set.

No. of exp.	Parameter			Surface roughness measurements [μm]			$R_{a\ med}$
	A	B	C	R_{a1}	R_{a2}	R_{a3}	
1	495	2	0.04	0.209	0.229	0.203	0.214
2	495	2	0.14	0.235	0.231	0.238	0.235
3	660	4	0.04	0.511	0.551	0.571	0.544
4	660	4	0.14	0.515	0.512	0.504	0.515
5	495	4	0.04	0.191	0.195	0.209	0.198
6	495	4	0.14	0.210	0.221	0.223	0.218
7	660	2	0.04	0.502	0.557	0.540	0.533
8	660	2	0.14	0.715	0.712	0.740	0.722

4 The determination of the process parameters contribution

To calculate the contribution of the cutting process parameters and their interaction, the measurements presented in table 3 are used. Using the Taguchi's method itinerary, the Variance analysis was adopted to determine the influence of the studied parameters exerted on the surface roughness. Using Minitab 17, in table 4 is presented this analysis.

Table 4. The Variance Analysis ANOVA (Minitab 17).

Source	DF (degree of freedom)	SS (sum of square)	Contribution
Regression	7	1.554	99.99%
A	1	1.404	90.35%
B	1	0.044	2.85%
C	1	0.000	0.00%
A x B	1	0.066	4.30%
A x C	1	0.013	0.85%
B x C	1	0.002	0.17%
A x B x C	1	0.023	1.48%
Error	1	0.000	0.01%
Total	8	1.554	100.00%

In table 4, it can be seen that 2 parameters and 3 interactions have a significant contribution with an influence exerted on the surface quality for the established experimental domain: the parameter A – cutting speed, the parameter B – cutting depth and the interactions between A x B – cutting speed with cutting depth and A x C – cutting speed with feed per tooth, and the interaction between all of three parameters A x B x C – cutting speed with cutting depth and feed per tooth.

5 The regression equation determination of the process surface roughness

Based on the table 4 results, the next step is to determine the regression equation of the surface roughness obtained under the cutting process parameters influence. For this, will be used only the parameters and the interactions previously set, that have the greatest influence on the surface quality. Using Minitab 17 software, the multiple linear regression model of the surface roughness for the Taguchi's method was determined, as it is shown in equation

1. This multiple linear regression model was obtained *without using the Box-Cox transformation*. The Variance analysis of this equation is presented in table 5.

$$RaM=0.000508 \cdot A-0.2499 \cdot B+0.000467 \cdot A \cdot B+0.00412 \cdot A \cdot C-0.00107 \cdot A \cdot B \cdot C \quad (1)$$

Table 5. The Variance Analysis ANOVA of the regression model without Box-Cox transformation.

Source	DF (degree of freedom)	SS (sum of square)	Contribution
Regression	5	1.526	98.14%
A	1	1.404	90.35%
B	1	0.044	2.85%
A x B	1	0.063	4.08%
A x C	1	0.005	0.37%
A x B x C	1	0.007	0.50%
Error	3	0.028	1.86%
Total	8	1.554	100.00%

The regression statistics of the equation (1) are:

- R squared: 98.14%;
- R squared adjusted: 95.04%;
- and R squared predicted: 87.76%

Now, to optimise this equation (1), it is necessary to use the *Box-Cox transformation with rounded* λ. The resulted regression model is presented in the equation (2):

$$lnRaM=-0.00198 \cdot A-1.330 \cdot B+0.002310 \cdot A \cdot B+0.0071 \cdot A \cdot C-0.00174 \cdot A \cdot B \cdot C \quad (2)$$

In table 6 is presented the ANOVA analysis of the regression model (2).

Table 6. The Variance Analysis ANOVA of the regression model with rounded λ.

Source	DF (degree of freedom)	SS (sum of square)	Contribution
Regression	5	1.362	96.55%
A	1	7.456	69.48%
B	1	0.720	6.71%
A x B	1	2.140	19.94%
A x C	1	0.024	0.23%
A x B x C	1	0.020	0.19%
Error	3	0.370	3.45%
Total	8	10.732	100.00%

The regression statistics of this equation (2) are:

- R squared: 96.55%;
- R squared adjusted: 90.80%;
- and R squared predicted: 76.67%

Another possible optimisation of the equation (1) can be made using *Box-Cox transformation with* $\lambda=0.5$. In this situation it is resulting the equation (3).

$$(RaM)^{0.5}=0.000942 \cdot A-0.1194 \cdot B+0.000227 \cdot A \cdot B+0.00269 \cdot A \cdot C-0.000681 \cdot A \cdot B \cdot C \quad (3)$$

In table 7 it is presented the ANOVA analysis of the regression model (3).

Table 7. The Variance Analysis ANOVA of the regression model with rounded λ=0.5.

Source	DF (degree of freedom)	SS (sum of square)	Contribution
Regression	5	3.171	99.76%
A	1	3.139	98.74%
B	1	0.013	0.43%
A x B	1	0.012	0.40%
A x C	1	0.002	0.09%
A x B x C	1	0.003	0.10%
Error	3	0.007	0.24%
Total	8	3.179	100%

The regression statistics of this equation (3) are:

- R squared: 99.76%;
- R squared adjusted: 99.37%;
- and R squared predicted: 98.40%

6 Conclusions

This research paper presents a methodology to determine the surface roughness obtained under the cutting process parameters influence. The studied cutting process parameters were: the cutting speed, the cutting depth and the feed per tooth. The research method used in this paper was the experiment and the Taguchi design of experiment. After applying the Taguchi method to calculate the influence of the studied cutting process parameters on the surface quality, it was found that the cutting speed is the factor with the greatest influence on the studied response - with 90.35%. According to the influence of the studied cutting process parameters and their interactions, a multiple linear regression model was realised in three different ways to optimise the prediction of the proposed equation:

• The first type of the regression model without transformation it is with R squared predicted 87.76%,

• The second type of the regression model using rounded λ transformation it is with R squared predicted 76.67%,

• The third type of the regression model using λ=0.5 transformation it is with R squared predicted 98.40%.

In this study the best prediction is given by the regression using rounded λ transformation, followed by the regression using λ=0.5 transformation. The lowest degree of prediction is given by the regression with rounded λ transformation. Based on the aspects mentioned above, we can concluded that it is possible to improve the prediction of the regression equation that describes the surface roughness in any point of the settled experimental field, using different type of regression and using the same measured values of the surface roughness.

References

1. A.B. Bontiu Pop, Scientific Bulletin, Serie C, Fascicle: Mechanics, Tribology, Machine Manufacturing Technology ISSN 1224-3264, **XXVIII** (2014)
2. A. B. Bonţiu Pop, M. Lobonţiu, Applied Mechanics and Materials, **808**, 34 (2015)
3. A.B. Pop, M.A. Ţîţu, IOP Conf. Series: Materials Science and Engineering **161**, (2016)

4. M.A. Ţîţu, A.B. Pop, IOP Conf. Series: Materials Science and Engineering **161**, (2016)
5. D.C. Montgomery, *Design and analysis of experiments*, 8th edition, Wiley (2013)
6. M. Ţîţu, C. Oprean, A. Boroiu, *Cercetarea experimentală aplicată în creşterea calităţii produselor şi serviciilor*. Bucureşti: AGIR (2011)
7. N. Medan, Magazine of Hydraulics, Pneumatics, Tribology, Ecology, Sensorics, Mechatronics, **3** (2016)
8. A. B. Bonţiu Pop, *Calitatea suprafeţelor la aşchierea aliajelor de aluminiu cu freze cilindro-frontale*, PhD. Thessis. University Baia Mare (2015)

MATEC Web of Conferences 137, 03012 (2017) DOI: 10.1051/matecconf/201713703012
MTeM - AMaTUC 2017

Multi-objective optimization of single pass turning using performance diagrams

Miroslav Radovanovic[1,*]

[1]University of Nis, Faculty of Mechanical Engineering, 18000 Nis, Serbia

Abstract. Maximization of productivity and minimization of cost with acceptable quality of the parts are the most important goals of the production. In machining multi-objective optimization is a real problem. For solving this problem modern optimization methods are used. In this paper is presented a multi-objective optimization of single-pass turning of stainless steel with coated carbide tool using performance diagrams.

1 Introduction

Optimization of cutting performances is one of the most important problems in machining processes. Multi-objective optimization provides optimal or near-optimal solution for two or more objectives in machining process. Number of researchers has studied multi-objective optimization in turning operations. Kubler F. et al. in [1] studied turning of 42CrMo4 using coated carbide cutting tool. They investigated the effect of depth of cut, feed rate and cutting speed on machining time, tool wear and process energy. For solving the multi-objective optimization problem with three objectives they applied GA based non dominated sorting algorithm-II (NSGA-II). Soni V. et al. in [2] studied turning of aluminum using carbide cutting tool. They investigated the effect of depth of cut, feed rate and cutting speed on material removal rate and surface roughness. For solving the multi-objective optimization problem with two objectives they applied Multi Objective Genetic Algorithm (MOGA). Abbas A. et al. in [3] studied turning of J-steel using uncoated carbide cutting tool. They investigated the effect of depth of cut, feed rate and cutting speed on machining time and surface roughness. For solving the multi-objective optimization problem with two objectives they applied Multi-objective Efficient Global Algorithm. Dhandapani K. et al. in [4] studied turning of AISI 4340 using uncoated carbide cutting tool. They investigated the effect of depth of cut, feed rate and cutting speed on material removal rate, flank wear and surface roughness. For solving the multi-objective optimization problem with three objectives they applied GA based non dominated sorting algorithm-II (NSGA-II). Soury E. et al. in [5] studied turning of St 37 using high speed steel cutting tool. They investigated the effect of depth of cut, feed rate, cutting speed, tool rake angle and nose radius on tool life and surface roughness. For solving the multi-objective optimization problem with two objectives they applied the optimization software program based on Visual Basic.

* Corresponding author: mirado@masfak.ni.ac.rs

In this paper is presented a multi-objective optimization of single-pass turning of stainless steel with coated carbide tool. For solving the multi-objective optimization problem with two objectives (material removal rate and machining cost), three factors (tool nose radius, feed rate and cutting speed) and nonlinear constraints (surface roughness, tool life and cutting power), performance diagrams are used.

2 Multi-objective optimization

Procedure of solving the multi-objective optimization problem has four phases. First phase is selecting objectives, factors and constraints. Second phase is determining optimization problem. Third phase is selection of method for solution the optimization problem. Fourth phase is solution the optimization problem. First phase in multi-objective optimization is selection of objectives, factors and constraints. Two objectives, material removal rate and machining cost, and three factors, tool nose radius, feed rate and cutting speed, are selected. For nonlinear constraints, surface roughness, tool life and cutting power are used.

Tool life is performance of the cutting tool from which depends continual work. Relation between the tool life and the cutting factors is expressed with the Taylor's formula.

$$T = \frac{C_T}{a_p^r f^m v_c^p} \tag{1}$$

Where: T (min) – tool life, a_p (mm) - depth of cut, f (mm/rev) - feed rate, v_c (m/min) – cutting speed, C_T, r, m and p – empirical constants relevant to a specific tool-workpiece combination.

Surface roughness is performance of the workpiece which must be satisfy. Ordinarily, average surface roughness is used.

$$R_a = 32\frac{f^2}{r_\varepsilon} \tag{2}$$

Where: R_a (μm) - surface roughness, f (mm/rev) - feed rate, r_ε (mm) - tool nose radius.

Cutting power is important in view of available machine tool power.

$$P_c = \frac{F_c v_c}{60000} = \frac{k_{c1.1} a_p f^{1-m_c} v_c}{60000(\sin\kappa)^{m_c}} \tag{3}$$

$$P_c \le \eta P_m \tag{4}$$

Where: P_c (kW) - cutting power, $k_{c1.1}$ (N/mm^2)- unit specific cutting force, m_c - exponent of specific cutting force, κ - cutting edge angle, η - efficiency, P_m - machine power (main motor power).

Machining time for single pass turning is:

$$t_m = \frac{L}{fn} = \frac{\pi DL}{1000 f v_c} \tag{5}$$

Where: t_m (min)- machining time, L (mm) - cutting length, f (mm/rev) - feed, n (rpm) - spindle speed, D (mm) - working diameter, v_c (m/min) – cutting speed.

Material removal rate is volume of removed material in unit time.

$$MRR = a_p f v_c \tag{6}$$

Where: MRR (cm^3/min) - material removal rate, a_p (mm) - depth of cut, f (mm/r) - feed, v_c (m/min) – cutting speed.

Machining cost for machining working diameter on cutting length is:

$$C = C_r t_n + C_r t_m + \frac{t_m}{T}(C_r t_d + C_a) \tag{7}$$

Where: C (EUR) – machining cost per piece, C_r (EUR) – labor plus overhead cost, t_n (min) – nonproductive time, t_m (min) – machining time, T (min) – tool life, t_d (min) – tool changing time, C_a (EUR) – tool cost per cutting edge.

Machining cost for turning, according to (1) and (3), is:

$$C = C_0 + C_1\, f^{-1} v_c^{-1} + C_2\, a_p^r\, f^{m-1} v_c^{n-1} \tag{8}$$

$$C_0 = C_r t_n,\ C_1 = \frac{\pi D L C_r}{1000},\ C_2 = \frac{\pi D L (C_r t_d + C_a)}{1000 C_T},\ C_a = \frac{C_{wp} n_p}{n_{tp}}\left(1 + \frac{z_b}{2}\right) + \frac{C_{wh}}{n_{th}} + \frac{C_{we}}{n_{te}} + C_{wv}$$

Where: C_{wp} (EUR) - cost of insert, n_p - number of cutting edges in machining, n_{tp} - number of useful insert cutting edges, z_b - factor of fractures of cutting edge, z_b=0.2-0.4, C_{wh} (EUR) - cost of tool holder, n_{th} – number of tool life (cutting edge) to endure one tool holder, C_{we} (EUR) - cost of tool holder parts, C_{we} =(0.2-0.3)C_{wh}, n_{te} - number of tool life (cutting edge) to endure tool holder parts, n_{te}=(0.15-0.30)n_{th}, C_{wv} (EUR) - cost of preparing tool.

The proposed mathematical model of optimization consists of two objectives, nonlinear constraints and bounds. The mathematical model of multi-objective optimization for single-pass turning has the next form:

- Objective functions:

$$\max, MRR = a_p f v_c$$

$$\min, C = C_0 + C_1\, f^{-1} v_c^{-1} + C_2\, a_p^r\, f^{m-1} v_c^{n-1}$$

- Constraints:

$R_a \le R_{a,max}$	(Constraint related to surface roughness)
$T \ge T_e$	(Constraint related to economical tool life)
$\frac{P_c}{\eta} \le P_m$	(Constraint related to available machine power)
$\pi D n_{min} \le v_c \le \pi D n_{max}$	(Constraint related to available spindle speed)
$v_{c,min} \le v_c \le v_{c,max}$	(Constraint related to tool performance)
$f_{1,min} \le f \le f_{1,max}$	(Constraint related to available feed on machine)
$f_{2,min} \le f \le f_{2,max}$	(Constraint related to tool performance)
$r_{\varepsilon,min} \le r_\varepsilon \le r_{\varepsilon,max}$	(Constraint related to tool performance)

One example of optimization is presented. Workpiece is a bar made from austenitic stainless steel X5CrNiMo18 10, diameter of D=75 mm, length of L=200 mm, unit specific cutting force of $k_{c1.1}$=2600 N/mm^2 and m_c=0.19. Depth of cut is a_p=3 mm. Machine tool is the universal lathe "Potisje PA-C30" with motor power of P_m=11 kW and efficiency of η=0.8. Spindle speed range is n=20-2000 rpm and feed range is f=0.04-9.16 mm/rev. Tool holder is PCLNR, cutting edge angle of κ=95° and rake angle of γ=-6°, with insert CNMG, tool nose radius of r_ε=[0.4 0.8 1.2 1.6 2.4] mm, grade of IC9025 (Iscar). Recommended levels of the cutting factors are: a_p=0.10-10.00 mm, f=0.05-0.80 mm/rev, $v_{c,max}$=290 m/min. Economical tool life is T_e=15 min. Surface roughness is R_a=12.5 μm. According to cutting tool performances (f=0.05-0.80 mm/rev, $n_{max}=1000v_{c,max}/D\pi=1000\cdot 290/75\pi=1231$ rpm), useful range of feed rate is f=[0.049 0.053 0.062 0.071 0.080 0.089 0.098 0.107 0.124 0.142 0.160 0.178 0.196 0.214 0.249 0.285 0.321 0.357 0.392 0.428 0.499 0.571 0.642 0.714 0.785] mm/rev, and useful range of spindle speed is n=[20 24 29 37 45 54 66 81 100 120 145 180 205 330 400 500 600 740 910 1100] rpm. Other data are: C_r=30 EUR/h, t_n=1 min, t_d=1 min, C_{wp}=4.5 EUR, n_p=1, n_{tp}=4, z_b=0.2, C_{wh}=50 EUR, n_{th}=300, C_{we}=15 EUR, n_{te}=200, C_{wv}=0 EUR.

Tool life equation for turning stainless steel with coated carbide tool is [6]:

$$T=\frac{21.873\cdot 10^3}{f^{0.081}v_c^{1.348}} \tag{9}$$

Mathematical model of multi-objective optimization for this example is:

- Objective functions:

$$\max, MRR = 3fv_c$$

$$\min, C = 0.5 + 23.55f^{-1}v_c^{-1} + 0.004\frac{v_c^{0.348}}{f^{0.919}}$$

- Constraints:

$$32\frac{f^2}{r_\varepsilon}\le 12.5$$

$$\frac{21.873\cdot 10^3}{f^{0.081}v_c^{1.348}}\ge 15$$

$$0.163f^{0.81}v_c\le 11$$

$$4.71\le v_c\le 290$$

$$0.049\le f\le 0.785$$

$$0.4\le r_\varepsilon\le 2.4$$

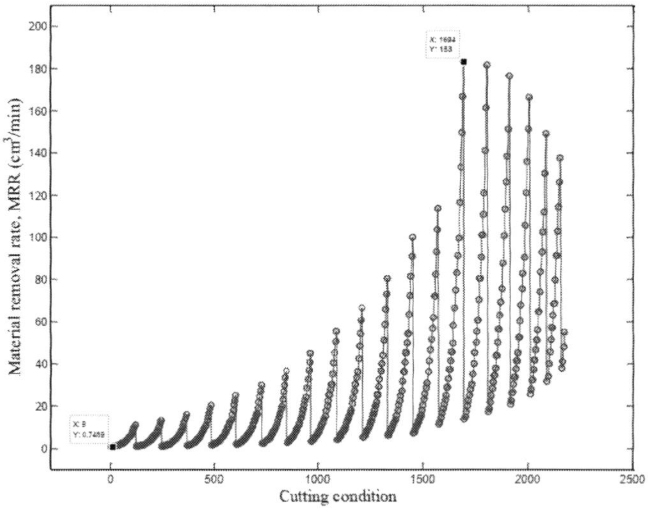

Fig. 1. Material removal rate vs cutting conditions.

For solving the optimization problem performance diagrams are used. Performance diagrams are drawn for all combinations of tool nose radius, feed rate and spindle speed, with respect to constraints of surface roughness, tool life and machine tool power. In this way 2500 different cutting conditions were obtained. In figure 1 is shown 2D diagram of material removal rate versus cutting conditions, and in figure 2 is shown 2D diagram of machining cost versus cutting conditions. Each point in 2D performance diagrams presents one particular cutting condition, i.e. combination of factor levels (r_ε,f,n).

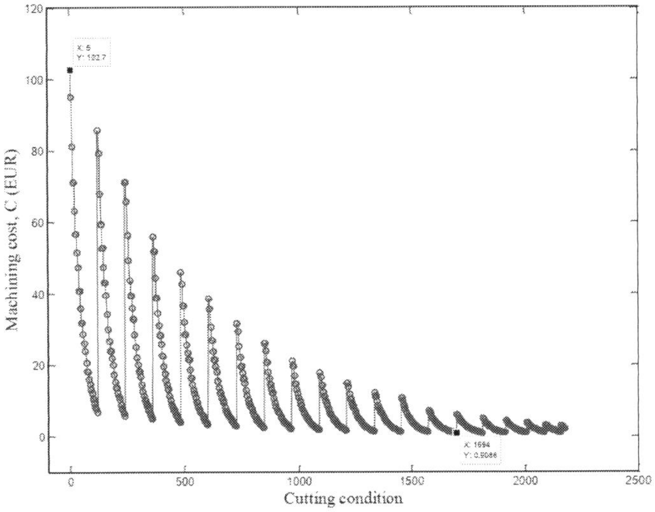

Fig. 2. Machining cost vs cutting conditions.

Based on the performance diagrams, the effects of factors on the objectives can be analyzed. From Figure 1 it can be seen that the point where the material removal rate has maximal value of 183 cm^3/min corresponds to combination 1694. From Figure 2 it can see that the point where the machining cost has minimal value of 0.9088 EUR corresponds to the same combination 1694.

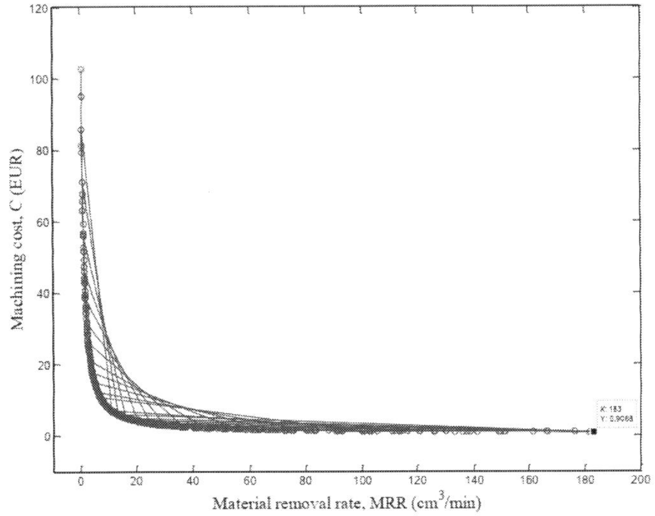

Fig. 3. Machining cost vs material removal rate.

In Figure 3 is shown the diagram of machining cost versus material removal rate. From Figure 3 it can see global optimum, i.e. the point positioned in the lower right corner, where the material removal rate has maximal value of 183 cm^3/min and machining cost has minimal value of 0.9088 EUR. This point corresponds to combination of factor levels 1694 with data: r_ε=2.4 mm, f=0.785 mm/rev and n=330 rpm (v_c=77.715 m/min). For these factor levels material removal rate is maximal MRR_{max}=183 cm^3/min, machining cost is minimal C_{min}=0.908 EUR, machining time is t_m=0.772 min, tool life is T=63 min, surface roughness is Ra=8.22 μm and cutting power is P_c=8.33 kW.

3 Conclusion

Performance diagrams were used for solving multi-objective optimization in single-pass turning of stainless steel with coated carbide tool. Cutting factors, tool nose radius, feed rate and cutting speed, were selected for maximal material removal rate and minimal machining cost. For factor levels r_ε=2.4 mm, f=0.785 mm/rev, and n=330 rpm (v_c=77.715 m/min) material removal rate is maximal MRR_{max}=183 cm^3/min and machining cost is minimal C_{min}=0.908 EUR. In this way the performance diagrams can be used as decision making tool in process planning for selection of cutting conditions.

The paper is a result of the technological project TR35034 which is supported by the Ministry of Education, Science and Technological Development of Republic of Serbia.

References

1. F. Kubler, *Procedia CIRP*, **29**, 822-827 (2015)
2. V. Soni, S. Mondal, B. Singh, *IJISET-International Journal of Innovative Science, Engineering & Technology* **1** (3), 418-423 (2014)
3. A. Abbas, K. Hamza, M. Aly, E. Al-Bakhali, Advances in Matterials Science and Engineering, http://dx.doi.org/10.1155/2016/6429160 (2016)
4. K. Dhandapani, P. Vasanthkumar, S. Nagarajan, Journal of Advanced Engineering Research **1** (2), 105-113 (2014)
5. E. Soury, E. Esfahany, A. Shahreza, Indian J Sci Res **1** (2), 54-63 (2014)
6. A. Bubakar, http://etheses.dur.ac.uk/5287/ (1996)
7. M. Madic, M. Radovanovic, P. Jankovic, D. Petkovic, S. Mladenovic, Journal of Production Engineering **19** (2), 1-6 (2016)
8. M. R. Radovanovic, M.M. Radovanovic, Tribological Journal Bultrib **6**, 85-90 (2016)
9. D. Petkovic, M. Radovanovic, Journal of Engineering Studies and Research **19** (1), 47-55 (2013)

MATEC Web of Conferences **137**, 03013 (2017) DOI: 10.1051/matecconf/201713703013
MTeM - AMaTUC 2017

Optimal production size with partially inspected batches

Arik Sadeh[1*] and *Cristina* Feniser[2]

[1]HIT, Holon Institute of Technology, Technology Management Faculty, 52 Golomb St. Holon, 58102, Israel

[2]Technical University of Cluj Napoca, Department of Management and Economic Engineering, Cluj Napoca, Romania

Abstract. While it is desirable to manufacture and ship perfect products, in many cases it is possible to ship products even though they do not stand up to all aspects of inspections. In this study a single machine is considered, that produces m products per day and there are n parameters that each product should pass. The number of passes (b) is a decision vector. A mathematical programming approach is used in order to model and optimize the number of products to be shipped under the constraints of failures. The results can be used for predetermined manufacturing, i.e. to set a series of tests less than m for each batch of production. In addition, aspects of calculation and computation are considered since the problem involves integer variables. Numerical examples and simulation runs of the suggested model are provided along with sensitivity runs.

1. Introduction

The practice of inspecting products before their shipment is a common procedure. While it is desirable to pass all inspections, there are many cases in which it is acceptable to ship products that may not pass all inspections. On the other hand, it could be of interest to avoid a certain number of inspections in order to ease operations and to reduce costs. The risk associated with this should then be taken into consideration.

A product or a batch of products either passes or fails a given inspection. There are cases in which not reaching full quality is applicable, yet a reasonable threshold for a maximum of failures is essential. One way to handle such a case is by using binary planning and zero-one programming.

This study intends to explore some aspects of partially inspected batches of products. It is of interest to provide a mechanism to help decision makers to handle the optimal set of rules. In the next section a literature review is provided, followed by a formal description of the problem. Then a numerical example is given along with an application section and a summary.

* Corresponding author: sadeh@hit.ac.il

1.1 Literature review

The problem of partially inspected batches can be treated as a binary optimization problem. Padberg [1] and Nemhauser and Wholsey [2] discussed thoroughly the usage and advantage of zero-one problems. Bilitzki and Sadeh [3] addressed several special cases using zero-one programming.

These types of problems are also addressed in the literature on treating missing data. Missing data values appear in surveys in many disciplines, such as natural sciences, engineering, social sciences and clinical research. The naïve approach is to discard a batch when there is a missing inspection, as in the case of missing values.

A counterpart to eliminating all observations with any missing value is to fill in the missing values with empirical indices such as the sample average, the sample median, etc. Raaijmakers [4] introduced the relative mean substitution as a procedure for replacing missing data based on sample means. Other categories of procedures are based on the correlations between variables with and variables without missing data, e.g. multiple regression imputation (Raymond [5]). Another example of the second type of procedure is given by Mason [6]. Three tests for examining and treating missing values as a result of attrition in clinical studies are described in that study.

Rubin [7] introduced the method of multiple imputation (MI) following the work by Dempster, Laird and Rubin [8], which provided the EM algorithm for efficient estimation when there are missing values in the data set. A comprehensive description of the MI method is provided in Schafer and Olson [9] along with practical tools to use that method.

This study intends to provide a useful method to deal with partially inspected batches of manufactured products. It takes into consideration the predetermined number of inspections needed for a complete inspection along with a predetermined number of inspections to be executed.

The suggested model is described first along with an illustration; then a few examples based on simulated data are used in order to show the usefulness of the mechanism. The paper concludes with a summary.

2. The model

2.1 General formulation

Consider a machine that produces n products per day and each product j (j=1,2…n) has to pass m inspections (i=1,2…,m).
Let's mark

$$A_{ij} = \begin{cases} 1, \text{if product j fails to pass inspection i} \\ 0, \text{otherwise} \end{cases} \quad (1)$$

That is, a failure to pass an inspection gives the value 1. Each column in matrix A is associated with inspection i, while each row in matrix A is associated with product j.

The decision is whether or not to consider manufactured product j as a suitable product for shipment. Let's define the decision variables x_j:

$$x_j = \begin{cases} 0, & \text{if product j to be discarded from the shipment set} \\ 1, & \text{otherwise} \end{cases} \quad (2)$$

For a given inspection i, d is the percent (proportion) of allowed failures. Consider the criterion for maximizing the number of manufactured products to be shipped:

$$\max \sum_{j=1}^{m} x_j$$
$$s.t. \quad \frac{\sum_{j=1}^{m} x_j A_{ij}}{\sum_{j=1}^{m} x_j} \le d_i \ , \ i = 1,2,,,n \quad (3)$$
$$x_j = \begin{cases} 0 \\ 1 \end{cases}$$

The constraint equations can be rewritten as

$$\sum_{j=1}^{m} x_j A_{ij} - d \sum_{j=1}^{m} x_j \le 0 \ , \ i = 1,2,,,n \quad (4)$$

This is an integer programming formulation with binary decision variables and the constraints mean that for each variable i, the ratio between the number of not shipped products and the total number shipped should be below a given threshold d_i. The value of d_i should be predetermined. When d_i is zero, only products that have fully passed inspection i can be considered for shipment. When d_i is 1, any product can be considered for shipment regardless of the result of inspection i.

When a portion of a product is applicable, e.g. j is a batch of many products, then the variables x_j can be bounded between zero and one and linear programming can be used.

Let's consider the parameter b_i as the number of allowed failures of inspection i, rather than the parameter d_i. Without loss of generality, the problem becomes:

$$\max \sum_{j=1}^{m} x_i$$
$$\text{s.t.} \quad \sum_{j=1}^{m} x_j A_{ij} \le b_i \ , \ i = 1,2,,,n \quad (5)$$
$$x_j = 0,1$$

2.2 Special common cases

Let's consider a case with p inspections where q inspections of p can be ignored (q<p). Now define matrix $B^{q,p}$ with p rows and $\binom{p}{p-q}$ columns. Each column in matrix $B^{p,q}$ has (p-q)

values of "1" and p values of "0". Note that each row of matrix $B^{p,q}$ has $\binom{p-1}{q-1}$ values of "1".

It would be easy to define index t (t=0,1,2...$\binom{p-1}{q-1}$ -1).

Now define the revised optimization problem to be:

$$\max \sum_{j=1}^{\binom{p}{q}} x_j$$
$$\text{s.t. } B^{pq}x^{pq} \leq t \tag{6}$$
$$x_j = 0, 1$$

where $\mathbf{x}^{p,q}$ is a vector of size $\binom{p-1}{p-q}$ and **t** is a vector of t and t can be any number between 0 and $\binom{p-1}{p-q}$.

When p is large, this is a big problem to be solved using binary programming. A way to ease the computation is by using linear relaxation, then using the solution to find the binary solution.

When the product $\left(t \cdot \frac{p}{q}\right)$ is an integer, the solution is not necessarily an integer, but the objective value is an integer. This could help to find the optimal solution of the binary problem. In practice, it is advised to choose parameters l, k and r such that this product is an integer.

These findings lead to the following suggested procedure.

Step 0: Define values for parameter p and parameter r (q<p).

Step 1: Compute the range values of t between 0 and $\binom{p-1}{p-q}$, choose t.

Step 2: Compute the term $\left(t \cdot \frac{p}{q}\right)$. If it is an integer, go to Step 3, if not, go to Step 1 and manipulate t.

Step 3: Build the matrix B^{pq}. It is a matrix of zeros and ones, with k rows and $\binom{p}{q}$ columns. Note that in each column there are q ones, and in each row of B^{pq} there are $\binom{p-1}{q-r}$ones.

Step 4: Solve the linear relaxation of Equation (6) and keep the objective value.

Step 5: Add a constraint that the objective value is bounded by the objective value of Step 4. Solve Equation (6) with this additional constraint.

The procedures described above are illustrated using numerical examples in the next section.

3. Numerical example

Consider a production line with p=9 inspections. If a batch passes any five of them, it is defined as a good batch. That is, q=5. The parameter t can be between 0 and 70, but for t=0,5,10,15…,70, the linear relaxation of Equation (6) will be an integer. For this illustration, the value t=20 is chosen. The term $\left(t \cdot \frac{p}{q}\right)$ is 36. Let's build the matrix $B^{9,5}$ and optimize Equation (6) without linear relaxation. The solution is a set of 36 variables (36 columns of the matrix $B^{9,5}$). Each column has 5 ones, and each row has 20 ones.

This gives the five inspections that each batch should undergo.

4. Conclusions

A mathematical programming is used to develop a model to choose which batches to consider after inspections. The model helps to design the set of inspections needed per batch. Several convenient working rules are determined in the study that can help to ease the computation in the case of a large number of batches and inspections.

The model is very useful for cases with a limited number of batches and many inspections. The contribution of the suggested model is when there are k inspections, and it is for the decision maker to decide both how many inspections to make and how many batches to inspect.

References

1. M. W. Padberg, Mathematical Programming **6**, 180 (1974)
2. G. L. Nemhauser, L.A. Wolsey, *Integer and Combinatorial Optimization* (John Wiley and Sons, NY, 1988)
3. A. Bilitzky, A. Sadeh, Journal of Combinatorial Optimization **10**, 227 (2005)
4. Q. A. Raaijmakers, Educational and Psychological Measurement **59**, 725 (1999)
5. M. R. Raymond, Evaluation and The Health Professions, **9**, 395 (1986)
6. J. M. Mason, Measurement and Evaluation in Counseling and Development **32**, 111 (1999)
7. D. B. Rubin, *Multiple imputation for nonresponse in survey* (J. Wiley & Sons, New York, 1987)
8. A. P. Dempster, N. M. Laird, D. B. Rubin, Journal of the Royal Statistical Society, Series B (Methodological) **39**, 1 (1977)
9. J. L. Schafer, K. M. Olsen, Multivariate Behavioral Research **33**(4), 545 (1998)

MATEC Web of Conferences **137**, 03014 (2017) DOI: 10.1051/matecconf/201713703014
MTeM - AMaTUC 2017

Using the factorial experiment method to analyze the corrosion protection process

Mihail Aurel Țîțu [1,*], *Constantin* Oprean[2], *Alina Bianca* Pop[3] *and Ștefan* Țîțu[4]

[1]Lucian Blaga University of Sibiu, Victoriei Street no. 10, code 550024, Sibiu, România
[2]Lucian Blaga University of Sibiu, Victoriei Street no. 10, code 550024, Sibiu, România
[3]SC TechnoCAD SA, Vasile Alecsandri no. 72, code 430351, Baia Mare, România
[4]Iuliu Hațieganu University of Medicine and Pharmacy, Victor Babeș Street no. 8, code 400012 Cluj-Napoca, România

Abstract. The organization functions are: research-development, production, commercial, financial-accounting, personnel and quality. In this paper the factorial experimental method will be applied, which is currently one of the most widespread methods used in the research-development departments of the organizations, due to its advantages and efficiency. The experiment was carried out at SC Coifer Impex SRL-Mirsa's metal structures factory. In this paper it is presented the factors modelling that exerts their influence on two objectives functions: the ensuring the nominal thickness of the rough-cast film and the consumption limiting. For data processing the STATISTICA 7 software was used which provides an accurately and effectively way to determine the influence degree of each variable on the settled objectives.

1 Introduction

The organization functions are: research - development, production, and commercial, financial - accounting, personnel and quality [1]. Currently, due to their advantages and their efficiency, the factorial experiment method is one of the most widespread methods used in the research and development departments of the organization [2]. The main advantage offered by the factorial experiments is that they are designed and developed according to an optimal strategy [3]. Other important advantages are:

• The use of the factorial experiment method allows to estimate the effect caused by an influence factor on several levels of the other factors, generating conclusions as variables for the entire range of experimental conditions;

• It is more efficient than the classic experiment, and its efficiency increases with the number of variables (influence factors);

• It is absolutely necessary to use this method when the interactions between the influence factors are present, to avoid the erroneous conclusions [4].

The input data to be processed are noted as x1, x2, ..., xk and they are called - influence factors (independent variables), and the outputs are noted as y1, y2, ..., ym and represent

* Corresponding author: mihail.titu@ulbsibiu.ro

the objective functions (responses) [5]. The factorial experiments are very efficient where the two or more factors influence exist on one or more objective functions [6]. The motivations of the factorial experiment modelling, are: highlighting the influence factors action on the researched system; verifying the hypotheses about internal system interactions; the researched system studying to obtain newer and more complete data; the system optimizing against different criteria [7].

2 Research method

The experiment was carried out on SC Coifer Impex SRL – Mirsa, a metal structures factory, and it presents the influence factors modeling of two main objectives:

- Ensuring the nominal thickness of the rough-cast film;
- The consumption limiting.

Ensuring the efficient and durable protection of the metal structures executed in the Coifer Impex factory, is accomplished by observing the operations that compose the technological process of anti-corrosive protection.

The technological process consists of:

- Verifying and preparing the surfaces;
- A good quality equipment and materials use;
- The provided times Keeping in the technical data sheets;
- Depositing / drying layers, re-coating intervals;
- The performing control of this phase.

The product on which the experiment was carried out is a two-component anti-corrosion rough-cast based on epoxy resin and EPOMID G3200S-FT zinc phosphate. The nominal film thickness is 50 [μm] and falls into a theoretical consumption of 160 [g/m^2] (no loss).

According to the technical product file of this rough-cast type, in order to obtain the optimal results, the following parameters must be applied:

- Spray application with AIRLESS method;
- Painting pressure: 140 - 160 [atm];
- Dilution: 3 - 5 [%];
- Nozzle diameter: 0.28 - 0.53 [mm];
- Spray distance: 250-800 [mm];
- Roughness: 50-75 [μm].

After the corrosion protection process measurements performing, the following input data were obtained for modeling using the factorial experiment method.

The input data were obtained by results measuring using appropriate devices for each type of the variable. The painting pressure [atm] is measured by means of a pressure-indicator regulator mounted on the used painting equipment. The dilution [%] represents the proportion of the added solvent to the amount of rough-cast. The nozzle diameter [mm] represents the hole through the rough-cast is pulverized, with a standard size, guaranteed by the manufacturer. The surface roughness [μm] is measured using a Rugosimeter. The dry film thickness [μm] is measured with Elcometer 456.

The consumption [g/m^2] represents the amount of rough-cast used to cover one square meter of rough-cast and it is obtained by dividing the amount of rough-cast used to paint the piece surface. The input data was processed using the statistical software STATISTICA 7, which enables the 2D and 3D graphics realization with a high accuracy and quality.

Table 1. Input data.

Piece no.	Paint pressure [atm]	Dilution [%]	Nozzle diameter [mm]	Spray distance [mm]	Roughness [µm]	Dry film thickness [µm]	Consumption [g/m²]
	X1	**X2**	**X3**	**X4**	**X5**	**Y1**	**Y2**
1	140	3	0.33	390	65	52.7	169.3
2	175	1	0.53	500	50	73	185.6
3	152	3.5	0.38	480	65	62.3	190
4	135	5	0.38	620	67	46.3	179.11
5	140	5	0.28	400	65	82.3	201.5
6	170	2	0.45	320	80	85.7	210
7	170	5	0.38	690	73	62	177
8	135	5	0.28	450	65	72	180
9	110	5	0.28	450	80	60.5	210
10	180	2	0.53	600	60	59.3	185
11	185	4	0.38	450	70	65.1	169.5
12	130	5	0.28	500	80	55	192
13	150	4.5	0.33	400	55	70.3	180
14	150	3	0.33	400	65	63.5	195
15	160	4.5	0.38	390	65	72.2	209
16	155	6	0.28	620	60	75.3	219.5
17	155	5.5	0.28	590	65	70.4	213
18	160	3	0.53	450	62	73.7	220.5
19	110	5	0.28	450	50	62.5	184
20	180	5	0.45	600	75	58.3	201
21	180	3	0.53	450	75	66	185.8
22	150	3	0.38	410	65	68.4	191.2
23	160	4.5	0.33	410	55	60.5	188
24	160	2	0.53	390	65	70.7	224
25	185	4	0.38	450	60	62.3	169.5
26	178	1	0.45	420	80	68.5	192
27	185	4	0.38	380	75	78.5	222.3
28	155	4.5	0.28	585	63	82.3	205.5
29	155	2	0.53	400	80	59.3	196.6
30	160	5.5	0.28	610	62	80.5	218
31	130	5	0.28	400	60	58.7	169
32	180	3	0.53	600	65	59	170

3 Results and discussions

After the data processing using STATISTICA 7, the most relevant aspects were graphically presented. First, based on histograms, the experimental data distribution was analysed. So about the variable X1, it was found that the pressure used to apply the rough-cast gets the ranges between 100 [atm] and 190 [atm] - Figure 1. The variable X2 indicates the applied dilution ratio, which was between 0.5 and 6 [%], the values are outside the tolerance field - Figure 2. The histogram of the X3 influence factor indicates that nozzles with diameters between 0.28 and 0.53 [mm] were used, according to the prescription of the used product - Figure 3. The histogram of the X4 influence factor indicates that the distance by the technical prescription of the rough-cast has been respected, i.e. the measured distances ranged from 250-700 [mm] - Figure 4. In the X5 variable situation the histogram indicates that the measured roughness ranges are between 40 and 80 [μm], it results that are values outside the tolerance range, but most of them are in tolerance - Figure 5. Also, from Y1 histogram we can see that we have values outside the tolerance field, which indicates that we have too thin coat parts that still require a layer coating and also to thick coat parts, parts which lead to a waste of material and energy - Figure 6. The Y2 histogram allows us to notice that the all the pieces painting and the prescribed rough-cast consumption was exceeded and most of the pieces are close to prescribed consumption, i.e. 165 [g/m^2] - Figure 7.

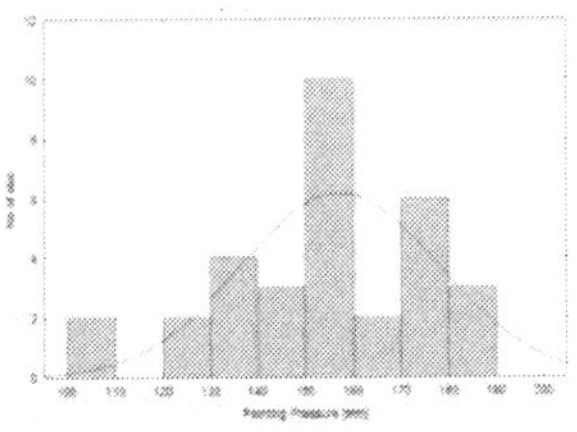

Fig. 1. Painting pressure (X1) histogram.

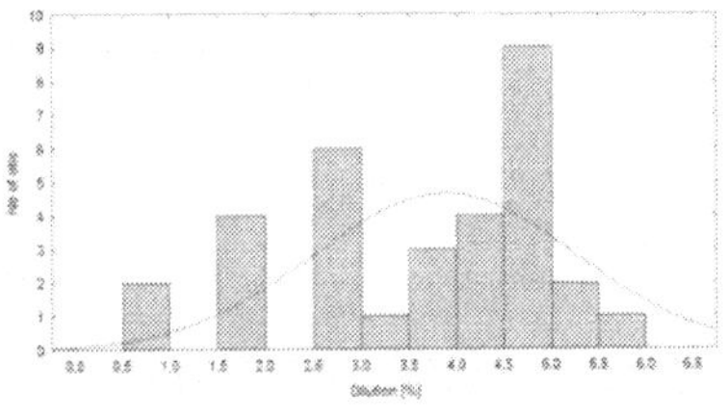

Fig. 2. Dilution (X2) histogram.

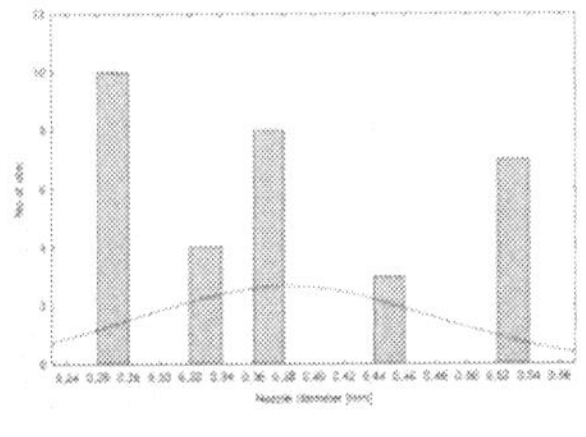

Fig. 3. Nozzle diameter (X3) histogram.

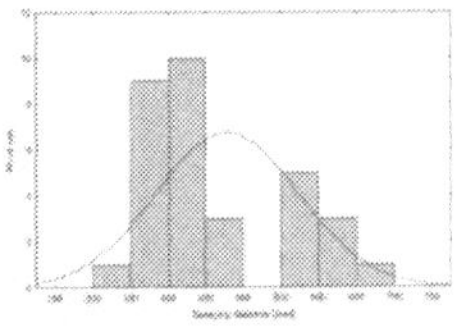

Fig. 4. Spraying distance (X4) histogram.

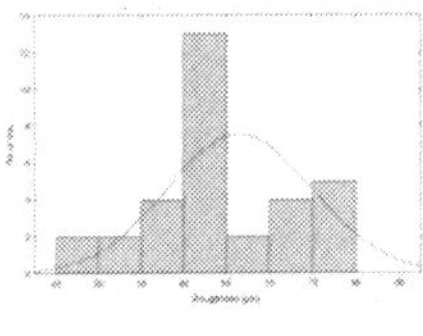

Fig. 5. Roughness (X5) histogram.

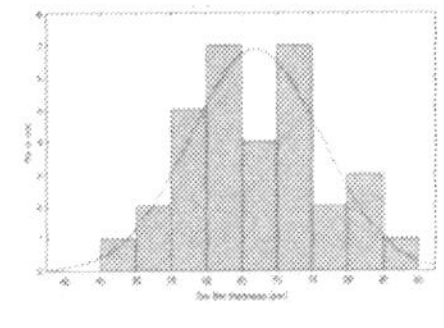

Fig. 6. Dry film thickness (YI) histogram.

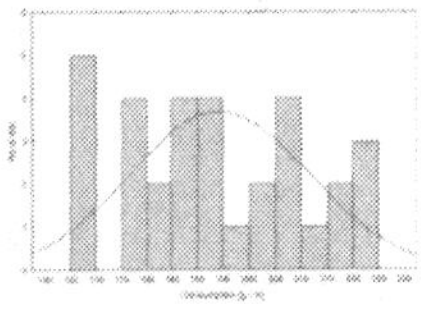

Fig. 7. Consumption (Y2) histogram.

From the presented histograms we can conclude that some values of the measured variables are outside of the tolerance field, so the objective functions are negatively influenced by the studied factors and the researched process results that it is not reliable.

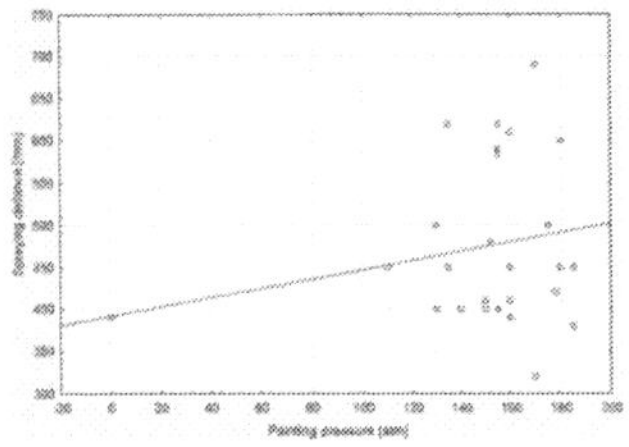

Fig. 8. Dispersion diagram between painting pressure and spraying distance correlation.

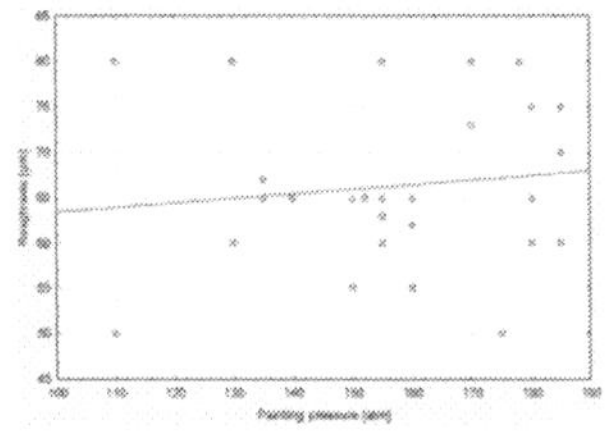

Fig. 9. Dispersion diagram between painting pressure and surface roughness correlation.

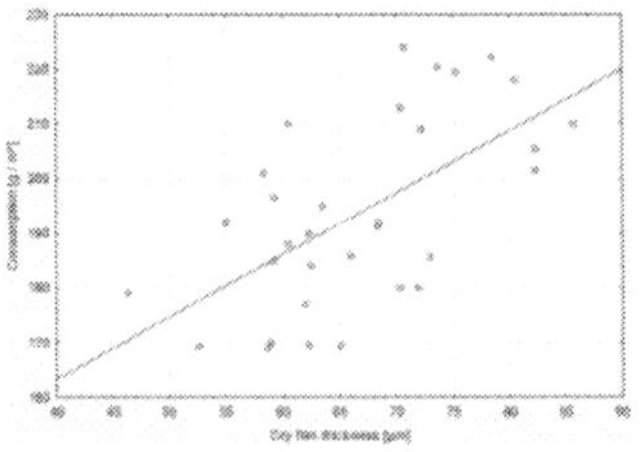

Fig. 10. Dispersion diagram between dry film thickness and rough-cast consumption correlation.

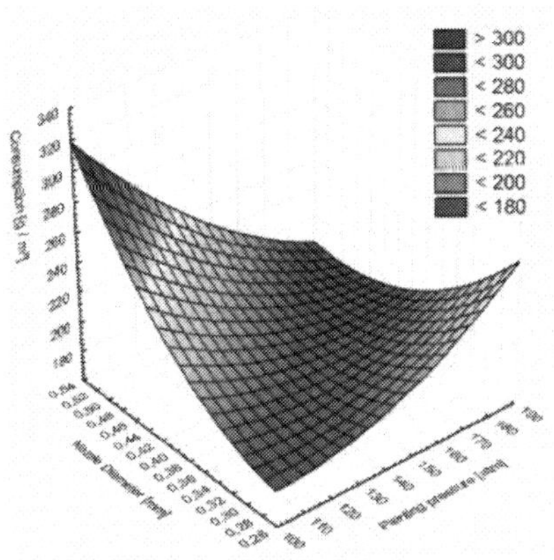

Fig. 11. The consumption 3D variation under the nozzle diameter and painting pressure influence.

Next, the data dispersion was also studied. In this situation it was found that the closest correlations are between painting pressure and spraying distance (Figure 8), between painting pressure and surface roughness (Figure 9) and between dry film thickness and rough-cast consumption (Figure 10).

The other correlations between: nozzle diameter and spray distance, between nozzle diameter and surface roughness, between nozzle diameter and rough-cast film thickness, between nozzle diameter and rough-cast consumption, between surface roughness and layer thickness and between surface roughness and consumption, have a very large dispersion. The Y1 "Layer Thickness" objective function is positively influenced by nozzle diameter increasing, by the spraying distance and the dilution reduction.

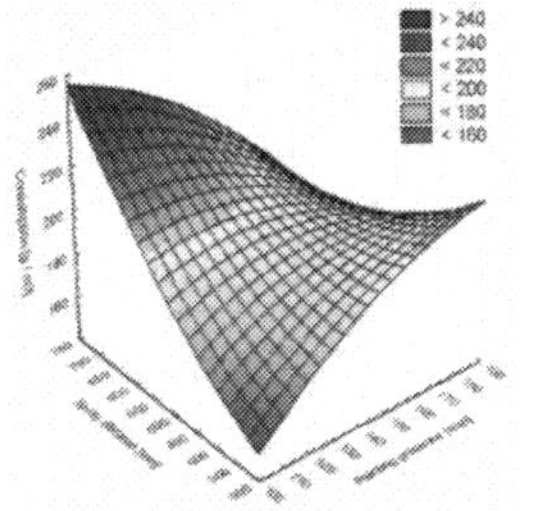

Fig. 12. The consumption 3D variation under the spray distance and painting pressure influence.

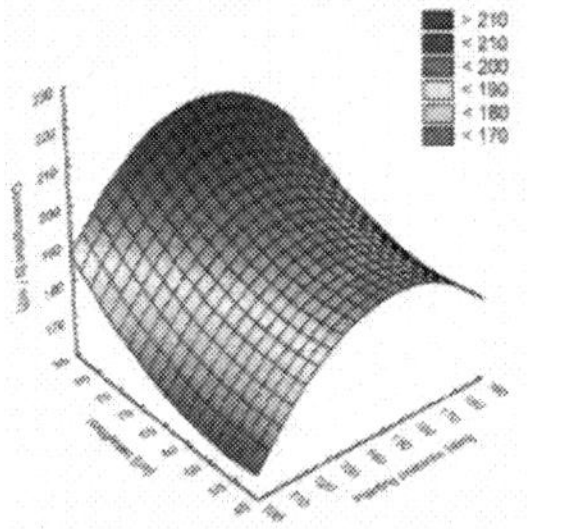

Fig. 13. The consumption 3D variation under the roughness and painting pressure influence.

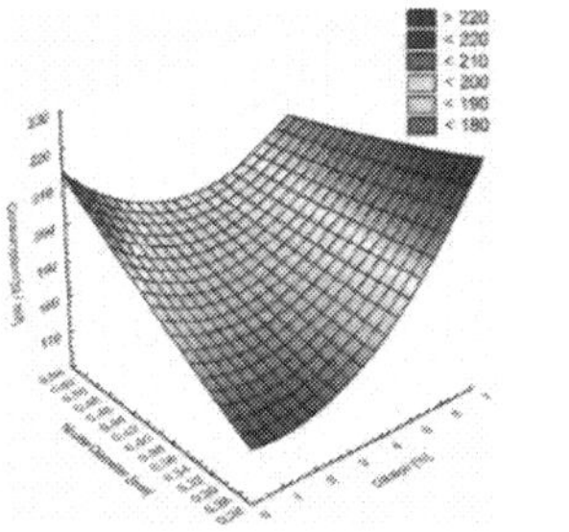

Fig. 14. The consumption 3D variation under the nozzle diameter and dilution influence.

 DOI: 10.1051/matecconf/201713703014

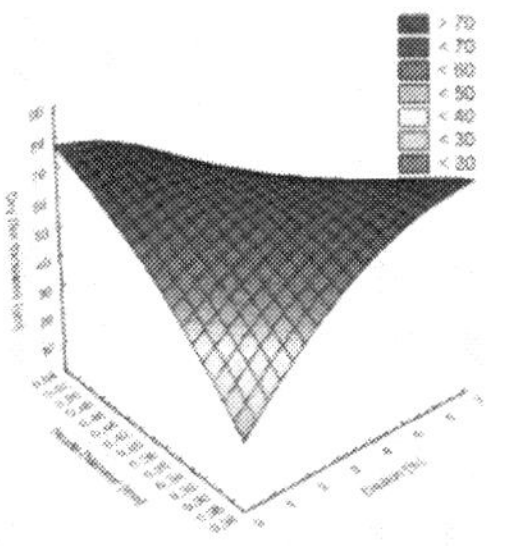

Fig. 15. The dry film thickness 3D variation under the nozzle diameter and dilution influence.

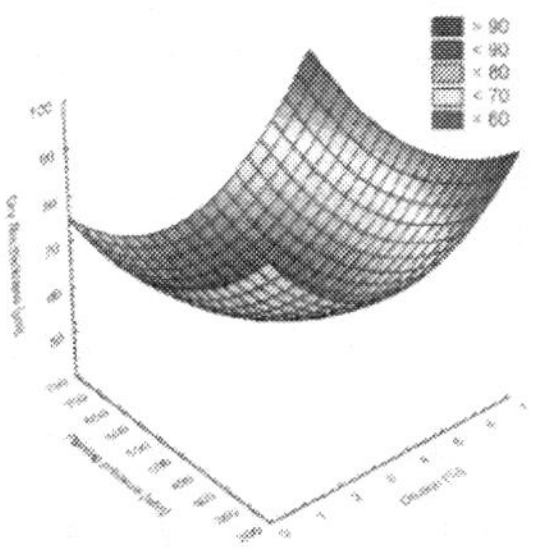

Fig. 16. The dry film thickness 3D variation under the painting pressure and dilution influence.

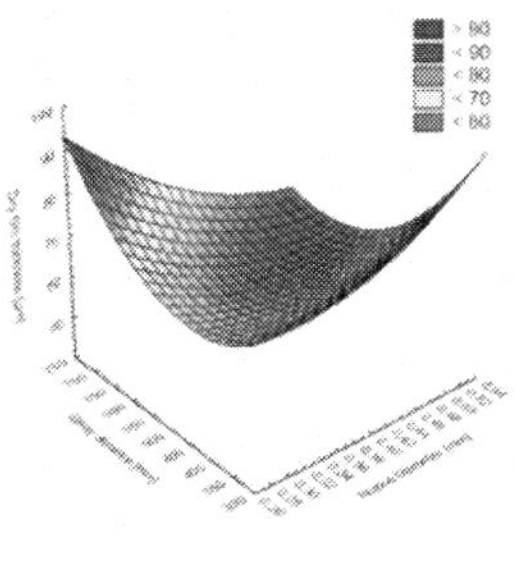

Fig. 17. The dry film thickness 3D variation under the spray distance and nozzle diameter influence.

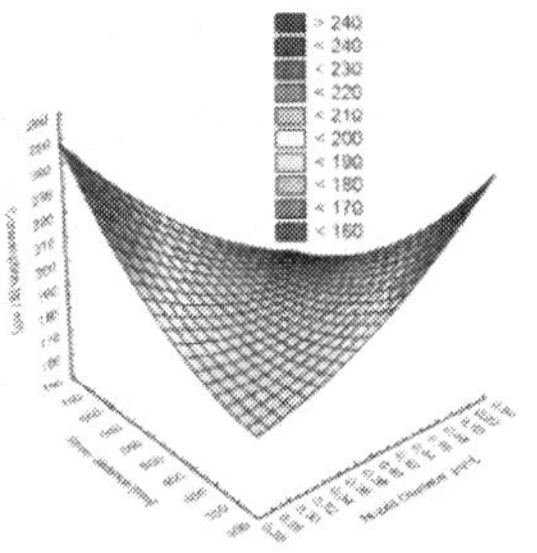

Fig. 18. The consumption 3D variation under the spray distance and nozzle diameter influence.

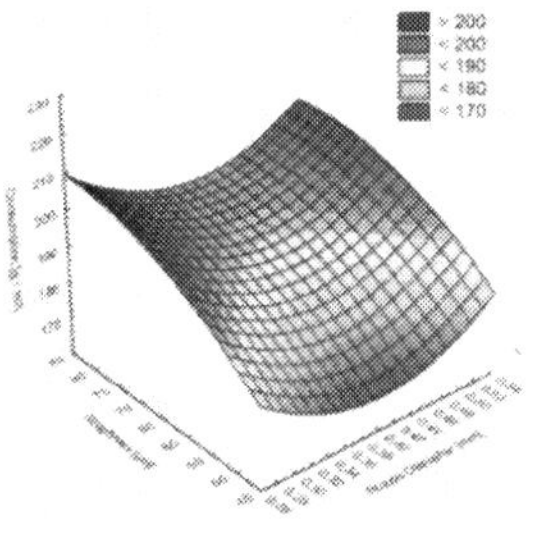

Fig. 19. The consumption 3D variation under the roughness and nozzle diameter influence.

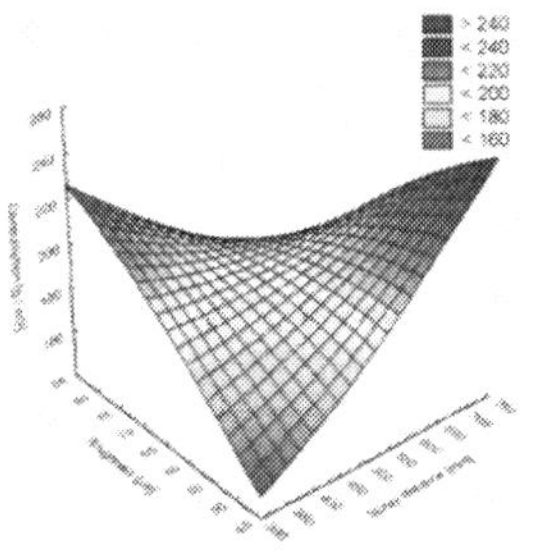

Fig. 20. The consumption 3D variation under the roughness and spray distance influence.

The Y2 objective function represented by rough-cast consumption is negatively influenced by spray distance, painting pressure, nozzle diameter and surface roughness increasing (Figure 11-20).

These observations were made after the three-dimensional (spatial) analysis made on the objective functions in order to track their variation under the influence of controllable factors.

4 Conclusions

The used factorial experiment is currently one of the most widespread methods to investigate the purposed research objective, in the form of inputs and outputs.

In this paper some obtained experimentally data (input data) were modelled by the factorial experiment method, and their processing was made using specialized software of statistical data processing, STATISTICA 7.

The experimental researches were aimed to define the most advantageous ways to obtain the protected surface roughness against the corrosive factors by applying the anticorrosive rough-cast by spraying and framing the estimated rough-cast consumption.

The input data are represented by the measurements made on the corrosion protection process of the company where the experiment was conducted and after which we could observe the influence types (negative or positive), as well as the influence degree of certain factors on the objective functions.

The data processing graphs mated with specialized software allowed to obtain essential information that can be used to optimize the investigated process. It also resulted that the STATISTICA 7 software used, provides a modality that accurately and operable determines the influence degree of each variable on the final values (objectives).

It also emerged that there are possibilities to improve the investigated process by variables controlling, by preparing an optimization plan of the anti-corrosive protection process. The process must be optimized in such a way that it becomes a focused and precise process, in other words - to be a reliable process. In the present paper, it was opportune to use the factorial experiment method of the investigated anticorrosive protective process due to the possibility of the process optimization.

References

1. A.M. Ţîţu, C. Oprean, *Managementul calitatii in economia si organizatia bazate pe cunostinte*, Editura Agir, Bucuresti, (2008)
2. D Montgomery, *Design and Analysis of Experiments*, Eighth Edition ed. Hoboken: John Wiley & Sons, Inc. (2013)
3. A.M. Ţîţu, C. Oprean, *Cercetarea experimentală şi prelucrarea datelor*, Partea I, Editura Universităţii Lucian Blaga, Sibiu (2006)
4. A.M. Ţîţu, C. Oprean, *Cercetarea experimentală şi prelucrarea datelor*, Partea a II-a, Editura Universităţii Lucian Blaga, Sibiu (2007)
5. A.M. Ţîţu, C. Oprean, I. Tomuţă, *Cercetarea experimentală şi prelucrarea datelor, Studii de caz*, Editura Universităţii Lucian Blaga, Sibiu (2007)
6. T. P. Ryan, J. P. Morgan, Journal of Statistical Theory and Practice **1** (3-4) (2007)
7. M. J. Anderson, P. J. Whitcomb, Kirk-Othmer Encyclopedia of Chemical Technology, DOI: 10.1002/0471238961.0405190908010814.a01.pub3 (2010)

The Emotiv EPOC interface paradigm in Human-Computer Interaction

Dorina Ancău[1], *Nicolae-Marius* Roman[1], and *Mircea* Ancău [2,*]

[1]Faculty of Electrical Engineering, Technical University of Cluj-Napoca, Romania
[2]Faculty of Machine Building, Technical University of Cluj-Napoca, Romania

Abstract. Numerous studies have suggested the use of decoded error potentials in the brain to improve human-computer communication. Together with state-of-the-art scientific equipment, experiments have also tested instruments with more limited performance for the time being, such as Emotiv EPOC. This study presents a review of these trials and a summary of the results obtained. However, the level of these results indicates a promising prospect for using this headset as a human-computer interface for error decoding.

1 The Emotiv EPOC interface

Emotiv EPOC is a new portable, multichannel Brain-Computer-Interface (BCI), with high resolution, intended for research practice in assistive technology. Somewhat, the evolution of this device capable of deciphering brain processes that can handle virtual or real objects, resembles that of the Wii Balance Board platform originally used as a controller for Nintendo video games and then as a medical tool to rehabilitate people with balance problems.

Although it has been designed for computer games, today the EPOC interface suggests numerous experiments on commanding an external device with a thoughtful intention.

Designed as a piece of equipment combining two parts, one hardware and one software, the EPOC interface is essentially capable of decoding the bioelectric brain activity, analyzing it using a specialized algorithm, and then transmitting a command to a computer.

The hard drive requires a headset (Figure 1) with 14 sensors capable of detecting and recording electrical potentials from scalp level in real time in the form of an electroencephalogram. The software

Fig.1. Emotiv EPOC.

* Corresponding author: mircea.ancau@tcm.utcluj.ro

attached to the device provides, some flexibility in, viewing the recorded data stream in EEGLAB format, monitoring the contact area quality, automatic or manual scaling of the desired channel, inserting timed markers that can be stored in the data stream, and analyzing of specific signals (FFT).

There are currently 76 companies building equipment and programs for BCI interfaces. Certainly, recent attempts to commercialize such viable devices (Intendix - Austria, 2012; Emotiv - USA, 2009; Cyberlink Brainfingers - USA; 1995; BioSemi "Pin-type"- Netherland, 1998) only support investment in the future development efforts of such systems.

The impact of this interface may be, primarily, clinical. Its effects would manifest itself directly on people with neuromuscular disabilities or neurodegenerative diseases, for which there is no other treatment. For these patients, it represents, in this way, an alternative to natural communication or by neuromuscular pathways, via an artificial bypass.

2 Emotiv epoc headset in human-computer interaction

The past few years have marked remarkable performance in the development of computer science applications in medicine. Among them, the new BCI interface for direct communication between the human brain and an external device has inspired numerous research into signal processing and computer learning by translating thought processes.

Experiments realized by Lievesley and others [1] in 2011, demonstrate that Emotiv EPOC can be used as a communication interface with a computer that, in turn, could control a neuroprotective device. There were, in fact, 2 projects that started from the following assumptions:

1. The EPOC interface transfer rate would be superior to that of a system of two head-operated switches attached to the headrest of an wheelchair which could be operated by left-right movements of the head (thus simulating the way of communication of a person with disabilities);
2. A virtual object could be moved on the computer screen using neurophysiological signals reflecting cognitive state.

The first hypothesis was tested by detecting facial expressions from the perception of specific tasks simulated on a computer. When using head-operated switches, the success rate was 100% while in case of the interface 73.3%. The same ratio was also obtained for the reaction time, 0.74 s versus 1.03 s with Emotive EPOC while a significantly higher one was displayed, in favour of the switches, in the deduction of the transfer rate of 1.38 b/s compared to 0.57b/s (0.44 - 2.66 b/s being the limits of the values obtained in other studies so far). Although the differences were not significant in all situations, the findings contradicted the initial hypothesis (provided the subjects were able to move his neck).

Not the same thing resulted from the run of the second experiment. Here, the participants used their thoughts to control, the image of a cube on a display, which could be moved "away" when the subject thought "push" or would remain in a static position during "neutral" EEG activity.

The success rate was about 79% in the absence of a threshold time and much higher, about 96%, in its presence. The data transfer was performed at a minimum rate of 0.65 mb/s and a maximum of 65 mb/s (the transfer rates identified in previous research averaged 60 mb/s for devices employing cortical potentials and 10 mb/s achieved by those applying sensorimotor rhythms).

Evaluating these results, it can be said that despite the lower transfer rates compared to other instruments for scientific use that measure EEG or cortical potentials (but which are, on the other hand, more impractical and also more expensive), and in the absence of a more accessible alternative, the Emotive interface EPOC shows a real potential for assistive technology.

The EPOC system has also been successfully used to measure the response of cortical response to auditory stimulation (evoked potential, P_3 component) [2, 3] and visual stimulation [3, 4]. The results indicated a high accuracy for ERP (event-related potential or evoked potential) peaks, ie. P_1, N_1, P_2, N_2 and P_3, but not for the MMN component (mismatch negativity).

In 2013, Badcock and others [5] adapted the Emotiv EPOC interface to the recording of evoked potentials by auditory stimuli (peaks P_1, N_1, P_2, N_2 and P_3) in adults and later, in 2014, in children (differences between these groups being given by the different levels of cortical and cognitive development). They have tested, in this way, the likeness of the waveforms of a professional system with those of an Emotive EPOC interface and have demonstrated through their results their similarity in morphology, amplitude and latency.

Taken together, the results of these studies suggest the validity and reliability of the EPOC interface to measure evoked potentials and EEG activity. It also opens new research paths, because it can, in principle, be used anywhere (for example, schools, homes, shopping centres, hospitals). The fast configuration procedure, portability, simple use and, last but not least, the price, gives the interface the status of a promising instrument for measuring electrical signals generated in the brain.

3 The evoked potentials

Environmental stimuli trigger distinct neuronal responses (evoked potentials or ERPs) in the human brain that can be measured, for example, by an electroencephalogram (EEG). An external action can be a stimulus for the subject who perceives it, triggering the triggering of evoked potentials. If, in a particular way, the action should, from the perspective of the subject, no longer continue or return to an earlier stage, the potential evoked takes the form of an error-related potential or ErrP.

The error potential (ErrP) is basically a particular case of evoked potential. The perception of an error can be recorded using the encephalogram as a distinct neural signal [6]. It appears in the brain automatically, involuntarily, without previous training of the subject. Although physiologically, the response mechanism of the brain to the perception of an error is complex and involves more brain areas, it has basically been demonstrated that the response to an error can be detected [7] in the anterior medial frontal cortex (pMFC).

Research of recent years highlights the fact that these error signals can be effectively used in human-computer communication and BCI applications [8, 9, 10]. Such a signal can also be used to reduce the possibility of returning an error through the system's ability to self-correct "by learning from its own fault" [11, 12].

Eric W. Sellers and others [13] present, in 2010, an interface that allows people with disabilities, including also serious cases to move a cursor on a display using the signals extracted from the sensory motor cortex.

In 2014, Chavarriaga and others [14] use a BCI interface that measures the response of the brain to a stimulus produced by the recognition of an error, to correct an action. The idea is taken over in 2015 by Ehrlich and Cheng [15] which validate an objective method of correcting an error by detecting signals resulting from the perception of the inaccuracy / correctness of a bivalent task performed by a robot. Their result confirms previous reports and demonstrates that robot action can be detected and modelled from EEG signals with an accuracy of approximately 70%. Frank and others, Hajcak and others [16, 17], in 2005, Taylor et al [18], in 2007, and then Chavarriaga and others [14], in 2014, describe the shape of these signals and their latency from the onset of the stimulus, and demonstrate that their amplitude depends on the following factors: the subject's grasp on the relevance of the task, the degree of involvement of the subject in the task (the amplitude of the signal is lower if the subject passively follows an external device than if it controlled this device) as well as the frequency of stimuli.

Ferrez and Millán [9], in 2008, Chavarriaga and Millán [19], in 2010, confirm that the ErrP signals are stable over time, that their shape is somewhat maintained for repeated measurements after an indefinite period of time - one year, for example - and then their triggering does not depend on the type of task but rather on the perception of incorrectness.

The form of error signals is also influenced by pathological and subject-specific conditions [20]. They reappear, demonstrated in another study, in the same form upon repetition of stimuli [21] regardless of the nature of the tasks [22].

An evoked potential triggered by a particular stimulus takes the form of the signal in Figure 2. It is characterized by a positive peak at about 100 ms from the onset of the stimulus, followed by a negative deviation at 260 ms (250 ms).

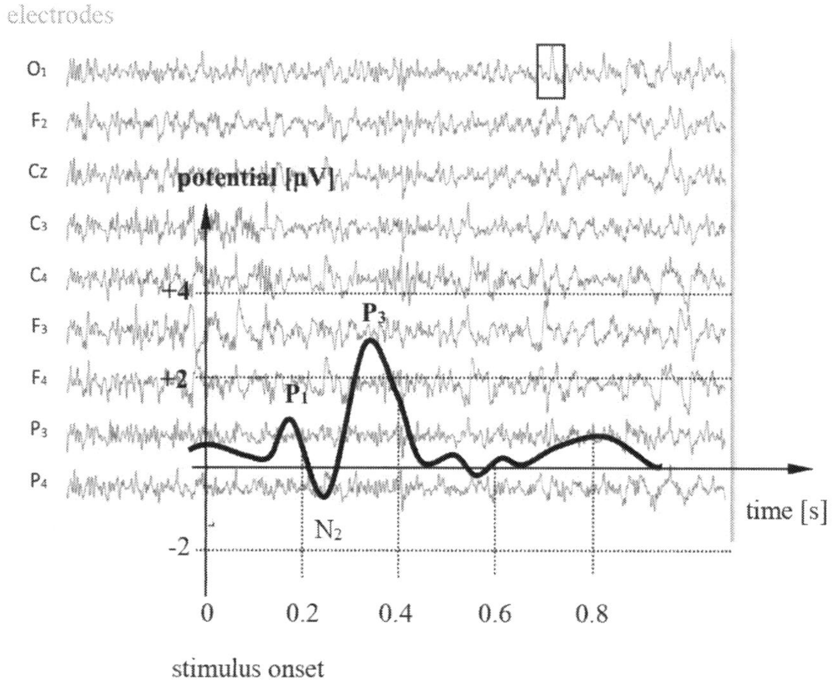

Fig. 2. Event-related potential (electrode O_1).

A second, more pronounced positive peak occurs at around 330 ms after the stimulus [19]. Although they come from the same sources as an EEG, the evoked potentials are much smaller than these (up to about 5 µV, while EEG waves can reach 50 µV). Therefore, they cannot be routinely distinguished over the background of an EEG record with the naked eye, being covered by surrounding, much larger signals. To emphasize the former, it is necessary to sum up, filter and mediate EEG signals so that the regularity of the signal background is distinguished.

ErrP is recognized in a record by negative signal deflection (ERN) immediately after triggering the stimulus in the alpha frequency in the frontal-centre area (Figure 3). The ERN component, consistently associated with conflict cases (such as incongruent stimuli), is followed by a positive deviation that reaches a maximum value at about 400 ms, called error positivity or P_e, and attributed by specialists to error awareness processing .

 DOI: 10.1051/matecconf/201713704001

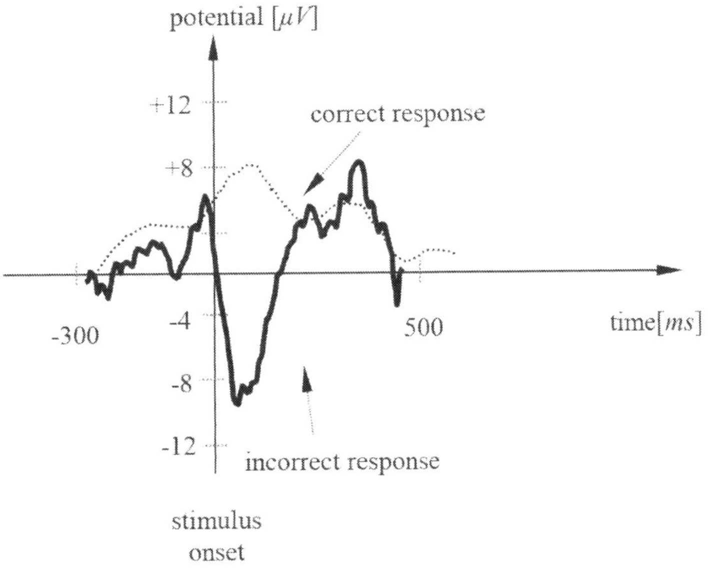

Fig. 3. The shape of the signal after the correct and incorrect response (electrode FC_z).

ErrP integration into human-computer interaction systems involves two key factors: detecting the error in single-trial basis and using this information to modify the online process. But practical applications of neurological measurements, pursuing the initial purpose of these researches, should take into account the performance of the system and its applicability to people with disabilities or neurodegenerative diseases.

4 Conclusions

The potential of the EPOC Emotive set for the acquisition and analysis of EEG signals has already been confirmed by numerous studies.

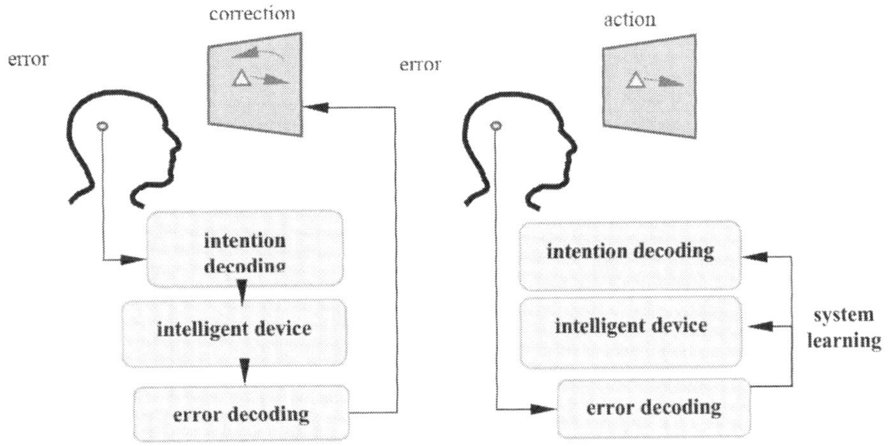

Fig. 4. Use of error-related potentials to improve BCI performance. Left: erroneous actions is used to trigger a corrective action; Right: error is used to update the BCI classifier and reinforce the learning.

While exhibiting lower transfer rates and accuracy, the results were still encouraging, particularly in view of the versatility and practicability of the device.

The work revised in this paper supports the feasibility of decoding potential errors using the Emotive EPOC interface and their use to improve the performance of human-computer communication systems. There is solid evidence that ErrPs can be successfully detected on a single-trial basis and that they can be effectively used in both HCI and BCI applications. The possibility of detecting potential errors can be used to deduce the optimal behaviour of performing bivalent tasks that will then allow the output data to be adapted to a controllable human-computer environment (Figure 4).

The use of BCI opens up many perspectives in a vast interdisciplinary field that brings together neurology, electrical engineering, prosthetics, psychology and computer science. Among these, the latest applications integrate, first of all, health and education.

References

1. R. Lievesley, M. Wozencroft, D. David Ewins, Journal of assistive technologies **2**, 67-82 (2011)
2. S. Debener, F. Minow, R. Emkes, K. Gandras, M. de Vos, Psychophysiology **49**, 1617-1621 (2012)
3. M. De Vos, M. Kroesen, R. Emkes, S. Debener, J. of Neu. Eng **11** (2014)
4. M. Duvinage, T. Castermans, M. Petieau, T. Hoellinger, G. Cheron, T. Dutoit, **12**, 56 (2013)
5. N.A. Badcock, K.A. Preece, B. de Wit, K. Glenn, N. Fieder, J. Thie, G. McArthur, *Validation of the Emotiv EPOC EEG system for research quality auditory event-related potentials in children*, (2015)
6. M. Falkenstein, J. Hoormann, S. Christ, J. Hohnsbein, Bio Psy **51**, 87–107 (2000)
7. M. Ullsperger, C. Danielmeier, G Jocham, Phy rev **94**, 35–79 (2014)
8. G. Schalk, J. Wolpaw, D.J. McFarland, G. Pfurtscheller, Clin Neurophy **111**, 2138–2144 (2010)
9. P.W. Ferrez, J.d.R. Millán, IEEE Trans. on Bio Eng **55**, 923–929 (2008)
10. B. Dal Seno, M. Matteucci, L. Mainardi, *On line detection of P300 and error potentials in a BCI speller*, Computational Intelligence and Neuroscience (2010)
11. X. Artusi, I.K. Niazi, M.F. Lucas, D. Farina, The IEEE J. on Em. and Sel. Topics in Circ and Sys **1**, 480–488 (2011)
12. A. Llera, M. van Gerven, V. Gómez, O. Jensen, H. Kappen, Neural Networks **24**, 1120–1127 (2011)
13. E.W. Sellers, D.J. McFarland, T.M. Vaughan, J.R. Wolpaw, *The Frontiers Collection*, Springer-Verlag Berlin, 97 - 111 (2010)
14. R. Chavarriaga, A. Sobolewski, J. del R. Millán, Frontiers in Neuroscience **8**, 208 (2014)
15. S. Ehrlich, G. Cheng, *A neuro-based method for detecting context-dependent erroneous robot action*, (2015)
16. M.J. Frank, B.S. Woroch, T. Curran, Neuron **47**, 495–501 (2005)
17. G. Hajcak, J.S. Moser, N. Yeung, R.F. Simons, **42**, 151–160 (2005)
18. S.F. Taylor, E.R. Stern, W.J. Gehring, Neuroscientist **13**, 160–172 (2007)
19. R. Chavarriaga, J.d.R. Millán, IEEE Trans on N Sys and Reh Eng **18**, 381–388 (2010)
20. D.M. Olvet, G. Hajcak, Clin l Psy Review **28**, 1343–1354 (2008)
21. D.M. Olvet, G. Hajcak, B Res **1284**, 89–99 (2009)
22. A. Riesel, A. Weinberg, T. Endrass, A. Meyer, G. Hajcak, Bio Psy **93**, 377–385 (2013)

Inference of power loss in spherical joints of the 6R̲SS parallel mechanism

Lucian Milica[1,*] and *Gabriel* Andrei[1]
[1]"Dunarea de Jos" University, Department Mechanical Engineering, 800201 Galati, Romania

Abstract. This paper presents an analysis of the power losses due to the friction between the components of the spherical joints, belonging to a 6R̲SS parallel mechanism. The determination of power losses in spherical joints has been done with an application where the characteristic point follows a closed space curve. The trajectory of the point on the convex semicuple in contact with the concave semicuple it has been determined by using facilities of the CATIA software.

1 Introduction

The paper presents a method for determining the power losses for a parallel mechanism by friction between the components of the spherical joints that form its six kinematic chains. Spherical joints are a special category of joints with the particularity that the contact surfaces are spherical, with or without lubricant between them.

The importance of spherical joints has led many researchers to carry out numerous studies to determine their influence on the dynamics of mechanical systems [1-4]. The contact forces acting in such joints are of a particular importance in the design and analysis of spherical joints [5-7]. It was also analyzed, based on contact forces, the energy dissipated in a spherical joint [8,9].

Recent studies have analyzed how the contact forces within the spherical joints that form space structures affect their stability and precision [10-12].

Liu et al. [13] have developed a mathematical model for approximation of the dimensional tolerance of a spherical joint which is based on distributed elastic forces.

2 Geometrical modeling of the spherical joint

Let us consider a kinematic chain of a 6R̲SS parallel mechanism (Figure 1). The kinematic chain is composed of rotational joints in the points O_k and spherical joints in the points A_k and B_k, respectively. The actuating arms $a_\mathrm{k} = O_\mathrm{k}A_\mathrm{k} = 254$ mm are rotated at an angle θ_k around the axis passing through the O_k point, by means of actuators mounted on the base plate. The position of the O_k point relative to the $\{R\}$ system, the length of the actuating arms k and the value of the angle θ_k are known. In Figure 2, has been shown the position of the O_k points.

* Corresponding author: milica.lucian@ugal.ro

Also, it has been noted with $e = 106$ mm the apothem of the base and with $g = 76$ mm the distance $O_k O_{k+1}$ which is the same distance as $B_k B_{k+1}$.

Let's look at one of the six kinematic chains of the 6<u>R</u>SS parallel mechanism (Figure 3). The degree of mobility of the mechanism is given by the Grübler-Kutzbach formula:

$$M = 6 \cdot m - \sum_{n=1}^{5} c \cdot k_n = 6 \cdot 13 - (6 \cdot 5 + 12 \cdot 3) = 12 \qquad (1)$$

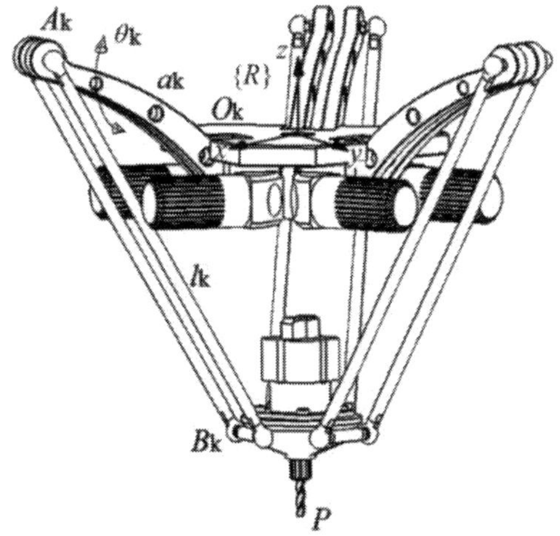

Fig. 1. Highlighting the O_k, A_k, r_k, θ_k parameters of the new 6<u>R</u>SS parallel manipulator.

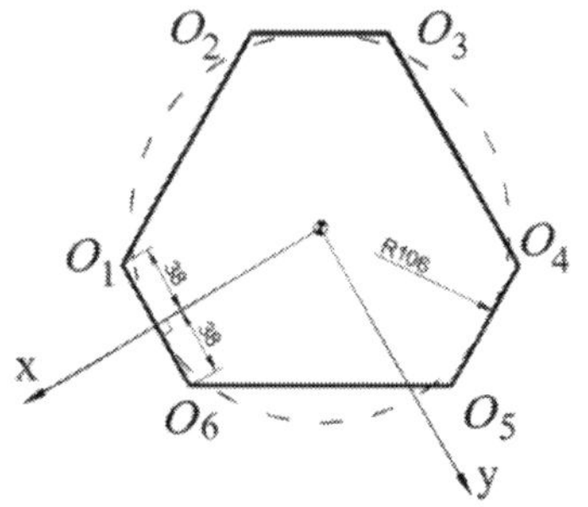

Fig. 2. The position of the O_k centers of the rotational joints.

Each actuating arm has a local mobility given by the rotation around the $A_k B_k$ axis, independent motion relative to the other possible motions of the mechanism. Eliminating these six isolated mobilities it results $M = 6$.

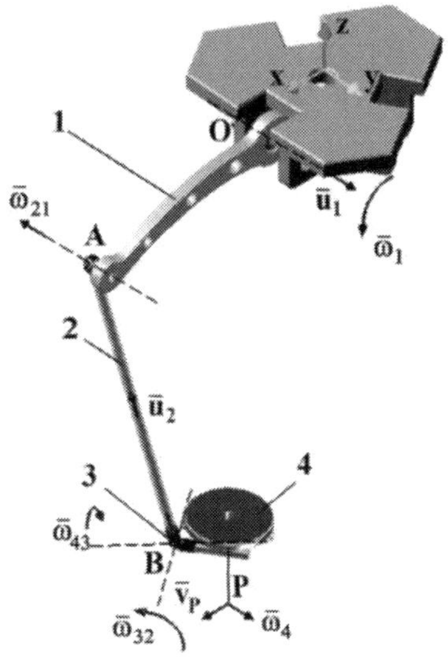

Fig. 3. Representation of a kinematic chain of the mechanism.

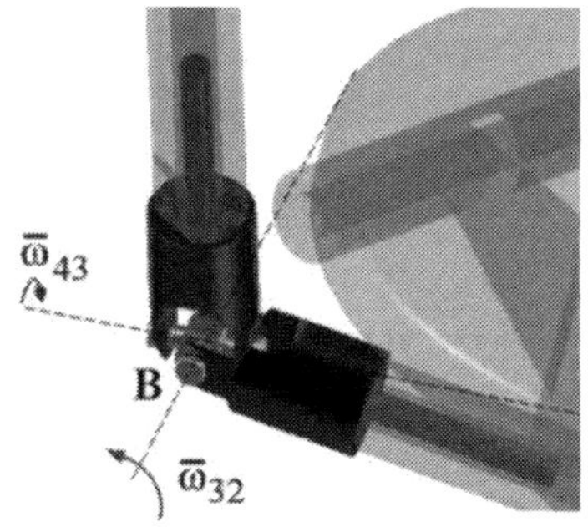

Fig. 4. Highlighting the 4th degree kinematic joint for the B point.

In the analysis of mobility it has been defined the following parameters:
- $m = n + 1$ (m represents the number of mobile kinematic elements);
- c is the class of the kinematic joints;
- k_n is the number of kinematic joints of class n.

For a more intuitive representation, it has been replaced the spherical joint from point B with a U-joint. This modification does not affect the degree of mobility of the mechanism. The direct kinematic model involves determining the speed of point P, based on six systems of equations, one for each kinematic chain (2):

$$S = \begin{cases} \boldsymbol{\omega}_1 \times \mathbf{OP} + \boldsymbol{\omega}_{21} \times \mathbf{AP} + \boldsymbol{\omega}_{32} \times \mathbf{BP} + \boldsymbol{\omega}_{43} \times \mathbf{BP} = \mathbf{v_P} \\ \boldsymbol{\omega}_1 + \boldsymbol{\omega}_{21} + \boldsymbol{\omega}_{32} + \boldsymbol{\omega}_{43} = \boldsymbol{\omega}_4 (\boldsymbol{\omega}_4 \equiv \boldsymbol{\omega}_P) \end{cases} \tag{2}$$

The variables in this case are: ω_1, ω_{32}, ω_{43}, $\boldsymbol{\omega}_{21}$, $\boldsymbol{\omega}_4$, $\mathbf{v_P}$.

Using the inverse kinematic model in which is known the linear speed ($\mathbf{v_P}$) and the angular speed ($\boldsymbol{\omega}_4 \equiv \boldsymbol{\omega}_P$) of point P, it can be determined six scalar variables: ω_1, ω_{32}, ω_{43}, ω_{21}^x, ω_{21}^y, ω_{21}^z.

3 Application for the inverse kinematic model

Let's look at the spherical joint of one of the points A_k.

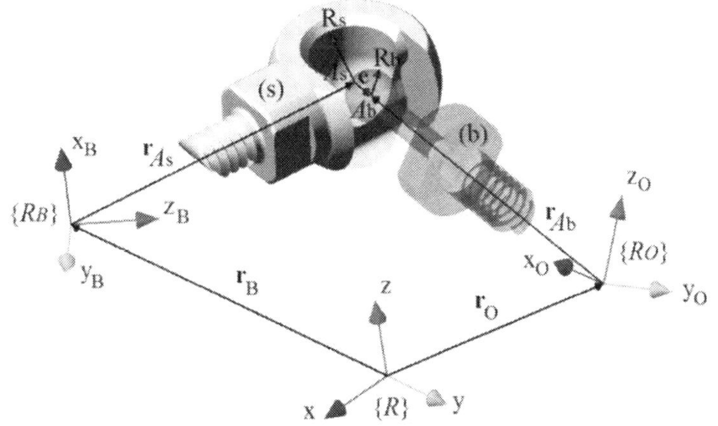

Fig. 5. Representing spherical joint with point position A_s and A_b.

Let's consider two bodies b (ball) and s (socket) that represent the spherical part and the concave part of a spherical joint from point A_1 (Figure 5).

The spherical part of the body b is inside the concave part of the body s. It will be noted with R_s and R_b the radius of the spherical body b and the concave body s. The difference between the radius R_s and R_b is the radial tolerance denoted by $j = R_s - R_b$ expressed in the above figure by the vector $\mathbf{e}$. The orientation of this vector coincides with the one of the axis passing through the points A_s and A_b, identical to the axis of the body b.

The position of the point A_s in the system $\{R_B\}$ is given by the position vector $\mathbf{r_{A_s}}$ and the position of the point A_b in the system $\{R_O\}$ by the position vector $\mathbf{r_{A_b}}$. Also, $\mathbf{r_B}$ and $\mathbf{r_O}$ are positional vectors of system origin $\{R_B\}$ and $\{R_O\}$ respectively. Norm of the vector $\mathbf{e}$ in the global system $\{R\}$ is:

$$|\mathbf{e}| = \sqrt{(x_A^b - x_A^s)^2 + (y_A^b - y_A^s)^2 + (z_A^b - z_A^s)^2} \tag{3}$$

It has been done an application in which was considered the characteristic point P in motion on a spatial curve of length L = 528.637 mm, when $\beta \neq \gamma \neq 0 = ct$ and α is variable (α, β, γ are the three orientation parameters of the mobile platform). In this application, the point P is in permanent contact with the curve. It has been stated that in the application done through the Catia software, the motion parameter is an arc of curve by equal length, dividing the curve into 72 equally distributed points on it, whose Cartesian coordinates are returned by the program.

The particularity of the application is that the inertia forces $\mathbf{F_i}$ are neglected, the motion being performed very slowly, therefore, there is only one axial force $\mathbf{F_{21}}$ acting in the **AB** direction.

Due to the imposed initial condition - the force acts in the **AB** direction - the eccentricity **e** is manifested in this direction throughout the motion. At the opposite pole of this eccentricity is the point of contact k between the spherical head and the concave part of the spherical joint (Figure 6).

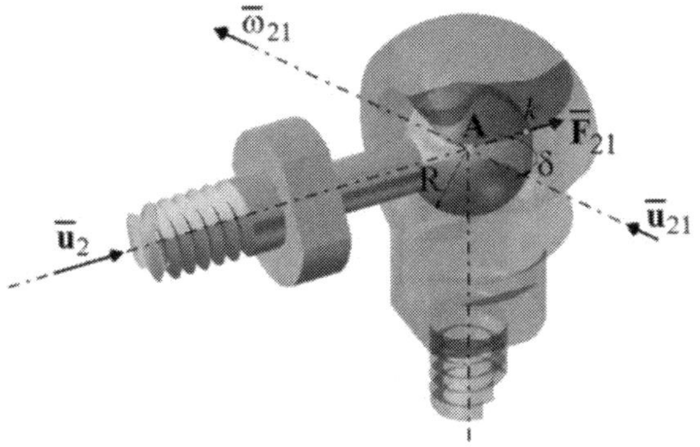

Fig. 6. Highlighting the point of contact k of the spherical joint A.

It has been noted with δ the angle between versor $\mathbf{u_2}$ and the versor $\mathbf{u_{21}}$ and with R the radius of the spherical head of the joint. Also $\boldsymbol{\omega}_{21}$ is the angular velocity of the contact point k.

Taking into account the imposed condition $\mathbf{F_i} = 0$, it can be written the following expression for the power loss $\mathbf{P_f}$ determinated by friction between the kinematic joint elements in point A:

$$\mathbf{P_f} = \mu \cdot \mathbf{F_{21}} \cdot \boldsymbol{\omega}_{21} \cdot \mathrm{R} \cdot \sin\delta \tag{4}$$

The mechanical work of friction forces is:

$$\mathbf{L_f} = \int_0^T \mathbf{P_f}\, \mathrm{dt} = \mu \cdot \mathbf{F_{21}} \cdot \mathrm{R} \int_0^T (\boldsymbol{\omega}_{21} \cdot \sin\delta)\, \mathrm{dt} \tag{5}$$

It can be considered the axial force $\mathbf{F_{21}}$ constant and unitary ($\mathbf{F_{21}} = \mathrm{ct} = 1$) throughout the entire during movement.

The value of the angle δ is given by the expression:

$$\cos\delta = \frac{\mathbf{u_2} \cdot \boldsymbol{\omega}_{21}}{|\mathbf{u_2}| \cdot |\boldsymbol{\omega}_{21}|} \rightarrow \sin\delta = \sqrt{1 - \cos^2\delta} \tag{6}$$

Based on of the Catia software, it can be determined the components of the vector $\boldsymbol{\omega}_{21}$, on the three directions (ω_{21}^x, ω_{21}^y, ω_{21}^z). Also, it can be determined the components (u_2^x, u_2^y, u_2^z) of the $\mathbf{u_2}$ versor of the spherical head axis.

From the six scalar components is resulting (7) and (8):

$$|\boldsymbol{\omega}_{21}| = \sqrt{{\omega_{21}^x}^2 + {\omega_{21}^y}^2 + {\omega_{21}^z}^2} \tag{7}$$

$$\mathbf{u_2} \cdot \boldsymbol{\omega}_{21} = \omega_{21}^x \cdot u_2^x + \omega_{21}^y \cdot u_2^y + \omega_{21}^z \cdot u_2^z \tag{8}$$

The integration of formula (5) was done numerically with the trapeze method, using the finite formula:

$$\mathbf{L_f} \approx \mu \cdot \mathbf{F_{21}} \cdot \mathrm{R} \sum_{i=1}^{72} (\boldsymbol{\omega}_{21} \cdot \sin\delta)_i \cdot \Delta t \quad (9)$$

where:

$$\Delta t = \frac{T}{72}\ \mathrm{s} \quad (10)$$

represents the time required to go through one of the 72 equal circle arcs.

Admitting a constant speed of motion of the point on the curve, $v = 10\ \mathrm{mm/s}$, and knowing its length (L), results the time T required for following the curve:

$$T = \frac{L}{v} = 52.8637\ \mathrm{s} \rightarrow \Delta t = 0.7342\ \mathrm{s} \quad (11)$$

In Figures 7 and 8 it has been represented the variation of vector components $\boldsymbol{\omega}_{21}$ and $\mathbf{u}_2$ depending on the position of the P_j point on the curve.

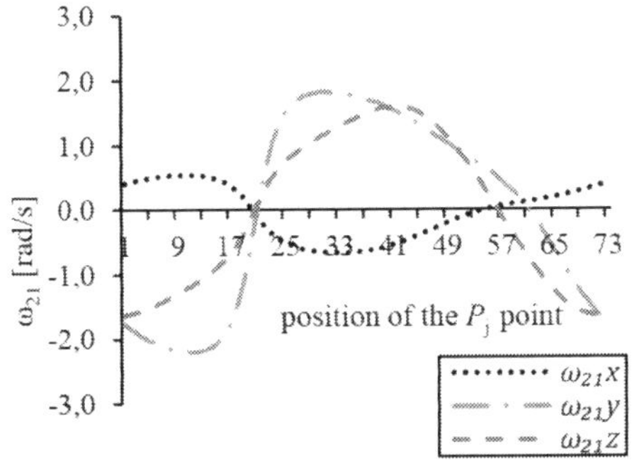

Fig. 7. Vector components $\boldsymbol{\omega}_{21}$ depending on the position of the P_j point.

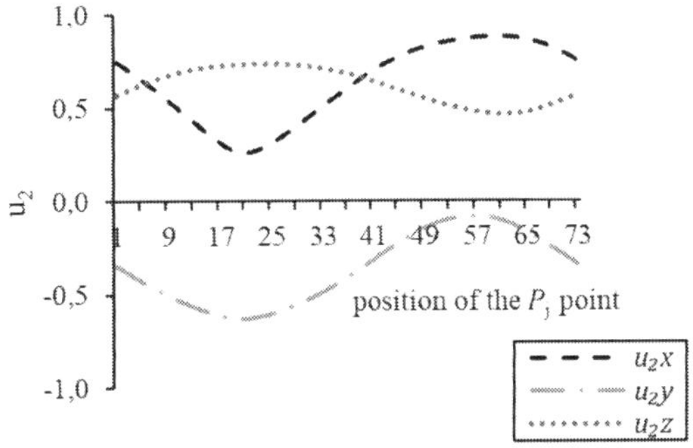

Fig. 8. Vector components $\mathbf{u}_2$, depending on the position of the P_j point.

By admitting a friction coefficient $\mu = 0.2$, from (4), (6), (7) and (8) it can be determined the lost power by friction $\mathbf{P_f}$ in the kinematic joint of point *A* for each position P_j of the characteristic point (Figure 9). By summing these values, it has been obtained the frictional power losses in the *A* kinematic joint, when the characteristic point follows the curve.

$$\mathbf{P_f} = \mu \cdot \mathrm{R} \sum_{i=1}^{72} (\boldsymbol{\omega}_{21} \cdot \sin\delta)_i = 0.13\ \mathrm{W} \quad (12)$$

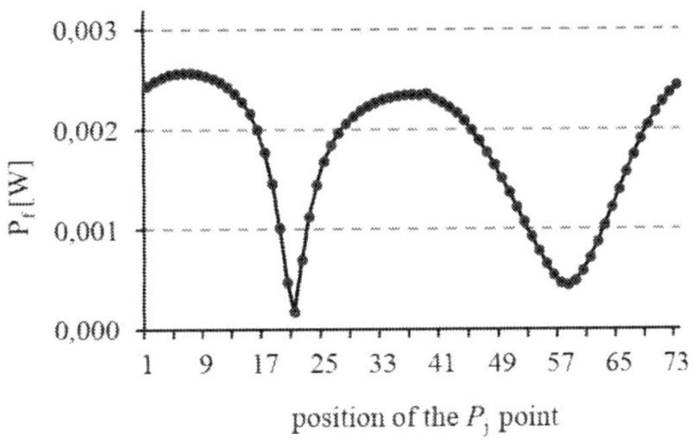

Fig. 9. Frictional power loss $\mathbf{P_f}$, for each position of the P_j characteristic point.

DOI: 10.1051/matecconf/201713704002

Also, it can be determined, using the functions of Catia software, the trajectory of a point k' belonging to the spherical head of the joint on its concave surface.

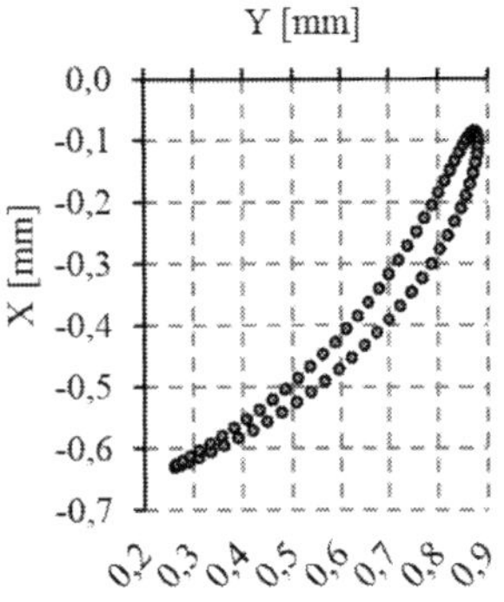

Fig. 10. Projection of the k' point trajectory on XY plane.

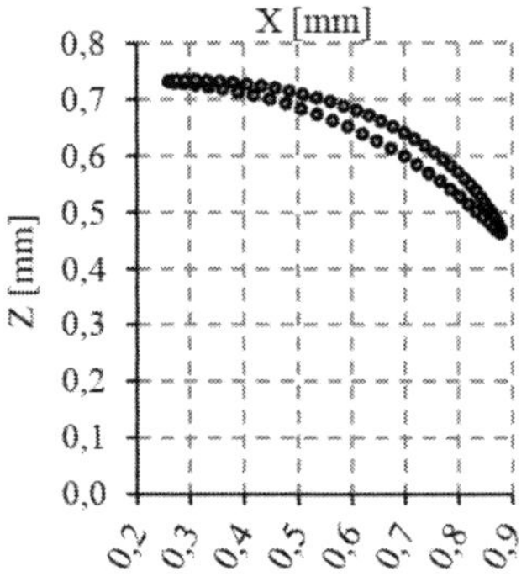

Fig. 11. Projection of the k' point trajectory on ZX plane.

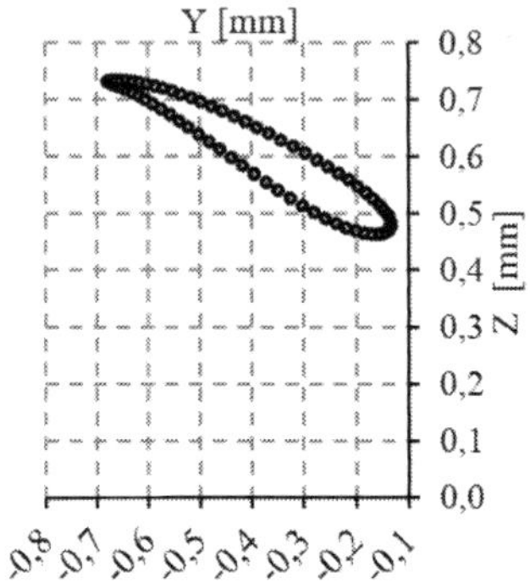

Fig. 12. Projection of the k' point trajectory on ZY plane.

Projecting the coordinates of these points on the three plans XY, XZ, ZY, it can be obtained the diagrams of Figures 10-12 which describe the trajectory of point k'.

4 Conclusions

This paper presents an original method for determining the power losses due to friction between the components of the spherical joints belonging to a 6<u>R</u>SS parallel mechanism.

The determination of these power losses in spherical joints was done in an application where, the characteristic point P follows a closed space curve, neglecting the inertial forces and the motion being very slow.

Using the Catia software, the variation graphs of the versor $\mathbf{u}_2$ of the direction **AB** and of the vector $\boldsymbol{\omega}_{21}$ have been determined depending on the position of the point P_j on the given spatial curve. The trajectory of a point belonging to the spherical head of the spherical kinematic joint of the parallel mechanism was obtained and the graphs of the traiectory's projections on the three plans XY, ZX, YZ were done.

References

1. S. Dubowsky, F. Freudenstein, ASME J. Eng. Ind., **93**, 305–309 (1971)
2. P. Flores, J. Ambrósio, J. P. Claro, H. M. Lankarani, C.S. Koshy, Mech. Mach. Theory, **41**, 247–261 (2006)
3. O.A. Bauchau, J. Rodriguez, Int. J. Solids Struct., **39**, 41–63 (2002)
4. P. Flores, Nonlinear Dyn., **61**, 633–653 (2010)
5. F. Farahanchi, S. W. Shaw, J. Sound Vib., **177**, 307–324 (1994)
6. G. S. Boltachev, N. B. Volkov and N. M. Zubarev, Int. J. Solids Struct., **49**, 2107-2114 (2012)
7. J. Rhee, A. Akay, Mech. Mach. Theory, **31**, 121–134 (1996)
8. P. Flores, J. Ambrósio, and J. P. Claro, Multibody Syst. Dyn., **12**, 47–74 (2004)
9. M. Lankarani, and P. E. Nikravesh, ASME J. Mech. Des., **112**, 369–376 (1990)
10. J. A. C. Ambrósio, NATO Adv. Sci. Inst., **2**, 15–33 (2002)
11. M. A. Brown, Adv. Space Res., **48**, 1747-1753 (2011)
12. Z. H. Qi, Y. S. Xu, X. M. Luo, and S. J. Yao, Multibody Syst. Dyn., **24**, 133-166 (2010)
13. C. S. Liu, K. Zhang, and L. Yang, J. Comput. Nonlin. Dyn., **1**, 160-167 (2006)

Using the modern CNC controllers capabilities for estimating the machining forces during the milling process

Radu-Eugen Breaz[1,*] and *Sever-Gabriel* Racz[1]

[1]Lucian Blaga University of Sibiu, Engineering Faculty, Emil Cioran 4, 550025 Sibiu, Romania

Abstract. Machining forces can nowadays be measured by using 3D dynamometers, which are usually very expensive devices and hardly available for most of the CNC machine-tools users. On the other hand, modern CNC controllers have nowadays the ability to display and save many outputs within the machining process, such as the currents or even the torques at the shaft's level for the feed motors on each axis. These outputs can be used for estimating the machining forces, but it is to be noticed that the above-mentioned currents and torques are proportional with the overall resistant forces, which includes not only technological forces, but also friction, inertial and pre-tensioning forces. This paper presents an approach for estimating the machining forces during a milling process, by using the outputs stored in the CNC controller and separating the effects of technological forces from the other forces involved in the process. The separation was made by running two sets of experiments, one set for dry-run regime and the other one for machining regime.

1 Introduction

Measuring the cutting forces is important with many respects. This measurement could be performed by using 3D dynamometers (Kistler), but this approach is suitable mostly for research facilities. Dynamometers are expensive and are reported to be easily damaged by cooling liquids. Consequently, alternative methods for measuring forces were developed. One of the most used method involves indirect measurements, by measuring the currents within feed and/or spindle motors using current sensors [1, 2]. Other approaches are based upon the development of special milling sensors [3, 4], systems with optical data transmission [5] or upon strain gauges [6]. There are also reported some solutions of non-commercial, special dynamometers developed by the research teams [7].

Modern CNC controllers can display and save many outputs within the machining process, such as the currents or even the torques at the shaft's level for the feed motors on each axis. These outputs can be used for estimating the machining forces, but it is to be noticed that the above-mentioned currents and torques are proportional with the overall resistant forces, which includes not only technological forces, but also friction, inertial and pre-tensioning forces.

* Corresponding author: radu.breaz@ulbsibiu.ro

2 Static torque and technological force

Usually, for a CNC machining centre, the feed drives on X and Y axes are composed of a rotary permanent magnet synchronous servomotor as actuation device and a recirculating ball screw as the solution for converting the rotary motion of the servomotor into linear slide motion.

The schematic diagram of a horizontal (for X and Y axes) feed drive is presented in figure 1.

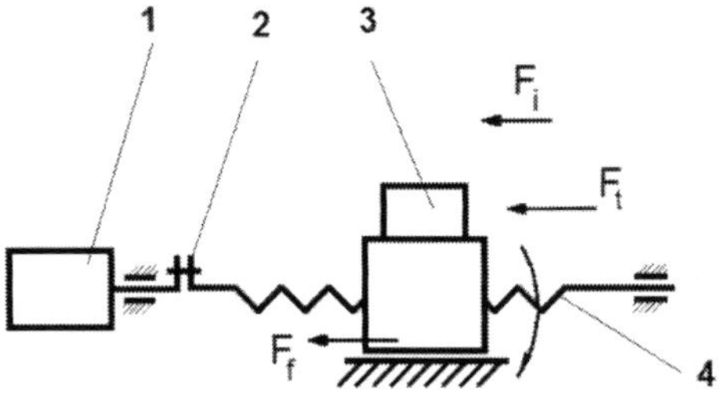

Fig. 1. Schematic loads diagram for a horizontal CNC feed drive.

The notations from figure 1 have the following significations:

1 – drive motor; 2 – axial coupling; 3 – horizontal slide and part; 4 – lead ball-screw.

The total static torque for a horizontal feed drive may be calculated with the following relation:

$$M_{st} = \frac{[F_t + F_f)] \cdot p}{2\pi \cdot i \cdot \eta_{tot}} + M_{pr} = \frac{[F_t + \mu \cdot m \cdot g)] \cdot p}{2\pi \cdot i \cdot \eta_{tot}} + M_{pr} \qquad (1)$$

where:

M_{st} – total static torque [Nm], F_t – technological force [N], F_f – friction force [N], F_i – inertial force [N], m – total mass (machine slide and workpiece) [kg], μ - friction coefficient, g – gravity acceleration [m/s^2], p – lead ball-screw step [mm], i – transfer ratio between rotational motor and machine slide, η_{tot} – total efficiency of the mechanical transmission between rotational motor and machine slide, M_{pr} – pre-tensioning torque of the nut of the main ball-screw system [N].

Taking into consideration the fact that, for the CNC machining centre used for this research the rotational motors for X and Y axes are directly coupled with the lead screws (i=1), the following relationships regarding the static torques on each axis may be written:

$$M_{stX,Y} = \frac{[F_{tX,Y} + \mu \cdot m_{X,Y} \cdot g] \cdot p}{2\pi \cdot \eta_s} + M_{prX,Y} \qquad (2)$$

Dynamic torques are also appearing during the milling process, on the acceleration and deceleration phase (fig. 2). If the data acquisition is done during the constant velocity (constant feed) phase, the effects of the dynamic torque may be neglected.

The notations from figure 2 have the following significations:

t_a – acceleration time (dynamic torque is positive non-zero);

t_{ct} – constant feed time (dynamic torque is zero);

t_d – deceleration time (dynamic torque is negative non-zero).

For the proposed approach, it was necessary to unfold the same movements cycle in no-load conditions, prior to unfolding it in load conditions, to be able to read and subtract the no-load and load-conditions torques, per the following relation:

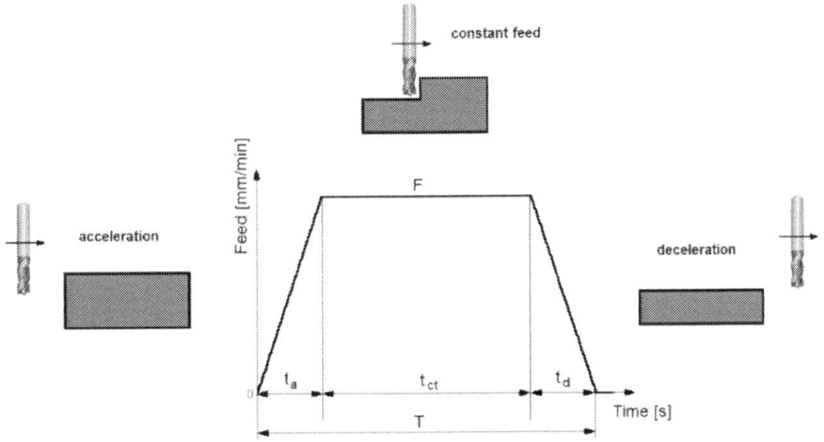

Fig. 2. Velocity profile for a milling cycle.

$$\Delta M_{st\,X,Y} = \frac{[F_{tX,Y} + \mu \cdot m_{X,Y} \cdot g)] \cdot p}{2\pi \cdot \eta_s} + M_{prX} - (\frac{[\mu \cdot m_{X,Y} \cdot g)] \cdot p}{2\pi \cdot \eta_s} + M_{prX,Y}) = \frac{F_{tX,Y} \cdot p}{2\pi \cdot \eta_s} \quad (3)$$

where:
$\Delta M_{X,Y}$ – difference between static torque in load and no-load conditions.
The technological forces can now be calculated using the relation:

$$F_{tX,Y} = \frac{2\pi \cdot \eta_s}{p} \cdot \Delta M_{st\,X,Y} \quad (4)$$

3 Experimental determinations

A vertical machining centre DMC v 635 ecoline by DMG Mori was used as experimental testbed. It is equipped with Sinumerik 840D CNC controller, which can measure, display and save many process outputs, such as torques on each axis, currents through feed motors and many others.

A single axis milling operation (on X-axis) with the following characteristics was unfolded:
- material: alloyed steel (33MoCr11), 151 HB
- cutting tool: flat-end mill from metallic carbide, diameter (d) 10 millimetres, 3 flutes;
- cutting regime: speed (n) 2000 rev/mins, cutting speed (v_c) 62.8 m/min, cutting feed (v_f) 180 mm/min, feed per tooth (f_z) 0.03 mm/tooth.

The TRACE function of the CNC controller was used for displaying the outputs.

The TRACE function allows the user to select and capture a selected output within the process for a given interval of time. The size of the interval was chosen of 30 seconds, while the sampling rate (T_s) was chosen of 12.5 milliseconds. The sampling rate was chosen to be the smallest one allowed by the size of internal memory of the CNC controller.

Straight lines of 30 millimetres length parallel to X axis were machined, with axial depth of cut (a_p) of 2 millimetres and radial width of cut (a_e) of 10 millimetres were machined (fig 3).

The movements cycle for machining the straight lines was programmed in no-load conditions (dry-run), where the tool was moved above the part, with no cutting process occurring and in load conditions, were the cutting process was unfolded.

 DOI: 10.1051/matecconf/201713704003

Fig. 3. Machined part.

The output chosen for the first approach was the real torque on X-axis. Figures 4 and 5 show a capture of these torques value in dry-run and in cutting conditions. The data were captured during the constant feed phase, to nullify the influence of the dynamic torque.

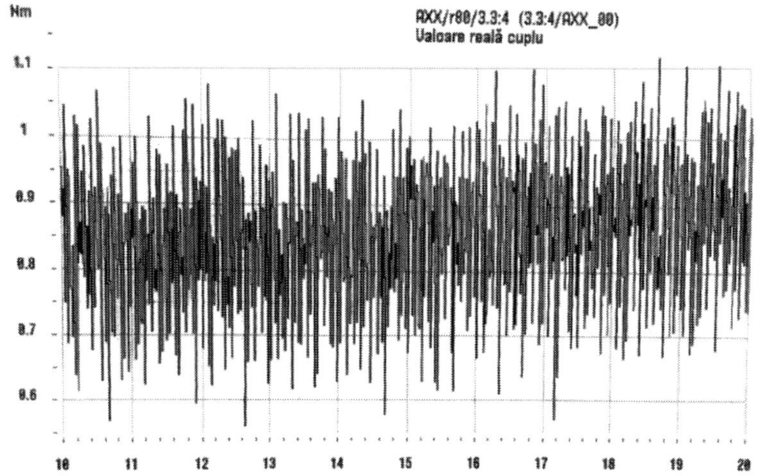

Fig. 4. Torque in dry-run (capture).

From figure 4 it is noticeable that the static torque in dry-run oscillates around a mean value of 0.85 Nm. Similarly, in figure 5 one can notice that the static torque in cutting conditions oscillates around the value of 1 Nm. Consequently, the value of torque defined by (3) can be expressed as:

$$\Delta M_{st\,X} = 1 - 0.85 = 0.15\ Nm \tag{5}$$

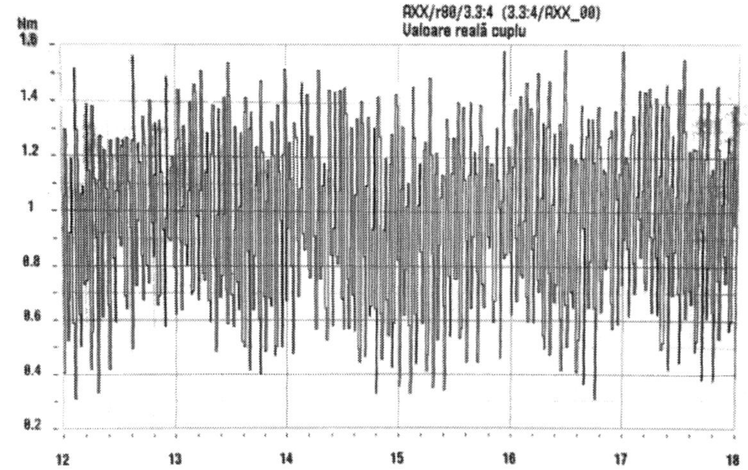

Fig. 5. Torque in cutting conditions (capture).

The technological force on X axes can now be calculated using relation (4):

$$F_{tX} = \frac{2\pi \cdot \eta_s}{p} \cdot \Delta M_{st\,X} = \frac{2\pi \cdot 0.9}{15 \cdot 10^{-3}} \cdot 0.15 = 56.55\ N \qquad (6)$$

A second approach involved the acquisition of the currents trough the feed motor of X-axis in dry-run and cutting conditions, presented in figures 6 and 7.

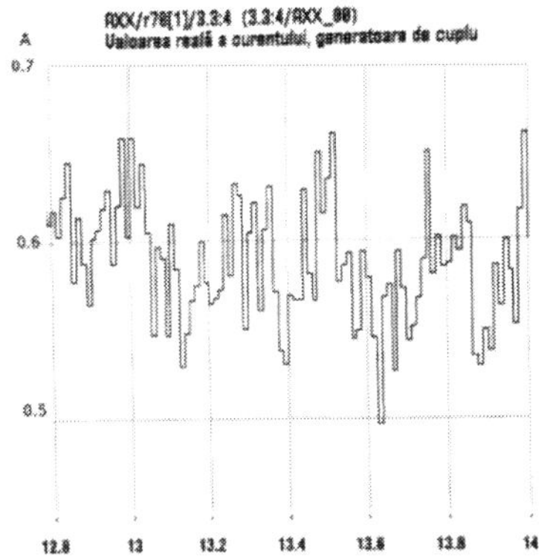

Fig. 6. Current through X-axis motor in dry-run (capture).

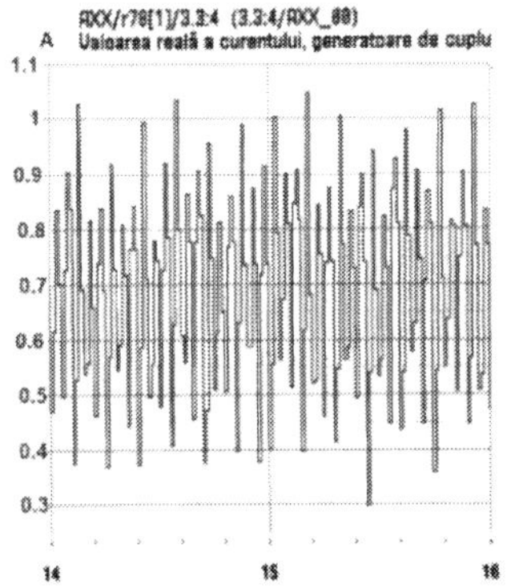

Fig. 7. Current through X-axis motor in cutting conditions (capture).

From figure 6 and 7 it is noticeable that the current in dry-run oscillates around a mean value of 0.6 A, while the current in cutting conditions oscillates around the value of 0.7 A. Consequently, the value of torque defined by (3) can be expressed as:

$$\Delta M_{st\,X} = (0.7 - 0.6) \cdot K_t = (0.7 - 0.6) \cdot 1.37 = 0.137\ Nm \qquad (7)$$

It can be noticed that the torque values determined from (5) and (7) are very close to each one. The K_t value form above (1.37 Nm/A) represents the torque constant of the 1FK7063-2AF71-1RH0 Simotics servomotor.

The cutting force for this milling operation was also determined by using the Kenametal on-line calculator [8] (figure 8).

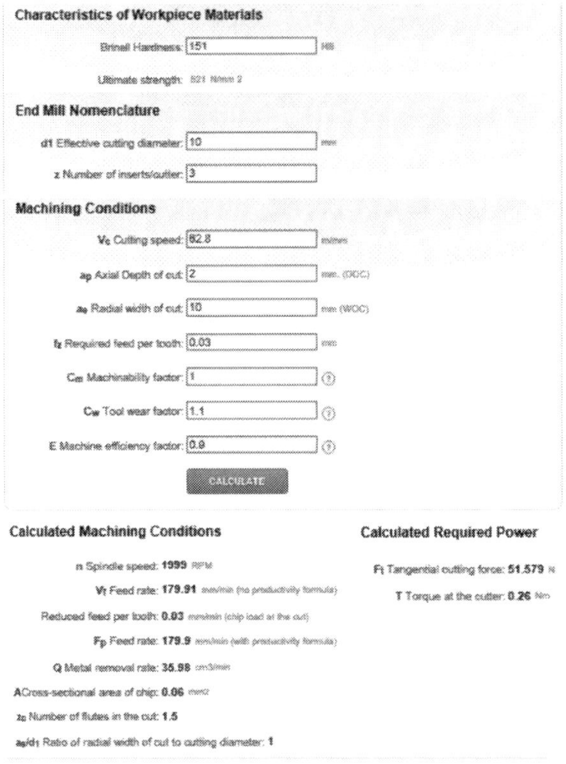

Fig. 8. On-line cutting force calculator

The calculated force F_t, per figure 8 has a value of 51.58 N which is very close to the experimentally determined one, which has a value of 56.55 N, per (6).

4 Conclusion

The proposed method allows the user to determine the cutting forces by using the facilities provided by the modern CNC controllers. The experimentally determined values followed very close the calculated ones. Further researches will be performed to tailor this method for multi-axis cutting operations.

Authors want to thank to Eng. Bogdan Prunel, CNC machining specialist, which provided valuable assistance during the experimental researches.

References

1. K. Dohyun, J. Doyoung, Precis. Eng. **35** , 143–152 (2011)
2. Y.-H. Jeong, D.-W. Cho, Int. J. Mach. Tools Manuf. **42**, 1559–1566 (2002)
3. C. Adolfsson, J-E. Stahl, Int. J. Mach. Tools Manuf. **35**, 1715 (1995)
4. C. Andersson, M. Andersson, J.-E. Stahl, Int. J. Mach. Tools Manuf. **51**, 67-76 (2011)
5. Y. Ikezaki, Y. Takeuchi, M. Sakamoto, Ann. CIRP, **33**, 61 (1984)
6. A.R.H. Shuaib, E. Garcia-Gardea, et al, J. Eng. Ind.-T. ASME, **103**(2), 235–240 (1981)
7. G. Totis , G. Wirtz, M. Sortino, D. Veselovac, E. Kuljanic, F. Klocke, Mech. Syst. Signal. Pr. **24,** 1844–1857 (2010)
8. *** https://www.kennametal.com/en/resources/engineering-calculators/end-milling-calculators/force-torque-and-power.html

Using Virtual Reality tools to support simulations of manufacturing instances in Process Simulate: The case of an iCIM 3000 system

Radovan Holubek[1,*], *Daynier Rolando* Delgado Sobrino[1], *Peter* Košťál[1], *Roman* Ružarovský[1] *and Karol* Velíšek[1]

[1] Slovak University of Technology in Bratislava, Faculty of Materials Science and Technology in Trnava, Institute of Production Technologies, 91701 Trnava, Slovakia.

Abstract. Nowadays the trends in manufacturing are aimed at exploding and using the full potential of production devices and systems. In this context, Simulation and VR tools play a key role in the evolution of such systems specially if used in conjunction. By using these tools, the present paper focused on the analyses of an iCIM 3000 system. The main goal of the paper lied on studying the possibilities of increasing the efficiency and effectivity of whole system and its individual devices. This was achieved by means of the creation of simulations and their comparison to the real system. At the same time, the results of the simulations themselves also became the basis for further and richer analyses by means of the use of VR tools that allowed a much deeper and detailed take of the whole system. The use of VR tools offered a better assessment and comprehension of the system under study what at the same time also worked as a feedback for the improvement of the initial simulations, and this by eliminating and/or improving unnecessary movements and collisions. The results of this study have a positive significance in the improvement of the system once these are implemented.

1 Introduction

The current trend of increasing competition in the marketplace results in a continuous improvement of manufacturing systems and processes and thus the need of implementing and deploying or put in into practice new production ideas and designs as quickly as possible. It is precisely this fast pace, a factor that implies huge challenges and complications that are to be corrected at each stage of the production process which one way or another always affects the final commissioning time of such systems. On the other hand, extending the deployment time and debugging of new equipment also results in an increase in investment to implement these into the production process. Because of these shortcomings and complications, the use of effective methods and tools become of

* Corresponding author: radovan.holubek@stuba.sk

tremendous importance in the reduction of the deployment times. Nowadays, most new production system projects starts from the same design tasks, i.e.: by creating CAD databases of production system models and all peripherals that are subsequently used as simulation objects in terms of designing operating production systems. This very first phase of creating accurate CAD models of production machines, industrial robots and other devices of a workplace is the most important stage in using such methods to design new or update/improve existing workplaces [1]. Despite a number of technology providers and system manufacturers offer their CAD databases on their devices, in many of the cases these models serve mostly as basic CAD model skeletons, often without much data for simulation (bindings, kinematic characteristics, etc.). On the other hand, it is only by using the exact CAD model of a system or process that is possible to design and properly analyze a reliable production system.

2 Using CAD models in simulations

Properly created CAD models can be loaded into various simulation softwares that allow to create simulations in technological and manipulation operations. The use of simulation methods in application software specifically designed for manufacturing equipment and systems are expected/designed to enable the most efficient projection of device layout solutions in terms of a proper deployment of these machines and devices. The same simulation of production and assembly processes reflects, as closely as possible, all the processes that take place in the selected case study in terms of the logistic flows of the individual equipment, the simulation of the technological and inter-operative processes, as well as the impact on the main and secondary production times.

As part of the research project "Excellent Teams of Young Researchers" - Grant STU, a project was solved with regard to the possibilities of using virtual simulation methods for the purpose of determining deviations between the virtual and the real environment. An iCIM 3000 flexible manufacturing system was chosen as a system suitable for real and virtual system comparisons; this system is located in our laboratories. The determined deviations could be first obtained in the simulation environment, then verified and better justified with Virtual Reality tools, and finally the findings and results can be implemented in the real environment/system.

First of all, it was necessary to correctly create and dismantle the CAD database of the layout according to an existing system Fig.1.

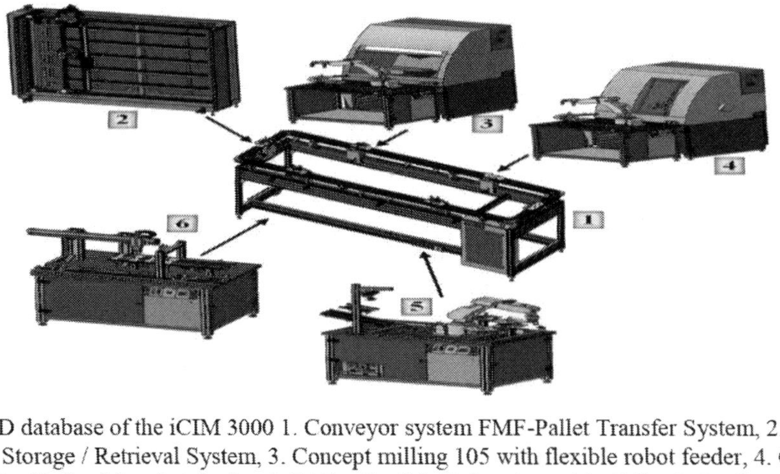

Fig. 1. CAD database of the iCIM 3000 1. Conveyor system FMF-Pallet Transfer System, 2. AS/RS Automatic Storage / Retrieval System, 3. Concept milling 105 with flexible robot feeder, 4. Concept turning 105 with flexible robot feeder, 5. FAC - Flexible Robot Assembly Cells, 6. Quality and Handling station QH 200.

Despite the iCIM 3000 system is to be understood as a whole, for simulation needs, it was necessary to dismantle and analyze y individual stations in order to maximize the use of each station and the equipment that makes up the system.

Individual stations had to be modeled and then put together/assembled in the CATIA CAD software. These CAD models had been created as static CAD models, thus it was then necessary to convert the CAD models into ".JT files", what was done in other CAD software able to open such CATIA models and able to export them with the needed format. Such ".JT files" are able to be loaded into the Process Simulate software that is the software chosen for the iCIM3000 simulations [2]. The Process Simulate software is designed to create mechanisms with integrated CAD databases and creating off-line simulations of various production and handling operations with the possibility of further development within manufacturing systems and processes. For simulation purposes, it was necessary to create the kinematics of individual clamping devices, robbers and linear guides. This was done taking into account the exact kinematics of the RV-2SDB industrial robot and its limits on the individual axes; creating a robot's kinematics is a necessary part of a successful simulation. Besides, in Process Simulate it is was necessary to first define kinematic links, fixed and moving parts (mother and son type of relationships), TCP tools,and last but not least to define the device type as a "robot". A robot defined as **"device"** can be integrated into the CAD model of the non-moving parts of the CAD system loaded into Process Simulate [3]. Based on the proper loading of the CAD model of the mounting station skeleton, the insertion of the gripper and robot with the properly defined kinematics Fig.2, it was possible to create individual robot trajectories with positioning.

Based on the simulation of this assembly process, it was possible to consider the optimization of individual robot paths. This mainly refers to the fact that in the real system the robot always returns to an initial position after each operation, which increases the assembly process time. This was the main purpose of the simulation of the assembly process of the iCIM assembly station. The modification of the robot's trajectories in the assembly process represents, in terms of time, considerable time savings compared to the real system.

This way, it was necessary to properly define each single device for the simulation of each single station in the iCIM 3000 system. In the creation of the simulation, the longest phase was the integration of the CAD files into usable formats of the type JT file.

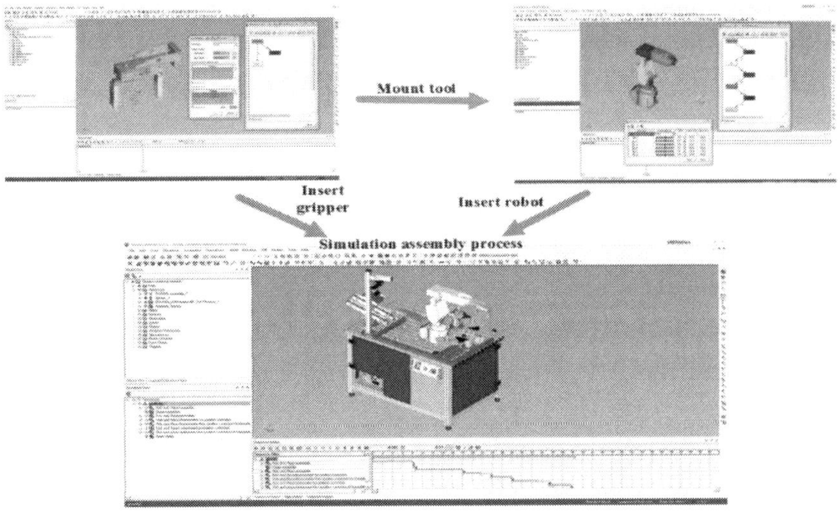

Fig. 2. Definition of device type and creation of kinematics in the assembly station.

Kinematic linkages, definition of the type of devices and peripherals also required lots of efforts and skills [4]. This procedure had to be followed for each device, parts, and element of the assembly cell. An important aspect is that after successfully creating the iCIM3000 system database and its components, such as individual stations, robots, manipulators, grippers, clamps, conveyors and the like, it is possible to use the entire database of Process Simulate software without having to define and modify them again.

Another complicated part while creating a suitable kinematic model for simulation purposes was the storage system Fig.3. Primarily, it is a Cartesian XYZ manipulator, but the axle, in addition to the translational movement, performs rotation around its axis by 180°. The creation of a suitable kinematic system and linkages was preceded by the definition of the movement parts depending on the kinematic structure of the manipulator. Since this is a specific device, the creation of the links and relationship analysis was more demanding. Unfortunately this is something that is not properly covered in the manuals for Process Simulate which mainly focus on general issues. Defining the kinematic structure of the XYZ keeping in mind the sometimes simultaneous translational and rotational movements was much more challenging, this specially in terms of specifying the conditions for each part of the CAD models of the assembly. A total of 50 positions were created for the manipulator. This at the same time presupposed much more work since each position included 4 other positions for the input - output of the pallets to and from the system.

In terms of the simulation and correction of the operating time in the storage system, we were able to reduce the storage system's operating times by up to 15% based on real storage system measurements in comparison with the simulated model. This results were based on time analyzes taking into account the speed limits of the individual axes of the real storage system. This information was also further verified by Virtual Reality simulations using an HTC Vive's VR glass set. Then, the creation of the simulation scenarios of the control station with the conveyor represented the next phase of simulation in the analyses and comparison of the iCIM3000 system, see Fig.4.

The control station has the task of controlling the diameter of the holes within the tolerance range using a strain gauge sensor. In trying to integrate the control station into the simulation environment, the kinematics of the gripper (for moving the pallets), manipulator and strain gauge sensor had to be also created. The part under control is a so-called "Baseplate", which is produced on a CNC milling machine of the same iCIM 3000 real system.

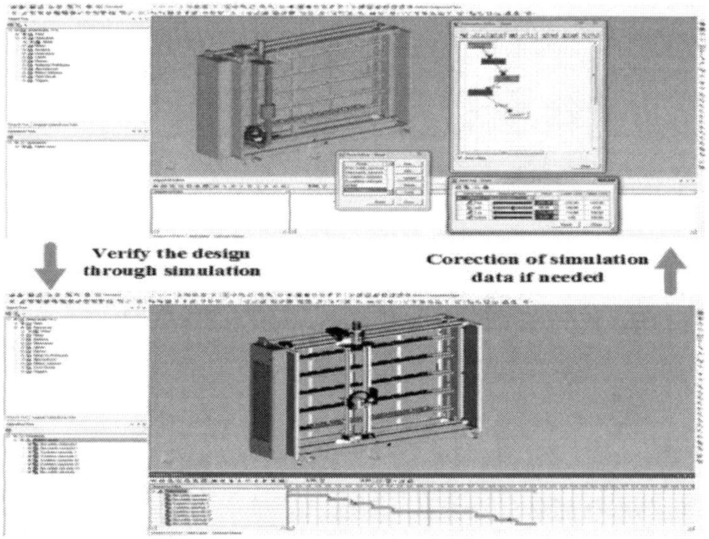

Fig. 3. Creating kinematics and positions of the storage system.

DOI: 10.1051/matecconf/201713704004

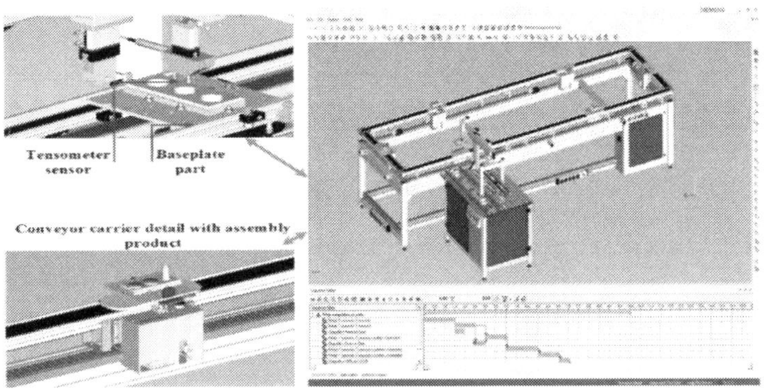

Fig. 4. Scenario of the quality control station and the conveyor.

Within the simulation scenarios, the conveyor was also integrated to connect all iCIM stations. The conveyor contains 8 pallet carriers with RFID chip identification. An important element in terms of the reducing time in comparison with the real system is that in a real system, when the conveyor stops at a station to supply a pallet, the other pallet carriers within the system stops as well till the operation, mounting for example, has been terminated. From the point of trying to reduce some of this time, this is to be eliminated by making the carriers to continue and then stop again just after the operation has been completed at a given station. This simulation verification appeared to be an appropriate elimination of the downtimes in the main and related sub-operation cycles in the comparison to the virtual and real system. This comparison of the real and simulated system with the incorporated changes was also verified by Virtual Reality simulation [5].

3 Virtual Reality tools to support simulations

The VR set used is the HTC Vive glasses headset. The VR includes two display panels with a resolution of 1,080 × 1,200 pixels. In order to create a virtual reality image, the video streams with a 90 Hz refresh rate. Two base station monitoring stations are used to accurately monitor the workspace and need to be placed in the corners of the room. Space monitors transmit laser beams to 32 headphones sensors and 48 sensors on handheld controllers to determine the exact position of the user in the pre-set motion space.

The tracking controls have a vibrational response that helps to intensify the virtual reality gain. The simulations we created in the Process Simulate software were used to visualize simulations in virtual environments using VR's support software. Enhancing Simulation in the Virtual Environment with HTC Vive Glasses allows for a much better visualization of the simulated process as the controls allow for 360 ° shooting, stop simulation at any given time, hide different CAD parties that can hide possible collisions in normal simulation Fig.5.

A very important thing within the virtual environment simulation is the possibility of "breaking" the CAD parts and thus getting into hidden parts of CAD models, which can be seen in a completely different view and dimension.

The knowledge gained from the virtual environment significantly helped to eliminate major and minor times that were not originally seen in desktop PC simulation. A very important factor influencing the best visualization in CoR is the most realistic rendering in real time. In a simplified form, we can say that CAD models and / Simulation of the production process, must be able to transmit the visual experience in real time to the user without undue delay [6, 7].

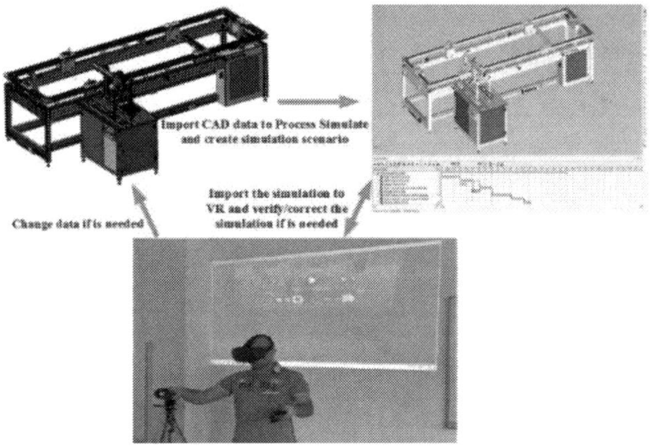

Fig. 5. Simulation scenario in virtual reality using Headset HTC Vive.

4 Conclusions and further research

The use of virtual reality in production systems has a huge importance in terms of safety, time-based deployment of the real system in production and the elimination of collisions, a very effective solution based on the exact verified facts within the results achieved. In order to use the VR for simulation of production systems, it is necessary to have a very good and powerful computing technique with a good graphics card. The use of VR is a very good tool because the acquired visualization is without the necessity of intervening in the real system and thus as much security as possible, especially during the incorporation of changes into the real system, increase the efficiency of comparison and overall deployment (comparison of desktop simulation, VR simulation Vs. real system), as well as the elimination of collisions, the possibility of damage to the system without prior knowledge. From the point of view of using the CoR in comparison with the real system, the use of such technologies is of great importance especially in incorporating changes in prototyping new machine and equipment products.

This research was supported by the STU Grant "Scheme for Support of Excellent Teams of Young Researchers – (1349 - Use of simulation methods in the context of the Virtual Commissioning as a tool for production systems in the Digital Factory).

References

1. K. Velíšek, R. Holubek, D.R. Delgado Sobrino, R. Ružarovský, N. Vetríková, *MATEC Web of Conf*, **94**, online ISSN 2261-236X (2017)
2. H. Yap, Z. Taha, S. Dawal, J. Zhejiang University - Science C, **12**, ISSN 1869-196X (Online) (2012)
3. L.V. Guerrero1, V. V. López, J. E. Mejía, *Computer-Aided Design and Applications*, **11,** S11-S19, DOI: 10.1080/16864360.2014.914400 (2014)
4. M. Lorenz, M. Spranger, T. Riedel, F. Pürzel, V. Wittstock, P. Klimant, *Research and Innov in Manuf: Proceedings of the 48th CIRP Conf on Manuf Sys*, **41**, 358–363 (2016)
5. M. Hincapi, M. Jess Ramrez, A. Valenzuela, J. Valdez, *Int J for Interactive Design and Manuf*, **8**, 1-22, DOI: 10.1007/s12008-014-0206-7 (2014)
6. Z. Tuma, J. Tuma, R. Knoflicek, P. Blecha, F. Bradac, *Procedia Engineering*, **69**, 1015-1020 (2014)
7. M. Eissele, O. Siemoneit, T. Ertl, *IEEE Conference on*, DOI: 10.1109/RAMECH.2006.252671 (2006)

 DOI: 10.1051/matecconf/201713704005

Novel trends in the assembly process as the results of human – the industrial robot collaboration

Radovan Holubek[1,*], *Roman* Ružarovský[1], *Daynier Rolando* Delgado Sobrino[1], *Peter* Košťál[1], *Adam* Švorc[1] and *Karol* Velíšek[1]

[1]Institute of Production Technologies, Faculty of Materials Science and Technology in Trnava, Slovak University of Technology in Bratislava, Bottova 25, 917 01 Trnava, Slovakia

Abstract. The contribution is focused on the creation of an idea proposal and simulation of the assembly system in cooperation of the human and the industrial robot. The aim of the research is to verify the feasibility of this cooperation between the human and the industrial robot on the basis of the created simulation in the assembly process. The important step of the design this collaboration is the determination of rules and safety of this cooperation. The paper also presents the method of working with the selected software and its functionalities and sequence of steps at the simulation creation. The objective of the research is the evaluation of the idea proposal of the collaborative assembly system on the basis of the created simulation. The analysis and evaluation of the simulation confirm the feasibility and safety of the cooperation of the man and robot and also verified the possibility of assembly made by man and robot from the disposition and dimension on point of view of the proposed workplace.

1 Introduction

Currently, it can be concluded that intelligent human and robot collaboration takes place in the foreground. This is mainly due to increasingly complex customer requirements, as well as efforts to gain the highest market competitiveness. The implementation of collaborative workplaces has its justification for shortening innovation cycles and a more flexible response to the more predictable market demands. In processes such as human with fast moving robot cooperation, it is necessary to take extra protection of persons [1].

The role of intelligent human cooperation with the robot can be considered to facilitate the recurrent, dangerous, and often very demanding work activities [2, 3]. The standard role of the human in the manufacturing working is separated from the industrial robot. The reason is preventing collisions and situations that directly threaten human health in the production process. The concept of human-robot collaboration can be seen as a common work in which these two action members complement in the manipulation, palletizing, surface treatment, assembly [4], disassembly [5] and other technological operations.

* Corresponding author: radovan.holubek@stuba.sk

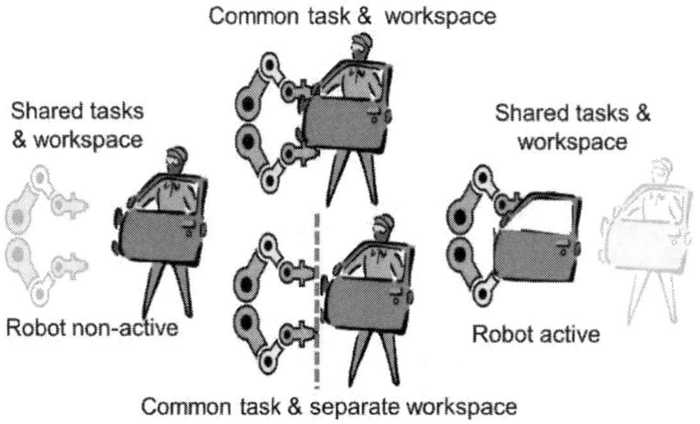

Fig. 1. Taxonomy of Human-Robot collaborative tasks and workspaces.

The human operators and robots are able to perform either single or cooperative tasks. The robot and the human operator could have a common task and workspace, a shared task and workspace, or a common task and a separate workspace (Fig. 1) [6].

The collaborative robot is also usable in control and tuning applications where its use is advantageous because it eliminates the potential failure of human factor. The robot is able to position itself in harder places, but at the same time it can achieve better accuracy even at higher speeds of movement. It is very advantageous to point out the fact that man can be exempted from monotonous activities and can devote himself to more valuable tasks and thought processes. The advantage of collaborative, also called cooperating workplaces is the fact that they combine the advantages of human's manual work and a fully automated workplace with an industrial robot [7]. The aim of the contribution is to verify the feasibility of this collaborative-collaboration based on the simulation of the conceptual design of the human-robot assembly workplace.

2 Objectives

The work deals with the idea of the assembly process, in which human will cooperate with a collaborative robot. The assembly process in the proposed collaborative workplace is simulated in the Siemens Tecnomatix Process Simulate software. An industrial robot performs predetermined tasks throughout the manufacturing process. The robot is capable of handling with adequate accuracy with hot, heavy but also sharp objects of manipulation. Compared to humans, however, it is not flexible in tasks in more complicated assembly processes. Therefore, if a human and a robot cooperate, it can also perform more complex tasks that would not be possible without their cooperation. Care must be taken to ensure its safety and health in operations where the working robot is near a human. These new safety standards have been defined by the American Association RIA - Robotic Industries Association in 2013 and have been approved by the US National Institute, designated ANSI / RIA R15.06-2012. This standard has been harmonized with the International Standard for Safety of Robots ISO 10218 [8]. Among the leading manufacturers of collaborative robots are the following companies: Universal Robots (UR3, UR5 and UR10), Rethink Robotics (Baxter, Sawyer), KUKA (LBR iiwa), ABB (Yumi, Roberta), F&P Robotics (P-Rob), Fanuc (CR-35i).

3 Methods

The research method is focused on the evaluation of idea design of a collaborative assembly workplace based on the simulation. Simulation allows to come to analyse the process itself before putting the real production system into operation and debug any errors or collisions [9-11]. The basis of the designed assembly workplace is the human – industrial robot cooperation. It was considered the subject of assembly in which we could connect the benefits of human work (such as the ability to manipulate in assembly) with the benefits of robot work (e.g. use of precision and repeatability). The simulation analysis will verify or our claim about the effectiveness of the collaboration is true or not. The first stage of the design was to determine the subject of the assembly.

3.1 Design of a collaborative assembly workplace

The first stage was mainly in the plane of thought processes, inspiration was sought in model-like, assembled workplaces. The second stage included the transfer of design from the idea plane to the software modelling environment and to give it the appropriate dimensions and shapes. Here it was adapted or modified the ideological proposals (retaining the essence of the idea proposal). It was achieved that the conceptual design of the cooperation workplace would be the most feasible.

The subject of the assembly is the engine block on which the robot applies the sealant, and then the man puts the lid on the engine block, assembles and screws the screws of the motor cover. The main idea of the assembly workplace is the industrial robot cooperated with the human operator. It is used the cooperative industrial robot from Universal Robots Company, the type UR5, Fig. 2. It was considering compliance with safety standard 10218, which specifies the rules of the collaborative workplace.

Fig. 2. Designed collaborative assembly workplace with robot UR5.

3.2 The simulation of the assembly process with human – the industrial robot collaboration

In this part of the paper is briefly described the creation of the simulation sequence. The creation of the simulation was actually a composition into a coherent time sequence with the logical continuity of the activities of the already created primary operations of individual elements and humans. The simulation has a duration of 150 seconds, starting with the movement of the carrier plate with the engine block from the starting point to the conveyor and ending at the end point at the end of the conveyor by assembling the entire engine block, the lid and the screw seal. The simulation course was divided into nine basic steps:

1. A delivery of the carrier plate with a motor block on the conveyor,
2. Carrier plate with engine block stopped by a stopper, set and clamped by a clamp, while man is coming to his workplace and the robot moves into the preparatory position, Fig. 3a.
3. The industrial robot is applying a sealant (Loctite 510), at the same time the human operator has taken over and holds the lid and controls the activity performed by the robot. Fig. 3b.
4. The robot UR 5 leaves the workspace after completing his activities, and when he leaves the working position, the human operator places the lid on the engine block (we've got enough screws with screws to prevent the screws from falling out of the lid), Fig. 3c.
5. The robot has reached the Home position, at the same time the human operator has taken the screwdriver and performs the action of screw-drivers on the motor lid, Fig.3d.
6. The human has complete the screwing operation and he has put the screwdriver on the defined place of storage on the side of the assembly desk.
7. The human moves to the designated point where he check a conveyance and at the same time is given the instruction for opening the support plate clamp with the assembled motor block.
8. The leaving of the carrier plate with a motor block on the conveyor and the preparing for the next operation.
9. Repeat the assembly process.

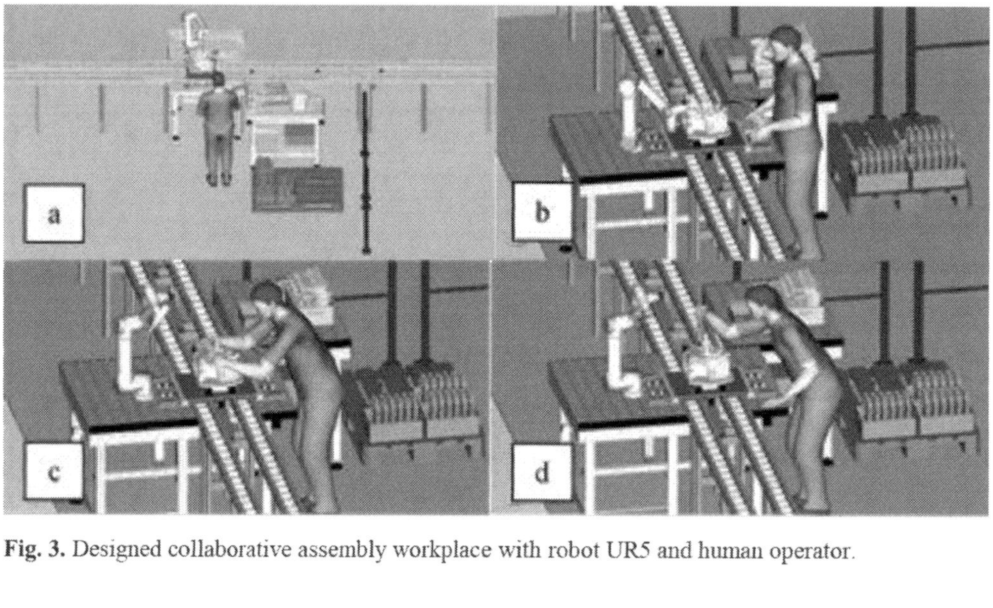

Fig. 3. Designed collaborative assembly workplace with robot UR5 and human operator.

4 Results from the simulation

The design of the assembly workplace was based on the idea of a hand-assembled workstation with a low level of automation. We have three goals in the proposal:

1. Human being the manipulation of less effort to accuracy, but also complicated with regard to the shape of the lid (we are considering several of its types).
2. Remove the human activities straining for its precision and concentricity (putting a seal around the perimeter of the engine block) and industrial cooperate robot being work (UR 5).
3. To preserve the original workspace (space efficiency, costs) while at the same time achieving the safety of the workplace in which these two members will work on a common subject.

The final point is obvious and necessary at the same time, the use of an industrial robot designed to work with a human. The workplace is not protected by fencing, optical gates or other systems. Human safety is only maintained by using a UR 5 robot that meets the conditions of collaboration.

The purpose of the simulation was to verify and evaluate the proposed dimensions of the workplace elements, its layout solution, With respect to the positioning of the robot with the effector and the necessary human availability in the assembly positions while maintaining ergonomic conditions. Based on observations during the simulation, it would like to give us found the design in which we in the subsequent correction made changes: changing the method of fixing and clamping the engine block onto the carrier plate, changing the method of clamping the support plate during assembly or to find and choose a faster way of mounting and clamping the support plate during assembly, by observing the movements and positions of the human operator when he screwing, it would be convenient to place the engine block at a lower height, Use the battery-powered screwdriver (or pneumatic actuator) on the balancer to reduce screwing stress, The end effector for the application of the seal is designed so as to meet the safety of the collaborative workplace (its surface is coated with softened PVC, like the robot surfacc).

The following figure, Fig. 4, it can be observed the robot and human workspace. It was verified that the robot does not interfere with a zone where a person is waiting to fit the lid on the engine block.

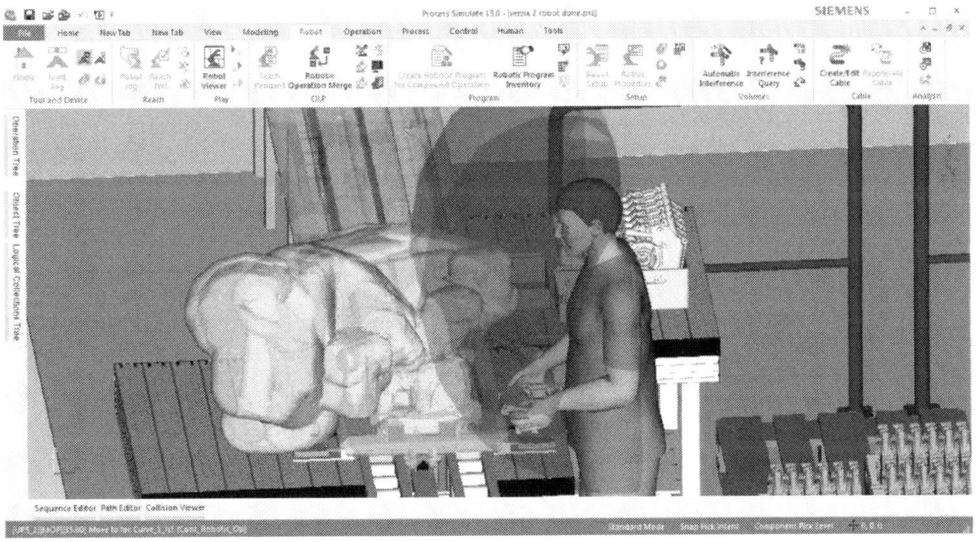

Fig. 4. The robot and human workspace, the Swept Volume function in Process Simulate.

5 Conclusion

Nowadays it can be seen growing interest and also the increasing activity in the field of innovation and innovative approach in different directions their competence, especially in Industrial Revolution in Automation - Industry 4.0. The human - robot cooperation definitely belong to this sphere. The aim was to create a simulation of the assembly process in the proposed workplace to verify the possibility of human interaction with the robot while preserving the required security. Followed it was analysed and evaluated the idea design of our workplace, it was mentioned the areas in which should be modified it and were made the additional corrections. The main contribution of this paper is the fact - verified by the simulation that collaboration of a human with a collaborative robot in the common space is feasible.

This research was supported by the STU Grant "Scheme for Support of Excellent Teams of Young Researchers – (1349 - Use of simulation methods in the context of the Virtual Commissioning as a tool for production systems in the Digital Factory).

References

1. J. Teiwes, T. Bänziger, A. Kunz, K. Wegener, ICAC (2016)
2. D. Antonelli, S. Astanin, Proc. CIRP, **41**, 352 (2016)
3. D. Schröter, P. Jaschewski, B. Kuhrke, A. Verl, Proc. CIRP, **55**, 2016, 12 (2016)
4. Å. Fast-Berglund, F. Palmkvist, P. Nyqvist, S. Ekered, M. Åkerman, Proc. CIRP, **44**, 175 (2016)
5. H. Cheng, W. Xu, Q. Ai, Q. Liu, Z. Zhou, D. Truong Pham, Proc. Man., **10**, 26 (2017)
6. G. Michalos, S. Makris, P. Tsarouchi, T. Guasch, D. Kontovrakis, G. Chryssolouris, , Proc. CIRP, **37**, 248 (2015)
7. P. Tsarouchi, A.-S. Matthaiakis, S. Makris, G. Chryssolouris, Int. Jour. of Comp. Int. Manuf., **30**, 6 (2017)
8. Ch. Vavra, Control Engineering Cz, **11**(4-83), 12 (2016)
9. K. Velíšek, R. Holubek, D. R. Delgado Sobrino, R. Ružarovský, N. Vetríková, MATEC Web of Conf. (2017)
10. K. Baizid, A. Meddahi, A. Yousnadj, S. Ćuković, R. Chellali, MATEC Web of Conf. (2016)
11. Z. Tuma, L. Kotek, J. Tuma, F. Bradac, MM Sci. Jour., **1**, 1, 1487 (2016)

Advanced engineering design capabilities applied for developing a technological device for automated assembly

Ionut Madalin Pista[1*], *Vasile* Merticaru, *Gheorghe* Nagit[1], and *Marius Ionut* Ripanu[1]

[1]"Gheorghe Asachi" Technical University of Iasi, Department of Machine Manufacturing Technology, Prof. D. Mangeron Street, no. 39A, Iasi, Romania

Abstract. The theoretical approach and the related case study presented in the paper are part of a larger research regarding the optimization of assembled products and assembly processes by enhancing the effectiveness and efficiency of design activities, including by embedment of DfX/DFMA principles. First of all, a conceptual model for the theoretical approach is proposed and discussed. The case study refers to the engineering design of an indexing mounting table for automated assembly of an electrical plug inlet. Some advanced principles and theories are discussed in relation to the technical solution design. Advanced engineering design capabilities provided by Solid Edge have been used for solving the 3D modelling task and for the other detailed design activities. Finally, some related conclusions are included, together with some directions of further researches.

1 Introduction. Problem statement

Within the NPD (New Product Development) process in general, an appropriated detailed establishment of product requirements and their translation into product specifications are of vital importance for aligning the best design solution to the client needs, [1]. One could consider that those product design activities cover mainly the Concept Design phase and System-Level Design phase from the design process, [2].

An advanced design philosophy, which we consider useful to be applied in the Concept Design phase and being able to facilitate the selection of the most adequate product concept solution among various others is the Axiomatic Design (AD), developed by Suh, [3].

For the System-Level Design phase, there can be considered as appropriated the involvement of DfX (Design for Excellence) design philosophy, [4], mainly DFMA (Design for Manufacturing and Assembly), [5], being important for an optimized development of assembled products and assembly processes.

The theoretical approach and the related case study presented in the paper are part of a larger research regarding the optimization of assembled products and assembly processes by enhancing the effectiveness and efficiency of design activities, including by embedment of some Axiomatic Design and DfX/DFMA principles.

* Corresponding author: ionutpista@yahoo.com

2 Conceptual model of the research approach

First of all, a conceptual model for the theoretical approach is proposed, as in Figure 1.

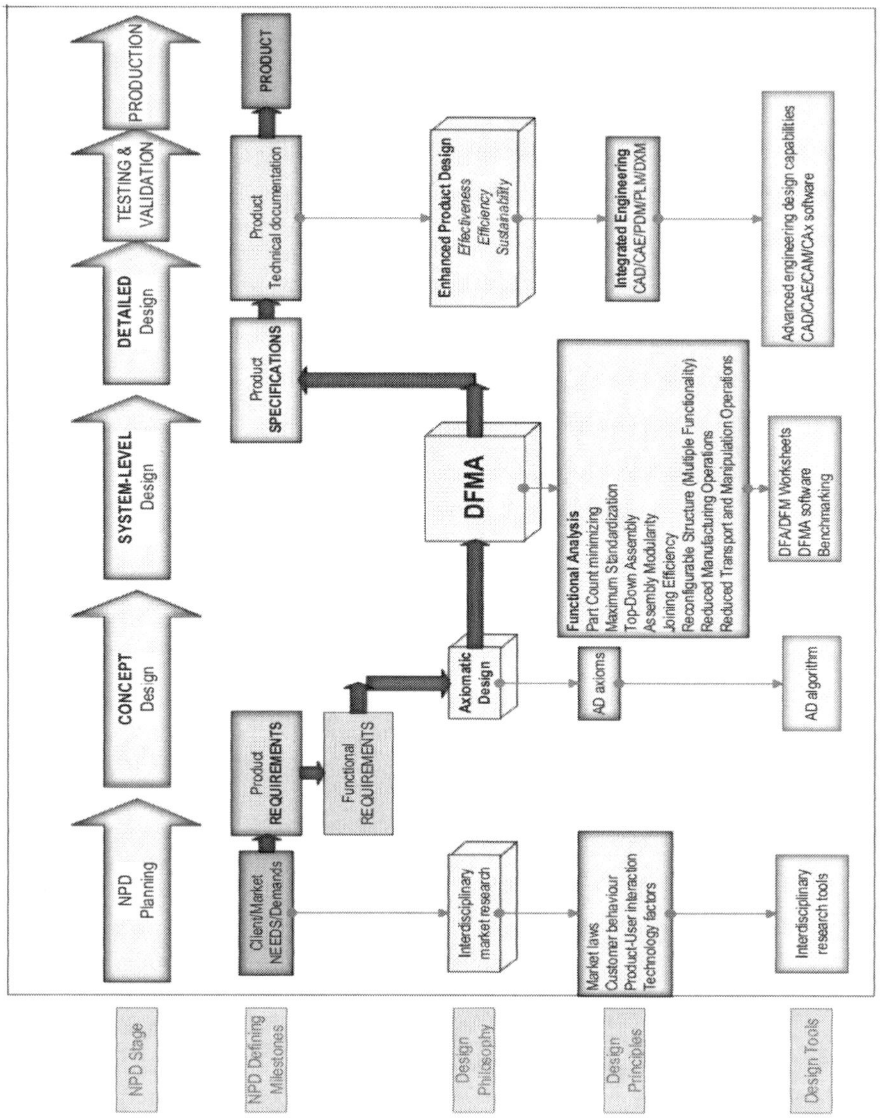

Fig. 1. Conceptual model of the research approach.

Within the model, an exhaustive interdisciplinary market research is recommended for defining the client/market needs/demands, which are translated into product requirements. Among those, the functional requirements have been mainly considered within the research approach, as input for the AD study, where the two AD axioms (axiom of *independence* and axiom of *information*) have been considered as design principles. The four domains of the theory are proposed to be covered within the AD study, for sequentially establishing the customer's requests, the functional requirements, the design structural parameters and respectively the process variables. Based on the recommendations of Stienstra, [6], the following principles are considered for the DFMA study: Functional Analysis; Part Count minimizing; Maximum Standardization; Top-Down Assembly; Assembly Modularity;

Joining Efficiency; Reconfigurable Structure (Multiple Functionality); Reduced Manufacturing Operations; Reduced Transport and Manipulation Operations.

For the Detailed Design phase, within the philosophy of Enhanced Product Design by effectiveness, efficiency and sustainability, the principles of Integrated Engineering, respectively those of CAD/CAE/PDM/PLM/DXM are considered [7,8], where: CAD – *Computer Aided Design*, CAE – *Computer Aided Engineering*, PDM – *Product Data Management*, PLM – *Product Lifecycle Management*, DXM – *Data Exchange Management*.

3 Case study

The case study refers to the engineering design of an indexing mounting table for automated assembly of an electrical plug inlet. For applying the algorithm and the principles highlighted in the previously presented conceptual model, the considered assembled product has been first of all analysed.

3.1 Assembled product description

The assembled product subjected to study, respectively an industrial electrical plug inlet, is shown in Figure 2, where the 3D view, the exploded view and the assembly scheme for it are presented.

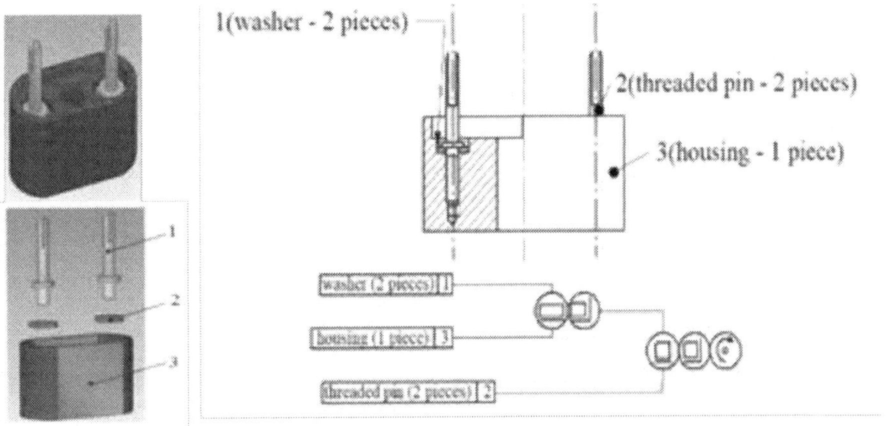

Fig. 2. 3D view, exploded view and assembly scheme for the studied assembled product.

3.2 Concept design for the indexing mounting table

Within the Axiomatic Design study, the following customer needs have been identified:
CN1: the device could be achieved within a laboratory for automated assembly, or within an assembling workshop from an electrical equipment factory;
CN2: constructive simplicity and component availability, in conditions of maximal productivity.

Further on, the main functional requirement has been established, as following:
FR0: design a technological device for automated assembly of an electrical plug inlet, adaptable for a laboratory for automated assembly, or for an assembling workshop from an electrical equipment factory, in conditions of cost effective manufacturing and assembly.

A set of second order functional requirements has been also established and those have been decomposed in third order functional requirements. In relation to the above mentioned

DOI: 10.1051/matecconf/201713704006

functional requirements, a set of design parameters were identified, corresponding to the constructive structure of the designed indexing mounting table, as shown in Figure 3.

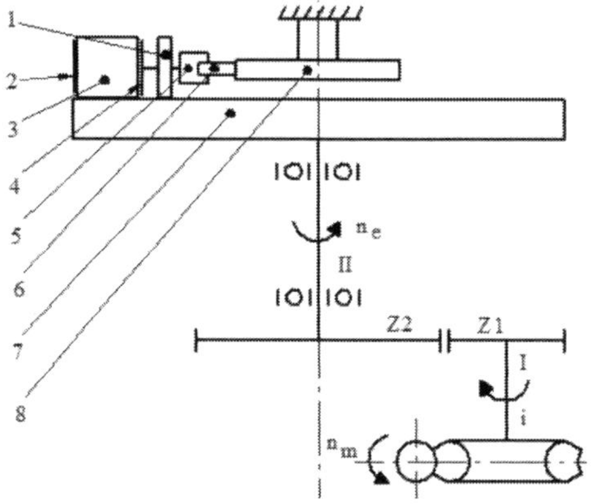

Fig. 3. Conceptual design scheme for the indexing mounting table. 1 – guiding element; 2 – orienting prism; 3 – plug inlet; 4 – fixing jaw; 5 – follower rod; 6 – roller; 7 – rotating plate; 8 – cam.

Assembly Name: Orienting-Fixing Device

Parts

	ID Number	Quantity	Part Description	Theoretical Min. Part	Practical Min. Parts	Not Standard or non shared part
1	1		Orienting subassembly			
2	1,1	1	Prism	Y	Y	N
3	1,2	2	Screw	N	N	Y
4	2		Follower			
5	2,1	1	Follower rod	Y	Y	N
6	2,2	1	Roller	Y	Y	N
7	2,3	2	Washer	N	N	N
8	2,4	1	Screw	N	N	Y
9	2,5	1	Nut	N	N	Y
10	3		Guiding subassembly			
11	3,1	1	Guiding element	Y	Y	N
12	3,2	2	Screw	N	N	Y
13	4	1	Fixing jaw	Y	Y	N
14	5	1	Spring 1	N	Y	Y
15	6	1	Spring 2	N	Y	Y
16	7	2	Screw	N	N	Y
17	8	2	Spring fixing plate	N	N	N
18						
19						
Total Parts		16	Totals	5	7	7
Sub-Assemblies		3	Categories T			

Assembly Name: MountingTableAssembly

Parts

	ID Number	Quantity	Part Description	Theoretical Min. Part	Not Standard or non shared part
	1	1	Base plate	Y	N
	2	1	Bottom subassembly		
	2,1	1	Bottom housing	Y	N
	2,2	1	Electrical motor	Y	Y
	2,3	1	Worm gearbox	Y	Y
	2,4	1	Flexible coupling	Y	Y
	2,5	1	Shaft	Y	N
	2,6	1	Pinion	Y	Y
	2,7	1	Angular contact ball bearing	Y	Y
	3	5	Screw	N	Y
	4	1	Upper subassemmbly		
	4,1	1	Upper housing	Y	N
	4,2	1	Coroana de sustinere	N	N
	4,3	5	Screw	N	Y
	4,4	1	Angular contact ball bearing	Y	Y
	4,5	1	Radial ball bearing	Y	Y
	4,6	1	Thrust ball bearing	Y	Y
	5	5	Screw	N	Y
	6	1	Rotary table		N
	6,1	1	Tubular Shaft	Y	N
	6,2	1	Bearing bushing	N	N
	6,3	1	Rotary plate	N	N
	6,4	5	Screw	N	Y
	6,5	1	Prism	Y	N
	6,6	2	Screw	N	Y
	6,7	1	Follower rod	Y	N
	6,8	1	Roller	Y	N
	6,9	2	Washer	N	N
	6,1	1	Screw	N	Y
	6,11	1	Nut	N	Y
	6,12	1	Guiding element	Y	N
	6,13	2	Screw	N	Y
	6,14	1	Fixing jaw	Y	N
	6,15	1	Spring 1	N	Y
	6,16	1	Spring 2	N	Y
	6,17	2	Screw	N	Y
	6,18	2	Spring fixing plate	N	N
	6,19	1	Radial ball bearing	Y	Y
	6,2	1	Roata dintata condusa	Y	Y
	7	1	Cam subassembly		
	7,1	1	Cam	Y	N
	7,2	1	Cam shaft	N	N
	7,3	1	Grower Washer	N	Y
	7,4	1	Screw	N	Y
	8	1	Grower Washer	N	Y
	9	1	Nut	N	Y
Total Parts		63	Totals	21	25
Sub-Assemblies		4	Categories T		

Fig. 4. DFA worksheets (partial view) for the Mounting Table assembly and for the Orienting-Fixing subassembly.

3.3 System level design for the indexing mounting table

By the embedment of DfX/DFMA principles within the research approach case study, a set of DFMA worksheets, [9], have been completed for the Mounting Table assembly and for the Orienting-Fixing subassembly.

3.4 Detailed design for the indexing mounting table. Results and discussion

Advanced engineering design capabilities provided by Solid Edge, [10], have been used for solving the 3D modelling task and for the other detailed design activities, [11].

Within the detailed design activities, the design model structuring has been considered as very important, so that the decomposition rules based on functional requirements above mentioned have been mainly targeted to be applied. For providing the Enhanced Product Design effectiveness, efficiency and sustainability, the assembling technology and its importance upon the product's functional qualities has been considered, together with the grouping of components in subassemblies, based on the functional role of the components, for optimizing the general functioning conditions. In developing the CAD models, the constraints imposed by components positioning in the assembly and the mobility degrees of the components in assembly have been also considered. A 3D view and components' identification for the Orienting-Fixing subassembly and its multiplied disposal on the mounting table (Solid Edge 3D parameterized model) are shown in Figure 5.

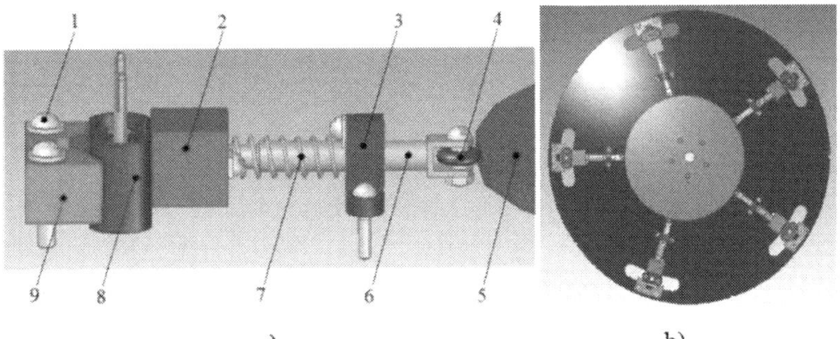

Fig. 5. Solid Edge 3D parameterized models. a) Orienting-fixing subassembly. b) Multiplied disposal of the orienting-fixing subassembly on the mounting table. 1 – screw; 2 – fixing jaw; 3 – guiding element; 4 – roller; 5 – cam; 6 – follower rod; 7 – spring; 8 – plug inlet; 9 – orienting prism.

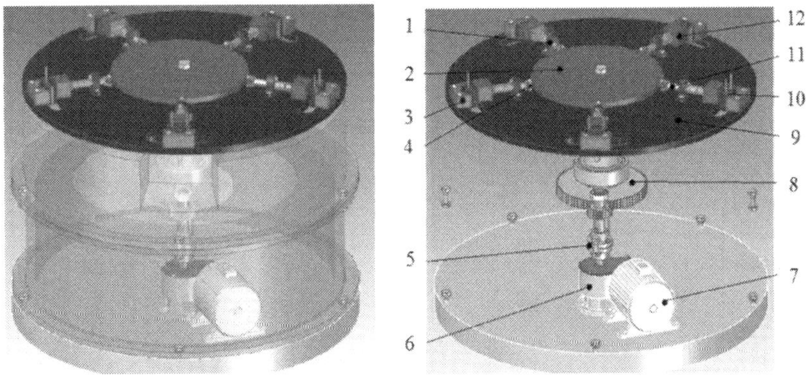

Fig. 6. Assembly 3D view and components' identification for the indexing mounting table (Solid Edge 3D parameterized model). 1 – guiding element; 2 – fixed cam; 3 – orienting prism; 4 – roller; 5 – flexible coupling; 6 – worm gearbox; 7 – electrical motor; 8 – spur gear; 9 – rotary plate; 10 – plug inlet; 11 - follower rod; 12 - fixing jaw.

The advanced capabilities of the product model dynamic generation and those of its embedded module entitled Engineering Reference, as extremely valuable CAE instrument, have been also used and referred within the presented case study.

An assembly 3D view and components' identification for the indexing mounting table (Solid Edge 3D parameterized model) are shown in Figure 6.

4 Conclusions

The proposed conceptual model for enhancing the effectiveness and efficiency of design activities, by embedment of Axiomatic Design and DfX/DFMA principles is consider as being valuable for defining the client or market needs or demands, which will be translated into product requirements and further on into product specifications.

We can further on conclude here that integrated CAD/CAE/PDM/PLM/DXM software tools like Solid Edge and Velocity Series from Siemens PLM provide enhanced capabilities to develop a product parameterized model characterized by flexibility and easiness in delivering of technical data for new or for modified product versions.

As long as the theoretical approach and the related case study presented in the paper are part of a larger research regarding the optimization of assembled products and assembly processes, we can identify as future research directions the approaching on some other aspects of DfX philosophy and principles within Product/Process Development and also on the Holistic Design philosophy and principles.

References

1. V. Merticaru, A.C. Paraschiv, M.I. Ripanu, MATEC Web of Conferences **112**, 03008 (2017)
2. M. Kressy, *Industrial Design* (Rhode Island School of Design, Designturn Inc., 2002)
3. N.P. Suh, *Axiomatic Design: Advances and Applications* (Oxford University Press, New York, 2001)
4. G.Q. Huang, *Design for X - First edition* (Springer Science, 1996)
5. G. Boothroyd, P. Dewhurst, W.A. Knight, *Product Design for Manufacture and Assembly* (Marcel Dekker, New York, 1994)
6. D. Stienstra, *Introduction to Design for (Cost Effective) Assembly and Manufacturing* (Georgia Institute of Technology, USA)
7. V. Merticaru, M.I. Ripanu, M.A. Mihalache, M.M. Cucos, Applied Mechanics and Materials **809-810**, 1492-1497 (2015)
8. V. Merticaru, M.I. Ripanu, Applied Mechanics and Materials **371**, 499-503 (2013)
9. *** Templatestaff, *DFA DFM Analysis Worksheet for Microsoft Excel*, (2013) http://templatestaff.com/dfa-dfm-analysis/
10. *** SIEMENS, *SolidEdge* (2017)
11. I.M. Pista, *Bachelor Thesis* (TUIASI, 2015)

DOI: 10.1051/matecconf/201713704007

Allowance treatment static designed couple and repeatable precision in assembly

Štefan Václav[1], *Jozef* Jurko, *PhD.*[2] and *Šimon* Lecký[1,*]

[1] Slovak University of Technology, Faculty of materials science and technology in Trnava, Institute of production technologies, Slovakia
[2] The Technical University of Košice Faculty of Manufacturing Technologies with a seat in Prešov, Slovakia

Abstract. In practice, we deal with static designed couples frequently. Construction engineers use them so they can upscale rigidity of product. In paper are described principles how to change after effects of problematic assembly in often used feed couples which use to be statically designed. Important problem in assembly is repeatable precision, mainly in construction of assembly lines. By imprecision of robot which inserting shank into the hole and by imprecision of position of hole in carrier shank won't be successfully inserted. In article are shown reasons and security approach of repeatable precision in assembly operations executing.

1 Introduction

Nowadays it is unlikely that any producer or manufacturer will produce their products without any assembly process. The more usual case is that a producer will have several components manufactured by themselves or suppliers and then assembled together [1]. The product designer is normally limited only by a few critical dimensions and has much scope in the design of individual components and assemblies [2]. For example, challenging precision boning is the interdependency between alignment and boning [3]. There are many good and bad examples, in the works of DFA. The good ones are statically determinate, the bad ones are over determinate [4]. In practice, the cases of statically predetermined couples are often encountered. The designers like using them to enhance rigidity of the product. A typical example is a ball bearing, which should theoretically have the maximum of three balls, but it has more of them in order to increase their toughness.

Figure 1 illustrates a hydraulic press that is statically predetermined. When the support moves on two cylinders, the couple is clearly statically predetermined, as shown in the Grübler - Valentovič equation, according to which the number of degrees of freedom is minus two.

There are more conditions for trouble-free assembly regarding the fact that the couples are not parallel but concurrent or even skewed. In such circumstances, the couple must have

* Corresponding author : simon.lecky@stuba.sk

gaps enabling smooth assembly even despite non-parallelism of the couple, which in turn partially decreases toughness.

It is often necessary to bond two elements (Figure 1) which apparently form a statically predetermined kinematic couple.

This concerns also a flanged bond, e.g. the bond of white and black flanges with four pins (Figure 2a). It is desirable to label both parts in order to enable assembly in any case (trouble-free assembly).

The above-mentioned suggests that there always should be pins inside the holes (Figure 2d). This can only be achieved by labelling the pins (Figure 2, pos. 2) and holes (Figure 2, pos. 1) separately, each from the same base. The same will be done with the trusses (Figure 2b). The reason for this method of labelling is due to the fact that manufacturing tolerances are not added up.

The depicted couples are very strongly statically predetermined structures, where the expected number of degrees of freedom equals one, in order to enable feed.

The formula in Figure 1c suggests that the number of degrees of freedom "is = -10", implying that the system is statically prearranged 9 times. The right labelling can assure trouble-free assembly.

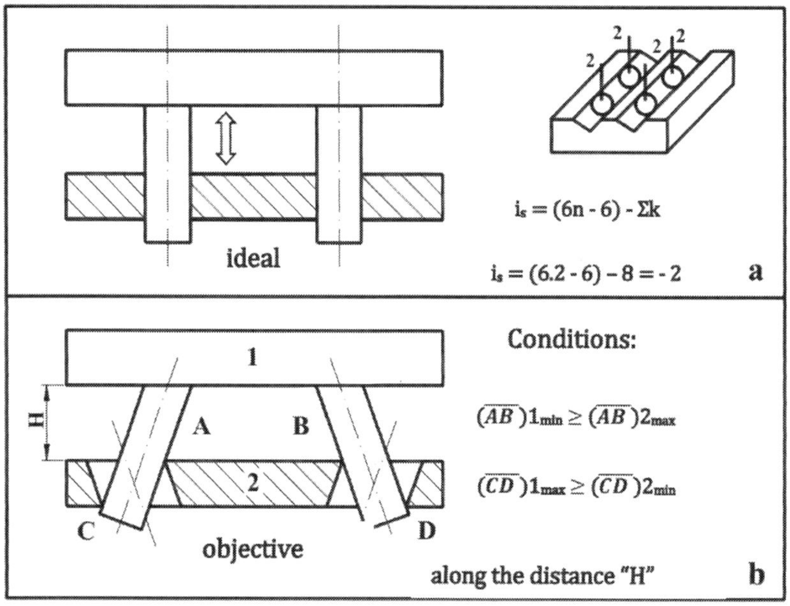

Fig. 1. Toleration treatment of statically redetermined movable couples.

This concerns also a flanged bond, e.g. the bond of white and black flanges with four pins (Figure 2a). It is desirable to label both parts in order to enable assembly in any case (trouble-free assembly).

The above-mentioned suggests that there always should be pins inside the holes (Figure 2d). This can only be achieved by labelling the pins (Figure 2, pos. 2) and holes (Figure 2, pos. 1) separately, each from the same base. The same will be done with the trusses (Figure 2b). The reason for this method of labelling is due to the fact that manufacturing tolerances are not added up.

The depicted couples are very strongly statically prearranged structures, where the expected number of degrees of freedom equals one, in order to enable feed.

The formula in Figure 2c suggests that the number of degrees of freedom "i_s = -10", implying that the system is statically prearranged 9 times. The right labelling can assure trouble-free assembly.

Such treatment enables smooth assembly of the components manufactured in large series.

Figure 3 shows a different method of mitigating the consequences of troublesome assembly of very frequent sliding couples which are generally statically prearranged. The couple depicted in Figure 3a is evidently statically prearranged; the consequences may be mitigated when first assembling the sliding workbench with rods and then connecting them to the base (Figure 3b).

Figure 3c is the best illustration of trouble-free assembly: there are two rotary-sliding couples (four-point) on one shaft, and a one-point couple on the other one (touches the left or the right side). Such a structure is statically prearranged but defined, which may assure completely trouble-free assembly.

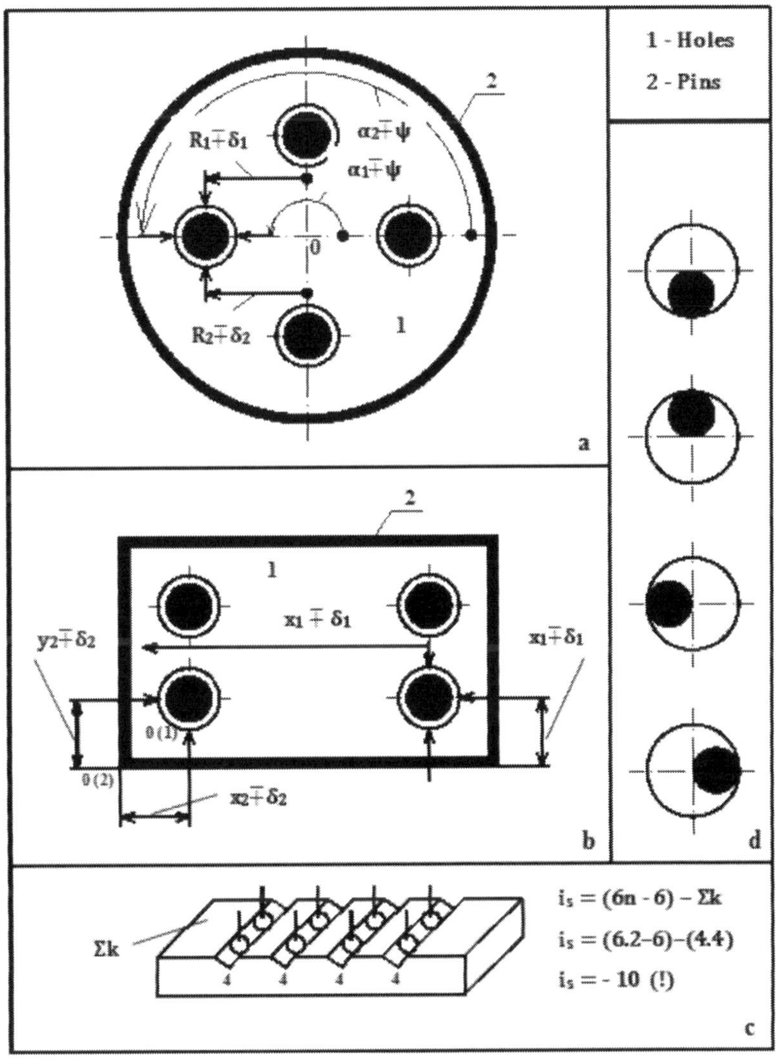

Fig. 2. Assembly ability of statically redetermined unmovable couples: a, b – dimension method, c – number of degrees of freedom, d – assemblability check.

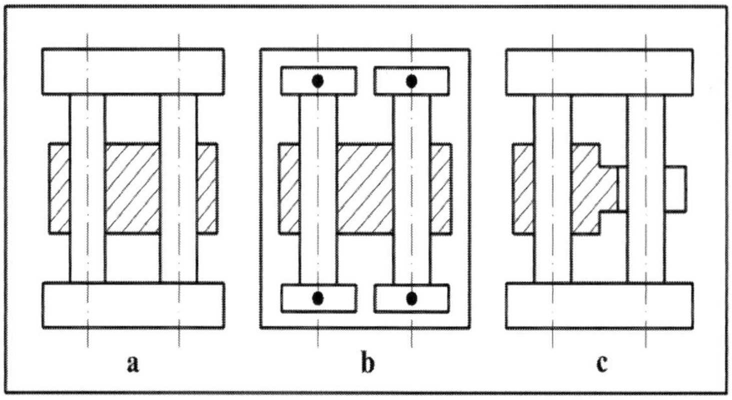

Fig. 3. Assembly ability of movable couples: a – statically prearranged structure, b – gradual assembly of statically prearranged structure, c – statically determined structure (trouble-free assembly).

If high accuracy in the product is not needed, so called extenders of gaps can be used. The principle is shown in Figure 4.

2 Repeated precision of assembly

A very important issue in assembly is repeated accuracy, especially when constructing assembly lines. Inaccuracy of the robot inserting the pin into the hole may cause the inaccuracy of the hole location in the gripper, which may prevent a successful insertion of the pin into the hole. This may be due to the reasons described below [4].

The first reason for the failure of assembly is that a robot collecting pins from the palette has a certain repetitive accuracy (uncertainty) of positioning, i.e. when inserting the pins, their spikes are not positioned to the same point (position), but make a circle of eccentricity "Ek" (Figure 4a).

Another reason for the failure of assembly is that the center of the hole is not in the same place with respect to the frame of the machine, but may be in a different place in each clamp, which is due to inaccuracy of the gripper and clamp, and sometimes also due to inaccurate contours of the part in the clamp (Figure 4a).

The issue poses a problem since the two causes may be combined (inaccuracies may increase by mutual summation) and it can be easily proved (Figure 4b) that to insert a pin by the radius R_{kmax} into the hole, the radius of the hole must be:

$$R_{dmin} = E_d + E_k + R_{kmax} \text{ [mm]} \quad (1)$$

where:

R_{dmin} – minimum radius of opening [mm],
E_d – eccentricity of opening [mm],
E_k – eccentricity of pin [mm],
R_{kmax} – maximum radius of pin [mm].

From experience it is known that such assembly requires a disproportionately large gap between the pin and the hole. This can be eliminated by using the maximum chamfer of both components (Figure 4c) and then trying to insert the pin into the hole by utilizing flexibility.

Another option is a little more complicated: the pin is placed next to the hole and then oscillatory motions are performed until success is achieved, under some pressure, to capture the pin into the hole (Figure 4d).

DOI: 10.1051/matecconf/201713704007

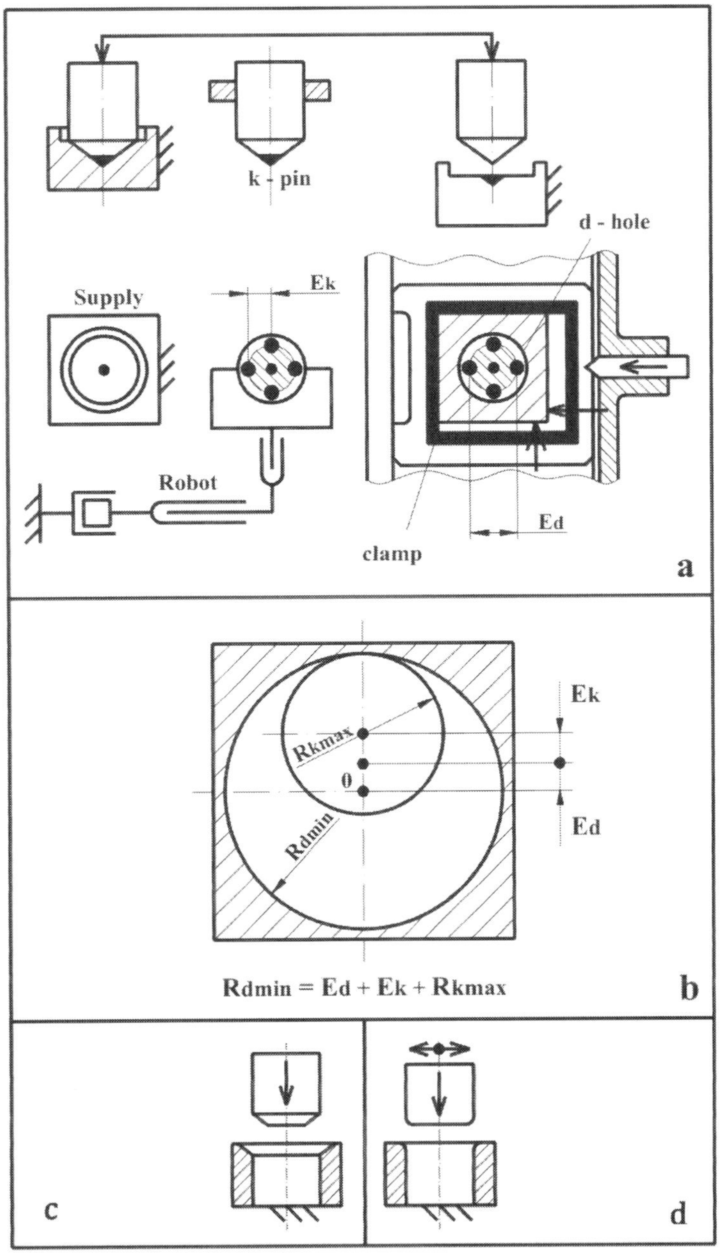

Fig. 4. Repeated accuracy of the robot and clamp increases the probability of the insertion of pin into hole, a – robot and clamp, b – conditions of insertion, c – chamfers improve probability of insertion, d – vibration improves probability of insertion.

3 Conclusion

The above-mentioned suggests that the shorter and of better quality the line, the lower the values of E_k a E_d. As illustrated in Figure 4b, assembly will be trouble-free under the following condition:

$$R_{dmin} = E_d + E_k + R_{kmax} \text{ [mm]} \quad (2)$$

or for the diameters:

$$\frac{\emptyset dmin}{2} = E_d + E_k + \frac{\emptyset kmax}{2}$$

$$\emptyset d_{min} = 2E_d + 2E_k + \emptyset_{kmax}\ [mm] \qquad (3)$$

After supplying particular values, e.g. $\emptyset_{kmax} = 40.1$ mm, $E_k = 0.2$ mm, $E_d = 0.1$mm:
$\emptyset_{dmin} = 0.2 + 0.4 + 40.1 = 40.7$ [mm]

Such a large gap (0.7 mm) cannot be accepted; the diameter of the hole will be therefore reduced. From experience, it is known that the gaps between pin and hole will be disproportionately large in such assembly.

The gap can be eliminated by using the maximum chamfer of both parts (Figure 4c) and then applying flexibility to insert the pin into the hole.

Another option is a little more complicated: the pin is placed next to the hole and then oscillatory motions are performed until success is achieved, under some pressure, to capture the pin into the hole (Figure 4d).

The machine to carry out such operations must be fitted with artificial sight and feeling. Such systems are very complex.

This contribution brings new insights into the tolerance treatment of statically assigned pairs for trouble-free mounting.

Another benefit of the article is the discovery of causes and the way to ensure the required repeat accuracy in carrying out assembly operations.

This article has been developed as part of the project: VEGA 1/0477/14 Examining the influence of selected machining process characteristics using HI-machining technology on the resulting quality of machined surfaces and seamless assembly.

References

1. H. A. Crostack, R. Refflinghaus, J. Liangsiri, *Defining Tolerances in Assembly Process with the Aid of Simulation*. In: S. Ratchev (eds) Precision Assembly Technologies and Systems. IFIP Advances in Information and Communication Technology **315**. Springer, Berlin, Heidelberg (2010)
2. C.S. Syan, K. G. Swift, *Design for manufacture*. In: C.S. Syan, U. Menon (eds) Concurrent Engineering. Springer, Dordrecht (1994)
3. T. Müller et al., *Robust Adhesive Precision Bonding in Automated Assembly Cells*. In: S. Ratchev (eds) Precision Assembly Technologies and Systems. IFIP Advances in Information and Communication Technology **435**. Springer, Berlin, Heidelberg (2014)
4. G. Boothroyd and P. Dewhurst, Product Design for Assembly, Wakenfield. USA (1987)
5. E. Valentovič. Assembly Automation **20** (3) 233 (2000)

Simulation as a support tool in assembly systems planning

Štefan Václav[1] , *Šimon* Lecký[1,*], *Katarína* Senderská[2] and *Albert* Mareš[2]

[1] Slovak University of Technology, Faculty of materials science and technology in Trnava, Institute of production technologies, Slovakia
[2] Technical University of Košice Faculty of Mechanical Engineering, Slovakia

Abstract. This paper deals with assumption that simulation is the only reliable method for manufacturing and assembly systems profiling. Simulation is essential software tool that improve design and planning of complex automated manufacturing and assembly systems. Systems with high level of complexity need to be tested even before they are constructed in real manufacturing plant. Elimination of errors in production before the first test production is initiated means huge savings in expenses and time. In this article are mentioned theoretical base of topic and example as a case study. For case study was used software Tecnomatix plant simulation from company SIEMENS.

1 Introduction

In the introduction is described the basic concept of modeling and simulation of product manufacturing and analysis of production processes. Below the model, is given a simplified picture of reality [1]. This term is used to designate a material or non-material imitation of an object, regardless of the purpose for which it is created [5].

Modeling is a theoretical and cognitive process that is based on abstracts, respectively. Ideally defined as recognizing objects through other objects that are mostly artificially created. Modeling imitates structures and behaviour of the real system [2]. Simulation of these systems is a specific form of cognitive process. The basic principle is to draw conclusions using experiments and simulations under a defined system with objects and their interactions. The basic classification of simulations can be seen in Figure 1 [4].

The simulation model can be characterized as a system that mimics the actual behavior of the simulated system. Simulation defines the artificial material objects that were created. An important element is that imitation of the system must maintain the same configuration changes over time. So, the sequence of events cannot be changed. It must be kept in simulations as it is in the original scheme [3].

"Every simulation and simulation program is now increasingly used in industrial enterprises. The pressure of competition and customer requirements are steadily increasing and therefore businesses need not to simulate production processes. Simulation programs are tools that test the effects of different model decisions and then evaluate production

* Corresponding author: simon.lecky@stuba.sk

DOI: 10.1051/matecconf/201713704008

capacities, duration of operations, and other production parameters. They also eliminate the possibility of introducing a change into a process that would not benefit the system" [5].

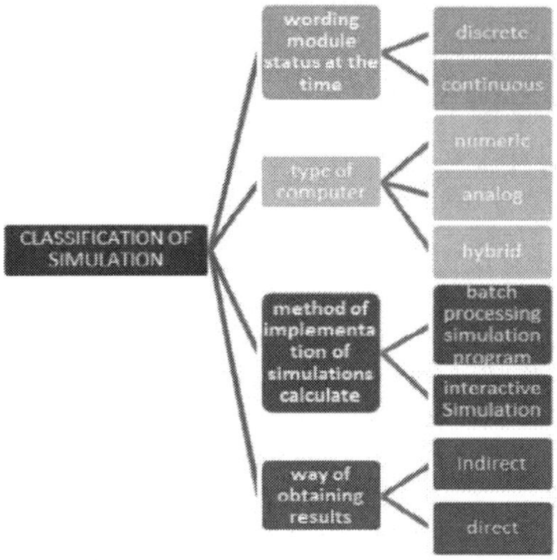

Fig. 1 Classification of simulation [12].

Simulation by Houzek R. is a numerical method that experiments with a special mathematical model [11]. Simulation is understood as a procedure, with the help of the process being studied. Respectively, its steps are generated over time, based on the parameters of the display system. The simulation model is a mathematical prescription to see these steps [10].

According to the authors of R. Debnar, J. Košturiak and I. Kurica, the main tasks of the simulation are:

• "Support for decision-making in system design and operation,
• Analysis and optimization,
• Predicting and Looking Forward,
• Replacement of the real system (training, testing, danger, etc.)" [9].

Simulation model means a mathematical technique that consists of the following elements:

• A set of logical and mathematical relationships that express the functional or other characteristics of the system,
• Include probable characteristics in the form of random effects on the model,
• Include the time for which changes are made to the model,
• There are also gradual calculations with different inputs that mimic real-time real-time behavior [8].

2 Simulation in assembly systems and workplaces

The simulation of workplaces and assembly systems helps in planning new systems or improving current systems (Figure 2). It is also often used in decision-making processes such as return of investment, cost of planned changes, project confirmation, production process analysis, and so on. Simulation in assembly systems facilitates the design phase and enhances the smoothness of project design and execution through its outputs. At present, simulations are up to 99% accurate. Of course, it depends on the input parameters and the validity of the model.

Benefits of simulation:
• Testing innovative strategies in risk-free virtual environments,
• Maximum use of production resources,
• Reducing investment risk through rapid simulation,
• Optimizing the size of systems and storage space,
• Rapid identification of sources of problems in logistics and production spheres,
• 20-60% reduction in inventory due to system size,
• 5 - 20% reduction in investment costs for the new system,
• Reduction of capacities for personnel and handling equipment,
• Quickly achieve positive results and identify impacts [7].

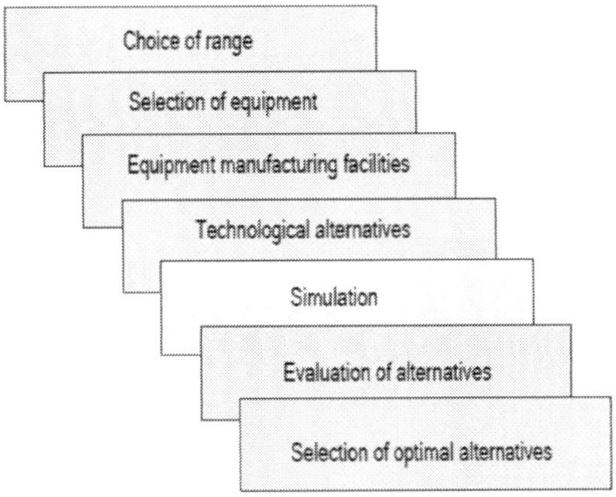

Fig. 2 Integration of simulation in designing of production cells [5].

Simulation of assembly sites is currently being used in robotic positions or in the area of cooperative stations. Similarly, software for simulating staff movements, examining workplace ergonomics, or testing the load on a worker's spine while working with loads is also widely used. Examples of such software can be Tecnomatix Process Simulate for simulation of robotic workplaces or Tecnomatix Jack for simulation of human factors and ergonomics.

3. Example of simulation and analysis in assembly system

This section discusses an example of a design and analysis of an assembly system consisting of three assembly workplaces, one control stand, one conveyor and one buffer (Figures 3 and 4). Simulation as a tool that helps analyse assembly systems allows assembly systems to improve without interruption of production. The Tecnomatix Plant Simulation software from SIEMENS was used to model and simulate the assembly system.

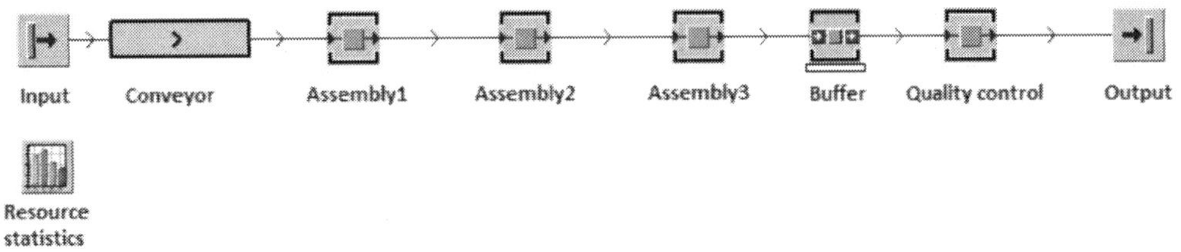

Fig. 3 Base model of assembly system.

DOI: 10.1051/matecconf/201713704008

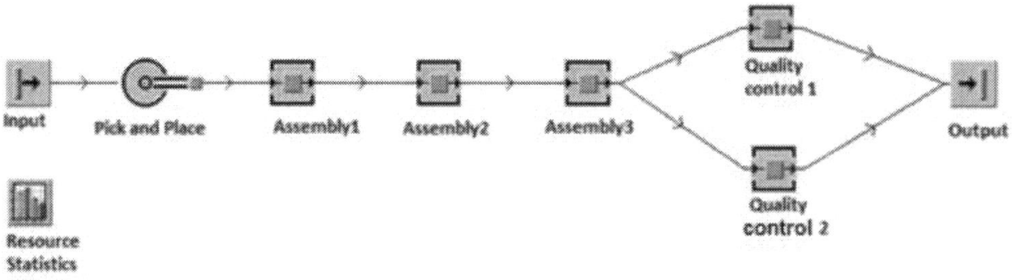

Fig. 4 Customized model of assembly system.

When analysing the assembly system, it was assumed based on the time pool of individual positions that the bulk will accumulate before the inspection position. After analyzing the model simulation results, it was found that 399 products were produced after eight hours of assembly time. Against the basic model where there were 149 products made after eight hours are 250 pieces more products. This means an increase in productivity by 267.78%.

Based on the source statistics, it is possible to see how the model of the assembly system differs before and after the adjustment, Figure 5. Mounting position number three is not blocked after the simulated model has been modified, and the addition of a further control opinion has reduced the control load and allowed for better system throughput.

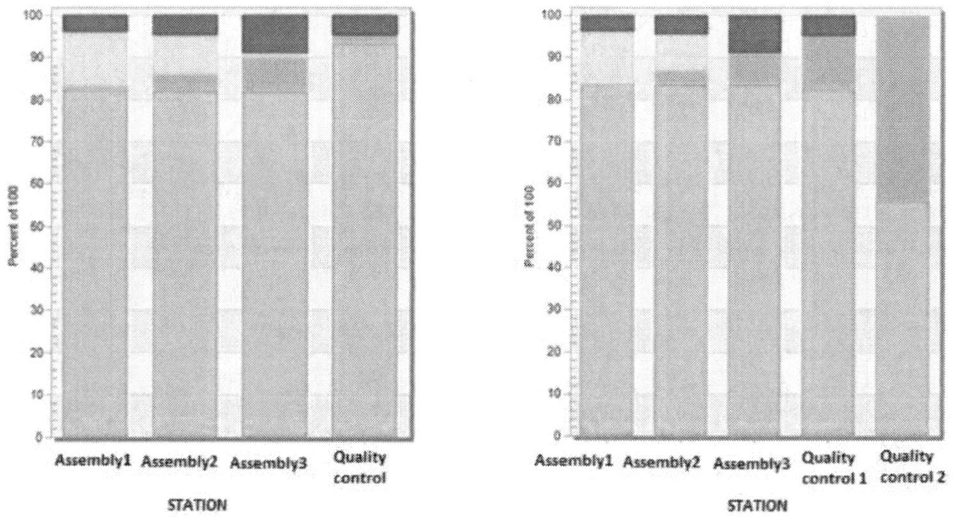

Fig. 5 Resource statistics – Base model (left), Customized model (right).

Legend: Work ▒ Blocked ☐ Failure ■

4. Conclusion

At present, simulation is increasingly used in assembling and especially in the development phase of assembly workplaces and systems. The use of simulations is mainly driven by the simulation benefits that are mentioned in the article. The main advantages of assembly simulation are to shorten development times, to eliminate design errors in assemblies before putting them into operation, to test several versions of assembly systems or workstations. An example simulation of the assembly system was done in the Tecnomatix

Plant Simulation software, and a proposal to improve the assembly system was introduced. Results shown that throughput of the assembly system was increased by 267.78%.

Gradually, as IT advances in the world, simulations are becoming increasingly prominent and assembly simulation can be expected to be one of the core tasks of development teams.

The article was written as part of the VEGA 1/0477/14 project "Research of influence of selected characteristics of machining process on achieved quality of machined surface and problem free assembly using high Technologies" supported by the scientific grant agency of the Ministry of Education of the Slovak Republic and of Slovak Academy of Sciences.

References

1. L. Réveszová, D. Paľová, Základy modelovania podnikových procesov, Košice, Technická univerzita v Košiciach (2009)
2. S. Klos, J. Patalas-Maliszewska, Management and Production Engineering Review (2013)
3. M. Gregor a kolektív, Simulácia systémov, Žilina, Vysoká škola dopravy a spojov v Žiline (1992)
4. D. Malindžák: Modelovanie a simulácia v GPSS, Košice (1988)
5. L. Pachniková, Ľ. Šidlovská, Simulation as support tool for design of manufacturing systems, Transfer inovácií 18 (2010)
6. M. Straka, D. Malindžák, Acta Montanistica Slovaca 14 (1) (2009)
7. TECNOMATIX Plant Simulation, Product overview, SIEMENS (2016)
8. J. Burieta, Simulácia. In Ipaslovakia (2007)
9. R. Debnár, J. Košturiak, I. Kuric, Simulácia ako nástroj pre zvyšovanie produktivity a zisku podniku (2016)
10. M. Houška, Simulačné modely I. Praha: Česká zemědělská univerzita v Praze. (2005)
11. T. H. Naylor, D. S. Burdick, K. Chu, *Computer Simulation Techniques*. New York (1968)
12. P. Trebuňa, M. Kliment, Edl, M., M. Petrik, Creation of simulation model of expansion of production in manufacturing companies. Modeling of mechanical and Mechatronic Systems MMaMS (2014).

MATEC Web of Conferences **137**, 05001 (2017) DOI: 10.1051/matecconf/201713705001
MTeM - AMaTUC 2017

Selected properties of laser cladding coatings shaped using Flow drill technology

Dariusz Bartkowski[1, *], *Waldemar* Matysiak[1] , and *Aneta* Bartkowska[2]

[1] Poznan University of Technology, Faculty of Mechanical Engineering and Management, Institute of Materials Technology, Poznan, Poland
[2] Poznan University of Technology, Faculty of Mechanical Engineering and Management, Institute of Materials Science and Engineering, Poznan, Poland

Abstract. The paper presents the investigations of selected chemical and mechanical properties as well as macro- and microstructure of materials formed using thermal drilling process (Flow drill). The aim of this study was to determine the microstructure of the coatings produced using laser cladding with powder technology. The coatings were produced on the low-carbon steel using 1 kW disc laser. After modification of surface, the thermal drilling process was applied. To produce all coatings, the pure copper powder was used. In this study the laser power equal of 500, 700 and 900 W were used. The microstructure, chemical composition (EDS) and microhardness were investigation. It was found that the surface modification of low carbon steel and next conducted thermal drilling process caused change the surface properties on the hole flange. It was found that surface modification of steel using laser cladding with cooper powder and next Flow drill process contributes to the change in microhardness and chemical composition on hole flange.

1 Introduction

Surface engineering technologies have influenced on the development of other production technologies and materials processing such as surface modification [1 - 13] as well as metal forming processing [14-22]. One of the metal forming processes is the hole flanging which is intended to increase the diameter of the previously cut hole and to form around it the conical or cylindrical flange. The traditional method is commonly known and consists of flanging through the cylindrical punch with a tapered end which is intended to pierce through the material pointwise and then to form a corresponding hole shape and a flange around it. Technological advances have strongly influenced the conventional edge drilling technology by Flow drill technology. Many characteristics of this process have benefited from it compared to the older known processing methods. There are several important advantages of this process [14, 15]. Firstly, in the thermal drilling process there is no material loss. Material is plasticized and then formed. Thanks to this, material is fully utilized to fabricate the collar and forming the sleeve. Additionally the Flowdrill technology allows the produces sleeve up to three times the height of the output material. Material

* Corresponding author: dariusz.bartkowski@put.poznan.pl

thickness may range from 0.5 to 10 mm. A slight increase in the recommended values of the machining parameters allows the edge thickness to increase by several percent. The thermal drilling is a one-step process, it does not require any change of tools or additional treatment before and after the process, and the time needed to produce the hole flange is small, just a few seconds. In addition, this process is very fast and easy to automate, making it suitable for serial or mass production.

This work is focused on the issue of the hole flanging using the thermal drilling technology (Flow drill). The process was used on materials with coating produced by the laser cladding process which is commonly used [1-4, 11]. Selected properties such as macro- and microstructure, coating thickness, microhardness, and chemical composition were investigated. The results were analysed and they are very promising.

2 Research Methodology

2.1 Materials and technology

The output material used for the laser cladding and thermal drilling processes was the general purpose structural S235JRH steel. It chemical composition of this material was shown in Table 1.

Table 1. Chemical composition of S235JRH steel used for testing.

Steel grade	Maximum weight share [%]				
	C	Mn	P	S	N
S235	0.7	1.40	0.045	0.045	0.009

Specimens of 30 x 24 x 20 mm with a wall thickness of 2 mm were cut from the profile. On the specimens were produced cooper coatings using laser cladding technology. On the series of specimens of S235JRH steel the copper coatings using laser beam have been made. The manufacturing process was carried out at the Industrial Institute of Agricultural Machines in Poznan. The TRUMPF laser model TruLaser Cell 3008 was used for produced specimens for research. The working area of this device includes the axes: X - 800 mm, Y - 500 mm and Z - 400 mm. The laser device includes components such as a working chamber, TruDisk 1000W disk laser equipped with a three-stream powder feeding system as well as cooling and filtering circuits. Prior to making the coatings, the specimens must be properly prepared: the surface of each was polished and cleaned. The laser cladding process was carried out in a closed chamber where powdered material was supplied coaxial to the laser beam by carrier gas as shown in Figure 1.

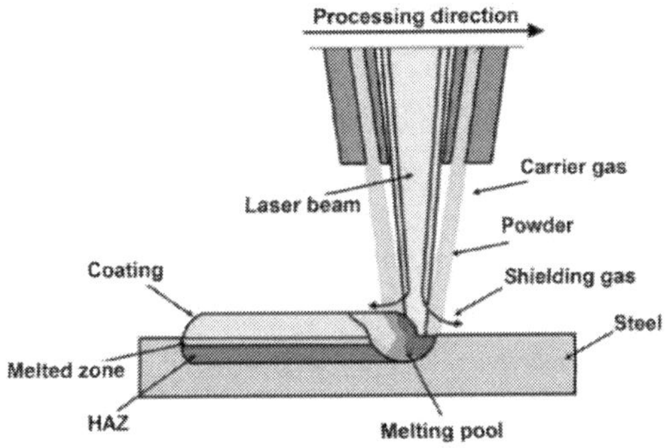

Fig. 1. Scheme of a laser cladding process with additive material in the form of powder.

The laser head contains three powder delivery nozzles and a central channel through which the shield gas and the laser beam are supplied. The coatings were prepared for each specimens using varying laser power parameter (700, 800 and 900 W). The process was carried out with constant parameters: feedrate 600 mm/min; laser beam diameter 2 mm; flow rate of gas supplying the powder (helium) 8 l/min; shield gas flow rateo (argon) 8 l/min; powder feeder rate 0.8 obr/min.

Specimens with laser cladded coating were subjected to a flow drill process. Holes and flanging for each specimens were made on a JET JMD-18VS drilling machine. The machine was equipped with a handle with cooling ring in which a drill bit of 7.3 mm was mounted. The tool was secured with a Flowdrill FDKS paste that was applied each time before the hole was made to protect the tool from high temperatures. Tools turnover amounted to 3000 rev/min, and feed rate was performed manually. Figure 2 shows a scheme of the hole flanging process in materials with coating produced using laser cladding with powder technology.

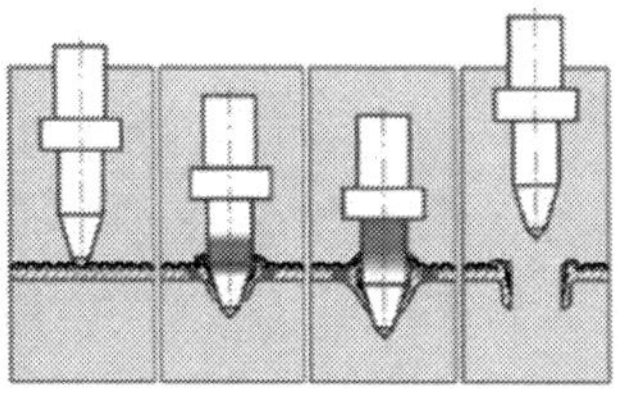

Fig. 2. Scheme of flanging using Flowdrill technology in materials with laser cladded coating.

2.2 Microstructure and microhardness study

Macroscopic observation of all specimens was performed using the magnifying glass. Color as well as presence or absence of cracks and porosity were determined. Photos were taken using the Nikon D5000. Microstructural observations were made using Huvitz HRM-300 optical microscope with digital image recording. For the tests, metallographic cross section were made. The specimens were cutting and next placing in conductive thermoset resin. To reveal the microstructure of specimens substrate and its surface, the mixture of nitric acid and ethanol was used. Microhardness tests were performed using the Vickers method on the Buehler microhardness tester.

2.3 Chemical composition study

The chemical composition of specimens was examined using the scanning electron microscope (SEM) Vega Tescan equipped with PGS's Prism 2000 EDS (Energy Dispersive Spectroscopy) system. EDS analyzers allow both surface and volumetric identification of the chemical elements contained in the test material. A linear analysis of the chemical composition was carried out for each specimen.

3 Results of study

3.1 Microstructure and microhardness results

In Figure 3 were shown a macroscopic images of three samples with coatings produced using different laser beam powers. For the first specimen (700 W), it can be seen that the resulting coating has a copper color without discoloration and other changes. Width of tracks are the same. No deformation and no cracks were observed. For the specimen produced using laser beam power equal 800 W, the part of surface have color of copper, but slightly darker than on first specimen. It was also found the dark area that results from excessive heating of the workpiece during the laser cladding process. In addition, no cracks, porosities or other surface damage were noted. In the case of specimen produced using laser beam power equal 900 W, the surface is mostly gray with one track of copper-colored track. Deformation of track was much greater than that preevious specimens. Based on macroscopic examination of three series kind of specimens produced using different laser beam power, it was found that the hole flanging process was successful and the form of holes and flange of all specimens were very similar. No geometric defects were observed in the form of cracks or other discontinuities of the material.

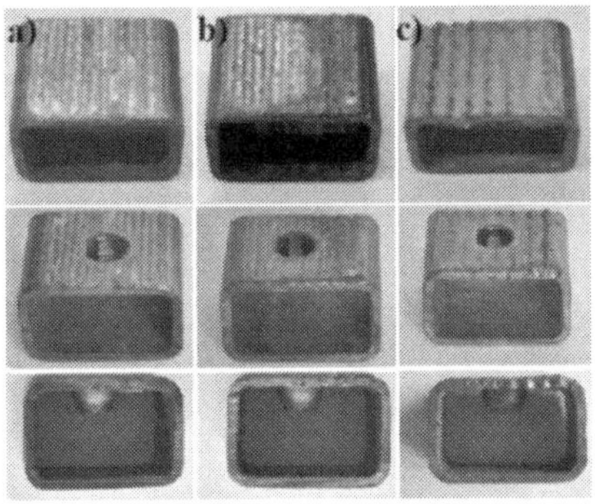

Fig. 3. Specimens with coating produced using laser cladding at laser beam power and after Flowdrill a) 700 W b) 800 W and c) 900 W.

On the basis of the conducted examinations of metallographic cross section of specimens the microstructure was determined. The results of microstructure were shown in Figure 4. In the case of specimens produced using laser beam power of 700 W, there are three distinct zones in the material: a distinctly uneven copper layer, a transition layer, and the substrate.

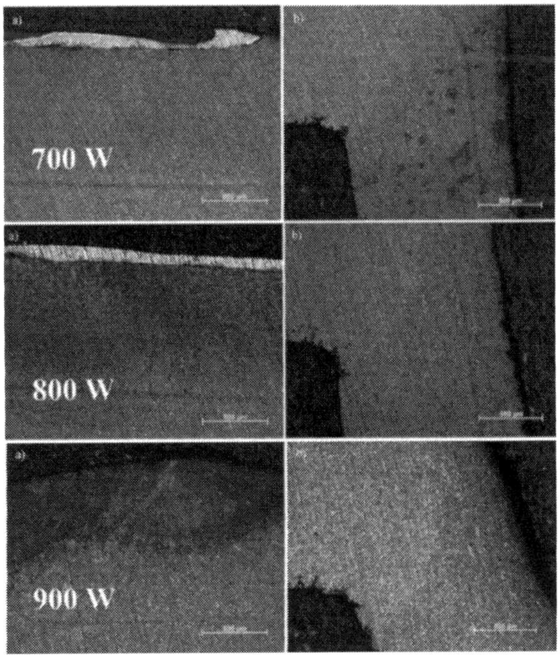

Fig. 4. Microstructure of a specimen with a laser cladded coating of cooper : a) coating, b) hole flange.

On the hole flange edge, it was found that the distinctive copper coating disappears, leaving only the transition zone. In the case of specimens produced using laser beam power of 800 W, there also three separate zones. This time the copper coating is spread evenly over the whole length, and it disappears on the edge of hole flange. The microstructure of specimen produced using laser beam power of 900 W, clearly differs from the previous. Higher temperature of laser claddind process caused the copper powder was mixed with the substrate material to a much greater extent than in the case of two previous kind of specimens. This results in a thicker outer coating without copper-colored and a difficult-to-find border of the transition zone. It is also possible to observe the absence of pores within the rim of the rolled edge. Microstructural studies additionally included a microhardness test of hole flange. For each specimens, five measurements were made in the cross section of the flange from the inner wall to the inside of the material. The results microhardness measurments for all specimens are present in table 2.

Table 2. Microhardness of specimens.

Measurement number	Microhardness [HV]		
	700W	**800W**	**900W**
1	255	236	221
2	220	199	176
3	196	161	213
4	274	213	171
5	220	193	171
Average:	233	200.4	190.4

The highest micro hardness value was achieved with the specimen produced using laser beam power equal to 700 W. This hardness was about 22% larger than the specimen produced at 900 W and about 16% larger than the specimens produced using 800 W. With the increase in laser beam power, the substrate was melted and mixed with the coating material, which is harder than substrate. Hence, at the smallest power of the laser beam, the surface of the flange is characterized by the highest microhardness. Increasing the power of the laser beam causes mixing the coating material with the soft substrate.

3.2 Chemical composition study

A linear analysis of the chemical composition of the specimens was made. Content of copper and iron was determined (Fig. 5). From the diagram (700 W), it can be determined that the produced coating have about 100 μm thickness in the material. The copper content on the surface varies considerably from about 75% to 40%, and beyond this 100 μm, a sharp decrease in the percentage of copper and a stabilization of about 5% was observed. In contrast to copper, the concentration of iron was change. For specimen produced using 800 W, the situation is similar to that of the specimen produced using 700 W. The coating containing copper concentration of about 50% reached 70 μm deep into the test material. The initial high content of 80% quickly declined, stabilizing between 45% and 65%. This time the iron content was interwoven with the copper in the top layer, where the rise and fall were recorded, but the value itself did not exceed 55%. As the graph shows, after a distance of 70 μm, the content of the elements has been stabilized after a visible drop of copper concentration to about 5% and a sharp increase in iron concentration to about 95%. In the case of specimen produced using 900 W, a large difference in the course of the concentration change can be observed compared to the other two kind of specimens. The copper content in the top of coating did not exceed 60% and to about 120 μm did not decrease below 50%. Iron content in the top of coating was the highest of all three specimens, its concentration ranged between 40 and 50%. Closer to the steel substrate there is a gradual decrease in copper content and an increase in iron content which, as in previous cases, has stabilized, reaching values of about 5% for copper and about 95% for iron approximately 260 μm from surface.

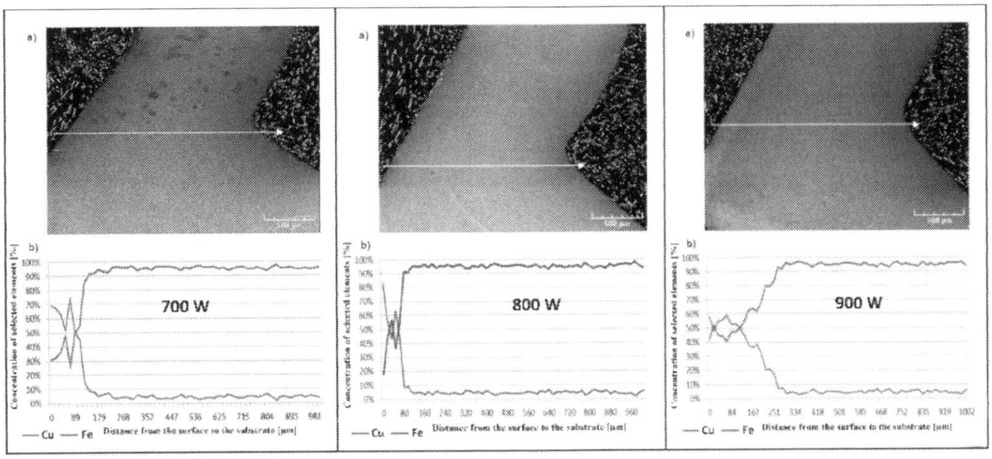

Fig. 5. Chemical composition of specimen with cooper coating produced using laser cladding: a) direction of linear composition analysis b) concentration of Fe and Cu.

4 Conclusions

Both technology presented in this work, laser cladding as well as Flowdrill are innovative technologies. Combining these two processes allows get high-quality connections in a very short time. Additionaly both of these methods are very quick and easy to automate. Laser cladding and Flowdrill processes are characterized by the fact that they are based on the treatment of elements with very high temperatures.

Based on this studies can be drawn the following conclusions. From the macrostructure study, it appears that the copper layer made with laser technology did not negatively affect the quality of the hole flange. For all three kind of specimens no discoloration, cracks or porosities were observed and the inner surface of the flange was good quality. Analysis of the chemical composition showed an increased copper concentration at the inner wall of flange. The process of thermal drilling caused the diffusion of coating material to the steel substrate. The method described may be used in the case of joining materials which have previously been subjected to surface treatment.

References

1. M. W. Steen, J. Mazumder, *Laser Material Processing*. Springer (2010)
2. D. Bartkowski, A. Bartkowska, IJRMHM **64** 20-26 (2017)
3. D. Bartkowski, G. Kinal, Int J Refract Met H **58**, 157-164 (2015)
4. D. Bartkowski, A. Młynarczak, A. Piasecki, B. Dudziak, M. Gościański, A. Bartkowska, Opt Laser Technol **68** 191-201 (2015)
5. A. Bartkowska, A. Pertek, M. Jankowiak, K. Jóźwiak, Archiv Metall Mater **57,** 211-214 (2012)
6. A. Bartkowska, A. Pertek, Surf Coat Tech **248**, 23-29 (2014)
7. W. Li, K. Yang, D. Zhang, X. Zhou, J Therm Spray Techn **25,** 1-2 (2016)
8. S-J. Dong, B. Song, G-S. Zhou, J Therm Spray Techn **22,** 7 (2013)

DOI: 10.1051/matecconf/201713705001

9. R. Gr. Maev, V. Leshchynsky, E. Strumban, J Therm Spray Techn **25**, 1-2 (2016)
10. M. Reza Bateni, P. Wei, O. Kesler, A. Petric, Mater Manuf Process **28** (2013)
11. A. R. Yan, Y-J. Li, Z-Y. Wang, Lasers in Eng **29**, 365-377 (2014)
12. V. Torabinejad, A. Sabour Rouhaghdam, M. Aliofkhazraei, Bulletin of Materials Science **39**, 3 (2016)
13. E. Abakay, S. Sen, U. Sen, Acta Physica Polonica **129** (2016)
14. W. Matysiak, L. Bernat, Metalurgija **54,** 235-238 (2015)
15. A.H. Streppel, H.J.J. Kals, CIRP Annals **32**, 167-171 (1983)
16. C. Bates, American Machinist **152,** 56-58 (2008)
17. S. F. Miller, J. Tao, A. J. Shih, Int J Mach Tool Manu **46,** 1526-1535 (2006)
18. L. Sobotová, J. Kmec, L. Bičejová, IJE, **IX** 1584-2673 (2011)
19. H.-M. Chow, S.-M. Lee, L.-D. Yang, J Mater Process Tech **207,** 180-186 (2008)
20. W. Matysiak, B. Barišić, I. Mamuzić, Metabk **49,** 13-17 (2010)
21. W. Matysiak, Metal forming **4**, 49-57 (2005)
22. S. Dobatkin, J. Zrnik, I. Mamuzic, Metalurgija **48,** 157-160 (2009)

The lack of uniformity of the deformations and the degree of deforming in case of orbital deformation

Sorin-Dumitru Grozav[1*], *Gabriel Nicodim* Ciuşcă[1] and *Vasile* Ceclan[1]

[1]Technical University of Cluj Napoca, Manufacturing Engineering Department, ROMANIA

Abstract. The degree of deformation and the lack of uniformity of the deformations are two important factors orbital deformity. The degrees of deformation in the case of orbital oscillating deformation between 40 and 70%, is much higher than in conventional methods. Two other important parameters that may change unevenness strains are axial force and advance. Their influence can be seen from the diagrams unevenness, and hence the higher the value of these parameters is reduced unevenness strains.

1 Introduction

Orbital deforming it is an unconventional forming technology of metals, having as a particular the fact that applying in a convenient way the forming pressure; it is obtained a considerable reduction of forming pressure for the same forming straight. The principle of orbital deforming was patented in 1910, but his extension began in the last 25 – 30 year, after the apparition of the machines able to realization of this operations. This method is based on the principle that the pressure is not applied on the whole surface of the work piece, and it is applied only a 20 – 30 of it. This it is show in the figure 1 [1], [2], [3].

The pressure application zone is variable in time, because the pressure is moving on the surface of the work piece with a constant frequency of 150 – 200 cycles/min. The most important traits of this method, having as result a sequential penetration of the pressure in every parts of the work piece surface are: the wobbling motion of the tool together with the superior tool reported to the vertical axes, and the four moves of that.

The determination of the tools track which is the most suitable for the work piece is important for metal flow in the tool. A similar influence has the choice of the wobbling limit angle has the same influence as forming pressure.

The pressure application zone is variable in time, because the pressure is moving on the surface of the work piece with a constant frequency of 150 – 200 cycles/min. The most important traits of this method, having as result a sequential penetration of the pressure in every parts of the work piece surface are: the wobbling motion of the tool together with the superior tool reported to the vertical axes, and the four moves of that [2], [3], [4], [5].

* Corresponding author: sorin.grozav@tcm.utluj.ro

 DOI: 10.1051/matecconf/201713705002

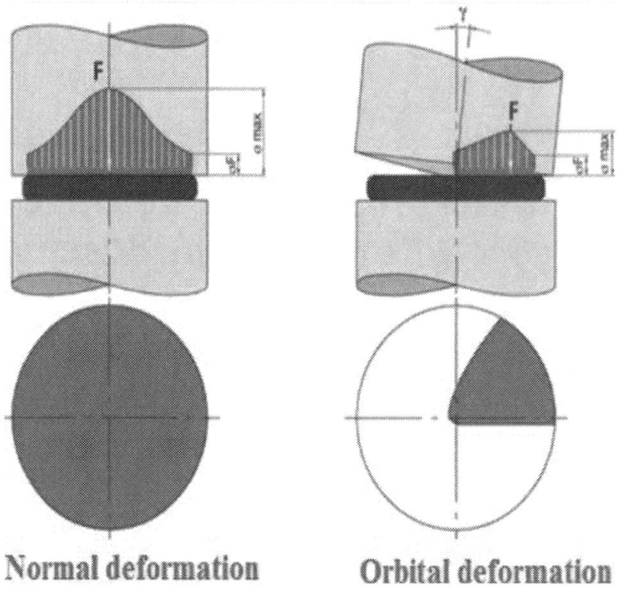

Fig. 1. Principle orbital deformation.

The determination of the tools track which is the most suitable for the work piece is important for metal flow in the tool. A similar influence has the choice of the wobbling limit angle has the same influence as forming pressure.

If the deformation of the conventional, inter-acting with the tool on its entire front surface of preform, when deformation orbital focus of deformation is located at any stage of the process (the shaded area in figure 1). The initial preform, having usually formed of a cylinder or ring, is seated in lower die which is vertically oriented axis of symmetry. The upper die having the conical shape of the roller is inclined from the vertical by an angle θ (figure 2) [6].

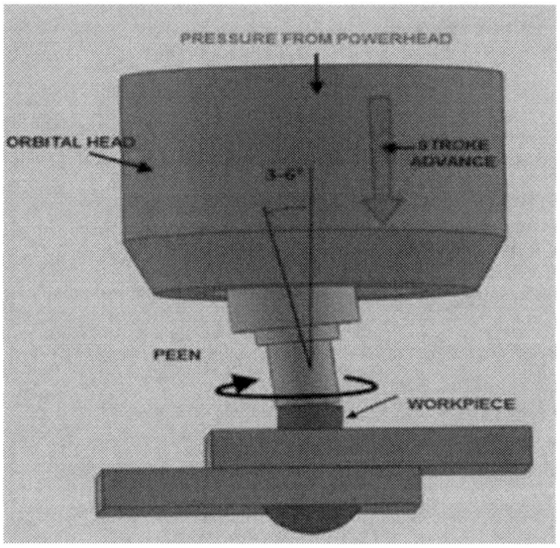

Fig. 2. Realization rivets by orbital deformation [6].

The upper die must be so positioned that the apex of the cone to be located always on the axis of symmetry of the lower die, which is the central axis of rotation. Changing the shape of the preform is carried out by oscillating the upper die around the central axis of rotation and the vertical movement of advance, either the upper die to the lower die, or vice versa. In any case, the tip of the cone, which is always fixed, becomes the pivot point of the upper die. Kinematic point of view it is a rigid fixed. Thus, at any given time a portion of the volume part is in the focus of deformation. Due to the oscillating movement of the upper mould and the advance due to axial deformation zone is dynamic. It moves the propeller downward sweep the front surface of the work-piece and approaching nearer to base the preform. Penetration depth of deformation on the preform height depends specifically track configuration and the actual amount of process specific parameters.

2 The lack of uniformity of the deformations and the degree of deforming

2.1 Experimental research

In order to be able to operate with maximum efficiency the facilities offered by this new and wider technology, it is necessary to study the flow of the material and the distribution of deformations in the volume of the orbital deformed blank, necessary for the design of a correct orbital deformation technology. [1], [2], [7], [8] For the study of the unevenness of the deformations at the orbital deformation we worked with lead cylindrical specimens with different ratios between height (H) and diameter (D): H / D = 0.2...2. We have drawn on the lateral surface a network of equidistant lines parallel to the base of the specimens, and on the front surface they have drawn equidistant concentric circles. For the modification of the parameters of the networks after the orbital deformation we could characterize the distribution of the deformations and the flow of the material.

The experiments were performed on a Polish orbital PXW100 deforming machine, which can develop a 1600 KN axial force (F_a) with an oscillating angle of the upper tool of $\gamma = 2^0$ at an oscillation frequency of 200 oscillations / min. This press carries out four types of movements. [1], [2], [4]

The recording of the run-time parameters was performed on a x-y inscriber, and the force was determined from the pressure of the hydraulic cylinder read on the manometer. The specimens were orbital deformed, with varying degrees of deformation, using multiple values x-y of deformation force and axial feed (a). The obtained results were compared with those from the conventional deformation of identical specimens. Following the modification of the geometric shape of the deformed specimens, it is observed that, unlike the classic deformation in which the specimen is the shape of the barrel, the orbital deformation takes the shape of the fungus, having the maximum diameter in the upper part and not in the middle of the height.

2.2 Experimental results

The degree of deforming is calculated with the equation:

$$\varphi_{h1} = \frac{\Delta h}{h} \tag{1}$$

where: Δh is the height variation after the repression and h is the initial height of the semi-finished product.

In the figure 3.a) you can see that for higher values of the rapport h/d, the more powerful deforming is situated at the superior part of the semi-finished product, while at the smaller values of the rapport, the inferior part of the semi-finished product suffers a more powerful deforming. This phenomenon is caused by the passage of the material from the side part to the lower one, fact that leads to the diminishing of the side network distances. The middle area of the semi-finished product is a little deformed and we can say that it doesn't depend on the rapport h/d.

The degree of radial deforming can be expressed by the equation:

$$\varphi_r \frac{\Delta R}{R} \tag{2}$$

where: ΔR is the changing of the cylinder radius and R is its initial radius.

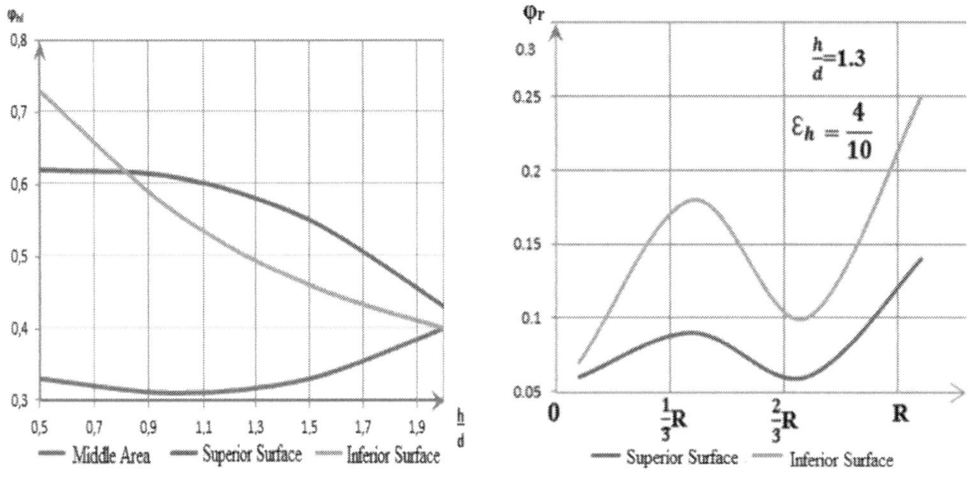

a) b)

Fig. 3. The variation of the deforming grade: a) in height; b) radial.

In the figure 3.b) there is represented the variation of the radial deforming degree. From the figure you can see that on the inferior surface of the semi-finished product the distribution of the radial deforming is similar to the one from the conventional repression, having a stationary middle area, respective an edge area with a maximum deforming. But, on the inferior surface, the repartition of the deforming has as a distinctive particularity the appearance of a medium deforming, even in the central area.

The knowing of the degree of unevenness of the deformations has a special importance and it can be express by the rapport:

$$\varphi_{du} = \frac{d_{msx}}{d_{\min}} \tag{3}$$

Where d_{max} and d_{min} are the maximum and the minimum diameters of the deformed semi-finished product (figure 4).

The dependence of the unevenness of the deformations according to the degree of deformation in height, for different rapports h/d is presented in the figure 4.a) and in the figure 4.b) there is presented the dependency of the unevenness degree towards the rapport h/d at the same deforming grade. From this diagram results that the unevenness of the deformations increases almost linear till the maximum value reached by the deforming degree between 0.4…0.5, afterwards decreasing slowly. The unevenness is more accentuated for higher values of the rapport h/d.

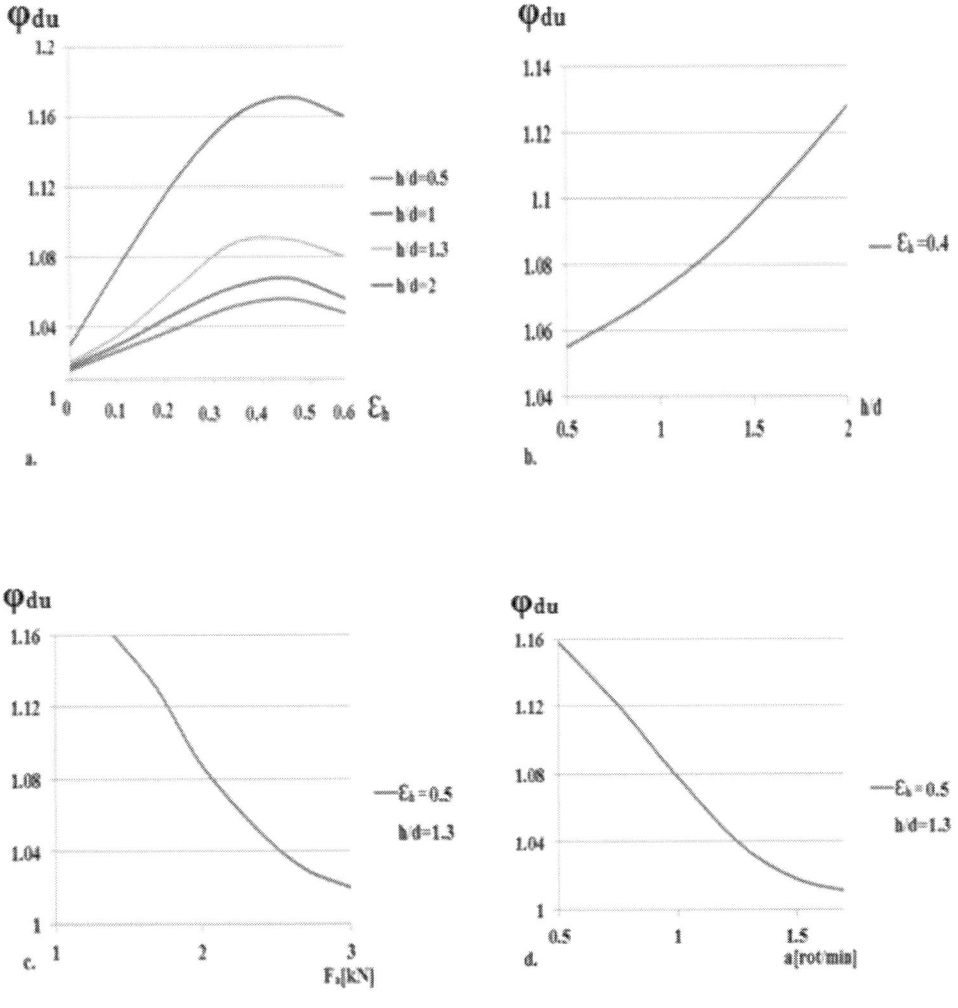

Fig. 4. The evolution of the unevenness degree of the deformations φ_{du}: a – according to the deforming degree on height; b – according to the rapport h/d; c – according to the axial force; d – according to the advance.

Other 2 important parameters which can modify the unevenness of the deformations are the axial force and the advance. Their influence upon the unevenness can be observed in the figure 4,c and d, from where results that, together with the increasing of these parameters, the unevenness of the deformations is reduced.

3 Conclusion

Following the analysis of the results presented in the diagrams presented, some conclusions are drawn which are very important for the establishment of a correct orbital deformation technology:

• at the orbital deformation of the pieces with different outer diameters, the maximum diameter will be located at the upper part because the unevenness of the deformations will help to fill the moulds;

• at large diameter differences the force and the axial feed will be reduced in order to increase the unevenness of the deformations;

• the parts with a constant outer diameter, which are desirable for the smallest unevenness of the deformations, will be deformed with a maximum axial force and advance provided by the deforming machine;

• degree of deformation in orbital deformation oscillating between 40 and 70%, is much higher than in conventional methods;

• geometrical format of the contact surfaces in orbital deformation directly the lack of uniformity of the deformations and the degree of deforming.

References

1. S. D. Grozav, *Studies and researches in process parameters variation in Rotary Forging,* PhD Thesis, University of Cluj-Napoca (1995)
2. S. D. Grozav, Orbital Deformation MEDIAMIRA Publishing House, Cluj-Napoca, Engineering Collection, ISBN 978-973-713-237 (2009)
3. P.M. Standrig, *Investigation into Rotary Forging*, Ph.D. Thesis University of Nottingham (1982)
4. S. D. Grozav, A. Crai, *Annals of the Oradea University*, ISSN 1583-0691 (2006)
5. S. D. Grozav, Gh. Gligor, *Annals of MTeM*, 175 -178, ISBN 973-9087-83-3 (2007)
6. *** http://msvs-dei.vlabs.ac.in/Orbital.php
7. T. Canta, I. Dragan, Gh. Achimas, *Constr. de Maşini*, **39**, 2-3, 101-104 (1987)
8 Drăgan, T. Canta, I. Nistor, Gh. Goron, D. Săbăduş, *Constr. de Maşini*, **41**, 2-3, 100-103 (1989)

Evaluation of deep drawing force under different friction conditions

Lucian Lăzărescu[1,*], and *Dorel* Banabic[1]

[1]Technical University of Cluj-Napoca, CERTETA Research Centre, B-dul Muncii nr. 103-105, Cluj-Napoca, Romania

Abstract. The purpose of this study is to investigate the variation of the required punch load during the deep drawing process under different friction conditions. In this regards, several deep-drawing tests of cylindrical cups were conducted under four friction conditions at the tool–blank interface. The first case was the dry deep-drawing, considered as a reference friction condition, while in the other three cases hydraulic oil, lithium-based grease and animal fat were used as lubricants. For each friction case, three levels of blank holding force were adopted, namely 10, 20 and 25 kN. The finite element simulation of the deep-drawing process was used to generate a set of calibration curves. By overlapping the experimental load-stroke curves on the calibration curves, the friction coefficient was estimated for each friction case.

1 Introduction

Deep-drawing is one of the most used forming methods with a wide range of applications in different fields such as automotive and aerospace industries. In this process a sheet metal is transformed into a hollow part by means of forming tools whose geometry is consistent with the shape of the part. There are several process variables that influence the success of a deep-drawing process. These include the blank-holding force (BHF) and the friction between tools and blank.

In recent years, some research works have been conducted to study the influence of friction conditions and BHF on the deep-drawing process. Ahmetoglu et al. [1] consider that BHF is the major factor affecting the occurrence of defects. Padmanabhan et al. [2] used FEM and Taguchi technique to determine the relative influence of die radius, blank-holding force and friction coefficient on the punch load. They have found that the die radius has a major influence followed by friction coefficient and BHF. Raju et al. [3], also conducted a study using the Taguchi method and their results have revealed that the die shoulder radius has a major influence on the thickness distribution of the deep-drawn cups made from AA 6061 aluminum alloy sheet, followed by BHF and punch nose radius. Browne et al. [4] have investigated the effects of many factors, including lubrication type and blank-holding pressure, on the punch load using Taguchi's L8(27) orthogonal array. Their results have shown that the friction is one of the parameters that have the greatest

* Corresponding author: lucian.lazarescu@tcm.utcluj.ro

influence on punch load. The authors of the paper [5] have studied the effect of die radius, surface roughness of the tools, drawing speed, blank-holding force, and lubrication type on dynamic friction coefficient between the flange and fillet regions of the tools and sheet metal by using the ANOVA method. Their results have proved that the lubricant type is an effective parameter for the flange and fillet regions.

Deep-drawing is also regarded as a laboratory test that can be used to assess the lubricant performance [6-9]. According to [7] the lower is the punch force, the better is the lubricant.

The above literature review shows that friction at the tool-blank interface and BHF are two important parameters and there are few studies that analyses the influence of friction conditions combined with the effect of BHF on the punch force.

The current study investigates the relationship between the punch force and friction conditions at the tool–blank interface in combination with the effect of bank-holding force in the deep-drawing process of cylindrical cups.

2 Experiments

2.1 Material and lubricants

A DC04 cold rolled steel sheet with the nominal thickness of 0.85 mm was used to perform the experiments in this study. The main mechanical properties of DC04 material are listed in Table 1. The values of the yield stress and the Lankford coefficient were obtained by tensile tests performed on specimens cut from the metal sheet at different orientations with respect to the rolling direction: 0, 45 and 90°. A tensile testing machine Zwick Roell Z150 was used to perform the tests. The parameters of Swift's hardening law were identified by fitting the experimental true stress–true plastic strain curve by equation 1. The data listed in Table 1 is used to describe the mechanical behavior of the blank in the finite element simulation of the deep-drawing process, as described in Chapter 3:

$$\sigma = K\left(\varepsilon_0 + \varepsilon_p\right)^n, \tag{1}$$

where σ is the flow stress and ε_p is the plastic strain.

The lubricants used in experiments were hydraulic oil (HLP 46), lithium-based grease (Repsol Grasa Mobilgras EP 2), and animal grease provided by the Erichsen Company, Germany.

Table 1. Mechanical properties of DC04 steel sheet (thickness 0.85 mm).

Parameter		Value at different specimen orientations		
		0°	45°	90°
Yield stress, $R_{p0.2}$ [MPa]		196	211	206
Lankford parameter, r [-]		1.96	1.30	2.19
Young's Modulus, E [MPa]		207000		
Poisson's ratio, ν [-]		0.28		
Mass density ρ [kg/m^3]		7800		
Coefficients of Swift's hardening law	Strength coefficient, K [MPa]	546.5		
	Strain hardening coefficient, n [-]	0.239		
	Initial yield strain, ε_0 [-]	0.009		

2.2 Experimental equipment

The deep-drawing of cylindrical cups was performed on a universal sheet metal testing machine (Erichsen type). Figure 1 shows the main dimensions of the tools and the components of the deep-drawing device.

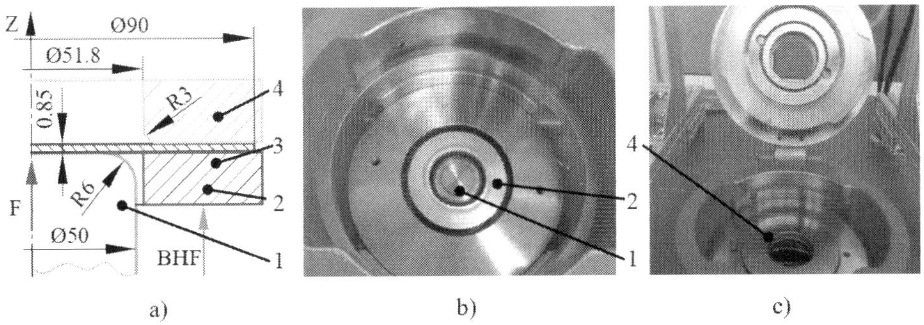

Fig. 1. Deep-drawing tools: a) – tool dimensions; b) and c) – universal sheet metal testing machine (Erichsen type); 1 – punch; 2 – blank holder; 3 – blank; 4 – die.

The punch is placed under the blank and is movable, while the die and the blank holder are stationary. The blank diameter is 90 mm; the punch and die diameters are 50 and 51.8 mm, respectively. The die and punch fillet radii of 3 and 6 mm, respectively, were selected to perform the experiments. The punch force and the punch stroke were continuously recorded by the machine software.

2.3 Experimental conditions

In order to investigate the evolution of the punch force during the deep-drawing process under different friction conditions (FC) at the tool–blank interface, four friction cases were considered: FC1 – deep-drawing without lubricant; FC2 - use of hydraulic oil; FC3 - use of a lithium-based grease, and FC4 - use of a lubricant based animal grease. These four cases of friction conditions were combined with three levels of the blank-holding force: 10, 20 and 25 kN, respectively. Before starting each individual test, the tools and the blank were cleaned using alcohol, after which a new lubricant was applied. Twelve different experimental conditions were considered, as a result of combining four friction conditions and three levels of the blank-holding force. Figure 2 shows cups drawn in different combinations of friction conditions and blank-holding forces.

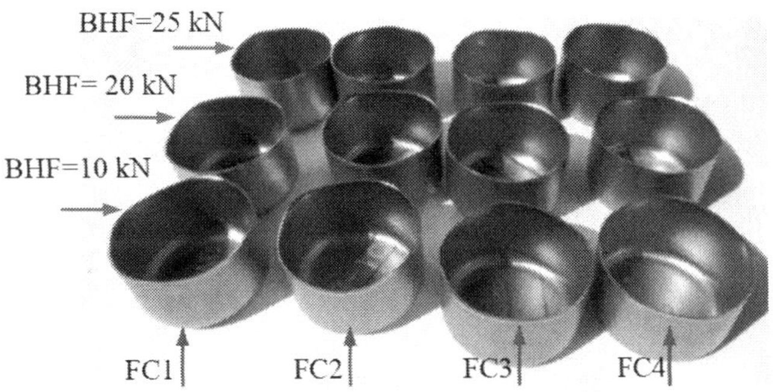

Fig. 2. Representative deep-drawn cups resulted from different combinations of friction conditions (FC) and blank-holding forces (BHF).

3 Finite element simulation of the deep-drawing process

The aim of the finite element simulation of the deep-drawing process was to obtain the calibration diagrams. A calibration diagram consists in a set of deep-drawing force – punch displacement curves obtained using different values of the friction coefficient. By overlapping the experimental deep-drawing force – punch displacement curves on the calibration diagram, the friction coefficient for each friction condition can be estimated. Three values of the friction coefficient (0.10, 0.15 and 0.20) and three levels of the blank-holding force (10, 20 and 25 kN) were considered to obtain the calibration diagrams.
The Dynaform finite element software [10] was used for simulating the deep-drawing process. Figure 3 shows the FE model before and after generating the mesh. The FE model includes: punch (1), blank holder (2), blank (3) and die (4). The blank was modeled as a deformable body using shell elements with an average size of 1.5 mm. The tools were modeled as rigid bodies. Swift's hardening law defined by (Eq. 1) was used to model the hardening of the blank and Barlat'89 yield criterion was chosen to describe the yield surface.

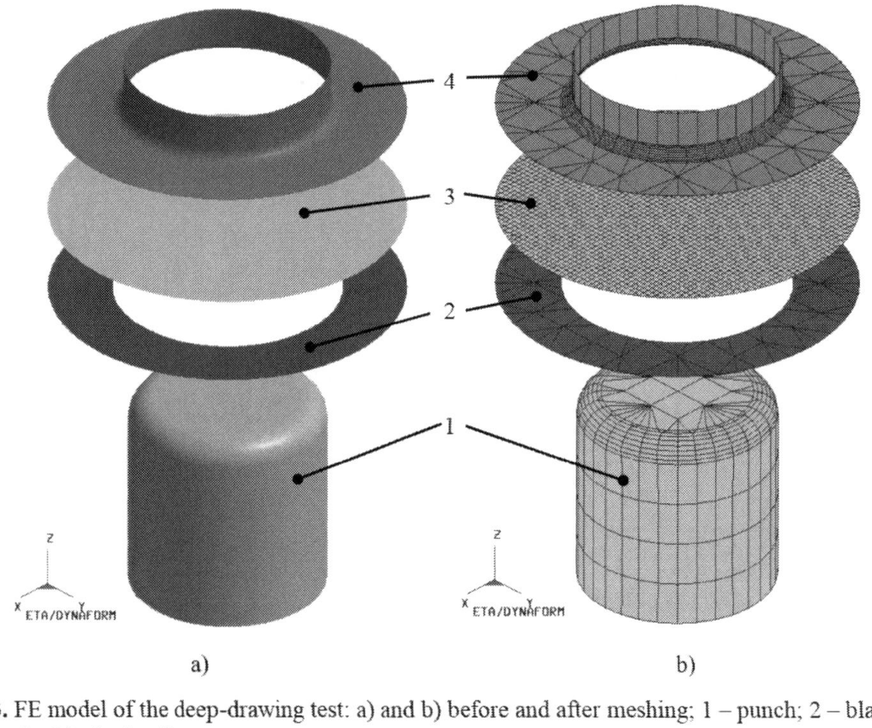

Fig. 3. FE model of the deep-drawing test: a) and b) before and after meshing; 1 – punch; 2 – blank holder; 3 – blank; 4 – die.

4 Results

4.1 Variation of the deep-drawing force under different friction conditions

Figure 4 shows the deep-drawing force versus punch stroke curves corresponding to four friction conditions at the tool–blank interface and three levels of the blank-holding force considered in this study. From this figure one may see that, in all cases, the deep-drawing force increases when the punch moves up to about 15 mm, where the maximum force is reached, after which the force begins to decrease.

DOI: 10.1051/matecconf/201713705003

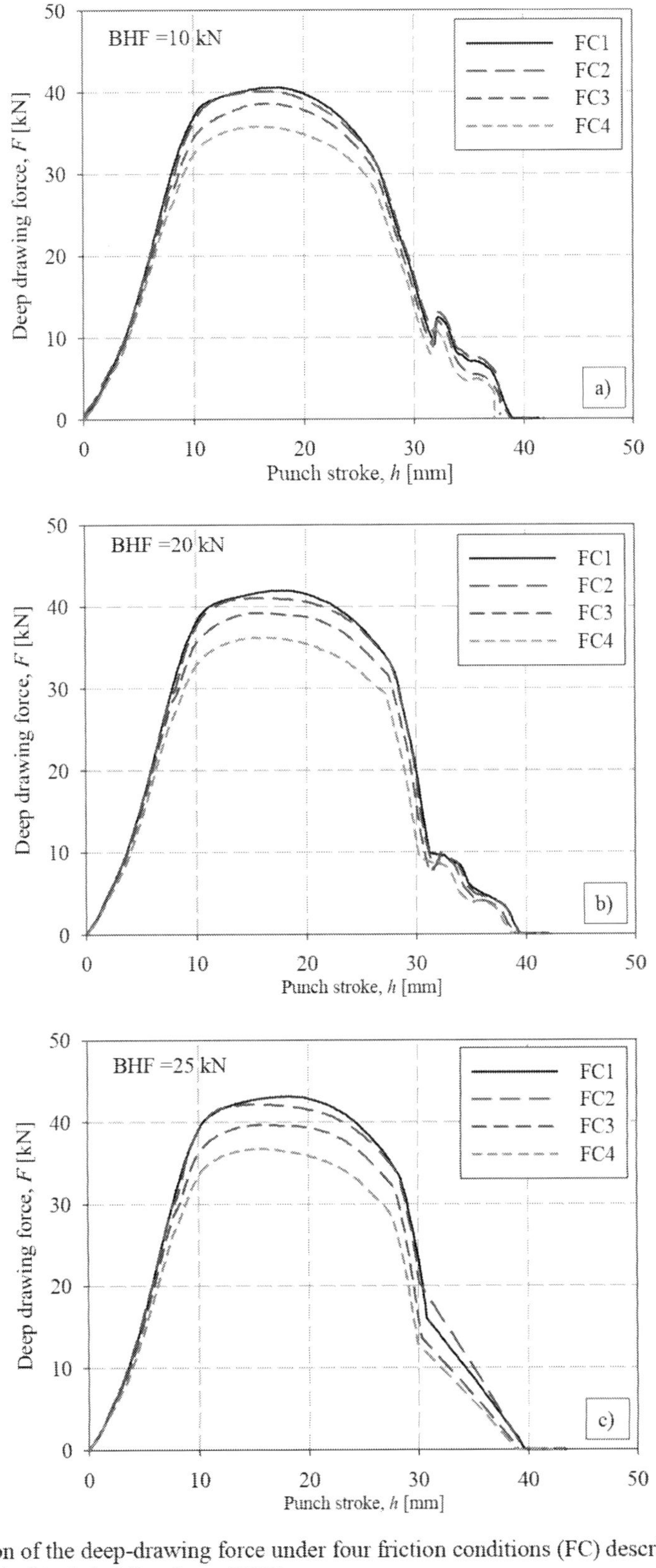

Fig. 4. Variation of the deep-drawing force under four friction conditions (FC) described in § 2.3 and three levels of the blank-holding force (BHF).

When the blank-holding force is low (BHF = 10 kN, Fig. 4a), the evolution of the deep-drawing force shows a slight increase towards the end of the punch stoke (at about 33 mm punch stoke). This increase can be caused by wrinkles that are unfolded when the drawn cup passes through the cylindrical cavity of the die. When the blank-holding force is high (BHF = 25 kN, Fig. 4c), wrinkles do not occur and the local increase in the deep-drawing force is not visible. This evolution of the deep-drawing force during the forming process is consistent with the data reported in the literature [11].

One may also observe in these diagrams that, for all levels of the blank-holding force, the higher deep-drawing force occurs in the case of friction condition FC1, when no lubricant at the blank/tool interface was used. The force decreases in the other friction conditions in the following order: FC2 (hydraulic oil), FC3 (lithium-base grease) and FC4 (animal fat).

In order to quantitatively study the variation of the deep-drawing force under different friction conditions, the maximum deep-drawing force (F_{max}) was plotted against the blank-holding force (BHF) for all four friction conditions, as shown in Figure 5.

One may observe in Figure 5 that the maximum deep-drawing force is most sensitive to the change in the blank-holding force, when no lubricant is used at the blank/tool interface (friction condition FC1). In this case, when the blank-holding force is increased from 10 to 25 kN, the maximum deep-drawing force increases from 40.55 to 43.09 kN; this means an increase of 2.53 kN in the maximum force. In the order of decreasing in sensitivity of the F_{max} to the change in the blank-holding force, follow the cases: FC2; FC3 and FC4. In FC2 and FC3 cases, when BHF increases from 10 to 25 kN, the maximum deep-drawing force increases with 2.05 and 1.08 kN, respectively.

F_{max} is less sensitive to the variation of the blank-holding force, when animal fat is used as lubricant (friction condition FC4). In this case, the increase of the blank-holding force from 10 to 25 kN implies an increase in the maximum deep-drawing force from 35.81 kN to 36.74 kN, which means an increase of 0.93 kN.

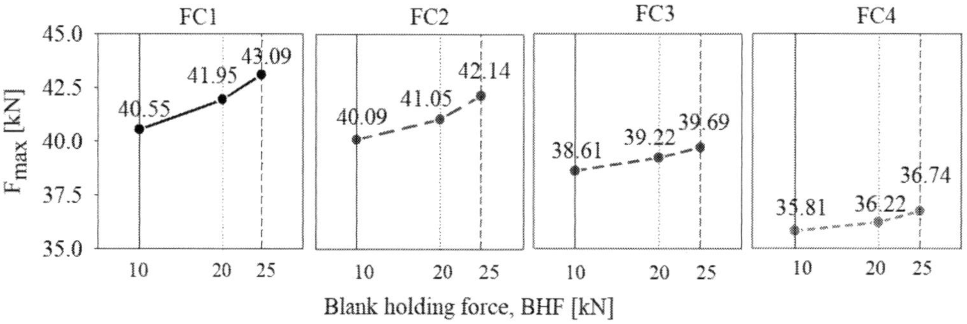

Fig. 5. Maximum deep-drawing force for three levels of the blank-holding force and different friction conditions (FC) described in § 2.3.

4.2 Estimation of the friction coefficient

Figure 6 shows the calibration curves generated by FE simulation of the deep-drawing process using three different values of the friction coefficient (μ) and three levels of the blank-holding force. The experimental deep-drawing force vs punch stroke curves overlapped on these diagrams allow estimating the friction coefficient corresponding to each friction condition. One may deduce from Figure 6 that in the case of deep-drawing without lubricant (case FC1) the friction coefficient exhibits a value greater than 0.2. A friction coefficient slightly above 0.2 was obtained in the case of hydraulic oil (case FC2). The friction coefficient for the lithium-based grease lubricant (case FC3) lies between 0.15 and 0.2. A friction coefficient between 0.1 and 0.15 was obtained for animal fat (case FC4).

 DOI: 10.1051/matecconf/201713705003

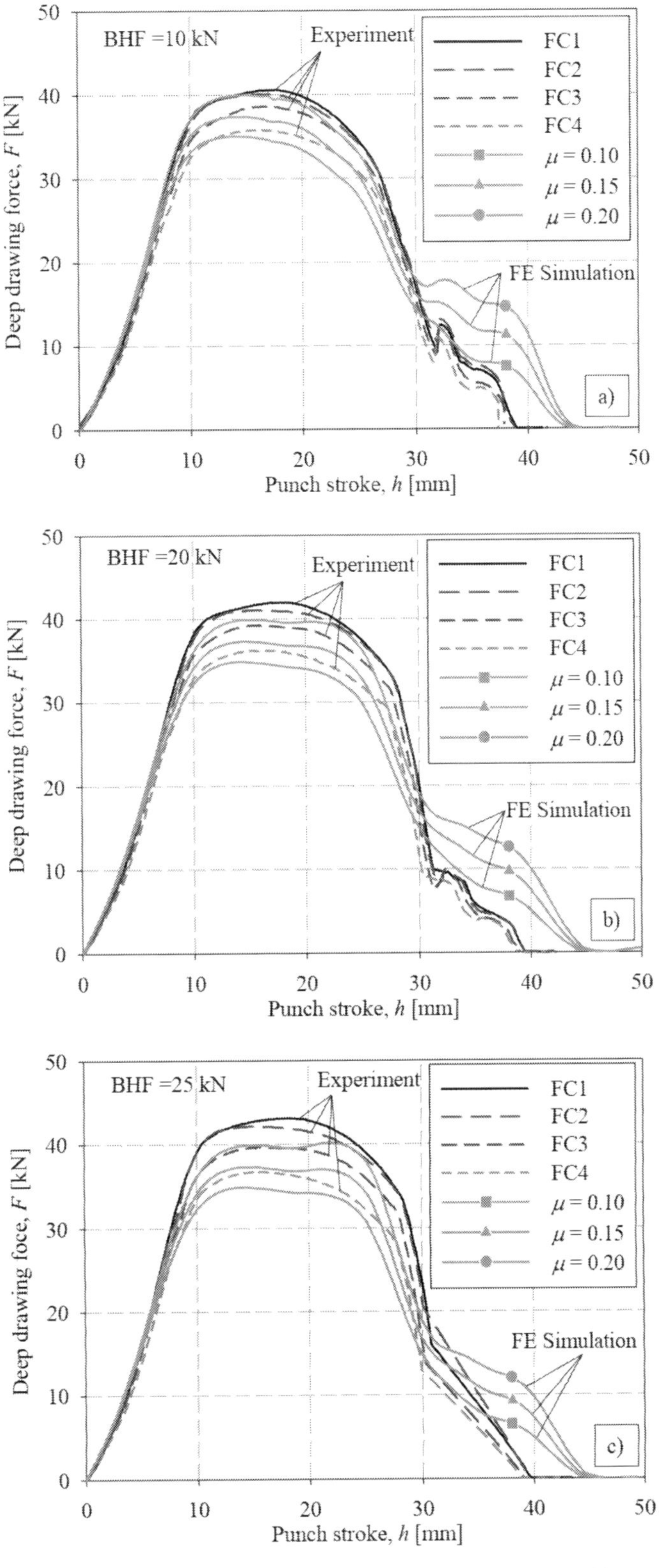

Fig. 6. Calibration of the friction coefficient by overlapping FE simulation with experimental results.

5 Conclusions

The effect of friction conditions combined with the intensity of blank-holding in the deep-drawing process of cylindrical cups was investigated by experiments and Finite Element simulation. Based on the results of this paper, the following conclusions can be formulated:

1) The drawing force is significantly affected by all friction conditions considered in this study. The drawing force increases in the following sequence of lubricants used: animal fat, lithium-base lubricant, hydraulic oil. As expected, the higher drawing force was found in the case of dry deep-drawing.

2) The drawing force is less sensitive to the changes in the blank-holding force when animal fat was used as a lubricant; the sensitivity increases when lithium-based lubricant, hydraulic oil and dry deep-drawing was used.

3) The estimated values of the friction coefficients are as follows: a friction coefficient between 0.10 and 0.15 corresponds to animal fat; the friction coefficient for the lithium-based lubricant lies between 0.15 and 0.20; a higher friction coefficient, very close to 0.20, was obtained for hydraulic oil; the case of dry deep-drawing corresponds to a friction coefficient higher than 0.20.

References

1. M. A. Ahmetoglu, A. Coremans, G. L. Kinzel, T. Altan, SAE Technical Paper No. 930287 (1993)
2. A. R. Padmanabhan, M.C. Oliveira, J.L. Alves, L.F. Menezes, Finite Elem. Anal. Des. **43**, 1062 (2007)
3. S. Raju, G. Ganesan, R. Karthikeyan, Trans. Nonferrous Met. Soc. China **20** 1856 (2010)
4. M. T. Browne, M. T. Hillery, J. Mater. Process. Technol. **136** 64 (2003)
5. M. Dilmec, M. Arap, Int. J. Adv. Manuf. Technol. **86** 747 (2016)
6. H. Kim, J. H. Sung, R. Sivakumar, T. Altan, Int. J. Mach. Tools Manuf. **47** 2120 (2007)
7. H. Kim, T. Altan, Q. Yan, J. Mater. Process. Technol. **209** 4122 (2009)
8. S. Subramonian, N. Kardes, Y. Demiralp, M. Jurich, T. Altan, J. Manuf. Sci. Eng. **133** 061001 (2011)
9. L. Ju, T. Mao, J. Malpica, T. Altan, J. Manuf. Sci. Eng. **133** 051010 (2015)
10. eta/DYNAFORM, *User's Manual* (Engineering Technology Associates, 2010)
11. L. Lăzărescu, I. Nicodim, D. Banabic, Key Eng. Mater. **639** 33 (2015)

Selected properties of galvanic coatings shaped using thermal drilling technology

Waldemar Matysiak[1,*], *Dariusz* Bartkowski[1] , and *Aneta* Bartkowska[2]

[1] Poznan University of Technology, Faculty of Mechanical Engineering and Management, Institute of Materials Technology, Poznan, Poland
[2] Poznan University of Technology, Faculty of Mechanical Engineering and Management, Institute of Materials Science and Engineering, Poznan, Poland

Abstract. The paper presents the investigations of mechanical properties and microstructure of materials formed by thermal drilling process (Flow drill). The aim of this study was to determine the microstructure of the surface layers produced using galvanic method on the low-carbon steel after formed this layer using thermal drilling technology. The coatings of copper and nickel thicknesses of 5, 10 and 15 μm were applied. Microstructure, chemical composition and microhardness was study. It was found the possibility of forming galvanic coatings using thermal drilling technology and change in chemical composition of treated material.

1 Introduction

At present, the requirements for the construction of tools and machine parts are increasing. Therefore, the search for new technological and material solutions is inevitable and becomes the goal of many researchers [1 - 7]. In order to make good connections, both detachable and inseparable, not only must the material have specific properties but also the processing technology used must provide the highest quality end product. To achieve specific requirements, it is advisable to improve the properties of the material using various enhancements. These can include, but are not limited to, surface modifications of the material by creating a new surface layer with unique properties that could not be obtained using the substrate material. Surface engineering technologies [4 - 8] also affect other technologies, such as plastic processing [3, 8 - 18]. One of the processes of plastic working is the hole flanging process, which is intended to increase the diameter of the previously cut hole and to form the material around as a conical or cylindrical flange. There are several ways to do this in different sources. One of them is flanging through the cylindrical punch with a tapered end which is supposed to pierce through the material by punch, and then form a corresponding hole shape and a flange around it. A major disadvantage of this method is the poor quality of the flange. There is also a technology for the hole flanging process through the thermal drilling (Flow drill). This is a method of producing flange holes without chip. Using this process, during contact of the tool with the workpiece, due to the

* Corresponding author: waldemar.matysiak@put.poznan.pl

high vertical force and the heat generated by the rotation and friction, the material is plasticised. This makes the conical part of the tool squeeze through the material and forming a hole. In the further part of the process, two fins are created: the top, from which the flange can be formed, and the bottom, which is shaped of sleeve. After shaping the flange of hole, the tool is retracted and the element cooled down in the air. The process takes place at very high speeds [8, 10, 12].

This work focuses on the flanging hole process of the elements with the galvanized coating. Flow drill technology is used in this research, which is significantly different from the classic methods of flanging. Selected features such as macro- and microstructure, coating thickness, microhardness, and chemical composition of the produced were investigated and results were analysed.

2 Research Methodology

2.1 Materials and technology

The starting material used for the hole flanging process was the general purpose structural S235JRH steel. The chemical composition of this material was shown in Table 1, and mechanical properties were shown in Table 2.

Table 1. Chemical composition of S235JRH steel used for testing.

Steel grade	Maximum weight share [%]				
	C	Mn	P	S	N
S235	0.17	1.40	0.045	0.045	0.009

Table 2. Selected properties of S235 steel used for testing.

Steel grade	Yield strength R_{eH} [N/mm^2]	Tensile strength R_m [N/mm^2]	Minimal elongation A_5 [%]
S235	235	360	20

Specimens of 30 x 24 x 20 mm with a wall thickness of 2 mm were cut from the profile, and next were galvanized. Nickel plating and galvanic copper plating were performed. Nickel plating was carried out by chemical deposition in a bath consisting of nickel chloride, sodium hypophosphite and sodium acetate. The bath temperature was equal to 60 °C. Copper plating was carried out in the acid bath (sulphate) composed of copper sulphate and sulphuric acid. The bath temperature was 38 °C, while the current density was 2 A/dm^2. These parameters were shown in Table 3.

Table 3. Parameters of galvanic treatment.

Galvanic treatment	**Nickel-plating**			**Copper-plating**		
Temperature [°C]	60			38		
Current density [A/dm^2]	-			2		
Time [h]	3	6	9	0.5	1	1.5
Average coating thickness [µm]	5	10	15	5	10	15

Specimens with galvanized coating were subjected to a flow drill process. Holes and flanging for each specimen were made on a JET JMD-18VS drilling machine. The machine was equipped with a handle with an aluminium cooling ring in which a drill bit of 7.3 mm was mounted. The tool was secured with a Flow drill FDKS paste that was applied each time before the hole was made to protect the tool from high temperatures. Tools turnover amounted to 3000 rev / min, and feed rate was performed manually. Figure 1 shows a scheme of the hole flanging process in materials with a galvanized coating.

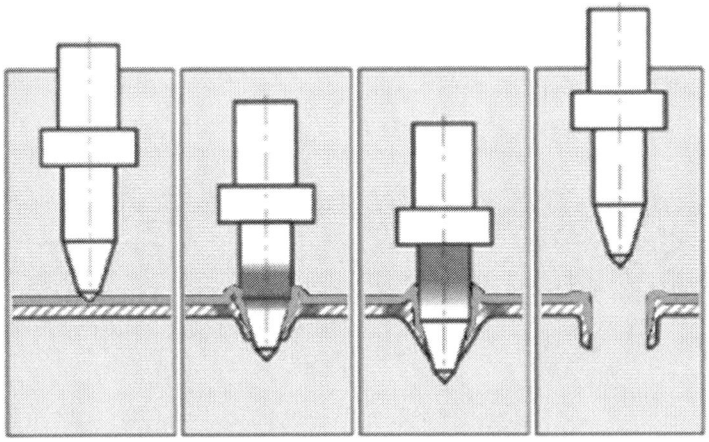

Fig. 1. Scheme of flanging using Flowdrill technology in materials with coating.

2.2 Microstructure and microhardness study

Macroscopic observation of individual specimens was performed using an unarmed eye and a magnifying glass. For each specimen, colour, cracks and porosity were determined. Photos were taken with the Nikon D3100. Microstructural observations were made using the Huvitz HRM-300 microscope. For the tests, metallographic cross section sewer made, which were prepared by cutting the specimens and placing them in Buehler's Conduktomet conductive thermoset resin. To reveal the microstructure of specimen's substrate and its surface, nital was digested, which is a mixture of nitric acid and ethanol. The Axio Vision microscope software program from Carl Zeiss was used to investigate the thickness of coatings. Microhardness measurement were performed using the Vickers method on the Buehler microhardness tester.

2.3 Chemical composition study

The chemical composition of specimens was examined using a scanning electron microscope (SEM) equipped with PGS's Prism 2000 EDS (Energy Dispersive Spectroscopy) system. EDS analysers allow both surface and volumetric identification of the chemical elements contained in the test material. A linear analysis of the chemical composition was carried out for each specimen.

3 Results of study

3.1 Microstructure and microhardness results

In Figure 2 was shown macroscopic images of specimens made after the galvanic copper plating process. The process was carried out in such a way that the final thickness coatings was obtained 5, 10 and 15 μm. Good surface quality has been found after the coppering process. Stains created during the process do not belong to technical drawbacks of galvanic processes. In Figure 5 was shown macroscopic images of specimens after nickel plating on 5, 10 and 15 μm thick. They were characterized by surface homogeneity.

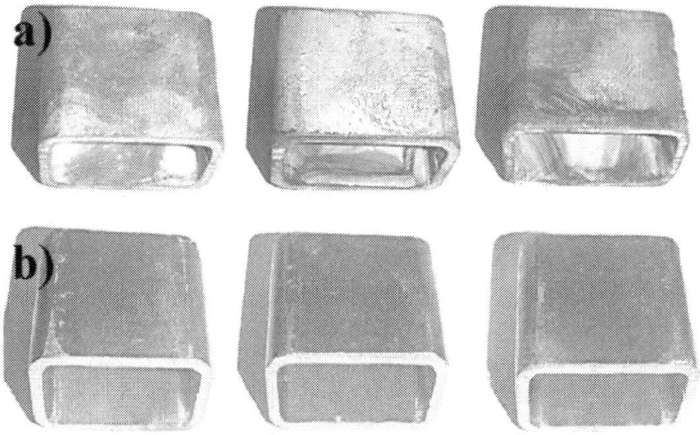

Fig. 2. Specimens after: a) galvanic copper plating, b) galvanic nickel plating with thickness of 5, 10 and 15 μm (left to right).

The microhardness of the copper coating is about 154 HV on average. During the study it was found that the application of the copper coating and its plastic shaping increased the microhardness of the inner side of the flange to over 300 HV. Microhardness measurement results are shown in Figure 3a. Microhardness of the nickel coating is about 400 HV on average. However, it has been shown during the tests that the application of the nickel coating and its plastic shaping does not affect the hardening of the inner side of the flange. The hardness remains the same as the steel substrate. Micro hardness measurement results were shown in Figure 3b.

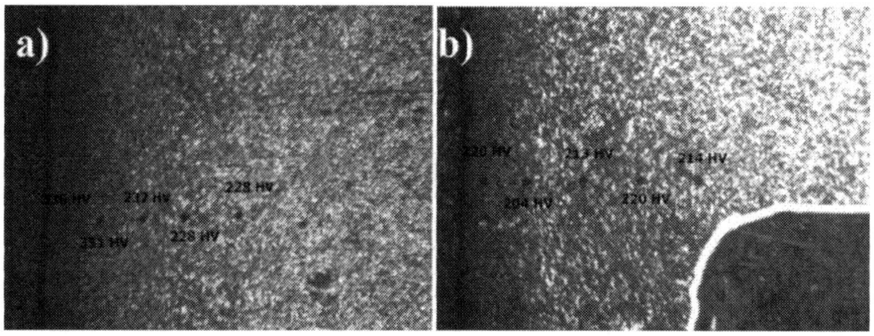

Fig. 3. Microhardness of the inner part of flange: a) after galvanized copper plating and Flow drill process, b) galvanized nickel plating and Flow drill process.

In Figures 4a-4c were shown the results of the microstructure study of galvanic copper coatings. It has been possible to shape copper galvanized products using Flow drill technology. The coating on the outside of the fold does not crack, but on the inside the copper modifies the substrate through diffusion processes. In Figures 5a-5c were shown the results of microstructural studies on nickel coatings. It has been possible to shape copper galvanized products using Flow drill technology. The coating on the outside of the fold did not crack at 5 and 10 μm thick, but at 15 μm thickness cracks were observed. On the inside, the nickel coating modifies the substrate through diffusion processes.

DOI: 10.1051/matecconf/201713705004

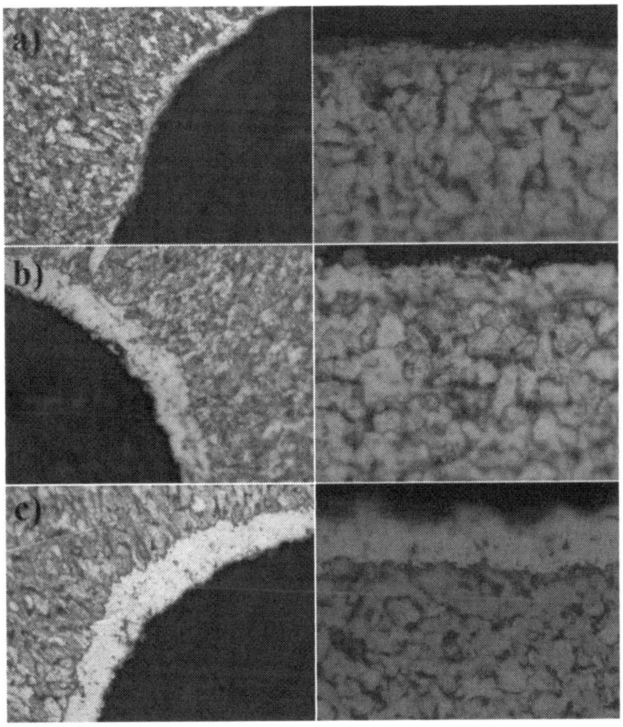

Fig. 4. Galvanic coating of copper having a thickness of : a) 5 μm, b) 10 μm, c) 15 μm after flanging of hole edge; magnification: figure left – 100x, figure right – 500x.

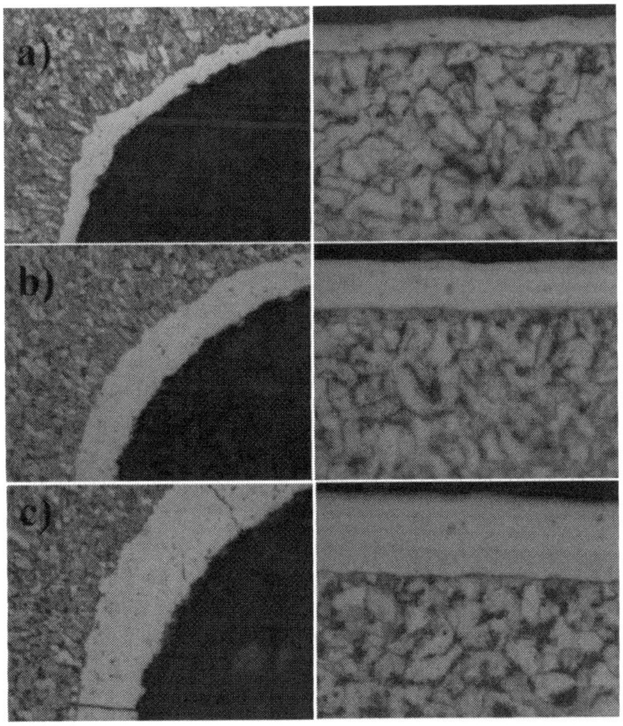

Fig. 5. Galvanic coating of nickel having a thickness of : a) 5 μm, b) 10 μm, c) 15 μm after flanging of hole edge; magnification: figure left – 100x, figure right – 500x.

 DOI: 10.1051/matecconf/201713705004

3.2 Chemical composition study

Figure 6 shows a linear distribution of the chemical composition on specimen after galvanic copper plating process on thickness of 15 µm and after Flow drill process. As can be seen, the copper coating during the thermoforming process has been mixed by diffusion with the steel substrate. It can be found that the coating was well adhered to the steel substrate and plastically shaped. Due to the increased temperature and friction the properties of the steel surface have been altered at the point of flanging on the inner side of the sleeve.

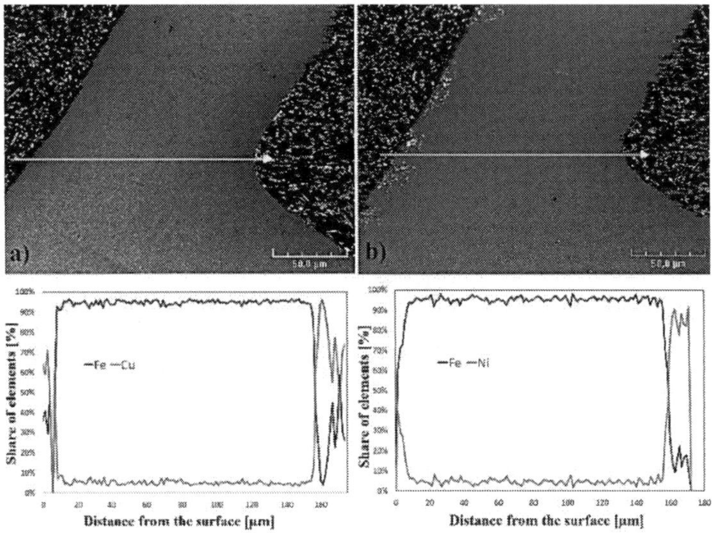

Fig. 6. Linear distribution of chemical composition after: a) galvanic copper and Flow drill process, b) galvanic nickel and Flow drill process.

Figure 15 shows a linear distribution of the chemical composition on a plated nickel plated sample with a 10 µm film thickness under Flow drill. As can be seen, the nickel coating during the thermoforming process has been mixed by diffusion with a steel substrate. It can be stated that the nickel coating has not been abraded, but has been plastically shaped. Due to elevated temperatures and friction, the steel surface properties of the rolled steel surface have been altered on the inside of the sleeve.

4 Conclusions

Based on the presented studies it was possible to perform the thermoforming process in materials with the produced surface layers in the form of galvanic coatings. General work requests are presented below:

• There is the possibility of plastic forming of steel with galvanic coatings using Flow drill technology

• The thickness of the surface layers produced significantly influences the possibility of their shaping by thermal drilling. Thicker coatings of nickel are more prone to cracking. Copper coatings do not tend to crack during shaping.

• Flow drill process on galvanic coatings modifies the steel surface. The chemical composition of the inner part of the hole flange was changing.

The method described in this paper may be used in the case of joining materials with coatings for example, if the galvanic coatings already existed on the jointed products, or if no other connection is possible e.g. by welding.

References

1. L. Sobotová, Transfer inovácií (in Slovak) **10,** 144-148 (2007)
2. L.Sobotová, E. Spišák, Automotive industry magazine (in Slovak) **2,** 62-64 (2009)
3. W. Matysiak, L. Bernat, Metalurgija **54,** 235-238 (2015)
4. D. Bartkowski, A. Młynarczak, A. Piasecki, B. Dudziak, M. Gościański, A. Bartkowska, Opt Laser Technol **68**, 191-201 (2015)
5. A. Bartkowska, A. Pertek, M. Jankowiak, K. Jóźwiak, Arch Metall Mater. **57,** 211-214 (2012)
6. A. Bartkowska, A. Pertek, Surf Coat Tech **248**, 23-29 (2014)
7. D. Bartkowski, A. Bartkowska, IJRMHM **64**, 20-26 (2017)
8. S. Zhou, X. Dai, H. Zheng, Opt Laser Technol **44,** 190-197 (2012)
9. T. Baidridge, G. Poling, E. Foroozmehr, V. Kadekar, M.C. Gupta, Opt. Lasers Eng. **51,** 180-184 (2013)
10. S. Zhou, Y. Huang, X. Zeng, Appl Surf Sci **254,** 3110-3119 (2008)
11. S.M. Zhu, Y.D. Zhang, Appl. Mech.Mater **556-562,** 189-192 (2014)
12. Q. Ma, Y. Li, J.Wang, K. Liu, Mater Des **92,** 897-905 (2016)
13. W. Matysiak. D. Bartkowski, M. Hatala, J. Zajac, S. Radhenko, R. Cep, Key Engineering Materials **669,** 87-94 (2016)
14. W. Matysiak, Metal forming **4**, 49-57 (2005)
15. W. Matysiak, B. Barišić, I. Mamuzić, Metabk **49,** 13-17 (2010)
16. S. Dobatkin, J. Zrnik, I. Mamuzic, Metalurgija **48,** 157-160 (2009)
17. H.-M. Chow, S.-M. Lee, L.-D. Yang, J Mater Process Tech **207,** 180-186 (2008)
18. S. F. Miller, J. Tao, A. J. Shih, Int J Mach Tool Manu **46,** 1526-1535 (2006)

Deformation prediction and Finite Element Analyses of precision seamless tubes during cold drawing

Martin Necpal[1,*], *Maroš* Martinkovič[1] and *Ivan* Buranský[1]

[1]Slovak University of Technology, Faculty of Materials Science and Technology in Trnava, Institute of Production Technologies, J. Bottu 25, 917 24, Trnava, Slovakia

Abstract. This paper deals with the deformation of precision seamless tubes during the cold drawing process, using Finite Element Methods. The value of strain obtain by stereology measurement of degree grain boundaries deformation are compared with simulation result for each drawing pass. The material chosen was the E235 and E355 steel grade, which is drawn from the initial dimension of Φ70 x 6.3 mm to the final dimension of Φ 50 x 3.75 mm by two passes. FEM Software DEFORM 2D with the Lagrangian method is used to numerical analyses of the cold drawing process.

1 Introduction

Cold drawing technology used for manufacturing of precision seamless tube depends on many factors. Cold forming process parameters, tool geometry initial and final tube dimension are selected to ensure a high quality of the tubes. In order to achieve the final diameter and wall thickness, tubes are reduced successively in several cold drawing steps. This can be done by either drawing the tube through a die and by adding a plug, which results in better defined wall thickness and inner surface quality. Cold drawing using a fixed plug is a versatile technology that is used usually in the Železiarne Podbrezová ironworks company. The experiment results of this company are compared with the numerical simulation described in this paper.

2 Experimental Material

In this work, steel grade E235 and E355 (ferritic-pearlitic carbon steel, see Table 1 [1]) was considered; as a feedstock for cold drawing, hot rolled tube with dimensions of Ø70 x 6.3 mm was selected. Mechanical properties of E235 steel are as follows: yield stress $R_e = (226 \div 245)$ MPa, tensile strength $R_m = (343 \div 441)$ MPa, ductility $A_5 = 24$ %, mechanical properties of E355 steel are as follows: yield stress $R_e = (335 \div 355)$ MPa, tensile strength $R_m = (490 \div 630)$ MPa, ductility $A_5 = 24$ % [2].

* Corresponding author: martin.necpal@stuba.sk

Table 1. Chemical composition of E235 and E355 steel grade in wt %.

	C	Mn	Ni	Cu	Mo	Sn	P	S	Si	Ti	Cr
E235	0.090	0.420	0.060	0.140	0.020	0.012	0.013	0.011	0.200	0.002	0.060
E355	0.180	1.220	0.110	0.170	0.030	0.017	0.025	0.012	0.270	0.002	0.100

2.1 Workpiece and tools of material modelling

As the temperature stays lower in cold drawing, the material properties do not vary and can therefore be supposed to by independent of the temperature variation. As the material behavior is considered temperature independent, the power law constitutive equation of material plastic properties can be used eq. 1. Where ε represent the equivalent effective plastic strain, $\dot{\varepsilon}$ represent the effective plastic strain rates, and c, n, m, y material parameters are determined by means of tensile tests [1, 5], they are listed in Table 2. The tube material was considered as plastic, the hardening is assumed as isotropic, and yield function type is set as Von Mises.

The geometry of the tube has been meshed. The eight elements are across the wall of the tube in first pass the six are in second pass Fig. 5. This mesh of workpiece is sufficient for accuracy and does not need to be re-mesh procedure during the calculation.

The forming tools, (mandrel and die) was considered as rigid bodies. The mesh of tools did not need to be generated for simulation.

$$\sigma = c\varepsilon^n \dot{\varepsilon}^m + y \tag{1}$$

Table 2. Mechanical properties of the stainless steel materials.

		E235 material	E355 material
c	Material constant (-)	178	404
n	Strain exponent (-)	0.61	0.63
m	Strain rate exponent (-)	0.02	0.02
y	Initial yield value (MPa)	235.5	345

2.2 Numerical simulation

DEFORM 2D and the Lagrangian calculation has been set for numerical simulation of cold drawing technology, according to the tool dimension given on Table 3. The initial axisymmetric geometric configuration has been set. The drawing velocity was 9 m.min^{-1} for all passes in experimental drawing and simulation boundary conditions. The friction model between the tools and the material was chosen to be a shear-type with the value of 0.08. The data interpolation functionality was used to set the second pass calculation. The final data from first pass was transfer, and to set second pass. This data transformation must be done, because drawing technology were performed without an interoperative annealing.

2.3 Tube Drawing Technology

The main production steps for tube cold drawing technology in Železiarne Podbrezová are as follows:

- feedstock pre-processing (hot rolled tube with dimensions of Ø70 x 6.3 mm).
- cold/hot pointing of tube ends (target diameter 40 mm).

- chemical treatment of the tube (pickling, phosphating, lubrication).
- cold drawing (see Tab. 3 for details).
- intermediate annealing and final annealing in protective atmosphere.
- final conditioning.
- surface inspection, packaging, rust-proofing.
- dispatch.

The final reduction of the tube area for two-pass technology (from Ø70 x 6. 3 mm to Ø50 x 3.75 mm) was 57.78 %; the value was divided rather uniformly among all passes (see Table 3). Calculation of tube cross-section (area) reduction:

$$Reduction = \frac{S_R}{S_0} \cdot 100\ [\%], \tag{2}$$

Where:
$S_R = S_0 - S$ [mm^2] and
S_0 – tube area before drawing [mm^2],
S – tube area after drawing [mm^2].

In the 1st pass the area reduction was 40.52 %, and in the 2nd pass the area reduction was 27.34 %.

Table 3. Parameters of tube drawing (O.D. – Outer Diameter, W.T. – Wall Thickness, L – Length).

	Feedstock dimensions [mm]			**Tube dimensions [mm]**			**Pointing [mm]**		**Reduction [%]**		
Pass No.	**O.D.**	**W.T.**	**L**	**O.D.**	**W.T.**	**L**	**L (tip)**	**Ø (tip)**	**Total**	**W.T.**	**O.D.**
1	70	6.30	4300	55	4.75	6725	300	40	40.52	25	21
2	55	4.75	6725	50	3.75	9255	300	40	27.34	21	9

Local plastic deformation of tube was analyzed. During subsequent drawing steps, the tube undergoes plastic deformation. Final geometry and properties of deformed tubes depends on strain value. It is possible to investigate change of material structure and predict plastic deformation.

The microstructure of each pas was observed in orthogonal tangential and longitudinal plane of deformed tube with about 500x magnification of light microscope and metallographic cut Fig1. Tangential metallographic cut is don three times. On the inside of the pipe surface "I", on the outside of the pipe surface "O" and on the middle "M" of the pipe wall Fig. 2. Due to plastic deformation, the structure anisotropy increased – grain boundaries orientation in deformed place of piece was observed. The local strain in analyzed place was obtained by stereological measurement of degree of grain boundaries orientation [4, 6]. The anisotropic microstructure was decomposed into isotropic and planar oriented components using quantitative metallography [3, 4] (Saltykov stereology methods with oriented test lines). Measured and calculated parameters of local plastic deformation in deformed zone are listed in Table 4.

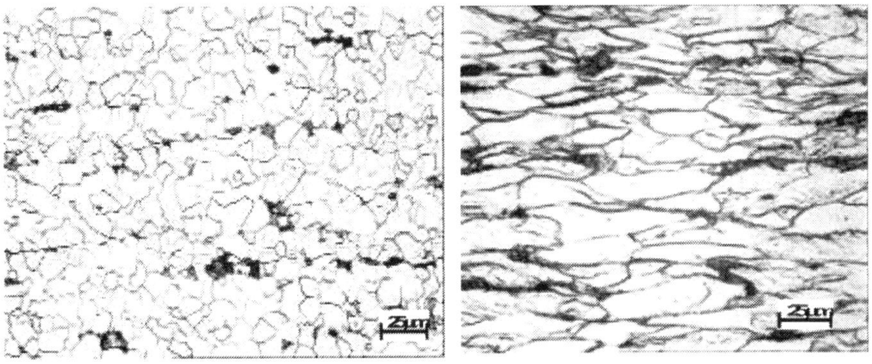

Fig. 1. Example of undeformed microstructure (left), deformation after 2 pass (right) [1].

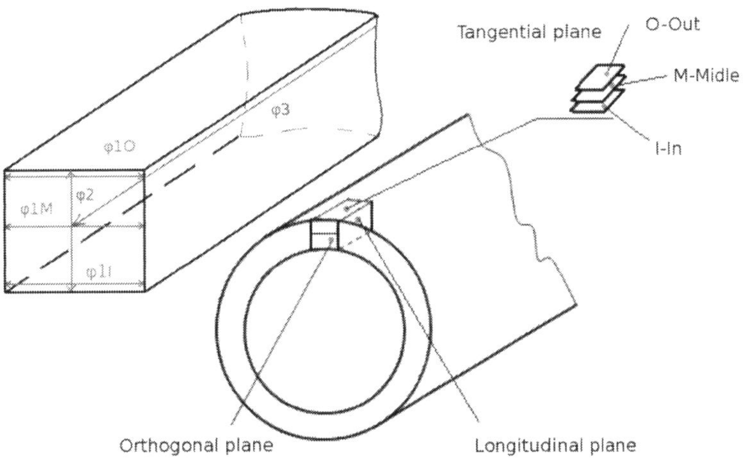

Fig. 2. Metallographic cuts and planes on the part of the deformed tube.

Table 4. Stereology methods and simulation strain values result.

		Values obtained by stereological measurement [1]					Values obtained by simulation				
		φ1O	φ1M	φ1I	φ2	φ3	φ1O	φ1M	φ1I	φ2	φ3
E235	Pass 1	-0,241	-0,237	-0,215	-0,282	0,520	-0,243	-0,239	-0,232	-0,298	0.537
	Pass 2	-0,336	-0,320	-0,301	-0,519	0,839	-0,339	-0,324	-0,305	-0,551	0,874
E355	Pass 1	-0,241	-0,237	-0,232	-0,282	0,520	-0,243	-0,239	-0,233	-0,297	0,536
	Pass 2	-0,336	-0,320	-0,300	-0,519	0,839	-0,339	-0.324	-0,305	-0,55	0,874

3 Results and discussion

After simulation, it was possible to compare the strain values obtained by stereological method and the values obtain by the simulation. The strain in all directions according Fig. 2, was analyzed. Theta φ1 strains are analyzed in three sections by the cross-section of the tube wall Fig. 2, 3. Strain φ1I is near the inner surface, strain φ1O is near the out surface, and

DOI: 10.1051/matecconf/201713705005

strain φ1M is in the middle of the tube wall. As shown in Fig. 3, after the 1 pass, the difference of strain between the E235 and E355 is minimal. On the inside, "A" of the tube is smaller negative strain, that on the outside "B" of the tube. The same result is after pass 1.

Calculated values of Radial φ2 strain for all passes and materials can be seen in Fig. 4. Due to high reduction 40.52 % after the first pass, there is observable positive φ2 strain in deformed area. After the second pass, where the reduction area is 27.34 %, there is only negative φ2 strain.

Calculated values of φ3 strain for all passes and materials can be seen in Fig. 5.

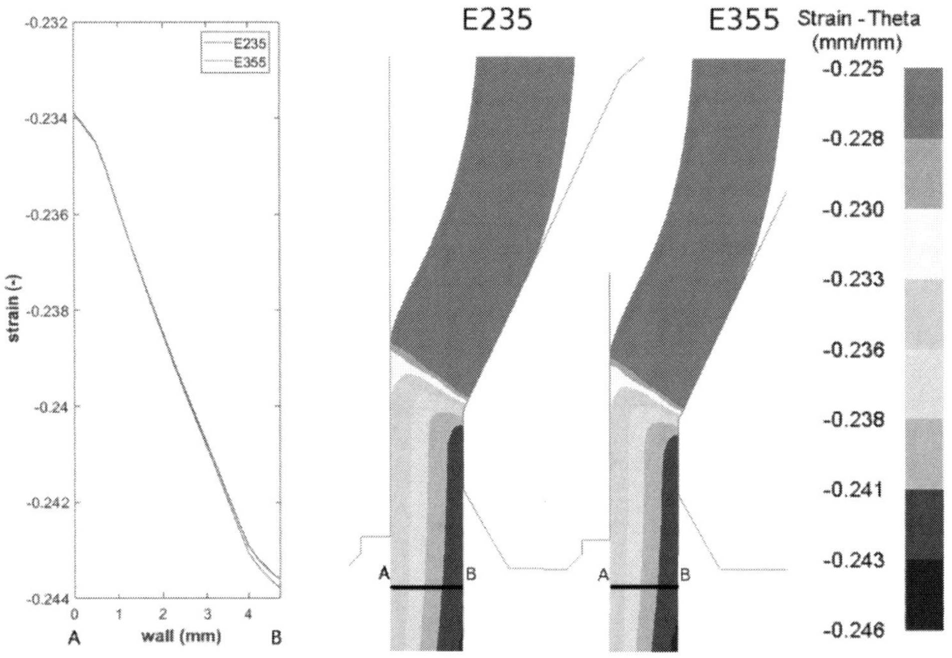

Fig. 3. Strain φ1 across the wall of the tube after 1 pass.

Initial strain of tube which is not deformed is 0. After first pass φ3 strain is approximately 0.54 for E235 and also E355 material Fig. 5 left. The data interpolation method was used for initialization of second pass. Initial strain for second pass is between 0.550 and 0.625 (Fig. 5, right).

 DOI: 10.1051/matecconf/201713705005

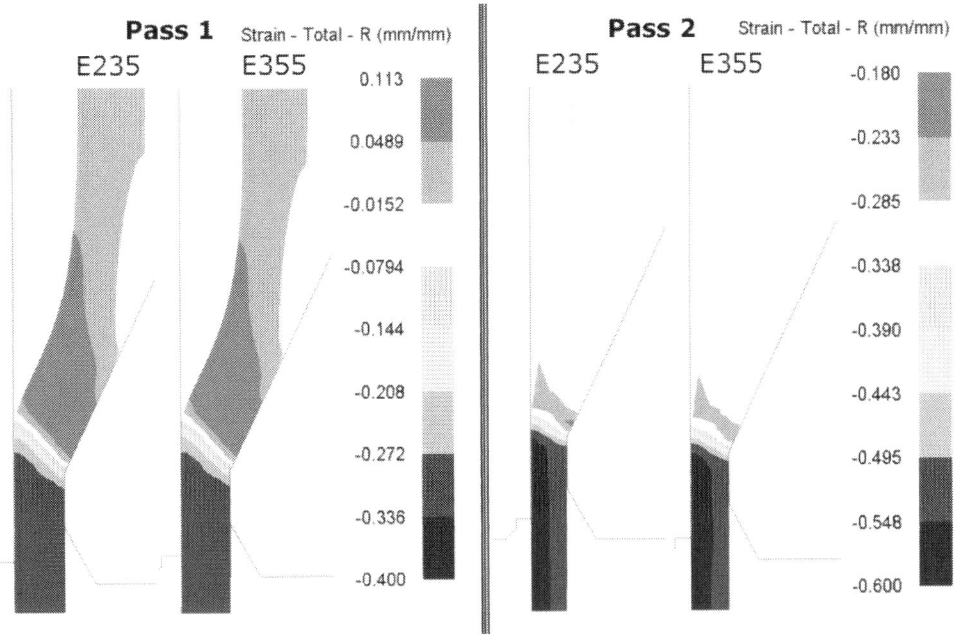

Fig. 4. Pass 1, and pass 2 strain φ2 with material E235 and E355.

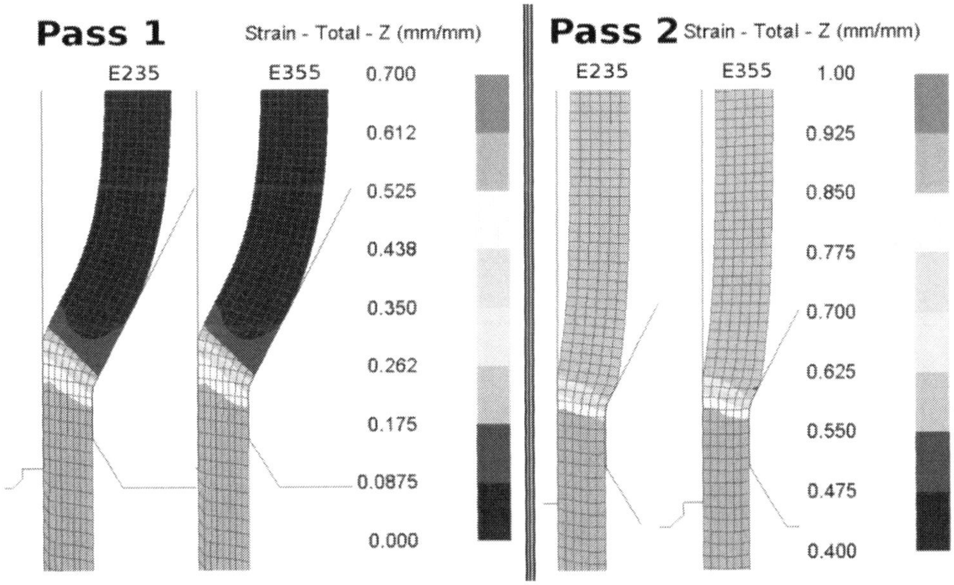

Fig. 5. Pass 1, and pass 2 strain φ3 with material E235 and E355.

4 Conclusion

The key topic of this paper was to compare S. A. Saltykov, Stereometric metallography method to investigation of deformation with result of numerical FEM simulation using DEFORM 2D software. The result can be summarized in:

 DOI: 10.1051/matecconf/201713705005

- Beside the experimental investigation using the stereology measurement and FEM simulation are showed very good agreement of final strain values
- By comparing the results of simulation of two different material models (E235 and E355) we can confirm that the material model has a minimal effect on the strain result.
- Simulation using software DEFORM 2D and axisymmetric Lagrangian method can be a valuable tool for prediction deformation strain in cold drawing process.

This work was supported by the Slovak Research and Development Agency under the contract No. APVV-15-0319

References

1. M. Ridzoň, *Research on technological parameters affecting the production and properties*, (PhD thesis) - State Technical University Bratislava, Trnava, (2010)
2. I. Fürbacher, K. Macek, K. Steidl, K. a kol, *Lexikon technických materiálů* [Lexicon of technical materials], **1,** Praha: Verlag Dashöfer, (1999)
3. M. Martinkovic, S. Vaclav, Engineering and Medicine, **5,** 585 – 588 (2013)
4. S. A. Saltykov, *Stereometric metallography*, **3**, Metallurgia, Moskva, (1970)
5. T. Holmquist, G. Johnson, J. de Physique **IV** Colloque, **01** (C3), C3-853-C3-860, (1991)
6. M. Martinkovic, *Ceramics-Silikáty*, **61.2**, 141-146 (2017)

Surface roughness obtained at stamping of bearing cages

Marius Ionuţ Rîpanu[1], *Gheorghe* Nagîţ[1], *Laurenţiu* Slătineanu[1], *Oana* Dodun[1,*] and *Andrei Marius* Mihalache[1]

[1] "Gheorghe Asachi" Technical University of Iaşi, Department of Machine Manufacturing Technology, Blvd. D. Mangeron, 59A, Iaşi, Romania

Abstract. Stamping of the bearing cages ensures a high enough productivity and consistency with the requirements of a large series manufacture. The roughness of stamped surfaces is one of the parameters for assessing the quality of the cages, which in its turn determines the behaviour in service of the bearings. A large number of input factors corresponding to the stamping process exert influence on roughness parameters. An experimental research has been conceived and developed to assess the influence exerted by the thickness of the blank and the clearance between the active elements of the die on the value of the surface roughness parameter *Ra*. The mathematical processing of experimental results allowed the identification of empirical mathematical models and graphical representations were designed to offer additional information concerning the influence of the two input factors in the stamping process on a quality parameter of the bearing cage.

1 Introduction

The bearings are complex machine elements used to support parts found in rotation or oscillation movements (shafts, gears, belt pulleys, cable rollers, rotary tables etc.). Essentially, a bearing includes an outer ring, an inner ring, rolling elements and bearing cage. The main function of the bearing cage is to avoid the direct contact between the rolling elements. In the case of roller bearings, the bearing cage ensures also the guiding of the rolls, while in the bearings with separable rings, the bearing cage retains the rolling elements so that they could not fall, when mounting or dismounting the bearing [1, 2].

The quality of the bearing cage could be connected with its functions; due to the dynamic solicitations, the materials for bearing cages should be characterized by a good mechanical strength. A low value of the friction coefficient could increase the efficiency of using bearing cages. On the bearing cage service behaviour, the accuracy, the surface roughness, the properties of the surface layer and the state of edges also exert influence.

Due to high quality requests specific to the bearings and due to their various fields of use, certain materials and manufacturing techniques were gradually selected for obtaining bearing cages. Thus, for the bearing of universal use and of small dimensions, the bearing

* Corresponding author: oanad@tcm.tuiasi.ro

DOI: 10.1051/matecconf/201713705006

cages are achieved from drawn sheets of carbon steel, by successive processes of cutting, punching and drilling. If the bearing cage includes two components, as the case of radial ball bearings is, the assembling is achieved by clinch.

In the case of stamped bearing cages, the main factors able to affect the bearing accuracy are the dimensions and the configuration of the bearing, inclusively the blank thickness, the type and the properties of bearing cage material, the state of the cutting tool edges, inclusively the size and the uniformity of their wear, the clearance between the tool components involved in the stamping process, the sheet flatness, the type and the constructive solution of the stamp, the way of clamping the blank, of limiting the blank feed movement, the existence of a lateral pressure exerted on the blank, the type and the state of the stamping machine etc.

The state of the cut surface state is affected by the same factors able to influence the stamping accuracy.

Over the years, the researchers developed investigations concerning the influence of distinct factors on the stamping accuracy and on the state of the cut surfaces. Usually, the cut surfaces show two distinct zones, one smooth along about 1/3 of the blank thickness and the other characterized by rupture aspects, along 2/3 of the blank thickness. Due to the presence of clearance between stamping tool components, the resulted surfaces could present a certain slope.

Saez de Argandoña et al. proved the negative influence of the clearance size between the stamp active components on the geometry of the part, on the cut surface state and on the bur size [3]. They consider that there is an optimal value of the clearance between the surfaces of the stamp active components, when the negative results are minimum.

If the clearance have a low size, one could show that the surface roughness parameter *Ra* could have value under 3.2 μm and even under 1.6 μm, the burs size does not exceed 0.01 mm, the thickness of the hardened surface layer is lower than 0.3 mm and the hardness could be with 40 pp– 80 % higher than the hardness of the blank base material [4, 5]. In order to obtain a low surface roughness, higher cutting speed and lower blank thickness could be used, when it is possible [6, 7].

The main objective of the research presented in this paper was to estimate the optimal values of the stamping process input parameters, to obtain a surfaces of good quality.

2 Process of cutting the windows in the bearing cages

Some considerations concerning the obtaining of cages of radial-oscillating bearings with coil rolls will be subsequently formulated. The technological process of manufacturing bearing cages includes operation of cutting, drawing, cutting and calibration of the bottom, perforation of the windows, turning of the external surface, deburring in drums, phosphating, quality control etc.

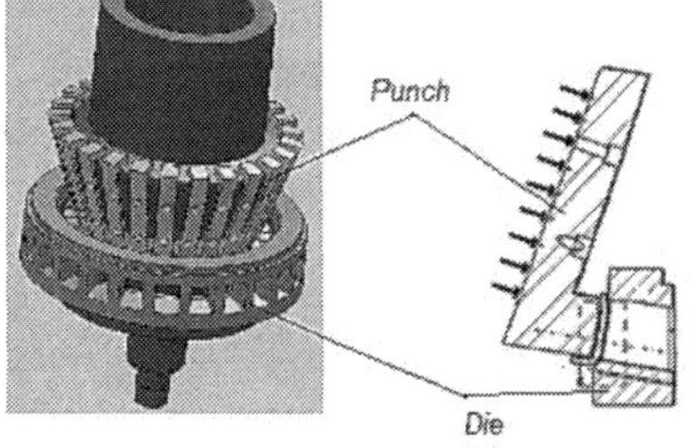

Fig. 1. Stamp components acting on the blank material during the process of cutting the windows in bearing cage and the way in which the punch detaches the window zone [2].

Generally, the cutting process involved in obtaining the cage windows needs a movement achieved by the punches along a direction perpendicular to the generatrix of the internal conical surface

DOI: 10.1051/matecconf/201713705006

of the cage (fig. 1). This movement is achieved under the action of the conical surface of the active zone of a central mandrel. As a result of force exerted by the punches sharpened edges, the blank material is pressed up to the moment when the first cracks appears; practically, a shearing process is gradually initiated. The propagation of the cracks and the continuation of the work movement achieved by punches, in a subsequent sequence, determines the removal of the blank material corresponding to windows. Since a certain clearance exists between the surfaces of active components of the stamping tool, a burr attached to the cage is also generated. As above mentioned, the height of the burr is dependent on the size of some process input factors.

The blank material has a not uniform movement under the action of the punch; on the other hand, the tool edge is not a simple smooth curve; at a micro level, certain irregularities of the tool edges could be observed. As a consequence, the cut surface will present some asperities which constitute the surface roughness.

Generally, the surface roughness includes the periodical and not periodical asperities characterized by a ratio between the wave length lower than 50 (sometimes, 40 is the value taken into consideration for this delimitation). There are many parameters applied to evaluate the surface roughness; one of the most used such parameters is the arithmetic mean deviation of the profile, symbolized by *Ra*.

One could consider that the main factors able to exert influence on the cut surfaces in stamping process are the clearance between stamp active components, the blank thickness, the length of the punch stroke, the cutting speed.

It is expected that if the clearance, the work stroke length and blank thickness are low, the surface roughness could be lower also.

A systemic analysis of cutting windows in bearing cages could highlight the main

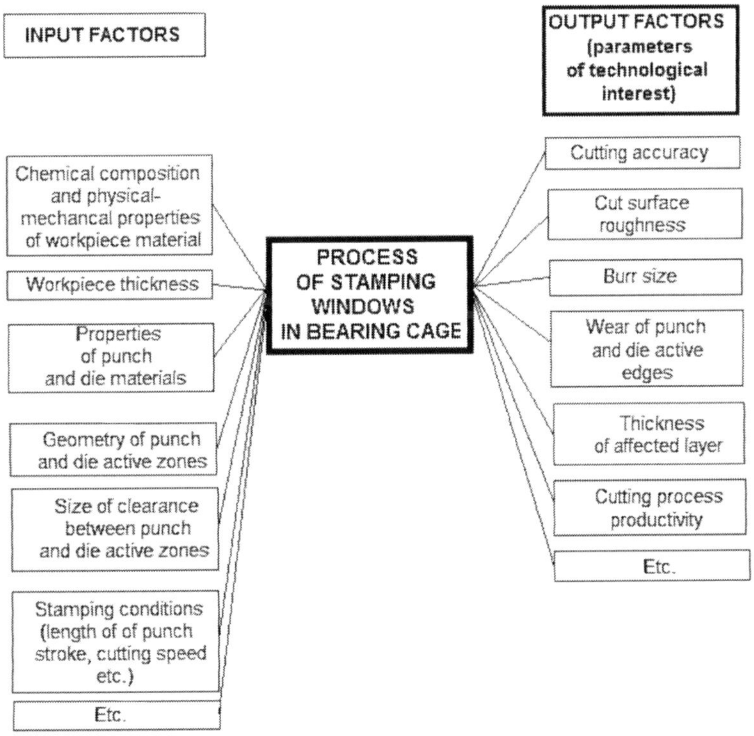

Fig. 2. Systemic analysis of factors able to affect the values of the parameters of technological interest at cutting of windows in bearing cages.

groups of factors able to affect the sizes of the parameters of technological interest (cutting accuracy, cut surface roughness, burs size, punch wear, thickness of hardened surface layer (fig. 2). One could see that the main such groups of factors are:
1. Chemical composition and physical – mechanical properties of blank material (friction coefficient, shear resistance, hardness, yield strength, formability etc.);
2. Properties of punch and die materials;
3. Geometry of punch and die active zones;
4. Size of clearance between punch and die;
5. Length of punch work stroke and cutting speed etc.

3 Conditions of experimental research

An experimental research was planned and materialized in to partially test the above-mentioned hypotheses concerning the factors able to affect the surface roughness at the cutting of windows in bearing cages. Metallic plates made of DD13 (1.0335, EN 10111) steel were used as test pieces.

The experimental research was developed on a hydraulic pressing machine type LVD 600, equipped with a stamp for perforating and calibrating windows in bearing cages [1]. The values of the surface roughness parameter *Ra* were measured by means of a Form Talysurf Series 2 equipment.

Experiments using distinct values for the blank thickness *t* and clearance *c* between the active components of the stamp were planned and achieved. Each experiment was achieved twice. Three values for the blank thickness were taken into consideration: t_1=3.5 mm, t_2=4 mm and t_3=5 mm. As one could see in Table 1, the clearance *c* had values between 0.11 and 0.18 mm.

Table 1. Influence exerted by the clearance *c* and blank thickness *t* on the surface roughness parameter *Ra* (length of work stroke *l*=6.5 mm).

Number of experiments set	Clearance, *c*, mm	Surface roughness parameter *Ra* (μm), for blank thickness, *t*, mm		
		3.5	4.0	5.0
1	0.11	0,43	0,47	0,47
2	0.13	0,50	0,56	0,62
3	0.14	0,57	0,63	0,68
4	0.15	0,60	0,67	0,73
5	0.16	0,62	0,70	0,74
6	0.17	0,63	0,70	0,76
7	0.18	0,70	0,77	0,81

The measured values of the surface roughness parameter *Ra* were inscribed in Table 1. The experimental results were mathematically processed by means of a specialized software based on the use of the method of least squares [9].

The software allows the selection of the most adequate empirical model by means of the Gauss's criterion; these criterion takes into consideration the sum of least squares corresponding to the differences between measured values and values allocated by considering a certain mathematical model, for the same experimental points. The lower is the value of the Gauss's criterion, the more adequate is a certain mathematical empirical

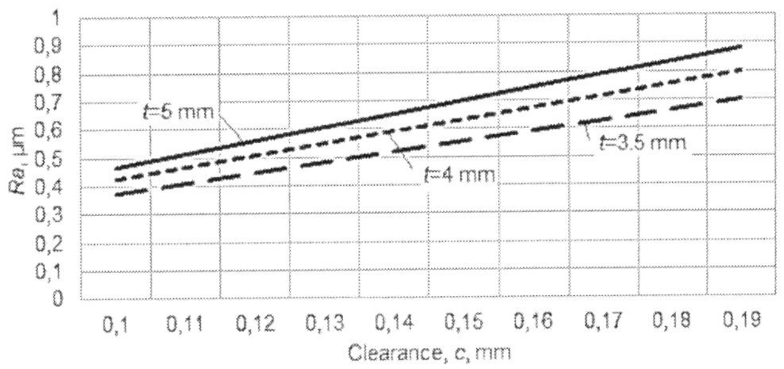

Fig. 3. Influence exerted by the clearance on the surface roughness parameter at cutting the windows in bearing cages (work stroke length l=6.5 mm).

model. The empirical models proposed are the polynomial of first and second degree, the power function, the exponential function, the hyperbolic function.

By mathematical processing of the experimental results, among the five possible empirical functions, the following was found as the most adequate to the experimental results:

$$Ra = -1.848 + 12.320c - 28.330c^2 + 0.549t - 0.0561t^2 \quad (1)$$

the value of the Gauss's criterion being S_G=0.0003344253. Since in machine building the power type function are used frequently, one established also such a function:

$$Ra = 2.176c^{0.985}t^{0.456}, \quad (2)$$

in this case, the Gauss's criterion having the value S_G=0.0006295096.

On the base of the mathematical empirical model constituted by the relation (2), the graphical representations from figures 3, 4 and 5 were elaborated.

The analysis of the mathematical empirical model (2) and of the graphical representations allowed the formulation of some conclusions. Thus, one notices that the clearance c exert an influence higher than the blank thickness t, since the value of the exponent attached to c in the mathematical empirical model (2) is higher than the value of the exponent attached to the blank thickness t (0.985>0.456). The increase of the both factors, clearance c and blank thickness t determines an increase of the surface roughness parameter Ra.

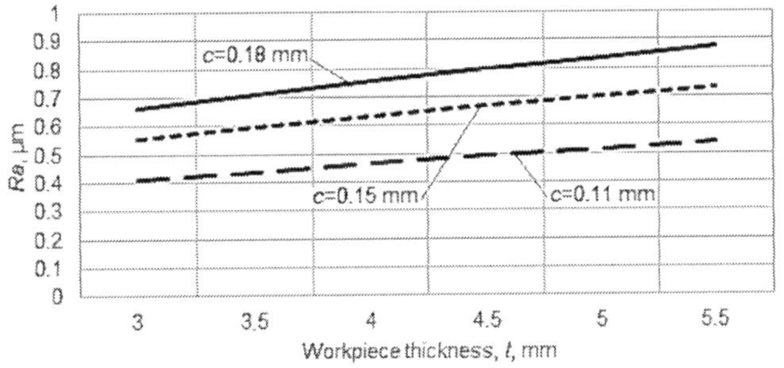

Fig. 4. Influence of the blank thickness t on the value of the surface roughness parameter Ra at cutting windows in bearing cages (stroke length l=6.5 mm).

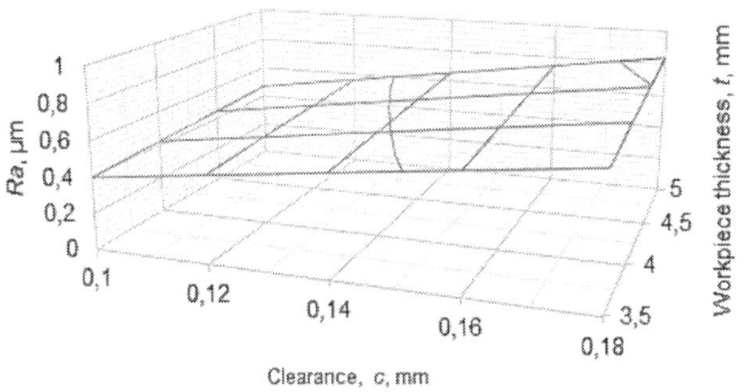

Fig. 5. Influence of clearance j and blank thickness g and on the value of the surface roughness parameter *Ra* at cutting windows in bearing cages (work stroke length l=6.5 mm).

4 Conclusions

One of the technological solutions applied to obtain windows in bearing cages is the application of a stamping process. The main factors able to affect the values of the surface roughness parameter *Ra* of the cut surfaces obtained by stamping process are the clearance between the active components of the stamp, the blank thickness and the cutting speed. In order to obtain a mathematical empirical model able to highlight the influence exerted by the clearance and the blank thickness on the value of the surface roughness parameter *Ra*, a theoretical analysis of the stamping process and an experimental research were planned and achieved. The experimental results were mathematically processed by means of a software based on the method of last squares. A mathematical empirical model power type was determined. The graphical representations elaborated on the base of the mathematical empirical model could be used to choose the right value of the clearance for a required value of the surface roughness. In the future, there is the intention to experimentally validate the mathematical empirical model and to extend the experimental research, to include the cutting speed in the empirical mathematical models.

References

1. M.I. Rîpanu, *Theoretical and experimental researches regarding the quality and precision of stamped bearings cages*, Doctoral thesis ("Gheorghe Asachi" Technical University of Iaşi, 2014)
2. FAG Rolling Bearings, *Ball bearings. Roller bearings. Housings. Accessories*, (Catalogue WL 41 520-3 EA), (1999)
3. E. Sáenz de Argandoña, A. Azoturia, C. García, N. Arana, A. Aguirre, P. Filaria, Robot. CIM-Int. Manuf. **24**, 773-779 (2008)
4. P. K. Rai, A. Mohammad, H. Z. Jafri, Int. J. of Eng. Res. and Appl. **3**, 511-515 (2013)
5. G. Angara, G. Imbroglio, et al, J. Clean. Prod., **29**, 255-268 (2012)
6. A. Florentino, E. Curette, A. Athanasios, L. Mazzini, C. Giardini, Int. J. Mater. Form., **2**, 805 (2009)
7. P.V.P. Macondo's, A.M. Eton, P.A.C.Beltrão, P.C. Borges, J. Mater. Process. Tech. **206**, 184-193 (2008)
8. P. Sartkulvanich, B. Kroenauer, R. Grolle, A. T. Alta, CIRP Ann.-Manuf. Tech. **59**, 279-282 (2010)
9. G. Creţu, Fundamentals of experimental research. Laboratory handbook ("Gheorghe Asachi" Technical University of Iaşi, 1992)

 DOI: 10.1051/matecconf/201713705007

Methodology supporting production control in a foundry applying modern DISAMATIC molding line

Robert Sika[1,*], and *Paweł* Popielarski[1]

[1]Poznan University of Technology, Institute of Material Technology, Piotrowo 3 Street, Poland

Abstract. The paper presents methodology of production control using statistical methods in foundry conditions, using the automatic DISAMATIC molding line. The authors were inspired by many years of experience in implementing IT tools for foundries. The authors noticed that there is a lack of basic IT tools dedicated to specific casting processes, that would greatly facilitate their oversight and thus improve the quality of manufactured products. More and more systems are installed in the ERP or CAx area, but they integrate processes only partially, mainly in the area of technology design and business management from finance and control. Monitoring of foundry processes can generate a large amount of process-related data. This is particularly noticeable in automated processes. An example is the modern DISAMATIC molding line, which integrates several casting processes, such as mold preparation, assembly, pouring or shake out. The authors proposed a methodology that supports the control of the above-mentioned foundry processes using statistical methods. Such an approach can be successfully used, for example, during periodic external audits. The mentioned methodology in the innovative DISAM-ProdC computer tool was implemented.

1 Introduction

In times of modern globalization and development of commerce, followed by a need of fulfilling more and more restrictive conditions of acceptance by a client, almost any significant manufacturing company must automate its production processes [1]. Modern and specialized machines and devices, installed widely, are usually sold together with a dedicated computer software. With a help of measurement devices, they are responsible for monitoring of the production processes, supplying a user with necessary information about a process state, to control it [2]. In the era of the "Industry 4.0" concept gaining more and more popularity, there is an assumption of a close relation between an industrial revolution, automation and data processing, but also with consideration of notions such as the cyber-physical systems [3] or Virtual Reality [4]. The companies are therefore equipped with proper tools to manufacture and supply products and services of high quality, maximally adjusted to needs and expectations of clients. On the other hand, technical and organizational aspects of company operation is also important, meaning

* Corresponding author: robert.sika@put.poznan.pl

supplying products in an acceptable time, shortening of production cycle, limiting of wastage and costs of production [5]. It is possible not only thanks to application of modern production devices, but also due to reasonable use of verified methods and tools (including the statistical ones) for aiding the decision making in the field of manufacturing, its organization and control [6].

Manufacturing processes in the foundry industry are among the most complex ones and automation is particularly observable in this industry in recent years [7]. This is mostly a result of process and organizational complexity of casting production. It results directly of diversification of materials used for casting production, but also out of course and relations between different stages of partial processes [8, 9]. The foundry processes, especially the fully automated ones, usually generate high amount of data related to course of processes. As an example, modern molding lines of DISAMATIC type, more and more frequently installed in foundries in the last two decades, are a good example here. They can integrate even several processes, such as mold preparation, its assembly, pouring, knocking out a casting, cooling and shot-blasting. Unfortunately, there is a problem with the mentioned amount of generated data assigned to particular parameters, often not much useful for on-going production control. Depending on a type of installed molding line, even several hundred parameters can be generated, while most of them are useful only to employees of maintenance departments.

The authors have proposed a new methodology, which aids control of foundry processes related to the DISAMATIC molding line, with use of statistical methods. Such an approach may be applied not only for the current production control, but also, for example, during periodical external audits, when it is important to show influence of specific parameters on final casting quality.

2 DISAMATIC Automatic Molding Lines

The Automatic Molding Lines (AML) have gained certain popularity in foundry for a few important reasons. They are the obtained processing effect, the level of efficiency and reliability of molding above 98% [10]. The process of automatic forming allows obtaining of a very high quality of a mold, of uniform and high degree of densification of the mold sand, as well as high dimensional accuracy. Prevalence of AML use in processes of casting manufacturing is also a result of economics of the molding process. Ratio of quality of an obtained mold and efficiency of the process to costs of manufacturing of molds is very beneficial, especially in the case of DISAMATIC working lines [11].

Development of AML is related to a continuous improvement in field of mechanization and automation. Modern solutions of building molding lines concentrate mostly on processes of automated control, diagnosis and operation with use of robots and manipulators. Adjustment of processes is based on an on-going control of casting processes [12]. Current development of AML strives towards building flexible structures, based on modular construction, what corresponds to needs of quick re-arming of a production line. Here it must be mentioned, that by the end of XX. century many foundries still used molding lines, where processes of core manufacturing, mold pouring and knocking out of castings were realized outside an automated cycle. New generations of automated molding lines concern full automation of many activities related to casting making [13]. Such lines, fully automated, commonly use microprocessors for control, diagnosis and data recording. In the solutions proposed nowadays, visualization systems are used for imaging of current course of casting processes, as well as for their control and adjustment [11-13].

Despite such an advancement, tremendous amount of recorded data may be overwhelming and – as results of authors' experience – usually is not useful for on-going control of casting processes. It is particularly visible in case of work of quality control departments, where it is obvious to look for reasons behind the most frequent defective products.

3 New concept of production control for the DISAMATIC lines

3.1 Research preparation. Data acquisition

The studies were conducted in an European foundry, producing cast iron castings, of mass between few and few dozen kilograms, mostly for automotive and machine construction industry. Major part of the casting production is carried out on an automated molding line of DISAMATIC 230 type, with vertical mold division. For the studies, core castings manufactured using this line were selected, of mass of approx. 20 kg. As the production is for the automotive industry, a client often personally receives and accepts a batch of castings, according to guidelines by a branch standard ISO-TS16949 [14].

The authors proposed a juxtaposition of the most significant process parameters, recorded on the DISAMATIC line and outside it. The parameters were selected by their influence on a most frequently occurring defect in a series (approx. 140 pieces of poured molds, 2 castings in a mold), which is shrinkage. As results of gathered literature [15, 16] this defect is frequently noted as a leading one in case of manufacturing castings using AML. The mentioned parameters were assigned to topography of data sources by the authors (Fig. 1).

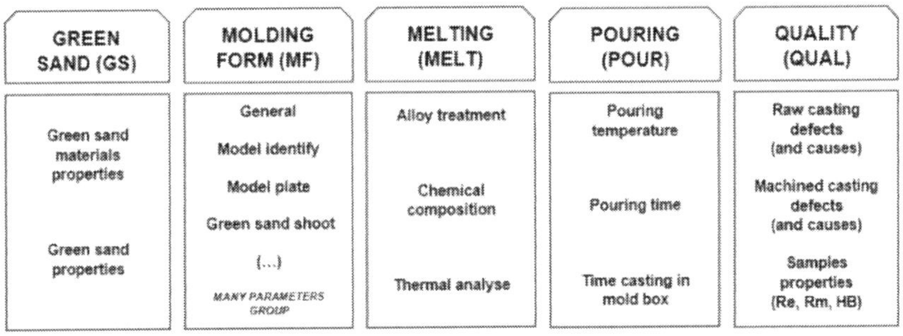

Fig. 1. Topography of data sources in the investigated foundry – simplified schema.

The most important parameters, in authors' opinion, related to a phenomenon of shrinkage defect forming, were put together (see Table 1). The parameters were divided into 3 groups:

• G1 – parameters recorded by the DISAMATIC modules (automated measurements),

• G2 – parameters recorded outside the DISAMATIC line (automated and semi-automated measurements),

• G3 – parameters recorded by a worker (manual measurements).

Table 1. Summary of parameters collected in the foundry (chosen data sources).

GROUP 1		GROUP 2		GROUP 3	
The most important from Molding Form	**Unit**	**Alloy treatment**	**Unit**	**Green sand parameters**	**Unit**
Number of model complet	n/a	Furnace comp. (metal, non-metal materials)	Kg	Moisture	%
Planned number of form	pcs.	Inoculator	Kg	Permeability	$\frac{m^2}{Pa * s * 10^8}$
Model plate A temp.	°C	Spheroidization	Kg	Compression	MPa
Model plate A temp.	°C			Compaction	%
Green sand shoot pressure	bar	**Chemical composition**	**Unit**	**Defects**	**Unit**
Green sand level in tray	%	CEQ, C, Si, Mn, Mg, Cu, S, Ni, Sn	%	Chosen raw casting defects (current defect, from one series or one day)	%
Pressure of green sand press	kG/m^2	**Pouring parameters (POUR)**	**Unit**		
Compressibility of molding form	%	Pouring temperature	°C		
Forcefulness of combining molding form	kG	Pouring time	sec.		

 DOI: 10.1051/matecconf/201713705007

Group 2 contains parameters recorded by measurement devices not directly coupled with the electronic control system of the molding line (automated measurements or semi-automated measurements with partial engagement of a human operator). Both group 2 and 3 contain very important parameters, that were not recorded by the DISAMATIC device at the time of conducting the studies by the authors.

3.2 Development of a methodology of production control for the DISAMATIC molding line

As mentioned in chapter 2, amount of data collected in automated foundries is very high. First of all, it should be limited to the most relevant parameters, which directly influence a phenomenon of forming of a specific defect. Frequency of gathering these parameters is also different. It is therefore necessary to assume, that juxtaposition of parameters on one time axis will require certain approximation of time relations between them, regarding a specific poured mold or series. However, in practice, such a full synchronization is not possible in each real case and it must be based upon certain averaging of values of the most frequently recorded parameters. It is therefore important to assume one parameter with a leading timeline (a reference time), to which all the other parameters will be associated. Usually it is a process being the most close to shaping a casting, meaning the process of mold pouring (a moment of pouring).

A solution proposed by the authors is a methodology, which consists in use of parameters coming from the DISAMATIC (GROUP 1) and outside this molding line (GROUP 2 and GROUP 3). The authors propose to select only certain parameters in scope of the first group. The parameters should be related mostly to process of mold preparation at a stage of its shaping with use of modelling plates and mold completing (assembling two half-molds, which eventually create a mold ready for casting). The presented methodology of production control assumes selection of parameters and efficient evaluation of their influence on final quality of castings in a foundry, which applies the DISAMATIC automation. A concept scheme of proceeding is shown in Figure 2.

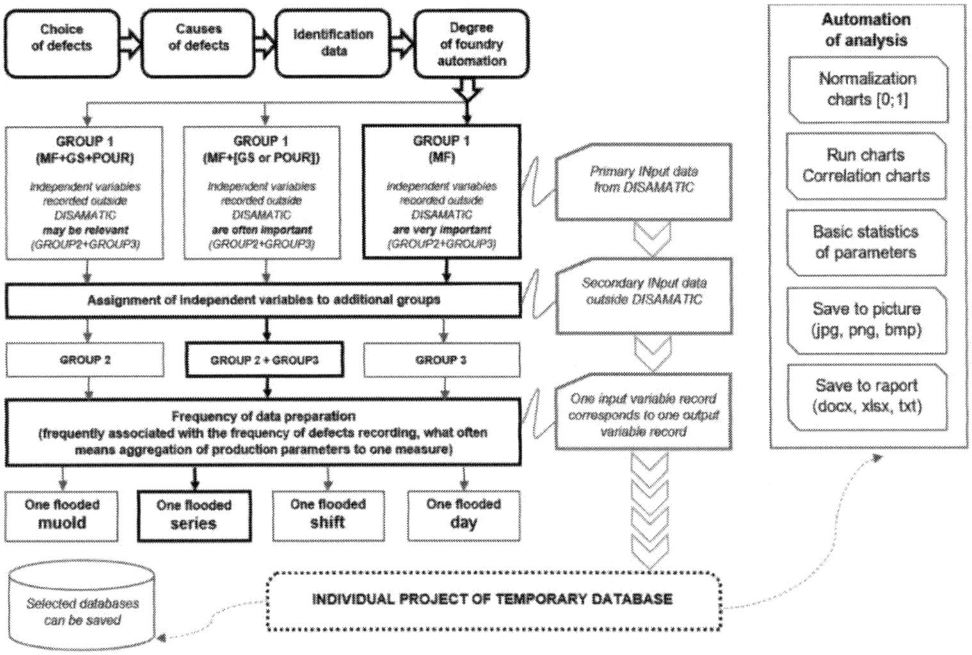

Fig. 2. Scheme of production control methodology for the DISAMATIC molding line.

3.3 Project of the DISAM-ProdC computer system

This section shortly presents the DISAM-ProdC (DISAMATIC Production Control) application. Its task is automation of activities described in the production control methodology. Two modules were proposed in the test phase: "automation of data acquisition" and "automation of analyses".

In the first case, the application generates a new database, on the basis of defined search criteria. The prepared database is analyzed in the "automation of analyses" module. Most of all, it was assumed to have a possibility of studying influence of one or several independent variables on a defect (dependent variable), presented to an employee in a clear and readable way. Figure 3 presents an exemplary screenshot of the "automation of data acquisition" module, while Figure 4 presents exemplary results of analyses – influence of selected parameters on the shrinkage type defect. In this case, the authors applied data scaling to a range between "0" and "1", where the first negative value means a minimal value of a set of a specific parameter, while a positive value means a maximal value of this set. An aim is to illustrate a graphical change of values of these parameters, in relation to percentage share of the defect (from 18.10.2016 to 21.10.2016 for each series per day: S1, S2…etc.). The real value is shown after clicking on a specific data point in a diagram with a mouse button (see "ID point" in Fig.4).

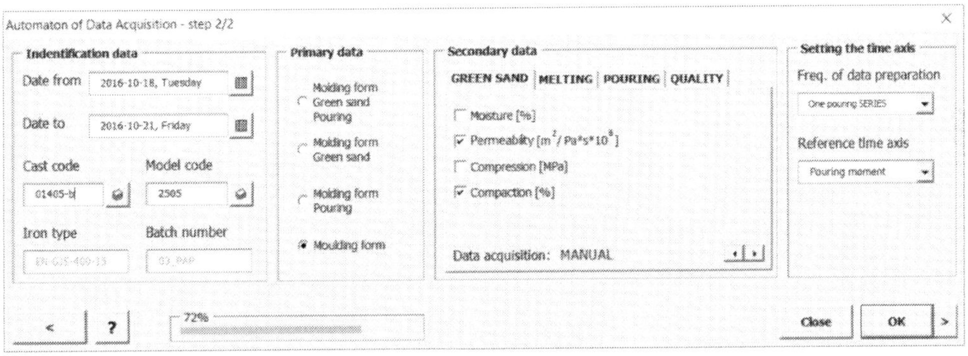

Fig. 3. The "Automation of Data Acquisition" module.

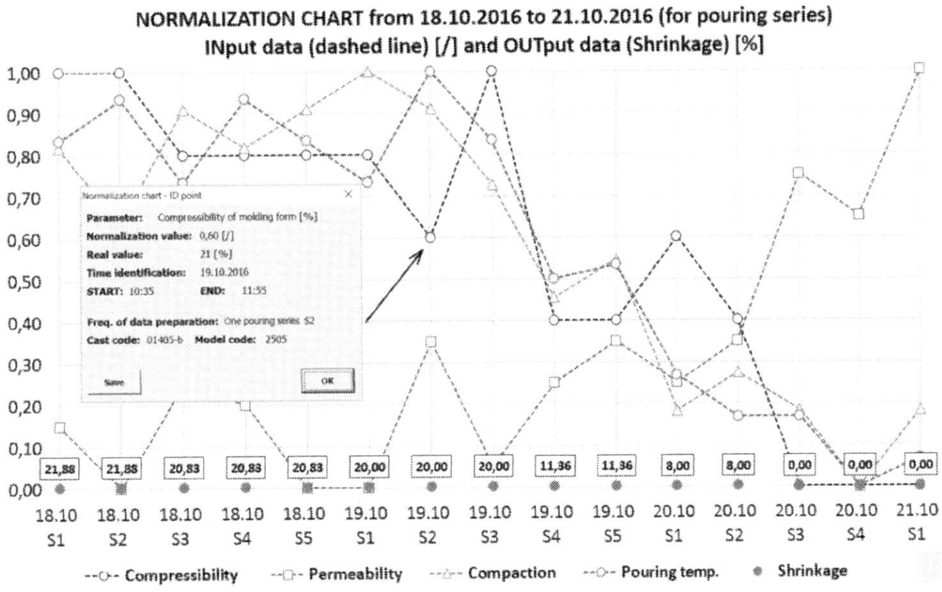

Fig. 4. The "Automation of Analyses" module.

4 Summary

The paper presents authors' own approach to production control in an automated foundry using a modern forming line DISAMATIC. The foundry is an example of a production system, where integration of many production processes responsible for receival of a final product (a casting) is visible. In these companies, modern machines and devices for work automation are installed more and more often and this contributes to generation of a huge amount of data, out of which not all are useful for the on-going production control.

The authors have proposed a modern methodology, that, by assumption, supports control of production processes with use of statistical methods. Above all, optimal selection of independent variables was assumed, considering a criterion of efficient conducting of basic statistical analyses.

The methodology was implemented in authors' own program DISAM-ProdC. It must be emphasized, that for the employees of the foundry, it is particularly important to perform quick analyses in a simple way. Excess functions, parameters and preliminary settings in computer programs aiding the production control may only discourage using them. Aim of the authors was, most of all, showing that among many independent variables, the most important ones can be selected to present their influence on final quality of castings, in form of simple analyses and statistic sheets. As results of authors' experience, such quick and practical analyses are very often useful for quality controllers, aiding current control of casting quality.

References

1. D. Schütz, A. Wannagat, C. Legat, IEEE Trans. On Ind., **9**(4), 2397, (2013)
2. K. J. Aström, B. Wiitenmark, *Adaptative Control. Second Edition*, (Lund Intitute of Technology, Mineola, New York, 2013)
3. J. Lee, B. Bagheri, H. Kao, Man. Lett., Elsevier, **3**, 18 (2015)
4. F. Górski, R. Wichniarek, P. Zawadzki, P. Buń, A. Hamrol, Proc. Comp. Sc., **75**, 137 (2015)
5. M.R.L. Varela, J. Trojanowska, S. Carmo-Silva, N.M.L. Costa, J. Machado, Manag. and Prod. Eng. Rev., **8**(2), 69 (2017)
6. A. Kujawińska, R. Rogalewicz, M. Diering, A.Hamrol, WorldCIST, Adv. in Int. Sys. and Comp., Springer, **571**, 499 (2017)
7. Z. Ignaszak, R. Sika, M. Perzyk, A. Kochański, J. Kozłowski, Arch. Of Found. Eng., **16(1)**, 5 (2016)
8. S. Saalem, M. Malik, S. Gottschling, J. Hartman, Proc. 71st World Foun. Cong. (WFC), Bilbao, Spain. **1** (2016)
9. T.R. Vijayaram, S. Sulajman, A.M.S. Hamouda, M.H.M Ahmad, J. of Mat. Proc. Techn., 178 (2006)
10. R. Banchhor, S. K. Ganguly, Proceedings in: *Interational Journal of Advanced Engineering Research and Studies*, **4**, 227 (2015)
11. M. Colditz, S.K. Kang, H.S. Kim, P. Larsen, Found. Trad. J. Int. (2015)
12. *** DISA, A Norican Technology, Moulding Solution, https://www.disagroup.com/ (2017)
13. J. Dańko, A. Fedoryszyn, Sold. Of Metals And Alloys, **1**(41), 34
14. ISO/TS 16949, Automotive Quality Management, (2009)
15. I. Hahn, J. C. Sturm, *Proceedings of the 69th World Foundry Congress,* (2010)
16. M. Sutaria, D. Joshi, M. Jagdishwar, B. Ravi, Trans. Amer. Found. Soc., **119** (2011)

The analysis of instantaneous tool displacements during precise ball end milling

Szymon Wojciechowski[1,*]

[1]Poznan University of Technology, Faculty of Mechanical Engineering and Management, Institute of Mechanical Technology, Poznan, Poland

Abstract. The paper presents the quantitative and qualitative analysis of monolithic ball end mill's instantaneous displacements generated during precise milling of inclined surfaces. The conducted experiment involves the measurements of tool's joining part displacements with the application of laser displacement sensor and cutting forces with piezoelectric dynamometer. The milling tests were carried out for the hardened alloy 55NiCrMoV6 steel in a range of variable feed per tooth and surface inclination angle values. The obtained results can be applied for the selection of effective milling parameters allowing the improvement of machined surface finish.

1 Introduction

The high machined surface quality is the fundamental requirement regarding the precise ball end milling process. It is usually depended on the selected milling parameters, as well as the specific machine-milling process interactions. The literature survey shows that geometrical product's specification (GPS) obtained after machining process is strongly affected by the cutting parameters [1], dynamical stability [2], thermal phenomena in the machining zone [3], as well as the ploughing mechanisms correlated with the minimum uncut chip thickness [4, 5]. Nevertheless, apart from the above-mentioned factors, the tool-workpiece vibrations (periodical displacements) can be also the critical quantity affecting the machined surface finish [6, 7].

The instantaneous tool displacements generated during milling process are usually induced by the geometrical errors (e.g. run out) and dynamic properties (e.g. damping, stiffness) of machine-toolholder-tool system [8 ,9]. Moreover, in case of a ball end milling process, the variations of surface inclination angle along the toolpath cause the changes in tool's effective cutting speed and thus significantly affect the process dynamics [10]. Therefore, the identification of the crucial factors influencing the tool's instantaneous displacements during milling is important in terms of machined surface quality improvement. However the accurate on-line measurement of tool's displacement during precise milling is a difficult task because of the inhibited access to the tool's working part, chip formation phenomenon and tool's rotational motion distorting the measured signals. Thus, the vibrations are usually measured with the application of indirect methods which

* Corresponding author: sjwojciechowski@o2.pl

involve the use of accelerometers fixed to the spindle head or workpiece. However the acquired signals require post-processing, including the filtration and integration, as well as they are not fully compatible with the real tool's displacement. Therefore, recently the novel, non-contact on-line measurement methods of tool's instantaneous displacements, based on laser vibrometry, capacitive gap sensors and laser displacement sensors have been developed.

Tatar and Gren [11] applied the scanning laser vibrometer (LDV) for the measurements of indexable 2-toothed mill's displacements during milling of AA7010 aluminum alloy. The similar approach has been adopted by Wojciechowski et al. [12] for the on-line measurements of 6-toothed end mill's joining part displacements during milling of hardened alloy steel. Miyaguchi et al. [13] carried out the measurements of ball end mill's instantaneous displacement with the application of capacitive gap sensor during cutting of hardened steel.

The state-of-the-art shows that despite a few researches, the problem of tool's instantaneous displacement evaluation during precise milling is still insufficiently recognized. Therefore this work focuses on the quantitative and qualitative analysis of ball end mill's instantaneous displacements generated during precise ball end milling of inclined surfaces. The carried out measurements involve the application of laser displacement sensor. The obtained results can contribute to the better understanding of the phenomena occurring during the GPS formation in milling process, and consequently to the selection of effective milling parameters enabling the improvement of surface finish.

2 Experiment

2.1 Workpiece and tool

The experiments have been carried out on hardened alloy steel 55NiCrMoV6 sample with average hardness of 52 HRC. The chemical composition of the work material is presented in table 1. The selected tool was monolithic ball end mill made of fine-grained sintered tungsten carbide (WC) with diameter D=10 mm and number of teeth z=2. The tool's geometry was as follows: $\gamma_o = -10°$, $\lambda_s = 30°$, $\alpha_o = 6°$. The ball end mill has been clamped in a HSK-63 shrink fit toolholder. The selection of this toolholder was dictated by its recommendation to High Speed Milling technology, because of high dynamical balancing. Experiments were carried out on 5-axes CNC milling workstation.

Table 1. The chemical composition of the 55NiCrMoV6 steel.

C	Si	Mn	Cr	Mo	Ni	V
0.5–0.6%	0.1–0.4%	0.65–0.95%	0.6–0.8%	0.25–0.35%	1.5–1.8%	0.07–0.12%

2.2 Research range and method

Milling tests included the measurement of ball end mill's instantaneous displacements and cutting force components, conducted in the range of variable surface inclination angles α and feed per tooth f_z. In order to obtain surface inclination during the milling process, tool's rotational axis was inclined against the workpiece in the range of various α angles (Fig. 1).

 DOI: 10.1051/matecconf/201713705008

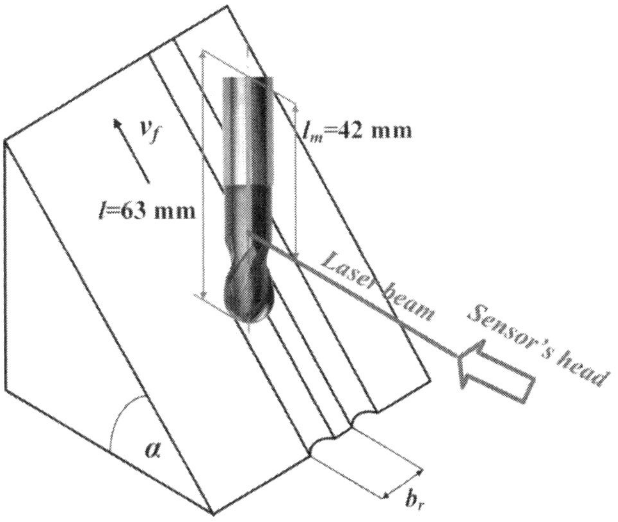

Fig. 1. Ball end milling kinematics and measurement of tool displacement.

In all investigated cases tool's effective diameter was lower than the value of pick feed – $D_{ef}<b_r$. The milling parameters applied in the research were depicted in table 2.

Table 2. Design of experiment.

Cutting speed v_c [m/min]	Depth of cut a_p [mm]	Rotational speed n [rev/min]	Inclination angle α [°]	Feed per tooth f_z [mm/tooth]	Overhang l [mm]
200	0.3	7351–18614	0 – 60 20 interval	0.05 – 0.2 0.05 interval	63

The cutting force components were measured in the three directions (F_x [N] –feed normal direction, F_y [N] - feed direction, F_z [N] - thrust direction) with the application of the 3-directional piezoelectric sensor (Fig. 2).

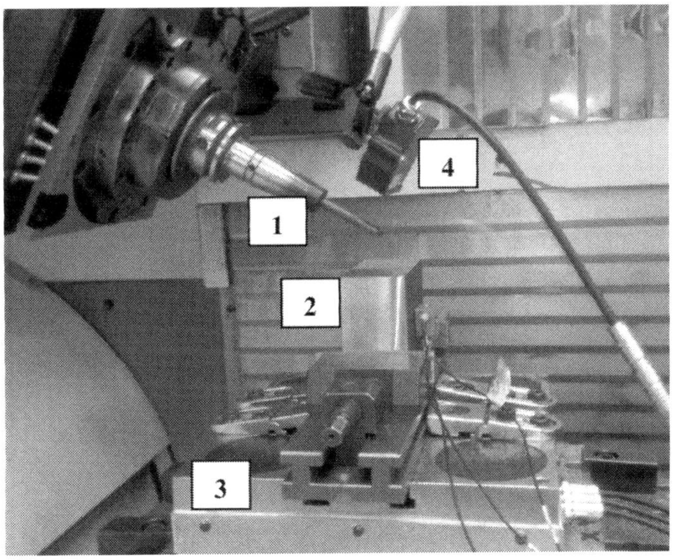

Fig. 2. The experimental set-up for the on-line measurement of forces and tool displacements (1 – ball end mill-toolholder system, 2 – workpiece, 3 – force dynamometer, 4 – laser displacement sensor).

 DOI: 10.1051/matecconf/201713705008

The force dynamometer was connected through amplifiers and band-pass filters to a data acquisition computer. The acquisition of signals was made with the application of dedicated software. This software was applied also for the calculation of signal's root mean square (RMS) values and filtration carried out with a low-pass filter.

The **optoNCDT Micro-Epsilon** laser sensor has been applied for the on-line measurements of ball end mill's instantaneous displacements (Fig. 2) in the direction perpendicular to the tool's rotational axis and collinear to the feed motion vector (Fig. 1). The laser's beam wavelength was $\lambda = 670$ nm, the measuring range was 10 mm and FSO (full-scale output) resolution at sampling frequency of 2500 Hz was equaled to 0.5 μm. The acquired signal has been filtered with the application of software low-pass Chebyshev's filter in order to eliminate the undesirable distortions. The cut off frequency f_c was determined on the basis of the following expression:

$$f_c \approx \frac{z \cdot n}{30} + 10\% \qquad (1)$$

3 Results and discussion

The figure 3 depicts the measured instantaneous values of ball end mill displacements in function of surface inclination angle α and feed per tooth f_z. During the precise milling, which is usually carried out with the relatively low depths of cut and feeds, tool's instantaneous displacements can be simultaneously affected by the geometrical errors of machine-toolholder-tool system (e.g. tool run out) and deflections induced by cutting and centrifugal forces.

Preliminary studies involving the measurement of tool displacements in conditions of the executed rotational motion of spindle-tool system revealed that tool's radial run out e_r is equal to 15μm. Nevertheless, in a majority of investigated cases the maximal measured displacement exceeded the e_r value. This observation confirms that displacement is affected also by the tool's elastic deformations induced by the effect of forces.

Figure 3a shows that for the slot milling conditions ($\alpha = 0$), the shape of the displacement time courses are similar to sine function, independently on the selected feed per tooth f_z value. However, for the upward ramping strategy ($\alpha > 0$) the displacement signals are less regular and consisting of components with higher frequencies (see – Figs. 3b, c). These discrepancies are induced by the distinct tool working angle values, affecting the shape of cutting force time course. During the slot milling, the periodically variable forces are found. On the other hand, in case of upward ramping in finish conditions, the active number of teeth is less than 1. Thus, the forces are pulsating which consequently affects the shape of tool displacement time course.

The investigated milling parameters (surface inclination angle α and feed per tooth f_z) affect also the tool displacements maximal values (see – Figs. 3 and 4).

It was observed that growth of feed per tooth f_z induces the linear increase of displacement peak values (Fig. 4) for the majority of the investigated surface inclinations. According to tool's dynamical deflection model, based on the differential motion equation, the tool deflection amplitudes are proportional to cutting force values, which in turn are usually linearly proportional to feed per tooth. In case of surface inclination angle α, the growth of this factor leads to the non-linear decrease of displacement peak values, independently on selected feed per tooth. Moreover, this relation is qualitatively similar to cutting force components curves in function of surface inclination angle (Fig. 5). Typically, the highest force values are obtained during the slot milling because of intense ploughing action induced by the very low cutting speeds located in the area of the tool tip. It should be also noted that in case of slot milling, the displacement amplitudes reach up to 140 μm and thus exceed the selected feed per tooth values (for f_z = 0.05 mm/tooth and f_z = 0.1 mm/tooth).

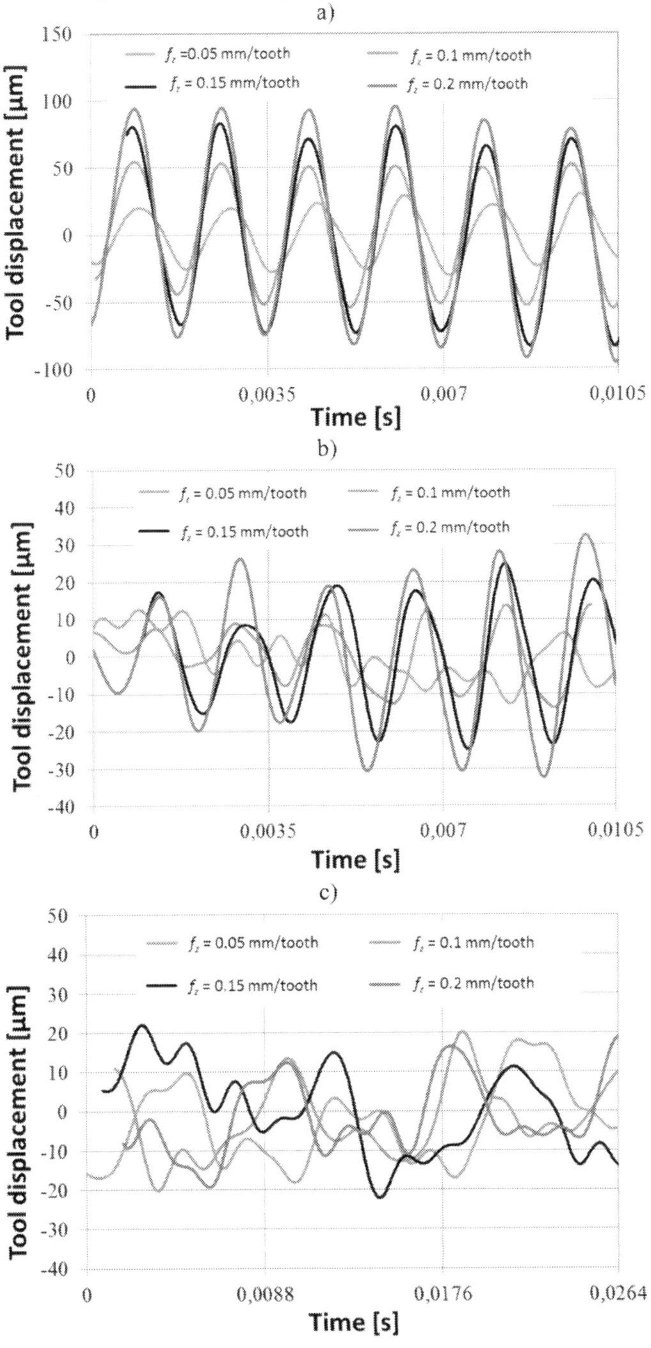

Fig. 3. Instantaneous tool displacements for: a) $\alpha = 0$; b) $\alpha = 20°$; c) $\alpha = 60°$.

Consequently, the tool can be pushed off from the material which is being currently cut, thereby contributing to the formation of considerable machining errors (e.g. the growth of surface roughness height and/or surface location error). In order to avoid this undesirable phenomenon, the milling process should be carried out at higher surface inclination angles.

DOI: 10.1051/matecconf/201713705008

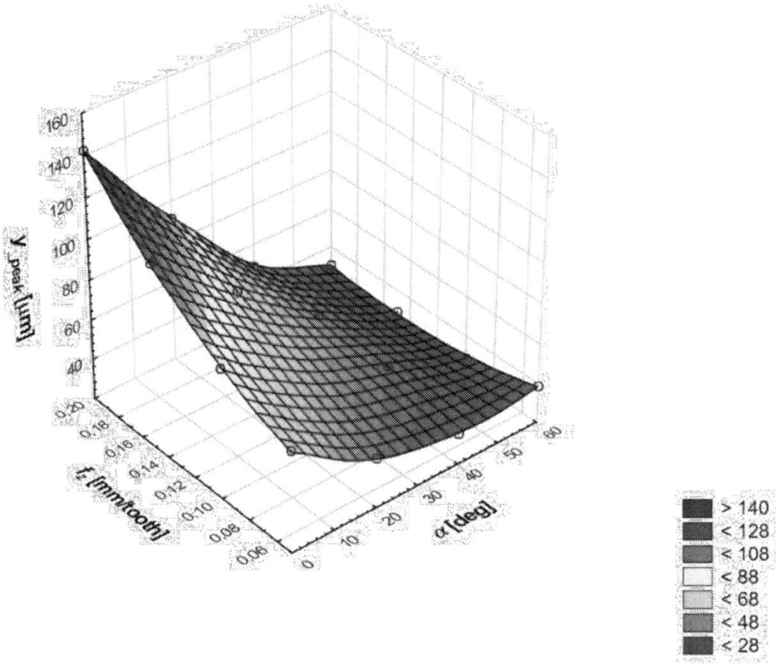

Fig. 4. Ball end mill's displacement peak values in function of surface inclination angle and feed per tooth.

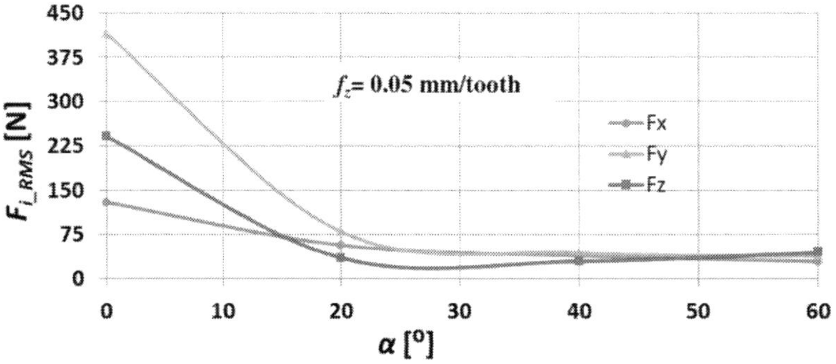

Fig. 5. Cutting force components in function of surface inclination angle.

4 Conclusions

On the basis of the carried out investigations, the following conclusions are formulated:

• The surface inclination angle affects the tool displacements qualitatively. In case of slot milling strategy ($\alpha = 0$) the shape of the displacement time course is similar to sine function. However, for the upward ramping strategy ($\alpha > 0$) the displacement curves are less regular and consist of components with higher frequencies;

• The growth of the α angle induces the non-linear decline in maximal displacement values, whereas the increase of f_z causes the linear growth of this quantity;

• In terms of tool displacement minimization, the ball end milling process of hardened 55NiCrMoV6 steel should be carried out with the constant surface inclination angle along the toolpath equaled to $\alpha = 60°$ and feed per tooth $f_z = 0.05$ mm/tooth. Thus, the obtained results can be implemented in industry during precise 5-axes ball end milling of a complex surfaces, enabling the minimization of surface roughness and surface location errors.

References

1. R.W. Maruda, G.M. Krolczyk, P. Nieslony, S. Wojciechowski, M. Michalski, S. Legutko, Journal of Manufacturing Processes, **24** 1 107–115 (2016)
2. R. Rusinek, M. Wiercigroch, P. Wahi, Procedia CIRP, **31**, 429–434 (2015)
3. D. Przestacki, P. Szymanski, S. Wojciechowski, Composites Part A: Applied Science and Manufacturing, **91** 1, 370–379 (2016)
4. S. Wojciechowski, R.W. Maruda, P. Nieslony, G.M. Krolczyk, International Journal of Mechanical Sciences, **119**, 360–369 (2016)
5. S. Wojciechowski, P. Twardowski, Procedia CIRP, **1**, 289 – 294, (2012)
6. L.N. López de Lacalle, A. Lamikiz, J.A. Sanchez, M.A. Salgado, International Journal of Advanced Manufacturing Technology, **24** 621–631 (2004)
7. S. Wojciechowski, P. Twardowski, M. Pelic, Procedia CIRP, **14** 113 – 118 (2014)
8. T.L. Schmitz, J. Couey, E. Marsh, N. Mauntler, D. Hughes, International Journal of Machine Tools & Manufacture **47**, 841–851 (2007)
9. S. Wojciechowski, Metrology and Measurement Systems, **18**, 3, 429 – 440 (2011)
10. X. Chen, J. Zhao, Y. Li, S. Han, Q. Cao, A. Li, Int J Adv Manuf Technol, **59** 885-898 (2012)
11. K. Tatar, P. Gren, International Journal of Machine Tools & Manufacture, **48**, p. 380 – 387 (2008)
12. S. Wojciechowski, P. Twardowski, M. Pelic, R.W. Maruda, S. Barrans, G. Krolczyk, Precision Engineering, **46** 158-165 (2016)
13. T. Miyaguchi, M. Masuda, E. Takeoka, H. Iwabe, Precision Engineering, **25**, 145–154 (2001)

MATEC Web of Conferences 137, 06001 (2017) DOI: 10.1051/matecconf/201713706001
MTeM - AMaTUC 2017

Low-cost industrial photogrammetry for rapid prototyping

Vlad Bocanet[1,*], *Marius* Bulgaru[1] , and *Vlad* Pop[1]

[1]Technical University of Cluj-Napoca, Faculty of Machine Building, Bvd. Muncii no. 103-105, Cluj-Napoca, 400641, Romania

Abstract. The field of extracting 3D information from photographs also known as photogrammetry has a wide range of applications, even in industry. It can be used to rapidly prototype a part with a decent precision. This paper analyses this possibility for a gas regulator that was cut in half. From the pictures taken of the part, a point-cloud can be generated that can be transformed in a mesh and subsequently in a CAD model. Alternatively, it can be compared with an existing CAD model to determine deviations.

1 Introduction

Being able to extract 3D information from an object is very important and leads to many applications especially in industry. In industrial applications, the extracted data can be integrated into a model that can be used to reverse engineer the original part or determine deviations of the physical part from an ideal CAD model.

There are many techniques used in measuring and reconstructing of 3D objects, that can be split into two types [1]:

- contact methods (for example, using coordinate measuring machines, calipers, rulers and/or bearings);
- non-contact methods (X-ray, photogrammetry, laser scanning).

Today, 3D models are mainly generated by use of non-contact methods based on light waves [1]. Geodetic measurement techniques such as spatial triangulation and optical alignment and levelling can be used [2]. One method, that shows increasing promise, is photogrammetry.

The word photogrammetry means measuring from a photograph [3]. This technique "allows one to reconstruct the position, orientation, shape and size of objects from pictures" [4].

When taking a picture of an object, we basically convert it from a 3D world into a 2D world. During this process, information, mainly related to depth, is lost. By using photogrammetry, we are able to convert 2D information back into 3D information. This process is based on triangulation, which consists of intersecting converging lines in space and determining the precise location of a point [5]. For this we need at least two photographs in which a certain point appears (Figure 1). Usually the camera moves around the object taking pictures that have a degree of overlap. The camera must move both in a

* Corresponding author: vlad.bocanet@tcm.utcluj.ro

horizontal plane as well as vertical, to "see" the object from all possible angles. Of course, this process is not perfect, because information is lost during the conversion stages.

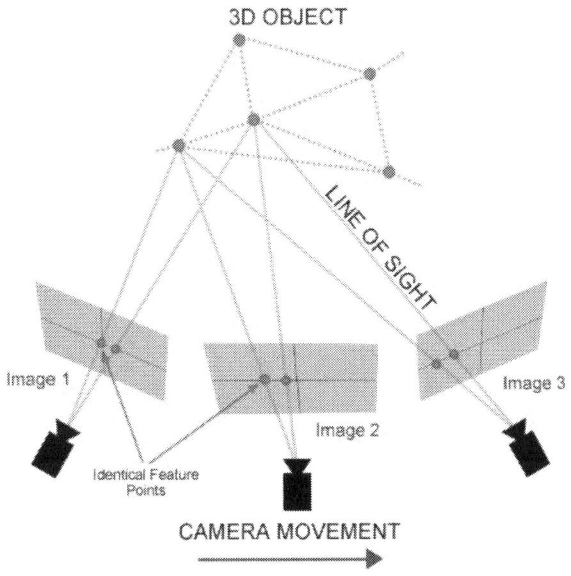

Fig. 1. Basic principle of photogrammetry [6].

The main application of photogrammetry is the production of topographic maps [4] but it has many other applications for big objects: mapping of cities or terrain, buildings, large equipment and even airplanes. Other applications include 3D modelling of agro-idustrial buildings [7] or a ship's hull [8] in the latter case even being preferred to laser scanning. In another study, it was used together with a neuro-fuzzy system to model and measure deformation of objects [9].

2 Experimental setup and methodology

The main purpose of this study was to determine the applicability of photogrammetry in the industrial environment for rapid prototyping of parts. To this end, a part was chosen as subjects and different variables were used to determine the factors that influence the most the accuracy of the resulting 3D model.

The experiment was conducted by using two different cameras: a 16 MPx camera (Nikon D5100) and a 12 MPx camera (Canon PowerShot SX20 IS). Their technical specifications are presented in Table 1.

Table 1. Specifications of camera used in the experiment [10,11].

Specs / Camera	Nikon D5100	Canon PowerShot SX20 IS
Image sensor	23.6 x 15.6 mm CMOS	6.17 x 4.55 mm CCD
Resolution	16.2 MPx	12.1 MPx
ISO range	100 – 6400	80-1600
Aperture range	f/3.5-5.6	f/2.8–5.7
Max. shutter speed	1/4000	1/3200
Focal length	18-55 mm	28-560 mm
Max. zoom	3x	20x

The software used for photo processing and extracting the 3D information was Agisoft Photoscan.

The subject part of the experiment was a gas regulator that was cut in half (Figure 2). The cast metal part is of medium size (207x135x102 mm) and has both inside and outside complex surfaces. One challenge with metal parts is the shine of machined sections that tends to interfere with the 3D reconstruction process. During the preparation phase, we noticed that a random patterned background helps in the identification of tie points.

Fig. 2. The part used in the experiment.

The first step in the reconstruction process is the alignment of photos. In this step the software detects points which are stable under viewpoint and lighting variations and generates a descriptor for each point. The approach is similar to the SIFT (Scale-Invariant Feature Transform) algorithm. The result is a cloud of tie points that were identified from the pictures (Figure 3).

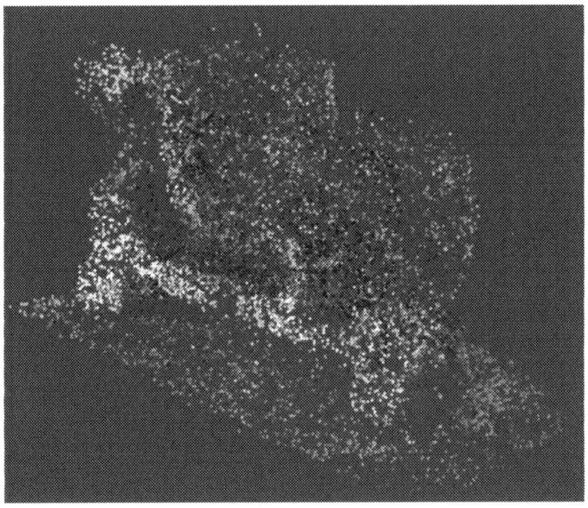

Fig. 3. Tie points resulted from the alignment of photos.

The dense cloud (Figure 4a.) is created by calculating depth information for each camera based on its estimated position. This step has two basic parameters that can be set: quality (the higher the quality the higher the processing time) and depth filtering (aggressive for lower detail and mild for higher detail). [12]

Next, the dense cloud data is used to build a polygonal mesh model (Figure 4b.). This approximation of the models' surface can be then used either in comparison with a CAD model or for reverse engineering. The mesh usually is not continuous, having holes where data is missing. It can be repaired (stitched) in a later step.

a) b)

Fig. 4. a. Dense cloud of points; b. the resulting mesh.

The last step is texturizing of the model (Figure 5). This is only to give the model a visual conformity with the original part. It doesn't have any useful geometric information.

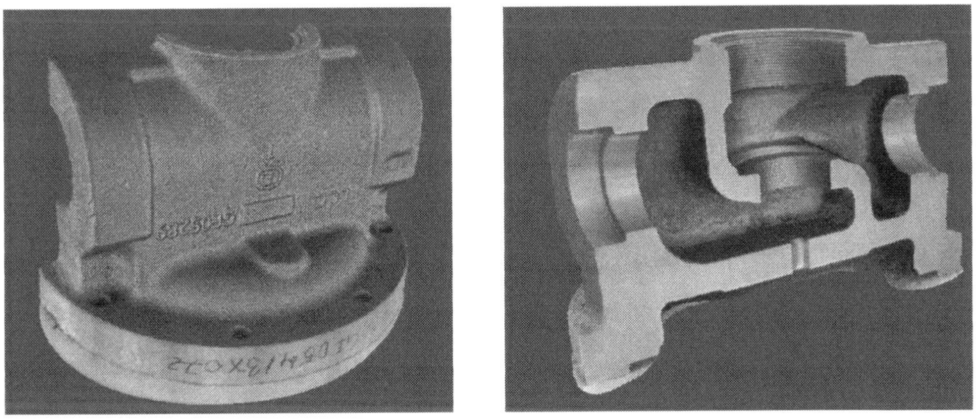

a) b)

Fig. 5. End result after texturizing: a. outside view; b. inside view.

In the general process described above some parameters were varied in order to see their influence on the end result.

3. Results and discussions

There are many factors affecting the reconstruction of 3D models from pictures through photogrammetry. In this study, we wanted to see the influence of: picture quality, image editing (masking), picture overlap and the usage of virtual markers. To this effect the photos where edited (cropped) either manually or by using Agisoft Photoscan. Then, virtual markers were used to scale the model after it was completed.

The camera was placed at about 30 cm from the subject. The settings that were used for the two cameras, are presented in Table 2. A number of 16 photos were taken on each horizontal plane which meant that there was a 22.5-degree angle between the photos. There were four vertical levels with an increment of about 15-degrees, resulting a total of 64 photos. The position was assured by using an adjustable camera mount.

Table 2. Camera parameters used in the experiment.

Parameter	Values for Nikon D5100	Values for Canon PowerShot
ISO	2500	1600
Focal length	40	17
F-stop	f/7.1	f/7.1
Shutter speed	1/200	1/100

Because the Canon is more suitable for long ranges, the results were not satisfactory and the experiments were continued by only using the Nikon camera.

The results of the three editing methods can be seen in Figure 6, with the gaps marked with blue.

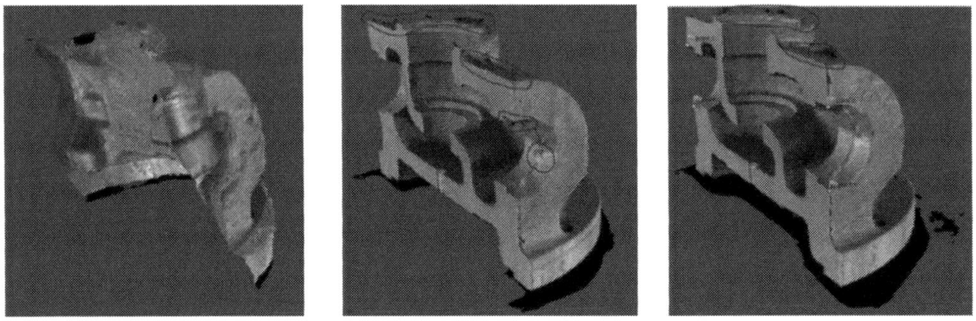

a) b) c)

Fig. 6. The resulting models: a) manually edited photos; b) photos edited in AgiSoft Photoscan; c) photos edited in AgiSoft Photoscan with scale markers.

The horizontal overlap of the pictures was given by the angle at which the pictures were taken. The tested horizontal angles were 22.5° and 45° (16 photos and 8 photos respectively). Reducing the number of photos by increasing the angle between them from 22.5° to 45° greatly reduced the quality of the model, creating gaps in the resulting mesh.

Some of the inside features of the regulator like the threaded part and the inside chamber required higher level of detail because they were constantly creating problems in the creation of the mesh. Thus, additional close-up pictures were taken to see if the quality of the resulting mesh would increase. Including these close-up pictures almost doubled their number, from 64 to 114, 111 of which were usable in the photo alignment stage. Adding virtual markers to the finished model, increased its accuracy because they could be used to scale the model. By defining a known distance between two points on the part, we could correct the scale of the model.

The best two results, the model without close-up pictures and the model with close-up and added virtual markers, were compared with a high precision scan of the part, in order to see their precision. The high precision scan was obtained with a GOM ATOS Core optical scanner and was used because there was no CAD model available for the part. The results of the two comparisons can be seen in Figure 7. The registration of the two clouds was made with CloudCompare using the Iterative Closest Point (ICP) method.

The greatest deviations were found on the bottom of the part where the part rested. In the inside part of the regulator the close-up pictures increased the precision of the model by reducing the distance of the mesh from the precision scan.

 DOI: 10.1051/matecconf/201713706001

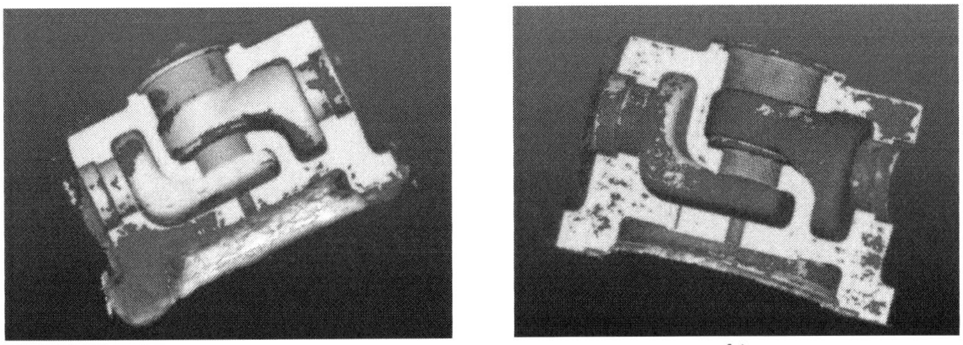

a) b)

Fig. 7. The deviation map of the two resulting meshes a) without close-up pictures; b) with close-up pictures and virtual markers.

The histograms of the distances from the precision scan can be seen in Figure 8. They have the same colour scale as in Figure 7. Adding the detail pictures and the virtual scale markers greatly reduces the deviations. If in the first case most deviations reach up to 6 mm, in the second case there are very few that go over 1.5 mm.

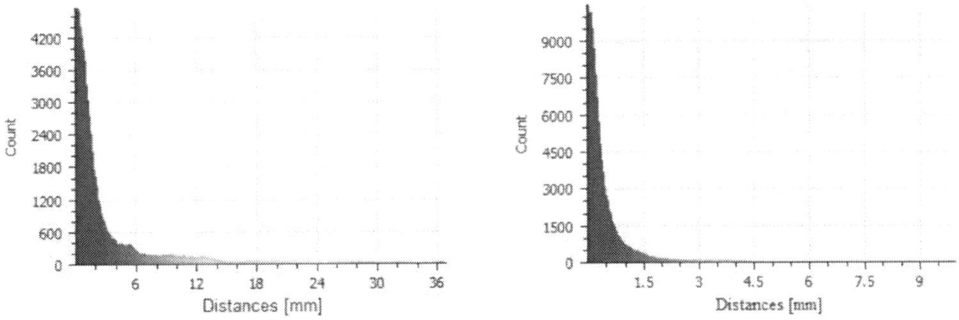

a) b)

Fig. 8. The deviation histograms for the meshes: a) without close-up pictures; b) with close-up pictures and markers.

4 Conclusions

The current research analyzed the possibility of using low-cost photogrammetry in an industrial environment to create a 3D mesh of a part.

This study shows that there is a minimum necessary degree of overlap to obtain a decent mesh. By using 16 photos in a horizontal plane and 4 levels in the vertical one, we managed to obtain a usable mesh. By supplementing the photos obtained through revolving around the subject with close-up photos of complex zones, the quality of the model can be increased.

There are many factors affecting the quality of the resulting model that haven't been included in this study, such as lighting (position, number of sources, colour), markers placed on the part, higher resolution cameras etc.

This low-cost method could be a rapid solution for rapidly prototyping a part, without the need of costly equipment or tools.

This work was supported by a grant of the Romanian National Authority for Scientific Research and Innovation, CNCS/CCCDI–UEFISCDI, project number PN-III-P2-2.1-BG-2016-0216, within

PNCDI III and Technical University of Cluj-Napoca through research Contract no. 2001/12.07.2017, Internal Competition CICDI-2017.

References

1. F. Remondino, S. El-Hakim, The Remote Sensing and Photogrammetry Society and John Wiley & Sons **21**(115), 269–291 (2006)
2. V. Uffenkamp, eConf C930928 **018** IWAA-1993-018 (1993)
3. K. K. Rampal *Handbook of Aerial Photography and Interpretation Concept* (Publishing Company, ISBN 817022649X, 9788170226499, 1999)
4. K. Kraus *Photogrammetry: Geometry from Images and Laser Scans, Volume 1*, De Gruyter Textbook, Walter de Gruyter, ISBN 3110190079, 9783110190076, (2007)
5. *** https://www.geodetic.com/v-stars/what-is-photogrammetry/, 14.08.2017
6. *** https://www.clemson.edu/centers-institutes/conservation/equipment/ photogrammetry.html, accessed 08.07.2017
7. P. Ariasa, C. Ordonez, H. Lorenzo, J. Herraez, J. Armesto, *Low-cost documentation of traditional agro-industrial buildings by close-range photogrammetry*, Building and Environment **42** 1817–1827, (2007)
8. H. J. Koelman, Computer-Aided Design **42**, 731–743, (2010)
9. F. F. Ahmadi, Measurement **79**, 94–100 (2016)
10. *** https://www.cnet.com/products/nikon-d5100/specs/ - accessed 09.07.2017
11. ***https://www.dpreview.com/products/ canon/compacts/canon_sx20is/specifications, accessed 09.07.2017
12. *** http://www.agisoft.com/index.php?id=34, accessed 10.07.2017

A new approach of the design process for replacing wooden parts of furniture

Cornel Ciupan[1*], *Emanuela* Pop[1], *Ioan* Filip[2], *Emilia* Ciupan[1], *Emilia* Câmpean[1], *Ioan* Cionca[2] and *Vasile* Heres[2]

[1]Technical University of Cluj-Napoca, Memorandumului street, nr. 28, 400114 Cluj-Napoca, Cluj, Romania
[2]TAPARO SA, 43562 Sat Borcut nr. 98, jud. Maramures, Romania

Abstract. Based on different design processes for innovative products, the paper presents a new approach of the process. The new method studies the problem of replacing the wooden components of the resistance structure of furniture with other materials. This is a reverse engineering process, that starts with the entire component, following a series of steps to the part drawings, ready for manufacturing. The method was validated by redesigning some parts of the upholstered products manufactured by TAPARO Company, by replacing the wooden parts of the resistance structure of sofas with composite material.

1 Introduction

Innovation is an important pillar in the new product development, in an dynamic and competitive market, because the products have to meet a wide range of needs related to quality, cost, shape, visual appearance and ability to express to the user the objectives for which it was designed [1].

Creativity is needed to create innovative products and most creative ideas in the new product development process occur in the conceptional design phases [2]. For long-term growth and to remain market competition, companies need to be able to develop continuously and come up with innovative solutions to meet some needs [3, 4].

In addition to generating ideas and innovative solutions, new product development also involves companies' ability to apply suitable problem-solving methods to solve design problems, technology issues, etc. and also to analyse them in detail. The techniques and methods applied to solve and analyse problems, as well as to generate solutions studied in [2, 5, 6] are Axiomatic Design (AD), Quality Function Deployment (QFD), Failure Mode Effect Analysis (FMEA) and Total Quality Management (TQM) for problems analysis and for idea generation, Brainstorming, Heuristics, Mindmapping and TRIZ [7, 8]. The paper [2] proposes a hybrid-compact design matrix model that integrates problem analysis as well as generating ideas into the early stages of new product development (NPD) process, into the conceptual design stage.

* Corresponding author: cornel.ciupan@muri.utcluj.ro

These aspects explain, on the other hand, why the NPD process is complex and involves a close collaboration between the industrial designer and engineering designer [9]. KwanMyung and Kun-pyo [9] presents four types of typical collaborative product design process, identificated after interviews with industrial and engineering designers from 6 consumer product manufacturers. Currently there are various methods of NPD, some of them approach design in terms of environmental impact, functions, design and aspect, product recyclability, etc.

The exponential growth of the consumer market forces companies to be as innovative as possible, but also consider the environmental impacts, the origin of raw materials, the energy involved in the production process, transport, the way in which products are used and what happens with them at the end of their life cycle [10, 11, 12]. A new product from the customer's point of view should look good, have a reasonable price, be qualitative, be environmentally friendly and recyclable, and from manufacturer vision it have to be achieved as soon as possible , with affordable technologies and minor resource losses.

All these aspects are planned from the early stages of the NPD process and require a systematic approach of the processes, involving several phases, needs identification, task clarification, conceptual design, part design and the assembly [13]. Darian [10] presents in his study how environmental performance of a product can be planned in early stages of NPD, in requirements identification and conceptual design phases.

A development direction presented in the paper is the replacement of structural materials, like metals, wood, glass with composite materials. This was possible a long time ago, with the invention of polymers, as thermoset and thermoplastic synthetic polymers, with comparative properties. In order to increase their strength, they have been reinforced, at first using glass fibre, later carbon fibre and then various natural fibres. Composite materials have become very interesting with their use in the aerospace industry. They have spread quickly in aeronautics, sports equipment's, like tennis rackets, golf sticks, sailing yachts, and also in furniture industry [14]. The first attempts to use in the furniture industry have been known since 1948, with the development of glass fibre-polyester armchairs. This type of material is lightness, colourfulness, easily shaped into complex profiles, accessible. At the "Synthetic by nature" exhibition in April 2015 at Design museum in Gent, Belgium, were exposed over 50 products from flax or hemp fibre reinforced composites [15].

2 General aspects of the design process

Designers have always been the ones who change the world. The change occurs as a result of projects related to products or services and/or to machines and technologies for their implementation. The design process consists of a succession of correlated activities designed to generate the conceptual basis for achieving a product. Pugh et. Al, [16] presents a integrated product design model that takes into account all the problems of product development: customer requirements, design specifications, conceptual design, technical design, manufacturing, sales, etc.

A new approach of the innovative products design process, with the main steps and a creativity method are presented in [17]. Starting from the premise that a product is designed to meet a social need, because its users must be satisfied by the product, the designers are the ones who conceive the product, the idea of the product based on its necessity and utility, to an imaginary form, then to a descriptive phase and finally to a model in perspective. We can make a difference between an "artistic" designer and a design engineer. "To create a beautiful, accepted and appreciated form at one point is more an art and less a science. Engineers and designers are called upon to produce beautiful products, both to create human working conditions for those who use them, and to participate forcefully in competition on an international market" [18]. The aspect of a product is a synthesis of the art - fashion - medium - technical factors, and for the designer it is a challenge because the product has to satisfy the four basic conditions, useful-beautiful-functional-economical. For these reasons, the approaches of designers and engineers, although different, must have a convergence area in which aesthetic, technical, functional and economic aspects meet. This also results from the constraints that the new product has to satisfy (Figure 1).

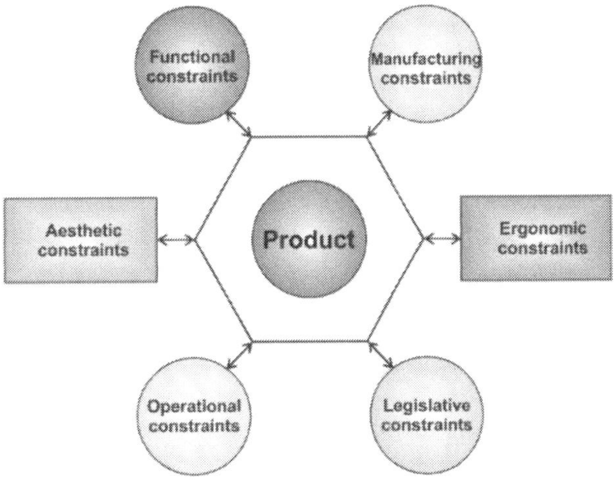

Fig. 1. The constraints on the design of a product.

By a new product we should not always see a "pioneer" product, a product with absolute novelty. The "renewal" of a product may consist in product modernization by changing shape, material, technology or by a new structural conception that gives it complementary or improving product features or socio-economic impact.

In the design of furniture products, the designer has to respect the constraints (requirements) shown in Figure 1. The constraints were established by a market research that addresses both the needs of the users and the existing products. Users' needs were determined by questioning a number of targeted consumers. When making questionnaires, a user-friendly language must be used, not a specialist language. The results by processing information from questionnaires must be transformed into constraints and design data.

Based on market research, product design specifications are developed. The design specifications will also contain legislative constraints or requirements imposed by the standards. The product concept is based on the design specifications resumed in Figure 2.

3 A method for redesigning components to replace wooden parts with composite materials

Replacing wood with composite materials involves redesigning the components. Redesigned parts should be analysed in the context of the assembly because their characteristics result from the functionality of the assembly. For every product, if we are going to change the material from which it was originally designed, we have to make a new project. The redesigned part will have to partially respect the shape and dimensions of the original part, at least at the level of the connection with the other components.

Then, in addition to size and shape constraints, the redesigned part must be suitable for manufacturing process by the new material, that is, to respect manufacturing constraints. For example, the shape of a wooden part, in assembly operations can not be preserved identically if it is made of plastics, injection or composite, by thermoforming.

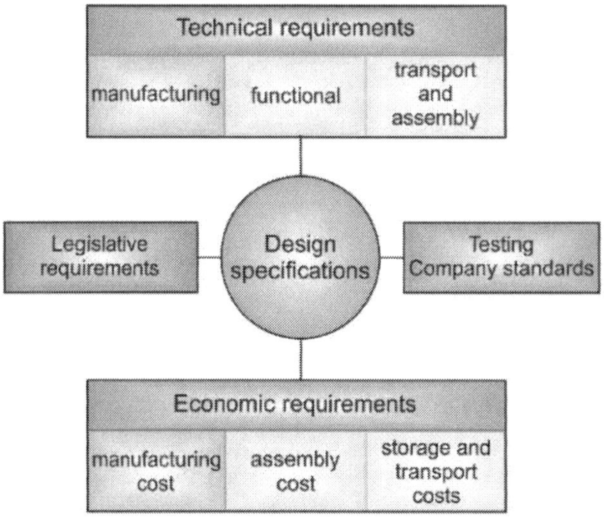

Fig. 2. Design specifications of a product.

The activities flow for furniture design, based on design specifications is shown in Figure 3.

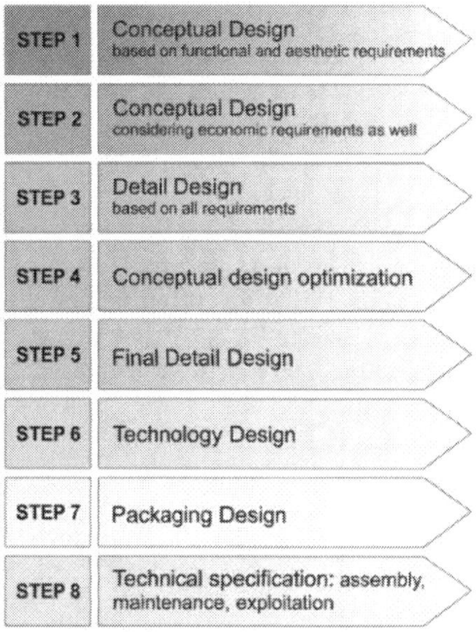

Fig. 3. The activities flow for furniture design.

A cause for which the shape change is needed is in the form of the matrix and the molding process.

There are many furniture products, especially those of the upholstered, which are successful on the market. Replacement of wood with composites in this product category should not be visible to the recipient, their external shape and functionality remaining unchanged. The parts of the resistance structure cannot be kept in their original form, especially due to manufacturing constraints. Then, functionally, the redesigned part must provide the characteristics of the original part. The study of replacing the wooden parts with composite materials based on vegetable fibres in furniture design was presented in paper [19].

A method of redesigning the parts by changing wood with composite materials, used for existing products is shown in figure 4.

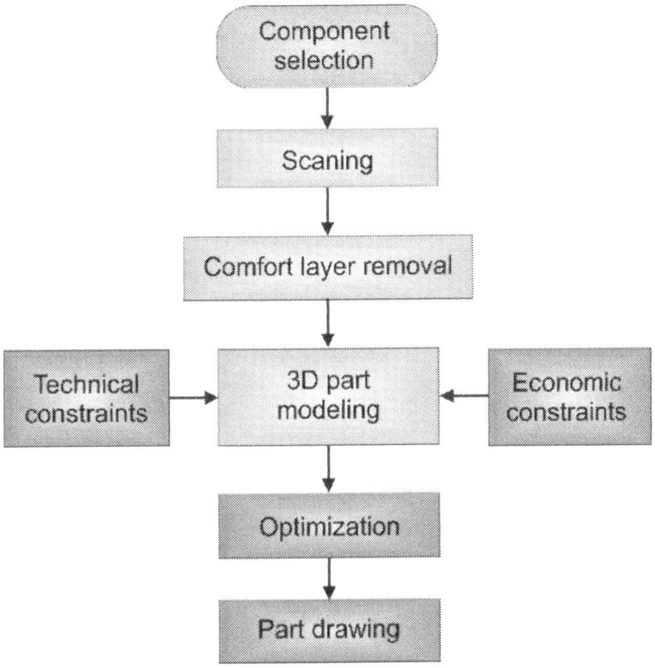

Fig. 4. The method of redesigning the parts.

Scan operation is required only if we do not have the 3D model of the assembly. By removing the comfort layer (generally known its thickness) it is obtain the volume and the outer surface of the part. Because the comfort layer is flexible (deformable), the outer surface of the workpiece should not be the same as that resulting from the removal of the comfort layer. These deviations from the resulting surface allow us to offer the piece a technological form. Thus, the new part will be designed to meet the following requirements:

- Dimensional and shape, to fit the determined volume;
- Functional, to resist the stresses imposed by the tests of the ensemble;
- Manufacturing, meaning to be manufactured under economic conditions.
- After the 3D modelling phase follows an optimization stage aimed at:
- Reducing material consumption by increasing the stiffness of the workpiece using ribs and various cut surfaces, as presented in paper [20]
- Increasing the quality of the parts and productivity through the design of highly feasible forms
- Reduce the manufacturing cost, on all possible ways.

4 Case study

Below is a case study to illustrate how the above method was applied (table 1). The selected furniture part is a sofa side, for which 3D shape can be obtained by scanning it. The next step is to remove the comfort layer, considering its known thickness, to obtain the outer volume of the part. The part volume has to be designed and optimized, to obtain the part drawings for manufacturing.

Table 1. Case study for a sofa side.

1. The selected part – sofa side	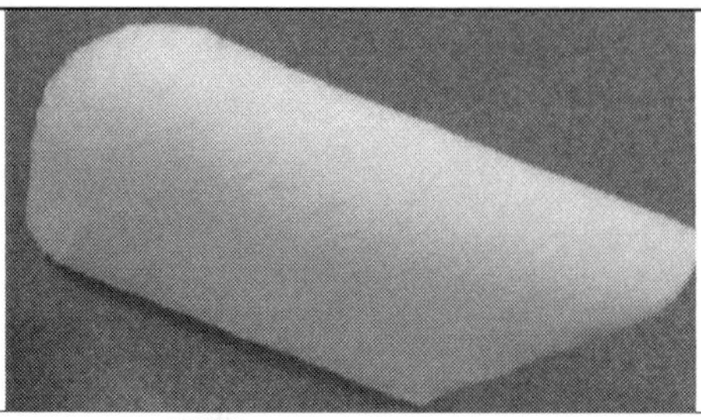
2. The 3D model of the sofa side	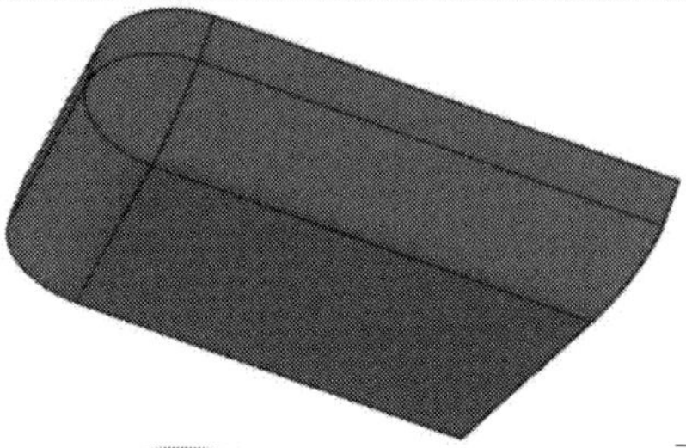
3. Comfort layer removal	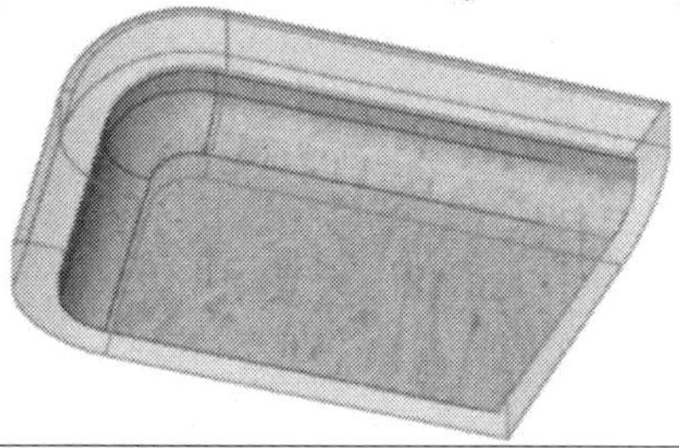
4. 3D Model	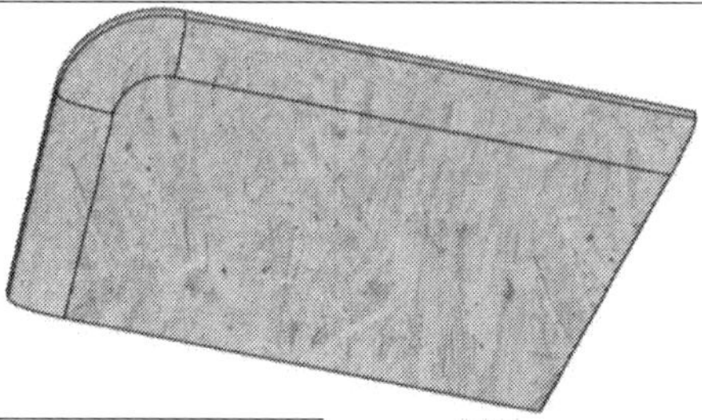
5. Part drawings	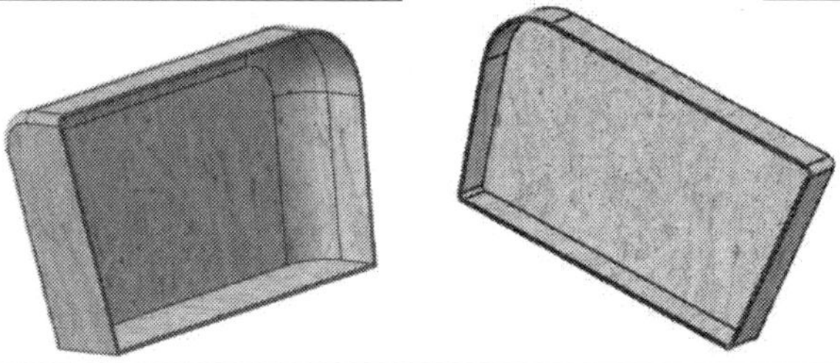

5 Conclusions

Based on another model for designing innovative products [17], the authors have developed a new method for redesigning the resistance structure by replacing the wooden parts of upholstered furniture. Generally, designers of upholstered products design, in a first stage, the external shape that gives the appearance and functionality of the furniture, and then deals with the design of the resistance structure and the technological aspects. The method has a general character and can be applied both to upholstery existing products, to which we want to change the material of the resistance structure by reverse engineering, as well as for new products.

The method is based on the principle of determining the outer volume of the resistance structure by removing the comfort layer from the total volume of the component. Then, starting from the determined volume, the component's resistance structure is projected from the desired material by subjecting it to the functional and testing constraints of the furniture. Optimization takes into account both technical and economic constraints that the designer considers relevant.

The method was validated by redesigning different wooden components of products manufactured by TAPARO (sofas, chairs and armchairs). The wood was replaced with composite materials that offers the functional requirements of the product as a whole, but can also be replaced by other materials.

References

1. J.A. Diego-Mas, J. Alcaide-Marzal, International Journal of Industrial Ergonomics **53**, 102-114, Spain, (2016)
2. Y. Ko, , Computers & Industrial Engineering, **107**, 345-359, Taiwan, (2017)
3. K.T. Ulrich, S.D. Eppinger, *Product design and development*, **2** (2000)
4. D. Cavallucci, P. Lutz, In first international Conference on axiomatic design, Proceedings of ICAD, 21–23 (2000)
5. C. Kahraman, T. Ertay, G. Büyüközkan, European Journal of Operational, **171** (2), 390–411 (2006)
6. A. Segismundo, P.A. Miguel, International Journal of Quality & Reliability Management, **25** (9), 899–912, (2008)
7. G. Altshuller, *And suddenly the inventor appeared TRIZ, the theory of inventive problem solving* (2nd ed.) (1996)
8. V. Kokotovich, Design Studies, **29** (1), 49–69 (2008)
9. K. Kim, K. Lee, Design Studies, **46**, 226-260 (2016)
10. D. Visotsky, A. Patel, J. Summers, The 24th CIRP Conference on Life Cycle Engineering USA, Procedia CIRP **61**, 69-74 (2017)
11. H. Wenzel, M.Z. Hauschild, L. Alting, *Methodology, Tools and Case Studies in Product Development*. Springer Science & Business Media, ***1***, (2000)
12. V. Rayate, J.D. Summers, Int. Des. Eng. Tech. Conf. Comput. Inf. Eng. Conf. Chicago (2012)
13. G. Pahl, W. Beitz, *Engineering Design: A Systematic Approach*, **11**, Springer Science & Business Media (2013)
14. L. Pil, F. Bensadoun, J. Pariset, I. Verpoest, Composites: Part A, **83**, 193-205, (2016)
15. *** http://www.Designmuseumgent.Be/ENG/Exhibitions-2015/Synthetic-By Nature Php (2015)
16. S. Pugh, *Creating Innovative Products Using Total Design*. Addison-Wesley Inc, (1996)
17. C. Ciupan, Proceedings of 5th Int. Conf. on the management of Technological Changes, Alexandroupolis, Greece, **2**, 283-288, ISBN 978960-8932-2-9 (2007)
18. A. Toma, *Cercetări şi contribuţii privind conexiunea formă, ergonomicitate, funcţionalitate în domeniul maşinilor unelte*. Teza de doctorat, UTCN (2012)
19. I. Filip, E. Ciupan, I. Cionca, M. Ciupan, E. Pop, E. Campean, V. Heres, F. Rat, C. Gherghel, Proceedings of the XXIth Int. Scientific Conf. "INVENTICA 2017, 38-45, (2017)
20. M. Steopan, C. Ciupan, E. Campean, E. Pop, I. Filip, L. Stelea, Proceedings of the XXIth Int. Scientific Conference "INVENTICA 2017", 95-103, (2017)

Software tool used for automated design of customizable product

Roxana Pescaru[1], *Panagiotis* Kyratsis[2], and *Gheorghe* Oancea[1,*]

[1] *Transilvania* University of Brasov, Manufacturing Engineering Department, Mihai Viteazu No. 5, Brasov, Romania
[2] Western Macedonia University of Applied Sciences, Department of Mechanical Engineering and Industrial Design, Kila Kozanis, Greece

Abstract. The paper presents an original software tool implemented for an automated design methodology associated to a customizable product from footwear industry. The software tool is written in Visual Basic for Application and is integrated in CATIA environment. It is an easy to use software tool and combines reverse engineering with parametric design of a product that has complex shapes with many curves. The final result is a solid CAD model obtained in an automatically mode.

1 Introduction

Nowadays the Reverse Engineering (RE) became a common technology used in different fields such as [1, 2]: automotive, aerospace, medical engineering, entertainment, clothing and footwear, architecture and arts. RE is defined as a reinvention technology which uses the processes of measuring, analysing, and testing to reconstruct an object [1].

In the RE process are used different commercial software systems for point cloud handling and object redesigning [1, 3-8]. Usually a specific methodology used by the designer in commercial software systems for the redesigning process and the product is obtained in a parameterised mode [3-5]. The specific methodology can be also implemented in specific software tools for obtaining the CAD model corresponding to a customizable product, in an automatically mode. The paper aims to present such a software tool based on the methodology developed by the authors for a customizable product from the area footwear [4].

2 Software tool presentation

The software tool has been developed using Visual Basic for Application (VBA) under CATIA commercial software system. The VBA language has been used in the process of software tool developing that can be launched from the CATIA environment. The interface shown in Fig. 1 (in editing mode), has been designed in the Visual Basic environment [9, 10] and the application which manages the components placed on it has been conceived to

* Corresponding author: gh.oancea@unitbv.ro

ensure the flow of information in the CATIA design environment.

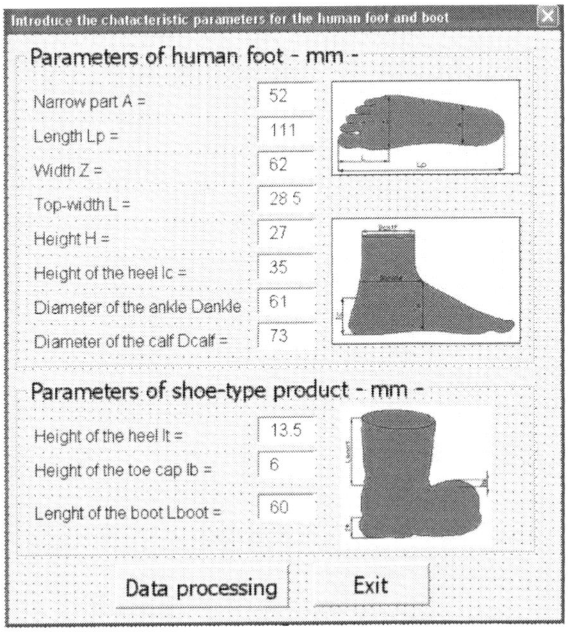

Fig. 1. Designed graphical user interface.

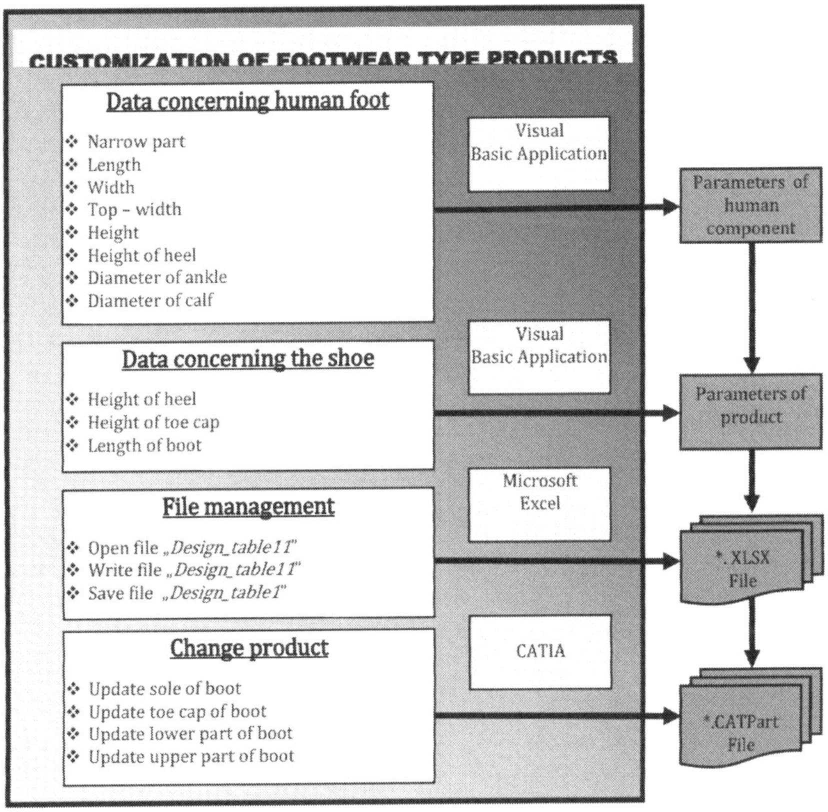

Fig. 2. Flow chart for customization of products of footwear type implemented in software tool.

On the interface are presented, in a graphical form, the product type and the corresponding human foot with the associated parameters.

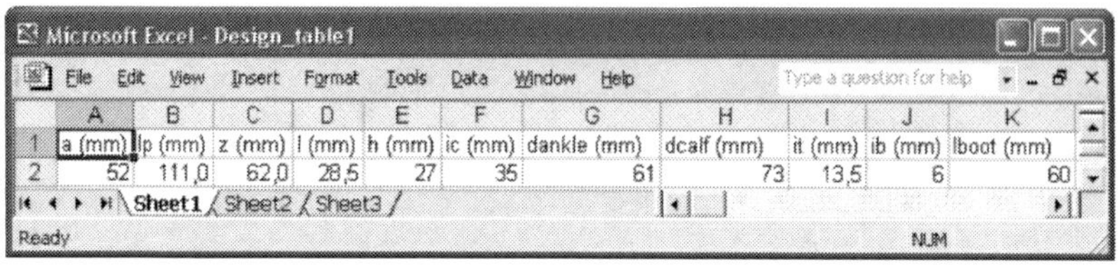
Microsoft Excel - Design_table1

	A	B	C	D	E	F	G	H	I	J	K
1	a (mm)	lp (mm)	z (mm)	l (mm)	h (mm)	ic (mm)	dankle (mm)	dcalf (mm)	it (mm)	ib (mm)	lboot (mm)
2	52	111,0	62,0	28,5	27	35	61	73	13,5	6	60

Fig. 3. Design table in Excel file.

According to the flow chart shown in Fig. 2, two interrelated modules are required in order to obtain the final product. The first module that consists of the VBA application has a role in the introduction of the data relating to the human component and is delivered as numeric values that represent the parameters required for the footwear customization. The second module is necessary in order to modify the product, using the CATIA environment in accordance with the values entered by means of the first module. The CAD model of the footwear product is modelled in a parametric mode using the RE methodology developed by the authors [4]. These two modules are interconnected in such a way that the user perceives the system as a whole without noticing when passing from one module to another.

In order to facilitate the introduction of the parameters necessary to carrying out the automated attainment of the curve/surfaces in the CATIA environment, the authors have chosen the design table method. The aim of the design table [11] is to control the parameters of a CAD model by means of external values. For any design table, the user has to associate the parameters of the model obtained in CATIA with the data stored in the table. By creating the associations, the user specifies which of the model parameters are set in correspondence with certain columns of the table. In this case the parameter values are stored as a Microsoft Excel table (Fig. 3).

The design table is defined using the Knowledge toolbar from the Shape module, by entering its name and most importantly entering the correct allocation of the parameters that compose the design table. The user can modify the values in the design table by editing and then saving the Microsoft Excel associated file. The CATIA environment automatically detects the changes and then updates the part according to the new values. In order to store the values of the parameters was used association data like in the case of the objects generated in virtual environments [12, 13].

3 Software tool testing

In order to test the software tool used in automated design for the products of custom footwear type, the user has to follow a series of steps from the moment the application is launched until the custom product is obtained like a solid model.

The real object for experimental product is shown in Fig. 4, and the steps that have to be followed are below presented.

The first step is to launch the CATIA software by Dassault Systems. The second is to open the"Parameterized_boot. CATPart" file generated using the algorithm developed by the authors [4]. The third step consists of data inputting in an window named "Introduce the characteristic parameters of the human foot and the boot", which is automatically displayed on screen (Fig. 5). The user has to enter both the data related to the foot for which product customization is desired and the parameters of the boot.

Fig. 4. Experimental footwear product.

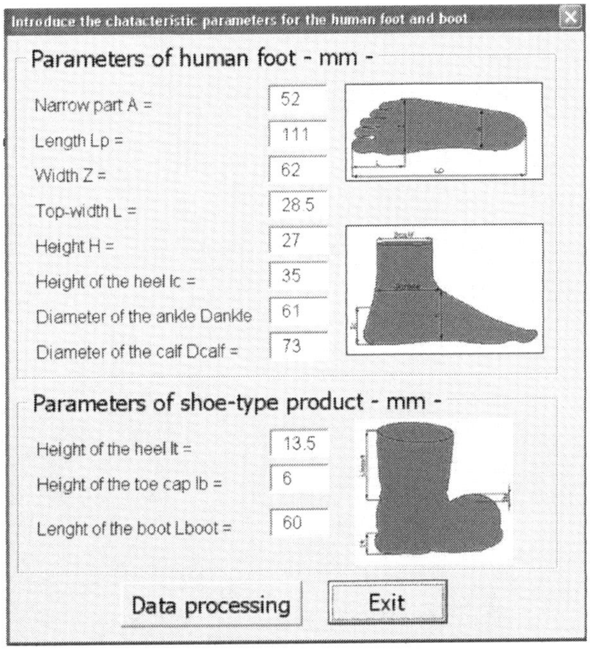

Fig. 5. Input parameters window.

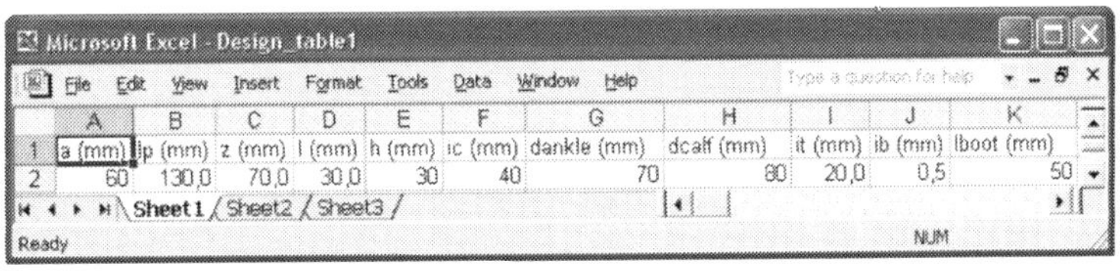

Fig. 6. Design_table1 file.

The fourth step, press the "Data processing" button and the data are automatically written in the "Design_table1" Excel file (Fig. 6), and then press the "Exit" button to close the window. Finally, the designer can notice how the components of the considered object are automatically changed according to the data entered in step 3.

All parameters have default values, being at the basis of the initial product presented in the Fig. 4. For a better illustration of the whole customization process of a footwear type product, table 1 shows both the default values of the parameters and those subsequently modified. Fig. 7 shows two products of the boot type that were obtained with data sets from the table 1. The final results are parametric CAD models obtained like a solid object.

DOI: 10.1051/matecconf/201713706003

Table 1. Default and modified values of parameters.

Parameter	Symbol	Default value [mm]	Modified value [mm]
Specific length for narrowing of foot	**A**	52	60
Length of foot	**Lp**	110,6	130
Specific length for width of foot	**Z**	61,4	70
Length from top of foot up to section where width of foot is known	**L**	28,5	30
Height of boot heel	**It**	13,5	20
Height of the foot	**H**	27	30
The height of the boot toe cap	**Ib**	6	0,5
The height of the human heel of the foot	**Ic**	35	40
Specific diameter for human ankle	**Dankle**	61	70
Height of boot	**Lboot**	60	50
Specific diameter the human foot's calf	**Dcalf**	73	80

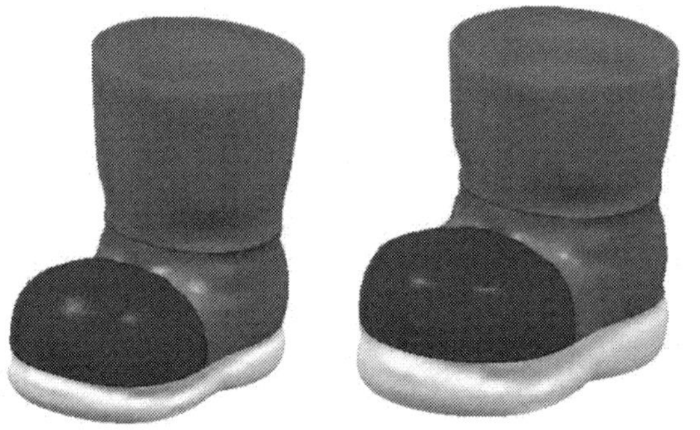

a) b)

Fig. 7. Final 3D CAD models based on automatically created solids: a) Default values of parameters; b) Modified values of parameters.

4 Conclusion

The paper presented a software tool developed by the authors for a footwear product type, using VBA and CATIA software system. In the tool is implemented an algorithm for the automated design of products with high level of customization for which, initially, no documentation was available. The paper also describes the steps for designing the products with a high level of customization, using the original software tool developed by the authors. The software tool is validated for a customized product which can be designed according to numeric data related to the human foot and product, entered by the user. Futures research will be focused on developing new case studies for different models of footwear products based on the methodology described by the authors.

This work supported by the Sectoral Operational Programme Human Resources Development (SOP HRD), financed from the European Social Fund and by the Romanian Government under the project number POSDRU/89/1.5/S/59323 and we hereby acknowledge the structural founds project PRO-DD (POS-CCE, O.2.2.1., ID 123, SMIS 2637, ctr. No 11/2009) for providing the infrastructure used in this work.

References

1. W. Wang, Reverse Engineering: *Technology of Reinvention* (Taylor & Francis Group, USA, 2010)
2. V. Raja and K. Fernandes, Reverse *Engineering an Industrial Perspective* (Springer Series in Advanced Manufacturing, UK, 2008)
3. G. Oancea, N.V. Ivan, R. Pescaru, AJME, **11** (4) (2013)
4. R. Pescaru, P. Kyratsis, G. Oancea, IOP Conf. Series: Materials Science and Engineering, **161** (2016)
5. R. Pescaru, G. Oancea, Advanced Materials Research, **7**, 591-593, (2012)
6. *** http://www.geomagic.com/en/products-landing-pages/designx. Accessed 12 July 2017
7. ***https://www.autodesk.com/products/powershape/features/reverse-engineering/reverse-engineering. Accessed 12 July 2017
8. *** https://www.3ds.com/products-services/catia/. Accessed 12 July 2017
9. C. Utley, A *Programmer's introduction to Visual Basic.NET*, (Sams Publishing, Indianapolis, USA 2001)
10. J. Walkenbach, *Excel VBA Programming for Dummies. 2nd Edition,* (Wiley Publishing, Indianapolis, USA, 2010)
11. I. Ghionea, *CATIA v5. Aplicații în inginerie mecanică* (Ed. BREN, Bucharest, Romania 2009)
12. G. Oancea, F. Girbacia , A. Nedelcu, *Software Module for Data Exchange Between AutoCAD and a Virtual Reality System, Product Engineering-Tools and Methods Based on Virtual Reality* (Springer Netherlands, 2008)
13. S.A. Haba, G. Oancea, Journal of Advanced Manufacturing Technology, **80** (5), (2015)

Comparison of process documentation generation in PTC Creo and NX systems

Martin Pollák[1,*], *Petr* Baron[1] , and *Marek* Kočiško[1]

[1]Technical University of Košice, Faculty of Manufacturing Technologies, 080 01 Prešov, Slovakia

Abstract. Process documentation is a specific type of progressive form for representation of the technological documentation which defines the sequence of the individual production steps on the chosen machine according to a pre-executed NC program. It is represented by a simplified form of automatically generated reports in a short time interval. The significant contribution of the creation is in the acceleration of communication between the designer and the technologist, as well as NC machine operators with the possibility of archiving and managing internal data. The article briefly describes and compares the creation of process documentation in the PTC Creo system and in the NX system environment.

1 Introduction

From the analysis of the total labour consumption accounts for 70 % to 80 % for the technological preparation of production. Thus, it can be said that this is the most time consuming stage in the production process. Requirements for creating technical documentation need to have a high level of technological preparation. The requirements make sure that the parts are manufactured correctly and the manufacturing steps in sequence. Technological preparation of production uses intuition, logical activity and the routine work mostly. Currently, due to the differences in the creation of technological documentation, it is necessary to divide the technological preparation of the production according to the area of its use and form [1].

Technological documentation is represented in form of [2]:

- technological process written verbally,
- technological process represented by images,
- NC code of program,
- simulation (visual presentation of the technological documentation).

The area of NC machining is a specific area in the production of technological processes. The main document for NC machining is the NC program. The NC program defines the sequence of the individual steps that are executed on the specified machine. Despite the amount of information that the NC program contains, it cannot be said that there is sufficient technology documentation.

The NC program consists of the basic features of the technological process, i.e. it specifies the basic machine setting and defines the precise tool paths as well as some

* Corresponding author: martin.pollak@tuke.sk

 DOI: 10.1051/matecconf/201713706004

auxiliary operations (tool change, lubrication, cooling...) [3]. The machine operator is missing additional data (base unit, number of pieces produced, material of the product and the like.)

For this reason, it is necessary to extend this documentation to its other forms. Today's CAD/CAM/CAE systems used to create NC programs offer us three solutions [1]:

• additional use of CAPP systems,
• implementation of complex PLM systems into the manufacturing company,
• possibility to use the automated generation of process documentation in the CAD/CAM/CAE system.

The article points to the creation of process documentation in the PTC Creo 3.0 system and in the NX system environment. The aim of creation is the elimination of lengthy processes during the step of technological preparation of production, using a simple and fast generation of process documentation directly in that system.

2 Creation of technological documentation and its automation

Mechanical work in the 20th century began to replace automated technical preparation of production, in order to save time and human forces. Manufacturing and machining equipment began to be produced, which were later replaced by NC machines. The most valuable aspect of today is the time which is the main reason why new technologies are introduced in the field of development and production. The main task of introducing new technologies into the manufacturing sphere is to shorten the design, production and other processes times to the minimum. The companies gradually began to use and implement a comprehensive CAD/CAM/CAE, CAPP and PLM systems.

The technical preparation of production includes all the activities that need to be carried out before the actual production of the product begins. All these activities are focused on the preparation and processing of product technical documentation (design, technology and projection). Variants of these solutions ensure maximum productivity and efficiency of production processes, such as shorter lead times, material savings, energy savings, labour intensity and production costs [4].

The manufacturing process provides the necessary manufacturing information, such as the times of individual manufacturing steps and intermediate steps, machines and tools used for the individual operations, movement of the tools, cutting speed, etc.

The information contained in the technological process can be divided into:

• the identification data: number of technological process, drawing number,
• the material information: designation of material quality, consumption amount of material per component, total amount of consumed material, initial state of the material,
• the operational data: number of operations, marking of the work, marking of the workplace, type of machine, tools, jigs, gauges, brief description of the operation, number of operational instruction cards, batch time, unit time, tariff class, total time on benefit, total cost of the operation,
• the production data: number of pieces in production batch, marking of the center designated for the production of parts, number of production orders, number of material warehouse.

3 The process of generating the process documentation

In the NC machining a technological documentation is based on the principle of a simplified process flow in the form of process documentation. There are already a lot of complex CAD/CAM/CAE systems that offer the possibility to use module for the automated generation of process documentation.

The process documentation represents a progressive form of automated generation of simplified technical documentation and is primarily intended for NC machining area. This option resulted from the requirements of the accelerated creation of technological documentation for the deployment of comprehensive CAD/CAM/CAE systems.

3.1 Generating process documentation in the PTC Creo 3.0 system

The PTC Creo 3.0 has this option available under the tab Application/Process Documentation. Automated generation is suitable for fast creation of workshop documentation, and its great advantage is the ability to archive in digital form HTML. Generated documentation is therefore possible, at any moment, to be customized, defined and used as a simplified form of representation of technological process. The Creo 3.0 provides two options for the creation of process documentation using predefined templates by:

- Process Documentation Defaults,
- Process Manager Defaults.

By selecting a template generation via Process Document Defaults are looking at a Report by operation choice of two types of display as follows: Process Main View and Tool View. The first type Process Main View contains a list of operation names and types of operations. In the second type, the Tool View is locating a system-preset list of tools and items such as:

- Name (name of operation),
- Tool_ID (tool name),
- Tool_Type,
- Cutter_Diam,
- Tool_Comment.

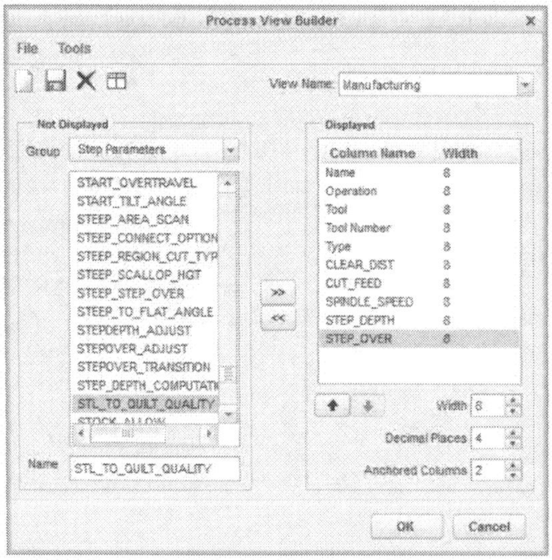

Fig. 1. Process View Builder.

Documentation generation content can be changed in accordance with the necessary data for production. The auxiliary window Process View Builder (Fig. 1) is divided into two parts that are not displayed and Displayed (not shown and displayed items) which determines the content of process documentation data. By gradually adding of items using the tool inserted, the user sets the appearance of process documentation. The selected data can then be sorted in the selected order of Displayed. In order to avoid of repeated and lengthy selection, the list of items can be saved separately.

After editing the appearance of the template settings are saved. Before the start of generation it is necessary to be able to select the desired look of generated process documentation. Process documentation may have a look as simple text or can be supplemented by graphic illustrations of operations and tools.

After checking a field Manufacturing Model Image, Step Image, eventually Tool and Solid Tool Image the choice in the generated documentation process reflected in the representations of the graphic representation of the machined areas, and also a graphic illustration of the use of tools. By the icon Create is launches the generation process (Fig. 2).

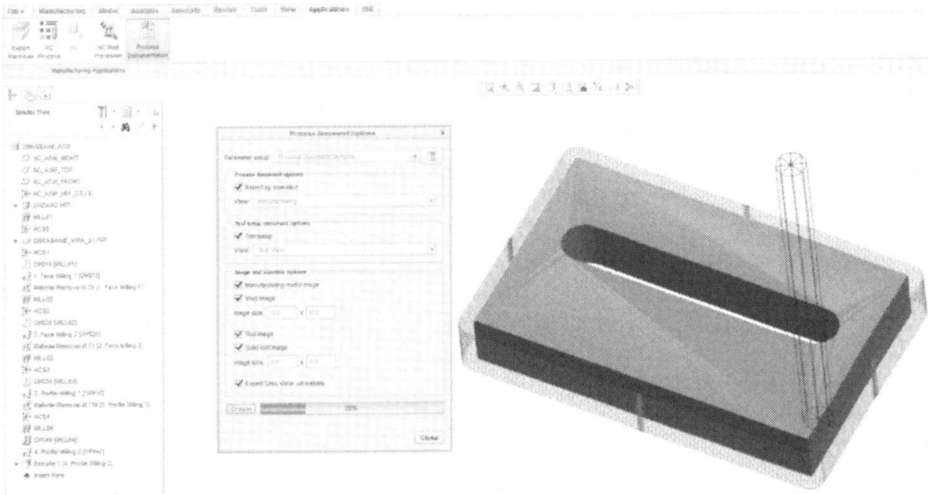

Fig. 2. The process of generating the process documentation.

After finishing the video demonstration, the system generates process documentation in HTML format (Fig. 3).

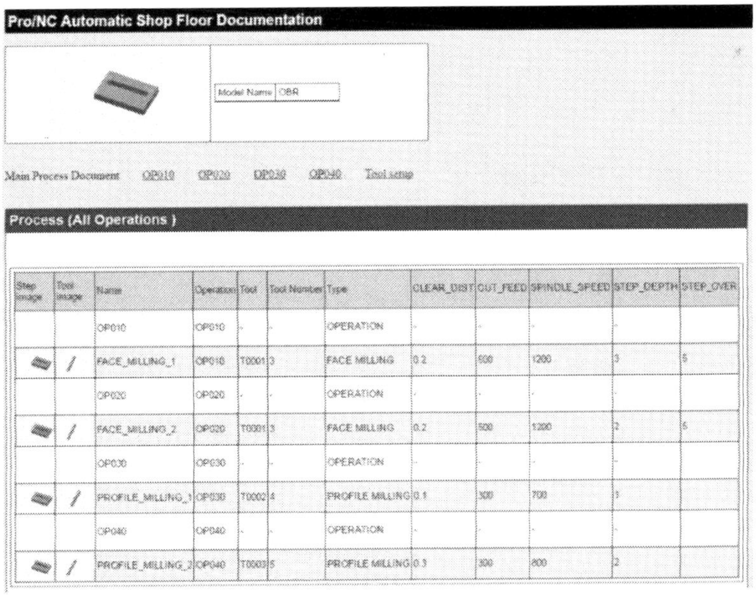

Pro/NC Automatic Shop Floor Documentation

Model Name	OBR

Main Process Document OP010 OP020 OP030 OP040 Tool setup

Process (All Operations)

Step image	Tool image	Name	Operation	Tool	Tool Number	Type	CLEAR_DIST	CUT_FEED	SPINDLE_SPEED	STEP_DEPTH	STEP_OVER
		OP010	OP010	-	-	OPERATION	-	-	-	-	
		FACE_MILLING_1	OP010	T0001	3	FACE MILLING	0.2	500	1200	3	5
		OP020	OP020	-	-	OPERATION	-	-	-	-	
		FACE_MILLING_2	OP020	T0001	3	FACE MILLING	0.2	500	1200	2	5
		OP030	OP030	-	-	OPERATION	-	-	-	-	
		PROFILE_MILLING_1	OP030	T0002	4	PROFILE MILLING	0.1	300	700	1	
		OP040	OP040	-	-	OPERATION	-	-	-	-	
		PROFILE_MILLING_2	OP040	T0003	5	PROFILE MILLING	0.3	300	800	2	

Fig. 3. Process documentation generated in the PTC Creo 3.0 system.

 DOI: 10.1051/matecconf/201713706004

By choosing of determined sequence of machining through graphical interpretation, the system displays more detailed magnified picture presenting view of the machining areas with exact specifications of set machining operation as well as information about the tool with which a given area is machined.

In the system PTC Creo 3.0 it is not possible in the generation of process documentation to change the look or format of documentation, which is the only disadvantage in the generating process. However, generated values in the table can be changed.

3.2 Generating process documentation in the NX system

Process documentation (in the NX system called "Shop Documentation") can be formed by four ways to view the machine sequences in terms of program, tool geometry and machining methods. These data can be viewed in a generated documentation using a set of objects that offers individual display options (machining sequences, geometry and component material, machining parameters, tool parameters, tool path information, tool parameter controls, control parameters...).

In the NX system, it is possible to create shop documentation using 4 predefined templates:

• Operation List Select (HTML/Excel) – generating a template in terms of machining operations in HTML format with editable templates in MS Excel (Fig. 4).

• Operation List Select (ASCII) – generating a template in terms of machining operations in the ASCII text format with the possibility of printing and distributing reports.

• Tool List Select (HTML/Excel) – generating a template in terms of used cutting tools with defined machining parameters in HTML with the ability to edit the template in the MS Excel editor. Tools are assigned to individual machining sequences. The tools are then, in the complete tool list, ordered by their assigned name.

• Tool List Select (ASCII) – generating a template in terms of used cutting tools with defined machining parameters in text format (without graphic elements and hyperlinks).

HTML (web) format presents data that can be browsed through a web browser and the ability to share documents on the web. In the created shop documentation, the graphs of the individual designs are displayed in the form of images with clear browsing through hyperlinks.

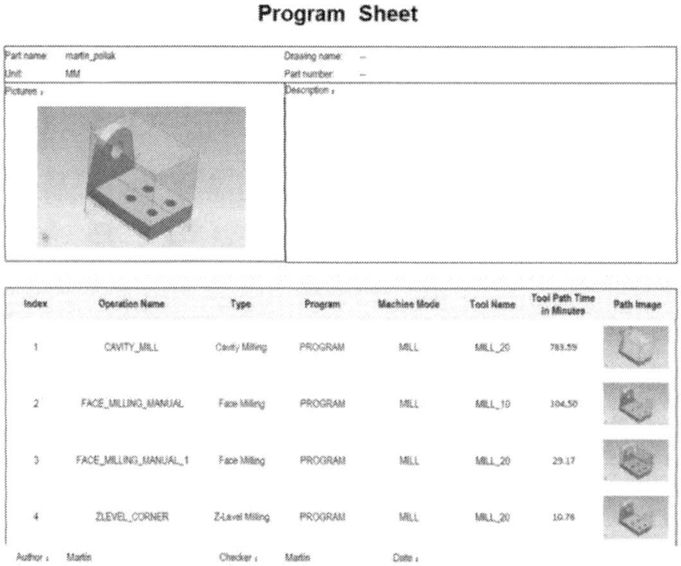
Program Sheet

Part name: martin_pollak   Drawing name: --
Unit: MM   Part number: --
Pictures :   Description :

Index	Operation Name	Type	Program	Machine Mode	Tool Name	Tool Path Time in Minutes	Path Image
1	CAVITY_MILL	Cavity Milling	PROGRAM	MILL	MILL_20	783.59	
2	FACE_MILLING_MANUAL	Face Milling	PROGRAM	MILL	MILL_10	104.50	
3	FACE_MILLING_MANUAL_1	Face Milling	PROGRAM	MILL	MILL_20	29.17	
4	ZLEVEL_CORNER	Z-Level Milling	PROGRAM	MILL	MILL_20	10.78	

Author : Martin   Checker : Martin   Date :

Fig. 4. Shop documentation generated in terms of machining operations in HTML format.

The template created in HTML format can be edited in the Microsoft Excel spreadsheet environment. Editing is done in the MS Excel worksheet file named shopdoc_template.xlsx (Fig. 5). Editing the file creates a new layout template according to the user's needs. Excel template file consists of two worksheets. The first worksheet includes a visual template page with the option of editing the layout, cell dimensions, data entry sequences and other editing options.

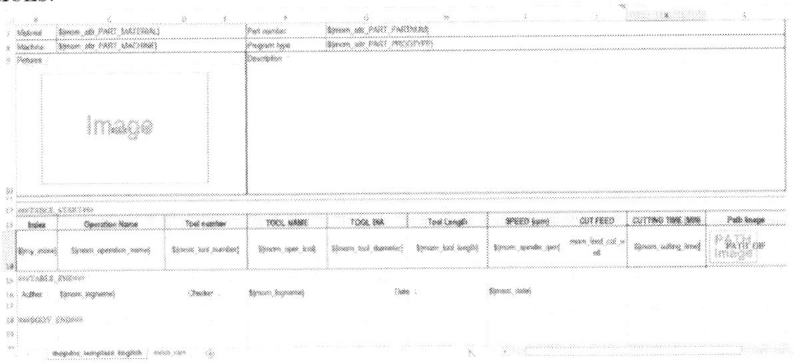

Fig. 5. Modification of shop documentation in MS Excel environment.

The second Excel worksheet contains an overview and description of the variables that are needed to produce changes in the first Excel worksheet, by simply copying and pasting the variables in the template cells (MOM). Variables are written to facilitate creation, without the need to study the linguistic expression statements of MOM variables.

4 Conclusion

Creation of process documentation is currently a required part of complex CAD/CAM/CAE systems. Therefore, the developers of these systems are continuously developing and incorporating this creation option into their products. Generating process documentation is relatively new and underdeveloped tool despite the efforts of the creators. Process documentation is simplified in the form of automatically generated reports containing the necessary parameters and graphical interpretation of specific machining cycles created in the system CAM environment of the given system.

The most important benefits of using automated generating of process documentation are:

• process documentation is a comprehensive, organized form of data that simplifies communication between the constructor, technology and NC machine operators,

• speed of processing of larger amount of information into a single document with the absence of the manual writing of technological documentation in the additional CAPP systems,

• outputs from these modules can also be used to archive and manage internal records, making it easier for management to produce productivity analyses,

• it allows quick upload of process documentation on the internet, because the output is directly in HTML, which improves communication between employees in different departments,

• process documentation can be used in training of new operators by providing information on the current project.

Each software maker has chosen a different approach to creating these templates. It should be noted that this option is not intended for normal system users. The user's simplest way to define a custom template is provided by PTC Creo but, on the other hand, the possibility of change is limited only to changing the individual displayed parameters in the output documentation (graphical form and visual appearance of documentation cannot be changed). The NX system provides wider possibilities for defining templates but there is a need to have knowledge of programming, technology and detailed knowledge of the system at the administrator's level. This system allows the creation of any graphics format of final documentation, which is a great advantage compared to the system PTC Creo 3.0.

The Ministry of Education, Science, Research and Sport of SR supported this work, contract VEGA No. 1/0492/16, and ITMS project 26220220125.

References

1. M. Staniszewski, S. Legutko, P. Raos, Tehnicki vjesnik-technical gazette, **21**, 1177-1181, ISSN 1330-3651 (2014)
2. I. Kuric, J. Kosturiak, A. Janac, J. Peterka, J. N. Marcincin, *Computer aided systems in mechanical engineering*, EDIS, 351, ISBN:80-7100-948-2 (2002)
3. P. Monka, K. Monkova, M. Balara, S. Hloch, J. Rehor, A. Andrej, M. Somsak, International journal of advanced manufacturing technology, **85**, 2325-2343, ISSN: 0268-3768 (2016)
4. S. Olejarova, J. Dobransky, J. Svetlik, M. Pituk, Measurement, **106**, 18-25, ISSN: 0263-2241 (2017)

Experimental studies on manufacturing of the complex shapes made from copper

Alina Ioana POPAN[1,*], *Bogdan* LUCA[1] and *Alexandru* CAREAN[1]

[1] Technical University of Cluj-Napoca, B-dul. Muncii, No. 103-105, 400114, Cluj-Napoca, Romania

Abstract. The paper presents the manufacturing process of a complex copper electrode. To process complex copper shapes is difficult because the material is soft and very sticky. The paper presents the steps for processing this copper electrode at a high dimensional accuracy and a good surface quality. For this process a special tooling solution is needed, optimal process parameters, smart CAD/CAM software and precise CNC equipment.

1 Introduction

CNC milling technology is one of the most important methods used in industry to machine materials [1, 2]. It has various distinct advantages over the other cutting technologies, such as: high machining versatility, high flexibility and has been proven to be an effective technology for processing various engineering materials [3-5].

Current trends in the manufacturing industry require parts to be processed in a short time, with a good surface quality, high dimensional accuracy and complex 3D shapes [6,7]. For manufacturing 3D complex parts are used a variety of CNC machines, such as: 3 or 5 axis milling machines, multitasking lathes, 6 axis robots, or even abrasive water jet centers [8-11].

To manufacture complex 3D parts is important to use CAD/CAM software. This kind of software must have various strategies for processing complex 3D shapes with fine details [12-14].

The electrodes which are used in electrical discharge machining are made from two materials graphite or copper. In this paper is presented the manufacturing process for a complex copper electrode. To obtain a good surface quality and a high dimensional accuracy for copper cutting are required dedicated tools for this type of application [15, 16].

Using wrong combination between tool geometry and process parameters, an accumulation of material against the rake face (built up edge - BUE) can appear. The built up edge can generate in this case a poor surface finish or tool brake [17].

* Corresponding author: alina.luca@tcm.utcluj.ro

DOI: 10.1051/matecconf/201713706005

2 The 3D model

The study was started with a 3D model, as in figure 1. This 3d model represents a copper electrode, designed for processing a complex mould.

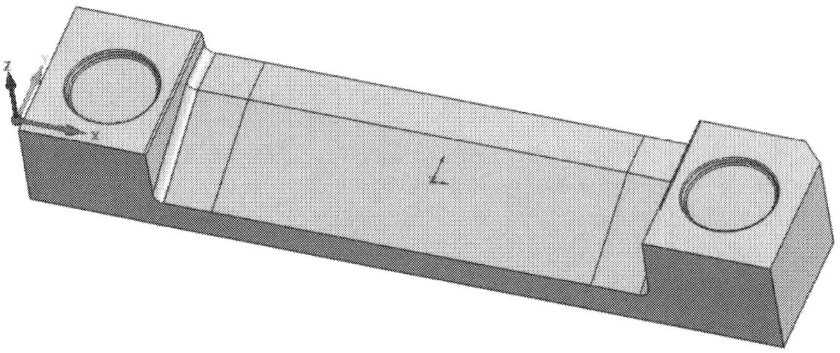

Fig. 1. The 3D model of the electrode.

3 Tools and process parameters

Copper alloys are difficult to machine by using normal tools, because they are very sticky. For process this copper electrode was chosen a special type of end mill made by Guhring Company.

To avoid the built up edge was selected a tool geometry with sharp edge and polished rake face. The end mills are made out of carbide, with 3 flutes and a bright surface finish with radial coolant ducts.

For roughing operations an end mill with the code 3367/12.00 was selected, as in figure 2.

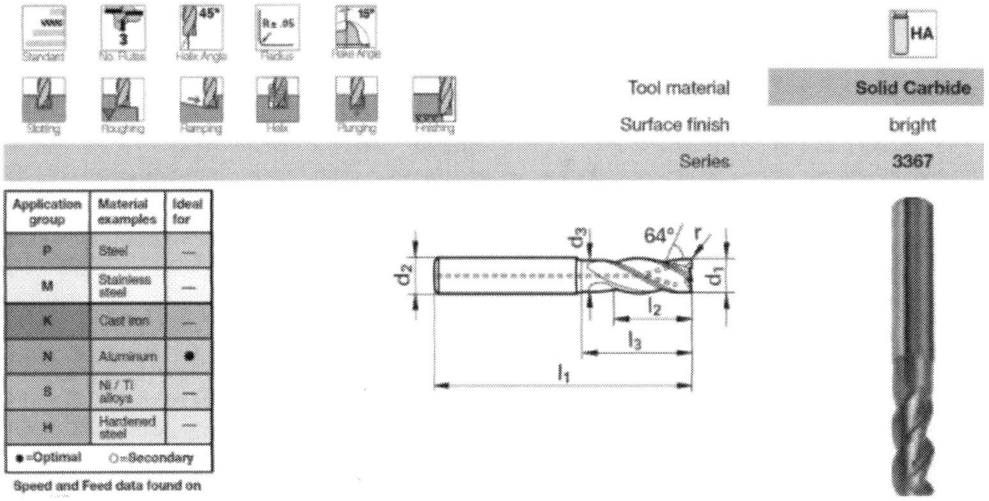

Fig. 2. The tool for the roughing operation.

The process parameters for roughing operation are:

- End mill 12 mm, code: 3367/12.000

DOI: 10.1051/matecconf/201713706005

Rotation speed: 3342 RPM; Feed Rate: 614 mm/min; Depth of cut: A_p=1 mm; Width of processing: A_e=12 mm.

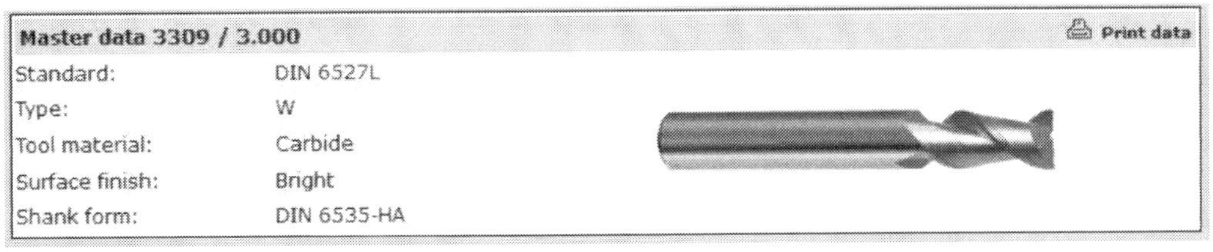

Fig. 3. The second tool for the roughing operation.

- End mill 3mm, code: 3309

Rotation speed: 10800 RPM; Feed Rate: 250 mm/min; Depth of cut: A_p=0.18 mm; Width of processing: A_e=0.05 mm.

For finishing operations was chosen a ball mill with code 3015, as in figure 4.

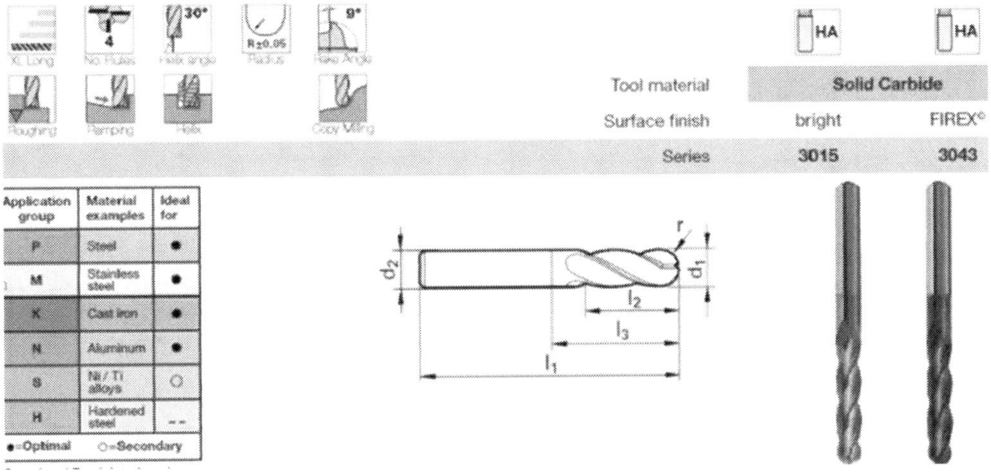

Fig. 4. The tool for the finishing operation.

The process parameters for finishing operation are: ball mill 3 mm, code: 3015; Rotation speed: 11400 RPM; Feed Rate: 950 mm/min; Depth of cut: A_p=0.18 mm;

The processing width, A_e is an important parameter for finishing of 3D complex surfaces. It has a direct influence on the surface quality.

To calculate this parameter the next formula can be used:

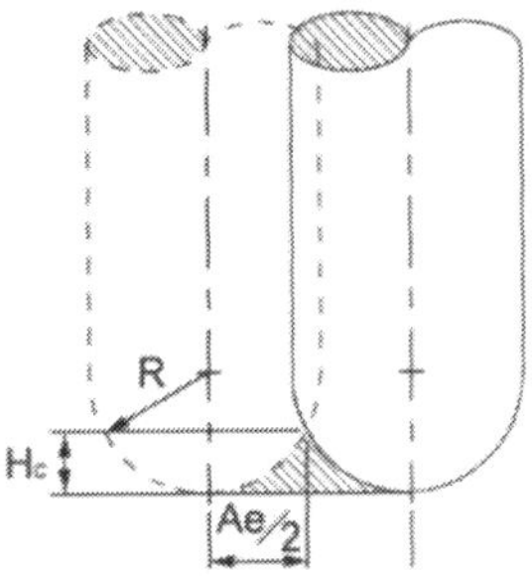

$$Hc = R - \sqrt{R^2 - \left(\frac{Ae}{2}\right)^2}$$

$$Ae = 2\sqrt{R^2 - (R - Hc)^2}$$ [18]

Fig. 5. The width of processing, A_e.

In this formula the H_c parameter represents the height of the surface roughness.

To convert H_c μm in the general parameter for surface quality R_a μm, table 1 can be used [18].

Table 1. To convert Hc μm in Ra μm.

H_C (μm)	0,2	0,4	0,7	1,25	2,2	4	8	12,5
R_A (μm)	0,03	0,05	0,1	0,2	0,4	0,8	1,6	3,2

In our study the width of processing was calculated at A_e=0.05 mm.

4 Experimental studies

For manufacturing the electrode was used a CAD/CAM software solution (SolidWorks and SolidCAM). The first step was roughing operation. For this was chosen the application *High Speed Machining* and *Contour Roughing* strategy. For finishing operation was selected the *Constant Z* strategy, in case of complex surfaces and *Linear strategy* processing for flat surfaces. The process simulation is presented in figure 6.

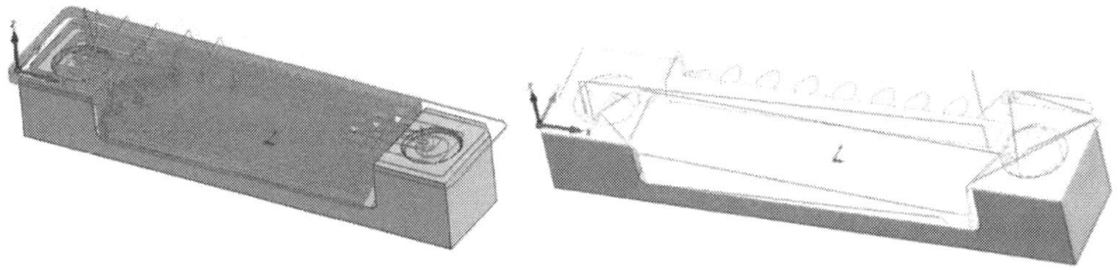

Fig. 6. The process simulation.

For this experimental study a 3 axis CNC Milling Centre was selected, HAAS VF2, with HAAS controller, from the Tehnical University of Cluj-Napoca. To fix the workpiece a special clamping device was developed, presented in figure 7.

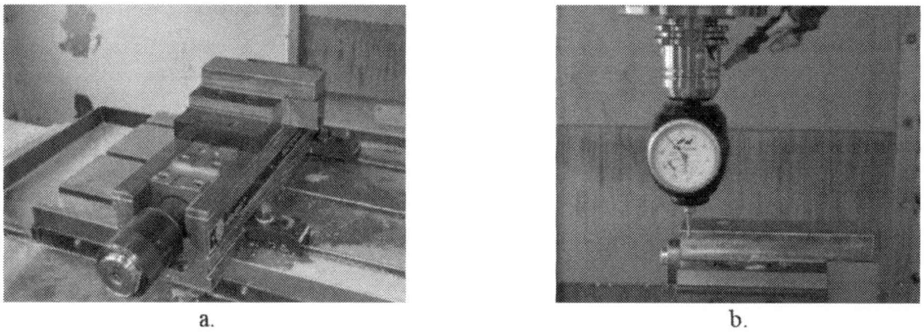

Fig. 7. a. The clamping system b. Defining the work offset.

Using a 3D Probing Device, the work offset of the part was taken and the required tools were measured (see figure 7).

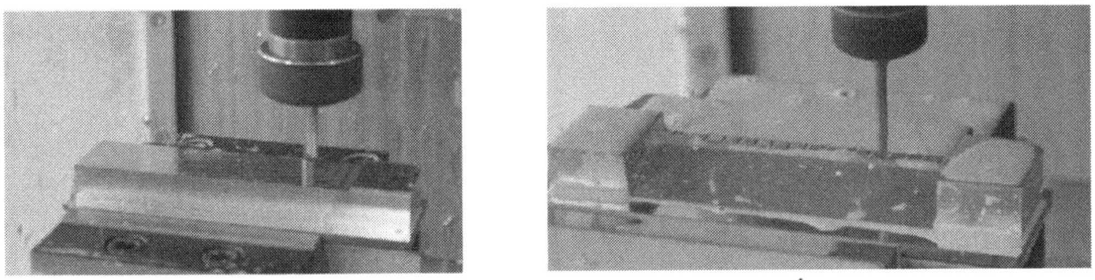

a. b.

Fig. 8. a. Roughing operation, b. Finishing operation.

For clamping the tools was used a tool holder with ER32 collet chuck. The cutting process was using internal and external cooling with fluid (for 12 mm end mill). In figure 9 are presented different phases of the machining process.

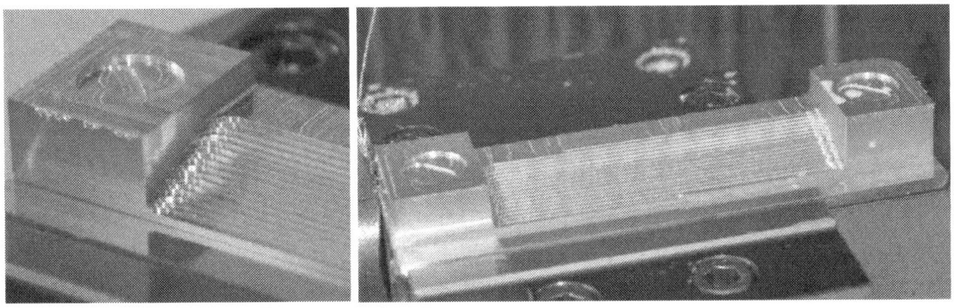

Fig. 9. Roughing operation with 12 mm end mill.

The entire roughing operation lasted 0.5 hour. The finishing phase of the process lasted 6.3 hours.

5 Investigation results

The manufactured electrode is illustrated in figure 10. Using a surface roughness tester Mitutoyo Surftest SJ-210 was measured the surface quality (R_a max 0.18 μm). The maximum dimensional accuracy, 0.03 mm was obtained during the measurements.

Fig. 10. The copper electrode.

6 Conclusions

This paper presents an experimental study for manufacturing a complex copper electrode.

The main problem of this application is built up edge. The built up edge must be avoided to obtain good surface quality and dimensional accuracy.

To manufacture complex 3D shapes out of copper one must have:

• CAD/CAM software with smart processing strategies;
• Special tools for processing copper;
• The proper cutting parameters;
• Cutting fluid is required.

This work was supported by a grant of the Romanian National Authority for Scientific Research and Innovation, CNCS/CCCDI–UEFISCDI, project number PN-III-P2-2.1-BG-2016-0216, within PNCDI III and the H2020 AMaTUC project (GA 691787).

References

1. N. Panc, V. Bocanet, M. Bulgaru, C. Beldean, Acta Technica Napocensis, Series Applied Mathematics and Mechanics, **57**(1) (2014)
2. I. A. Popan, A. I. Popan, S. C. Cosma, A. Carean, CoSME'16, doi: 10.1051/matecconf/20179402011 (2017)
3. I. Vuscan, N. Panc, Ed. Scoala Ardeleana, Cluj-Napoca, (2015)
4. M. Mera, S. Comsa, M. Damian, O. Mera, Conference of Technics and Engineering, ISBN 973-8254-07-8, 645 (2001)
5. I. A. Popan, N. Balc, A. I. Popan, D. Fratila, A. Trif, Applied Mechanics and Materials **808**, 54 (2015)
6. A. Carean, I. A. Popan, M. Carean, AJME, Vol. 11, Issue 3, ISSN 1583-7904, 50 (2013)
7. A. Trif, M. Borzan, D. Fratila, V. Ceclan, , A. Popescu, Conference - Automation in Production, Planning and Manufacturing, ISBN 978-80-89276-41-7, 170 (2014)
8. R. Curta, N. Balc, A. Carean, DAAAM,1155 (2010)
9. I. A. Popan, N. Balc, B. Luca, A. I. Popan, A., Applied Mechanics and Materials Vol **808**, 339 (2015)
10. N. S. Reddy, P. V. Rao, Int. J. Adv. Manuf. Technol. **26**, 1202 (2005)
11. N. Balc, AJME, Vol. X, Issue 1, ISSN 1583-7904, **12,** (2012)
12. I. A. Popan, A. Carean, I. A. Popan, V. Ceclan, N. Balc, AJME, **11**, 42 (2013)
13. J. Paulo Davim, V.N. Gaitonde, S.R. Karnik, Journal of materials processing technology **205** 16 (2008)
14. V. A. Ceclan, I. A. Popan, S. D. Grozav, C. Miron-Borzan, I. Kuric, Applied Mechanics and Materials **808**, 286 (2015)
15. Guhring Tool Catalogue (2017)
16. I. A. Popan, N. Balc, B. Luca, A. I. Popan, Acta Technica Napocensis, Series Applied Mathematics and Mechanics, **66,** 153 (2015)
17. A.B. Junior, A.E. Diniz, F.T. Filho, The International Journal of Advanced Manufacturing Technology, **43**, 756 (2009)
18. *** www.mstar.ro (Dormer)

MATEC Web of Conferences **137**, 06006 (2017) DOI: 10.1051/matecconf/201713706006
MTeM - AMaTUC 2017

Studies on predictive virtual models based on finite element analysis of the behaviour of geomembranes

Petre Vasiluta[1], *Nicolae Florin* Cofaru[1,*], *Ileana Ioana* Cofaru[2]

[1]Lucian Blaga' University of Sibiu, Industrial Engineering and Management Department, 550025, Emil Cioran Street, No. 4, Sibiu, Romania
[2]Lucian Blaga' University of Sibiu, Computer Science and Electrical Engineering Department, 550025, Emil Cioran Street, No. 4, Sibiu, Romania

Abstract. The study shown in this paper presents the behaviour of geomembranes used at the ecological landfills. For this goal was used a method designed to elaborate the virtual models of 3D geomembrane with one ply or 2 plies disposed at 30°, 45° and 90° and based on this model there were developed a number of 20 particular cases. For each situation it was realized the nonlinear analyses for all the developed cases with different vertical pressures representing different loads for the waste layer (1m, 2m, 10m, 15m, 20m). The main results are obtained in graphical form and represent the Maximum tensions Von Mises and total displacements of the geomembrane.

1 Introduction

The apparition and evolution of the computing system allowed the automation of the conception, designing and research processes.

In this context, the FEA Concept (Finite Elements Analysis) was born from the necessity to develop some strong computing systems of resistance and data stream or fields met in engineering.

Generally, the steps of approaching a simulation based on the method of finite elements are [11]:

• Generating the tridimensional model using its own 3D modulus of definition or importing it in a CAD specialized program.
• Defining the restraints, the forces, the pressures, the sources of heat, the radiations and the laws that govern them when the analysis is ruled dynamically.
• Dividing the initial model in finite elements using different techniques in order to obtain the wanted model in the Mesh structure.
• Ruling the analysis with the aim to obtain the statistic maps of tensions, siftings and temperatures and so on.

* Corresponding author: nicolae.cofaru@ulbsibiu.ro

DOI: 10.1051/matecconf/201713706006

Generally, the resulted maps regarding the static analysis and the films and the dynamic analysis express [11]:

• The stress obtains by the algorithm von Mises offering information on the charging of finite elements compared to the outside force.

• The strain of the mechanical system which reveals the lengths.

• The strain energy which offers information on the energetic consume needed to obtain the deformation status.

• The displacement which offers information on the variation of the defined nods.

In Romania the use of geosyntetic materials for construction works is regulated by MTCTNP 075/2002. The works of storage of waste and the use of geomembranes and geocomposits is regulated by Governmental Decision 162/2002 and the Technical Norms of the MAPM approved by Order 1147/2002.

In the entire world a lot of tests and studies were realized [1, 2, 3, 6, 9, 10] on the subject of the use of geomembranes. In a work [4] the use of protection materials installed on the geomembrane trough the experimental study was described.

In another study presented in the paper [5] the tension of encounter made in the system of the ecological deposit was analysed. The tension of encounter is due mainly to the weight of the body part elements of the ecological deposit and it varies. The phenomena of encounter between layers was also studied.

Another direction was given in a paper [7], which presented the classical lab tests analysing the puncturing effect which appears due to the sand layer currently placed on top of the geomembrane. In the tests, the liquid pressure was also taken into consideration.

In a paper [8], it was evaluated what was the mechanical answer of two polymeric geomembranes of PVC and polyethylene of 1,5 mm of high density, describing the phenomena in a proper mathematics model.

After analysing this worldwide published studies, it resulted the need to study the geomembrane actions when assembling and developing an ecological storage.

In this context, the current paper presents different studies using the finite element method which synthesizes the behaviour of the geomembrane submitted to different charges given by different thickness of waste pushing on it. A special attention has been offered to the plies that inevitably occur when mounting the geomembrane. These plies could have important consequences regarding the functioning and impermeability of the ecological landfill.

2 Virtual shaping and analysis with solid element of the geomembrane

The virtual model of the geomembrane was conceived in the Design Modeller Application, application that is a predecessor of the AnysWorkbench Application.

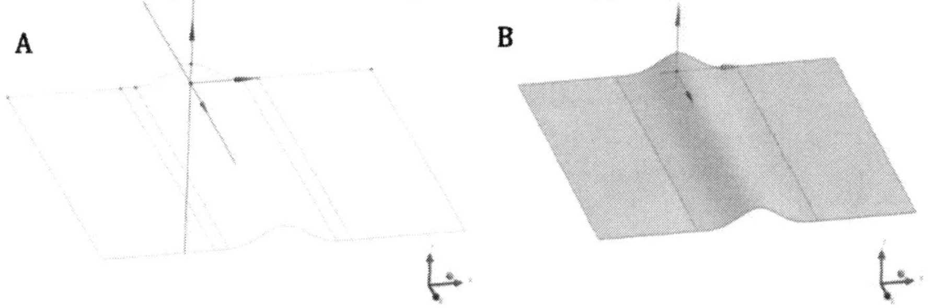

Fig. 1. A - Making the membrane surface; B - Making solid model of the membrane with a single ply.

In order to realize the geomembrane model with one ply, it was used a sketch and then every 2 plies models used different realization methods.

In order to achieve the 3D model, the first step is to draw in the coordinates system XY the 2D model Sketch, a curve line with a height of 57 mm (height of the ply). The sketch (figure 1A) realized in the pre-processor Design Modeller represents the contouring shape of the geomembrane ply.

.

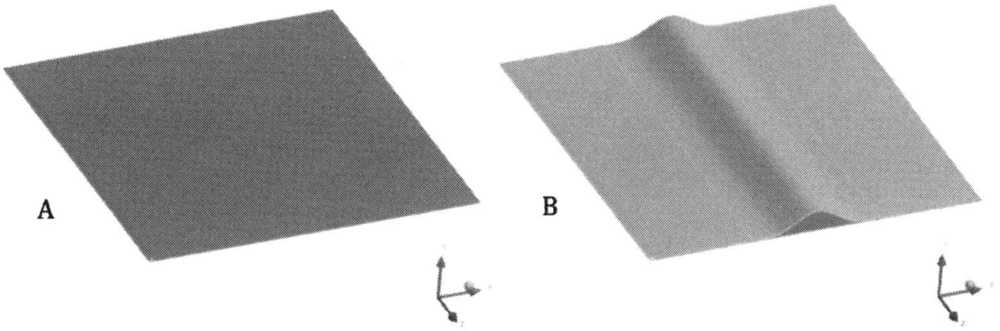

Fig. 2. A - „*Ground*" component; B- Virtual model geomembran1 (1ply) – „*Ground*".

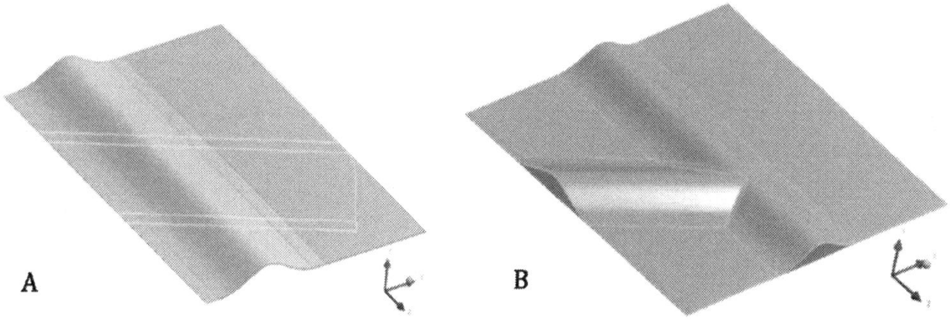

Fig. 3. A – The rotation of the ply at - 30°; B - The model of the geomembrane with 2 plies arranged at 30° and "Ground".

The surface of the geomembrane was conceived using the commands Line from Sketch and Extrude and then defining in solid by extrusion at 2 mm (figure 1.B). A contact on the surface of the geomembrane was imposed in order to make the analysis in such a manner that it won't go in one Direction-Y. This component, which was a simple plate with the same exterior dimensions like geomembrane, will be called Ground (figure 2A). The 2 components can be distinguished by their colour and position in figure 2.B. It is important to mention that the surface of the model is $1m^2$ (1m x 1m) and the thickness both of the geomembrane and of the Ground component is 2 mm. In this way the surface of the ply geomembrane solid model was realised (figure 2B).

For achieving the virtual models for the geomembranes with 2 plies arranged at 30°, 45° and 90°, special functions were used:

- Move – for rotation of the plies at different angle.
- Substract – Boolean operation.
- Fblend – for rounded of the new components.
- Repair – for a detailed checking of the consistence of the components.

In figure 3.A the rotation of the ply at the 30° is shown and in figure 3.B it is presented the model of the geomembrane with 2 plies together with "Ground" element.

In the same way, the others 3D models of the geomembranes with 2 plies disposed at 30°, 45° and 90° were achieved. In order to achieve a nonlinear static analysis with shell elements these 3D models used were suitable.

3 Nonlinear analysis of the geomembrane with 2 plies

In the following, the elaboration of the nonlinear static analysis for the geometric model of the 2 plies 30° geomembrane with shell elements will be presented. The 2 components were placed in a global XYZ system in order to correctly place the degrees of freedom of the 2 components.

The discretization of the geometric model in nods and elements was realized with 186 Solid hexahedric elements and 187 Solid tetraedic elements, both solid elements with middle nods. The middle nods are necessary in order to better approximate the results.

The Sweep method was used in order to discretize the elements for the Ground component, automatically controlling the formation direction of the source elements to the interest area.

For the encounter zones of ply surfaces, due to the fact that they are hard to discretize, advanced discretization methods were used, like HexDominant and PatchConforming, those being methods used to encounter the hexaedric and tetraedic elements. In order to obtain an efficient discretization the 10 mm and 20 mm dimensions were used.

Following the solving of the nods and elements network, 42. 731 nods and 15. 125 elements resulted (figure 4).

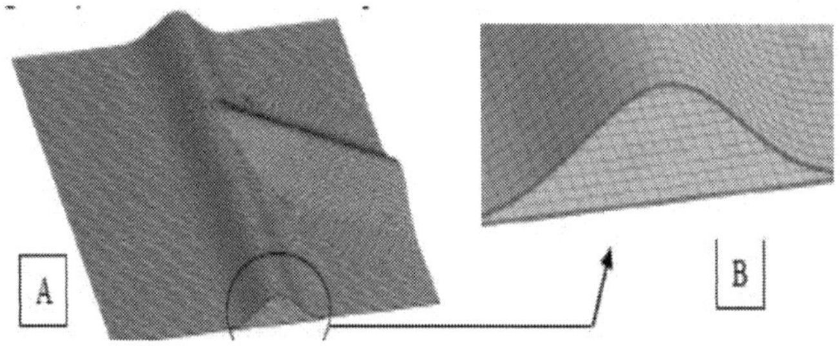

Fig. 4. A - Network modes and items obtained for model membrane 2 folds arranged at 30° – "Ground"; B - Picture fold local area.

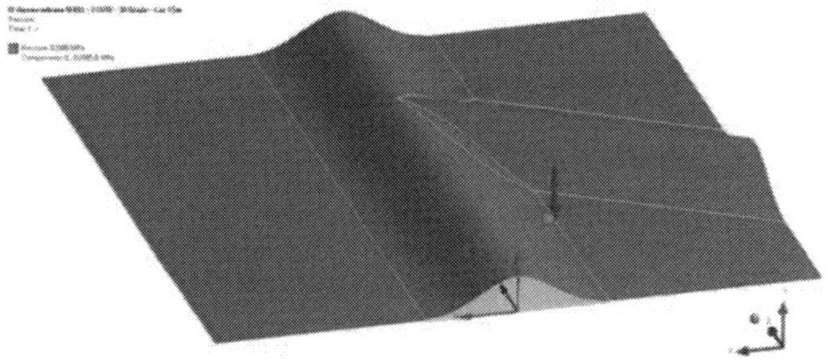

Fig. 5. Applying pressure evenly distributed on the upper surface of the membrane 2 folds arranged at 30°.

In order to realize this analysis, the limit conditions used were:

• Virtual time for run of the analysis was set at 1 sec.

• The solver was set as one being the iterative type using PCG method („Preconditioned Conjugate Gradient") level 2.

• Constraining the "ground component" using the commands „Fixed Support" on the bases surfaces, DOF = 0 (degree of freedom).

•Applying a Normal uniform distributed pressure on the superior surfaces of the geomembrane with a value dependent of the height of the waste (1m, 2m, 10m, 15m, 20m) (figure 5).

• Constraining the geomembrane on the exterior lines (parallel with the ply) using also the command Fixed Support.
• Constraining the lateral surfaces of the geomembrane with the support „Frictionless Support", blocking in this way the movement on the perpendicular direction.

The characteristics of the isotropic material used for the geomembrane are shown in Figure 6. The image is the caption from simulation module „Data Engineering".

Properties of Outline Row 3: Polyethylene

	A	B	C
1	Property	Value	Unit
2	Density	1020	kg m^-3
3	Isotropic Secant Coefficient of Thermal Expansion		
4	Coefficient of Thermal Expansion	0.00176	C^-1
5	Reference Temperature	22	C
6	Isotropic Elasticity		
7	Derive from	Young's Modulus and Poisson's Ratio	
8	Young's Modulus	2000	MPa
9	Poisson's Ratio	0.394	
10	Bulk Modulus	3.1447E+09	Pa
11	Shear Modulus	7.1736E+08	Pa
12	Tensile Yield Strength	63	MPa
13	Compressive Yield Strength	0	MPa
14	Tensile Ultimate Strength	33	MPa
15	Compressive Ultimate Strength	0	MPa
16	Isotropic Thermal Conductivity	0.2256	W m^-1 C^-1
17	Specific Heat	1386	J kg^-1 C^-1

Fig. 6. The characteristics of the isotropic material used for the geomembrane.

For the results set the following values were extracted: Total displacement, Equivalent total strain, Maximum equivalent Von Mises Stress, Maximum principal stress, Minimum principal stress.

In figure 7 it is shown the map of the Maximum equivalent Von Mises Stress.

The area of interest and importance is the maximum stress Von Mises and total displacement of the membrane. Due to the pressure imposed by 0.0192 MPa, total displacement is 25.38 mm and the maximum value calculated for maximum tension membrane is equivalent to 29.82 MPa.

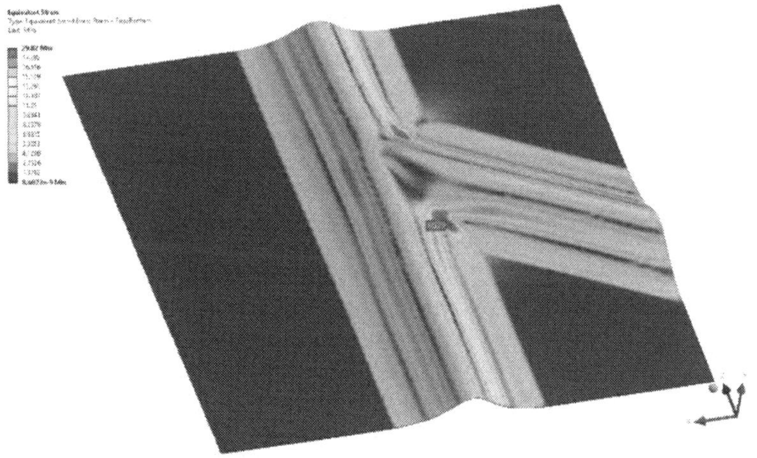

Fig. 7. Maximum equivalent Von Mises Stress.

These values were obtained at the analysis of the 30°, 2 plies geomembrane model with Shell elements for the application of the tension given by the waste layer with the 1m length. The 0.192 MPa tension is applied on a 1m^2 surface of the geomembrane. The values obtained for the 5 analyses with the respective tensions and waste layers of 1m, 5m, 10m, 15m, and 20m for the total deformation, Equivalent Total Strain and the Equivalent Total Strain can be found in Table 1.

Table 1. Values obtained for the 5 cases developed membrane with 2 plies arranged at 30° analysed with shell element.

H *[m]*	**PESD** *[MPa]*	**DT** *[mm]*	**DET** *[mm/mm]*	**TME** *[MPa]*
1	0,0192	25,38	0,0375	29,82
5	0,0961	35,99	0,0599	39,73
10	0,1923	41,13	0,9973	57,01
15	0,2885	45,78	0,1279	60,14
20	0,3846	47,69	0,1413	66,14

H - height layer of waste;
ESDP - pressure from the layer of waste;
DT - total displacement;
DET - total equivalent strain;
TME - the maximum stress equivalent von Mises

Table 2. Values obtained for the 20 cases developed membrane analysed with shell elements.

H *[m]*	**PESD** *[MPa]*	**Geomembrane 1 PLY – Shell**			**Geomembrane 2 PLIES – 30° - Shell**		
		DT *[mm]*	**DET** *[mm/mm]*	**TME** *[MPa]*	**DT** *[mm]*	**DET** *[mm/mm]*	**TME** *[MPa]*
1	0,0192	29,44	0,0249	23,68	25,38	0,0375	29,82
5	0,0961	41,78	0,0450	32,05	35,99	0,0599	39,73
10	0,1923	45,64	0,0672	50,54	41,13	0,9973	57,01
15	0,2885	47,35	0,0775	59,23	45,78	0,1279	60,14
20	0,3846	49,72	0,0963	64,18	47,69	0,1413	66,14
—		**Geomembrane 2 PLIES – 45°- Shell**			**Geomembrane 2 PLIES – 90° - Shell**		
1	0,0192	27,14	0,0379	31,11	32,81	0,0384	32,10
5	0,0961	37,03	0,0507	43,19	38,17	0,0406	45,12
10	0,1923	42,15	0,0671	59,02	41,71	0,0852	59,43
15	0,2885	46,22	0,0993	63,79	45,61	0,0929	64,11
20	0,3846	51,31	0,1129	68,23	49,17	0,1213	68,63

Using the same approach, the virtual models of 3D geomembrane with one ply or 2 plies disposed at 30°, 45° and 90° were achieved (20 particular cases).

The obtained values for different tensions from the waste layers are presented in table 2. The abbreviations used in Table 2 are the same as those used in Table 1.

It can be observed that by applying a pressure of 0.0192 MPa (1 meter height layer waste), the maximum total displacement is 32.81mm and the maximum stress equivalent von Mises is 32.10 MPa. This values were obtained for the geomembrane with 2 plies disposed at 90°. By applying a maximum pressure of 0.0348 MPa (20 meter height layer waste), the maximum total displacement is 51.31mm obtained for the geomembrane with 2 plies disposed at 45°, whereas the maximum stress equivalent von Mises was reached for the geomembrane with 2 plies disposed at 90°.

4 Conclusions

The main goal of this paper is to analyse the geomembrane behaviour, using the finite element method, in different situations. The main original contributions are the following. It was realised a numerical simulation of the geomembrane by the 4 developed models each and every one with 5 different loading cases. It was realized a static nonlinear analysis for all the developed cases with different vertical tensions representing different heights for the waste layer (1m, 2m, 10m, 15m, and 20m). In the paper are presented the analysis conditions for the simulations, the limit conditions and the material characteristics. There were extracted images for the maximum tensions Von Mises and for the total deformations of the geomembrane. By undergoing a comparative analysis with these cases, it resulted that:

- All evaluated parameters in all cases increase with the height of waste layer.
- For the maximum value of the waste layer (20 m), the total displacement is minimum for 2 plies disposed at 30° and maximum for 2 plies disposed at 45°, the total equivalent strain is minimum for 1 ply and maximum for 2 plies disposed at 30° and the maximum stress equivalent von Mises is minimum for 1 ply and maximum for 2 plies disposed at 90°

Because the appearance of the plies is inevitable at the mounting of the geomembrane it would be better to manage the disposition of these plies. According with this study it's important to avoid the situations when the values of the studied parameters are maximal because in these positions of the plies it is possible to damage the geomembrane with bad consequences on the good functioning and on the non-permeability of the ecological landfill.

We consider that the study is very important in order to achieve a good and correct mounting of the geomembrane.

References

1. D. Barbulescu, *Research on membrane protection dams against infiltration of local materials*, PhD Thesis, Technical University of Civil Engineering, Faculty of Hydrology, Bucharest, (2011).
2. S. J. E. Dickinson, *Physical Response of Composite Geomembrane/Geosinthetic Clay Liners under Simulated Landfill Conditions*, PhD Thesis, Queen's University, Kingston, Ontario, Canada (2008).
3. J. P. Giroud, K. Badu-Tweneboah, K.L. Soderman, **6**, 1019-1048 (1995).
4. P.J. Fox, J.D. Ross, J.M. Sura, R.S. Thiel, Geosyntetics International **5** (18), 272-279 (2011).
5. Y.L. Kwang, G.C. Chin, J.H Taik, H.Y. Seong, *The Stress-Strain Behaviour of Geotextiles under Puncture Loads in a Liner System*, Proceedings of The Twelfth International Offshore and Polar Engineering Conference, Kitakyushu, Japan (2002).
6. S.M Merry, J.D Bray, Geosynthetics International **3** (4), 517-536 (1996).
7. W. Haimin, S. Yiming, D. Linjun, T. Zhaoming, *Mechanical Behaviour of Interface between Composite Geomembrane and Permeable Cushion Material*, Hindawi Publishing Corporation, (2014).
8. A.Z. Katarzyna, D. Neil, F. Gary, D. Russel, V. Jones, B.Zhang, Geotextiles and Geomembranes, 224-235, (2014).

9. K.V.N. Raviteja, B. Munwar, S. Basha, *Variability Associated with interface friction between geomembrane and soil*. 50th Indian Geotechnical Conference, Pune, India, (2015).
10. S. Seeger, H. Böhm, G. Söhring, W. Müller, *Long term testing of geomembranes and geotextiles under shear stress*, Federal Institute of Materials Research and Testing (BAM), Berlin, Germany, (2015)
11. V. Oleksik, A. Pascu, *Proiectarea optimală a maşinilor şi utilajelor*, Editura Universităţii „Lucian Blaga" din Sibiu, ISBN 978-973-739-431-6, (2007)

An innovation diffusion model for new mobile technologies acceptance

Nadi Barkoczia[1], *Elena Simina* Lakatos[1, 2, *] and *Laura Andrada* Bacali[1]

[1]Faculty of Machine Building, Technical University of Cluj-Napoca, 400641 Cluj-Napoca, Romania
[2]Center for Innovation and Organizational Sustainability, Cluj-Napoca 400609, Romania

Abstract. This paper aims to approach the diffusion model developed in 1960 by Frank Bass has been utilized to study the distribution of different types of new products and services. The Bass Model helps by describing the process in which new products are adopted in a market. This model is a useful tool for predicting the first purchase of an innovative product for which there are competing alternatives on the market. It also provides the innovator with information regarding the size of customers and the adoption time for the product. The second part of the paper is dedicated to a monographic study of specific conceptual correlations between the diffusion of technology and marketing management that emphasizes technological uncertainty and market uncertainty as major risks to innovative projects. In the final section, the results of empirical research conducted in Baia-Mare, Romania will be presented in a way that uses diffusion Bass model to estimate the adoption period for new mobile technologies.

1 Introduction

The present study aims to emphasize the interferences of technology diffusion and marketing management by using one part of the mathematical Bass model. The research questions of the study are: How does interpersonal relations influence new technology adopting speed? Depending on the time needed for the new technology to be adopted, is it possible to predict the lifecycle of new product?

According to recent studies, the percentage of innovation failure varies from 5% to 90% depending on the nature of innovation and target customers [1]. Additional sources assume that 97% of new ideas and prototypes for new products enter the market and more than half of those fail within two years [2]. Anticipating success of the new product or service is therefore one of the most demanding tasks in critical management [3].

Currently, a significant number of empirical studies have been devoted to the diffusion of innovation. Impressive results have been achieved both in the theoretical and practical fields. Creating models for marketing purposes is an essential part of the so-called new marketing strategy [4].

* Corresponding author: simina.lakatos@mis.utcluj.ro

Information and communication technology facilitates the collection and recording of data. Analytical software enables fast and efficient data processing. However, new products forecast diffusion in comparison to well implemented processes that have certain limits to [3]: The limited availability of historical data on sales; Lack of knowledge about new markets which causes a high risk for producers and other members of the distribution channel. Section 2 of the paper is dedicated to a literature review of international technology and the diffusion of technology. It approaches interferences between the diffusion of technology and sourcing of marketing. Section 3 of the paper contains the methodology for empirical research and the data analysis followed by a presentation of the results Section 4. These sections are dedicated to the measurement of the first purchase of an innovative product for which competing alternatives exist on the market. It provides the innovator with information regarding the size of customers and the adoption time for the product. The last section provides the conclusions and theoretical and practical implications of the study. Technology continues to become a competitive instrument that is increasingly important for many organizations [5].

2 Current stage of research regarding diffusion in marketing

The research in the diffusion field seeks to understand innovation distribution through their life cycle modelling and from a consumers' interaction point of view [6]. Danciu [7] emphasizes the fact that both adaptation, such as permanent and progressive innovations of marketing and business, will be placed in the center of solutions that must be adopted as a consequence of the impact of evolutions and tendencies of the environment. One of the distinctive tendencies is the substantial shortening of the product life cycle. The shortest life cycle suggests a faster and faster entering of products on the global market and the disappearance of products that maintain for a long time their position on the market [8].

Each type of technology is followed by the apparition of new products. Specifically, products that rapidly progress through their life cycle [9]. Due to rapidly recorded technological progress, some specialists consider conventional theories about a products life cycle somehow obsolete. This is the result of shortened life cycles, consumers' change in lifestyle, and a rapid process of continuous change that characterizes the industry, such as the entire economy [10]. Beginning with the 90's, diffusion modelling in specialized literature has seeked to extend the Bass framework with the aim of reflecting on the complexity of a new product. Recent contributions to diffusion modelling research has reexamined this interpretation to identify and discuss about other social types of interaction [11, 12]. Clark and Goldsmith [13] examine the relationship between innovation and social influence (interpersonal) that can manifest either as normative influence or informational influence. The variable factors were analyzed separtely, although in the marketing research they are considered part of a global concept named consumer susceptibility to interpersonal influence.

3 Methodology

3.1 Bass Model Description

The Bass model was developed by Frank M. Bass [14] and has been widely used for new product forecasting. The main point in the Bass model is about adoption and diffusion of innovative products or technology. The Bass Curve formula is approximated by the following equation in discrete time:

$$n(t) = p * [M - N(t-1)] + \frac{q}{M} * N(t-1) * [M - N(t-1)] \quad (1)$$

n (t) = number of new buyers of the product or service concerned when t

N (t - 1) = cumulative number of buyers at time t - 1

M = the potential market, i.e., the total number of potential buyers of the final product that should be set to launch.

p = the coefficient of innovation or external influence that characterizes the probability that a person who has not yet used the product begins to use it due to external influence in the process of dissemination: advertising, price, product awareness in the media, personal needs and ease of use.

q = the coefficient of imitation or internal influence characterizes the probability that a person who has not yet used the product begins to use it due to the influence of those who are already using the product. With a total of three points available for parameter estimation, equation 1 reduces to:

$$n(t) = \alpha_1 + \alpha_2 * N(t-1) + \alpha_3 * N^2(t-1) \quad (2)$$

where: $\alpha_1 = p * m$, p and q $\alpha_2 = \alpha_3 = q / m$.

When q is greater than p, sales of innovation increases to a peak and then declines. When q is less than p, sales decrease monotonically. When q is 0, the model reduces to an exponential low. It is interesting to note that the exponential distribution is a variant of the Bass diffusion model by taking the parameter q = 0, i.e. considering only the category of innovators to identify new buyers. When p = 0, identical to the model Bass Mansfield.

To estimate the values of p, q and m of equation (1) Bass model we can find time T when adopting its peak. It is given by [15]:

$$T(peak) = \frac{1}{(p+q) \times \ln\frac{q}{p}} \text{ v} \quad (3)$$

Equations (1) and (3) are the most relevant. They allow us to anticipate how long it takes to adopt a new technology which is at the peak of its adoption.

3.2 Data analysis

We conducted a study based on the survey research method with its instrument, the questionnaire, to gather information from users about mobile technology with the aim of estimating the adoption period of new mobile products with the use of the Bass Model. Through the use of ordinary least square and the nonlinear regression method based on historical data, we estimate the parameters p, q and m. We also intended to investigate the changes of the model forecasting under different analogous products. We chose our subjects applying a non-random sampling, based on affordability.

The survey focuses on tracking the process of the diffusion of new technologies in the mobile market. The questionnaire provides information about consumers' behavior on the mobile phone market. Also, it includes personal information (gender, age, profession, income) requited to identify the target group. We applied a total number of 328 questionnaires taking into consideration three age categories (16-21, 22-30 and 30+). For this research we chose people between 22 and 30 years of age which are considered the most representative for applying the Bass model of technology diffusion, totaling 164 respondents. The questionnaire was administrated and implemented by the authors. The implementation period was between March 1, 2014 and August 31, 2014. The target group is represented by the consumers in the mobile phone market from Baia-Mare, Romania.

DOI: 10.1051/matecconf/201713707001

4 Results

According to the questionnaire on the technology diffusion for mobile phone products, we used the Bass model in terms of determining the number of new adopters of mobile products and the period of time required for adoption in the context of respondents between 22 and 30 years, separately for women and men. We considered all 164 respondents as potential adopters (Table 1). The research we made is divided into three parts:

a. Predicting the adoption of new technologies for M = 70 women, respectively M=94 men aged 22 to 30 years of age;
b. Forecasting the period for all 164 respondents to adopt;
c. Conducting a comparative study of the male and female adopters aged between 22 and 30 years of age.

Table 1. Estimating the potential market M, the innovation coefficients p and the imitation coefficients q.

Potential adopters	M	p	q
Women aged 22 to 30 years	70	0,09	0,50
Men aged 22 to 30 years	94	0,15	0,37

In the beginning, we estimated the values for the p and q parameters. These values were estimated using the answers given by the respondents to question 12 in the questionnaire, "If you were interested in purchasing, what information sources would you use?", whose answers allowed us to divide the respondents into two categories of adopters: innovators and imitators. The following step was to estimate the two coefficients for innovation (p) and imitation (q). Thus, the data needed for implementing the Bass mathematical model is found in the table below. Regarding the new male adopters, of the 94 potential adopters at time t = 1, we have a number of 14 new adopters, representing innovators, and at time t = 2, we have a number of 16 new adopters of which 12 are innovators and 4 imitators, and so on. As it can be seen from the data given in the 2 tables during the period of adoption, initially we only have innovators and only external influence factors on potential adopters are present.

Table 2. Determining the number of new females and respectively male adopters at the time t.

t	**Women**					**Men**				
	M - N(t-1)	**N(t-1)**	**Innovators**	**Imitators**	**n(t)**	**M - N(t-1)**	**N(t-1)**	**Innovators**	**Imitators**	**n(t)**
0	70	0				94	0			
1	64	6	6	0	6	80	14	14	0	14
2	55	15	6	3	9	63	31	12	4	16
3	44	26	5	6	11	46	48	10	8	17
4	32	38	4	8	12	31	63	7	9	16
5	21	49	3	9	12	18	76	5	8	12
6	11	59	2	7	9	10	84	3	5	8
7	6	64	1	5	6	5	89	2	3	5
8	3	67	1	3	3	3	91	1	2	3
9	1	69	0	1	1	1	93	0	1	1
10	0	70	0	1	1	1	93	0	0	1

Once the new technology was brought to the attention of potential adopters by innovators, imitators appear and thus along the diffusion process of new technologies, the number of innovators gradually decreases while the number of imitators increases. The women adoption period is 10 weeks and for men it is the same.

In Figure 1, we plotted the cumulative adoption of new technologies in time and the period required for adoption of the 70 women respondents and 94 men respondents, both categories within the interval of 22 and 30 years of age.

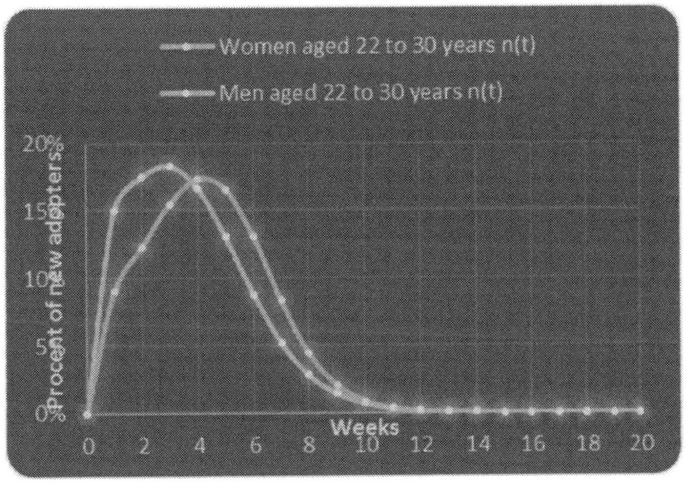

Fig. 1. The adoption period for women and respectively men.

Also, the number of imitators is higher among women, the coefficient of imitation is q = 0,50 compared to 0,37 in men. Regarding the number of new adopters n(t) at time t = 1, it is 6 for females compared to 14 for male adopters.

5 Conclusions and discussions

In this paper, we conducted a monographic study of specific conceptual correlations between the diffusion of technology and industrial marketing management. On one hand, and an empirical analysis of adopters' behavior regarding a new technology on the mobile phone market using Bass technology diffusion model. We cannot approach the product life cycle without an analysis of consumer's behavior. In the introduction and growth phases, the most innovative consumers will adopt the product. In the maturity phase, more and more cautious consumers will buy the product until a new product with more performance based features will appear and the consumers will move towards the new product. The fundamental problem for marketing specialists is the assessment of maturity phase duration. Regarding the empirical analysis we concluded that:

• Male consumers are more under the influence of external factors than women. This is explained by the greater value of the innovation coefficient (p). On the mobile phone market, they are more open to new technologies.

• The internal factors have a greater influence on women. The imitation coefficient (q) is higher on women.

• The number of imitators within the category of age 22 to 30 years is zero at first, then increases rapidly to reach a peak and then decreases when there are fewer people who have not yet bought the new product.

 DOI: 10.1051/matecconf/201713707001

• The coefficient of innovators (p) for men is higher than for women, which is due to the external influence, being stronger on men than on women. It can also be noticed that the imitation coefficient (q) has a higher value for women than for men which explain that women are more opened to word-of-mouth communication.

This study requires more investigation into the further implementation of the Bass mathematical model of diffusion in new technologies by addressing the lifecycle of products on the mobile market and notice the differences in adopting new technologies by the different categories of age and gender, but also other individual characteristics such as culture, financial affordability, level of education, etc. The comparative analysis of the results obtained will be useful for the formulation of future marketing strategies.

"This work was undertaken through the Partnerships in Priority Domains: Programmer -PN II, developed with the support of MEN-UEFISCDI, Project no.337/2014 - UNIINOI".

References

1. V. Trommsdorff, *Marketing inovací*. Nakladatelství CH Beck (2009)
2. N. Barkoczi, M. Lobonţiu, L. Bacali, Review of Management & Economic Engineering, **14** (2), 297 (2015)
3. M. Vasiľová, *Conceptual Framework and Research Propositions for Modeling the Diffusion of Technological Innovation*. Marketing-trends-congress (2012)
4. P. S. Leeflang, D. R. Wittink, International journal of research in marketing, **17** (2), 105 (2000)
5. L. Dovleac, *Launching Efforts Needed For A High-Tech Product*. Management & Marketing-Craiova, (1), 149 (2011)
6. R. Peres, E. Muller, V. Mahajan, International Journal of Research in Marketing, **27** (2), 91 (2010)
7. V. Danciu, *Marketingul viitorului, răspunsul adecvat la schimbările mediului*. Economie teoretică şi aplicată, 27 (2013)
8. B. Kim, Managing the transition of technology life cycle. Technovation, **23** (5), 371 (2003)
9. E. Viardot, *Successful marketing strategy for high-tech firms*. Artech House, (2004)
10. M. C. Demetrescu, *Metode de analiză în marketing*. Teora, (2001)
11. C. I. Ivanov, S. Avasilcăi, Procedia - Social and Behavioral Sciences, **124**, 397 (2014)
12. L.I. Cioca, and M. Cioca, WSEAS Transactions on Information Science and Applications **4** (2), 303 (2007)
13. R. A. Clark, R. E. Goldsmith, Journal of Marketing Theory and Practice, **14** (4), 275 (2006)
14. F. M. Bass, Management Sci **15** (5), 215 (1969)
15. D. R. Firth, C. Lawrence, S. F. Clouse, Interdisciplinary Journal of Information, Knowledge, and Management **1** (1), 1 (2006)

MATEC Web of Conferences **137**, 07002 (2017)
MTeM - AMaTUC 2017
DOI: 10.1051/matecconf/201713707002

The study of postural workload in assembly of furniture upholstery

Andrzej Marek Lasota[1,*] and *Krzysztof* Hankiewicz[2]

[1] University of Zielona Gora, Institute of Safety Engineering & Work Science, Prof. Z.Szafrana 4 Street, 65-516 Zielona Gora, Poland
[2] Poznan University of Technology, Chair of Management and Computing Systems, 11 Strzelecka Street, 61-845 Poznan, Poland

Abstract. The productivity of the workers is affected by the Work-related Musculoskeletal Disorders (WRMSDs) which common cause of health problems, sick leave and it can result in decreased quality of work and increased absenteeism. The objective of this study is to evaluate and investigate the postural workload of sewing machine operators in the assembly of upholstery in furniture factory by using the Ovako Working Posture Analysing System (OWAS) with sampling. The results indicated that posture code 2111 (back code: 2 – bent forward; arms code: 1 – both below the shoulder joint; legs code: 1 – sitting position; load code: – 1 less than 10 kg) was the most common working posture rating 38.1%; 63.9% of positions displayed non-neutral back postures and 52% received harmful action categories. The performed assembly tasks have an influence on harmless and harmful action categories. This study is crucial on assembly, and in the future work allows develop a framework for assessment the physical risk of WRMSDs in assembly.

1 Introduction

The ergonomics and environment factors have been the core issue for industry for many years and its profiles are rising. To ensure an ergonomics work environment, it is possible to require specific attention in carry out industry processes [1]. The needs that are the basis of the ability to work in a particular environment are dealt with in terms of working conditions [2]. The productivity of the workers is affected by the Work-related Musculoskeletal Disorders (WRMSDs) which common cause of health problems, sick leave and it can result in decreased quality of work and increased absenteeism in many professions. For example, in the United States, in 2015 musculoskeletal disorders accounted for 32% of all absence from work due to sickness [3]. In the Great Britain, in 2015/2016 WRMSDs accounted 41% of all work related illnesses and it is estimated 8.8 million working days were lost due to WRMSDs what represent 34% of all days lost due to work related ill health [4].

* Corresponding author: am.lasota@o2.pl

In Poland, in 2015 they cause over 33 million days of absence, accounting for 14.7% of all absences [5]. Thus, WRMSDs are a serious problem in many industry sectors [6, 7]. The main ergonomic risk factors for WRMSDs are: awkward posture, repetition, excessive physical load, forceful exertion, and duration of movement [8, 9]. Moreover, arrangement of workstations, tasks and working methods, tools and anthropometric characteristics of workers have an influence on working postures [10]. The other problem is the adaptation of work stands to the elderly. Tools should be designed according to their capabilities [11]. Additionally general fatigue during the work should be analyze in complex way and it is possible to develop fatigue management system [12].

Sewing machine operators usually perform piecework jobs and assemble parts of components to create a final product. Their job requires precision, is highly repetitive, and usually involves both upper limbs, is visually demanding and requires a high degree of concentration [13, 14]. Studies have shown the occurrence of WRMSDs in sewing machine operators in the textile and garment manufacturing industries [13, 15]. According to the authors' knowledge, there are not studies focused on sewing assembly process and furniture industry. Therefore the objective of this study was to evaluate and investigate the postural workload of sewing machine operators in the assembly of upholstery in a furniture factory.

2 Methodology

The case study was conducted in the upholstery sewing department of a furniture factory in Poland. The factory produces upholstered furniture (e.g. armchairs, sofas, etc.). The sample consisted of five women sewing machine operators aged from 29 to 51 years. Employees were informed about the study and participated in it as volunteers. The whole assembly process is divided into 5 tasks: sewing seats, sides, backs, on decorations and all parts together. Operators checked the dimensions of the fabric cut out with the dimensions in the project and began sewing. At the decoration sewing position, the employee first sewed them manually and then sewed them again on the machine. Assembly of all parts together began with checking of the correctness of parts sewn on the previous positions.

To collect data have been used: "pen and paper" observation, face-to-face interviews, task analysis and posture evaluation. Observations of the tasks performed by employees were preceded by face-to-face interview with supervisors and employees. The evaluations were recorded directly onto the prepared evaluation worksheet. Postures were selected with time sampling (30 s interval). This time interval allowed working postures to be assessed freely and a high density of samples to be recorded. In the entire assembly process 355 work postures adopted by the operators were evaluated; 71 postures for each of five tasks.

For the evaluation the OWAS method [16] was applied; it is one of the simpler observation techniques and allows a quick assessment. This method is based on a classification of different positions for the back, arms, legs and the force/load used in the work. The OWAS method uses a four-digit code to describe various postures and force/load combinations. The codes include four back postures (1 - straight, 2 - bent forward, 3 - twisted, 4 - bent and twisted); three arm postures (1 - both below the shoulder joint, 2 - one above the shoulder joint, 3 - both above the shoulder joint), seven leg postures (1 - sitting position, 2 - standing with straight legs, 3 - standing with one leg extended, 4 - standing with legs bent, 5 - standing with one leg bent, 6 - kneeling on one or both knees, 7 - walking); and three variants of load (1 - less than 10 kg, 2 - from 10 to 20 kg, 3 - over 20 kg). These four-digit codes from different body parts in a specific position combined with the estimated load provide information about each postural load. For example, 1241 indicates that the worker's back is straight (back code: 1), working with one arm above the shoulder joint (arms code: 2), standing with legs bent (legs code: 4), and handling a load weighing less than 10 kg (load code: 1). The combination creates categories describing the risk of exposure to WRMSDs and action categories (AC) necessary to improve the working conditions: AC1 – No risk: normal posture, with no particular adverse effect on the musculoskeletal system. Intervention is not required; AC2 – Low risk: working posture has a slight detrimental effect on the musculoskeletal system, there is a light load, immediate intervention is not required, but the ergonomic adjustment should be taken into account in future actions; AC3 – Medium risk: working posture has a significant detrimental effect on the musculoskeletal system; ergonomic intervention should be carried out as soon as possible; AC4 – High risk: working posture has a very high detrimental effect on the musculoskeletal system; ergonomic intervention is required immediately [17].

The Pearson's Chi-squared test was used to assess differences in harmless postures (AC1) with harmful postures (AC2, AC3, AC4). The Wilcoxon signed-rank test was conducted for comparing the AC results of evaluated tasks. Statistically significant differences were accepted at the 5% level of probability ($p < 0.05$).

3 Results

The frequency of posture codes in the following tasks: sewing seats, sewing sides, sewing backs, sewing on decorations, sewing all parts together and complete assembly of all parts in the final product were various (Table 1).

Table 1. Frequency (%) of posture code in tasks and entire process.

Posture code	Seats	Sides	Backs	Decorations	All parts	Total	Action Categories
1111	42.3	25.4	22.5	50.7	9.9	30.1	AC1
1121	-	1.4	-	-	-	0.3	AC1
1131	-	-	2.8	-	-	0.6	AC1
1211	-	-	-	4.2	-	0.8	AC1
1221	-	5.6	-	-	-	1.1	AC1
1311	-	-	4.2	-	-	0.8	AC1
1331	-	-	5.6	-	-	1.1	AC1
2111	45.0	35.3	26.9	36.6	46.4	38.1	AC2
2121	-	4.2	-	-	-	0.8	AC2
2131	-	-	8.5	-	-	1.7	AC2
2221	-	1.4	-	-	-	0.3	AC2
2311	-	-	4.2	-	-	0.8	AC3
2321	-	9.9	4.2	-	-	2.8	AC2
2331	-	-	5.6	-	11.3	3.4	AC3
3111	12.7	5.6	5.6	8.5	-	6.5	AC1
3121	-	-	-	-	1.4	0.3	AC1
3151	-	-	9.9	-	-	2.0	AC4
3221	-	-	-	-	9.9	2.0	AC1
3321	-	-	-	-	21.1	4.2	AC1
4111	-	7.0	-	-	-	1.4	AC2
4211	-	1.4	-	-	-	0.3	AC3
4221	-	2.8	-	-	-	0.6	AC2
Total	100	100	100	100	100	100	

In total, for the sewing process the most common posture code was 2111 (2 – back bent forward, 1 – arms both below the shoulder joint, 1 – legs, sitting position, 1 – load less than 10 kg) and amounted to 38.1%; whereas second most common was posture code: 1111 (30.1%). A total of 47.8% of postures classified as AC1, 45.7% as AC2, AC3 – 4.5% and AC4 – 2.0 %. At the seat sewing position, the most frequent results were: posture code 2111 (45.0%), followed by posture code 1111 (42.3%) and 3111 (12.7%). A total of 55.0%

of postures were classified into AC1 and for AC2 – 45%. For sewing sides 11 various postures were observed. The most common postures were: 2111 (35.3%), 1111 (25.4%) and 2321 (9.9%); the resulting action categories were AC1 – 38.0%, AC2 – 60.6%, AC3 – 1.4%. At the back sewing position 11 various postures were observed; the most common were: 2111 (26.9%), 1111 (22.5%) and 3151 (9.9%). There were: AC1 – 40.8%, AC2 – 39.4%, AC3 – 9.9%, AC4 – 9.9%. At the position of sewing on decorations 4 different postures were recorded. The most common postures were: 1111 (50.7%), 2111 (36.6%). In terms of action categories, at AC1 were 63.4% of cases and at AC2 – 36.6%. Finally, at the last task – sewing all parts together – 6 different postures were observed and the most common were: 2111 (46.4%), 3321 (21.1%), 2331 (11.3) and 1111 (9.9%). At AC1 were 42.3% cases, AC2 – 46.4%, AC3 – 11.3%.

Table 2. Proportion of OWAS working postures (%) during performed tasks.

Body parts			Seats	Sides	Backs	Decorations	All parts	Total
Back	1	straight	42.3	32.4	40.8	54.9	9.9	36.0
	2	bent forward	45.1	49.3	49.3	36.6	57.7	47.6
	3	twisted	12.6	5.6	9.9	8.5	11.3	9.6
	4	bent and twisted	0.0	12.7	0.0	0.0	21.1	6.8
Arms	1	both below the shoulder joint	100.0	80.3	76.1	95.8	57.7	82.0
	2	one above the shoulder joint	0.0	11.3	0.0	4.2	9.9	5.1
	3	both above the shoulder joint	0.0	8.5	23.9	0.0	32.4	12.9
Legs	1	sitting position	100.0	76.1	63.4	100.0	56.3	79.1
	2	standing with straight legs	0.0	23.9	4.2	0.0	32.4	12.1
	3	standing with one leg extended	0.0	0.0	22.5	0.0	11.3	6.8
	4	standing with legs bent	0.0	0.0	0.0	0.0	0.0	0.0
	5	standing with one leg bent	0.0	0.0	9.9	0.0	0.0	2.0
	6	kneeling on one or both knees	0.0	0.0	0.0	0.0	0.0	0.0
	7	walking	0.0	0.0	0.0	0.0	0.0	0.0
Load	1	< 10kg	100.0	100.0	100.0	100.0	100.0	100.0
	2	10 – 20kg	0.0	0.0	0.0	0.0	0.0	0.0
	3	>20kg	0.0	0.0	0.0	0.0	0.0	0.0

The proportion of OWAS work postures (%) during performed tasks (Table 2) shows that in total for this assembly process, the back in 36.0% of cases was straight, 47.6% – bent forward, 9.6% – twisted and 6.8% – bent and twisted. Arms were in 82.0% of tasks both below the shoulder joint, 5.1% – one above the shoulder joint and 12.9% – both above the shoulder joint. Results for the legs were: 79.1% – sitting position, 12.1% – standing with straight legs, 6.8% – standing with one leg extended and only 2.0% – standing with one leg bent. Load in all cases was less than 10 kg.

Table 3. Harmless (AC1) vs. harmful (AC2, AC3, AC4) action categories (%), Pearson's chi-square test, df(4).

Tasks	Harmless	Harmful	X^2	p
Seats	55	45	18.39	0.001
Sides	38	62		
Backs	41	59		
Decorations	63	37		
All parts	42	58		
Total	48	52	-	-

The results show differences in % of harmless (AC1) and harmful (AC2, AC3, AC4) action categories for each task and for the whole assembly process (Table 3). In total it was noticed more harmful action categories (52%) than harmless (48%) and it was observed a strong association between the tasks and the harmless and harmful action categories (X^2=18.39, $p < 0.001$). The Wilcoxon signed rank test shows statistically significant differences of AC between tasks performed by operators: "All parts" vs. "Decoration" ($p<0.005$), "All parts" vs. "Seats" ($p<0.05$), "Backs" vs. "Decoration" ($p<0.001$), "Backs" vs. "Seats" ($p<0.01$), "Decoration" vs. "Sides" ($p<0.001$) and "Seats" vs. "Sides" ($p<0.05$).

4 Discussion

This case study identified that the most common work posture for the overall process of assembly was 2111. Additionally the bent forward trunk was the most critical back posture that should be eliminated or reduced for sewing machine operators. It was only in the case of sewing on decorations that the most common posture was 1111 – neutral. These differences are caused by the fact that the decoration worker performs the task in two phases. In the first phase she sews them by hand and in the second stage attaches them by sewing on a machine. In three tasks out of five it was noted the most strenuous poor work postures (AC3, AC4). For the sewing of sides, only one posture was coded at 4211 (1.4%, AC3), which was observed when workers were reaching for scissors. At sewing of the backs, 3 postures were coded 2311 (4.2%, AC3), which was noticed during the exchange and reloading of spools of thread. Four postures were recorded as 2331 (5.6%, AC3) – in standing position, when worker rotating element and visual inspection; and 7 postures – 3151 (9.9%, AC4) during placement in the container. At the final assembly stage: 2331 (11.3%, AC3) – while rotating the entire element and conducting visual inspection. Over the entire assembly process only 6.5% of tasks at AC3 & AC4 are noticed.

The next authors' findings indicate that the proportion of back posture shows wide variation among tasks. However, the dominant postures are: bent forward and straight. The most common arm posture is both below the shoulder joint. However, in the task of sewing all parts together it was observed that approximately 1/3 arms postures were both above the shoulder joint. This can be explained in that the joining by sewing of all parts together and visual control of sewed parts requires manual handling, rotating of parts sometimes taking place in a standing position because of their large size.

The greater proportion of legs shows that the process was made in the sitting position but at certain workplaces: sewing sides, backs and assembly of all parts some operations required a standing position (e.g. rotating, placement, visual inspection of parts).

The next authors' discovery shows a strong association between performed tasks and the harmless and harmful action categories ($p < 0.001$). In this process of sewing it was observed more harmful AC (52%) than harmless (48%) postures. Moreover, the results indicates statistical significant differences in medians of action categories between "All parts" vs. "Decoration", "All parts vs. Seats", "Backs" vs. "Decoration", "Backs" vs. "Seats", "Decoration" vs. "Sides" and "Seats" vs. "Sides". This means that there are differences in harmful postures. This knowledge should help in the development of a job rotation schedule.

In this study the observations covered the whole assembly process. This approach has some advantages. Firstly, it makes it possible to identify the tasks that involve a high risk of exposure to poor postures. Secondly, this approach can recognize postures with a distinctly or extremely harmful effect on the musculoskeletal system for the identified tasks. Thirdly, this approach can recognize weaknesses in relation to postures in the process of assembly. Finally, this knowledge allows the management and control of risk on performed tasks in assembly before they develop into serious musculoskeletal disorders and allows the development of a job rotation schedule. There are, however, also some limitations to this study. Postures were evaluated at 30 seconds intervals and some postures occurring within the sampling period could have been missed. 71 samples are not cover enough of all possible postures at the task do not allow a full picture of working postures to be obtained.

5 Conclusions

The results of this study indicate a negligible workload that requires soon or immediate ergonomic intervention, however, a deeper analysis indicates excessive back workload that may lead to WRMSDs. The knowledge from this study should be used to identify and control risks associated with the performed tasks and to develop a job rotation system, eliminating adverse exposure before it is able to develop into serious musculoskeletal disorders. This study is crucial on assembly, and in the future work allows develop a framework for assessment the physical risk of WRMSDs in assembly.

Based on the results of this study the following are recommended:

- Rearrangement of equipment,
- Develop and implement a job rotation schedule,
- Develop training on correct working practices (e.g. posture during operation, proper usage of tools and workstations), rest.

References

1. A. Górny, MATEC Web of Conferences, **94**, 04011 (2017)
2. A. Górny, Procedia Manufacturing, **3**, 4700 (2015)
3. Bureau of Labor Statistics 2016, *Nonfatal Occupational Injuries and Illnesses Requiring Days Away from Work*, http://www.bls.gov/news.release/pdf/osh2.pdf (2015)
4. HSE, *Health and Safety Executive, Work Related Musculoskeletal Disorder Statistics* (WRMSDs) in Great Britain 2016, Available from: http://www.hse.gov.uk/Statistics/causdis/musculoskeletal/msd.pdf?pdf=msd (2016)
5. ZUS, *Absencja chorobowa w 2015 roku. ZUS*, Warszawa, Available from: http://www.zus.pl/documents/10182/39590/Absencja+chorobowa+w+2015+roku.pdf/c36a1432-adcf-4f6f-8d72-5d1108e7ab8a?version=1.0 (in Polish) (2016)
6. A. M. Lasota, K. Hankiewicz, *Evaluation of ergonomic risk in the production line of frozen food products*, Yue, X.G., Duarte, N.J.R. (Eds.), *International Conference on Economics and Management Innovations – CEMI,* ACSR-Advances in Computer Science Research. Atlantis Press, Paris, France, **272** (2016)
7. J. M. Lu, L. J. Twu, M. J. J. Wang, Int J Ind Ergonom, **52**, 40 (2016)
8. A. M. Lasota, K. Hankiewicz, *Assessment of risk to work-related musculoskeletal disorders of upper limbs at welding stations*, Arezes, P.M., Baptista, J.S., Barroso, M.P. Carneiro, P., Cordeiro,P., Costa, N., Melo, R.B., Miguel, A.S., Perestrelo, G. (Eds.), *International Symposium on Occupational Safety and Hygiene - SHO 2016*; proceedings book. Guimaraes, Portugal, **138** (2016)
9. P. Coenen, M. Douwes, S. van den Heuvel, T. Bosch, Ergonomics, **59** (9), 1182 (2016)
10. E. R. Vieira, S. Kumar, J Occup Rehabil, **14** (2), 143 (2004)
11. M. Butlewski, Applied Mechanics and Materials **657**, 1061 (2014)
12. M. Butlewski, G. Dahlke, M. Drzewiecka, L. Pacholski, Procedia Manufacturing, **3**, 4732 (2015)
13. N. J. Delleman, J. Dul, Int J Ind Ergonom, **30** (6), 341 (2002)
14. N. Öztürk, M.N. Esin, Int J Ind Ergonom, **41** (6), 585 (2011)
15. I. Dianat, M. Kord, P. Yahyazade, M. Karimi, A. Stedmon, Appl Ergon, **51**, 180 (2015)
16. O. Karhu, P. Kansi, I. Kuorinka, Appl Ergon, **8** (4), 199 (1977)
17. A. M. Lasota, K. Hankiewicz, *Working Postures of Spot Welding Machine Operators*, Arezes PM. et al. (Eds.), Occupational safety and hygiene IV, Taylor & Francis Group, London, **261** (2016)

Full Factorial DOE to determine and optimize the equation of impact forces produced by water jet used in sewer cleaning

Nicolae Medan[1,*], *Mihai* Banica[1], and *Sandor* Ravai-Nagy[1]

[1]Technical University of Cluj-Napoca, Department of Engineering and Management of Technology, Victor Babes 62A, Baia Mare, Romania

Abstract. The functioning of the equipment to cleaning sewer is dependent on certain process parameters, causing variations of the impact forces. The aim of this paper is to define and optimise an equation that determines the impact forces, forces produced when the water jets is used to clean the sewer. The research method to measure the impact forces is the experiment using full factorial design of experiment. To can made the experiments was used a stand to generate the pressure waterjets and a device to measurement the values of impact forces. In the first part of paper was determine the equation of impact forces produced by water jet and then was realised a multiple linear regression model in three different ways to optimise the prediction of the proposed equation.

1 Introduction

Industry water jet technology is frequently used in a lot of areas such: concrete hydrodemolition, jet cutting for different type of materials, mechanical processing of minerals, medical applications, rock fragmentation, surface preparation for protective coatings [1]. Industrial cleaning is a classic application of water jets technology. In the late 1950s, when reliable high-pressure pumps were built, the usage of water jets spread widely in the field of pipes and sewerage cleaning.

A series of complex phenomena appear in the cleaning jets. Most research in the field is experimental. So, Adler [2] studying the mechanisms that appear when a water jet hit a surface. Other works have studied the pressure variation in water jets such as Leach et al [3], Leu et al [4] and Guha et al [5].

To clean the sewer networks is used equipment that use high pressure water jets. The parameters of water jets adjusted by this equipment are determined by the process parameters [1]. Variation of these parameters produces variations of the impact forces.

To determine an equation who described the values of impact forces in concordance with the process parameters is necessary to realised practical experiments to determine the values of impact forces for different set-up values of process parameters [6,7].

* Corresponding author: nicolae.medan@cunbm.utcluj.ro

2 Apparatus used and methodology

To realize the experiments, was used a stand for generate water jets and a device to for determining the values of impact forces [6].

2.1 Stand to generate pressure water jets

In figure 1 is presented the diagram for the stand to generate water jet.

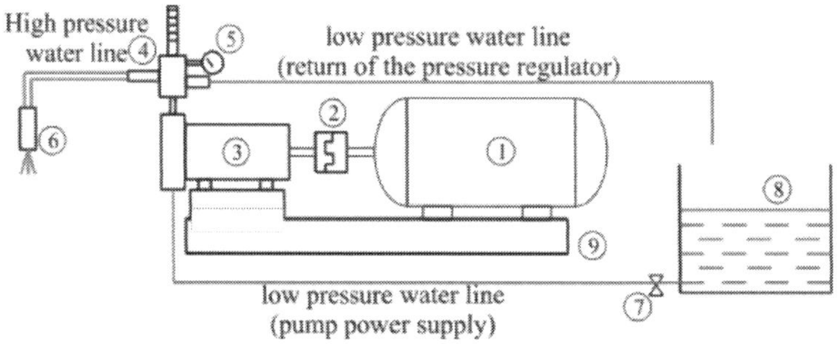

Fig. 1. Diagram of the stand to generate water jet.

Component parts of stand: (1) electric motor (2) flexible coupling; (3) high pressure pump, 4) pressure regulator, 5) pressure gauge, 6) nozzle, 7) tap water, 8) water tank, 9) chassis. From the high-pressure pump (3) water coming out and goes to the pressure regulator (4). This pressure regulator allows to adjust the working pressure to the desired value.

2.2 Device to measure the impact forces

In figure 2 is presented the diagram of the device used to measure the impact force produced by water jet when hit a surface.

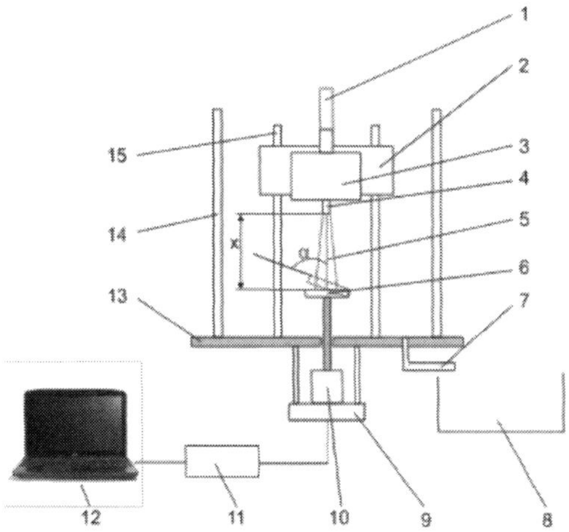

Fig. 2. Diagram of the device for measure the impact force.

The principle of working for the device is the following. From high pressure water hose (1) come water at a desired pressure p. At the outlet of nozzle is generated a water jet (5). This water jet (5) hit the target plate (6), This target plat (6) can be located at a certain distance x in front of the nozzle. The water jet (5) generates the impact force when he meets target plate (6). The impact force produces axial movement of target plate, axial movement converted into an electric signal by the piezoelectric sensor (10). The electrical signal is collected by data acquisition board (11), which forward data to a computer (12) and using processing software results the values of impact force.

2.3 Methodology of the measurements

The research method of this study is the experiment. To determine the values of impact forces produced by water jets is used the full factorial method. To determine the impact forces, it is necessary to set up the experimental domain.

In the water jet cleaning process, a series of parameters are involved [1]. These parameters can be divided into two major groups, namely: 1) target parameters which shall be defined according to the contact area between the water jet and the surface to be cleaned and 2) process parameters. In the measurement of the impact forces of a stationary water jet and flat and rigid surface the process parameters are involved.

2.4 Setting up the experimental domain.

The process parameters are:

1) D nozzle diameter, [mm];
2) P water pressure, [bars];
3) x distance between the nozzle and impact surface, [mm];
4) α impact angle (angle formed by the jet and impact surface), [^{0}].

In table 1 are presented the values of process parameters and their levels according with full factorial method.

Table 1. The values of the parameters for full factorial method.

Abbreviation	Parameter	Values
A	Nozzle diameter D [mm]	1, 1.5, 2
B	Pressure p [bar]	100, 120,140,160,180,200
C	Impact angle α [^{0}]	60, 75, 90
D	Distance x [mm]	25, 50,75,100,125,150,175,200

Corresponding to the 4 parameters and second-degree interactions of the parameters, for full factorial method result a plan of experiments who contain a number of 144 experiments.

Based on the parameters established in accordance with Table 1 the impact forces were determined. For each experiment was performed three measures of impact force and still it has worked with F_{med}, representing the arithmetic average Results a total of 432 measurements.

For brief of this paper, we dare not to present the values of F_{med} obtained.

3 Results

For determine the equation of impact forces, in the first step it is necessary to determine the contribution of each parameter and their interactions.

3.1 Determining the contributions of the parameters and their interactions

To calculate the contributions of parameters and their interactions was used the values of F_{med} obtained in concordance with design of experiments. Using the full factorial method has conducted an analysis the variance to determine the influence of each parameter and their interactions on the impact force. In table 2 is presented the analysis of the variance, using Minitab 17.

Table 2. Analysis of Variance (Minitab 17).

Source	DF (degree of freedom)	SS (sum of square)	Contribution
Regression	10	162559	99.14%
Diameter	1	116939	71.32%
Pressure	1	29592	18.05%
Angle	1	5959	3.63%
Distance	1	389	0.24%
Diameter*Pressure	1	7596	4.63%
Diameter*Angle	1	1523	0.93%
Diameter*Distance	1	148	0.09%
Pressure*Angle	1	385	0.24%
Pressure*Distance	1	20	0.01%
Angle*Distance	1	5	0.00%
Error	421	1406	0.86%
Total	431	163965	100%

3.2 Determining the regression equations of impact forces

According to the results obtain in table 2, the next step is to determine the equation of the impact forces using only the parameters and interactions previously set.

Using Minitab 17, the multiple linear regression model of impact forces was determined (equation 1):

$$F_{med} = 22.91 - 32.97 \cdot A - 0.2088 \cdot B - 0.26 \cdot C + 0.30072 \cdot A \cdot B + 0.3755 \cdot A \cdot C \quad (1)$$

This multiple linear regression model was obtaining without Box-Cox transformation.
In table 3 is presented the analysis of variance of equation (1).

Table 3. Analysis of Variance for equation 1.

Source	DF (degree of freedom)	SS (sum of square)	Contribution
Regression	5	161610	98.56%
Diameter	1	116939	71.32%
Pressure	1	29592	18.05%
Angle	1	5959	3.63%
Diameter*Pressure	1	7596	4.63%
Diameter*Angle	1	1523	0.93%
Error	426	2355	1.44%
Total	431	163965	100%

The regression statistics of equation (1) are: R squared: 98.56%, R squared adjusted: 98.55% and R squared predicted: 98.52%.

To optimise the equation (1), the next step is to be made a regression using Box-Cox transformation with rounded λ. Result the equation (2):

$$\ln\left(F_{med}\right) = -0.618 + 1.3611 \cdot A + 0.007137 \cdot B + 0.00969 \cdot C \quad (2)$$

In table 4 is presented the analysis of variance of equation (2).

Table 4. Analysis of Variance for equation 2.

Source	DF (degree of freedom)	SS (sum of square)	Contribution
Regression	5	178.455	97.56%
Diameter	1	143.163	78.27%
Pressure	1	29.325	16.03%
Angle	1	5.959	3.26%
Diameter*Pressure	1	0.009	0.00%
Diameter*Angle	1	0.00	0.00%
Error	426	4.462	2.44%
Total	431	182.917	100%

The regression statistics of equation (2) are: R squared: 97.56%, R squared adjusted: 97.53% and R squared predicted: 97.50%.

Another possible optimisation of equation (1) is realised made a regression using Box-Cox transformation with λ=0.5. Result the equation (3):

$$\left(F_{med}\right)^{0.5} = 0.097 + 0.216 \bullet A - 0.00082 \bullet B - 0.0002 \bullet C + 0.01428 \bullet A \bullet B + 0.0173 \bullet A \bullet C \quad (3)$$

In table 5 is presented the analysis of variance of equation (3).

Table 5. Analysis of Variance for equation 3.

Source	DF (degree of freedom)	SS (sum of square)	Contribution
Regression	5	1240.51	98.98%
Diameter	1	963.07	76.84%
Pressure	1	214.07	17.08%
Angle	1	42.99	3.43%
Diameter*Pressure	1	17.15	1.37%
Diameter*Angle	1	3.23	0.26%
Error	426	12.79	1.02%
Total	431	1253.30	100%

The regression statistics of equation (3) are: R squared: 98.98%, R squared adjusted: 98.97% and R squared predicted: 98.95%.

In table 6 are presented the regression statistics for all of three equations determined.

Table 6. Regression statistics for all 3 equations.

Type of regression	Equation no.	R squared	R squared adjusted	R squared predicted
without transformation	(1)	98.56%	98.55%	98.52%
rounded λ transformation	(2)	97.56%	97.53%	97.50%
λ=0.5 transformation	(3)	98.98%	98.97%	98.95%

4 Verification of determined equations

For verifying the three previously determined equations was performing a new set of experiments. For each experiment was performed three measures of impact force and still it has worked with F_{exp}, representing the arithmetic average.

Table 7 lists the default values for D, p, and α input parameters. The distance x was set for all experiments at 100 mm.

In order to compare the obtained results, the relative error between F_{exp} and F_{cal} was calculated for each of the three determined equations.

The relative error is given by the relationship:

$$\varepsilon_x = \frac{|F_{cal} - F_{\exp}|}{F_{\exp}} \tag{4}$$

were F_{cal} represents the impact force calculated with one of the three previously determined equations and F_{exp} represents the impact force measured experimentally.

Table 7. The relative error for the three determined equations.

Exp no				F_{exp} [N]	F_{cal} [N]			Relative error ε [%]		
	D [mm]	P [bar]	α [^{0}]		F_1	F_2	F_3	ε_1	ε_2	ε_3
1	1.25	130	70	16.85	18.08	15.41	16.66	7.30	8.55	1.13
2	1.75	170	85	19.87	21.22	17.78	19.39	6.79	10.52	2.42
3	1.25	130	70	22.88	24.76	20.84	22.69	8.22	8.92	0.83
4	1.75	170	85	26.25	27.9	24.06	25.86	6.29	8.34	1.49
5	1.25	130	70	33.26	34.28	31.09	32.77	3.07	6.52	1.47
6	1.75	170	85	39.31	40.24	35.9	38.14	2.37	8.67	2.98
7	1.25	130	70	43.85	46.98	42.32	44.78	7.14	3.49	2.12
8	1.75	170	85	51.74	52.94	48.87	51.02	2.32	5.55	1.39

From table 7, it can be seen from the point of view of the relative error that the most accurate is equation 3 followed by equation 1, and the latter equation 2.

This is in accordance with the regression statistics for all of three equations determined above (table 6).

5 Conclusions

1) In this work, it is presented a methodology to determine the impact forces produced by water jets used in sewer cleaning. The impact forces dependent on certain process parameters.

2) The research method used is full factorial design of experiment. After applying full factorial itinerary for calculating the percentage of influence of parameters and their interactions of the impact forces, it is found: nozzle diameter D is the largest influence, with a percentage of 71.32%. In the second place is pressure P with a value of 18.05%, follow by interaction between nozzle diameter D and pressure p with a value of 4.63% and the impact angle α with a value of 3.63%. The lowest value is given by interaction between nozzle diameter D and impact angle α with a value of 0.93%.

3) For the experimental domain, the parameter distance x has a percentage of influence of only 0.24% for impact forces, practically insignificant.

4) According with the influence of parameters and their interactions, was realised a multiple linear regression model in three different ways to optimise the prediction of the proposed equation:

• first type regression without transformation, with R squared predicted 98.52%;
• second type regression using rounded λ transformation, with R squared predicted 97.50%;
- third type regression using λ=0.5 transformation, with R squared predicted 98.95%.

5) The best prediction is given by the regression using λ=0.5 transformation, followed by the regression without transformation. The lowest degree of prediction is given by the regression using rounded λ transformation.

6) It is possible to improve the prediction of the equation who describe the impact forces only using different type of regression and the same measured values of the impact forces.

7) After completing a new set of experiments and calculating the relative error for the three equations, the hierarchy for regression statistics for all 3 equations is confirmed.

References

1. A. W. Momber, *Hydroblasting and Coating of Stell structures*, Oxford: Elsevier Ltd, (2003)
2. W. F. Adler, The Mechanics of Liquid Impact, TMST, 127-183 (1979)
3. S. J. Leach, G. L. Walker, Some Aspects of Rock cutting by High Speed Water Jets. PTRSL, Series A, **260**, 295-308 (1966)
4. M. C. Leu, P. Meng, E. S. Geskin, L. Tismeneskiy, Mathematical modelling and experimental verification of stationary waterjet cleaning process, JMSE, **120**(3), 71-579 (1998)
5. A. Guha, R. M. Barron, R. Balachandar, An Experimental and Numerical Study of Water Jet Cleaning Process. JMPT, 610-618 (2011)
6. N. Medan, S. Ravai Nagy, Determining the Equation of the Impact Forces Produced by Water Jets Used in Sewer Cleaning, IMANEE 2015, AMM **809-810**, 1579-1584 (2015)
7. D.C. Montgomery, *Design and analysis of experiments*, 8th edition, Wiley, (2013)

 DOI: 10.1051/matecconf/201713707004

Analytical inventory of process variables for sustainable development of a small business for integrated production of wood pellets

Vasile Merticaru[1,*], *Bogdan* Rusu[2], and *Romeo-Mihai* Ciobanu[2]

[1]Gheorghe Asachi, Technical University of Iasi, Department of Machine Manufacturing Technology, Prof. D. Mangeron Street, no. 39A, Iasi, Romania
[2]Gheorghe Asachi, Technical University of Iasi, Department of Engineering and Management, Prof. D. Mangeron Street, no. 28, Iasi, Romania

Abstract. The paper presents some theoretical considerations and the related deliverable results obtained within a research approach developed for analytically inventorying the process variables for a wood pellets micro-production activity integrated with self-providing crops of energetic willow. The study is part of a larger research approach intended to accurately define the appropriate specifications for developing a sustainable small business in this area. Within the particular research approach being discussed in the paper, some conceptual models have been structured and are proposed, based on process systematic analysis, as following: a model of the research general hypotheses; a model of the general conceptual research frame; a process flow model for the considered integrated processes, respectively wood pellets micro-production and energetic willow crops; a model for sustainability groups of indicators to be considered; two models for process variables inventorying and classification, one for each of the two considered integrated processes. After the research hypotheses are simplified by considering some particular implementing conditions, a final model of the investigated variables is submitted to discussion. Some adequate conclusions revealed by the research approach, together with some directions of further research development are finally identified and presented in the paper.

1 Introduction. Research problem identification

Within the actual global context, the problem of renewable energy resources is of vital interest. Biomass pellets and more specifically wood pellets gain more and more importance in this direction, as long as published statistical data show that in EU the consumption of pellets have increased from 3.8 mill. tons in 2005 to 9.8 mill. tons in 2010 and is expected to reach about 24 mil. tons in 2020, where about a half of it is imported, [1].

Among the biomass solutions for pellets production, energetic willow (Salix viminalis) outstands as a very performant wood type resource, if considering and balancing the

* Corresponding author: merticaru@tcm.tuiasi.ro

 DOI: 10.1051/matecconf/201713707004

positive and the negative arguments, [2]. More than that, looking from a holistic point of view, willow crops systems and the associated value chains can be considered as having a multifunctional role, as along with producing sustainable energy they bring other value-added benefits to the related ecosystems and communities, [3].

In the particular situation of countries like Romania or even for more specific particularly delimited locations where a severe forest depletion by intensive wood mass exploiting is registered and where the local communities are highly dependent on the wood mass as heating energy resource, the sustainable development of some small businesses for integrating wood pellets micro-production and energetic willow crops appears as a viable solution. In order to provide sustainability for such projects the problem that arises is to identify the success/failure factors, to hierarchize and quantify them through a set of technical and economical specifications which the business process should be organized on.

Further on, the paper presents some theoretical considerations and the related deliverable results obtained within a research approach developed for analytically inventorying the process variables for a wood pellets micro-production activity integrated with self-providing crops of energetic willow. The study is part of a larger research approach intended to accurately define the appropriate specifications for developing a sustainable small business in this area.

2 General conceptual research frame

Within the above mentioned research approach, the conceptual frame and the related theories and principles have been clarified and first of all the research general hypothesis have been defined through a set of relations of association and of dependence and/or determination, as it is shown in the model presented in Figure 1.

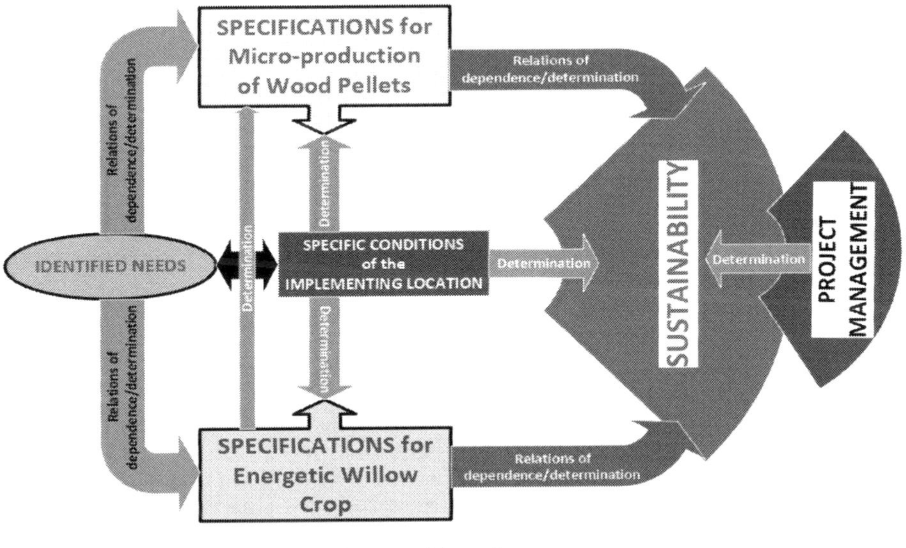

Fig. 1. Conceptual model of the research general hypotheses.

The particular needs targeted by the research project intention have been investigated and defined and the research general hypotheses have been further on customized and simplified by strictly reporting to the factors deriving from the specific conditions of the considered implementing location and to the particularities identified for the energetic willow cropping and for the wood pellets production processes, considered to be integrated. A model of the general conceptual research frame resulted from that phase, as in Figure 2.

 DOI: 10.1051/matecconf/201713707004

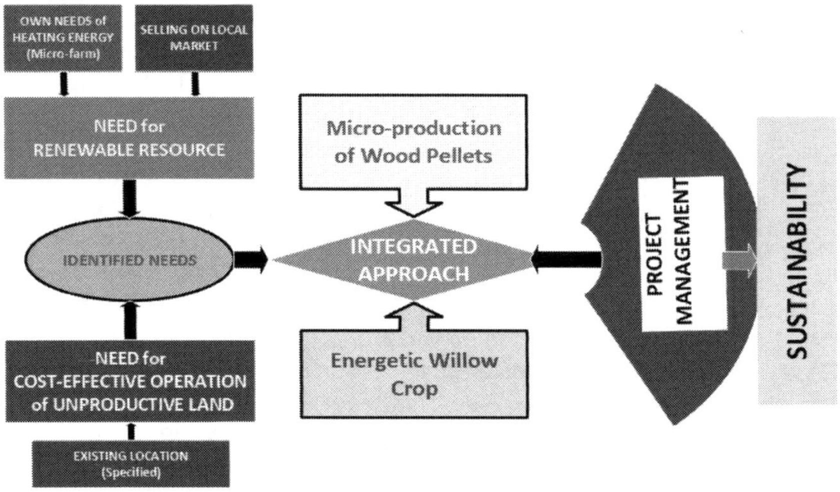

Fig. 2. Model of the general conceptual research frame.

3 Systematic analysis of the integrated processes

For analytically inventorying the process variables for providing the sustainable development of the particular process or project subjected to research study, a process systematic analysis has been carried out, starting with the functional and structural analysis of the two considered integrated processes, respectively wood pellets micro-production and energetic willow crops, in relation to the particularly targeted needs of renewable resources, respectively to satisfy some own needs of heating energy for a micro-farm and to be able to respond to the opportunity of selling wood pellets on local market, together with that of cost-effective operation of an existing unproductive land in specified location.

A comprehensive process workflow model has been targeted to be obtained as deliverable result, as it is shown in Figure 3. The functional and structural analysis of the energetic willow crop process have considered the technical recommendations of SUNY-ESF, [3], while that for the wood pellets micro-production process have considered technical data published by Piraglia et al., [4]. Not at last, some specific system/process/function decomposition rules have been considered in developing the right above mentioned process model.

Further on, a simple model as it is shown in Figure 4, has been elaborated for sustainability groups of indicators to be considered within the enlarged research approach, based on the state-of-the-art study and information related to LCSA – Life Cycle Sustainability Assessment, published by Kloepffer, [5]. All the three components of the Sustainable Development, respectively the ecological, economic and social impacts have been considered through some specific categories of indicators.

The detailed systematic analysis of the considered integrated processes have allowed the identification and classification of a large number of involved factors, which further on have been organized as process variables within two fishbone diagram models, one for each of the two investigated business processes, as it is shown in Figure 5.

As long as practically is very difficult to consider all the identified process variables in defining some relations of association and of dependence and/or determination between the input set of technical and economical specifications and the output set of qualitative and quantitative sustainability indicators, some further simplifying hypotheses have been considered, in relation to the particular implementing location and to the identified needs.

 DOI: 10.1051/matecconf/201713707004

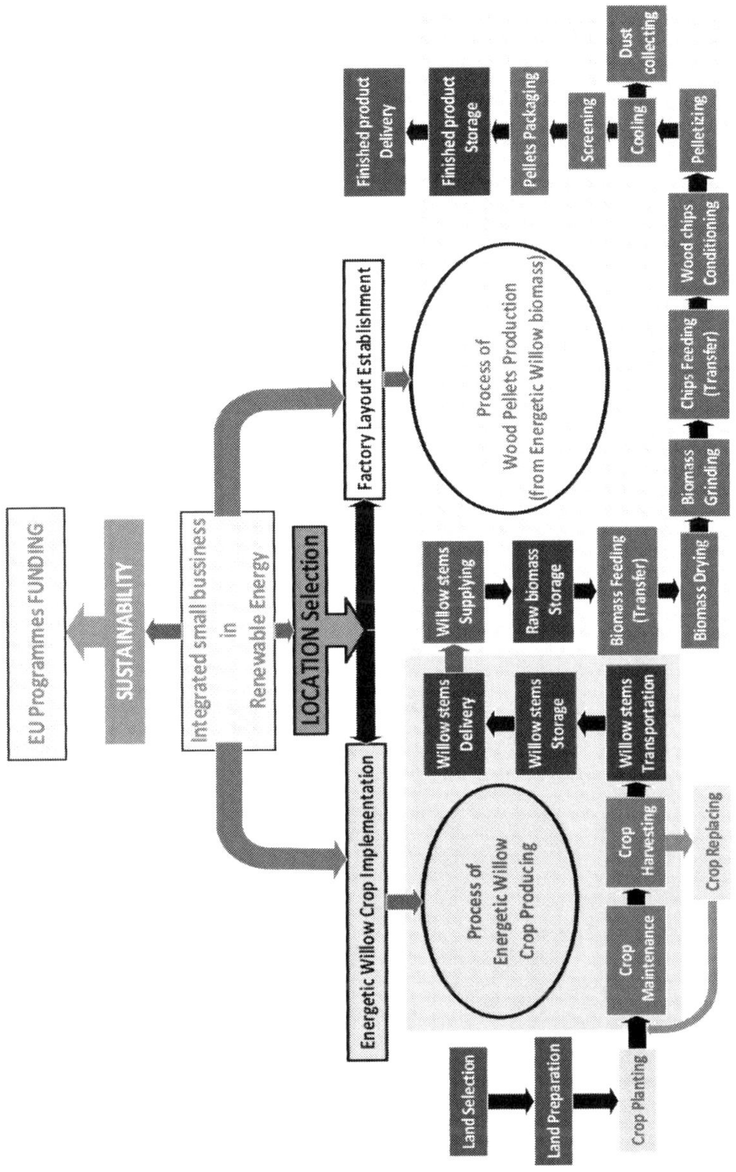

Fig. 3. Process flow model for the considered integrated processes.

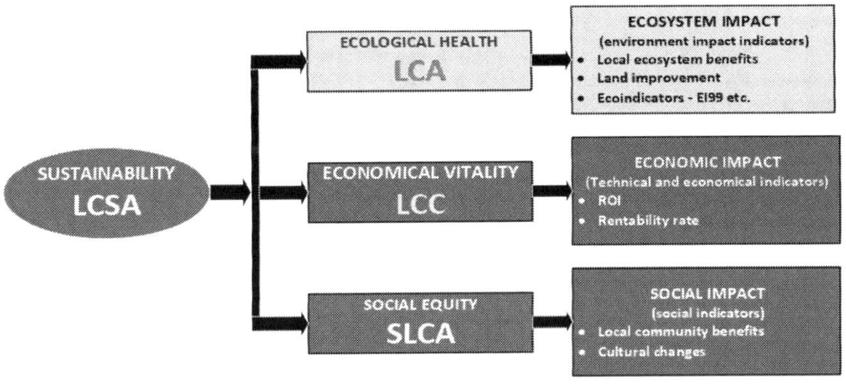

Fig. 4. Model for sustainability groups of indicators to be considered.

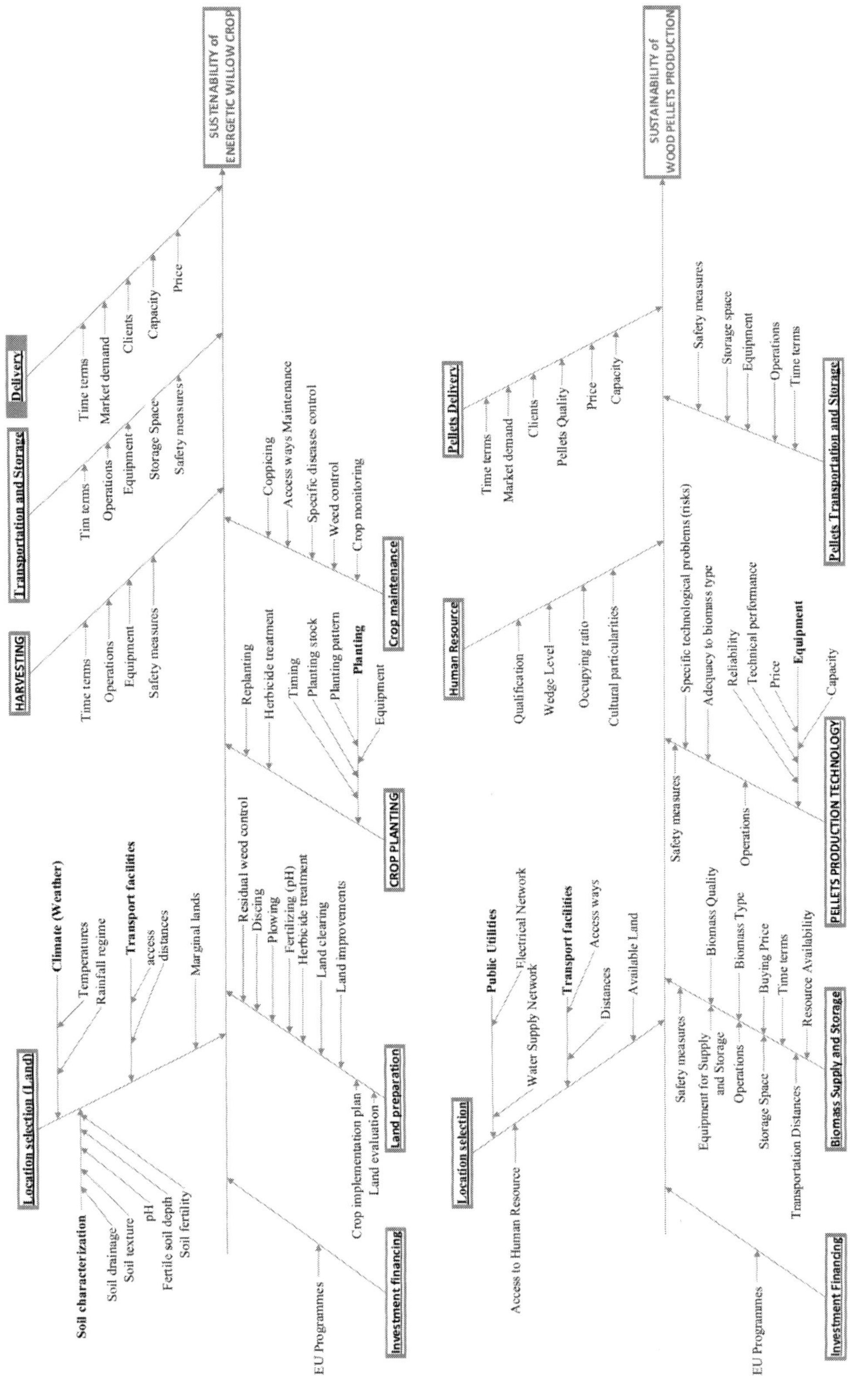

Fig. 5. Models for process variables inventorying and classification.

4 Selection of key process variables

After the research hypotheses have been simplified by considering the set of the particular implementing conditions, a final model of the investigated variables has been elaborated as presented in Figure 6.

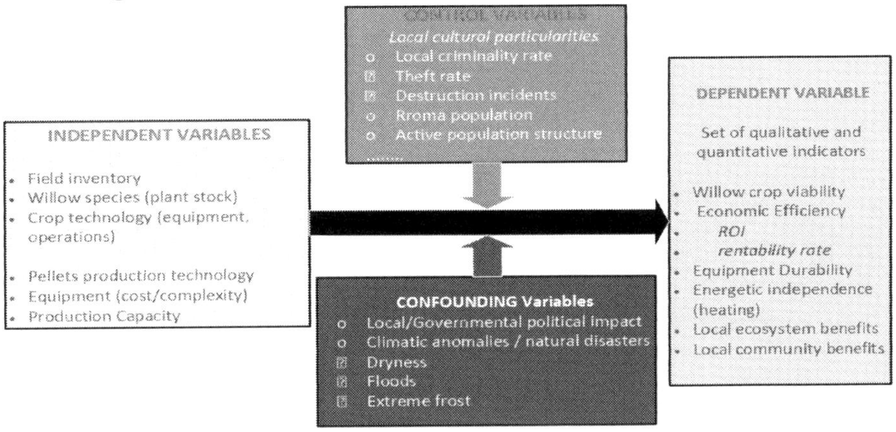

Fig. 6. Model of the research investigated variables.

5 Conclusions

In relation to the above presented considerations and models, there can be concluded that process systematic analysis based on process workflows models, on functional requirements analysis and, not at last, on system/process/function decomposition rules, can be an effective approach in analytically inventorying process variables for the sustainable development of a particular process or project.

As another conclusion of the above presented approach, some directions of further research development can be finally identified and presented here. First of all, there can be considered as useful and of interest to investigate the adapting of Axiomatic Design principles and tools, [6], in analytically inventorying, classifying and selecting the process variables to be considered in translating process requirements into process specifications for providing sustainability of a project or of a small business as in the discussed case study. Of real interest seem to be to further investigate the appropriateness of involving some Lean tools, [7], as well as some simple Excel-based inventory models, to identify the key process input variables.

References

1. C. Verhoest, Y. Ryckmans, *Industrial Wood Pellets Report* PellCert (2012)
2. A. Olejniczak, A. Cyganiuk, A. Kucińska, J.P. Łukaszewicz, *Energetic Willow (Salix viminalis) – Unconventional Applications*, Chapt. 9 in Sustainable Growth and Applications in Renewable Energy Sources, Ed. M. Nayeripour (2011)
3. SUNY-ESF, *Willow Biomass Producers Handbook*, (2017)
4. A. Piraglia, R. Gonzalez, D. Saloni, BioResources **5**(4), 2374 (2010)
5. W. Kloepffer, Int J LCA **13**(2), 89 (2008)
6. N.P. Suh, *Axiomatic Design: Advances and Applications*, Oxford University Press, New York, (2001)
7. J.W. Martin, *Lean Six Sigma to Reduce Excess and Obsolete Inventory*, iSixSigma (2010)

Open innovation in IT brands in Cluj-Napoca

Ligia Maria Nan[1, 2 *], *Roxana Lavinia* Păcurariu[1] and *Laura* Bacali[1, 2]

[1]Faculty of Machine Building, Technical University of Cluj-Napoca, 400641 Cluj-Napoca, Romania
[2]Center for Innovation and Organizational Sustainability, Cluj-Napoca 400609, Romania

Abstract. This article presents the concept of open innovation in IT brands, an industry that is always evolving. We will present the results of a preliminary study regarding the benefits of open innovation on Cluj-Napoca IT brands, the evolution of the number of employees from large companies in the industry and the future direction of research. The sample of the study contains 37 firms selected from a total of 1,490 companies registered in Cluj-Napoca.

1 Introduction

This study aims to analyze how the concept of how open innovation helps and develops the IT industry. Due to the accelerated growth and development of this industry, traditional innovation is no longer enough to withstand on the market, therefore the firms have to approach and implement open innovation. The study is part of early-stage research in the city of Cluj-Napoca, regarding how the online communication channel is being used, specifically social networking sites. It offers future prospects for research and aims to enrich the existing literature.The concept of open innovation can be interesting for organizations and it encourages the development and improvement of websites to obtain a better firm image. In the first part of the study, we will present the theoretical part of the concept of open innovation for its importance in branding and in the IT industry. Then we will present the way the study was carried out, which was possible by analyzing the content of the chosen companies' websites. To investigate how IT firms present the concept of open innovation on their websites, we will try to answer to the following questions in defining the research: How noticeable is the communication of the concept of open innovation on specific firms' websites? and To what extent do they communicate this concept? In the last part, we will present the findings and conclusions of this study, as well as the future directions of research. The growing interest for open innovation indicates the fact that this is becoming a necessity for firms, in order to be able to grow, to be active in innovation and to remain competitive. The concept of open innovation began to be promoted by Chesbrough, thus he created a new paradigm, which is accepted worldwide, being implemented in companies and studied in schools. The concept of open innovation encourages companies and organizations to use both external and internal ideas for innovation and to share their own ideas with other firms wanting to grow [1].

* Corresponding author: ligianan@yahoo.com

2 The concept of open innovation

The concept of open innovation has been greeted with great receptivity from the beginning and has been adopted in a multitude of areas. It has brought major changes to the innovation process, it has led organizations to be more open and it has managed to create new opportunities and to demolish barriers [2]. In other words, open innovation supports the collaboration between different people, R&D institutions, startups, SMEs, large companies, universities and other organizations, aiming at innovation and sharing both risks and gains [3, 4]. The main benefits of implementing the concept of open innovation in a firm are [5]: lower-cost profitability for R&D activities, customer involvement from the early stages of development, competitive advantage and improved productivity of the R&D process. During the implementation process, risks and challenges automatically occur. The main disadvantages identified are [6]: the likelihood of disclosing too much information that is not intended for disclosure, potential issues related to intellectual property ownership and the likelihood of losing the competitive advantage due to the disclosure of too much information. Open innovation brings benefits and creates opportunities for entrepreneurs and small companies. It can help them develop and introduce new brands on the market. This can be accomplished by outsourcing the design, involving the end user in all stages of the development process [7]. The benefits of engaging in open innovation have impacts on the long term process. We can identify the following benefits: abundant ideas and knowledge, opportunities to create value, the improvement of reputation, building relationships, discovering talents that can be potential innovators or employees and having a shared customer base [8]. Herzog [9] identifies the following as the most common forms of open innovation: internal research, patenting, joint R&D agreements, innovation competitions, joint ventures and acquisitions. The crowdsourcing is considered to be a form of open innovation by which consumers become producers. It plays a major role in the process of creating, organizing and sharing knowledge, making it easy to use online platforms and social networks [10, 11]. The process of open innovation contains a phase of exploration and knowledge creation, both internally and externally [12, 13, 14], a phase of knowledge retention [15] and a phase of knowledge transfer [16]. The internal exploration of knowledge represents the generation and development of new ideas within the company, and the external one is about bringing ideas from outside the firm. The knowledge retention phase is necessary to keep the knowledge and relationships established for a long time. The final phase is all about applying the ideas of internal innovations, as well as sharing them with the company's external environment [17]. The influence of the brand on open innovation can be seen in a study by Füller and Matzler [18]. Through this study, they wanted to see whether the brand is one of the factors that influence consumers in participating in open innovation projects of a company and if consumers have the desire to share their ideas and knowledge with the firm [19]. The findings show that the willingness to participate in open innovation projects is heavily influenced by the consumer's capacity, such as motivation, skills and creativeness. Then, the consumer can also be influenced by the trust he has in the brand. Nørskov, Antorini and Jensenh [20], show that partner and consumer attitude towards the brand determines their involvement in the firm's projects.

Brand loyalty influences the desire to share ideas and strengthens collaboration on open innovation. In order to maximize the involvement of all stakeholders, the firm must be the one to take the first step and to create communities in which innovation is encouraged [21]. The companies have to invest in relationships, to share their innovations and to create opportunities for interaction on open innovation activities. In order to involve easily the stakeholders of a firm brand in open innovation activities, they can turn to social networks. It is important and beneficial for brands to interact with all stakeholders, users, consumers and suppliers. They want to interact with a brand they can trust and identify with. By maintaining relationships, brands create experiences, which can be both useful and entertaining, bringing satisfaction to both. In this way, the firms gain brand support, which improves brand identity, brand positioning and brand reputation [22]. Social networks have a major influence in the open innovation process, because through them communication costs are reduced and they bring multiple possibilities for interaction and active participation in the user innovation process [23, 24]. In the IT industry, the concern for open innovation began with the open source trend. Today, the largest companies place the concept of open innovation at strategic level. Open innovation process starts from simple outsourcing transactions in order to reduce overload, costs and risks [25]. We can identify several factors that make it difficult to apply open innovation to the IT industry. These factors relate to: privacy, human resources, brand, image, resources, costs, management, culture, organization, market, partnerships and technology [26]. In recent years, an improved concept was developed, called open innovation 2.0. The essence of the improvements is due to multidisciplinary experiments. This new paradigm is based on the creation of common values and common vision, integrated collaboration, accelerated collaborative research, co-creation of shared value, virtual innovation ecosystems and co-experimentation and user involvement [27, 28, 29].

3 Methodology

To analyze the existence of the concept of open innovation in the IT industry, we have chosen a database provided by the Chamber of Commerce and Industry of Cluj-Napoca. This database contains 1490 IT firms from Cluj-Napoca, registered under the CAEN codes 62 and 63. We selected for this study only those firms that meet the conditions of being a large company: firms with more than 50 employees in 2016 and a turnover of more than 500,000 euro (equivalent to 2,225,000 RON) in 2016. Following this selection, 37 firms remain to analyze. In this analysis, we decided to use only these 37 large companies and to research their websites to find out information about open innovation. We searched for any information, involvement or projects linked to the concept of open innovation.

Being a growing industry, we wanted to observe the evolution of the number of employees in the large companies in Cluj-Napoca. We have made a graph to show the evolution of the number of employees from the 37 large IT firms in Cluj-Napoca in the last five years.

This study is exploratory and descriptive. The study includes 37 large companies; out of a total of 1,490 IT firms in Cluj-Napoca.

4 Preliminary results

4.1 The analysis of the concept of open innovation within the IT brands in Cluj-Napoca

The study implies analyzing the websites of the largest companies in Cluj-Napoca in order to see which of them have submitted information about open innovation. The analysis of the relevant websites was carried out, in the period between 05 and 30 June 2017. From a total of 1,490 companies, only 247 have a Facebook social network page.

The analysis showed that out of 37 firms used in our study, only three had an open innovation section and these were subject to further analysis. These three firms (8%) literally specify the term open innovation and they have initiated projects and programs that encourage the development of the concept. These are: NTT Data, NEUSOFT and ALTRAN. As for the other firms, they do not mention this concept directly.

For 21 companies (57%) we can assume that they have information that indirectly refers to this concept, even if the term is not literally specified on their websites. We can assume this from the various activities that we are undertaking to develop into everything related to open innovation.

We recall some of these activities: partnerships with the education system and other areas different from the main field of activity, startup collaborations, participation in conferences on open innovation topics, development of professional platforms and forums for users, offering open source solutions and infrastructure. Then, there are the remaining 13 firms (35%), who have no information about this concept. The findings of the analysis are summarized in Table 1. NTT Data has a separate website dedicated to the open innovation concept, detailing the three ways it addresses and implements the concept of open innovation. These are: open innovation forum called "From the Port of Toyosu", open innovation business contest, open innovation assistance program called DCAP (Digital Corporate Acceleration Program).

Table 1. Research results.

	Number of firms	%
Sample size	1490	
Firms registered on Facebook	247	16,6%
Firms with over 50 employees	37	2,5%
Firms with a turnover over 2,225,000 RON	37	2,5%
Characteristics of open innovation concept		
Mentioned directly	3	8%
Mentioned indirectly	21	57%
Not at all mentioned	13	35%

NEUSOFT focuses on sustainable improvement through open innovation. The mission statement of the company declares its commitment to alliances and open innovation. ALTRAN organizes the Open Innovation Conference in Italy and it has opened the Open Innovation Institute in France. Innovative Product Development is the way the company applies the concept of open innovation, thereby encouraging outsourcing of the industrialization, production or development processes.

4.2 The evolution of the number of employees in the IT industry in the last five years in the largest companies in Cluj-Napoca

Softvision is the largest company in 2016, based on the number of employees, having 1,160 employees, with 464 more than in 2012, as can be seen from Figure 1. NTT Data is followed by 646 employees, with 406 employees in addition to 2012, and in the third place is Fortech, having 550 employees, with an increase of 311 employees. Arobs Transilvania Software has 457 employees, with 180 employees in addition to 2012. Evozon Systems has 437 employees, with 290 more than in 2012. The largest evolution regarding the number of employees can be seen in seven companies (19%), with an increase of more than 200 employees in the last five years. These companies are: Sofvision, NTT Data, Fortech, Evozon Systems, Access IT Systems, Betfair Romania Development and 3Pillar Global. Six companies (16%) registered an increase in the number of employees between 100 and 200 and the remaining 22 companies (60%) registered an increase in the number of employees between 1 and 100. Only two companies (5%) experienced a significant decline in the number of employees, Neusoft EDC, which dropped by 104 employees and Active Power Solutions dropped by 50 employees. Over 40% companies have doubled their workforce over these five years.

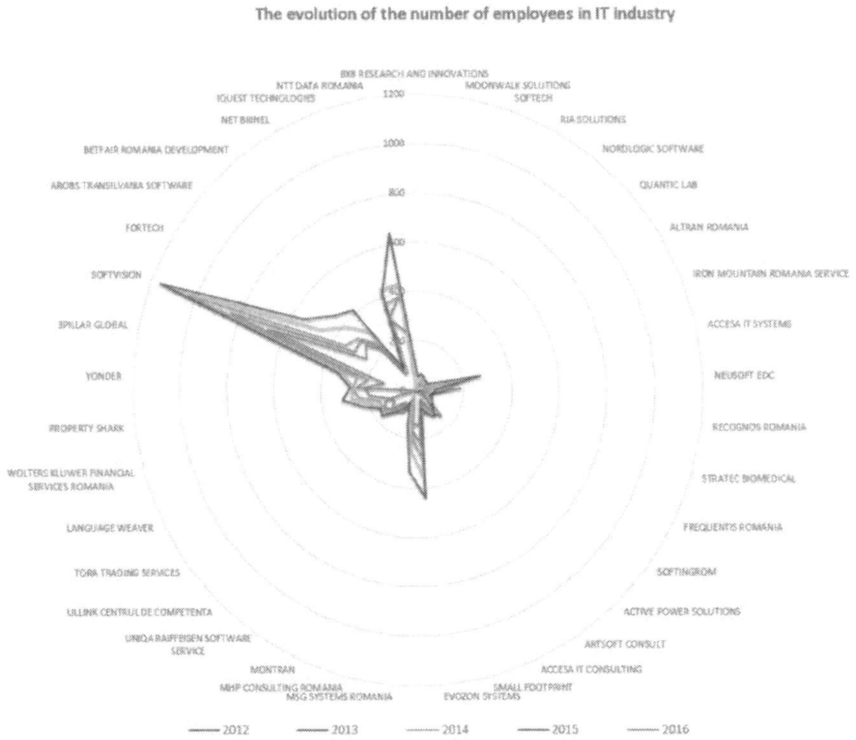

Fig. 1. The evolution of the number of employees in IT industry.

5 Conclusions and following research direction

The IT industry is always concerned and focused on innovation. It is open to new ideas and always applies the latest concepts. The findings show that the number of large IT companies in Cluj-Napoca with information about open innovation on their websites is very low. Companies that specifically mentioned the concept of open innovation have an orientation for it and have initiated projects and programs to promote and develop the concept. Most of the companies did not specifically mention the term open innovation, but they began to take a few steps towards assimilation, by attending conferences on open innovation topics, or using some of the forms of open innovation. The recognition of the concept of open innovation is largely influenced by the reputation of the brand that promotes the concept. The IT industry has a great impact on the concept by the fact that this industry is founded on innovation. In the future, we will expand the research by studying the role of social networks in branding and the extent to which companies are open to the new concepts of open innovation. We will deepen our understanding on the influence and the impact that social networks have on the branding concept in the IT industry.

"This work was undertaken through the Partnerships in Priority Domains: Programme-PN II, developed with the support of MEN-UEFISCDI, Project no.337/2014- UNIINOI".

References

1. H. Chesbrough, *Open Innovation: The New Imperative for Creating And Profiting from Technology,* Harvard Business School Press, Boston, Massachusetts, (2003)

2. J. West, A. Salter, W. Vanhaverbeke, H. Chesbrough, Research Policy **43** (5), 805 (2014)
3. K. Şimşek, N. Yildirim, Procedia - Social and Behavioral Sciences, **235** (24), 719 (2016)
4. O. Granstrand, *The economics of IP in the context of a shifting innovation paradigm*, Innovation Report 2011: Shifting Innovation Paradigms and the Role of Intellectual Property. WIPO, Geneva (2011)
5. S.J. Marais, C.S.L. Schutte, *The Development of Open Innovation Models to Assist the Innovation Process,* University of Stellenbosch, South Africa (2010)
6. P. Trott, D. Hartmann, International Journal of Innovation Management, **13** (4), 715 (2009)
7. K. Hamid, Int. J. Entrepreneurship and Small Business, **7** (4), 457 (2009)
8. H. Chesbrough, *The Era of Open Innovation,* MIT Sloan Management Review, **44** (2003)
9. P. Herzog, *Open and closed innovation: different cultures for different strategies*, Gabler Verlag, (2008)
10. R. B. Bauer, T. Gegenhuber, *Crowdsourcing: Global search and the twisted roles of consumers and producers,* Institute of Organization and Global Management Studies, Johannes Kepler University Linz, **22** (5), 661 (2015)
11. M.N. Wexler, *Reconfiguring the sociology of the crowd: exploring crowdsourcing*, International Journal of Sociology and Social Policy, **31**, (2011)
12. K.G. Smith, C.J. Collins, K.D. Clark, Academy of Management Journal, **48** (2), 346 (2005)
13. P.J. Lane, B.R. Koka, S. Pathak, Academy of Management Review, **31** (4), 833 (2006)
14. B. Cassiman, R. Veugelers, Management Science, **52**, 68 (2006)
15. L. Argote, B. McEvily, R. Reagans, Management Science, **49**, 571 (2003)
16. W. C. Bogner, P. Bansal, Journal of Management Studies, **44** (1), 165 (2007)
17. U. Lichtenthaler, E. Lichtenthaler, Journal of Management Studies, **46**, (2009)
18. J. Füller, K. Matzler, M. Hoppe, The Journal of Product Innovation Management, **25,** (2008)
19. J. Füller, Advances in Consumer Research Volume, **33**, (2006)
20. S. Nørskov, Y. M. Antorini, M. B. Jensen, International Journal of Innovation Management, **20**, (2016)
21. M. R. Martínez-Torres, Expert Systems with Applications, **7**, (2013)
22. J. Prell, R. Schleich, *Social Open Innovation in Online Brand Communities with Particular Regard to the Social Exchange Theory and Brand Management,* Lund University, School of Economics and Management, (2015)
23. J. Bughin, A. Hung Byers, M. Chui, *How social technologies are extending the organization*, McKinsey Quarterly, (2011)
24. E. S. Lakatos, L. Bacali, O. B. Bercea, C. M. Muresan, A. Moldovan, *The benefits of IT tools in innovation process for SME sustainability*, In International Conference in Advances in Management, Economics and Social Sciences, (2015)
25. O. Gassman, E. Enkel, H. Chesbrough, *The future of open innovation,* R&D Management, **40**, (2010)
26. W. Krause, C. Schutte, N. du Preez, *Open Innovation in South African Small and Medium-Sized Enterprises*, CIE42 Proceedings, Cape Town, South Africa, **26**, (2012)
27. M. Curley, B. Salmelin, *Open innovation 2.0: A new paradigm*, Cambridge, MA: Academic Press, (2013)
28. M. Curley, *Twelve principles for open innovation 2.0*, Nature, (2016)
29. E, Motoiu, O. A. Pavel, E. S. Lakatos, Review of Applied Socio-Economic Research, **11**, (2016)

MATEC Web of Conferences 137, 07006 (2017) DOI: 10.1051/matecconf/201713707006
MTeM - AMaTUC 2017

Impact of environmental conditions on the reliability of MEMS components from optical applications

Marius Pustan[1,*], *Corina* Birleanu[1] and *Florina* Serdean[1]

[1] Micro & Nano Systems Laboratory, Mechanical Systems Department, Technical University of Cluj-Napoca, 400641 Muncii Street, Romania

Abstract. The reliability design and the lifetime of MEMS from optical applications is strongly dependent on the operating conditions. In optical devices an additional stress is introduced by the temperature gradient appearing during operation. The thermal stress introduced in the structure modifies the component response and its lifetime. The investigated sample is a micromirror fabricated from gold using the MEMS technologies and tested by an atomic force microscope. The scope is to analyse the temperature and relative humidity influence on the mechanical and tribological behaviours. Using a mechanical force given by the bending deflection and stiffness of the AFM probe, the dependence between the applied force and the micromirror deformation is monitored as a function of temperature. By increasing the temperature of structures the stiffness decreases based on the thermal relaxation of the material. If the relative humidity increases the adhesion between the micromirror and substrate is improved based on the capillarity effect. The results obtained are useful to MEMS designers who can predict the micromirror response as a function of the operating conditions with influence on the actuation energy, the accuracy in response and the system lifetime.

1 Introduction

Micro-opto-electro-mechanical Systems (MOEMS) integrate optical, mechanical and electrical components on a single chip and allow the use of various optical elements in a compact and low-cost device. Some principle commercial applications of MOEMS are: optical switches (for telecommunication routeing), displays (head up displays, vehicle displays), shutter arrays (switching), tuneable filters (broadband systems), waveguides (biomedical/ diagnostic), environmental pollution detection (multisensory arrays with optical detection) [1]. The main mechanical part of MOEMS is a moving component as microbridges, microcantilevers or a more complicated structure such as a micromembrane. These flexible components are integrated in MOEMS as mirrors, shutters, filters and attenuators [2]. The fabrication technology of MOEMS is based on micromachining

* Corresponding author: Marius.Pustan@omt.utcluj.ro

techniques combined with integrated circuits (IC) processing methods. The integration of the components in a complex MOEMS device can be performed in a monolithically method or a hybrid one. The micromachining technology together with IC- based processing techniques is used in the fabrication of complex opto-electronic integrated circuits and micro-electromechanical alignment devices in production quantities [1].

The failure mechanism of Micro-electro-mechanical Systems (MEMS) is more complicated than in microelectronics. For instance, the MEMS devices interact with different environmental conditions such as temperature or humidity. The reliability design of MEMS must consider the environmental effect on the device response and the stresses that are induced by the environment [3]. The Physics of Failure approach is one method to improve the reliability of MEMS. This approach can be used to determine the effect of environmental stress of MEMS sensitivity [3].

In order to be able to perform experimental tests on MEMS components as a function of environmental conditions the testing equipment must be integrated into an environmental chamber with the possibility to change the temperature, the humidity levels and the medium pressure. A thermal gradient introduces softening due to Young's modulus - temperature relation and a thermal relaxation, which changes the rigidity of material [4, 5]. The lifetime of optical MEMS devices can also be influenced by temperature. The temperature changes the stiffness of the investigated structure with influence on the restoring force of components from substrate after stiction. Stiction is a very common reliability problem in MEMS devices. Stiction occurs when the adhesion force between contacting surfaces exceeds the restoring force that stiffness is dependent. Adhesion depends on the capillary condensation, which is affected by the moisture and surface contamination as well as the van der Waals forces that are influenced by the surface roughness. The medium humidity increases the adhesion based on the capillary forces between contact surfaces.

The next sections present the experimental tests performed on a MEMS component from optical application fabricated from gold as a reflective material. The sample is experimentally investigated in an environmental chamber with the possibility of temperature control as well as the environmental relative humidity (RH).

2 Experimental procedure

2.1 Sample description

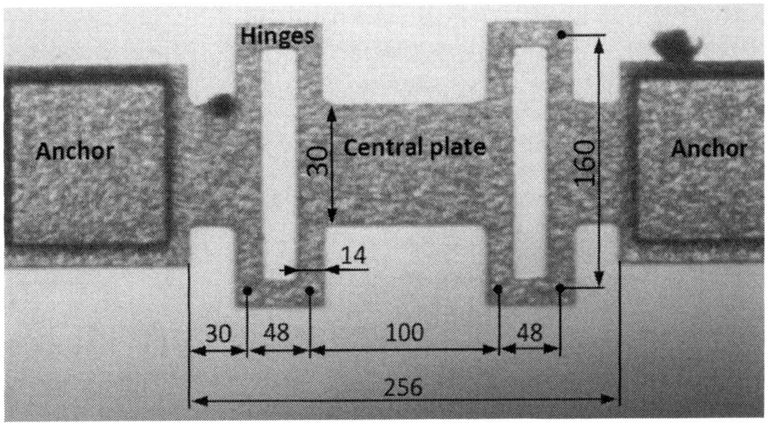

Fig. 1. Micromirror used in experimental investigations and the geometrical dimensions in µm.

The optical MEMS component used in the experimental investigation is a micromirror (Fig.1) fabricated using MEMS technology from a reflective material (gold) on a silicon substrate [4]. The gold material is the most used material in optical applications. The central plate of micromirror is supported by two rectangular hinges connected to the anchors. The rectangular hinges of structure are introduced in order to compensate the thermal expansion of the structure under the action of a thermal gradient. The central plate of structure can perform out-of-the plane motion. The thickness of flexible plate is 3μm and the gap between it and the silicon substrate is 2μm. The others in plane geometrical dimensions of the flexible part (in μm), measured by using an optical microscope, are presented in Fig.1.

2.2 Atomic force microscopy procedure

In order, to determine the environmental conditions effect on the micromirror behaviour, an Atomic Force Microscope (AFM) XE70 was used. The spectroscopy-in-point operating mode is used to deflect the sample from the initial position toward the substrate under different testing conditions (temperature and humidity). Using this AFM module, the vertical displacement of the scanning head together with the AFM probe is controlled by the AFM software. The AFM probe used in the experiment (TD21464) has a stiffness equal by 156N/m and the resonant frequency is 54kHz. In order to investigate the temperatures influence on the mechanical response of the AFM probe, it vibrates close to the thermal stage and the resonant frequency is monitored during increasing temperature. The resonant frequency of AFM probe is not modified if the medium temperature increases from 20°C to 140°C.

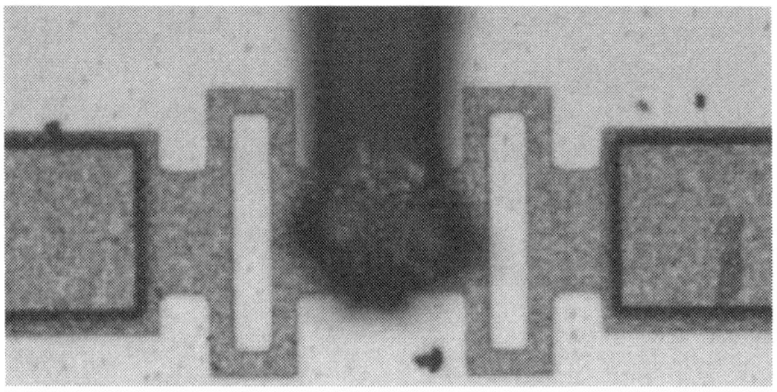

Fig. 2. AFM probe in contact with the investigated micromirror.

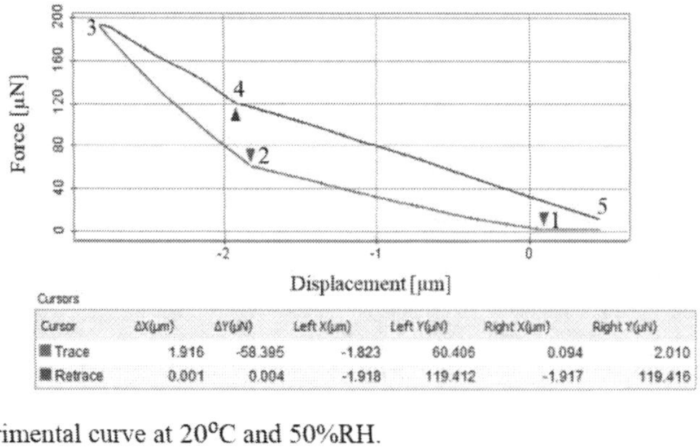

Fig. 3. AFM experimental curve at 20°C and 50%RH.

Firstly, the AFM probe comes in contact with the investigated sample in the mid-position of central plate (Fig. 2). Secondly, the vertical displacement of the AFM probe together with the scanning head is controlled. The sample is fixed on the AFM table and the probe is mounted in the scanning head. In this step, the bending deflection of AFM probe is optically determined while the AFM probe bends together with the sample. The stiffness of the AFM probe is known. The detected deflection of AFM probe multiply with its stiffness give the force applied on structure.

Figure 3 presents the AFM experimental dependence between the vertical displacement of AFM probe and the applied force. There are two different curves, each of them with two slopes. The loading curve corresponds to the part 1-2-3 (Fig. 3) and the unloading is given by distance between 3 and 5 with an intermediate position in 4. The first part of curve 1-2 corresponds to the bending of AFM probe together with investigated sample. The sample touches the substrate in position 2, after that is only the bending deflection of AFM probe. The unloading process starts in position 3, there is only unbending deflection of AFM probe between 3 and 4, in position 4 the sample is detached from substrate and comes together with the AFM probe to the initial position 5. The first part 1-2 is used to determine the sample stiffness and the adhesion between the central plate and substrate can be observed in the position 4 on the experimental AFM curve. The difference between controlled vertical displacement of the AFM probe (and the scanning head) and the detected deflection of AFM probe gives the displacement of investigated sample. The applied force divided by the sample displacement means the sample stiffness. The tests are performed under different temperatures and for different humidity of the testing environment conditions. For accuracy measurements, the tests were repeated five times and the average results were considered.

3 Temperature effect on the optical MEMS component response

The temperature applied on the investigated structure additionally introduces a thermal stress gradient. The thermal gradient deforms the micromirror based on thermal dilatation of structure. Moreover, the modulus of elasticity decreases under the thermal relaxation effect of material with influence on the mechanical response of system [4]. The mechanical response of system depends on the out-of-the plane stiffness that is directly proportional with the modulus of elasticity of material. As a consequence, by increasing the operating temperature of structure the stiffness decreases as well as the mechanical response of micromirror.

Experimentally, the temperature is applied directly to the sample anchors between 20^0C and 140^0 by using a thermal controlled stage. After, the spectroscopy-in point of AFM is used to bend the flexible plate of micromirror and the stiffness is determined. Figure 4 presents the experimental out-of-the plane stiffness of the structure as a function of temperature.

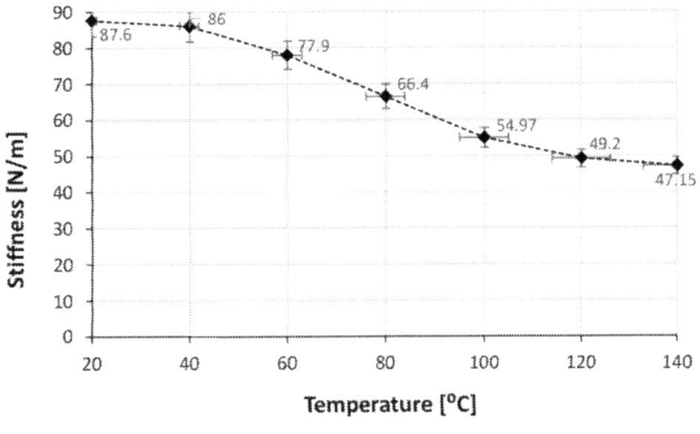

Fig. 4. Experimental out-of-the plane stiffness of micromirror versus temperature.

The out-of-the plane stiffness decreases as temperature increases, respectively. A nonlinear dependence stiffness versus temperature is experimentally observed. The stiffness decreases with about 46% if the operating temperature increases from 20°C to 140°C. At the beginning, a small decrease of the stiffness is determined, in the temperature range between 20°C and 40°C. This is based on the relaxation of material provided by the residual stress appearing in the structure during fabrication process [6]. A strong dependence between stiffness and temperature is determined in the temperature range between 40°C and 140°C based on thermal relaxation of the modulus of elasticity of the material.

4 Humidity influence on the optical MEMS components response

In order, to determine the humidity effect on the behaviour of investigated micromirror, a humidity controlled chamber is used. The humidity is modified between 50%RH and 90%RH. The humidity has not a significant influence on the micromirror stiffness. A small difference in stiffness of micromirror is experimentally determined if the humidity increases from 50% to 90% based on the damping effect given by the surrounding medium.

The spectroscopy-in-point of AFM is used to bend the central plate of micromirror until substrate under different controlled humidity. During unloading, the detachment of investigated micromirror from substrate (the position 4 from Fig. 3) is delayed due to the adhesion effect that gives a jump in the AFM probe deflection (Fig. 5). The jump size is directly proportional to the strength of adhesion between contacted surfaces. The tests performed on the adhesion between the flexible plate and substrate at room conditions (50%RH) provides small adhesion force, around 4nN (Fig. 3). When the humidity is increased, the adhesion effect between contact surfaces increases significantly. The adhesion force is given by the difference between the cursors 4′ and 4″on the unloading AFM curve.

Figure 5 presents the adhesion force for two different humidities (55%RH and 90%RH). The value of the adhesion force is given by ΔY from the retrace values provided in the table under diagrams.

he relative humidity is step-by-step increasing between 50% and 90% and the adhesion force is measured. Figure 6 presents the humidity effect on the adhesion force between the investigated micromirror and its substrate. The adhesion effect increases as the medium humidity is increased, with a maximum measured force of 17.16 μN at 90%RH.

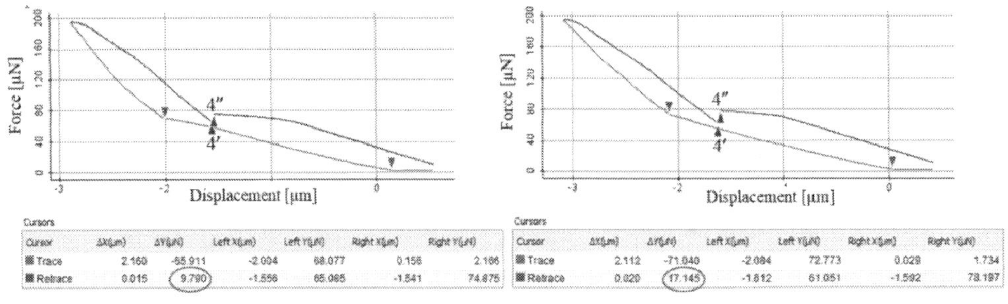

(a) (b)

Fig. 5. Adhesion force between micromirror and substrate for: (a) 55%RH and (b) 90%RH.

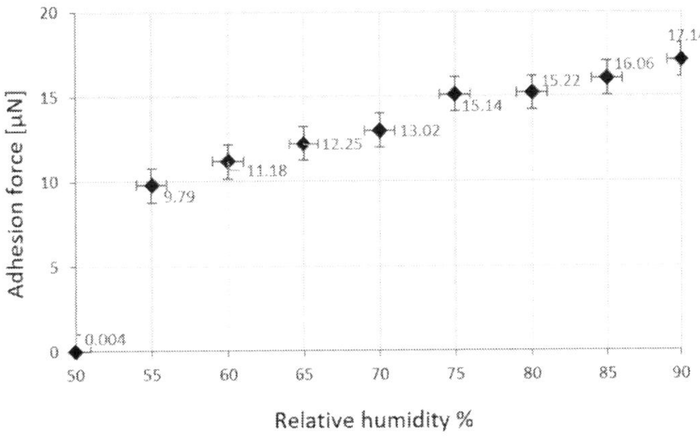

Fig. 6. Experimental adhesion force between micromirror and substrate versus relative humidity.

5 Conclusions

The environmental conditions change the mechanical and tribological behaviours of microcomponent from MEMS applications with influence on the accuracy in response and the system lifetime. In optical applications a temperature gradient occurs and changes the stiffness of flexible structures. A micromirror supported by rectangular hinges was chosen to investigate the environmental effect on the structure response. Out-of-the plane stiffness of microcomponent under a mechanical force was determined as well as the adhesion force. For adhesion analysis the flexible plate of microcomponent was deflected until substrate. The temperature was applied directly on the anchor using a controlled thermal stage. As the temperature increases, the stiffness decreases with influence on the system response (force and/or displacement). The stiffness decreases with about 46% if the testing temperature increases from 20°C to 140°C. The temperature modifies the modulus of elasticity based on the thermal relaxation of material. Moreover, the stiffness has influence on adhesion effect based on the restorative force of microstructure from substrate.

There are some MEMS applications that operate in ambient conditions as the devices used in the environmental condition monitoring.

The humidity has a strong influence on adhesion effect based on the capillary forces. As the humidity increases, the capillarity is increased providing higher adhesion force between movable structure and substrate. Using a humidity controlled chamber, the relative humidity during testing was modified. For investigated MEMS structure, the adhesion force between the central plate and substrate significantly increases if the relative humidity is increased from 50%RH to 90%RH.

The research work and the results presented in this paper are useful to MEMS designers in order to develop reliable MEMS structures operating under different environmental conditions.

This work was supported by a grant of the Romanian National Authority for Scientific Research and Innovation, CNCS/CCCDI UEFISCDI, project number PN-III-P2-2.1-PED-2016-1727, PNCDI III.

References

1. A. El-Fatatry, *MEMS Aerospace Applications, MOEMS Applications* (RTO-EN-AVT-105 Lecture Series, NATO) (2015)

2. I. Stanimirović, Z. Stanimirović, *Optical MEMS for Telecommunications: Some Reliability Issues* (In Advances in Micro/Nano Electromechanical Systems and Fabrication Technologies, InTech) (2013)
3. P. Schmitt, F. Pressecq, X. Lafontan, P. Pons, J. M. Nicot, C. Oudea, D. Estève, H. Camon, J.Y. Fourniols, *MEMS behavioral simulation: a potential use for physics of failure (PoF) modeling* (4th Round Table on Micro/Nano Technologies for Space) (2003)
4. M. Pustan, C. Dudescu, C. Birleanu, Phys. Statutus Solidi C **12**, 1322-1327 (2015)
5. M. A. Tadayon, H. Sayyaadi, N. Jazar, J. of Physics: Conference Series **34**, 89–94 (2006)
6. M. Pustan, V. Rochus, J.C. Golinval, Microsyst. Technol. **18**, 247-256 (2012)

Silver Nanoparticles (AgNP) impregnated filters in drinking water disinfection

Alexandru Rus[1,*], *Vasile-Dănuț* Leordean[1] and *Petru* Berce[1]

[1]Technical University of Cluj-Napoca, Department of Manufacturing Engineering, B-dul Muncii 103-105, 400641, Cluj-Napoca, Romania

Abstract. This paper describes how simple portable devices could eliminate water pathogens by using Silver Nanoparticles, based on their antimicrobial properties. Recent studies indicated that silver nanoparticles can achieve up to 100% antibacterial activity removal. Results are showing that Silver Nanoparticles retention in the filter structure, E. coli bacteria removal, water quality and water flow rate must be evaluated as main efficiency indicators of the designed filters, in order to obtain the optimal filter. To apply the antimicrobial property of Silver in drinking-water treatment, a filter is produced using Additive Manufacturing techniques and coated with different concentrations of silver solutions.

1 Introduction

Today world is facing real challenges in meeting increasing demands of drinking water as the available supplies of freshwater are exhausting due to lack of rain, population growth and more stringent health based regulations. Clean water is vital to human health, while in countries such as India, 80% of the diseases are due to bacterial contamination of drinking water.

The most common techniques used for water treatment are using mostly chemical and physical agents such as chlorine and its derivatives, bromine (Br), water sediment filters (fiber and ceramic), activated carbon, reverse osmosis etc. but the use of halogens has many problems because of their high toxicity and vapour pressure [1]. Nanoscience is currently widely investigated to be used in water purification. This new interdisciplinary subject is referring to the study of chemical and physical changes at nanoscale level (1–100 nm). Nanotechnology is currently used for developing new materials and devices [2] and it is also intensively investigated in water purification.

This paper is referring to the use of nanotechnology, specifically the use of silver nanoparticles (AgNP) in water purification purposes. It describes the properties and methods used for obtaining AgNP, the types of substrates used in general for AgNP filters and proposes a new filter structure.

* Corresponding author: alexandru.rus86@gmail.com

2 Methods for obtaining the Silver Nanoparticles (AgNP)

Silver is a white metal characterized by its shimmering, soft proprieties, being a good electric and thermal conductor. Silver is also a therapeutic agent used in medicine for treating infectious diseases and surgical infections. [3] Silver may be found in different form/silver compounds, such as Silver Nitride ($AgNO_3$), Silver Chloride (AgCl), Silver fulminate (AgONC), etc. The most popular compound is Silver Nitrate ($AgNO_3$) used as the starting for synthesis other silver products and painting area. Silver Chloride (AgCl) is used for potentiometric and PH testing equipment [4].

It is well known that noble metals are very resistant to severe conditions and exhibit unique properties when they are produced at nanoscale. [5] Metal nanoparticles like silver are particles raging from 1nm and 100nm with very good electrical, optical and thermal properties. [6] Silver nanoparticles (AgNP) have a good conductivity, are chemically stable, their catalytic activity has been proved as being a very good one and are used due to exceptional antimicrobial activity [7].

AgNP is one of the most popular nanoparticles due to its many applications and relatively low manufacturing costs. Some of the AgNP applications are described below Figure 1:

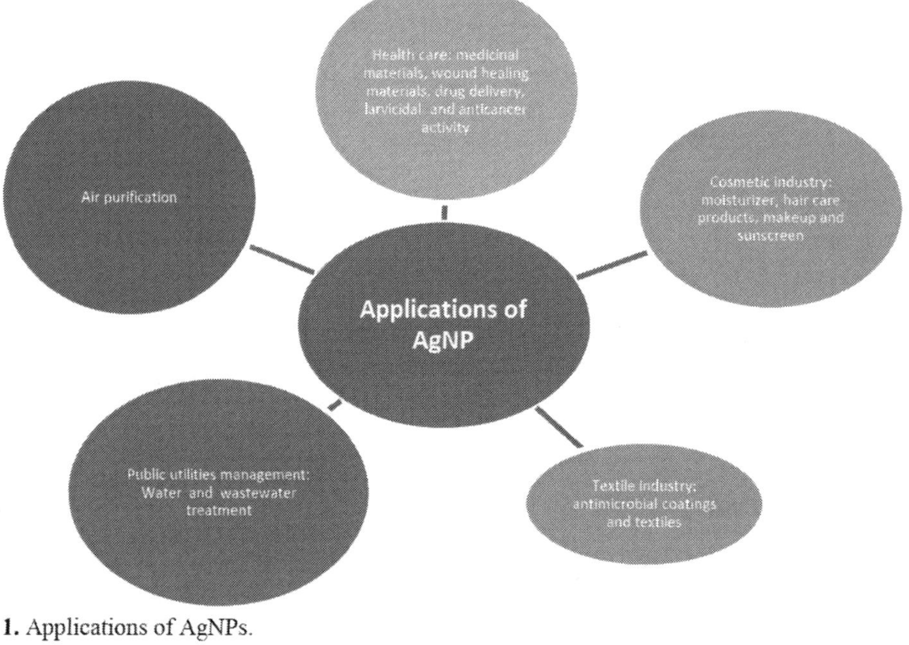

Fig. 1. Applications of AgNPs.

Currently, there are a lot of researches addressing the potential of AgNP for water treatment, mostly combined with filtration. Silver is either impregnated on domestic water filters either used in conjunction with copper ionization as a preventative measure against colonization of Legionella spp. A recent application is the usage of silver ions combined with copper and chlorine for swimming pools sanitation systems. [8] Among the advantages of this technology are: no odor, corrosion and no formation of unwanted oxidizing compounds. [9]

Over the time, different methods have been used to synthesize AgNP: biological, physical, and chemical. The more common methods are chemical and physical methods that produce pure and well-defined nanoparticles, but in general, the chemicals used in the synthesis are toxic, energy consuming and expensive [3].

2.1. Chemical methods

Chemical methods refer to the synthesis of AgNPs by using organic and inorganic reducing agents. An advantage of the chemical methods is the simplicity of preparing AgNPs in aqueous solution. The most common method of producing AgNP is the chemical reduction of AgNO3 dissolved in water with a reducing compound such as sodium citrate, borohydride, ascorbate, glucose, hydrazine, etc. [10] These agents have the role of reducing Ag+ to metallic silver (Ag0). During this process, it is important to use protective agents to stabilize the dispersion of the nanoparticles and avoiding their agglomeration. There are also other several methods of producing AgNPs such as: microemulsion techniques, UV-initiated photoreduction, electrochemical reduction, solution irradiation, microwave-assisted synthesis or cryochemical method [10].

2.2. Physical methods

Evaporation-condensation and laser ablation are the most common physical approaches for AgNPs synthesis having advantages such as the absence of solvent contamination in the prepared thin films and the uniformity of nanoparticle distribution. Thus, the result is pure and uncontaminated metal colloids.

However, these methods are also well known since the devices usually occupy large spaces and are expensive, they consume a great amount of energy while achieving the adequate temperature around the source material and require a lot of time to achieve thermal stability.

When using laser ablation in solution for AgNPs synthesis several factors such as laser wavelength, the duration of the laser pulses, the laser fluency, the ablation time duration and the effective liquid medium, with or without the presence of surfactants, should be taken into consideration [10]. Other physical methods for are referring to are thermal decomposition method, ceramic heating process, arc discharge technique [7].

2.3. Biological methods

There is a growing need to develop environmentally and economically friendly processes, which do not use toxic chemicals in the synthesis protocols [11]. In general these methods use: bacteria (*Pseudomonas stutzeri AG259, Lactobacillus* strains, etc.), fungi (*Fusarium oxysporum, Aspergillus flavus*), algae (*Lyngbya majuscule, Spirulina subsalsa, Rhizoclonium heiroglyphicum, Chlorella vulgaris*), plants (green tea *(Camellia sinensis)*, alfalfa (*Medicago sativa*), lemongrass (*Cymbopogon flexuosus*), geranium (*Pelargonium graveolens*)). Synthesis of NPs using plants is very cost effective, and thus can be used as an economic and valuable alternative for the large-scale production of NPs

This method depends on a series of factors such as: the nature of the organisms, appropriate circumstances for cell growth and enzymatic activity, optimal reaction media, and appropriate choice of biocatalyst suitable to obtain the desired nanoparticle [10].

3 Type of supports for AgNP filters

Several researchers have shown that using different methods in preparing silver nanoparticles can lead to very good bacterial activity removal. Materials development is enhancing silver nanoparticles deposition on solid materials for the deactivation of microorganisms in water treatment [12]. In the case of drinking-water treatment, various forms of silver nanoparticles coated on materials/supports were investigated during the years:

Sand impregnated with AgNP is less expensive, effective and can be self-constructed with limited local skills [13].

Zeolites as cation exchangers in water treatment, have increased in usage due to their availability, low cost, high surface area and sorption capacity, chemical inertness and low or null toxicity for human [14].

Fiberglass needs only a small amount of Ag for a strong bactericidal response against E. coli. The evidence suggests that the adhered Ag nanoparticles to the fiberglass substrate are active against bacteria in a filtration application. Falling desorption rates suggest the adsorption of Ag by the E. coli [15].

Activated Carbon filter was used as support for Silver Nano-particles from silver nitrate solutions using glucose as a reducing agent. The batch method showed that with the increase in contact time and the decrease in the bacterial count, complete inhibition was achieved. In column test, the prepared AgNPs/AC showed a complete bacterial reduction in synthetic water and actual polluted water samples. Data revealed a direct relation between Ag-NPs concentration and contact time toward bactericidal activity against E. coli which represented as an indicator for microorganism's pollution, whereas the concentration of AgNPs and contact time increased, bacterial counts of E. coli decreased and inhibition zone increased for all bacterial concentrations used [16].

Blotter paper allows microorganisms to meet the biocide, but attachment to the fiber surfaces limits the levels of silver in the effluent water. The primary purification mechanism is not the removal of bacteria from the effluent by filtration but rather the deactivation of bacteria as they percolate through the AgNP paper structure. Consequently, the filter effluent contains dead bacteria. The large pore size in the filter paper allows for reasonably rapid flow by gravity, without the need for pressure or suction [17].

Cellulose filters: Cellulose filter paper was coated with silver nanoparticles using a chemical reduction method, with sodium borohydride: silver nitrate ratios of 2:1 and 10:1. The efficiency this cellulose filter in E. coli removal ranged from 99% to 100%. Moreover, the performance of the silver nanoparticle-coated cellulose filter paper was evaluated in terms of silver in the effluent and compliance with the drinking-water quality standards. Although further research is still necessary, the performance of silver nanoparticle-coated cellulose filter paper and its short and easy preparation method suggests its suitability as an antibacterial water filter during emergency situations [18].

Ceramic filters impregnated with silver nanoparticles were investigated in the laboratory and in the field in Guatemala. The addition of silver nanoparticles to the ceramic filters improved the performance for all mass percentages of sawdust relative to filter media without nanoparticle treatment. Filters with higher porosity achieved higher bacteria removal than those with lower porosity. After laboratory testing, ceramic filters were manufactured and distributed to 62 households. Over the course of the study, the average percent reduction in total coliforms and E. coli was around 90% and the average effluent concentration of ionic silver was 0:02 mg=L (below the U.S. EPA standard of 0:1 mg=L) [19].

Polyurethane Foams, can be used for Silver nanoparticles coating in diverse forms. This material can be washed, dried, and stored for extended periods without the loss of nanoparticles. The performance of the material as an antibacterial water filter was checked and no bacterium was detected in the output water when the input water had a bacterial load of 1 105 –1 106 CFU/mL. The antibacterial action was also checked online for a flow rate of 0.5 L/min and no bacterium detected, which suggests that domestic use of this technology is possible. The results are in line with the WHO requirements for drinking water. The experiments suggest the possibility of the use of this material in drinking water purification, air filtration, domestic and industrial air quality management, antibacterial packaging, etc. The chemicals involved in the synthesis of nanoparticles are commonly available, cheap, and non-toxic [20].

4 Design of AM produced bottle cap filter

The filter consists of a double threaded bottle cap, allowing contaminated water to flow from the support bottle into the receiver bottle. This is perforated by a 19,4-mm diameter cylindrical filter impregnated with silver nanoparticles. The AM techniques allow to produce a controlled porosity structure, which is influencing the efficiency of the filter in terms of silver retention in the filter, water flow rate and E. coli bacteria removal.

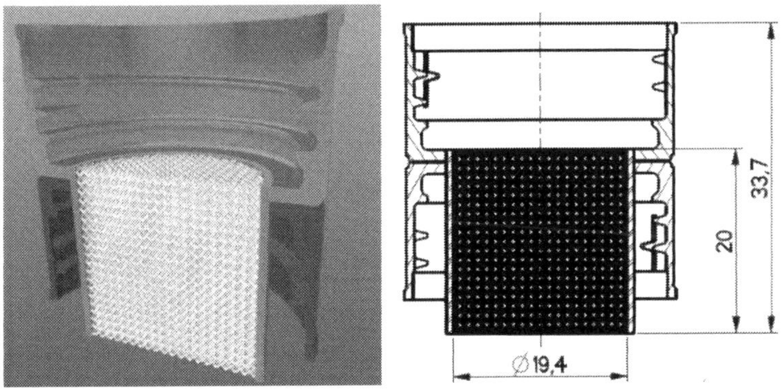

Fig. 2. AgNP Filtering bottle cap.

Dosoky et al. evaluated the bactericidal efficiency of AgNPs by its application in different concentrations to water sample, and allowed to interact with bacteria for different duration. Then, the bactericidal efficiency of AgNPs was determined by comparing the counted bacteria before and after the treatments resulting that the highest concentration of AgNP exhibited highest bactericidal efficiency in total bacteria count (TBC), where, after two hours, 0.1, 0.05 and 0.01 mg/L AgNP was found to be sufficient to inhibit 91.85, 89.14 and 74.92% TBC from surface water [21].

Therefore, the AgNP coating of the AM filter will be done using 0.1 mg/L concentration of AgNP. Water flow simulations through the designed filters must be evaluated to obtain indications on the optimal filter dimension and structure, to optimize the filter manufacturing process and the efficiency of filtration.

5 Conclusions

The research presented shows that different constructions of filters impregnated with silver nanoparticles can significantly improve water quality. The main intention is to analyse and develop an innovative filter with the optimal coating of AgNP which could completely remove pathogenic bacteria from test water. Moreover, further researches are needed to fully understand the behaviour of the designed filter structure in terms of porosity, AgNP coating, silver concentration in the effluent and E. coli removal from test water.

This research was supported by the AM-CIR project, PN-II-RU-TE-2014-4-1157, no. 37/01.10.2015 financed from the UEFISCDI by the Romanian Government.

References

1. D. K. Tiwari, J. Behari, P. Sen, World Applied Sciences J, **3**, 417-433 (2008)
2. *** https://www.sciencelearn.org.nz/resources/1640-nanoscience-explained, accessed on 4th of July 2017
3. J. W. Alexander, **10**, 3, 289–292 (2009)
4. *** http://ssnano.com/ag___silver, accessed on 4th of July 2017
5. Mü. Sökmen, S.Y. Alomar, C. Albay, Göü. Serdar, J. of Alloys and Compounds, **3** (2017)
6. *** http://www.mknano.com/Nanoparticles/Elements/Silver-Nanoparticles/, accessed on 28th of June 2017
7. K. Alaqad, T.A. Saleh, J. of Env. & Anal. Tox., **6**, 2-3 (2016)
8. L. Fewtrell, Silver: Water disinfection and toxicity, **2**, 9 (2014)
9. ***http://www.watertechonline.com/silver-copper-have-anti-bacterial-effects/, accessed on 28th of June 2017
10. S. Iravani, H. Korbekandi, S.V. Mirmohammadi, B. Zolfaghari, Research in Pharmaceutical Sciences, **9** (6), 385–406, (2014)
11. Popa, I.M. Nicolai, V. Ceara, C. Boboc, C.A Danes, ICPE, EEA, **64**, 2, 104-112 (2016)
12. L. Mpenyana-Monyatsi, L. H. Mthombeni, M.S. Onyango, M. N. B. Momba, Int J Environ Res Public Health, **9**, 1, 244–271 (2012)
13. S.N. Mahmood, S. Naeem, N. Basit, T.H. Usmani, Envir. Tech., **14**, 151–157 (1993)
14. Y Matsumura, K. Yoshikata, S.I. Kunisaki., T. Tsuchido, Appl. Environ. Microbiol. **69**, 4278–4281 (2003)
15. G. Nangmenyi, Z. Yue, S. Mehrabi, E. Mintz, J. Economy, Nanotechnology, **20**, 495–505 (2009)
16. A.H.M. El-Aassar, M.M. Said, M. Abdel-Gawad, H.A. Shawky, Australian J of Basic and Applied Sciences, **7** (1), 422–432 (2013)
17. T. A. Dankovich, D. G. Gray, Envir. Sci Tech., **45**, 1992–1998 (2011)
18. S.M. Praveena, L. S. Han, L.T.L. Than, A.Z. Aris, J of Experimental Nanoscience, **11**, 17, 1307-1319 (2016)
19. E. N. Kallman, V. A. Oyanedel-Craver, J. A. Smith, J. Environ. Eng., **137**, 407-415 (2011)
20. J. Prashant, T. Pradeep, *Potential of Silver Nanoparticle-Coated Polyurethane Foam as an Antibacterial Water Filter*, Wiley InterScience, DOI: 10.1002/bit.20368 (2005)
21. R. Dosoky, S. Kotb, M. Farghali, J. Adv. Vet. Anim. Res., **2** (2), 175-184 (2015)

Carbon epoxy front hood for an electrical city vehicle

Paul Bere[1], *Calin* Neamtu[2,*], *Mircea* Dudescu[3], *Radu* Comes[2] and *Sergiu* Solcan[2]

[1] Technical University of Cluj-Napoca, Manufacturing Engineering, Bd. Muncii 103-105, Romania
[2] Technical University of Cluj-Napoca, Design Engineering and Robotics, Bd. Muncii 103-105, Romania
[3] Technical University of Cluj-Napoca, Mechanical Engineering Department, Bd. Muncii 103-105, Romania

Abstract. In the last decade fiber-reinforced polymer (FRP) had a very impressive development. Due to its physical and mechanical properties, this material is used in many high-end domains such as: aerospace, aviation, automotive, medical, engineering or building constructions. In the last period FRP are being intensely used in the automotive industry especially for the chassis manufacturing and other vehicle structural components. In this paper, the authors present the model of a carbon epoxy front hood of a two-passenger electrical car which is specially designed in urban area and which makes use of advanced FRP manufacturing.

1 Introduction

In recent decades, there have been significant changes in the use of materials in different fields across the world. Fiber reinforced polymer composite (FRP) materials represent a unique category of advanced materials with an outstanding scientific and industrial interest.

Today, major manufacturing companies are switching from traditional thermal-electric cars to electric one. Analysis predicts that the total cost of ownership of electric cars will dip below those with internal combustion engines in 2022. Electric cars are mostly intended to be used in the urban environment and the electric car research is focused on improving the deficiencies of the current vehicle systems. The use of composite materials in the construction process of the electric road vehicles will have a major impact to decisively reduce the mass of the car in order to reduce the energy consumption while also offer possibilities to increase the autonomy of the car.

Different researchers presented the current problems that may occur due to the use of these materials [1-3]. Whether it is constituents, micro or macro mechanics, theory of breakage of these materials, mechanical behavior, or general notions about the technology of obtaining them, these researchers have approached these subjects. The authors aim to achieve the best performance in terms of weight, global material cost and the best stiffness.

Different structures of reinforced composite materials are presented in the work of [4-8]. These researches are focused on either the mechanical characteristics or on reducing

* Corresponding author: calin.neamtu@muri.utcluj.ro

their mass by using hybrid solutions of sandwich structures. The applications of these materials in the automotive field create the prerequisites for obtaining complex structures that have superior characteristics to the traditional materials regarding their mass reduction. The material behavior under high temperature conditions of composite materials is different from that of traditional materials. This is influenced by the type of matrix and the thermal treatment applied to its polymerization [9-13].

Today the manufacturers and designers of composite material products are not addressing the problem correctly at the design stage. They try to reproduce metal parts and make them from composite materials which represent a totally uninspired approach that greatly reduces the composite materials benefits. This approach is known as "Black Metal Pieces" and the result is a black carbon fiber piece made in the same shape as the metal parts. This approach doesn't make use of the composite material benefits, thus, there is a need to implement a new concept from the product design phase, namely the specific design of the composite parts.

In this paper, the authors present the design problems of creating a front hood for a small electric car designed for urban traffic. Starting from the typical static loads cases presented in the literature to determine the hood stiffness, the authors propose innovative solutions for the development of the material and its structure. In the case of a carbon fiber / epoxy reinforced composite material, the hood becomes stiffer using a sandwich structure. The ANSYS Workbench software solutions [14] are used to model the composite material architecture and to evaluate the stiffness behavior of the part.

2 E-car concept 96BG

The 96BG concept represents a small electric car designed to run in crowded cities. The car is intended to be used for various day-to-day activities within the city and it has four seats in order to accommodate a wide range of users. Design of the car is illustrated in Figure 1 and is adopted for the FRP material solutions.

Fig. 1. E-car concept 96BG.

Figure 2 presents the designing procedure of the classic metal hood (on top of the figure) and the composite developed hood (on the bottom of the figure). For the composite hood, the ribs reinforcements on the back side used in the metallic design are eliminated and they are replaced by a composite structure.

The border of the hood is a like a frame that continues with an inner area having a different layup of the composite structure, in order to increase the stiffness of the hood without adding some separate glued parts. The cross-sectional view of the hood made of carbon/epoxy with respect to a Cartesian system is presented in the Figure 3. that illustrates the above described concept of the composite architectural layout.

 DOI: 10.1051/matecconf/201713708001

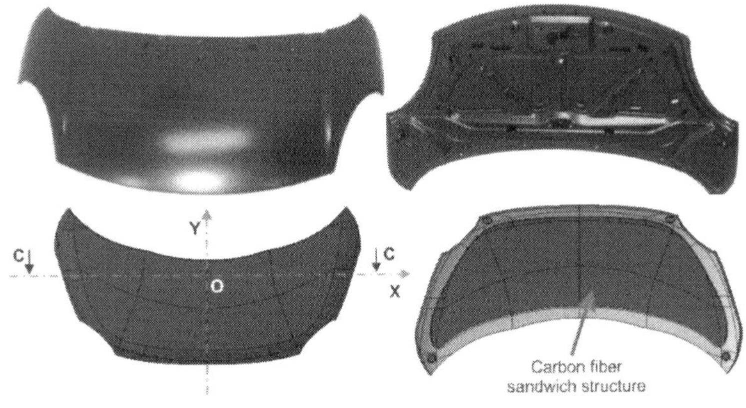

Fig. 2. Front hood design: up sheet metal, down FPR.

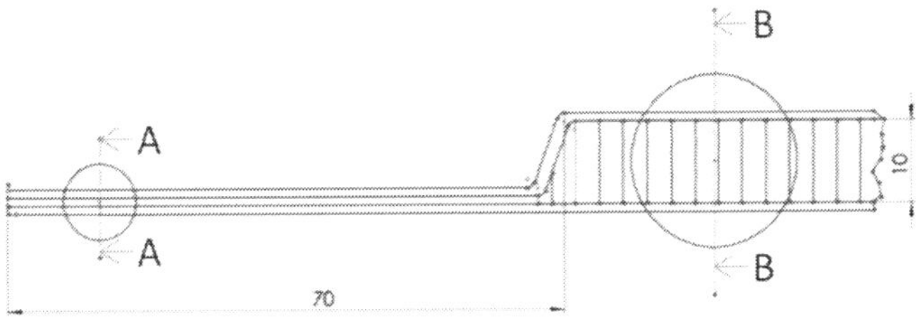

Fig. 3. Section view of the layers on the X axis.

3 FRP material designs

The edge of the car hood consists of three types of laminated structure along a length of 70 mm. The laminated structures architectural layout is illustrated in detail in the A-A section. In the Figure 4 are presented the layers distribution of the proposal composite laminates. For the composite laminate denoted with ***1*** in Figure 4, the following architectural layers were chosen: [90/0±45_2/90/0]. For a 0/90 ply a Twill structure type of 200 g/m^2 with a 0.25 mm thickness/layer and for the ±45 a biaxial fabric of 300 g/m^2 with a thickness of 0.3 mm/layer have been used. Another possible architectural arrangement for laminate denoted by ***1*** could be the following: [90/0±452/90/0]. For both 0/90 and ±45 plies a Plain fabric of 90 g/m^2 of 0.1 mm thickness / layer was used and these layers are the first one arranged in the die and they represent the outer part of the bonnet.

On the interior, both on the border and the entire circumference of the bonnet the layers denoted in Figure 4.a. by ***2*** is applied. The layers configuration is: [±45/0-902/±45]. For the stratified layer denoted ***3*** in the Figure 4.a.-b., the following architectural layout was used: [±45/0-90]. For the 0/90 a Twill type of 200 g/m^2 of 0.25 mm thickness/layer was used and for the ±45 ply a biaxial fabric of 300 g/m^2 with a thickness of 0.3 mm/layer. Another architectural arrangement for layers denoted by ***3*** could be: [90/0±45/90/0]. For 0/90 and ±45 plies a plain fabric of 90 g/m^2 with 0.1 mm thickness was chosen.

In Figure 4.b. a detail is presented for the center of the sandwich structure of the bonnet. The laminated marked as ***1*** and ***3*** remain unchanged, and for layer ***4*** a Nomex honey comb with a thickness (height) of 10 mm and a diameter of 5 mm is used. For the points of interest (fastening of the hinges, the fasteners, and the support points, layers were superimposed by the biaxial fabric of 300 g/m^2, 8 layers, in this area the sandwich structure was eliminated.

DOI: 10.1051/matecconf/201713708001

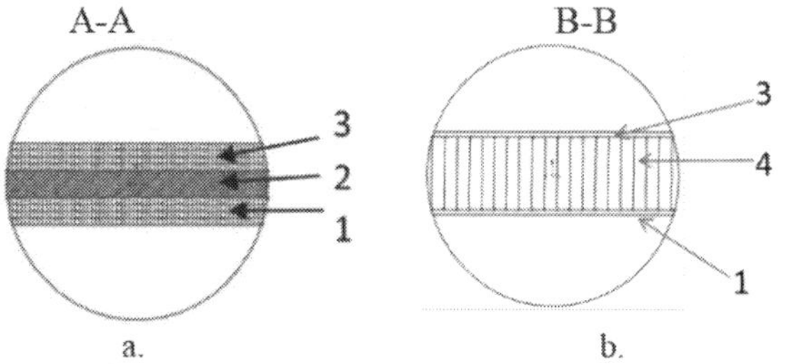

a. b.

Fig. 4. Details of the composite layers.

The staking sequence of the layers and the type of the reinforcement layers are selected to ensure a balanced structure to have on X-Y direction closed mechanical coefficients. In the same time, the proposed laminate is selected to avoid the material contraction in different directions. The fibers are applying to 0-90° and ±45°.

4 Numerical simulations

The finite element model contains all geometrical components of the above described hood. The simulation was realized using ANSYS Workbench software and the ACP tool for layup modelling of the composite materials. The composite material used in the simulation was defined as in the manufacturing process. The orthotropic material proprieties of the fabric epoxy carbon woven being those provided by the software material database (Epoxy Carbon Woven (230 GPa prepreg). The two analyzed cases differ only in stacking sequences and their thickness for the layer 1 and 3 as presented in figure 3. The shell elements are used where the number of the elements was 2800. The bases for the analysis are the established static load cases [15-18] to determine the longitudinal, transversal and torsional stiffness. The hood is bended in cases of the lateral and transversal stiffness tests and distorted in the torsional stiffness test. The loads are applied in vertical direction and have for each load case the intensities shown in Fig. 5.

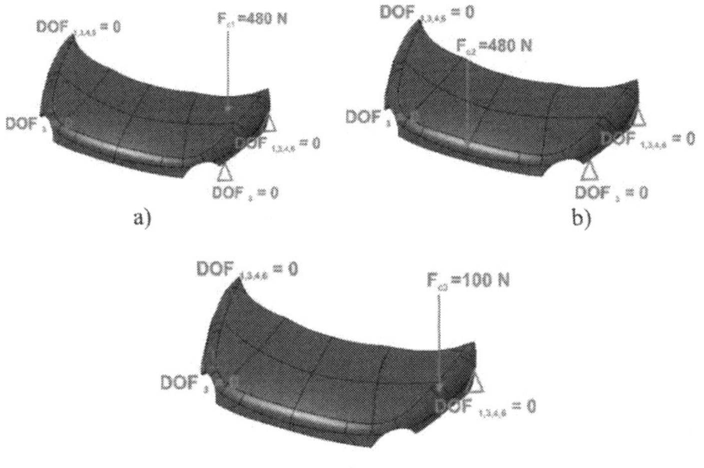

c)

Fig. 5. Load cases for: a) lateral stiffness, b) transversal stiffness and c) torsional stiffness.

In all three tests, the hood is mounted in design position and constrained in the supporting points, the hinges in the rear hood area and the buffer points in the front area, respectively. The assessment criterion is the total deflection of the load application point on the outer panel. Fig. 6 presents the total displacement distribution for investigated load cases.

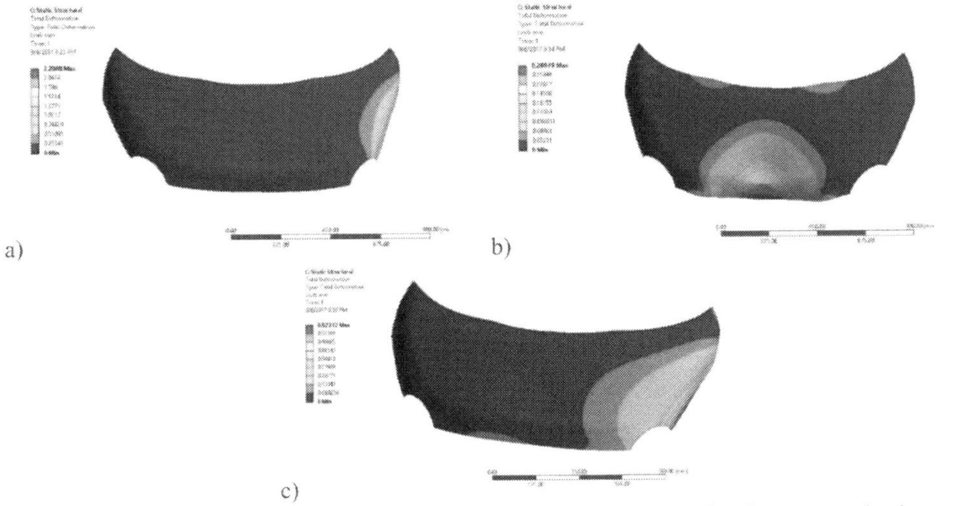

Fig. 6. Displacement distribution along the force direction obtained for first composite lay-up: a) lateral stiffness, b) transversal stiffness and c) torsional stiffness.

The obtained values give an indication about the hood stiffness in comparison with a similar one made from steel and ensure in this stage the proper composite and structural design. The stiffness results for these three cases of material are presented in table 1.

Table 1. The numerical stiffness of the hood in case of different materials.

Stiffness K=F/δ [N/mm]	Lateral [N/mm]	Transversal [N/mm]	Torsional [N/mm]	Mass [kg]
Case I Carbon layers by Twill **200g/m²**	209,6	1655,1	160,5	2,38
Case II Carbon layers by Plain **90g/m²**	80,1	861,3	66,6	1,41
Case III from **Steel**	39,5	406,9	23	5,6

A similar steel hood (Figure 7) made of structural steel sheets of 0.5 mm thickness, has been investigated to check and validate if the of the proposed design of the composite structure is a proper one in terms of strength and stiffness.

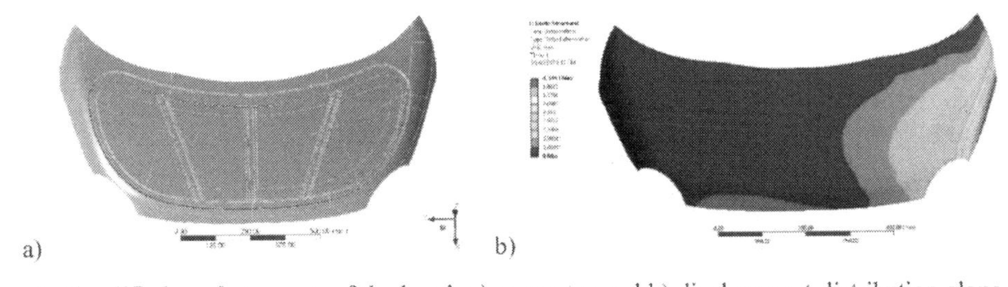

Fig. 7. Simplified steel structure of the hood: a) geometry and b) displacement distribution along the force direction for the load case torsional stiffness.

Comparison between experimental and numerical stiffness of the structure is necessary to be performed in the next step and consists future work.

Comparatively with the stiffness values calculated for a similar steel hood (lateral – 39.5 N/mm; transversal – 406.9 N/mm and torsional – 23 N/mm) we obtained higher values even for the lighter structure containing Plain fabric with $90g/m^2$ (lateral-80.1 N/mm;

transversal - 861 N/mm and torsional - 67 N/mm). Thus, the composite structure proved to be two times minimum stiffer than the steel structure.

5 Conclusions

For the analyzed hood having different composite structural design (Twill 200g/m^2 vs. Plain 90g/m^2), a decrease in lateral stiffness of 60%, in transversal stiffness of 45% and in torsional stiffness of 56% was obtained. If we analyses the weight of the hood in the two cases we obtained a decrease of 41%. The mass decrease of 2.4 times comparing with the steel hood. The first structure Twill 200g/m^2 is stiffer compared to the second one (Plain 90g/m^2), for both cases the obtained stiffness values are higher than the stiffness of a similar steel hood. This validates the proposed architecture of the composite material designed to fabricate a light structure of a car bonnet. The front hood solution adopted is new and specific for FRP design. Both configuration layers studied have been numerically analyzed under typical loads in order to determine the lateral, transversal and torsional stiffness.

This work was supported by a grant of the Romanian National Authority for Scientific Research and Innovation, CNCS/CCCDI–UEFISCDI, project number PN-III-P2-2.1-96BG-2016-0210, within PNCDI III.

References

1. Ch. Decolon, Structures composites, Hermes, Paris, (2000)
2. L.P. Kollar, G.S. Springer, Mechanics of Composite Structures, Springer, Cambridge, (2003)
3. I.A., Popan, A.I. Popan, Acta Tehnica Napocensis, **60**, II, 251-254 (2017)
4. J. Zhang, K. Chaisombat, el al, Materials and Design **36**, 75–80 (2012)
5. P. Beardmore, C.F. Johnson, Compos Sci Technol, **26**, 251–81 (1986)
6. H.A. Al-Qureshi, J. Mater Process Technol, **118**, 58–61 (2001)
7. P. Hreha, A. Radvanska, et al, Metrol. Meas. Syst., **22**(2), 315-326 (2015)
8. N.F. Betzler, C. Slater, M. Strangwood, Sports Eng., **14**(1), 27–37 (2011)
9. Y.I. Tsai, E.J. Bosze,et al., Compos Sci Technol, **69**(3–4), 432-437 (2009)
10. P.J. Liotier, A. Vautrin, J.M. Beraud, Compos A Appl Sci Manuf, **42**(4), 425–437 (2011)
11. G.M. Krolczyk, P. Nieslony, Krolczyk J.B., Samardzic I., Legutko S., Hloch S., Barrans S., R.W. Maruda, J. Measurement, **70**, 203-213 (2015)
12. I.A. Popan, N. Balc, A.I. Popan, MSE, MATEC Web Conf., **121** (2017)
13. E.C. Botelho, L.C. Pardini, M.C. Rezende, J Mater Sci., **41**(21) (2006)
14. ANSYS Workbench, Release 16.0, ANSYS, Inc., Canonsburg, USA (2015)
15. M. Hamacher, R. Wohlecker, L. Ickert, Abaqus Users' Conference, May 19-22, Newport, Rhode Island, USA, 1-16 (2008)
16. C.E. Torricelli, L. Splendi and M. Pettazzoni, in Proceedings of the World Congress on Engineering WCE 2011, July 6 - 8, London, U.K (2011)
17. N. Bhaskar, P. Rayudu, International Journal of Current Engineering and Technology, **5** (2015)
18. S. M. Darwish, S. M. Elseufy, A. Ahmad, Journal of Management & Engineering Integration, **7**(62) (2014)

Humidity influence on the adhesion of SU-8 polymer from MEMS applications

Corina Birleanu[1,*], *Marius* Pustan[1], *Rodica* Voicu[2], *Florina* Serdean[1] and *Violeta* Merie[1]

[1]Technical University from Cluj-Napoca, Micro and Nano System Laboratory, 400641, B-dul Muncii 103-105, Romania
[2]Laboratory of Modeling, Simulation and CAD, National Institute for R&D in Microtechnologies - IMT Bucharest, 126A Erou Iancu Nicolae Street, 077190, Bucharest, Romania

Abstract. In this paper, the adhesion behaviors of SU-8 polymer thin film from MEMS application were investigated as a function of relative humidity. The adhesion test between the AFM tip and SU-8 polymer have been extensively studied using the atomic force microscope (AFM), for a relative humidity (RH) varying between 20 and 90%. The samples for tests are SU-8 polymers hard baked at different temperatures. The hard bake temperature changes the tribo-mechanical properties of polymers. The paper reports the measurements and the modeling of adhesion forces versus humidity in controlled ranges between 20 to 90%RH. To investigate the effect of relative humidity on adhesion for SU-8 polymer hard baked we used an analytical method which encompasses the effect of capillarity as well as the solid-to-solid interaction. While the capillary force expression is considered to be the sum of the superficial tension and the Laplace force for the solid-solid interaction is expressed by the Derjagin, Muller and Toropov (DMT) model of solids adhesion. The analytical results obtained are in accordance with those obtained experimentally.

1 Introduction

Microsystems field is one of the most interesting and exciting fields that present interest in science and technology today. Many physical phenomena that are neglected at macro scale acquire dominant characteristics at micro and nano scale. These phenomena exploited in an intelligent way lead to the manufacture of Micro-Electro-Mechanical-Systems (MEMS) devices with operation principles completely different than the macro scale counterpart system [1].

These micro-scale devices are commonly made from silicon (Si) or some cross-linked polymers such as SU-8. Though Si has many attractive properties related to hardness, stiffness, fatigue and the ease of mass-fabrication by the semiconductor fabrication technology, Si is very poor in tribological performances [2]. In a dynamic state when silicon is in contact with itself or other similar material, it exhibits high friction, high initial and dynamic friction, and considerable wear [2]. The main failure mechanisms in MEMS are stiction, friction and wear. When two surfaces are close to each other, the attractive

* Corresponding author: Corina.Barleanu@omt.utcluj.ro

surface forces (capillary, van der Waals and electrostatic forces) play an important role in device performance [2, 3].

SU-8 is an epoxy-based negative photoresist material near UV photoresist based on EPON SU-8 epoxy resin that has been first developed and patented it compositions by IBM as far back as 1989 [4, 5]. Besides the many advantages of SU-8 photoresists over available UV photoresists, an import advantage is its use in the fabrication of MEMS. Because of its aromatic functionality and highly cross-linked matrix, the SU-8 resist is thermally stable and chemically very inert [6].

Moreover, the SU-8 polymer is insoluble in water, biocompatible and after polymerization has chemical durability. Also SU-8 polymer allows to obtain surfaces with the very best aspects of roughness, which is a basic need for designing reliable functional MEMS mechanical structures with good mechanical and tribological properties [7]. The SU-8 polymer parameters and data for the mechanical and tribological properties published in the literature vary substantially as a function of the processing conditions, testing method as well as the operating parameters [8].

In this paper, using an AFM equipped with an environment chamber, the nano adhesion behaviors of SU-8 polymer hard baked at different temperatures were systematically studied as functions of relative humidity, between 20 - 90%RH and the corresponding mechanisms were discussed. A full understanding of the adhesion behaviors of SU-8 polymer is an essential issue and this investigation can provide further insights into minimizing the potential failure of MEMS in humid environment. The experimental tests are performed under different testing condition in order to investigate the humidity effect on the adhesion phenomena of such MEMS material.

2 Materials and methods

2.1 Thin-film preparation

The SU-8 material used in this paper as testing material is a negative, epoxy-type, near-UV (350-400 nm) photo-resist in which eight epoxy groups are attached to a single molecule.

Silicon (100) wafers were used as substrates for the SU-8 polymer films.

SU-8 2015 polymer (Micro-Chem) was spin on the wafers using a spinner with thicknesses of around 10 μm. The SU-8 samples were hard baked at different temperatures to complete the crosslinking of the polymer. Four samples of SU-8 polymers were selected for experimental characterizations with different hard bake temperature such as: 125°C, 165°C, 195°C and 215°C. The final film thickness was about 10 μm. These samples are experimentally investigated under different testing humidity in order to determine their adhesion behavior.

The experimental result for contact angle θ of water vapor on the SU-8 surface was 65° ± 4° at 22°C ambient conditions. Contact angle measurements were made with a Theta Lite Optical Tensiometer TL 101 with a USB 2 digital camera. Contact angles were determined with OneAttension software by sessile drop method using Young-Laplace. Five measurements were made on surface under investigation.

2.2 Atomic force microscopy

The structural and surface morphology and adhesion forces between SU-8 samples and AFM (Atomic Force Microscopy) cantilever were determined at different relative humidity varying from 20 to 90%.

Surface roughness measurements were conducted by atomic force microscopy using AFM - XE 70 in tapping mode, to avoid the scratch and deterioration of the polymeric surface. The AFM probe used in the surfaces characterization is the PPP-NCHR AFM probe with unique features like: highly doped silicon to dissipate static charge, Al coating on detector side of cantilever, with the following parameters: the thickness is 4μm, the width is equal to 30μm, the length is 125μm, the force constant is 42N/m and the resonant frequency is 330 kHz.

The adhesive effect between the AFM probe (Si_3N_4) and the samples were evaluated using the force spectroscopy mode of the apparatus, the so-called force-displacement measurements, in a controlled environment (clean room). The tip used was Si_3N_4 AFM tip HQ - NSC 35/Hard/Al BS with a nominal value of the spring constant k = 16 N/m and radius < 20nm. The load applied for all measurements was 50 nN. The set point was 10 nN while the scan rate was 0.75 Hz. Measurements were performed over a range of processing RH between 10 to 90% and a contact time up to a few seconds.

Such measurements provide correct information about the surface energy and the adhesion only if the contact between the AFM tip and the cantilever surface remains in the elastic range. Literature indicates that at 20°C and 35%RH, when the load is varied between 20 nN and 250 nN, there is no significant change in the adhesion force [9].

For all experimental investigations, the AFM was enclosed in an Environmental Chamber (Glove Box). Relative humidity within the glovebox was controlled by a Humidity Control System. The water used to humidify the chamber was distilled and deionized. Humidity levels were monitored continuously by a radiosonde. Prior to measurement, the humidity was allowed to reach equilibrium as indicated by a constant value. After that, going from low to high humidity and vice versa to dry conditions was achieved in less than 10 min. The relative humidity is changed, from 10% up to 90% in steps of 10 %. The air flow did not significantly affect AFM operation and all the common adjustments were readily accessible via the glove ports. Operating over extended humidity of 80% RH was avoided because of the possible damage to the AFM scanner. Each measurement was repeated many times in order to improve the accuracy of the experimental results and the mean value was computed.

3. Results and discussions

The paper is motivated by tribological problems observed in MEMS / NEMS devices. It is postulated that a fundamental understanding of nanoscale tribological behaviour can create guidelines for the design and manufacture of reliable MEMS devices.

3.1 Topography

To analyse the adhesion force between SU-8 surface and AFM tip it is important to measure the roughness value of the contact surfaces.

For evaluating the surface topography of the samples was utilized the Non-Contact AFM mode (NC-AFM) based on the attractive inter-atomic force between the tip and the sample surface. In AFM, a sharp tip at the end of a cantilever is vibrated near the surface of a sample and was scanned over 3μm × 3μm surface area.

Surface roughness was evaluated by flattening the images (second order). A second-order polynomial plane correction was carried out in order to reduce waviness due to scanner bow and variations of the film thickness. Then a line-wise leveling was performed to reduce a possible line-by-line repetition error. As surface roughness parameters we qualified the root mean square (RMS or R_q) surface roughness provided by the XEI software. R_q is more sensitive to occasional highs and lows, making it a valuable complement to the arithmetic average height of roughness-component irregularities from the mean line R_a. For instance, the acquirements of AFM images of morphologies surface and roughness surface for upper side of SU-8 sample surface hard baked at 125°C is the R_q = 0.409 nm and R_a = 0.325 nm.

The AFM images for each of the contact surfaces, together with the statistical parameters provided for roughness by the software used for interpreting the experimental data (XEI Image Processing Tool for SPM) are given in Fig. 1 (a) and (b).

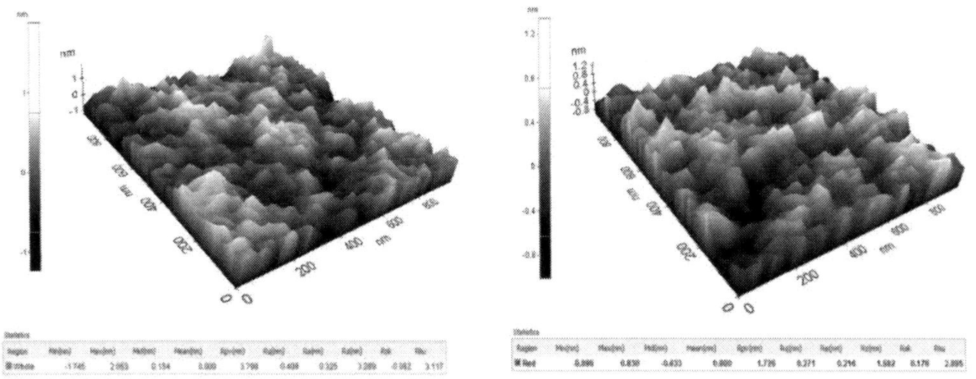

Fig. 1. Topographic images in NC-AFM mode of (a) SU-8 hard baked at 125°C; (b) SU-8 hard baked at 215°C.

The roughness parameters of investigated of SU-8 polymers decrease as the hard bake temperature is increased from 125°C to 215°C. The R_a roughness is 0.325nm for SU-8 polymer hard baked at 125°C, it decreases to 0.296 nm for 165°C hard baked temperature, 0.268 for a hard bake temperature of 195°C and 0.216 for the polymer baked at 215°C.

Reproducible roughness measurements require a well-defined set of operating conditions and parameters such as the AFM peak radius, pixel size, scanning speed, and width. In order to compare the results obtained between different batches of samples or different laboratories, it is important to know these parameters.

3.2 Adhesion force

3.2.1 Experimental evaluation of adhesion force.

The experimental determination of the pull-off forces between the AFM tip and SU-8 surface at different hard bake temperature was conducted using the spectroscopy in point mode of the AFM.

Figure 2 shows the pull-off forces measured for different humidity values for the SU-8 thin film with at different hard bake temperature at ambient temperature, 22°C. Each reported value is the average of five measurements taken at different regions on the surface of each sample.

Based on such experimental investigations as those in figure 2, the values of the adhesion force for the SU-8 hard baked at 125°C and 215°C investigated at different humidity are shown in figure 3.

The increase of the pull-off forces with the RH is more pronounced in the 30% to 70% range. For a RH less than 30% the thickness of the water film on the SU-8 surface is too thin to form a capillary neck with the tip asperity. At around 70% RH, the adhesion force shows a maximum value, but if the RH is further increased the pull-off force shows a decrease. When the water film reaches a minimum required thickness at 30% RH, a water bridge forms between the asperity and the surface, thus leading to a sudden increase of the capillary force. Binggeli and Mate [10] explained that the decrease in adhesion force at high relative humidity as a result of the interactions between capillary forces and forces related to the chemical bonding of the water molecules [11]. As a result, one can distinguish between three regimes of adhesion force versus humidity: a regime dominate by solid-to-solid interaction, van der Walls regime at low RH (RH < 30%), a regime dominated by the capillary force (30% <RH< 70%) and a high humidity regime (RH > 70%).

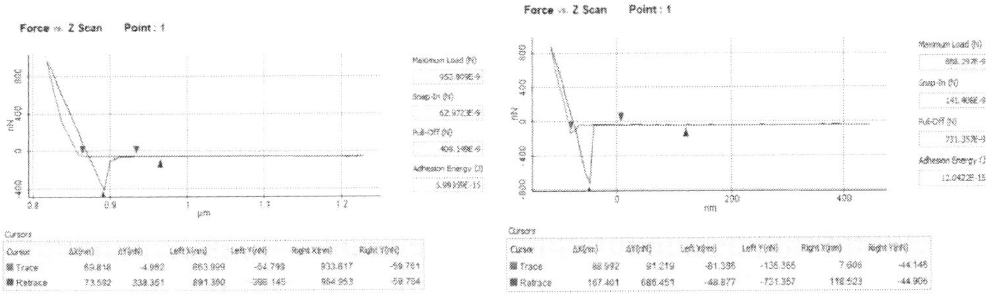

Fig. 2. Pull-off force between AFM tip and SU-8 hard baked at 125oC and test at 22oC and different RH values: 20 % and 60 %.

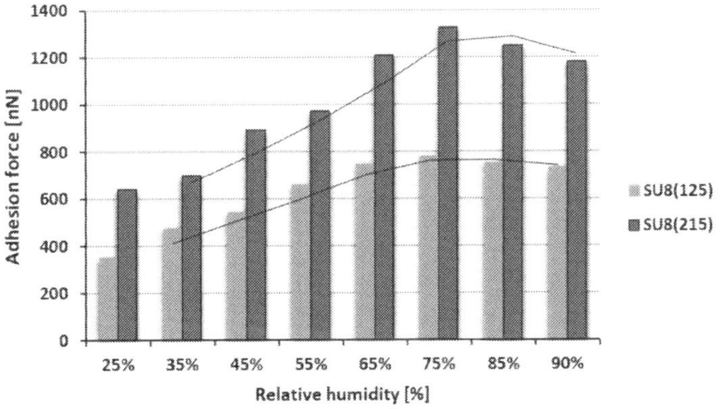

Fig. 3. Adhesion forces versus relative humidity [RH%] measured between the SU-8 hard baked at 125°C and 215 °C and AFM tip.

Increasing humidity generally promotes increases water adsorption leading to increased contributions from capillary bridges between the tip and SU-8 thin film surface towards the pull-off force. During measurements of the pull-off force no deteriorations in surface samples were observed and there was no significant change in the radius on the AFM tip. Therefore, we can say that solid adhesion component is the main parameter contributing to the pull-off force.

3.2.2 Analytical interpretation of adhesion force

In the presence of an annulus of pure liquid the adhesion force P_{adh} can be written taking into account the influence of four attractive forces acting, which are van der Waals force, electrostatic force, capillary force and others.

$$P_{adh} = P_{vdw}(d) + P_{cap}(\Delta p) + P_{ten}(\lambda_{LV}) + P_{others} \quad (1)$$

Where: $P_{vdw}(d)$ is the force due to van der Walls interaction with separation d between surfaces; $P_{cap}(\Delta p)$ is the capillary force based on Laplace pressure (Δp); $P_{ten}(\gamma_{LV})$ is the force due to liquid-vapor surface tension (γ_{LV}) and other interactions P_{others}. In most practical cases, when a capillarity bridge was formed the capillary force can be expressed as:

$$P_{cap}(\Delta p) = 4\pi R\gamma_{LV} \quad (2)$$

Most experimental results indicate that predominant mechanism of adhesion is given by van der Walls forces, solid-to-solid interactions. Recent research shows that capillary forces are a factor that contributes to the phenomenon of adhesion. Capillary forces created by natural air humidity further increase the adhesion force [12].

The van der Waals forces (P_{vdw}) which act between two surfaces and result from the interaction between the instantaneous dipole moments of atom are given by [13, 14]:

$$P_{vdw}(d) = \frac{AR}{6d^2} = 4\pi R\gamma_{SL} \quad (3)$$

where: A is the Hamaker constant of the liquid, d is the minimum distance between surfaces and γ_{SL} is the solid-liquid surface tension.

The force due to the other interactions P_{others} can be neglected.

Assuming that both asperities of the contact surfaces are spherical, a single asperity contact occurs when the adhesive force acting on the asperity deforms it by at least the RMS roughness and is close to the DMT model. Therefore, we applied the DMT theory for the contact between solids. The surfaces in contact are assumed to be rigid and the adhesion is estimated based on folowing equation:

$$P_{adh}^{DMT} = -2\pi W_{adh} R \quad (4)$$

In this equation R is the reduced radius of contact, which is calculated as $R = (1/R_1 + 1/R_2)^{-1}$; R_1 and R_2 are radii of contacting surfaces; R_1 = 18 nm and R_2 = 0.304 nm. Because the differences between the two rays are significant, more than 18 times in the calculations R are considered to be equal with AFM probe radius R_1 (18 nm). The work of adhesion of two flat surfaces W_{adh} can be calculated using the following equation [15]:

$$W_{adh} = -\frac{A}{12\pi d^2} \quad (5)$$

A is Hamakar constant which depends on the properties of the two materials and the intervening medium. Typical value for A for SU-8 and AFM tip typically made of silicon nitride are $A_{air} = 6.5\text{x}10^{-21}$ J in the air [16] and $A_{water} = 1.8\text{x}10^{-21}$ J in the water [17]. In this case when two surfaces are in contact in a humidity atmosphere the absorbed and condensed water molecules are present between these surfaces separated at a gap with thickness of molecular distance.

The work of adhesion of the two surfaces in contact separated by an atomic distance $d \approx$ 0.4 nm [15] is approximately equal to –43.1 mJ/m^2 in the air and -11.93 mJ/m^2 in the water.

Assuming tip radius R is 18 nm, the DMT adhesion forces in the air and the water for a single contact based on (6) and (16) are $P_{DMT}^{air} = 487.45nN$ and $P_{DMT}^{water} = 134.92nN$, respectively. If the humidity increases from 0 to 100% the DMT adhesion force will take a value between P_{DMT}^{air} and P_{DMT}^{water}. For an intermediary humidity will be used the Wan et al.

[18] approximation method. The total adhesion force based on analytically interpretation is calculated as the sum of DMT and capillary adhesion.

To take into account the analytical capillarity contributions to pull-off forces for SU-8 samples, the contact angler for the SU-8 surface and AFM tip is set as a value of $\theta = 65°$. The contact radius $R = 18$ nm, the ambient temperature $T = 22°C$, the surface tension $\gamma_{LV} = 73$ mJ/m^2 and molecular volume of water $V = 1.8x10^{-5}$ m^3/mol [14].

In figure 4 is presented the variation of pull-off forces, capillary forces, surface tension forces and solid-to-solid forces based on DMT model in relation with relative humidity, RH.

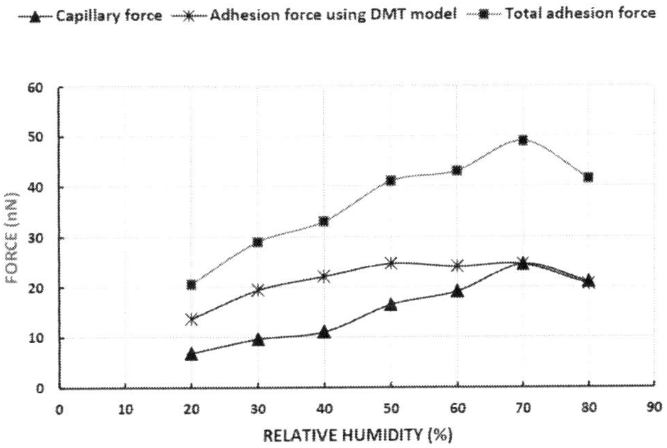

Fig. 4. The variation of the capillary force, the adhesion force based on DMT model force and the total adhesion force with the relative humidity for a single contact between the SU-8 hard baked at 125°C and AFM tip.

The total capillary force increases to the 24.5 nN at 70% relative humidity. The DMT adhesion force decreases with an increase a relative humidity. The DMT adhesion force occupies a large value in total adhesion. The total adhesion force increases slowly to 49 nN at a value of relative humidity of 70%.

Due to the force applied at the thin film from SU-8, the contact between the tip and the surface is along a line segment. Studying the contact line segment, it was established that there were 15 contact points which were modelled as the contact between the sphere and a plane due to the roughness of SU-8 and AFM tip.

Then the total adhesion was computed according to [19] and the results are presented in Fig. 5.

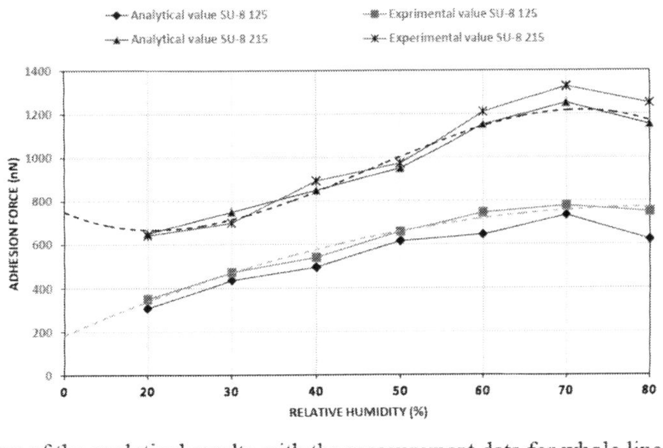

Fig. 5. Comparison of the analytical results with the measurement data for whole line contact between the AFM tip and SU-8 hard baked at 125°C and 215°C, respectively.

4. Conclusions

This paper presents the influence of the relative humidity on the values of the pull off forces under normal loading based on analytical and experimental studies in the case SU-8 polymers hard baked at different temperatures. The salient conclusions arising from this work are summarized below:

Accurate determination of pull-off forces at the molecular level leads to a significant insight into the areas of interest such as SU-8 material.

In order to investigate the effect of capillarity due to the presence of humidity range on adhesion (pull-off forces), we considered the capillarity force consisting of the Laplace force and the surface tension force and the solid-to-solid interaction by DMT theory, starting from a single contact. In [20] is suggested that single asperity contact occurs when the adhesive force acting on the asperity deforms it by at least the RMS roughness. Then, using the model proposed by the authors in [19] the results were extrapolated to the entire length of contact between AFM tip and samples. In interpreting the data presented in the ambient temperature at different values of relative humidity it was necessary to assume a value for Kelvin radius as 0.4 nm and to establish the equilibrium separation.

In conclusion, we used the AFM investigation to determine the pull-off forces on a nanostructure as SU-8 samples. The capillarity forces theory and analytical model used for adhesion forces components gives results which are in good agreement with the experimental data.

This work was supported by a grant of the Romanian National Authority for Scientific Research and Innovation, CCCDI - UEFISCDI, project number 22/2016, within PNCDI III, ERA-MANUNET-II-Robogrip project "Microgrippers as end-effectors with integrated sensors for microrobotic applications (ROBOGRIP)".

References

1. M. Pustan, G. Ekwinski, Z. Rymuza, Int J Mater Res, **98** (5) (2007)
2. N.B. Tay, M. Minn, S.K. Sinha, Tribol Lett, 2011. **44** (2) (2011)
3. W.M. Van Spengen, R. Puers, I. De Wolf, J Adhes Sci Technol, **17** (4) (2003)
4. J. Gelorme, R. Cox, S. Gutierrez, US Patent No. 4,882,245. Washington, DC: US Patent and Trademark Office (1989)
5. R. Feng, R.J. Farris, J Micromech Microeng **13** (1) (2002)
6. R. Martinez-Duarte, M.J. Madou, *Microfluidics and Nanofluidics Handbook* (2011)
7. J Hammacher, A Fuelle, J Flaemig, J Saupe, Microsyst Technol, **14** (9-11) (2008)
8. J Lee, H Shin, S Kim, S Hong, J Chung, Jpn J Appl Phys 2003. **42** (10A) (2003)
9. Y. Ando, Wear, **238** (1) (2000)
10. M. Binggeli, C. Mate, Appl Phys Lett, **65** (4) (1994)
11. M Farshchi-Tabrizi, M Kappl, Y Cheng, J Gutmann, Langmuir, **22** (5) (2006)
12. T.W. Kim, B. Bhushan, J R Soc Interface, **5**(20), (2008)
13. A Torii, M Sasaki, K Hane, *Proceedings An Investigation of Micro Structures, Sensors, Actuators, Machines and Systems, IEEE* (1993)
14. D. Xu, K.M. Liechti, K. Ravi-Chandar, J Colloid Interface Sci, **315** (2) (2007)
15. J.N. Israelachvili, *Intermolecular and surface forces*, Academic press (2011)
16. S. Matope, A. Van der Merwe, Y. Rabinovich, S. Afr. j. ind. eng, **24** (2) (2013)
17. H.-J. Butt, B. Cappella, M. Kappl, Surf Sci Rep, **59** (1) (2005)
18. K.T. Wan, D.T. Smith, B.R. Lawn, J Am Ceram Soc, **75** (3) (1992)
19. M. Pustan, C. Birleanu, C. Dudescu, Appl Surf Sci (2016)
20. D. Tabor, J Colloid Interface Sci, **58** (1) (1977)

Characterization of a thermoforming composite material made from hemp fibers and polypropylene

Emilia Ciupan[1,*], *Lucian* Lăzărescu[1], *Ioan* Filip[2], Cornel Ciupan[1], *Emilia* Câmpean[1], *Ioan* Cionca[2] and *Emanuela* Pop[1]

[1]Technical University of Cluj-Napoca, Memorandumului Street, No. 28, 400114 Cluj-Napoca, Cluj County, Romania
[2]TAPARO SA, 43562 Village Borcut, No. 98, Maramures County, Romania

Abstract. The paper refers to a composite material developed for manufacturing thermoformed products with applications in furniture making, automotive industry etc., a method and machinery for manufacturing the material in unwoven form. From this material, Research and Development Department of TAPARO SA has designed and built a series of furniture components. The composite material made of a thermoplastic fibrous component and hemp fibre component, the way of obtaining and the properties of the thermoformed material presented in the paper are necessary in the process of designing and optimizing the parts.

1 Introduction

The purpose of the paper is to present the mechanical properties of a composite material made by TAPARO SA Company and used for thermoforming of different furniture parts. The majority of the padded products contain a resistance structure in the form of a timber frame. Wood is an excellent material in terms of functional, environmental and aesthetic but its overexploitation has an impact on the environment, and most of the countries have adopted legislation regarding the forestry exploitation [1, 2]. Therefore manufacturers that have large series of articles containing pieces of wood, including furniture manufacturers, seek alternatives of wood with other recyclable products that provide productivity and overall cost benefits of the product. In this regard were developed a series of composite materials made of natural fibers and thermoplastics, which enable the thermoforming manufacture of the parts previously made of wood. Replacing wooden products involves redesigning components, but the optimization of the project requires a series of physical and mechanical properties of the product: tensile strength, elastic modulus, shear modulus, mass density, Poisson' ratio etc. These properties vary depending on the component fibrous mixture and the thermoforming process, but the literature offers no concrete values.

Panaitescu [3] studied the thermal stability of hemp fibers when treated with different silane treatments. Polypropylene copolymer BJ380MO was mixed with maleic anhydride

* Corresponding author: emilia.ciupan@mis.utcluj.ro

grafted polypropylene, Polybond 3200, and also with short cut hemp fibers with the length of about 2 mm. Using an electrically heated press (Dr. Collin, Germany) were moulded at the dimension of 100 x 100 x 1 mm and then compressed, at a temperature of 180° C.

The samples were noted as PP/HF (polypropylene/ hemp fibers) for the composite with untreated fibers and PP/HF–APS, PP/HF–MPS, PP/HF–GPS and PP/HF–KP for the composites containing HF modified with APS (γ-Aminopropyltriethoxysilane), MPS (γ-Methacryloxypropyltrimethoxysilane), GPS (γ-Glycidoxypropyltrimethoxysilane) and KP (potassium permanganate), respectively. Using a Universal Testing Machine Instron 3382 tensile properties of the composites were determined for five specimens of each sample. The results concluded that when the amount of HF–MPS grown from 0 wt% to 40 wt% the PP reinforced with HF–MPS has an increase of modulus compared to the composites containing untreated fibers. The increase was with 67% in PP/40 wt% HF–MPS and with only 30% when PP was reinforced with the same amount of untreated fibers.
Also, nanoindentation tests were performed. Using the Oliver–Pharr method, hardness and elastic modulus were determined. The results shown different hardening effect of treated HF, better for KP, MPS and MAPP (maleated polypropylene) treatments, intermediate for APS and worse for GPS, similar to macro-level mechanical results. Also, the results showed that TM (tensile modulus of elasticity) increased with 47% and 31% for the composites with MAPP and KP treatments, and with 26% and 37% for NM (nanoindentation modulus).

Z.L. Yan [4] studied the difference in morphology, composition and thermal resistance of noil fibre for the fabrication of injection moulded NFRC (natural fibre reinforced composites). The results were compared with scutched hemp fibre (SHF). The tensile, flexural and impact strengths of composites with both kinds of fibre are compared with different fibre and maleic anhydride polypropylene (MAPP) addition content. Theresa Sullins [5] studied also the effects of the fiber chemical treatment and/or polymer matrix treatment with coupling agent, on the mechanical properties of the hemp fiber polypropylene matrix composites. The study used different fiber length during the processing. Another study that used hemp as raw material was the study of M.R. Bambach [6] how studied the intrinsic compression behaviour of natural fibre composites. For the study, composite plates from flax, jute and hemp fibre-resin were fabricated having the width of 150 mm and 450 mm length, with 2,3 or 4 fiber layers for the flax and jute channels, and 3, 4 and 6 fibre layers for the hemp channels.

Although the literature offers articles on the thermoforming of fibrous composite materials, it does not offer concrete values related to this process. The mechanical properties of this composite material, like specific stiffness and strength, have not been studied sufficiently and the producers of such materials do not provide concrete values.

2 Composite materials and the manufacturing process

2.1 Composite material

The majority of upholstered products have a structure in the form of a wood frame. For ecological reason, a series of composite materials made of natural and thermoplastic fibers have been developed, materials which can be thermoformed so as to replace products made of wood.

Patent KR970008215, "Thermoplastic composite material reinforced with hemp fibers", refers to a composite material made of a thermoplastic reinforced with hemp fibers and filler represented by wood. The thermoplastic is represented by polypropylene, polyethylene, a copolymer of ethylene and polypropylene, a copolymer of acrylonitrile -

butadiene-styrene or simply nylon. The composite is manufactured as sheets used in die-cutting or pellets used in injection moulding.

Another patent FR2781492 „Composite thermoplastic material for use in production of various moulded articles includes hemp fibers of specified dimensions and humidity" refers to a thermoplastic composite which includes hemp fibers of sizes and humidity fit for moulded products. The composite material is formed from thermoplastic material, with a maximum melting point of 200°C, and hemp fibers shorter than 2 mm with a diameter smaller than or equal to 0.2 mm. The patent describes a method of manufacturing the material consisting in melting the thermoplastic and mixing hemp fibers into it.

The disadvantage of the material obtained by the patented method consists of the fact that it has small strength due to the short fibers and is recommended to be used in injection moulding and less for thermoforming.

The composite material for thermoforming, which will be characterized in this work, is made of a thermoplastic fibrous component consisting of 20-60 mm long and 7-16 DEN fine polypropylene fibers representing 40% to 50% of the total material weight, and a plant fiber component which can be hemp who has 70-80 DEN fine and 60 to 100 mm in length and represents 50% of the total material weight.

2.2 The manufacturing process and the machinery

The manufacturing process of the composite material consists of the following operations:
a. taking the plant fibers from the bale and cutting them to lengths between 5 and 100 mm, using a rotating blade chopping machine;
b. simultaneous weighing of the plant fibers resulted from the previous phase in order to obtain a mix for the composite material of which the plant fibers represent 50-60% of the total mass;
c. coarse mixing of the plant and polypropylene fibers and defibering them with the help of a fiber opener with nails;
d. mixing and finely shredding the materials;
e. interlacing the material and reducing its thickness by a factor of 4 to 5
f. pulling and rolling the material in order to make a fabric with consolidated fibers.

The machinery for producing the composite material (Fig. 1) consists of at least two feeding modules, one for the thermoplastic fibers (1) and the other for the plant fibers (2), one module (3) which weighs and feeds correct proportions of each type of fiber, one module (4) for the primary mixing and the coarse defibering, one module (5) for the fine mixing, one module (6) for defibering, one module (7) for interlacing and one module (8) for pulling and rolling the material.

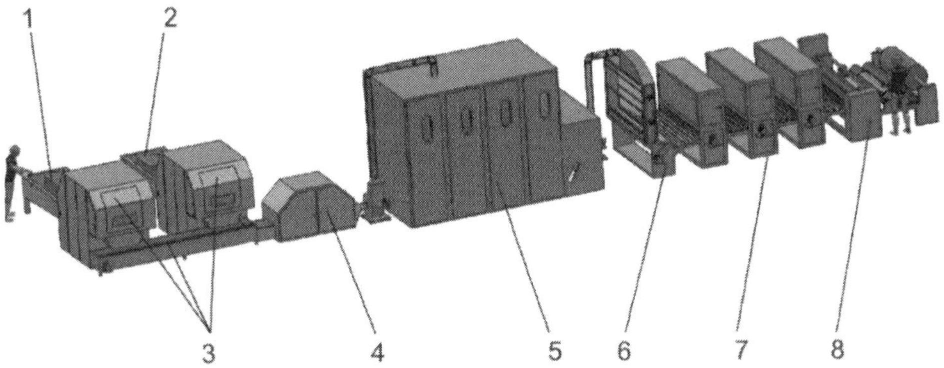

Fig. 1. The modular structure of the machinery for the manufacturing of the composite material.

2.3 The samples

From the nonwoven material, plates of 350x350 mm were cut and then packages were made, as shown in Table 1. With L, the feed direction of the material to the interlacing machines was noted, and with T, a direction perpendicular to it. From the materials presented in Table 1, plates were thermoformed. The thermoformed process was carried out in two stages:

a. Heat the material at 220 °C using a heated plate press.
b. Simultaneous pressing of 5 plates on a 600kN press.

The plates were numbered according to the number of layers and layer placement direction.

Table 1. Making the sheet.

No.	No. of layers	Layout	No. of sheet
1.	3	L-L-L	5
2.	3	L-T-L	5
3.	4	L-L-L-L	5
4.	4	L-T-L-T	5

From the heat-treated plates (Table 1), with a guillotine, 110 samples with the dimensions of 250x25mm were cut (Fig. 2, Table 2).

3 Performing the tests and composite material characterization

3.1 Testing machine

The Zwick Roell Z150 was used to perform the tensile tests. The samples and test methods were performed according to the standard D3039, „Test Method for Tensile Properties of Polymer Matrix Composite Materials".

The machine used is an electromechanical one, having a kinematic axis with a leading screw.

The machine control, the data acquisition, as well as the processing and export of results are done using the computer through a program called testXpert II.

The test was done having the following machine settings:

• grip to grip separation at the start position 150 mm;
• pre-load -0,4 MPa;
• test speed -2 mm/min.

As said before, from the thermoformed plates (Table 1) 110 samples with the dimensions of 250x25mm were cut (Table 2). All the samples from Table 2 were labelled, as shown in Figure 2.

Table 2. Test samples.

No.	No. of layers	Layout of	No. of
1.	3 (L-L-L)	L	10
2.		T	10
3.		D	10
4.	3 (L-T-L)	L	10
5.		T	10
6.		D	10
7.	4 (L-L-L-L)	L	10
8.		T	10
9.		D	10
10.	4 (L-T-L-T)	L	10
11.		D	10

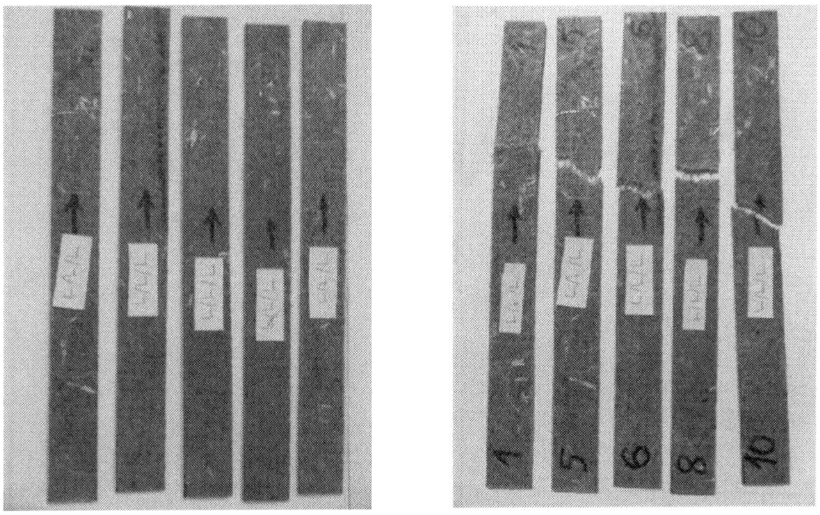

Fig. 2. Samples before and after the tests.

3.2 Results

The results of the tests are shown graphically in the figures from Table 3. In Table 3 the following notations were used: σM - tensile strength, εM – strain at tensile strength, σB – stress at break, εB – strain at break.

Table 3. The stress-strain diagrams

<table>
<tr><th>No.</th><th>No. of layers</th><th>Layout of samples</th><th>Tensile stress versus strain curves</th></tr>
<tr><td>1.</td><td rowspan="2">3 (L-L-L)</td><td>L</td><td>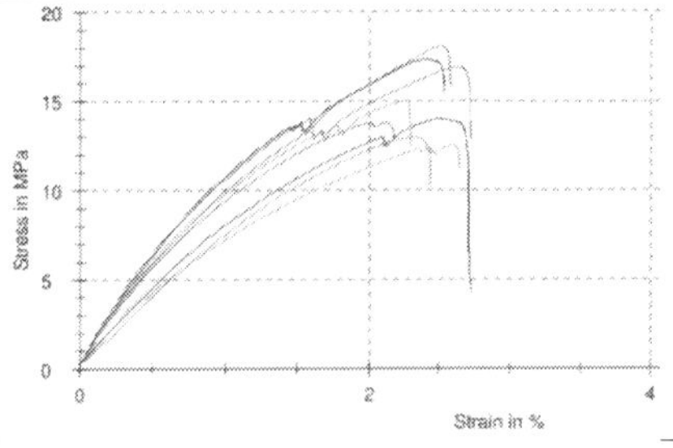
</td></tr>
<tr><td>2.</td><td>T</td><td>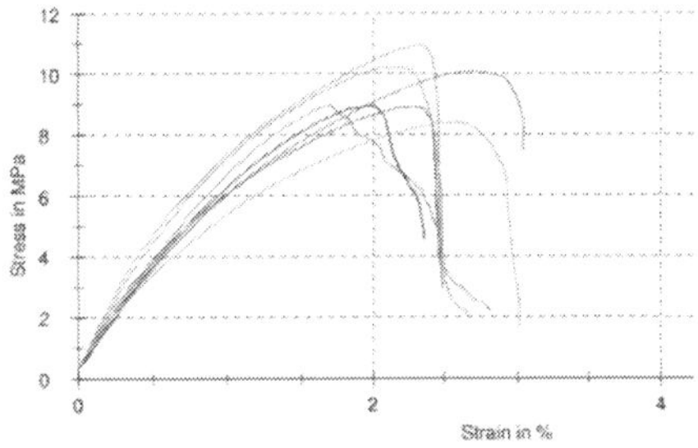
</td></tr>
</table>

3.		D	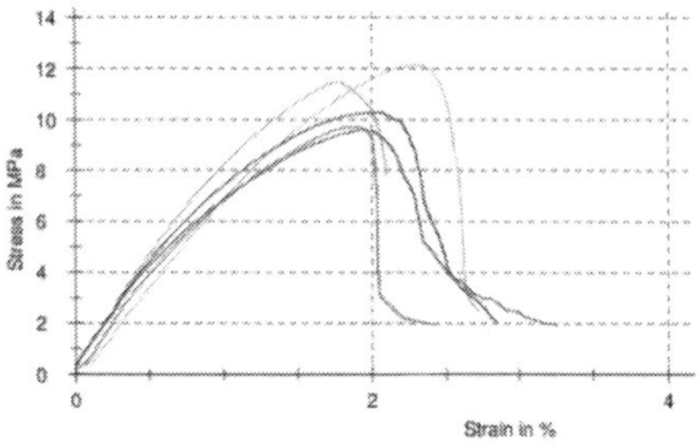
4.	3 (L-T-L)	L	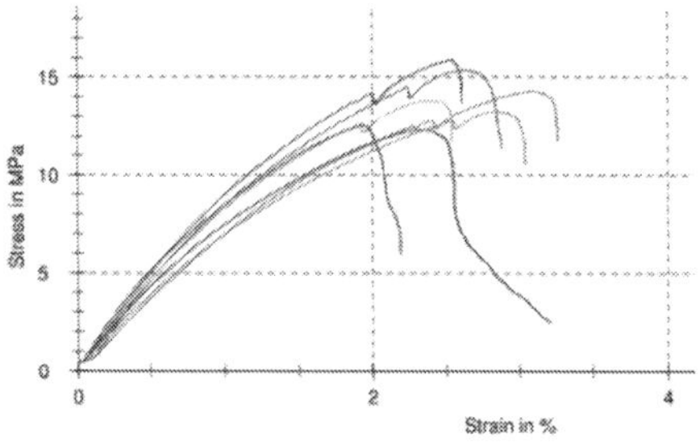
5.		T	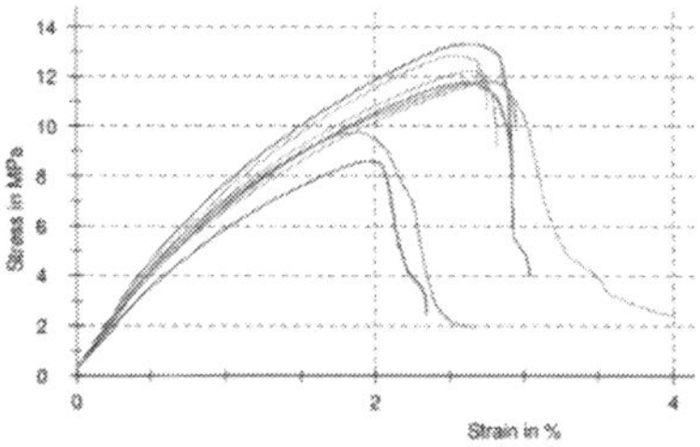
6.		D	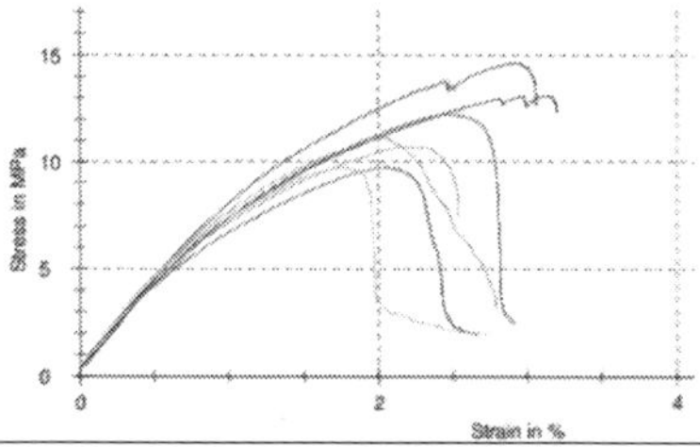
7.	4 (L-L-L-L)	L	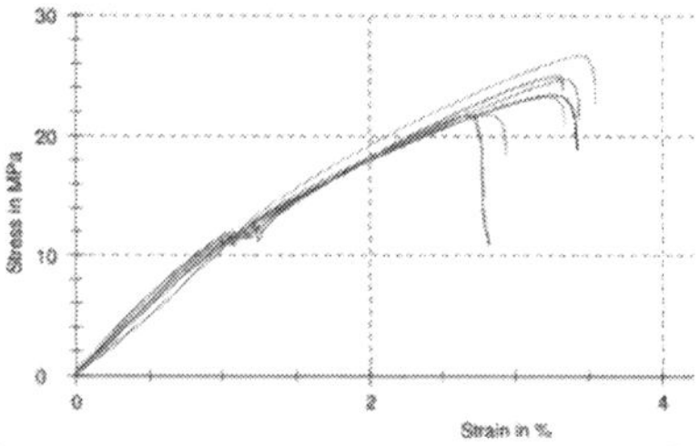
8.		T	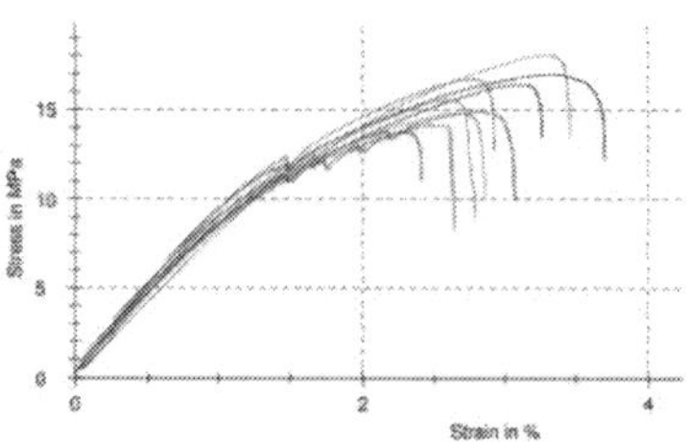

9.		D	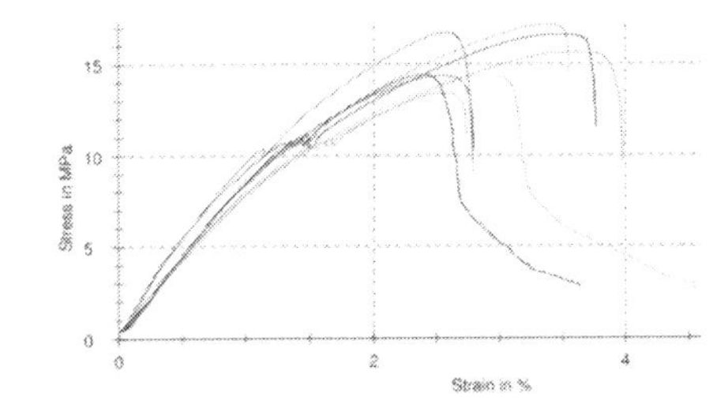
10.	4 (L-T-L-T)	L	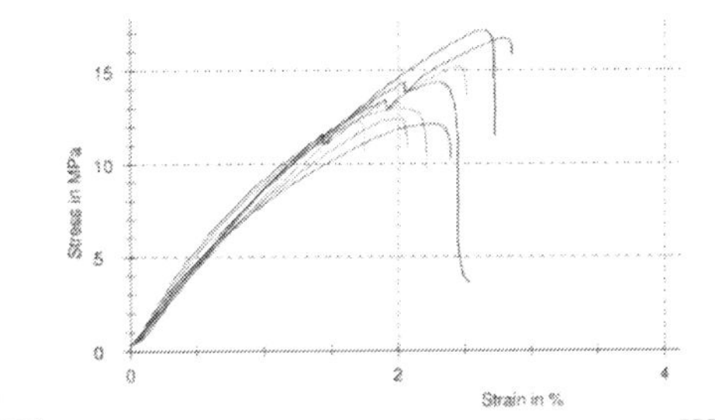
11.		D	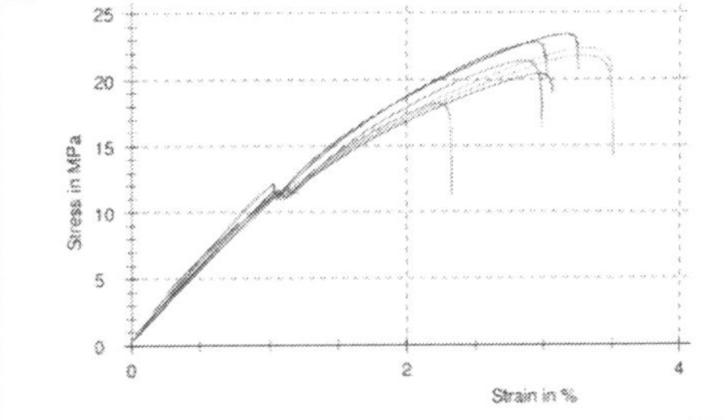

The mean values of the tests are summarized in Table 4. After performing the tests, a smoothing of the results was made by removing samples that gave extreme values. In columns 3-6 are presented the mean values of a set of 7 samples for each layout of samples (L - longitudinal, T - transversal, D - diagonal), and in columns 7-10 were calculated the mean values for all samples Layout of layers from column 1.

Table 4. The tensile properties of thermoformed composite material.

Layout of layers	Layout of samples	Average values - Layout of samples				Average values - Layout of layers			
		σ_M [MPa]	ε_M [%]	σ_B [MPa]	ε_B [%]	σ_M [MPa]	ε_M [%]	σ_B [MPa]	ε_B [%]
1	2	3	4	5	6	7	8	9	10
LLL	L	15.1	2.4	15.1	2.4	11.77	2.20	9.53	2.27
	T	9.5	2.2	7.58	2.4				
	D	10.7	2	5.91	2				
LTL	L	13.9	2.5	12.5	2.6	12.33	2.47	10.04	2.60
	T	11.5	2.5	9.36	2.5				
	D	11.6	2.4	8.26	2.7				
LLLL	L	23.8	3.2	23.8	3.2	18.30	2.97	17.77	3.00
	T	15.4	2.9	13.8	3				
	D	15.7	2.8	15.7	2.8				
LTLT	L	16.7	2.9	16.7	2.9	19.10	2.95	19.10	2.95
	D	21.5	3	21.5	3				

 DOI: 10.1051/matecconf/201713708003

4 Conclusions

The following conclusions are drawn from the conducted tests regarding the mechanical characteristics of the thermoformed material:

a. The thermoformed material made by overlapping multiple layers of reinforced nonwoven provides a higher σ_M mechanical strength. This is due to a better homogenize of the plant fibers and of the thermoplastic mixture. Thermoformed samples obtained by overlaying four layers give 55% more resistance than samples obtained by overlaying three layers.
b. Tensile strength varies depending on the layering of nonwoven fabrics. Thus, the best resistance is obtained in the direction of the feed direction (L) and the smallest in the perpendicular direction to it (T). At cross-sectional stress, the residual decreases by approximately 40% for the samples obtained by overlapping layers in the same direction (L).
c. The samples cut diagonally (D) at 45° to the L or T directions have higher resistance than the transverse ones, but close to it for the LLL and LLLL layers. When laying LTLT layers, the diagonal stress presents the best resistance (σ_M=21.5 MPa).
d. Cross-overlay of non-woven (LTL; LTL) layers increases the minimum strength (T-direction) and medium resistance by approximately 5%.
e. Tensile strength at break σ_B increases with the number of layers and is approaching the maximum tensile strength σ_M.
f. Strain at tensile strength has approximately the same value in the maximum breaking point.

References

1. I. Filip, E. Ciupan, I. Cionca, M. Ciupan, E. Pop, E. Campean, V. Heres, F. Raţ, C. Gherghel, The 21th International Conference of Inventics, Iasi, 38-45 (2017)
2. M. Steopan, C. Ciupan, E. Campean, E. Pop, I. Filip, L Stelea. The 21th International Conference of Inventics, Iasi, 95 - 103 (2017)
3. D. M. Panaitescu, C. A. Nicolae, Z. Vuluga, C. Vitelaru, C. G. Sanporean, C. Zaharia, D. Florea, G. Vasilievici, J. of Industrial and Engineering Chemistry, **37**, 137–146 (2016)
4. Z.L. Yan, H. Wanga, K.T. Lau, S. Pather, J.C. Zhang, G. Lin, Y. Ding, *Composites: Part B*, **46**, 221–226 (2013)
5. T. Sullins, S. Pillay, A. Komus, H. Ning, *Composites Part B*, **114**, 15-22 (2017)
6. M.R. Bambach. *Compression strength of natural fibre composite plates and sections of flax, jute and hemp. Thin-Walled Structures*, **119**, 103–113 (2017).

Proposal for a composite structure and graphic design for a parking barrier

Liana Hancu[1,*], *Gabriel* Marc[1], *Adrian* Popescu[1], *Paul* Bere[1] and *Simona* Rodean

[1]Technical University of Cluj-Napoca, Manufacturing Engineering Department, ROMANIA

Abstract. In the paper there is presented a case study of a cabinet barrier for a car parking. The materials used for manufacture are changed with polymeric composites. A design is proposed for the cabinet and a manufacturing technology is considered. The composite material is designed, realized and tested with tensile test ant three bending test and the results are presented.

1 Introduction

A composite is a material made from two or more components with very different physical or chemical properties. The new material obtained has different characteristics from the individual components which remain separate and distinct within the finished structure. The new material may be preferred because they are stronger, lighter, or less expensive when compared to traditional materials.

More recently, as presented in [1-3], researchers have also begun to actively include sensing, actuation, computation and communication into composites, which are known as Robotic Materials.

A composite material is made of the matrix, the reinforcement and the additives (coupling agents, catalysts and accelerators, pigments and fillers) [4-5].

Polymeric composites are combinations of two materials where the matrix is a polymeric material and the other one, called reinforcing phase, is in the form of fibers, sheets or particles and is embedded in the first one. Matrix materials are usually thermoplastics or thermosets like polyester, epoxy, silicon, phenolic, and so on. Fibers commonly used are: glass, graphite, boron and aramids.

The advantages such as light weight, low density, corrosion resistance, high stiffness, high heat resistance, fatigue resistance and high strength made them very attractive for industry.

There are many ways to arrange the reinforcement and the differences in the way the fibers are laid out give different characteristics to the material. Reinforcement usually makes the material more rigid and delays crack propagation. The fibers can greatly improve the composite's properties if they are properly select and arrange [6-11].

Fiber-reinforced composite materials are usually of two categories: short fiber-reinforced materials and continuous fiber-reinforced materials. Continuous reinforced

* Corresponding author: Liana.Hancu@tcm.utcluj.ro

materials are usually manufactured in layered or laminated structures which are available in a variety of forms.

Many researchers studied the mechanical behavior of a glass fiber reinforced composite and presented their connection with the fiber strength and modulus, fiber weight fraction, orientation, type, chemical stability, matrix strength and the interface bonding between the fiber and the matrix [12-18].

Considering that the strength and stiffness is not the same in all directions of a multilayered composite, the use of such a material needs a very accurate manufacture but only after the mechanical proprieties are determined.

2 Case study considerations

2.1 Product

A lot of products from the market, made of plastics are nowadays changed by similar ones but made of composites. The new technology that is needed is very different from the initial one. Such a product was a challenge for the authors who consider that the barrier cage used for automatic parking needs to be improved.

The proposal for the upper part of the product is presented in figure 1. The SolidWorks 2016 program was used to build the 3D model.

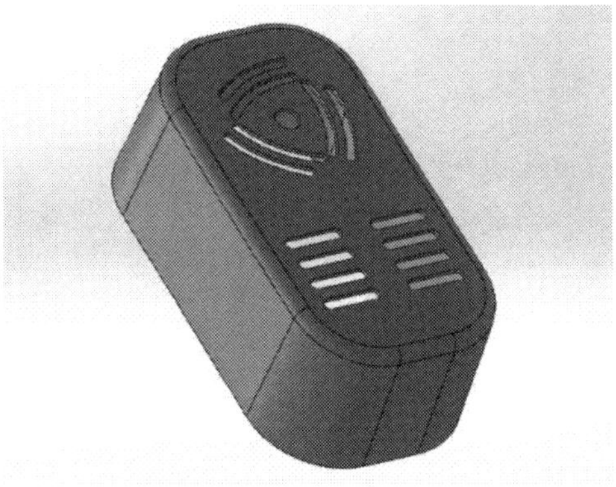

Fig. 1. Upper part of the barrier cabinet.

The barrier cabinet is designed to replace the old models currently available on the market. The upper part of the cabinet will be made by resin transfer moulding, and the material used for the reinforcement is a bidirectional glass fiber fabric. The dimensions of the upper part of the barrier are as follows: 1000 x 550 x 350 mm.

2.2 Manufacturing

The manufacturing technology proposed for the part is Resin Transfer Moulding (RTM) which is a low pressure moulding process in a closed mould for moderate volume production quantities.

Metered mixing equipment is used to control resin and catalyst quantities that are mixed through a mixer and injected into the mould port.

The main advantages that make the technology proper for our product are a uniform thickness, good quality surface for both sides of the component and low emissions. For optimum surface finish, a gel coat would be applied to the mould surface prior to moulding.

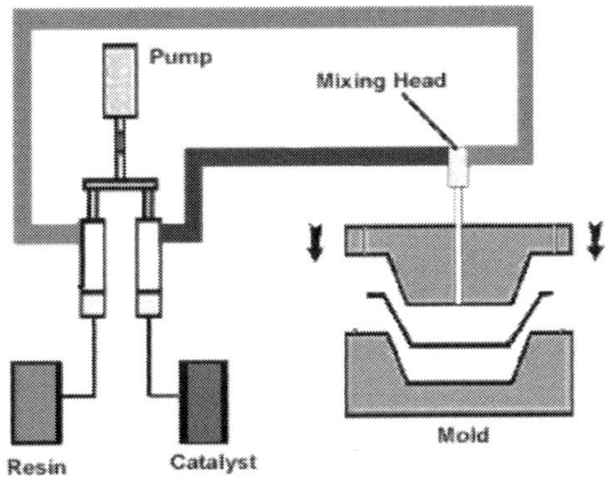

Fig. 2. Resin transfer moulding.

Fabrics or mats are usually used and laid upon the bottom (female) part of the mould. Preformed glass reinforcements are often used for complex mould shapes. The upper mould tool is then clamped over the first, and the resin is injected into the cavity. The technology is in some cases improved with the usage of vacuum which is applied to the mould cavity to assist resin into the fabrics. This is known as Vacuum Assisted Resin Injection (VARI). After all the fabric is wet out, the laminate is allowed to cure at ambient or elevated temperature.

2.3 Samples

Experimental research was performed in order to determine the mechanical characteristics for the proposed structure of the barrier upper part in order to be strength and also flexible as the usually demands for such a product are. The necessary experiments are tensile test and three point bending test.

In order to obtain the samples, a plate with initial dimensions of 250 x 250 x 2 mm was obtained using Resin transfer moulding technology.

An epoxy resin, type Epiphen RE 4020 / DE 4020, was used for the matrix. The mixing ratio of the parts (in weight) is 100:30, according to the manufacturer recommendations.

For reinforcement 4 layers of bidirectional balanced glass fabric by 300 g/m^2 was used and the calculated fiber volume fraction is 62.5%, which is very close to the maximum value of 65% that can be realized with this technology. The whole composite system was cured at 20°C for 24 hours.

From the plate obtained in this way, six samples for both tensile test and for three point bending test were cut.

The samples dimensions are made according to the prescriptions from the standards EN ISO 527-4 and ISO 527-5-2000 for tensile test and ISO 1425:2013 for bending tests.

The cutting of the samples was made with the water cutting machine.

3 Experimental results

The samples were tested on a Zwick/Roel Materials testing machine Z150 with test Control PC and the software testXpert II was used.

The results from the tensile tests are presented in figure 3 and the values registered by the machine are presented in table 1. It can be seen that all the graphics from sample 1 to 6 have the same slope, almost linear and this means that the plate was manufactured properly. The deformation is almost linear till the maximum force, when the specimens breaks, and is recorded the maximum force and deformation.

The medium value for the 6 samples, "x", the dispersion of the results, "s" and the dispersal coefficient, "v", are calculated by the machine software and presented in table 2. They all indicate that the manufacturing through Resin Transfer Moulding offers good characteristics all over the surface and this is very important for a part that is supposed to be hit during usage.

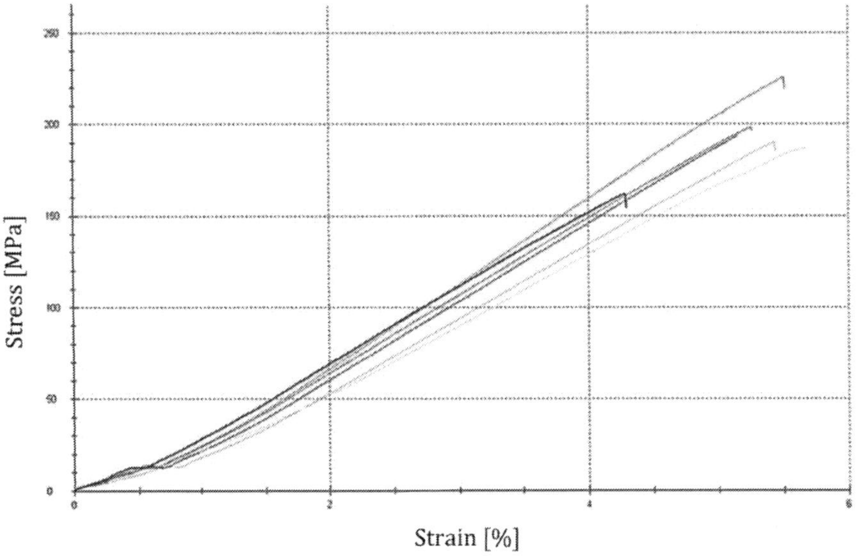

Fig. 3. Results of tensile tests.

Table 1. Individual values for the samples.

Nr	Date	σ_{low} MPa	σ_{high} MPa	E_t MPa	σ_{x1} MPa	σ_Y MPa	ε_Y %	σ_M MPa	ε_M %	ε_{tM} %	σ_B MPa	ε_B %	ε_{tB} %	b mm	h mm	A_0 mm²
2	6/6/2017	1.727	5.952	2080	-	-	-	198	5.3	5.3	198	5.3	5.3	25	2	50.00
3	6/6/2017	1.521	4.348	1400	-	-	-	190	5.4	5.4	190	5.4	5.4	25	2	50.00
4	6/6/2017	1.670	5.309	1830	-	-	-	193	5.1	5.1	193	5.1	5.1	25	2	50.00
5	6/6/2017	1.563	5.584	2020	-	-	-	187	5.7	5.7	187	5.7	5.7	25	2	50.00
6	6/6/2017	1.960	5.947	1980	-	-	-	225	5.5	5.5	225	5.5	5.5	25	2	50.00
7	6/6/2017	1.933	6.710	2390	-	-	-	162	4.3	4.3	162	4.3	4.3	25	2	50.00

Table 2. Medium values for the series of 6 samples.

Series n = 6	σ_{low} MPa	σ_{high} MPa	E_t MPa	σ_{x1} MPa	σ_Y MPa	ε_Y %	σ_M MPa	ε_M %	ε_{tM} %	σ_B MPa	ε_B %	ε_{tB} %	b mm	h mm	A_0 mm²
x	1.729	5.642	1950	-	-	-	193	5.2	5.2	193	5.2	5.2	25	2	50.00
s	0.184	0.790	328	-	-	-	20.4	0.49	0.49	20.4	0.49	0.49	0.000	0.000	0.00
v	10.64	14.00	16.83	-	-	-	10.58	9.44	9.44	10.58	9.44	9.44	0.00	0.00	0.00

The results from the bending tests are presented in figure 4 and the values registered by the machine are presented in table 3. It can be seen that all the graphics from sample 1 to 6 have the same shape and this means that the material has the same characteristics all over the plate.

The medium value for the 6 samples, "x", the dispersion of the results, "s" and the dispersal coefficient, "v", are calculated by the machine software and presented in table 4. They all indicate that the manufacturing through Resin Transfer Moulding offers good characteristics all over the surface and this is very important for a part that is supposed to be hit during usage.

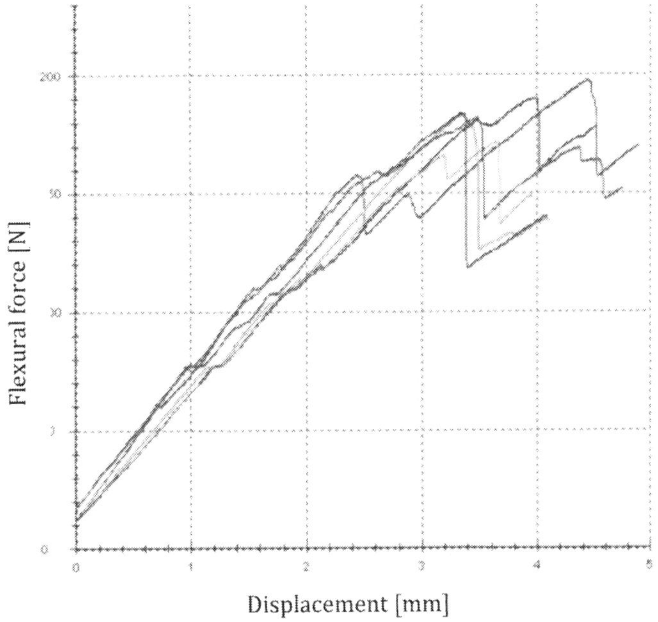

Fig. 4. Results of three points bending test.

Table 3. Individual values for the samples.

Nr	Height mm	Diameter d0 mm	Fmax. kN
106	272.43	20	0.20
108	272.47	20	0.18
109	272.63	20	0.19
110	272.70	20	0.17
111	272.61	20	0.18
112	272.62	20	0.18

Table 4. Medium values for the series of 6 samples.

Series n = 6	Height mm	Diameter d0 mm	Fmax. kN
$\bar{x}$	272.58	20	0.18
s	0.10	0.000	0.01
v	0.04	0.00	4.72

Before starting the test, the machine is set to the test parameters. As seen in figure 4, the behavior of the composite has four stages during the test. The first stage, almost linear, explains the elastic deformation of the composite. The second stage begins after the load reaches a peak value, when diminish of about 20–50 % of the peak load was observed in composite laminate structures.

This load diminish appears because of the fiber cracking and delamination, or some defects in the structure of the composite. In the third stage, because only the reinforcement carried the load, the specimen grew again but never exceeded the previous peak load.

 DOI: 10.1051/matecconf/201713708004

4 Conclusions

The case study presented shows that using a polymeric based composite material the product is more resistant to pressure than plastics without reinforcement. The technology proposed has the advantage, compared to other technologies used for composite parts, that assures a good quality for the part both in geometry and in quality surface. The design of the composite structure is important for obtaining high strength and also good flexural proprieties. The composite material was realized and tested and the results show a good behavior at traction but some delamination at bending.

The appearance and propagation of delamination between layers is of great importance and the prediction of the failure of the composite structures should be considered for study. One possible causes of delamination in our case can be the uniform structure used for the part as all the layers were in the same direction and the penetration of the resin between them was, in this way, difficult.

This work was supported by a grant of the Romanian National Authority for Scientific Research and Innovation, CNCS/CCCDI–UEFISCDI, project number PN-III-P2-2.1-96BG-2016-0210, within PNCDI III.

References

1. M.A. McEvoy and N. Correll. *Materials that couple sensing, actuation, computation, and communication*. Science 347(6228) (2015)
2. T. Sathishkumar, S. Satheeshkumar, J. Naveen, J. Reinf. Plast. compos. **33**, 1258-1275 (2014)
3. J.M. Stickel, M. Nagarajan, Int. J. Appl. Glass Sci., 3122-3136 (2012)
4. C. Florea, H. Iancau, A. Popescu, M. Simon, P. Bere, COMAT (2010)
5. A. Azzam and W.Li,, IOP Conf. Ser. Mater. Sci. Eng. **6**, 2012016, Global Conference on Polymer and Composite Materials (2014)
6. G. L. Enachescu, M. F. Stefanescu, J Plastic Mater **53**, 198-201 (2016).
7. G.M. Krolczyk., P. Nieslony, J.B. Krolczyk, I. Samardzic, S. Legutko, S. Hloch, S. Barrans, R.W. Maruda, J. Measurement **7**, 2003-2013 (2015)
8. R. Udroiu, A. Nedelcu, B. Deaky, Environmental Engineering and Management Journal **10**(9), 1387-1394 (2011)
9. W. Plowucha and W. Jakubiec, Quality and Innovation in Engineering and Management 183-186 (2011)
10. P. Bere, C. Neamtu, Central European Journal of Engineering, 1-7 (2014)
11. I. Campbell, et al., Rapid Prototyping Journal, **14**(5), 271-279 (2008)
12. B. Liu, R.I. Campbell, and E. Pei, Real-time integration of prototypes in the product development process. Assembly Automation, **33**(1), 22-28 (2013)
13. M.F. Ştefănescu, Mat. Plast. **35**(4), 231-238 (1998)
14. M.F. Ştefănescu, Mat. Plast. **43**(3), 254-257 (2006)
15. J.B. Krolczyk, Transactions of FAMENA, **37**(4), 43-54 (2014)
16. G. Krolczyk, J.B. Krolczyk, S. Legutko, A. Hunjet, Tehnički vjesnik/Technical Gazette, **21**(2), 447-450 (2014)
17. I.A. Popan, N. Balc, A.I. Popan, MSE, MATEC Web Conf., **121** (2017)
18. M. Shirinbayan, J. Fitoussi, M. Bocquet, F. Meraghni, B. Surowiec, A. Tcharkhtchi, Composites Part.B. Engineering, **115**(15), 3-13 (2017)

Characterization polyethylene terephthalate nanocomposites mixing with nano-silica and titanium oxide

Mircea A. Rusu[1,*], *Dan* Leordean[1], *Cosmin* Cosma[1], *Miuta* Filip[2], *Marioara* Moldovan[2], *Laura* Silaghi-Dumitrescu[2] and *Olga Hilda* Orasan[3]

[1]Technical University of Cluj-Napoca, Faculty of Machine Building, Cluj-Napoca, Romania
[2]Babes Bloyai University - Raluca Ripan Institute of Chemistry Research, Cluj-Napoca, Romania
[3]University of Medicine and Pharmacy "Iuliu Hatieganu", Cluj-Napoca, Romania

Abstract. Polyethylene terephthalate (PET) based nanocomposites containing nano-silica (Aerosil (Degusa)) and titanium oxide (TiO_2 (Merk)) were prepared by melt compounding. Influence of nano-silica and titanium oxide on properties of the resulting nanocomposites was investigated by scanning electron microscopy (SEM), Fourier transform infrared spectroscopy (FTIR) and atomic force microscopy (AFM). The possible interaction between nano-silica and titanium oxide particles with PET functional groups at bulk and surface was elucidated by transmission of FTIR-ATR spectroscopy. AFM studies of the resulting nanocomposites showed an increased surface roughness compared to pure PET. SEM images illustrated that nano-silica particles have tendency to migrate to the surface of the PET matrix much more than titanium oxide powder.

1 Introduction

The presence of large quantities of mixed plastic waste, impurities and contamination are the main challenge for efficient recycling of plastic materials from packaging. In the last years the efficient management of different waste streams (selective collection, automatic separation) has enabled the recovery of large volumes of relatively clean and homogeneous polymeric elements that are viable for mechanical recycling. Polymers are the most widely used materials in various fields due to their valuable characteristics such as good mechanical properties, low density, rather low cost, and also easy to process [1-4]. The total production of plastics is more than 230 million tonnes per year and to alleviate this waste, many solutions have been proposed for the management of plastics waste, such as recycling, incineration and degradation of materials.

The most common types of polymers are: PET (polyethylene terephthalate), HDPE (high density polyethylene), PVC (polyvinyl chloride), LDPE (low density polyethylene), PP (polypropylene), PS (polystyrene). The efficient treatment of polymers waste is still a difficult challenge. The traditional methods such as combustion or burying underground

* Corresponding author: mcrusu2206@yahoo.com

show a negative effect on the environment like formation of dust, fumes and toxic gases in the air, and the pollution of underground water and other resources. We can obtain the substantial enhancements of mechanical, thermal, optical and barrier properties using the nanofillers [5-6]. The nano-additives increase the interphase surface of components and superficial area/volume ratio leading to improvement of overall performances. If a comparison is made between blending recycled polymers with nanomaterials [7, 8] some studies have found that a small load is sufficient to develop excellent properties without greatly increasing density and cost. It has been found that they have acted as a nuclear agent and accelerated the rate of crystallization of the polymer, increasing the melting resistance and viscosity of the polymers [9-14].The aim of this study is to develop, characterize and investigate, according to the filler content, blends of recycled PET with nano-silica and titanium oxide, in order to obtain materials used in the manufacture of parts used for design verification, shape verification, models for casting and medical application processes.

2 Materials & Methods

The PET samples used in this study come from numerous commercial waste packaging materials (cups, glasses, bottles, etc.) made from these polymers. Nano-silica particles (Aerosil) were supplied by the German, Degussa Company and titanium oxide by Merck. PET samples with nano-silica particles and titanium oxide were mixed and dried, prior to the melting process. Each nanocomposites was prepared by mixing and was kept at 210^{0}C for 30 minutes, in order to ensure proper viscosity for the mixing process. The content of nano-silica and titanium particles in the matrix was ensured to be 10% by weight (wt%). Specimens to be tested were further processed by the aid of a laboratory press at a pressure of 220 psi(Fig.1.). The temperatures of both upper and lower plates of the press were kept at 220^{0}C. The chemical structure of the waste plastics PET, samples PET/with nano-silica and PET/titanium was confirmed by recording their IR spectra, SEM Microscopy and AFM. Characterization of the resultant nanocomposites was carried out using a FTIR spectrophotometer (JASCO, 610), equipped with a reflection ATR accessory for reflection mode. All the ATR spectra of PET/nano-silica and PET/titanium nanocomposites were comparated with the spectra of pure PET. The recorded wavenumber range was from 450 to 4000 cm^{-1}. Surface morphology analyses of the nanocomposite were carried out by the aid of a scanning electron microscope (SEM, FEI Inspect S, Netherlands) and atomic force microscopy (AFM, Veeco, Multimode type.).

Fig. 1. Laboratory press with plates.

3 Results and discussions

The recording of the spectra on the recycled samples may change depending on the morphology and the sample processing history, and the relative intensities of the low-to-medium intensity bands can be obtained. The chemical structure of the (PET) polymers, melted with silica and TiO_2 was confirmed by recording their IR-ATR spectra. The Fourier transform infrared (IR-ATR) results of samples in absorption mode are shown in table 1.

The infrared spectrum is known to contain information of the microstructure. The microstructure is a representation of the characteristics of the material, whether it is physical, chemical or mechanical. It may be noted that breadth and shape of the infrared bands are representation of the manner in which molecules relax from the vibrational excited state reach upon the absorption of an (infrared) photon.

Table 1. Infrared spectroscopic data of waste pet bottles (cm^{-1}).

Band Name/ samples	ArCH Wagg	ArCH *Out of plane*	O-C-C Asymm	C-C-O Asym m	C-H def. in alkane	C-H bending	Ar C-C bond	C=O Ester	C-H Asym m
PET	724	873	1091 1128	1238	1340 1402	1453	1501	1709	2960

The FT-IR spectra of PET confirms the presence of a carbonyl group at 1709 cm^{-1}. The peak at 1238 cm^{-1} is due to the asymmetric C-C-O stretching involving the carbon in aromatic ring. The aromatic C-H wagging appears at 724 cm^{-1}. The peak at 873 cm^{-1} corresponds to aromatic C-H out of plane bending. The O-C-C asymmetric stretching is split at 1128 and 1091 cm^{-1}. There are peaks at 1453 cm^{-1} (C-H bending), 1501 cm^{-1} (aromatic C-C stretching) 2960 cm^{-1} (C-H asymmetric stretching). Peak at 1402 cm^{-1} and 1340 cm^{-1} is due to deformation C-H alkane. Infrared spectra of the SiO_2 used in this study indicates the presence of Si–O–Si band, well reflected in nanosilica particles, as a strong bending band at about 1012-1200 cm^{-1}. The infrared spectra for composites PET/SiO_2 showed the weak C–H stretching, C–H bending and moderately intense C–C out of plane bending vibrations of the benzene rings in polyester appear at 3054, 793, 848 and 721 cm^{-1} respectively. The infrared spectra of the PET/TiO_2 nanocomposites indicate 1710 cm^{-1} band characteristic of the vibration of the carbonyl (C=O) of the ester group present in the polymer.

In the present study, the morphology of the surfaces of the blends was analyzed with SEM. The particles were very fine, the morphology and size of the powder were determined using SEM images, which revealed spherical particles and a uniform dispersion in the polymer (Fig.2).

Figure 2 (a-e) shows SEM micrographs of the surfaces of pure PET together with various PET/SiO_2, PET/TiO_2 composites. As can be seen, pure PET has a relatively smooth and uniform surface. Most of the crystallites are irregularly shaped and they seem to interlock with each other. The radial texture can be observed at some locations. However, the boundaries between the crystallites are not clear. As the polymer chains next to the nanoparticles are likely to be tethered onto the particules surfaces, a strong interaction may exist between the polimer and crystallites that prevents the boundary areas from being etched away. The strategy of using nanoparticles to modify structure and bulk properties of performance materials is well known. In general, the ability of additives to migrate to the surface is defined by several factors such as size, composition, mobility, functionalities and molecular architecture.

DOI: 10.1051/matecconf/201713708005

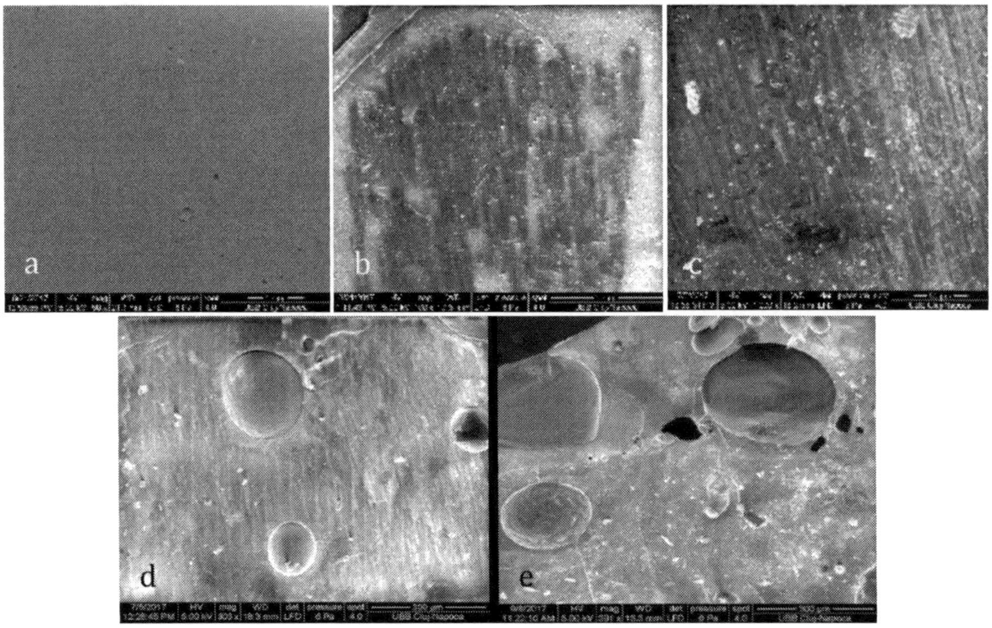

Fig. 2. Scanning electron images of various composites: (a) pure PET, (b) PET/ 10% nano-silica, c) PET/10% titanium oxide; Images in the section: (d) PET/10% hydrophilic nano-silica, (e) PET/10% titanium oxide.

As mentioned previously, the FTIR analysis of the samples illustrate that PET, SiO_2 and TiO_2 particles have some interfacial interactions within the PET matrix. However, it seems that these interactions are not strong enough to enable good dispersion of the particles within the PET matrix.

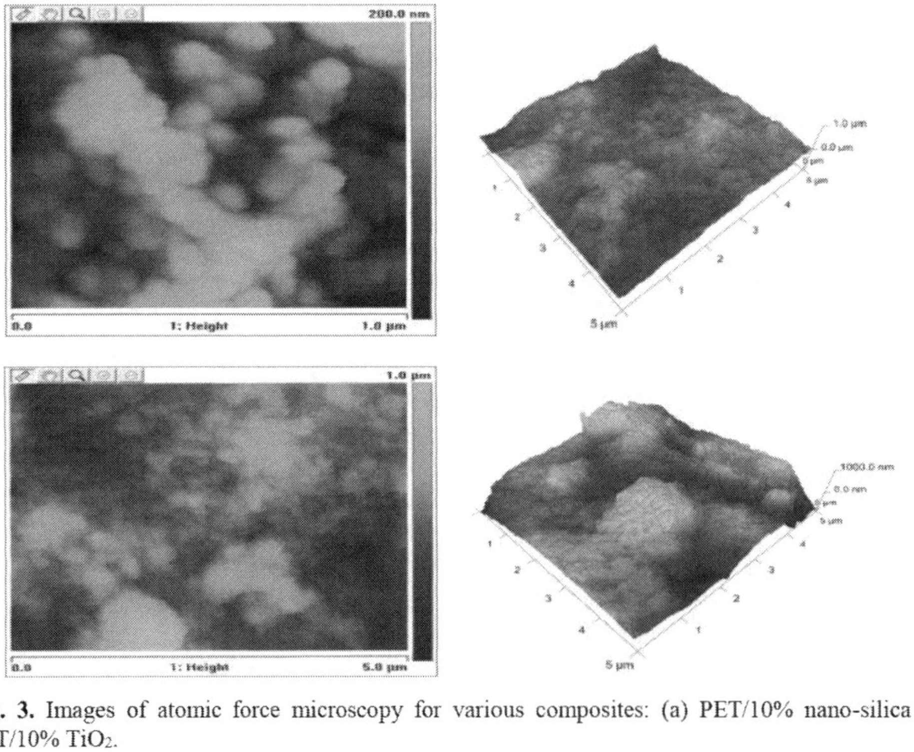

Fig. 3. Images of atomic force microscopy for various composites: (a) PET/10% nano-silica (b) PET/10% TiO_2.

Other authors have confirmed this phenomenon in cases where nano-silica and TiO_2 was incorporated into multilayer organic coatings [15-22].

Addition of nano particles to the PET matrix leads to various chain interactions enabling formations of aggregates and agglomerates on the surface of PET composites. The presence of these particles results in increased surface roughness which is in good agreement with the AFM images. Such agglomerations are less evident for SiO_2/PET composite causing smoother surfaces comparing with TiO_2/PET composites. Figure 3 shows the presence of agglomerations on the surface of the nanocomposites investigated.

Plastics besides polymers also contain additives, dyes and copolymer mixtures that can cause problems and are difficult to separate. By recycling them, after several processing cycles, the structure of polymer is degraded introducing poorer mechanical properties than those of a virgin one [10, 11] obtaining inferior quality products.

In addition to the chemical contaminants and degradation products generated during the first processing and use, impurities, such as microscopic particles, increase the frequency of defects in the polymer resulting in the quality degradation in the recycling flow. Because sorting, separation and cleaning processes are the most important and most difficult to achieve for the recycled polymer, small amounts of nano powder were added to the polymer, which can be an efficient recycling technique for polymers. Various nanoparticles such as nano-silica and TiO_2, have been used for recycling which should be developed for all polymer waste.

However, with the goal of successful and economic recycling, more research needs to be done on the whole aspect of the recycling process to enhance competitiveness and optimization of the melt mixing process in these systems. The first step could be to improve interfacial interaction and adhesion in nanocomposites, particle surface modification procedures, which should lead to a better compatibility with the surface of the polymer.

4 Conclusions

Products prepared by the addition of nanoparticles to recycled polymers can offer unexpected combinations of improved properties, low weight, easy processing and low costs that are not readily available and concurrently with other industrial products. Effects of the nanoparticle type on surface properties nanocomposites were studied

Obtained results suggest that there are more and larger silica and titanium oxide particles present at the PET/nanocomposite, causing increased surface scattering and roughness. FTIR-ATR studies have shown interfacial interactions between in the PET matrix with SiO_2 and TiO_2 nanoparticles, but are not strong enough to allow good dispersion of the particles in the PET matrix. This result is incidentally consistent with our other results obtained from AFM, and SEM analyses.

This research was supported by the AM-CIR project, PN-II-RU-TE-2014-4-1157, no. 37/01.10.2015 financed from the UEFISCDI by the Romanian Government and also by the HORIZON 2020 AMaTUC project (GA 691787) financed from EU.

References

1. A. Leszczy´ nska, J. Njuguna, K. Pielichowski, J.R. Banerjee, Thermochim. Acta, **453**, 75–96 (2007)
2. S. Sinha Ray, M. Okamoto, Prog. Polym. Sci., **28**, 1539–1641 (2003)
3. Ramezani-Dakhel H, Garmabi H. Journal of Applied Polymer Science, **118**, 969-979 (2010)
4. S.S. Pesetskii, S.P. Bogdanovich, N.K. Myshkin, J. Fric, Wear, **28**, 457–475 (2007)

5. J. Njuguna, K. Pielichowski, S. Desai, Polym. Advan. Technol., **19**, 947–959 (2008)
6. J. Ma, J. Xu, J.H. Ren, Z.Z. Yu, Y.W. Mai, Polymer, **44**, 4619–4624 (2003)
7. D. Burgentzle´, J. Duchet, J.F. Ge´ rard, A. Jupin, B. Fillon, J. Colloid Interf. Sci., **278**, 26–39 (2004)
8. W.Li, J.Karger-Kocsis, R.Thomann: Journal of Polymer Science Part B: Polymer Physics, **47**, 1616–1624 (2009)
9. S.C. Chung, W.G. Hahm, S.S. Im, Macromol. Res., **10**, 221–229 (2002)
10. L. M.Robeson, Polymer blends: A comprehensive review. Hanser, Munich (2007).
11. W. Liu, X. Tian, P. Cui, Y. Li, K. Zhang, Y. Yang, J. Appl. Polym. Sci., **91**, 1229–1232 (2004)
12. Y. Zare, H. Garmabi , Journal of Applied Polymer Science., **123**, 2309-2319 (2012)
13. S. H. Jafari1, A. Kalati-vahid, H. A. Khonakdar, A. Asadinezhad, U. Wagenknecht, D. Jehnichen, eXPRESS Polymer Letters, **6**, 148–158 (2012)
14. S.I.S Shahabadi, H. Garmabi. Polymer Letters, **6**, 657–671 (2012)
15. M. Goto, M. Sasaki, T. Hirose, Journal of Materials Science, **41**, 1509-1515 (2006)
16. T.Y Tsai, C.H Li, C.H Chang, W.H. Cheng, C.L Hwang, R.J. Wu, Adv Mater, **17**, 1769-1773 (2005)
17. S.C. Chung, W.G. Hahm, and S.S. Im, Macromolecular Research, **10**, 221–229 (2002)
18. O.H Lin., H.M Akil., Z.A.M Ishak, Polym. Compos., **30**,1693–1700 (2009)
19. J. Vega-Baudrit, M. Sibaja-Ballestero, P. Va´zquez, R. Torregrosa-Macia, J.M. Martı´n-Martı´nez, Int. J. Adhes., **27**, 469–479 (2007)
20. Y.G. Lee, C. Oh, D.W. Kim, Y.D. Jun, S.G. O, J. Ceram. Process. Res., **9**, 302–306 (2008)
21. C.S. Sipaut, N. Ahmad, R. Adnan, I.A.B. Rahman, M.A. Bakar, J. Ismail, C.K. Chee, J. Appl. Polym. Sci., **7**, 27–34 (2007)
22. K. Han, M. Yu, J. Appl. Polym. Sci., **100**, 1588–1593 (2006)

Mechanical behavior of composite materials using the finite element analysis

Emilia Sabau[1,*], *Adrian* Popescu[1] and *Cristian* Vilau[1]

[1]Technical University of Cluj-Napoca, Manufacturing Engineering Department, 103-105 Muncii Bvd. Cluj-Napoca, Romania

Abstract. A finite element analysis software has been used in order to determine several mechanical properties for samples made from unidirectional composite materials. Plates from reinforced unidirectional fiber glass polymer composites obtained by compression hand lay-up process were studied.Tensile tests were performed upon samples with different number and orientation of the composite ply layup.

1 Introduction

Composite materials represent a class of engineering materials with great scientific and technical interest. They are not only a perfect replacement for ferrous or non-ferrous materials, but also materials to solving technical problems in the whole range of different industries. Currently its properties and quality, polymer composite materials have a wide applicability in a number of areas such as building and civil engineering, electrical engineering, electronics, road, rail and marine, aeronautical and aerospace technique etc. [1-3].

The major advantage of the composites consists in the possibility of modeling the properties and obtaining a wide range of materials. In most cases, the composite comprises a base material, the matrix, in which a complementary material is dispersed in the form of fibers or particles, and the main properties that are intended to be obtained in an improved form are: tensile strength, wear resistance, density, high temperature resistance, superficial hardness, dimensional stability, vibration damping capacity.

Engineering calculation is an essential part of the design process, be it constructive or technological design. The great variety of issues that the engineer has to solve is reflected in the many computational methods currently used. These methods have gradually developed, as the theoretical and technological knowledge accumulates, in the last half century [4-7].

The finite element method is currently the most widely used numerical solution for engineering problems. The Finite Element Analyses (FEA) is a numerical solution of the problems that can be expressed in mathematical language in the form of partial derivative equation systems [8].

* Corresponding author: emilia.sabau@tcm.utcluj.ro

When applying FEA, the occupied domain of the physical system that we analyse, is meshed in finite subdomains bounded by rectilinear or curvilinear boundaries. Through this operation, the real system is replaced with a network of so-called finite elements.

The differential equations that describe the system behavior will be verified for each element. The mathematical construction of the finite elements ensures a certain degree of approximation continuity at crossing of the boundary between the elements. Continuity is realised using remarkable points associated with the elements (known as nodes). In fact, the approximation of the exact solution problem results as a function of the unknown values in the respective nodes. Applying FEA, we obtain a system of equations that is numerically solved in relation to nodal unknown values.

The mesh uses thousands or even tens of thousands of finite elements.

In the last years, a series of finite element computational programs have appeared and function as integrated modules in CAD systems. One example is SolidWorks Simulation.

The paper analyse unidirectional composite samples using SolidWorks Simulation Software. The program can optimize material selection, the number and orientation of the composite ply layup to ensure product quality, performance and factor of safety.

Using SolidWorks Simulation Software to analyse the composite parts we can reduce the need for costly prototypes and save time and development costs. Each layer can be defined by a unique set of material properties and orientation, giving the designer maximum control to find the optimum layup and material for maximum product performance [9-15].

The paper will analyze the mechanical characteristics for different polymer composite structures, both by finite element simulations and by experimental determinations. It will identify all factors influencing the quality, productivity and competitiveness of products made and will study all the phenomena that accompany the process of forming parts.

The results will be used to obtain new reinforced polymer composites at lower costs, technical and mechanical properties as high in order to use them in more and more fields.

2 Method and materials

The material used is unidirectional fibers glass tissue 225 g/m^2 and polyester resin, Poliester Norpol 440-M750 type, as a matrix.

The polyester resin is orthophthalic thixotropized. The technical properties of the resin are: Brookfield viscosity at 25 °C: 250-350 mPa.s; Density at 25 °C: 1350 kg/m^3; Gel time with 2% MEC (methyl-ethyl-ketone) in 100g resin: 8-15 min; Exothermic peak: 160-190 °C; Exothermic peak time: 15-23 min; Colour: transparent pink.

The manufacturing process was achieved at ambient temperature of 20 °C. The technology applied was compression hand lay-up technology, [16-19]. Different plates with different reinforced grade were obtained. Separate composite plates from each of reinforced type were made. The samples were cut from the obtained plates on three angles directions of the fibers, like 0°, 90° and 45 ° relative to direction of stress.

To determine the mechanical resistances of composite plates was done experimental trials at tensile tests.

The samples dimensions obtained from unidirectional tissue and tested on longitudinal direction were cut at 15x1x250 mm dimensions according to ISO 527-5 A, and the samples dimensions obtained from unidirectional tissue and tested on transversal direction were cut at 25x2x250 mm dimensions, according to ISO 527-5 B.

In order to highlight the influence of pressure on the mechanical properties at tensile tests of the lamine groups, unidirectional composite plates were made by compression hand lay-up process using a pressure of 0,16 MPa. Before polymerization, after the layers were achieved, the mold that consists of two metal plates (300 x 300 mm), is compressed with a power-press to homogenize the structure, to remove the exceeded resin and to increase the reinforcement degree. The compression of composite plates was achieved on the tensile-compression test machine, which is provided with calibrated dial indicating the value of applied force.

Tensile tests were performed upon samples with fibers tilted at 0°, 45° and 90° angles relative to direction of stress.

The results of the tensile tests for the samples that use unidirectional fabric are shown in Table 1, where: M_f – reinforced grade, $A=bxh$ – section aria of the sample, E – elastic modulus, F_{max} – maximum broken force, σ_r – tensile strength.

Table 1. UD 225 g/m² / Polyester Norpol 440-M750 composite

Nr. crt.	M_f [%]	Angle	$A=bxh$ [mm²]	E [MPa]	F_{max} [N]	σ_r [MPa]
1.	65	0°	15 x 1 = 15	10895	7023	468,2
2.		45°	25 x 2 = 50	3864	2810	56,2
3.		90°		2650	1940	38,8

It is noted that in the case of the UD 225 g/m² / Polyester Norpol 440-M750 composite, the value of the breaking strength at 45° fiber reinforcement angle decreases with 87,99% compared to the reinforcement at 0° and with 91,71% in the case of reinforcement at 90°.

3 Finite element analyses

The Finite Element Analyses is a numerical solution of the problems that can be expressed in mathematical language in the form of partial derivative equation systems. At the FEA base exists the meshing concept. In very intuitive terms, the meshing operation is a division of the spatial domain occupied by the analyzed body into regions of limited sizes known as finite elements.

This paper analyses samples using the SolidWorks Simulation Software. The analysis proposes as the first objective to determine the stress state and the deformation of the samples. A mesh with 3731 nods and 1732 elements has been generated, Figure 1.

Fig. 1. Finite element network and restrictions.

The right side of the sample is rigid fixed at the end, and at the left side we have a imposed displacement of 4,50 mm, presented in Figure 2. The samples realized from unidirectional tissue with polyester matrix was made from 3 plies with the thickness of 0,33 mm at a reinforced angle by 0°.

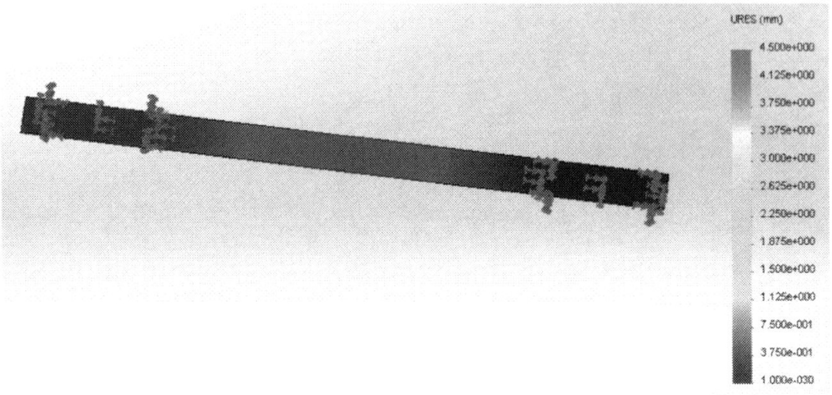

Fig. 2. Displacement distribution.

The SolidWorks Simulation finite elements program is used to point out the applicability of failure samples. The deformation of samples at tensile testing was analyzed. The visualization and interpretation of the simulation's results have been made with the help of SolidWorks Simulation module. This realizes the graphical representation of the numerical results. Thus, Figure 3 presents the distribution of von Mises equivalent stress in configuration of failure sample. It is observed that the maximum value for unidirectional tissue is 436,383 N/mm^2.

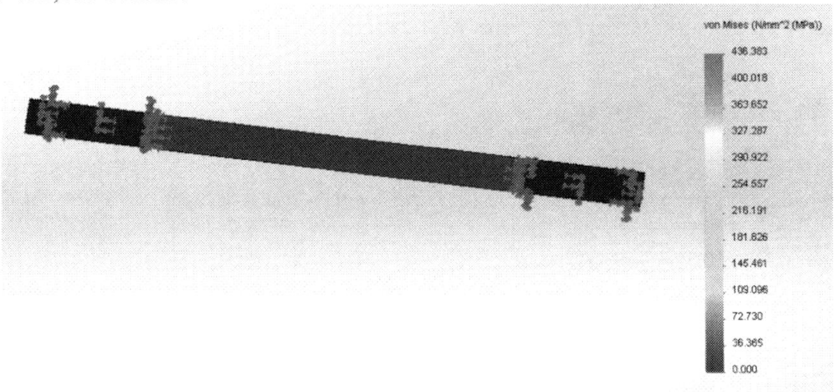

Fig. 3. Distribution of von Mises equivalent stress, reinforced angle by 0°.

We can conclude that the composite that use in the structure unidirectional tissue obtained similar values for the result forces, both in the experimental part and in the FEA simulation (Table 1, Figure 4).

Reaction Force (N)

Component	Selection	Entire Model
Sum X:	0.	-0.055176
Sum Y:	0.	7.189E-012
Sum Z:	0.	-0.40556
Resultant:	0.	0.4093

Reaction Moment (N-m)

Component	Selection	Entire Model
Sum X:	0.	7.4725E-013
Sum Y:	0.	0.
Sum Z:	0.	9.2271E-013
Resultant:	0.	1.1873E-012

Fig. 4. Result force for UD 225 g/m^2 / Polyester Norpol 440-M750 composite.

Also, two other particular cases were analyzed, when the sample was realized from 6 plies with the thickness of 0,33 mm at a reinforced angle by 45° and 90°. In these cases, the distribution of von Mises equivalent stress in configuration of failure sample is shown in

Figure 5, respectively Figure 6. It is observed that, the maximum value for unidirectional tissue with a reinforced angle of 45° is 86,735 N/mm^2.

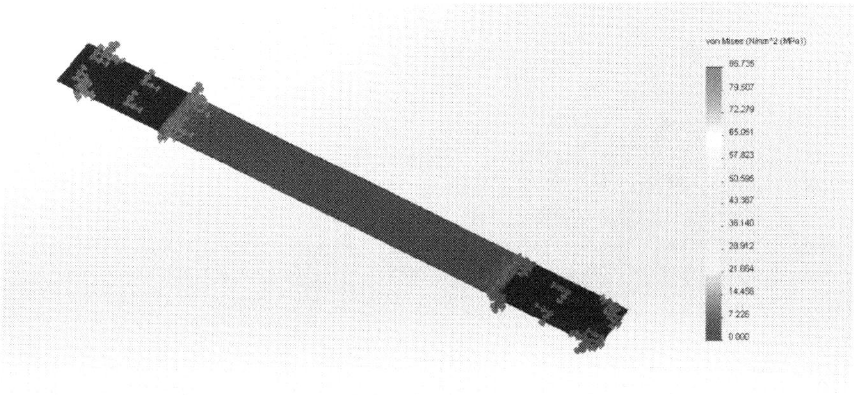

Fig. 5. Distribution of von Mises equivalent stress, reinforced angle by 45°.

The maximum value for unidirectional tissue with a reinforced angle of 90° is 54,548 N/mm^2.

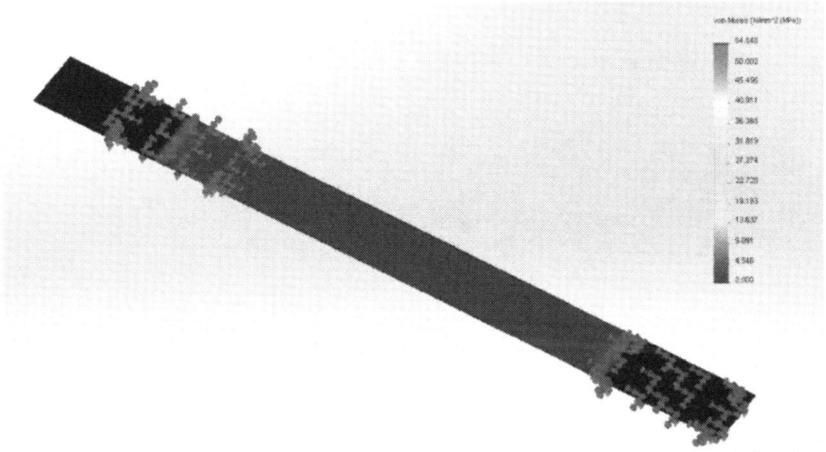

Fig. 6. Distribution of von Mises equivalent stress, reinforced angle by 90°.

The presented examples demonstrate the possibility of implementing the proposed model in the structure of finite element analysis programs. We can observe, that the FEA results are very close to the test results.

4 Conclusions

The paper analyzes the mechanical characteristics for unidirectional glass polyester composite structures, both by finite element simulations and by experimental determinations.

The mechanical behavior of the polymer-composite materials depends by the nature and by the reinforcing material, and by the technological procedure. The manufacturing technology for the unidirectional composite plates was compression hand lay-up processes. The experiments consist in tensile tests, fibers being tilted at 0º, 45º and 90º angles, relative to direction of stress. Several variants of composite samples were designed, by taking into account the reinforcement degree, the number and the orientation of the plies, the thickness ply, Young modulus, maximum broken force, tensile strength. The structure of the reinforcing material will influence the mechanical characteristics.

Applying a pressure to the composite plate during the hand lay-up process is beneficial because by the removal of exceeded resin the reinforcement degree increases which leads to a growth of mechanical strength characteristics.

By using SolidWorks Simulation Software to analyse the composite parts we reduce the need for costly prototypes and save time and development costs. Each layer can be defined by a unique set of material properties and orientation, giving the designer maximum control to find the optimum layup and material for maximum product performance.

We conclude, that the composite, that use in the structure unidirectional tissue, obtained similar values for the result forces, both in the experimental part and in the FEA simulation.

Making a comparison between results, it is obvious that the FEA results are very close to the test results. This means that the FEA is validated by the test results.

This work was supported by a grant of the Romanian National Authority for Scientific Research and Innovation, CNCS/CCCDI–UEFISCDI, project number PN-III-P2-2.1-BG-2016-0216, within PNCDI III and the H2020 AMaTUC project (GA 691787).

References

1. E.J. Barbero, *Introduction to Composite Materials Design* (West Virginia University, USA, 1998)
2. D. Gay, *Matériaux Composites* (Hermes, Paris, 1997)
3. M. Berthelot, *Matériaux Composites. Comportement mécanique et analyse des structures* (EDP TEC&DOC, Paris, 1999)
4. R.M. Jones, *Mechanics of Composite Materials* (Taylor and Francis, Philadelphia, 1999)
5. L.P. Kollar, G.S. Springer, *Mechanics of Composite Structures* (Cambridge University Press, Cambridge, 2003)
6. V.V. Vasiliev, E.V. Morozov, *Mechanics and Analysis of Composite Materials* (Elsevier Science, London, 2001)
7. E.E. Gdoutos, *Fracture mechanics. An introduction. Second Edition* (Springer, USA, 2005)
8. D.V. Hutton, *Fundamentals of finite element analysis* (Elizabeth A. Jones Publishers, USA, 2004)
9. E.J. Barbero, D.H. Cortes, Comp. Part B: Eng. **41**, 124 (2010)
10. R.H. Lopez, M.A. Luersen, E.S. Cursi, Comp. Part B: Eng. **40**, 731 (2009)
11. N.M. Hassan, R.C. Batra, Comp. Part B: Eng. **39**, 66 (2008)
12. W.K. Lim, W.K. Jeong, E.K. Tschegg, Comp. Part B: Eng. **41**, 94 (2010)
13. K. Bruyere, Revue des composites et des materiaux avonces, **76**, 69 (1997)
14. D. Gay, J. Gambelin, *Une aproche simple du calcul des structures par la methode des elements finis* (Hermes, Paris, 1991)
15. O.O. Ochoa, J.N. Reddy, *Finite element analysis of composite laminates* (Kluwer Academic Publishers, USA, 1992)
16. P. Bere, M. C. Dudescu, N. Balc, P. Berce, O. Nemes, A. M. Iurian, Mat. Plast. **51**, 2 (2014)
17. A. Popescu, L. Hancu, P. Bere, Appl. Mec. Mat. **371**, 394 (2013)
18. E. Sabău, N. Bâlc, P. Bere, AJME **9**, 4 (2011)
19. P. Bere, P. Berce, O. Nemeş, Comp. Part B: Eng. Int. J. **43**, 2237 (2012)

AUTHOR INDEX

AUTHOR INDEX